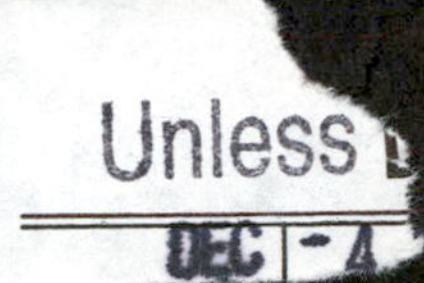

Fundamentals of Computer Aided Geometric Design

Fundamentals of Computer Aided Geometric Design

Josef Hoschek
Technische Hochschule Darmstadt
Darmstadt, Germany

Dieter Lasser
Universität Kaiserslautern
Kaiserslautern, Germany

Translated by
Larry L. Schumaker
Vanderbilt University
Nashville, Tennessee

A K Peters
Wellesley, Massachusetts

Editorial, Sales, and Customer Service Office

A K Peters, Ltd.
289 Linden Street
Wellesley, MA 02181

Originally published in 1989 by B. G Teubner, Stuttgart,
under the title Grundlagen der geometrischen Datenverarbeitung.

Library of Congress Cataloging-in-Publication Data

Hoschek, Josef.
[Grundlagen der geometrischen Datenverarbeitung. English]
Fundamentals of computer aided geometric design / Josef Hoschek and Dieter Lasser.
p. cm.
Translation of: Grundlagen der geometrischen Datenverarbeitung.
Includes bibliographic references and index.
ISBN 1-56881-007-5
1. Computer graphics. I. Lasser, Dieter, 1954- . II. Title.
T385.H69613 1993b
516'.6'0285—dc20 93-16733
CIP

Printed in the United States of America
97 96 95 94 93 10 9 8 7 6 5 4 3 2 1

Contents

Translator's Note

This volume is a direct translation of the second edition of *Grundlagen der geometrischen Datenverarbeitung* by Josef Hoschek and Dieter Lasser, published in the fall of 1992 by B. G. Teubner, Stuttgart, Germany.

My interest in undertaking this project began with a reading of the original lecture notes in 1986, and a closer study of the first edition of the book (published in 1989), which convinced me that an English version would be a valuable edition to the CAGD literature. In translating the text, I have tried to stay as close to the original German as possible. All of the figures, mathematics, and notation are exactly the same as in the second German edition.

I would like to express my special thanks to the authors for their help in choosing English terminology, for their careful reading of the various drafts, and for their work on the index. I would also like to thank my PhD student Greg Fasshauer for his help in comparing the translation to the original, and with proofreading. While I am responsible for all of the TEX errors, my thanks are due to Ruby Moore of the Department of Mathematics at Vanderbilt University for typesetting some of the displayed equations. Finally, I want to thank my wife Gerda for her constant support, and for valuable assistance with all stages of the project.

Nashville, October, 1992 — Larry L. Schumaker

Preface

Computer aided geometric design (CAGD) provides the mathematical basis for dealing with geometric data. There has been a rapid expansion in CAGD since powerful computers have become available. Some applications include:

- representation of large data sets,
- visualizing products,
- automatically producing sectional drawings,
- designing pipe systems, e.g. in chemical plants,
- modelling surfaces arising in the construction of cars, ships and airplanes,
- production and quality control, e.g. in the sewing machine, textile and shoe industries,
- drawing marine charts and city and relief maps in cartography,
- planning and controlling surgery,
- creating images in the advertising, television, and film industries,
- describing robot paths and controlling their movement,
- controlling milling machines used in manufacturing.

This list could go on and on, and there will undoubtedly be many new applications of CAGD in the future.

This book presents the mathematical foundations of geometric data processing, usually referred to as *computer aided geometric design*. For the everyday user, a CAGD system is a "black box". In this book we look at the "inside" of this box, i.e., we discuss the mathematical methods which underlie CAGD. Our aim is to provide mathematical techniques which will enable the reader to develop his own software for generating, describing, and modifying free form curves, surfaces, and volumes. We do not discuss fundamentals of computer science, such as hardware and software technology, graphics software systems, etc.

The book consists of 16 chapters, an extensive bibliography, and an index. Chapter 1 is concerned with the theory of projection, Chapter 2 gives both the geometric and the numerical basis for the representation of curves and surfaces, and Chapter 3 treats various types of splines. Our discussion in Chapters 4 and 6 of Bézier and B-spline methods for the generation of free form curves, surfaces, and volumes constitutes the core of the book. We treat both tensor-product and triangular surfaces, as well as general parameter domains. Chapters 5 and 7 deal with the question of geometric smoothness conditions between pieces of curves and surfaces. They also contain a description of various types of generalized splines such as β-, γ-, ν- and τ-splines. Surface representations are discussed in Chapter 8 (Gordon-Coons surfaces) and in Chapter 9 (scattered data surfaces). The latter chapter also includes the subject of surfaces defined on surfaces. Chapter 10 is devoted to the problem of converting between different curve and surface descriptions. This problem is important for data transfer. In Chapter 11 we discuss the subject of multivariate representations, along with various applications, such as the definition and design of algebraic curves and surfaces, and so-called free-form deformations. In Chapter 12 we present algorithms for the intersection of curves and surfaces, and in Chapter 13 smoothing methods for curves and surfaces are discussed. Blending methods are surveyed in Chapter 14, offset curves and surfaces in Chapter 15, and finally, in Chapter 16 we give an introduction to methods for controlling milling machines. Proofs are presented in detail only when they provide useful insights; otherwise we refer to the bibliography. The book also includes a discussion and references to a variety of applications, such as parallel curves and surfaces, the connection with finite element methods, geometrical criteria to recognize unwanted curve and surface regions, parametrization of sets of points to generate curves and surfaces, visibility tests, ray tracing, blending of curves and surfaces, etc.

The field of computer aided geometric design has been developing very rapidly in the past few years. We have tried to take into account the latest results. Our extensive bibliography supplements the text, and is designed to allow the reader to delve into areas which could only be mentioned briefly in the book. We have listed books first, then articles. To help the reader identify the nature of cited references, in the text we refer to books using author codes with all capital letters, e.g., [HOS 91], while for all other publications we capitalize only the first letter, e.g., [Hos 91].

This book grew out of lectures which we have given over the past several years at the Technical University of Darmstadt, and at the Universities of Karlsruhe and Kaiserslautern. Lecture notes were also prepared for courses for CAGD users in industry, and for a sequence of correspondence courses which we developed for the University of Hagen. The book does not include exercises — readers interested in exercises should refer to the above-mentioned correspondence course notes, see [HOS 87].

This volume is an English translation of *Grundlagen der geometrischen Datenverarbeitung* published by B. G. Teubner, Stuttgart, Germany, 1992. We would like to express our appreciation to Larry Schumaker for his interest in our work, and for his willingness to undertake the translation. We enjoyed working with him, and would like to express our thanks for his suggestions and comments which served to improve the quality of the book.

We are indebted to the editors at AKPeters Publishers for their professional guidance and support in the preparation of the English edition, and especially for undertaking the difficult task of inserting the numerous figures. Finally, we are thankful to our draftsman, Mrs. Elke Kniffki, for her skillful work in preparing all of the art work.

Darmstadt, October, 1992 Josef Hoschek and Dieter Lasser

1 Transformations and Projections

1.1. Introduction

In order to produce a good picture of a two or three dimensional object on a computer screen or on a plotter, we need to have some method for projecting the object onto a selected plane, and for placing it properly in the field of view. An artist can solve this problem with the help of intuition and experience, and an accomplished photographer can choose the best angles, distance, and field of view. On a computer, these human abilities have to be replaced by mathematical tools for describing the object, determining a projection mapping, and setting up transformations such as magnification, translation, and rotation of the image.

To develop the necessary tools, we use results from *linear algebra*, *analytic geometry* and *analysis* in $\mathbb{R}^2$ and $\mathbb{R}^3$. We shall describe objects mathematically in terms of points, lines, curves, surfaces, and volumes. Transformations of these objects will be obtained from corresponding mappings of the points, lines, curves, etc. The geometric background for this chapter is provided by descriptive geometry; for a computationally oriented treatment of this subject, see [HARTM 88] and [ADA 88].

We begin this chapter by giving mathematical descriptions of *translation*, *scaling*, and *rotation*. We shall see that *homogeneous* coordinates provide a good analytical way of treating a series of transformations to be carried out in succession. Objects in $\mathbb{R}^3$ can be projected onto a plane in various ways, including *parallel projection* (which corresponds to the shadow of an object), and *central projection* (which corresponds to taking a photograph). We shall give criteria for evaluating an image, along with methods for extracting three dimensional information from flat scenes (such as *stereographic images* and *anaglyphs*). In order to produce a good image of an object, it is also extremely important to be able to describe which parts of it are visible to the viewer. We also discuss the use of *shading* to improve the three dimensional appearance of the image.

1.2. Coordinate Transformations

Coordinate transformations in $\mathbb{R}^2$ and $\mathbb{R}^3$ are absolutely essential for creating images on a computer screen or plotter, see e.g., [ANG 81], [SCHUL 87]. In general, we have to distinguish the coordinate system in which the object is described from the coordinate system of the screen or plotter. The coordinate system of the object is frequently determined by geometric properties (such as special directions, symmetry, etc.). The coordinate system of the screen and the size of the field of view are usually determined by the properties of the device being used (for example, the origin might have to be in the upper left hand corner, and the x and y axes parallel to the edges of the screen). To transform the object system into the device system, we use coordinate transformations such as *translation*, *scaling*, and *rotation*. In general, we assume that we are working with orthonormal (Cartesian) coordinate systems.

1.2.1. Coordinate Transformations in the Plane

Suppose S and S' are systems with coordinates $(O; x_1, x_2)$ and $(O'; x'_1, x'_2)$, respectively. We can think of these as the *device system* and the *object system*. We begin by discussing the simplest transformation between the two systems, a *translation*, where we assume for now that the corresponding axes in the two systems are parallel to each other, see Fig. 1.1.

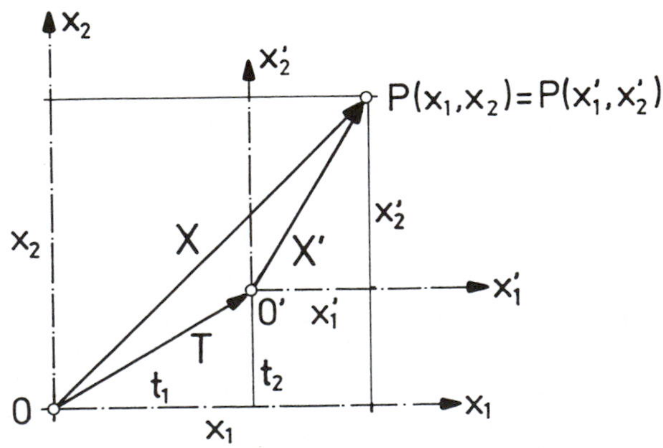

Fig. 1.1. Translation.

From Fig. 1.1, we can immediately read off the transformation equations

$$\begin{aligned} x_1 &= t_1 + x_1' \\ x_2 &= t_2 + x_2', \end{aligned} \tag{1.1}$$

where $\boldsymbol{T} := (t_1, t_2)^T$ represents the coordinates of the origin of the system S' with respect to the system S. Introducing the vectors $\boldsymbol{X} := (x_1, x_2)^T$ and $\boldsymbol{X}' := (x_1', x_2')^T$, it follows from (1.1) that

$$\boldsymbol{X} = \boldsymbol{T} + \boldsymbol{X}'. \tag{1.1'}$$

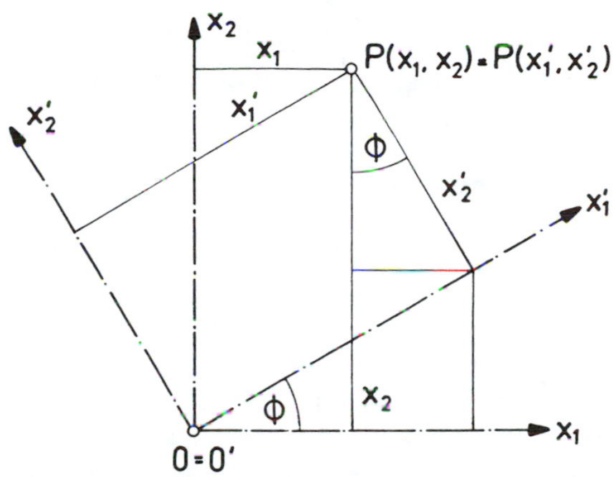

Fig. 1.2. Rotation by an angle ϕ.

If we rotate the system S' with respect to the system S by an angle of ϕ about the common origin $O = O'$, it follows from Fig. 1.2 that

$$\begin{aligned} x_1 &= x_1' \cos\phi - x_2' \sin\phi, \\ x_2 &= x_1' \sin\phi + x_2' \cos\phi. \end{aligned} \tag{1.2}$$

In terms of the (orthogonal) *rotation matrix*

$$\boldsymbol{R}(\phi) = \begin{pmatrix} \cos\phi & -\sin\phi \\ \sin\phi & \cos\phi \end{pmatrix},$$

(1.2) becomes

$$\boldsymbol{X} = \boldsymbol{R}\boldsymbol{X}'. \tag{1.2'}$$

Equation (1.2′) describes the transformation of points in the new system into points in the original system.

Sometimes it is necessary to stretch or shrink the system S'. Such *scaling* can be described by

$$x_1 = \lambda_1 x_1', \qquad x_2 = \lambda_2 x_2', \qquad\qquad \lambda_i \in \mathbb{R}. \tag{1.3}$$

In terms of the *scaling matrix*

$$\widetilde{\boldsymbol{S}} := \begin{pmatrix} \lambda_1 & 0 \\ 0 & \lambda_2 \end{pmatrix},$$

(1.3) can be rewritten in matrix form as

$$\boldsymbol{X} = \widetilde{\boldsymbol{S}}\boldsymbol{X}'. \tag{1.3'}$$

If we want to carry out a translation, a rotation and a scaling (one after the other), we can use the transformation

$$\boldsymbol{X} = \widetilde{\boldsymbol{S}}(\boldsymbol{T} + \boldsymbol{R}\boldsymbol{X}'). \tag{1.4}$$

The following pairs of transformations commute with each other: 1) translation with translation, 2) scaling with scaling, 3) rotation with rotation, and 4) rotation with scaling, provided it has a common scale factor. However, transformations of the form (1.4) are in general not commutative because of the addition in the expression in (1.4). Since in practice we often have to carry out just such transformations, we now introduce the idea of *extended coordinate vectors* (where vectors in $\mathbb{R}^2$ are mapped into vectors in $\mathbb{R}^3$):

$$\boldsymbol{X} := \begin{pmatrix} 1 \\ x_1 \\ x_2 \end{pmatrix}.$$

Now if we extend the matrix $\boldsymbol{R}$ in an analogous way to

$$\boldsymbol{A} := \begin{pmatrix} 1 & 0 & 0 \\ t_1 & \cos\phi & -\sin\phi \\ t_2 & \sin\phi & \cos\phi \end{pmatrix} = \begin{pmatrix} 1 & \boldsymbol{0} \\ \boldsymbol{T} & \boldsymbol{R} \end{pmatrix}, \tag{1.5}$$

and similarly extend $\widetilde{\boldsymbol{S}}$ to $\boldsymbol{S}$, we can then rewrite the mapping (1.4) as

$$\boldsymbol{X} = \boldsymbol{S}\boldsymbol{A}\boldsymbol{X}'. \tag{1.4'}$$

These matrices do not commute; *i.e.*, if we scale first, then rotate, and then translate, we get

$$\boldsymbol{X} = \boldsymbol{A}\boldsymbol{S}\boldsymbol{X}'.$$

Extended coordinates are a special case of homogeneous coordinates, see e.g., [GRO 57]. In terms of homogenous coordinates, each point $(x_1, x_2) \in \mathbb{R}^2$ is represented as a vector

$$X = \begin{pmatrix} w \\ wx_1 \\ wx_2 \end{pmatrix} =: \begin{pmatrix} u_0 \\ u_1 \\ u_2 \end{pmatrix}, \tag{1.6}$$

where w is the *homogenizing factor*. Clearly,

$$x_1 = \frac{u_1}{u_0}, \qquad x_2 = \frac{u_2}{u_0}, \tag{1.6'}$$

where usually u_0 will be nonzero. The case $u_0 = 0$ is allowed, however, and permits the analytic description of the *vanishing point* (or point at infinity) corresponding to the direction (x_1, x_2). This point at infinity will be of importance in later considerations. Here we should point out that often the homogenizing factor is taken to be the last coordinate in the vector $\boldsymbol{X}$, in which case (1.6) implies that

$$\boldsymbol{X} = (u_1, u_2, u_3)^T,$$

with u_3 the homogenizing factor; see e.g., [ENC 88].

Remark. A geometric description of homogeneous coordinates can be found in Section 4.3.6 below where rational Bézier curves are treated.

1.2.2. Coordinate Transformations in $\mathbb{R}^3$

In three dimensional space with a Cartesian basis, we can describe a coordinate transformation in analogy to the bivariate case by

$$\boldsymbol{X} = \boldsymbol{T} + \boldsymbol{R}\boldsymbol{X}', \tag{1.7}$$

where $\boldsymbol{T}$ is the translation vector, and the rotation matrix $\boldsymbol{R}$ is an orthogonal 3×3 matrix. Let

$$\boldsymbol{T} = \begin{pmatrix} t_1 \\ t_2 \\ t_3 \end{pmatrix}, \quad \boldsymbol{X} = \begin{pmatrix} x_1 \\ x_2 \\ x_3 \end{pmatrix}, \quad \boldsymbol{X}' = \begin{pmatrix} x'_1 \\ x'_2 \\ x'_3 \end{pmatrix}, \quad \boldsymbol{R} = (a_{ik}),$$

with

$$a_{ik} = \cos(\langle x_i, x'_k \rangle) = \boldsymbol{e}_i \cdot \boldsymbol{e}'_k.$$

Here $\boldsymbol{e}_i$ and $\boldsymbol{e}'_k$ are the orthonormal basis vectors of the associated Cartesian coordinate systems, respectively (see Fig. 1.3).

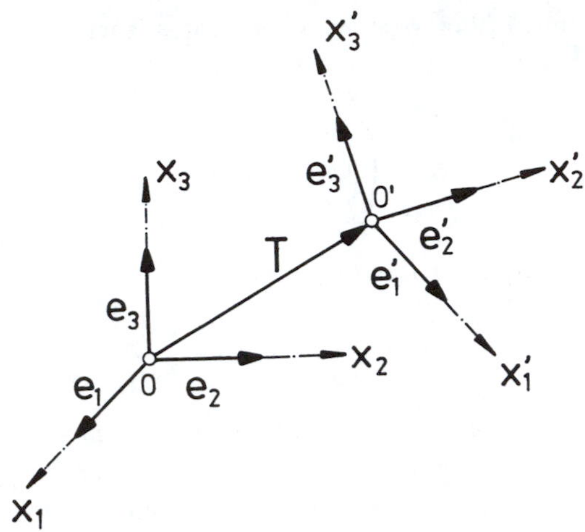

Fig. 1.3. Coordinate Systems.

In $\mathbb{R}^3$ we can also work with extended or *homogeneous coordinates*, defined by

$$\boldsymbol{X} = (1, x_1, x_2, x_3)^T.$$

Relative to these coordinates, the transformation (1.7) takes the form

$$\begin{pmatrix} 1 \\ x_1 \\ x_2 \\ x_3 \end{pmatrix} = \begin{pmatrix} 1 & \boldsymbol{0} \\ \boldsymbol{T} & \boldsymbol{R} \end{pmatrix} \begin{pmatrix} 1 \\ x_1' \\ x_2' \\ x_3' \end{pmatrix} \quad \text{or} \quad \boldsymbol{X} = \boldsymbol{A}\boldsymbol{X}'. \tag{1.8}$$

The scaling matrix now becomes

$$\boldsymbol{S} = \begin{pmatrix} 1 & & & 0 \\ & \lambda_1 & & \\ & & \lambda_2 & \\ 0 & & & \lambda_3 \end{pmatrix}, \tag{1.9}$$

and it follows that a general transformation is given by the formula

$$\boldsymbol{X} = \boldsymbol{S}\boldsymbol{A}\boldsymbol{X}'. \tag{1.8'}$$

Since $\boldsymbol{R}$ is an orthogonal transformation, $\boldsymbol{R}^{-1} = \boldsymbol{R}^T$, and the inverse transformations can be written as

$$\boldsymbol{A}^{-1} = \begin{pmatrix} 1 & \boldsymbol{0} \\ -\boldsymbol{R}^T\boldsymbol{T} & \boldsymbol{R}^T \end{pmatrix}, \qquad \boldsymbol{S}^{-1} = \begin{pmatrix} 1 & & & 0 \\ & \frac{1}{\lambda_1} & & \\ & & \frac{1}{\lambda_2} & \\ 0 & & & \frac{1}{\lambda_3} \end{pmatrix}.$$

The rotation produced by the matrix $\boldsymbol{R}$ can also be decomposed into rotations *about each of the three coordinate axes* as follows:

1) rotation about the x_3-axis by an angle α:

$$\boldsymbol{Z}(\alpha) = \begin{pmatrix} \cos\alpha & -\sin\alpha & 0 \\ \sin\alpha & \cos\alpha & 0 \\ 0 & 0 & 1 \end{pmatrix}, \tag{1.10a}$$

2) rotation about the x_2-axis by an angle β:

$$\boldsymbol{Y}(\beta) = \begin{pmatrix} \cos\beta & 0 & \sin\beta \\ 0 & 1 & 0 \\ -\sin\beta & 0 & \cos\beta \end{pmatrix}, \tag{1.10b}$$

3) rotation about the x_1-axis by an angle γ:

$$\boldsymbol{X}(\gamma) = \begin{pmatrix} 1 & 0 & 0 \\ 0 & \cos\gamma & -\sin\gamma \\ 0 & \sin\gamma & \cos\gamma \end{pmatrix}. \tag{1.10c}$$

Here the rotation angles α, β, and γ are always taken to be positive with respect to the right-hand system (x_1, x_2, x_3). Then, for example, for $\boldsymbol{R}$ we can write

$$\boldsymbol{R} = \boldsymbol{X}(\gamma)\boldsymbol{Y}(\beta)\boldsymbol{Z}(\alpha), \tag{1.11}$$

where first we rotate about the x_3-axis, then about the x_2-axis, and finally about the x_1-axis. The rotation can, of course, be done in a different order, but then we get a different representation of the rotation matrix $\boldsymbol{R}$, that is, for a given matrix $\boldsymbol{R}$, we get different rotation angles α, β, γ.

Another way to find the entries of the rotation matrix $\boldsymbol{R}$ is to use the *Euler angles*. The position of the new coordinate system relative to the old one is described by three angles:

a) the *nutation angle* $0 \le \theta < \pi$ between the positive x_3-axis and the positive x_3'-axis,

b) the *precession angle* $0 \le \psi < 2\pi$ between the x_1-axis and the line of intersection OX of the (x_1, x_2)-plane with the (x_1', x_2')-plane, see Fig. 1.4,

c) the *rotation angle* $0 \le \phi < 2\pi$ between the line OX and the x_1'-axis.

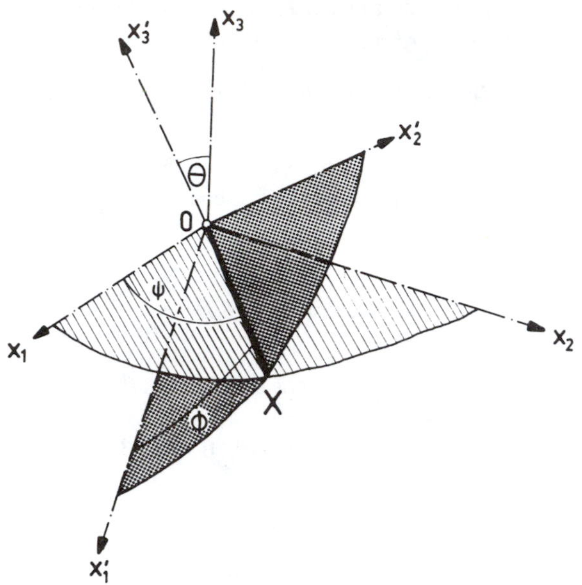

Fig. 1.4. Euler angles.

Introducing the abbreviations

$$c_1 := \cos\theta, \quad c_2 := \cos\psi, \quad c_3 := \cos\phi$$
$$s_1 := \sin\theta, \quad s_2 := \sin\psi, \quad s_3 := \sin\phi,$$

we can now write the rotation matrix $\boldsymbol{R}$ in terms of the Euler angles as

$$\boldsymbol{R} = \begin{pmatrix} c_2c_3 - c_1s_2s_3 & -c_2s_3 - c_1s_2c_3 & s_1s_2 \\ s_2c_3 + c_1c_2s_3 & -s_2s_3 + c_1c_2c_3 & -s_1c_2 \\ s_1s_3 & s_1c_3 & c_1 \end{pmatrix}. \tag{1.12a}$$

As an example, we consider the sequence of transformations shown in Fig. 1.5, where

original system $\boldsymbol{X}$	$\rightarrow$	intermediate system $\boldsymbol{X}' = \boldsymbol{AX}$ (after translation in the x-direction by a)
intermediate system $\boldsymbol{X}'$	$\rightarrow$	intermediate system $\boldsymbol{X}'' = \boldsymbol{BX}' = \boldsymbol{BAX}$ (after rotation about the x-axis by α)
intermediate system $\boldsymbol{X}''$	$\rightarrow$	final system $\boldsymbol{X}''' = \boldsymbol{CX}'' = \boldsymbol{CBAX} = \boldsymbol{DX}$ (after rotation about the z-axis by β).

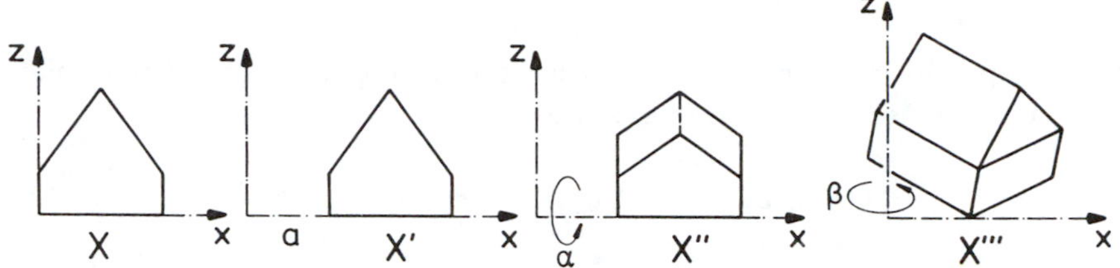

Fig. 1.5. Example of a Transformation.

The associated mapping matrix in this example is

$$D := \begin{pmatrix} 1 & 0 & 0 & 0 \\ 0 & \cos\beta & \sin\beta & 0 \\ 0 & -\sin\beta & \cos\beta & 0 \\ 0 & 0 & 0 & 1 \end{pmatrix} \begin{pmatrix} 1 & 0 & 0 & 0 \\ 0 & 1 & 0 & 0 \\ 0 & 0 & \cos\alpha & \sin\alpha \\ 0 & 0 & -\sin\alpha & \cos\alpha \end{pmatrix} \begin{pmatrix} 1 & 0 & 0 & 0 \\ -a & 1 & 0 & 0 \\ 0 & 0 & 1 & 0 \\ 0 & 0 & 0 & 1 \end{pmatrix}.$$

In practice (cf. e.g., [ENC 75]), the vectors of coordinates of points whose images one wants to find are inserted as columns in a so-called *object matrix*, leading to the following equation for the mapping of an object:

$$\begin{pmatrix} 1 & 1 & \\ \xi_1 & \eta_1 & \\ \xi_2 & \eta_2 & \dots \\ \xi_3 & \eta_3 & \end{pmatrix} = \begin{pmatrix} 1 & \mathbf{0} \\ \boldsymbol{T} & \boldsymbol{R} \end{pmatrix} \begin{pmatrix} 1 & 1 & \\ x_1 & y_1 & \\ x_2 & y_2 & \dots \\ x_3 & y_3 & \end{pmatrix}.$$

Rotation of an object in three space can also be described by rotation around a fixed axis $\boldsymbol{u} = (u_x, u_y, u_z)^T$ by an angle β, where we assume $|\boldsymbol{u}| = 1$. In this case, the associated rotation matrix is given by

$$\boldsymbol{R} = \begin{pmatrix} (1-c\beta)u_x^2 + c\beta & (1-c\beta)u_x u_y - u_z s\beta & (1-c\beta)u_x u_z + u_y s\beta \\ (1-c\beta)u_y u_x + u_z s\beta & (1-c\beta)u_y^2 + c\beta & (1-c\beta)u_y u_z - u_x s\beta \\ (1-c\beta)u_z u_x - u_y s\beta & (1-c\beta)u_z u_y + u_x s\beta & (1-c\beta)u_z^2 + c\beta \end{pmatrix} \tag{1.12b}$$

with $c\beta = \cos\beta$, $s\beta = \sin\beta$. Conversely, if we are given an orthogonal matrix $\boldsymbol{R} = \{a_{ik}\}$, then it follows that the angle of rotation is

$$\beta = \arccos\left(\frac{a_{11} + a_{22} + a_{33} - 1}{2}\right),$$

and the axis of rotation is

$$\boldsymbol{u} = \frac{1}{2\sin\beta}(a_{32} - a_{23}, a_{13} - a_{31}, a_{21} - a_{12})^T. \tag{1.12c}$$

It is also possible to describe rotations using normed *quaternions*. Quaternions are vectors in $\mathbb{R}^4$ with the representation

$$\boldsymbol{Q} = q_0\boldsymbol{e}_0 + q_1\boldsymbol{e}_1 + q_2\boldsymbol{e}_2 + q_3\boldsymbol{e}_3, \qquad q_i \in \mathbb{R}, \tag{1.13a}$$

where multiplication of the basis vectors $\boldsymbol{e}_i$ satisfies the following conditions [BLA 60]:

$$\boldsymbol{e}_0\boldsymbol{e}_j = \boldsymbol{e}_j\boldsymbol{e}_0 = \boldsymbol{e}_j, \quad \boldsymbol{e}_k\boldsymbol{e}_k = -\boldsymbol{e}_0, \quad k = 1, 2, 3$$

$$\begin{array}{l} \boldsymbol{e}_j\boldsymbol{e}_k = -\boldsymbol{e}_k\boldsymbol{e}_j = \boldsymbol{e}_l \\ j, k, l \text{ cyclical } 1,2,3 \end{array} \quad \Rightarrow \quad \begin{array}{cccc} \boldsymbol{e}_0 & \boldsymbol{e}_1 & \boldsymbol{e}_2 & \boldsymbol{e}_3 \\ \boldsymbol{e}_1 & -\boldsymbol{e}_0 & \boldsymbol{e}_3 & -\boldsymbol{e}_2 \\ \boldsymbol{e}_2 & -\boldsymbol{e}_3 & -\boldsymbol{e}_0 & \boldsymbol{e}_1 \\ \boldsymbol{e}_3 & \boldsymbol{e}_2 & -\boldsymbol{e}_1 & -\boldsymbol{e}_0 \end{array}$$

A quaternion is called *normed* provided that $|\boldsymbol{Q}| = 1$, and in this case, we can write

$$\boldsymbol{Q} = \left(\cos\frac{\beta}{2}, \boldsymbol{u}\sin\frac{\beta}{2}\right), \qquad |\boldsymbol{u}| = 1, \tag{1.13b}$$

where $\boldsymbol{u}$ is the direction of the axis of rotation, and β is the associated angle of rotation.

The vector $\boldsymbol{r}_2$ obtained by rotating a given vector $\boldsymbol{r}_1$ by an angle β about the axis $\boldsymbol{u}$ can now be simply written as

$$\boldsymbol{r}_2 = \boldsymbol{Q}\boldsymbol{r}_1\widetilde{\boldsymbol{Q}}, \tag{1.13c}$$

where $\widetilde{\boldsymbol{Q}} = (\cos\frac{\beta}{2}, -\boldsymbol{u}\sin\frac{\beta}{2})$ is the quaternion conjugate to $\boldsymbol{Q}$. Various applications of quaternions to problems in computer graphics are discussed in [Ple 89] and [Sei 90].

1.3. Projections

In order to display a three dimensional object on a computer screen, we need a projection (*linear mapping*) of the object onto the image plane. The choice of this projection will determine the quality of the display. Mathematically, a projection can be described by a mapping matrix. This mapping matrix will be determined by geometric requirements on the mapping: the given object (coordinate system $(O; x_1, x_2, x_3)$) can be projected onto the prescribed image plane ϵ (image coordinate system $(O'; \xi_1, \xi_2, \xi_3)$ with the ξ_3-axis perpendicular to ϵ) by

- parallel projection in the direction $\boldsymbol{p}$ (Fig. 1.6a),
- central projection with view-point $\boldsymbol{Z}$ (Fig. 1.6b).

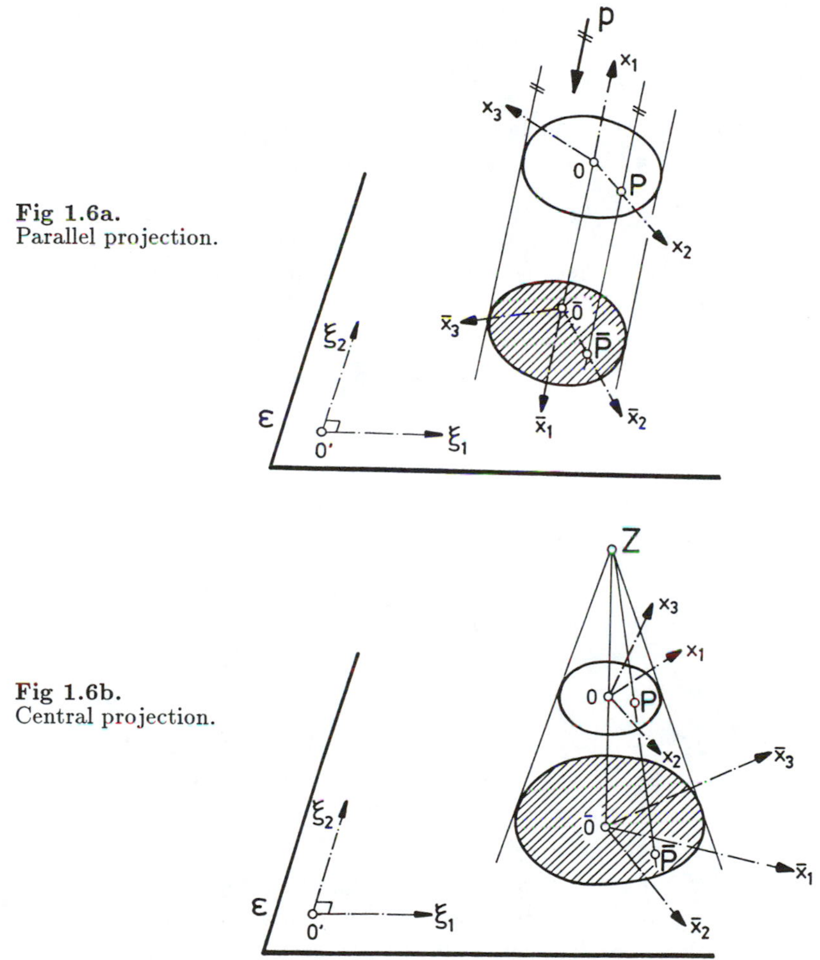

Fig 1.6a.
Parallel projection.

Fig 1.6b.
Central projection.

The choice of the projection $(\epsilon, \boldsymbol{p})$ or $(\epsilon, \boldsymbol{Z})$ effects the mapping. To compare projections, we can look at their effects on a unit cube.

1.3.1. Parallel Projection

Parallel projection corresponds to mapping the points of the object along projection lines ("light rays") parallel to a prescribed direction $\boldsymbol{p}$. The images of the points of the object are the points where the projection lines intersect the image plane ϵ. We differentiate between

- *orthographic* parallel projection ($\boldsymbol{p}$ perpendicular to ϵ),
- *oblique* parallel projection ($\boldsymbol{p}$ not perpendicular to ϵ).

In order to give a simpler mathematical description of *parallel projection*, we assume that before carrying out the parallel projection, the object coordinate system is transformed $(O; x_1, x_2, x_3) \to (\widetilde{O}; \widetilde{x}_1, \widetilde{x}_2, \widetilde{x}_3)$ so that

- the origin O' of the image system $(O'; \xi_1, \xi_2)$ coincides with the image $\bar{O}$ of the origin $\widetilde{O}$ of the object coordinate system,
- the image $\bar{x}_3$ of the $\widetilde{x}_3$-axis of the object coordinate system falls on the axis ξ_2 of the image system. The system $(\widetilde{O}; \widetilde{x}_1, \widetilde{x}_2, \widetilde{x}_3)$ is mapped onto $(\bar{O}; \bar{x}_1, \bar{x}_2, \bar{x}_3)$. Thus, we get the situation shown in Fig. 1.7 (cf. also Figs. 1.6a, 1.6b).

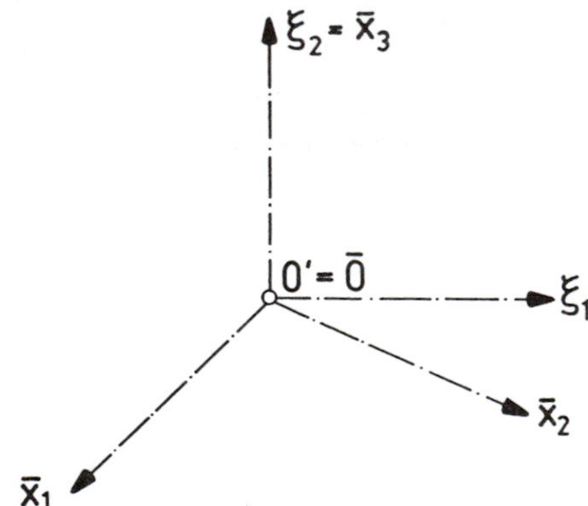

Fig. 1.7. Coordinate system in the image plane.

Now in order to find the image $\bar{\boldsymbol{P}}(\bar{x}_1, \bar{x}_2, \bar{x}_3)$ of an object point $\boldsymbol{P}(x_1, x_2, x_3)$ in the image coordinate system shown in Fig. 1.7, we consider the mapping of one of the object coordinate axes x_i. Under parallel projection in the direction $\boldsymbol{p}$, we find that its image $\bar{x}_i$ (see Fig. 1.8) is given by

$$\bar{x}_i = v_i x_i, \qquad i = 1, 2, 3, \tag{1.14a}$$

where v_i is a scalar distortion factor which is completely determined by the projection direction $\boldsymbol{p}$. From Fig. 1.8, we see that, in general, a parallel projection can be determined by two equivalent approaches:

- first the projection direction $\boldsymbol{p}$ is prescribed, and then the distortion factors v_i are calculated,
- first the distortion factors v_i are chosen, and then the image $(\bar{O}; \bar{x}_1, \bar{x}_2, \bar{x}_3)$ of the object coordinate system is "appropriately" selected.

The mathematical background for this assertion is the *Theorem of Pohlke* from constructive geometry, cf. e.g., [BRA 86], [HAW 62], and [HOW 51].

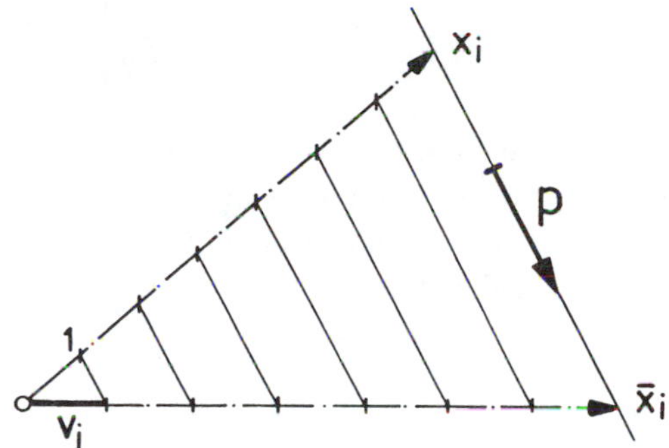

Fig. 1.8. Distortion under parallel projection.

1.3.2. Selection of the Distortion Factors

If the distortion factors v_i and the angles α (between $\bar{x}_1$ and ξ_1), and β (between $\bar{x}_2$ and ξ_2) are given, then by Fig. 1.9 and (1.14a), the coordinates $\bar{\boldsymbol{P}}(\xi_1, \xi_2)$ of the image $\bar{\boldsymbol{P}}(\bar{x}_1, \bar{x}_2, \bar{x}_3)$ of an object point $\boldsymbol{P}(x_1, x_2, x_3)$ can be computed as

$$\begin{aligned} \xi_1 &= v_1 x_1 \cos\alpha + v_2 x_2 \cos\beta, \\ \xi_2 &= -v_1 x_1 \sin\alpha - v_2 x_2 \sin\beta + v_3 x_3, \end{aligned} \tag{1.14b}$$

or in matrix form

$$\begin{pmatrix} \xi_1 \\ \xi_2 \\ 0 \end{pmatrix} = \begin{pmatrix} v_1 \cos\alpha & v_2 \cos\beta & 0 \\ -v_1 \sin\alpha & -v_2 \sin\beta & v_3 \\ 0 & 0 & 0 \end{pmatrix} \begin{pmatrix} x_1 \\ x_2 \\ x_3 \end{pmatrix}. \tag{1.14c}$$

From the position of the images $(\bar{x}_1, \bar{x}_2)$ of the corresponding coordinate axes, we can also determine the "visibility" of O, i.e., whether the observer is looking in the direction $\boldsymbol{p}$ into the positive octant of the object system

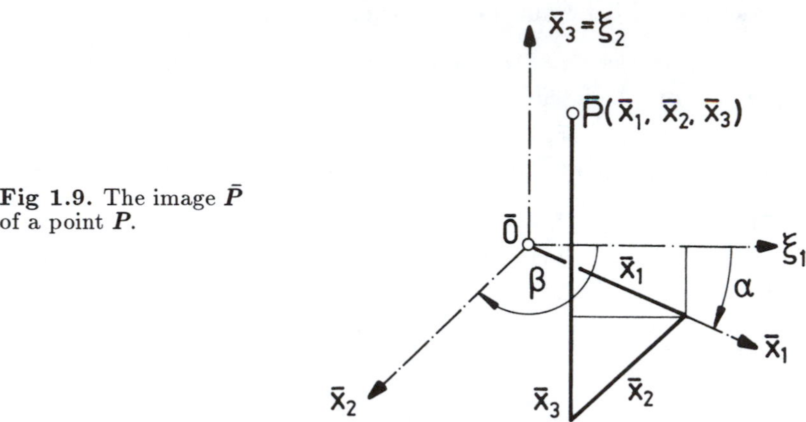

Fig 1.9. The image $\bar{P}$ of a point P.

(Case I), or into the negative octant of the object system (Case II). Here it is essential that we are using a right-handed Cartesian coordinate system. For a cube in the positive octant this means that in Case II, the origin is visible, while in Case I it is not, see Fig. 1.10.

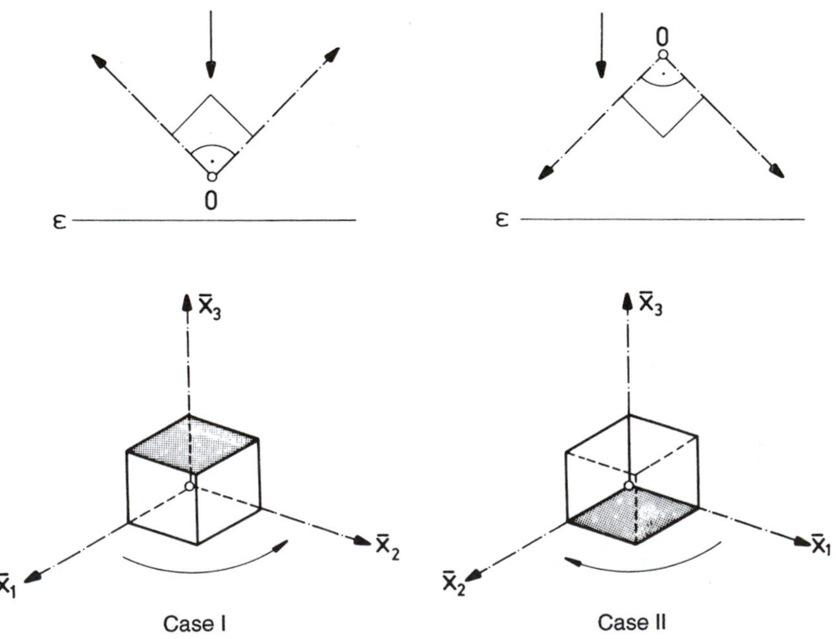

Fig. 1.10. Orientation of the images of the axes.

The distortion factors v_i (see [HOW 51], [WUN 76])

- can be chosen freely for an oblique parallel projection,
- depend on the chosen axes for orthographic parallel projection.

Appropriate images of axes under oblique parallel projection (called *oblique axonometry* in constructive geometry) are shown in Figs. 1.11, 1.12. Here we should note that a coordinate plane of the object remains undistorted if it is parallel to the image plane. The images of the associated coordinate axes form a right angle. The quality of the image is generally better if no coordinate plane is parallel to the image plane, cf. Fig. 1.13 with Figs. 1.11, 1.12; see [BRA 77].

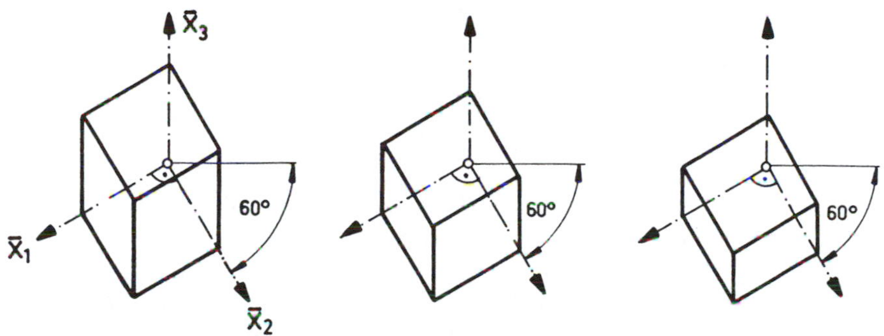

Fig. 1.11. Plan obliques (bird's-eye perspective).

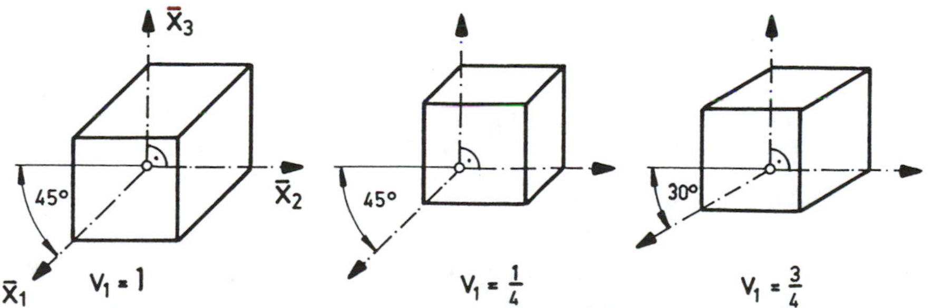

Fig. 1.12. Elevation obliques (cavalier perspective).

For orthographic parallel projection (*orthographic axonometry*), the distortion factors can be found from the angles between the images of the axes, see Fig. 1.14.

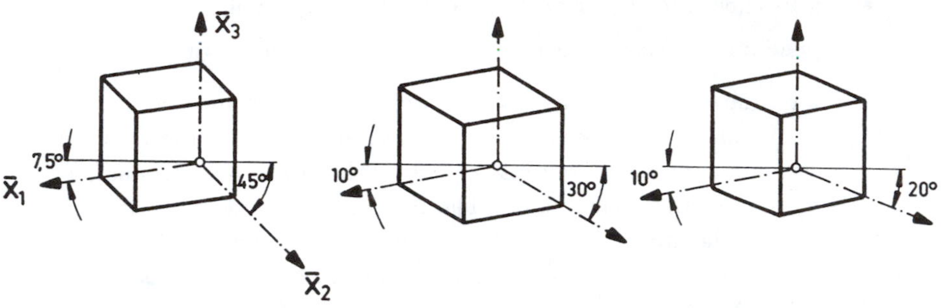

Fig. 1.13. Favorable oblique projections.

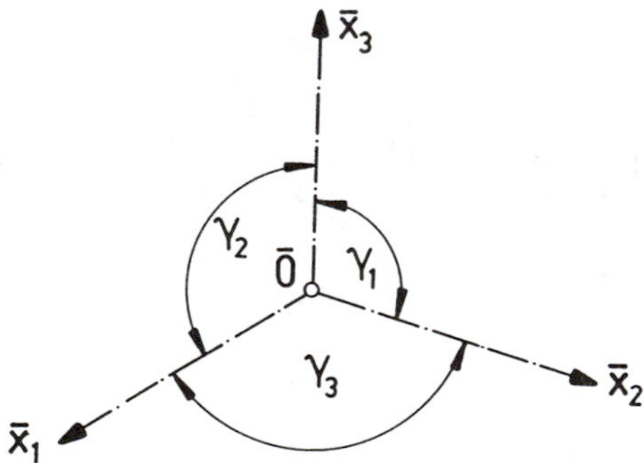

Fig. 1.14. Angles between the coordinate axes under orthographic projection.

We have (see [WUN 69])

$$v_1 = \sqrt{\frac{-\cos\gamma_1}{\sin\gamma_2 \sin\gamma_3}}, \quad v_2 = \sqrt{\frac{-\cos\gamma_2}{\sin\gamma_1 \sin\gamma_3}}, \quad v_3 = \sqrt{\frac{-\cos\gamma_3}{\sin\gamma_1 \sin\gamma_2}}.$$

Favorable axis images result from the choice $\gamma_1 = \gamma_2 = \gamma_3 = 120°$ which leads to $v_1 : v_2 : v_3 = 1 : 1 : 1$ (*isometric projection*), and for the *engineering projection* corresponding to $v_1 : v_2 : v_3 = 2 : 1 : 2$ (cf. Fig. 1.15).

The axis images for engineering axonometry are determined planimetrically from the short sides and the height of the triangle shown in Fig. 1.15 with side ratios 2 : 2 : 3.

Now in order to display images using one of these coordinate systems, we can find the mapping equations (1.14) by reading off the distortion factor and the angle as in Fig. 1.9 from the corresponding axis images.

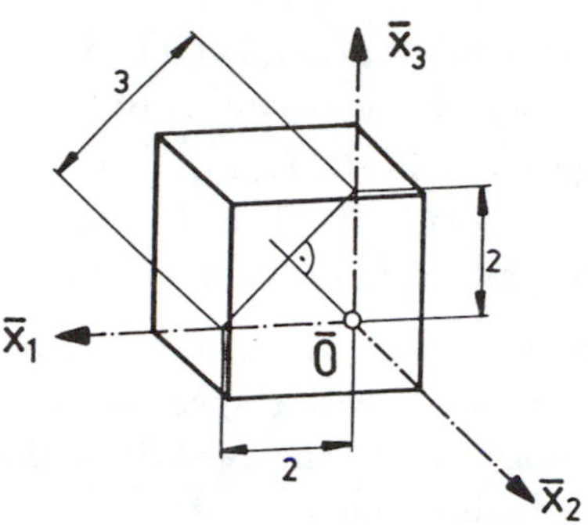

Fig. 1.15. Engineering axonometry.

Another way to determine the mapping equation was suggested by [Pau 88]. We interpret *parallel projection* of points (x_1, x_2, x_3) as an *affine mapping* using the transformation equation

$$\begin{aligned} \xi_1 &= a_0 + a_1 x_1 + a_2 x_2 + a_3 x_3, \\ \xi_2 &= b_0 + b_1 x_1 + b_2 x_2 + b_3 x_3. \end{aligned} \tag{1.15}$$

Now we can compute the images $\bar{\boldsymbol{P}}_i$ in the image plane of four vertices $\boldsymbol{E}_i$ of a unit cube, and from these find (a_i, b_i). The choice

$$\begin{array}{llllll} \boldsymbol{E}_0(0,0,0) & \rightarrow & \bar{\boldsymbol{P}}_0(0,0), & \text{yielding} & a_0 = 0, & b_0 = 0, \\ \boldsymbol{E}_1(1,0,0) & \rightarrow & \bar{\boldsymbol{P}}_1(\alpha_1,\beta_1), & \text{yielding} & a_1 = \alpha_1, & b_1 = \beta_1, \\ \boldsymbol{E}_2(0,1,0) & \rightarrow & \bar{\boldsymbol{P}}_2(\alpha_2,\beta_2), & \text{yielding} & a_2 = \alpha_2, & b_2 = \beta_2, \\ \boldsymbol{E}_3(0,0,1) & \rightarrow & \bar{\boldsymbol{P}}_3(\alpha_3,\beta_3), & \text{yielding} & a_3 = \alpha_3, & b_3 = \beta_3, \end{array}$$

is especially appropriate, and the coefficients of (1.15) are exactly the image coordinates of the corners of the cube!

1.3.3. Selection of the Projection Direction

The mapping matrix (1.14b) can be considerably simplified in the case of orthographic parallel projection (i.e. the projection direction $\boldsymbol{p}$ is perpendicular to the image plane ϵ) when the (x_1, x_2)-plane of the object system is parallel to ϵ. In (1.14b) we then have $v_1 = v_2 = 1$, $v_3 = 0$, and, furthermore, $\alpha = 0$ and $\beta = -\pi/2$ can be selected so that (1.14b) reduces to

$$\xi_1 = x_1, \qquad \xi_2 = x_2. \tag{1.16}$$

This is not generally possible for skew parallel projection (where $\boldsymbol{p}$ is not perpendicular to ϵ). Thus we start by

- rotating the object system $(O; x_1, x_2, x_3)$ into a new system $(O; \widetilde{x}_1, \widetilde{x}_2, \widetilde{x}_3)$ so that, e.g., the $(\widetilde{x}_1, \widetilde{x}_2)$–plane is parallel to the image plane ϵ,
- translating the object system $(O; \widetilde{x}_1, \widetilde{x}_2, \widetilde{x}_3)$ so that the origin O' of the image system coincides with O. The coordinate plane $(\bar{x}_1, \bar{x}_2)$ of the (new) object system then coincides with the image plane.

We can define this kind of mapping as follows: If $\boldsymbol{N} = (a, b, c)^T$ with $|N| = 1$ is the normal to the image plane ϵ with respect to the object system, then we require that $\boldsymbol{N}$ be transformed into the $\widetilde{x}_3$–axis of the transformed system. Then we can choose the $\widetilde{x}_2$ direction as

$$\boldsymbol{Y} := \frac{1}{\sqrt{a^2 + b^2}} (b, -a, 0)^T,$$

and it follows that the direction of the $\widetilde{x}_1$ axis is $\boldsymbol{X} = \boldsymbol{Y} \times \boldsymbol{N}$. We can thus describe this rotation by

$$\begin{pmatrix} \widetilde{x}_1 \\ \widetilde{x}_2 \\ \widetilde{x}_3 \end{pmatrix} = \begin{pmatrix} \boldsymbol{X}^T \\ \boldsymbol{Y}^T \\ \boldsymbol{N}^T \end{pmatrix} \begin{pmatrix} x_1 \\ x_2 \\ x_3 \end{pmatrix}. \tag{1.17}$$

Once these mappings have been performed, i.e.,

- the x_1-x_2 plane of the object system is rotated to be parallel to the image plane,
- the object system is moved to the origin of the image system,

then the object point $\boldsymbol{P}(x_1, x_2, x_3)$ will have coordinates $\boldsymbol{P}(a_1, a_2, a_3)$, and the transformed projection direction will have coordinates $\boldsymbol{p}(p_1, p_2, p_3)$.

Now if the coordinates of the image $\bar{\boldsymbol{P}}$ of $\boldsymbol{P}$ are described by $\boldsymbol{P}(\bar{x}_1, \bar{x}_2, 0)$, it follows from Fig. 1.16 that the mapping equation becomes

$$\bar{\boldsymbol{P}} = \boldsymbol{P} + \lambda \boldsymbol{p}, \qquad \lambda \in \mathbb{R}.$$

Since, in general, $p_3 \neq 0$ (for otherwise the projection is parallel to the image plane ϵ), λ can be computed from the third coordinate as

$$\lambda = -\frac{a_3}{p_3}.$$

Thus the mapping equation is

$$\begin{pmatrix} \bar{x}_1 \\ \bar{x}_2 \\ 0 \end{pmatrix} = \begin{pmatrix} 1 & 0 & -\frac{p_1}{p_3} \\ 0 & 1 & -\frac{p_2}{p_3} \\ 0 & 0 & 0 \end{pmatrix} \begin{pmatrix} a_1 \\ a_2 \\ a_3 \end{pmatrix}. \tag{1.18}$$

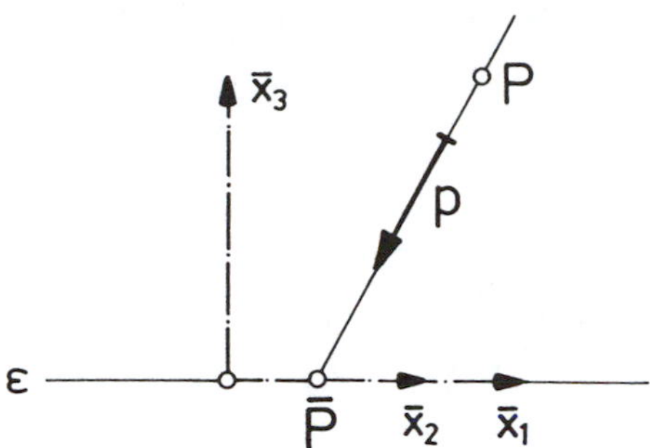

Fig. 1.16. The image $\bar{P}$ of a point P under parallel projection.

Remark. The projection direction can be found from $\boldsymbol{A}\cdot\boldsymbol{p} = 0$, where $\boldsymbol{A}$ is the mapping matrix in (1.18). This is a homogeneous linear system of equations (describing the kernel of $\boldsymbol{A}$).

Remark. $\boldsymbol{A}$ becomes trivial (see the remarks at the beginning of this chapter) if the projection direction $\boldsymbol{p}$ is parallel to x_3. Then $p_1 = p_2 = 0$, and (1.16) follows.

1.3.4. Central Projection

For central (perspective) projection, an object Φ is projected on an image plane ϵ through a *center of vision* $\boldsymbol{Z}$, see Fig. 1.17 and compare with Fig. 1.6b. Central projection can be regarded as a mathematical model of the processes of vision and photography. Under parallel projection, all parts of an object Φ are uniformly distorted. For central projection, however, the parts of the object which lie in front of the image plane ϵ (i.e., between ϵ and $\boldsymbol{Z}$) are enlarged, while those lying behind the image plane are reduced. Parts of the object which lie in the image plane remain undistorted, see [Sch 88].

When looking toward the image plane in a perpendicular direction, the eye can only see those objects which lie in a cone of vision with apex at $\boldsymbol{Z}$ and axis aligned with the line perpendicular to ϵ through $\boldsymbol{Z}$. We denote the point in ϵ at the foot of this perpendicular by $\boldsymbol{H}$, and denote the distance from $\boldsymbol{Z}$ to $\boldsymbol{H}$ by d. $\boldsymbol{H}$ is called the *center of perspective*, and is the midpoint of the *circle of vision* defined by the intersection of the cone of vision with the image plane ϵ. Experience shows that the radius of the circle of vision is approximately $d/2$, see e.g., [How 51], [MORE 52]. Now if the eye can be moved, then it covers a circle of radius approximately d, the so-called *distance circle*. Images inside the circle of visibility appear normal, those outside the

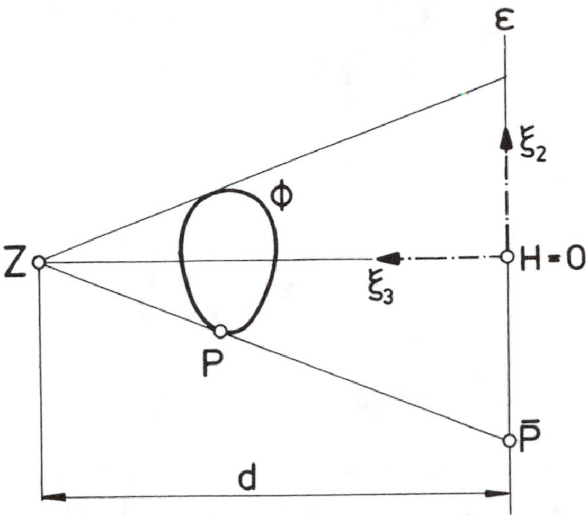

Fig. 1.17. Central Projection.

circle of visibility but inside the distance circle are slightly exaggerated, and those outside of the distance circle are strongly distorted, see Fig. 1.18.

In order to give a simple mathematical description of central projection, we introduce a special coordinate system $(O;\xi_1,\xi_2,\xi_3)$ with origin at the point $\boldsymbol{H}$, whose ξ_3 axis goes through $\boldsymbol{Z}$, see Fig. 1.17. Now we have to transform the object system $(O;x_1,x_2,x_3)$ into the image system.

By Fig. 1.17, the image $\bar{\boldsymbol{P}}$ of an object point $\boldsymbol{P}$ is given by

$$\bar{\boldsymbol{P}} = \boldsymbol{Z} + \mu(\boldsymbol{P} - \boldsymbol{Z}), \qquad \mu \in \mathbb{R}$$

or

$$\begin{pmatrix} \bar{\xi}_1 \\ \bar{\xi}_2 \\ 0 \end{pmatrix} = \begin{pmatrix} 0 \\ 0 \\ d \end{pmatrix} + \mu \begin{pmatrix} \xi_1 \\ \xi_2 \\ \xi_3 - d \end{pmatrix}.$$

Comparing the third components, we have

$$\mu = \frac{d}{d - \xi_3},$$

and thus the coordinates of the image point $\bar{\boldsymbol{P}}$ are given by

$$\bar{\xi}_1 = \frac{d\xi_1}{d - \xi_3}, \qquad \bar{\xi}_2 = \frac{d\xi_2}{d - \xi_3}. \tag{1.19}$$

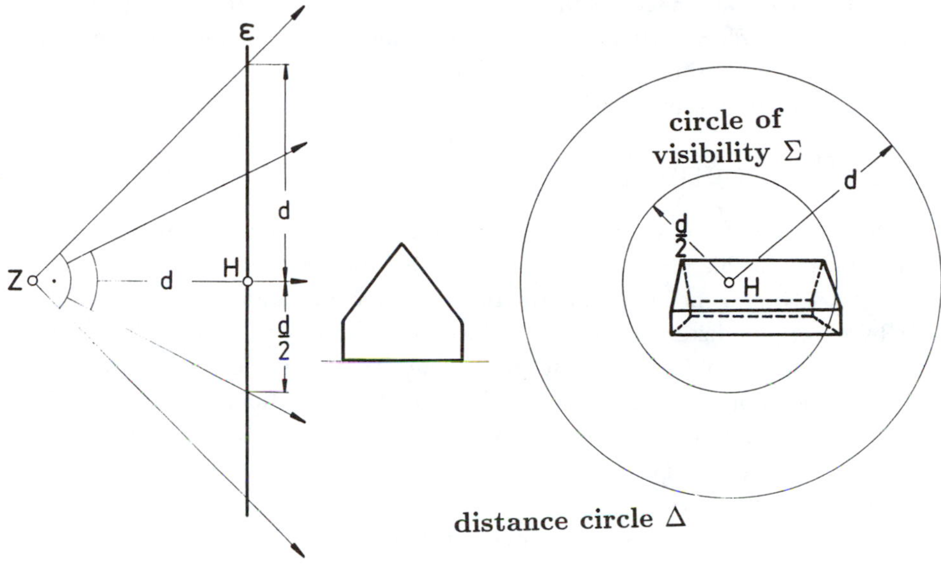

Fig. 1.18. Visual effect under central projection.

Introducing homogeneous coordinates via

$$\xi_i =: \frac{\eta_i}{\eta_0}, \qquad \bar{\xi}_i =: \frac{\bar{\eta}_i}{\bar{\eta}_0},$$

we can write the mapping equation (1.19) in matrix form as

$$\begin{pmatrix} \bar{\eta}_0 \\ \bar{\eta}_1 \\ \bar{\eta}_2 \\ 0 \end{pmatrix} = \begin{pmatrix} d & 0 & 0 & -1 \\ 0 & d & 0 & 0 \\ 0 & 0 & d & 0 \\ 0 & 0 & 0 & 0 \end{pmatrix} \begin{pmatrix} \eta_0 \\ \eta_1 \\ \eta_2 \\ \eta_3 \end{pmatrix}. \tag{1.19'}$$

From (1.19) we see that points in the plane $\xi_3 = d$ cannot be depicted. This plane through $\boldsymbol{Z}$ and parallel to the image plane ϵ is called the *vanishing plane*. Images of points on the vanishing plane lie at infinity, and thus, the images of figures which intersect the vanishing plane divide into two parts. If the point $\boldsymbol{Z}$ lies at infinity, central projection reduces to parallel projection.

The mapping equations for central projection can also be found by looking at the image of a cube, provided we interpret the central projection of the points $(O; x_0, x_1, x_2, x_3)$ as a *projective mapping*, see [Pau 88]: We consider the so-called *central axonometry*, which corresponds to the rational mapping

$$\xi = \frac{a_0 + a_1 x + a_2 y + a_3 z}{1 + c_1 x + c_2 y + c_3 z}, \qquad \eta = \frac{b_0 + b_1 x + b_2 y + b_3 z}{1 + c_1 x + c_2 y + c_3 z}. \tag{$*$}$$

If we compute the images of the corner points $\boldsymbol{E}_0 = (0,0,0)$, $\boldsymbol{E}_1 = (1,0,0)$, $\boldsymbol{E}_2 = (0,1,0)$, $\boldsymbol{E}_3 = (0,0,1)$, $\boldsymbol{E}_4 = (1,1,0)$, along with one additional point, say $\boldsymbol{E}_5 = (1,0,1)$, then $(*)$ leads to an easily solvable system of linear equations for the unknown coefficients.

1.4. Stereographic Images and Anaglyphs

With the methods developed so far, we are able to produce planar images of 3D objects. In this process 3D information is lost, although often the viewer can reconstruct some of this information by looking at additional projections or by observing distortions to determine what is close and what is far away. For complicated objects, this 3D reconstruction is frequently very difficult if not impossible, see e.g., [Scha 88]. In order to understand 3D objects well (e.g., complicated surfaces arising in analysis and differential geometry), plane projections alone are not adequate. For this purpose, *stereographic pairs* and *anaglyphs* can be useful, see e.g., [Bart 80], [Hau 77].

The technique for producing stereographic photographs has been known for some time (using a camera with two lenses). It is used, for example, in stereographic photogrammetry and in the study of the surface of the earth using satellites. It is also of use in medicine, for example in performing complicated operations (where stereographic X-rays are used), or in producing 3D images of organs using CAT scans. In these applications, polarized light is used. Anaglyph images (usually produced using red and green on a dark background) are often used in textbooks on descriptive geometry, see e.g., [EHR 64], [SCHÖ 77]. More recently, this technique has also found applications to television.

The idea behind a stereographic image is that the mind reconstructs the 3D image from two separate images, one for each eye. The given object Φ is mapped onto an image plane ϵ from two view-points $\boldsymbol{Z}_1$ and $\boldsymbol{Z}_2$ at a distance $2b$ apart and a distance d from ϵ, see Fig. 1.19. Projection through the view-point $\boldsymbol{Z}_1$ leads to an image B_1, and projection through $\boldsymbol{Z}_2$ leads to an image B_2.

In order to find the mapping equations for the stereographic pair B_1, B_2, we introduce a special coordinate system as in Section 1.3.4 whose (ξ_1, ξ_2) plane coincides with the image plane, and whose ξ_3 axis goes through the midpoint on the line segment $\boldsymbol{Z}_1\boldsymbol{Z}_2$, see Fig. 1.19.

Assuming as above that the view-points $\boldsymbol{Z}_1$ and $\boldsymbol{Z}_2$ are separated by a distance $2b$ and lie at a distance d from the image plane, and assuming that $\boldsymbol{Z}_1 \cup \boldsymbol{Z}_2$ is parallel to the ξ_2 axis, then their coordinates are given by

$$\boldsymbol{Z}_1 = (0, b, d), \qquad \boldsymbol{Z}_2 = (0, -b, d). \tag{1.20}$$

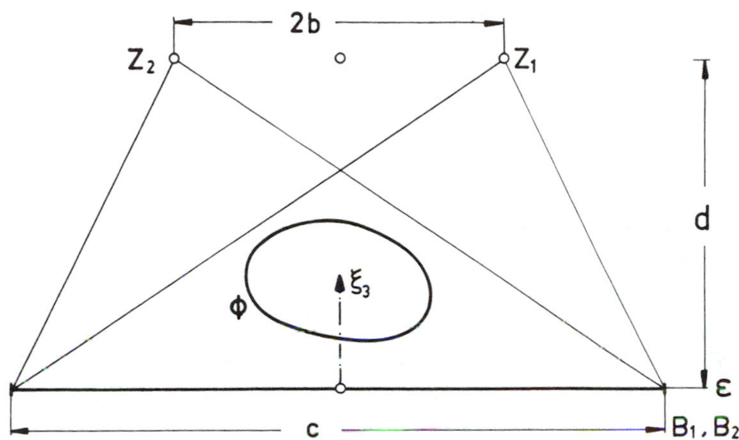

Fig. 1.19. Construction of a stereographic pair B_1, B_2 for an object Φ.

Now if in (1.20) we formally set $a_2 := \pm b$, then by analogy with (1.19), we get the following mapping equations for a point $\boldsymbol{P}(\xi_1, \xi_2, \xi_3)$:

$$\bar{\xi}_1 = \frac{\xi_1 d}{d - \xi_3}, \qquad \bar{\xi}_2 = \frac{\xi_2 d - \xi_3 a_2}{d - \xi_3}, \tag{1.21}$$

or in matrix form with homogeneous coordinates,

$$\begin{pmatrix} \bar{\eta}_0 \\ \bar{\eta}_1 \\ \bar{\eta}_2 \\ 0 \end{pmatrix} = \begin{pmatrix} d & 0 & 0 & -1 \\ 0 & d & 0 & 0 \\ 0 & 0 & d & -a_2 \\ 0 & 0 & 0 & 0 \end{pmatrix} \begin{pmatrix} \eta_0 \\ \eta_1 \\ \eta_2 \\ \eta_3 \end{pmatrix},$$

where the homogeneous coordinates are defined via

$$\xi_i = \frac{\eta_i}{\eta_0}, \quad i = 1, 2, 3, \qquad \bar{\xi}_j = \frac{\bar{\eta}_j}{\bar{\eta}_0}, \quad j = 1, 2.$$

The pair of images constructed by (1.21) may have to be reduced or enlarged.

Instead of working with the mapping equations corresponding to two view-points $\boldsymbol{Z}_1, \boldsymbol{Z}_2$, we also could have used directly the mapping equations (1.19) for central projection: The image B_1 is constructed by projection through $\boldsymbol{Z}$ onto ϵ, while the image B_2 is constructed similarly after translating the object by $2b$ along a line parallel to the image plane ϵ.

In viewing a stereographic pair, we have to be careful that each eye can see only one of the images, i.e., the fields of view of the two eyes must be (at least approximately) separated.

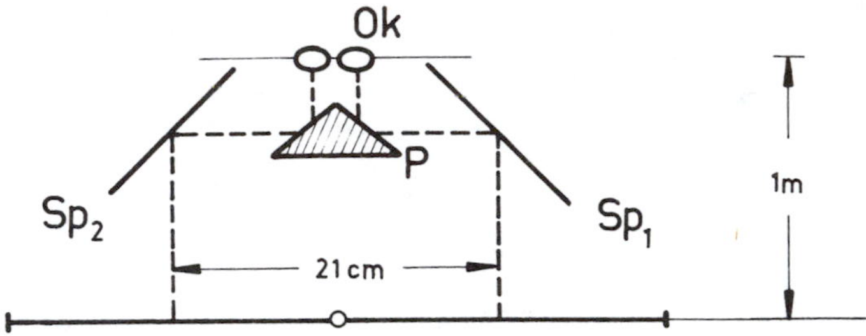

Fig. 1.20. Stereo viewer.

In using a stereo viewer (see Fig. 1.20), the two images are observed through two lenses O_k. Typically this is accomplished with the help of a prism P and two mirrors Sp_1, Sp_2. This allows the viewing of a pair of images of fairly large size (up to size $c = 21$ cm spaced 21 cm apart).

Mini-stereographic pairs (see Fig. 1.21 and e.g., [Bart 80], [Hau 77]) can be viewed directly through a pair of lenses. But if the eyes are separated by a distance of $2b = 6.5$ cm, for example, the images can be of size at most $c = 6.5$ cm (when the distance from the view-points to the image plane is $d = 30$ cm).

If several people want to view a stereographic image simultaneously, then we can use polarized light and a special pair of glasses. A recently developed viewing method (*wedge stereoscopy* [Kre 72], [Unb 85]) uses a stereographic pair of images positioned one above the other, and is viewed through a pair of glasses fitted with two prisms.

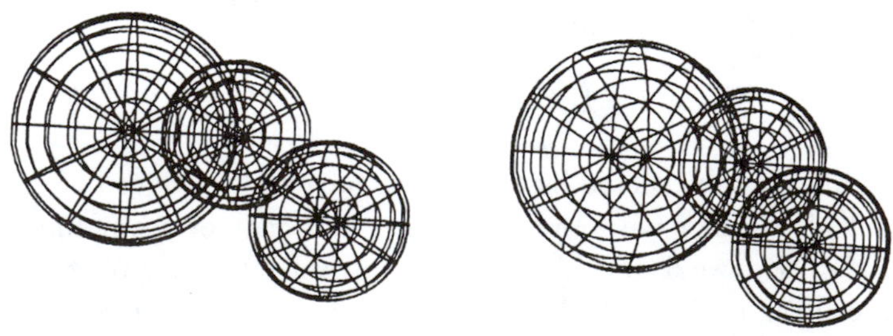

Fig. 1.21. Mini-stereographic image of three spheres.

Three dimensional images can also be produced on a video screen using polarized light, for example, by displaying left and right polarized images alternately, and viewing them through appropriately polarized glasses.

With some practice, one can also develop stereographic vision without the help of technical tricks. To do this, look at a pair of stereographic images lying side by side from a distance of approximately 60 cm. Now first focus on a point (e.g., the tip of a pencil) at a distance of about 30 cm with both eyes, where the right eye looks over the pencil point at the far vertex of the left figure, while the left eye looks at the far vertex of the right figure. At this point, one sees three blurred images. The middle image is the one of interest, and after some time (and a little practice), it can be made to appear as a sharp 3D image, cf. [Hau 77], [SCHW 76]. A practiced observer can vary these distances arbitrarily; it is only necessary to look for the 3D image at the right place. Indeed, an advanced group of observers can even view stereographic images shown on a screen using an overhead projector. In viewing stereographic images this way, both images have to be reversed in order to maintain the correct orientation (see Fig. 1.22).

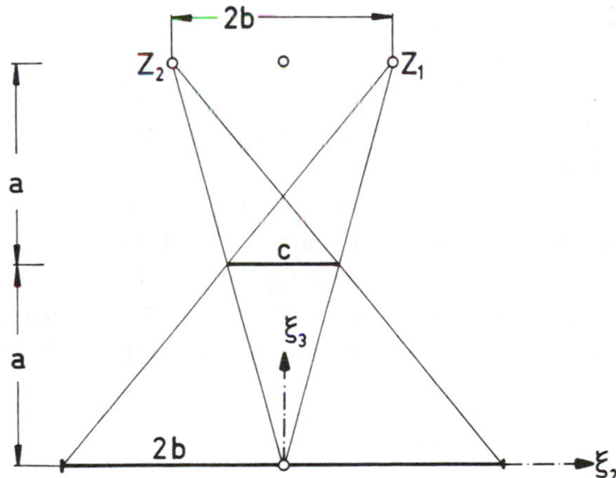

Fig. 1.22. Unaided stereographic vision reverses orientation.

Unaided stereographic viewing can be facilitated if we remove the two prisms in a wedge stereoscope, and position them before the eyes in such a way that the two mini-stereographic images merge.

To produce 3D images using *anaglyphs*, the two images can be printed on top of each other, since the separation of the two images is accomplished using complementary colors. Here there is in principle no limit on the size

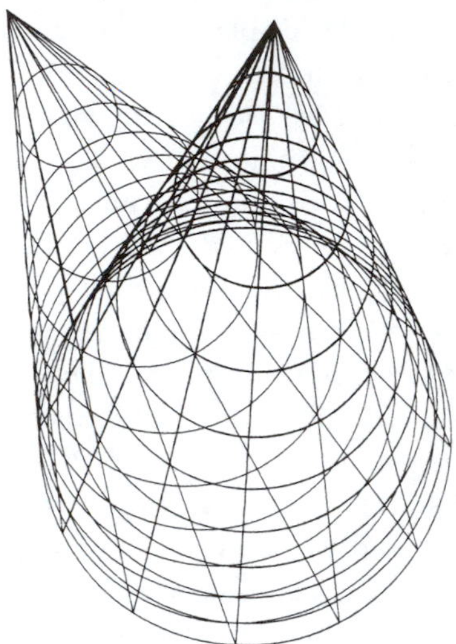

Fig. 1.23. Simulated anaglyphic image of a cone.

of the images. Fig. 1.23 simulates an anaglyph, where because of printing limitations, we have not used color.

Anaglyph images can also be shown to a larger group. In order to do this, we make separate plots of the anaglyph images, and mount them on slides which include a border. The anaglyph images can now be shown using two overhead projectors, where the slides are positioned so that the borders of the two images coincide. One set of slides should be made with red film, the other with green. Observing the resulting images with a pair of glasses with one red and one green lens produces the 3D image. For further discussion and explanation of various 3D viewing techniques, see [OKO 76], [Mcco 92].

1.5. Visibility Methods

In this section we discuss methods for determining which parts (such as vertices, edges, and surface patches) of the projected image of a 3D object are visible. Removing these "hidden" parts from the image greatly enhances its 3D effect. For the best results, it is crucial to determine as accurately as possible which parts of the scene are visible from the view-point. If this is done incorrectly, the result can be a completely false image of the object.

While Fig. 1.24 shows that incorrect visibility information can convey the impression of an impossible 3D object, Fig. 1.25 shows that different visibility decisions lead to different orderings of the object in a scene. Indeed, without

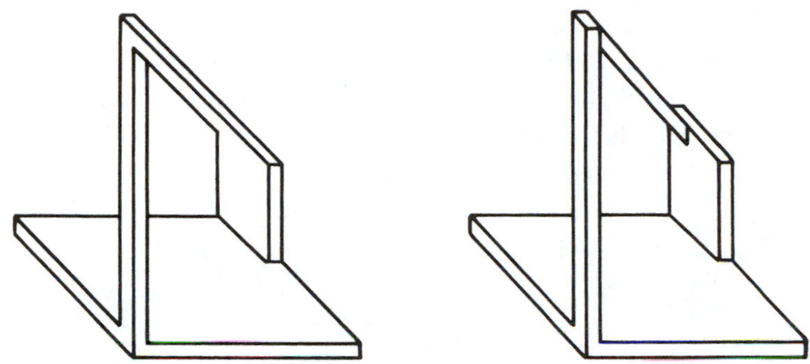

Fig. 1.24. An impossible object and the correct view.

deciding visibility, it is usually impossible to determine the space ordering of 3D objects.

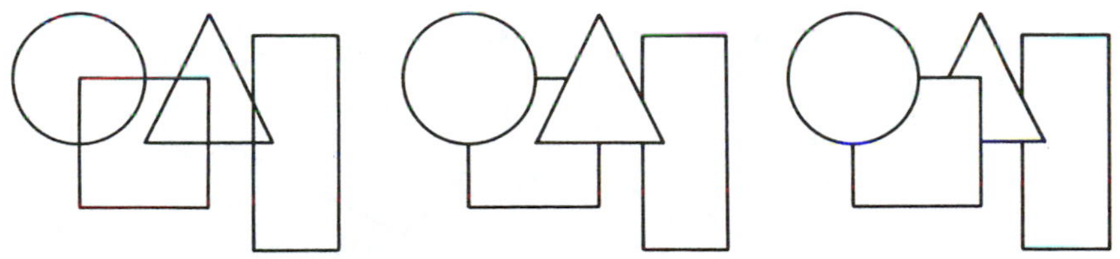

Fig. 1.25. Visibility and order.

Deciding visibility, i.e., eliminating hidden vertices, edges, and surfaces is one of the most challenging problems of computer graphics. There are still many open problems in this area, and much remains to be done, cf. [ENC 75], [FOLE 82]. In general, it does not pay to develop general purpose programs, as they are often inefficient – it is usually advisable to choose the method which works best for the situation at hand, cf. [MYE 83], [Witt 81], and [Sab 85a]. Visibility methods can be systematically divided into

- surface tests,
- point tests,
- point/surface tests.

In a *surface test*, the basic element which is to be tested for visibility is a piece of a surface. Here we only test whether a given surface patch on a convex body is hidden by the body itself.

For this test, we orient the outside surface of the object using an outward pointing normal vector, and compute the scalar product of the normal vector at a given point with the vector $\boldsymbol{p}$ describing the viewing direction. For parallel projection, $\boldsymbol{p}$ is determined by the direction in which we are projecting, while for central projection, it is determined by the line passing through the point of interest and the view-point $\boldsymbol{Z}$. A point is visible to the observer if and only if

$$\boldsymbol{N} \cdot \boldsymbol{p} < 0. \tag{1.22}$$

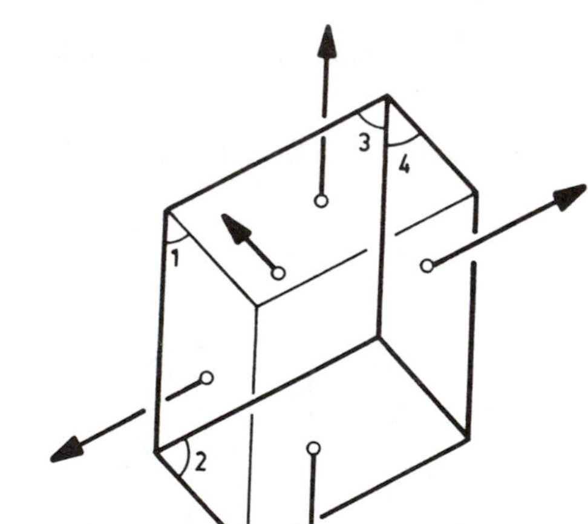

Fig 1.26. Visibility of a rectangular parallelepiped.

Fig. 1.26 shows a convex body bounded by planes (a rectangular parallelepiped) with outward oriented normals. Only the sides 2, 3, and 4 are visible. Fig. 1.27 shows the image resulting from using the surface test for a convex, continuously curved surface under parallel projection. The visible and hidden parts of the body are separated by silhouette curves on the surface defined by

$$\boldsymbol{N} \cdot \boldsymbol{p} = 0. \tag{1.23}$$

The points of these curves describe the shadow of the object in the plane ϵ, cf. Fig. 1.27.

If a convex surface patch is given by a function or a relation, then (1.23) is a nonlinear equation whose zeros are to be determined.

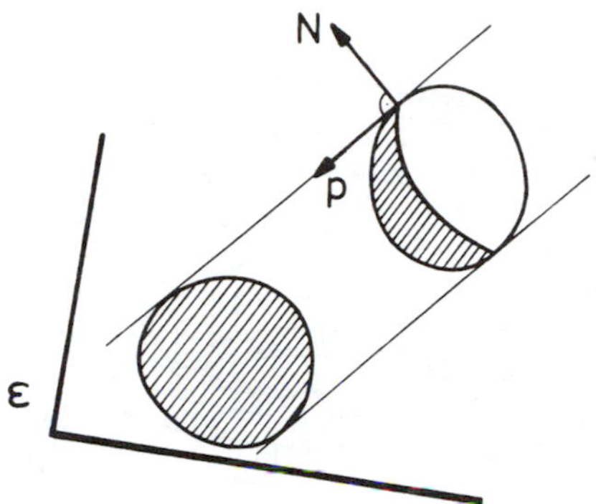

Fig. 1.27. Projections of the silhouette of a convex body.

As an example, consider the silhouettes of a torus written in parametric form as

$$\boldsymbol{X} = (r\cos v + \rho\cos v\cos u,\; r\sin v + \rho\sin v\cos u,\; \rho\sin u)^T,$$
$$\text{with } \rho < r \text{ constant}, \quad u, v \in [0, 2\pi]$$

under parallel projection with $\boldsymbol{p} = (0, -\cos\alpha, -\sin\alpha)^T$. By (2.8), the outward oriented normals are

$$\boldsymbol{N} = (\cos u\cos v,\; \cos u\sin v,\; \sin u)^T.$$

From the silhouette condition (1.23), we get

$$\boldsymbol{N}\cdot\boldsymbol{p} = \cos u\sin v\cos\alpha + \sin u\sin\alpha = 0,$$

whose solution is

$$u = \arctan(-\sin v\cot\alpha), \qquad v = \arcsin(-\tan u\tan\alpha).$$

From this we conclude that the lines $v = v_0 =$ constant and u lying between

$$u_1 = \arctan(-\sin v_0\cot\alpha) \quad \text{and} \quad u_2 = u_1 + \pi$$

are visible, while the lines $u = u_0 =$ constant are visible for v lying between

$$v_1 = \arcsin(-\tan u_0\tan\alpha) \quad \text{and} \quad v_2 = v_1 + \pi.$$

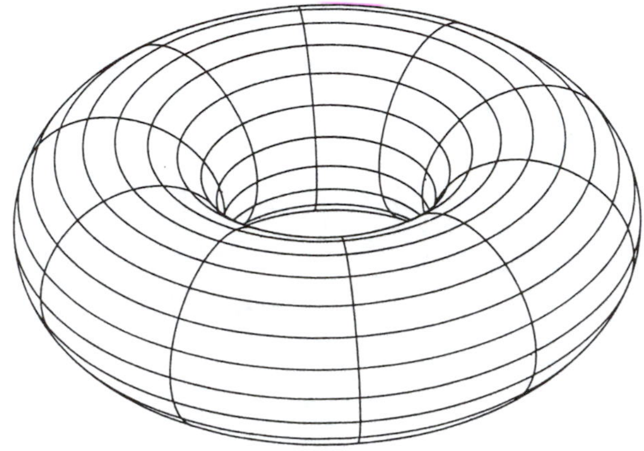

Fig. 1.27a. Silhouettes of a torus.

However, here we must take into account the fact that this holds only for $|-\tan u \tan \alpha| < 1$. If this condition is not satisfied, then the corresponding parameter lines are displayed completely, see Fig. 1.27a.

Recently developed visibility algorithms first find the sets in which (1.22) holds, i.e., which are possibly visible. The corresponding silhouettes determine domains in the parameter plane, and it remains only to test their boundaries against each other [Horn 85], [Elb 90].

For a *point test*, the observed lines or pieces of curves are divided into segments, and each segment is identified with a test point, e.g., the midpoint. The segment will be drawn provided that this test point is not hidden by any other part of the object, i.e., if no other point of the object lies on the line between the view-point and the test point. Thus, we first construct a line in the viewing direction passing through the test point $\boldsymbol{P}_i$, and check whether this line intersects the object or not. The quality of this method depends on the length of the segments being used. It can happen that a hidden edge does not end exactly at the boundary of visibility, or that a segment extends beyond the visibility boundary, see Fig. 1.28.

The most effective visibility algorithms combine both the point and surface tests, and attempt to accelerate the process by some additional considerations. An overview (in part with source programs) of such visibility algorithms (frequently also referred to as *hidden-line* or *hidden-surface* methods) can be found e.g., in [ENC 75, 88].

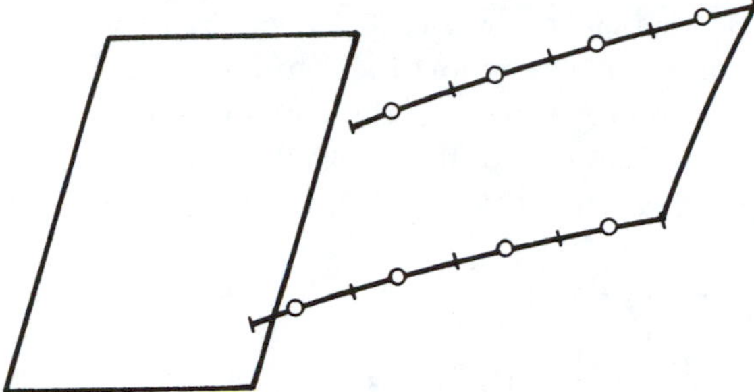

Fig. 1.28. Incorrect image using the point test.

As an example, we now describe in detail the *covering method*, cf. [ENC 75, 88]. First, each surface patch of the object is covered with a (u, v)-raster of size $\Delta u, \Delta v$, see Fig. 1.29. Each point $\boldsymbol{P}$ on the object is then determined by

- its cartesian coordinates (x, y, z),
- the corresponding (u, v)-parameters.

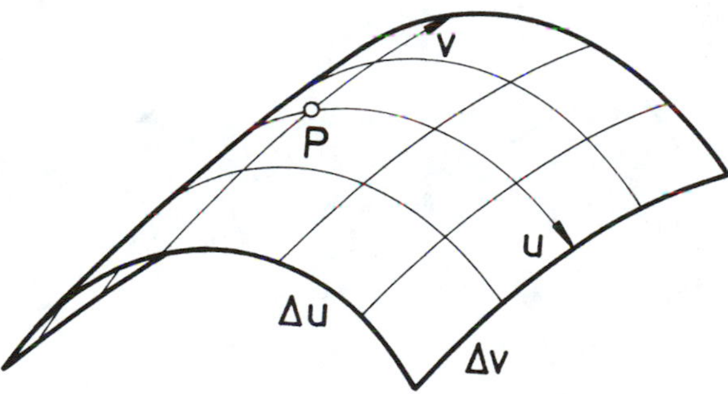

Fig. 1.29. (u, v)-parameters of a surface point.

Now the object is projected onto the image plane, which is taken to be the (x, y) plane, and which we cover with a raster of size Δ, see Fig. 1.30. To test a grid point $\boldsymbol{P}$ for visibility, we first locate the subrectangle in which it lies. Next we test all surface patches of the object which contain points whose projections onto the image plane fall into this rectangle. If the number of these is too large, then we can refine the image plane grid. In order to determine the visibility of $\boldsymbol{P}$ more precisely, we approximate the four-sided surface patches of interest by pairs of plane triangles (see Fig. 1.31), and remove the triangles which do not contain $\boldsymbol{P}$. Assuming the coordinates of $\boldsymbol{P}$ on the image plane are (x, y), we now use the vertices of the corresponding triangular elements to compute the approximate z coordinates of all points $\boldsymbol{P}_\ell$ on the object mapping to $\boldsymbol{P}$. Then the object point $\boldsymbol{P}_\ell$ with largest z coordinate is visible.

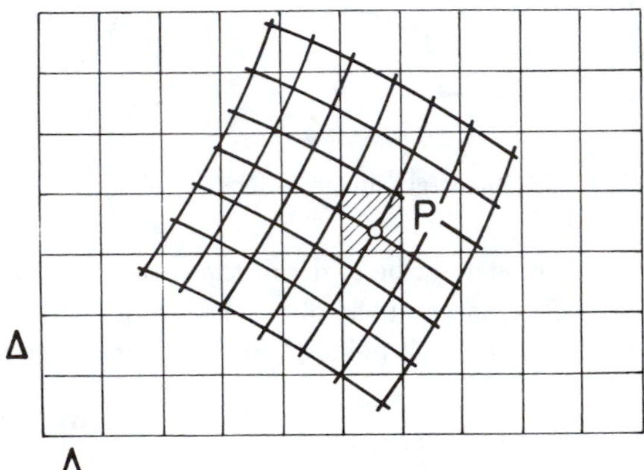

Fig. 1.30. A raster in the image plane.

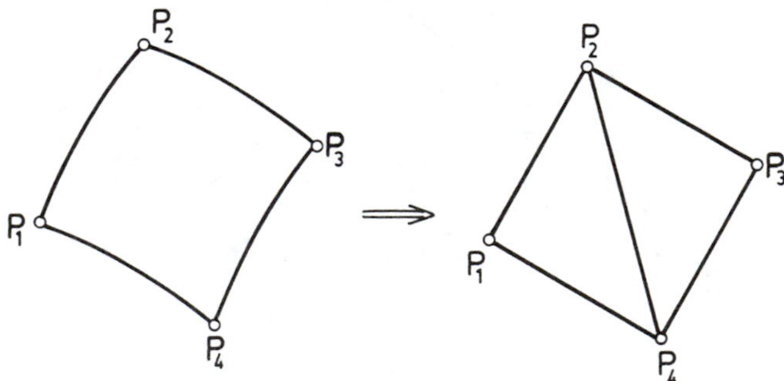

Fig. 1.31. Decomposition of a $(\Delta u, \Delta v)$ patch.

This method allows us to decide the visibility of all grid points, and it remains to check which of the connecting edges can be seen. If two neighboring grid points are both visible, then so is the line connecting them. If a grid point is visible and its neighbor is not, then we have to introduce one or more additional test points on the connecting edge, and the test procedure must be repeated. For example, the new test point can be chosen to be the intersection of the edge with the (u, v) parameter grid. The size of the original $(\Delta u, \Delta v)$ grid determines the accuracy of the method, cf. Fig. 1.32.

Fig 1.32. Refinement of the visibility test.

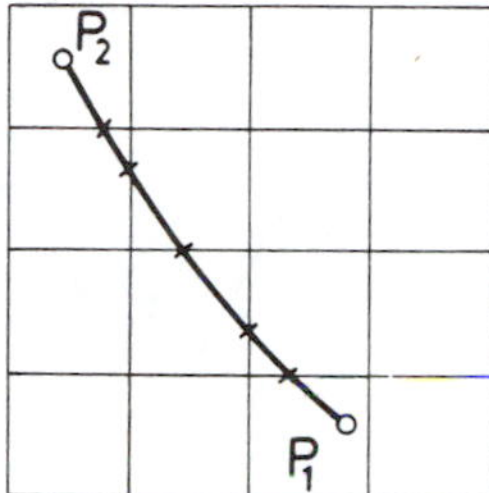

Fig. 1.33 shows the visible part of a relatively complex object given in parametric form (see Chap. 2). The figure was kindly provided by Giering and Seybold of the Institute for Geometry at the Technical University of Munich.

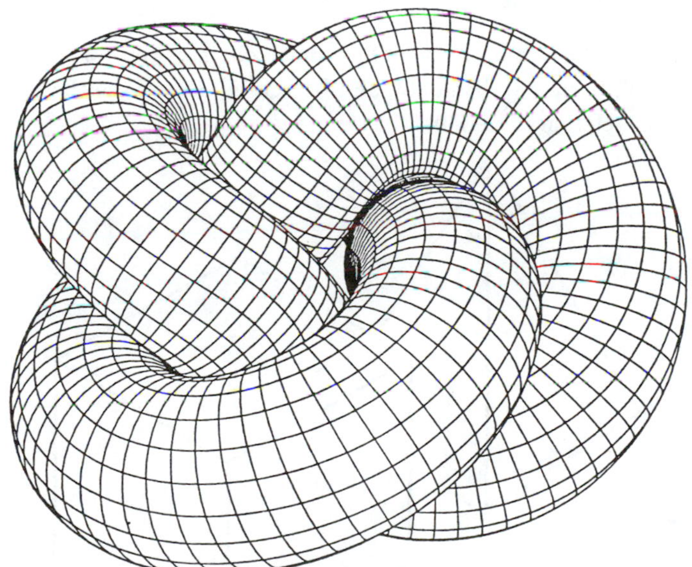

Fig. 1.33. Hidden line drawing of a complex object.

When drawing objects on a video screen, there is a simple method for deciding visibility based on commands for dealing with pixels. If each time a pixel is drawn its previous value is erased, then we can display a surface

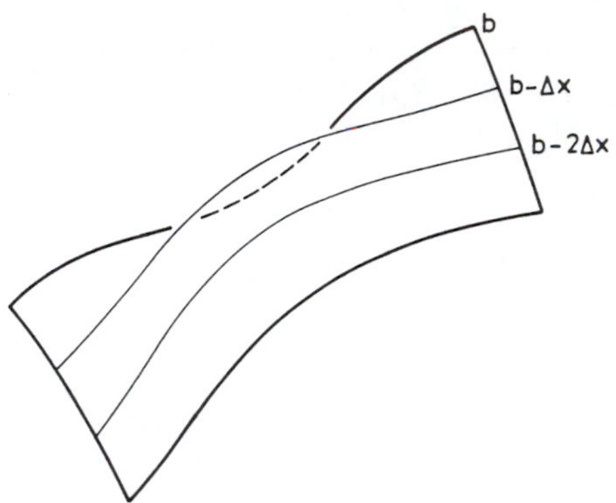

Fig. 1.34. Video output where the dotted line is erased.

described by $z = f(x, y)$ for $a \leq x \leq b$, $c \leq y \leq d$, as follows: we begin with the furthest edge $x = b$, and then draw the sequence of curves corrresponding to $x = b - i\Delta$ $(\Delta > 0)$; see Fig. 1.34 and [MYE 83], [Butl 79]. This method can also be applied to appropriate objects consisting of several parts (cf. Fig. 1.35).

For more information on visibility methods, see [Suth 74], [Grif 78, 79, 81], [MÜL 80], [Krip 85], [Bon 86], and [ROG 85], as well as the bibliography [Grif 78] and references therein.

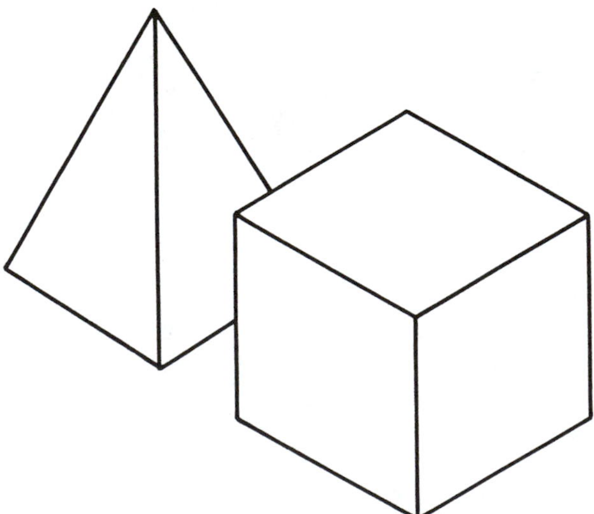

Fig. 1.35. Visibility of several objects on a video screen.

1.6. Shading and Reflection

One way of enhancing the 3D effect of a planar image is by removing hidden lines and surfaces using the visibility methods discussed in the previous section. In this section we discuss the use of *shading* as another tool for enhancing the display of 3D images. The idea is to use different shades of gray or color on the various parts of the image to create a more realistic impression, cf. [FOLE 82], [NEW 79], [Pho 75], [ROG 85], and [FEL 88]. Shading methods are becoming increasingly important not only in judging the quality of surfaces (see Chap. 13, [Pös 84]), but also in many other areas such as the design of buildings, the choice of building lots [Peck 85], and the analysis of energy use (e.g., the placement of solar cells).

The foundation of shading methods is provided by *geometric lighting models*, which provide a way of dealing with the brightness of surface elements. These models must be based on certain physical assumptions, which in general must come from experimentation.

There are two basic concepts of importance when dealing with shading of curved surfaces: the *isophotes* and the *isophengs* [Lang 84], [Rös 37]. Isophotes are *lines of equal brightness*, and are defined by

$$h_1 = C_1 \cos \lambda, \tag{1.24}$$

where C_1 is an appropriately selected constant, and λ is the angle between the surface normal $\boldsymbol{N}$ to the surface at the point $\boldsymbol{P}$ and direction $\boldsymbol{p}$ of the incident light rays, with $|N| = |p| = 1$, see Fig. 1.36.

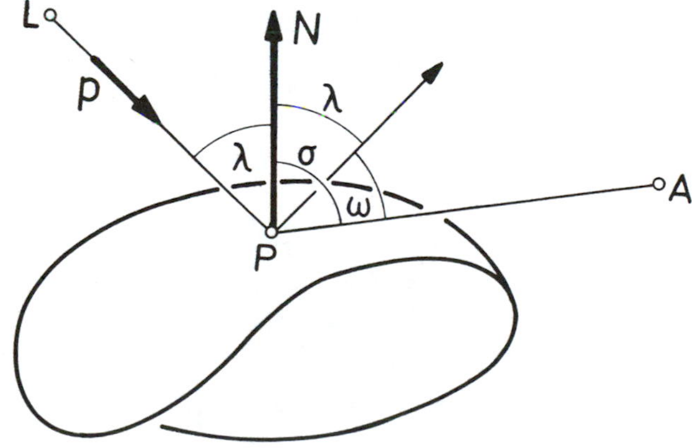

Fig. 1.36. Incident and reflected ray, light source $\boldsymbol{L}$, eye $\boldsymbol{A}$.

It should be pointed out that this definition is not completely accurate since it does not take into account the view-point of the observer. Thus, we introduce the concept of *isophengs*, the lines of equal *apparent* brightness, defined by

$$h_2 = C_2 \cos\lambda \cos\sigma, \tag{1.25}$$

where C_2 is an appropriately selected constant, and σ is the angle between the line of vision from the observer to the surface point $\boldsymbol{P}$ and the normal vector at that point, see Fig. 1.36. The use of isophengs overemphasizes the importance of the view-point of the observer, and results in shadings of surfaces which are not realistic. Particularly bothersome is the fact that using this model for the brightness, the outlines of the surface are darkest (since $h_2 = 0$ when $\sigma = \frac{\pi}{2}$). This has led to the development of several concepts of brightness which are combinations of the two above, see [FOLE 82], [NEW 79].

In choosing the constants for the above methods, it is important to take into account the fact that the absolute light intensity is proportional to R^{-2}, where R is the distance between the eye and the light source. On the other hand, the relative intensity is nearly independent of R, since the apparent size of the object also depends on R in the same way. Moreover, in selecting the constants, we should also take note of

- the nature of the surface (dull, shiny, etc.),
- the nature of the light (diffuse, point source, etc.).

To describe the nature of the surface, we introduce

- the *reflection coefficient* R_p, which describes how much light from a point light source will be reflected ($R_p \in [0,1]$),
- the *reflection coefficient* R_d, which describes how much light from a diffuse light source will be reflected ($R_d \in [0,1]$).

A surface appears colored when it has differing reflection coefficients for light of differing wavelengths. The nature of the light is described by

- I_p, the intensity of the incident light from a point light source,
- I_d, the intensity of the incident light from a diffuse light source.

Based on these concepts, we can describe various *shading models* using the various definitions of intensity. For example, for *diffuse reflection* from a dull surface, using a point light source and the isophote model, we have

$$I = I_p R_p \frac{\cos\lambda}{R^2}, \tag{1.26}$$

where λ is the angle between the incident light and the normal to the surface at the point $\boldsymbol{P}$.

In order to develop a more natural lighting model, we still need to take into account the fact that diffuse light is also present. Another problem with (1.26) is that for parallel illumination, $R \to \infty$. Even for central illumination, R^{-2} can range over a large set of values if, for example, the view-point is near the object. That would lead to a very large variation in brightness for surface points with the same angle of incidence λ. In view of this, [FOLE 82] has suggested replacing the denominator R^2 by $r + k$, where k is a constant and r is the distance between the surface and the view-point. This leads to

$$I = I_d R_d + I_p R_p \frac{\cos \lambda}{k + r}.$$

If the surface is "shiny" rather than dull, then a point light source leads to local bright spots, so-called *highlights*. [Pho 75] has attempted to capture this effect by introducing a *specular reflection coefficient* $W(\lambda)$ (see Fig. 1.37), which leads to the intensity formula (Phong-shading)

$$I = (R_p \cos \lambda + W(\lambda) \cos^n \omega) \frac{I_p}{k + r} + I_d R_d, \tag{1.27}$$

where ω is the angle between the line of sight (from the eye) and the reflected light (see Fig. 1.36), and $n \in [1, 10]$ is a measure of the *shininess* of the surface. Thus, a metal surface, which has very small bright spots has a high value of n, while a surface made of paper has a small value of n.

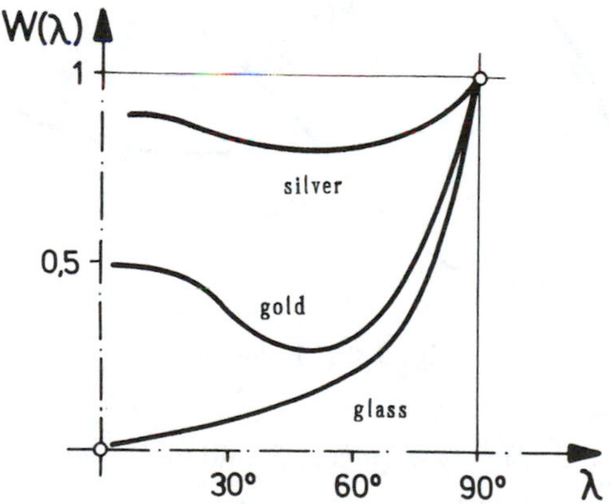

Fig. 1.37. Specular reflection coefficient (cf. [NEW 79]).

[Cook 81] has improved the Phong shading model by taking account of all the physical components of shading, resulting in image shading which is very similar to photography.

In order to shade a smooth surface, we can construct an appropriate set of points on the surface, and compute the intensity at each grid point by one of the above shading models. We then record these intensities at each grid point and use them to interpolate intensity values for all points on the model. This can be done by approximating the smooth surface by a polyhedral surface, computing normals at the vertices of the polyhedra by averaging the normals in the surrounding facets, using these to find the intensities at the vertices, and finally averaging over these values for points in each facet, cf. e.g., [FOLE 82].

In order to exactly locate highlights which appear in a shading model, we need to determine which reflected light rays pass through the view-point. To this end, let $\boldsymbol{L}$ be the light source, $\boldsymbol{P}$ the reflection point, $\boldsymbol{N}$ ($|N| = 1$) the normal vector at $\boldsymbol{P}$, and $\boldsymbol{A}$ the view point. Then the incident ray has direction $\boldsymbol{b} = \boldsymbol{PL}$ (see Fig. 1.38), and the reflected ray has direction $\boldsymbol{a} = \boldsymbol{PA}$.

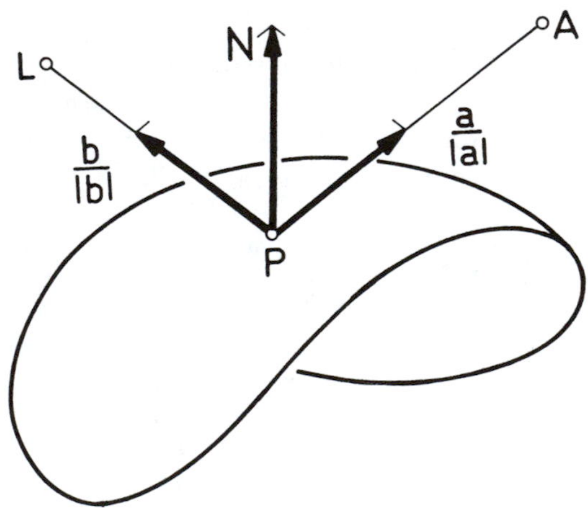

Fig. 1.38. Reflection.

Then by the reflection law (incident and reflected rays lie in the same plane with the normal vector)

$$\frac{\boldsymbol{a}}{|\boldsymbol{a}|} + \frac{\boldsymbol{b}}{|\boldsymbol{b}|} = \mu N, \tag{1.28}$$

where μ is a still to be determined scalar factor. Taking account of the fact that for reflection, the angle of incidence equals the angle of reflection, we have

$$\frac{\boldsymbol{a} \cdot \boldsymbol{N}}{|\boldsymbol{a}|} = \frac{\boldsymbol{b} \cdot \boldsymbol{N}}{|\boldsymbol{b}|}. \tag{1.29}$$

Taking the scalar product of (1.28) with $\boldsymbol{N}$, it follows from (1.29) that

$$\mu = 2\frac{\boldsymbol{b} \cdot \boldsymbol{N}}{|\boldsymbol{b}|}.$$

Now, in order to compute highlights corresponding to a given light source $\boldsymbol{L}$ and view-point $\boldsymbol{A}$, using (1.28), we are led to the following (in general nonlinear) equation for the unknown point $\boldsymbol{P}$:

$$\frac{\boldsymbol{L} - \boldsymbol{P}}{|\boldsymbol{L} - \boldsymbol{P}|} + \frac{\boldsymbol{A} - \boldsymbol{P}}{|\boldsymbol{A} - \boldsymbol{P}|} = 2\frac{(\boldsymbol{A} - \boldsymbol{P}) \cdot \boldsymbol{N}}{|\boldsymbol{A} - \boldsymbol{P}|}\boldsymbol{N}. \tag{1.30}$$

Bright spots of curves and surfaces were already of importance in several applications predating computer graphics, cf. e.g., [Wun 50].

A method which optimally simulates the physics of illumination and simultaneously deals with the visibility problem is *ray tracing*. In this method, the exact contribution of every (specularly reflected, diffusely reflected, transmitted, or possibly refracted) ray is traced back for every point in the image, see e.g., [Rot 82], [Kaj 82], [ROG 85], [Sny 87], [Yan 87, 87a], [MÜLL 88], and [LEI 91]. This process requires, of course, a considerable amount of computation, since the ray tracing has to be carried out for every pixel in the image. This ray-tracing process can be done independently for each image point, however, and so can be done efficiently with parallel processing, see e.g., [Plu 85].

2 Basics From Geometry and Numerical Analysis

2.1. Differential Geometry of Curves and Surfaces

2.1.1. Parametrization of Curves

The image of an open, closed, half open, finite, or infinite interval I under a continuous, locally injective mapping into $\mathbb{R}^2$ or $\mathbb{R}^3$ is called a *curve*. If the image lies in $\mathbb{R}^2$, it is a plane curve, while if it is in $\mathbb{R}^3$, in general it will be a space curve. A curve can be considered as a set of points $\boldsymbol{P}$, with respect to a given origin O. These points can be regarded as vectors $\boldsymbol{P}$ which are the values of a locally one–to–one vector-valued function $\boldsymbol{X} = \boldsymbol{X}(t)$ of a parameter t defined on an interval I. The function $\boldsymbol{X}(t)$ is called the *parametrization* of the curve.

The parametrization

$$\boldsymbol{X}(t) = (\cos 3t \cos t,\, \cos 3t \sin t)^T, \qquad t \in [0, \pi],$$

provides an example of a plane curve called a *trochoidal curve*; see Fig. 2.1.

Given real numbers r and h and an integer n, the parametrization

$$\boldsymbol{X}(t) = (r \cos t,\, r \sin t,\, ht)^T, \qquad t \in [0, 2n\pi],$$

describes a space curve called a *helix*; see Fig. 2.2.

A parametrization $\boldsymbol{X}(t)$ can be transformed into a new parametrization $\boldsymbol{X}^*(t^*)$ via a *parameter transformation* $t = t(t^*)$. The two parametrizations are said to be *equivalent* provided that $t = t(t^*)$ is an invertible function. A parametrization is said to be *regular* provided that it is at least continuously differentiable, and

$$|\dot{\boldsymbol{X}}(t)| \neq 0 \qquad \text{for all} \quad t \in I.$$

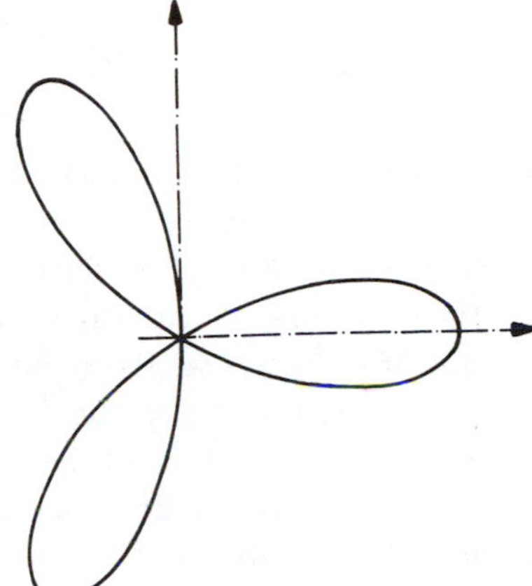

Fig 2.1. Trochoidal curve.

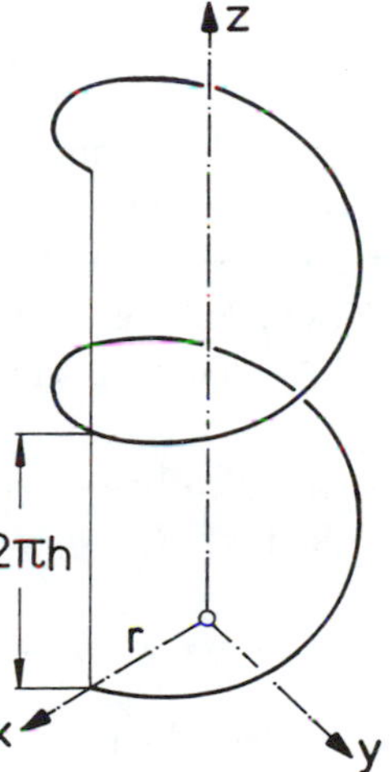

Fig. 2.2. Helix.

If $|\dot{X}(t)|$ has a zero at a point t_0, then in general the curve has a singularity (a cusp) at this point. The length of a curve can be computed as

$$L = \int_{t_0}^{t_1} |\dot{X}(t)| \, dt. \tag{2.1}$$

In particular, if $|\dot{X}(t)| = 1$, then the parameter t can be interpreted as the *arc length* s of the curve $\boldsymbol{X}$.

The deviation of a curve (which is at least twice continuously differentiable) from a straight line can be measured by its *curvature*

$$\kappa(t) = \frac{|\dot{\boldsymbol{X}}(t) \times \ddot{\boldsymbol{X}}(t)|}{|\dot{\boldsymbol{X}}(t)|^3}, \tag{2.2}$$

where $\boldsymbol{a} \times \boldsymbol{b}$ is the vector product (cross product) of the vectors $\boldsymbol{a}, \boldsymbol{b}$. If $\kappa = 0$ for all $t \in I$, then the curve reduces to a straight line; if $\kappa(t) = 0$ holds locally, then the curve has a point of inflection or a flat point. Since for an arbitrary curve, the curvature of a circle of radius r is given by $\kappa = 1/r$, we call $r = 1/\kappa$ the *radius of curvature* at the point $t = t_0$. The radius of curvature can be interpreted geometrically as the radius of the circle whose first and second derivatives agree with those of the curve $\boldsymbol{X}(t)$ at the point $t = t_0$.

We can measure the deviation of a curve (which is at least three times continuously differentiable) from being planar by its *torsion* τ, given by

$$\tau(t) = \frac{(\dot{\boldsymbol{X}}(t), \ddot{\boldsymbol{X}}(t), \dddot{\boldsymbol{X}}(t))}{|\dot{\boldsymbol{X}}(t) \times \ddot{\boldsymbol{X}}(t)|^2}, \tag{2.3}$$

where $(\cdot,\cdot,\cdot)$ denotes the determinant of the matrix formed by the three vector arguments. The torsion τ vanishes for planar curves whenever $\kappa \neq 0$. The functions κ and τ are *invariants* of a curve, *i.e.*, their values are independent of the parametrization. Further, $\kappa(t)$ and $\tau(t)$ (with $\kappa \neq 0$) form a *complete invariant system* [LAU 65], *i.e.*, they uniquely determine a curve (up to its position in $\mathbb{R}^3$).

In $\mathbb{R}^2$, curves can also be described as functions $y = f(x)$, or implicitly by an equation of the form $f(x, y) = 0$. For example, a line can be described by

$$y = mx + b,$$

and an ellipse can be described by

$$\frac{x^2}{a^2} + \frac{y^2}{b^2} = 1. \tag{$*$}$$

In $\mathbb{R}^2$, the appropriate functional form or implicit form can be found from the parametrization by elimination of parameters. For example, starting with the parametrization

$$x = a\cos t, \quad y = b\sin t, \qquad t \in [0, 2\pi],$$

of an ellipse, eliminating t leads to the implicit relationship $(*)$.

In the following, we will generally prefer the parametrization of curves because it gives us greater flexibility. Thus for us, a curve will be a vector-valued function $\boldsymbol{X}(t)$ of a parameter t, *i.e.*, the components of $\boldsymbol{X}$ will be real-valued functions of the parameter t.

2.1.2. Parametrization of Surfaces

Suppose G is a domain in the plane with parameters (u, v), for example, of the form

$$a \le u \le b, \qquad c \le v \le d,$$

and suppose we have a continuously differentiable and locally injective mapping $G \to \Phi$ which takes points (u, v) in G into $\mathbb{R}^3$. Then every point in the image set Φ can be described by a vector function $\boldsymbol{X}(u, v)$. $\boldsymbol{X}(u, v)$ is called the *parametrization* of the surface Φ, and u, v are called the parameters of this representation, see Fig. 2.3. The lines on the surface Φ corresponding to constant values of u and v are called the *parametric net*.

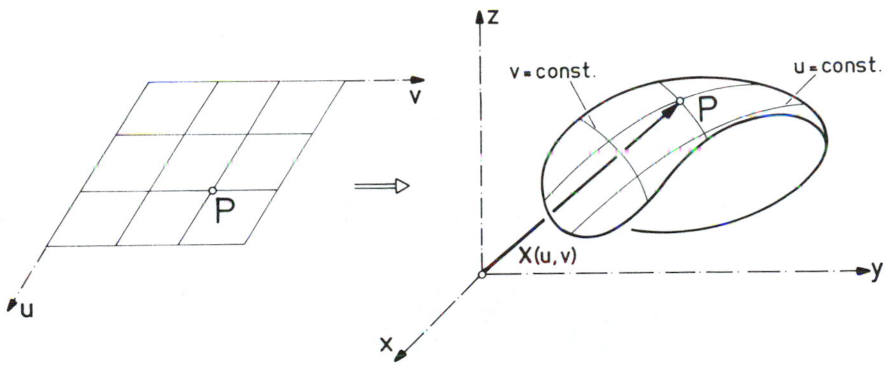

Fig. 2.3. Parametrization.

A parametrization is said to be *regular* provided that at every point of Φ, the normal vector is defined; i.e.,

$$\left| \frac{\partial \boldsymbol{X}}{\partial u} \times \frac{\partial \boldsymbol{X}}{\partial v} \right| \neq 0. \tag{2.4}$$

If $\left| \frac{\partial \boldsymbol{X}}{\partial u} \times \frac{\partial \boldsymbol{X}}{\partial v} \right|$ has a zero at the point $\boldsymbol{P}(u_0, v_0)$, then, in general, the surface has a *singularity* at $\boldsymbol{P}$ (a ridge or a cusp).

For surfaces, singularities may occur in their parametrization even when the surface itself has no singularity. This is the case, for example, at the poles of a sphere using the parametrization (2.6a). Such parametric singularities can be removed by reparametrization. We can reparametrize a surface using a transformation of the form

$$u = u(u^*, v^*), \qquad v = v(u^*, v^*). \tag{2.5}$$

Such a *parametric transformation* is called *admissible*, provided that it is invertible, *i.e.*, the associated functional determinant does not vanish on Φ.

We now give examples of parametrizations of three well-known surfaces:

1) *Sphere* (radius 1):

$$\boldsymbol{X} = (\cos u \cos v,\ \cos u \sin v,\ \sin u)^T, \qquad -\frac{\pi}{2} \le u \le \frac{\pi}{2}, \quad 0 \le v < 2\pi, \tag{2.6a}$$

2) *Cone*:

$$\boldsymbol{X} = (u \cos v,\ u \sin v,\ cu)^T, \qquad c = \text{const.}, \quad u \in \mathbb{R}, \quad 0 \le v < 2\pi, \tag{2.6b}$$

3) *Torus* (cf. Fig. 2.4):

$$\boldsymbol{X} = (r \cos v + \rho \cos v \cos u,\ r \sin v + \rho \sin v \cos u,\ \rho \sin u)^T,$$
$$\rho < r \text{ (const.)} \qquad u, v \in [0, 2\pi]. \tag{2.6c}$$

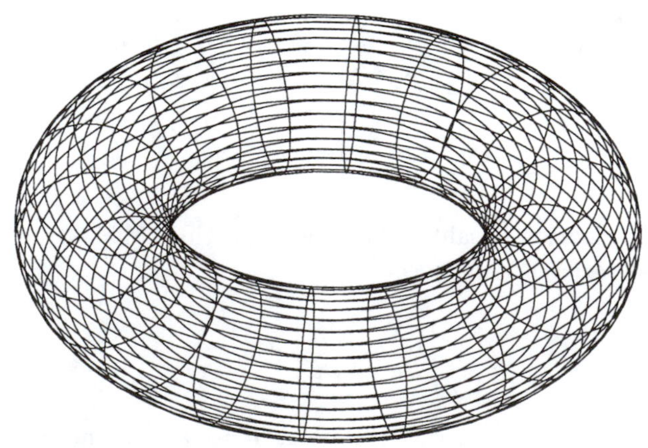

Fig. 2.4. Parallel projection of a torus.

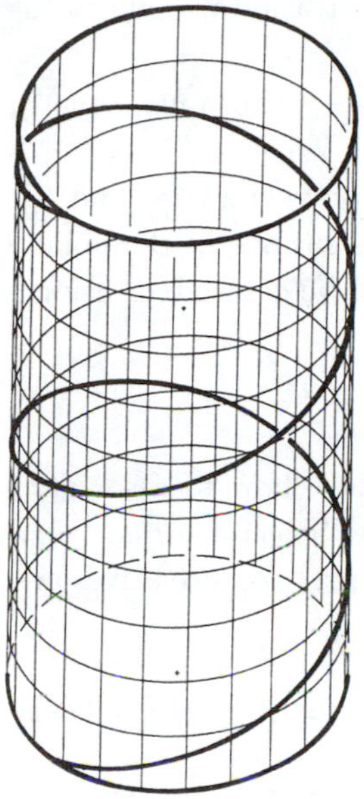

Fig 2.5. A helix as a curve on the surface of a cylinder.

Curves on a surface can be described by a function between the parameters; e.g.,

$$u = f(v) \qquad \text{or} \qquad u = u(t), \quad v = v(t).$$

For example, the equation $v = cu$ describes a *helix* on the surface of the circular cylinder of radius r given by

$$\boldsymbol{X} = (r\cos u,\ r\sin u, v)^T,$$

cf. Figs. 2.2 and 2.5.

Of course, surfaces can also be described as functions $z = z(x, y)$ or in implicit form as $f(x, y, z) = 0$. For example, we have

plane:	$ax + by + cz = d,$	*sphere:*	$x^2 + y^2 + z^2 = r^2,$
cone:	$c^2(x^2 + y^2) = z^2,$	*hyperboloid:*	$x^2 + y^2 - z^2 = 1.$

2.1.3. Curves on Surfaces and Curvature

The *tangent vector* to a surface curve $\boldsymbol{X}(u(t), v(t))$ can be computed as

$$\dot{\boldsymbol{X}} = \frac{\partial \boldsymbol{X}}{\partial u}\dot{u} + \frac{\partial \boldsymbol{X}}{\partial v}\dot{v}. \tag{2.7a}$$

In particular, the tangents to the parametric curves are given by

$$\boldsymbol{X}_u := \frac{\partial \boldsymbol{X}}{\partial u} \quad \text{(lines } v = \text{const.)}, \qquad \boldsymbol{X}_v := \frac{\partial \boldsymbol{X}}{\partial v} \quad \text{(lines } u = \text{const.)}. \tag{2.7b}$$

The two vectors $\frac{\partial \boldsymbol{X}}{\partial u}$ and $\frac{\partial \boldsymbol{X}}{\partial v}$ uniquely determine the plane which is tangent to the surface at the point $\boldsymbol{P}(u, v)$. The (normed) *normal vector* to the surface can be computed, using the vector product, as

$$\boldsymbol{N} = \frac{\boldsymbol{X}_u \times \boldsymbol{X}_v}{|\boldsymbol{X}_u \times \boldsymbol{X}_v|}. \tag{2.8}$$

We now show how to associate curvature values with each point $\boldsymbol{P}(u, v)$ on the surface. Select a surface curve $\boldsymbol{X}(s)$ which is parameterized by arc length. Then its first two derivatives are given by

$$\begin{aligned} \boldsymbol{X}'(s) &= \boldsymbol{X}_u u' + \boldsymbol{X}_v v', \\ \boldsymbol{X}''(s) &= \boldsymbol{X}_{uu}(u')^2 + 2\boldsymbol{X}_{uv} u'v' + \boldsymbol{X}_{vv}(v')^2 + \boldsymbol{X}_u u'' + \boldsymbol{X}_v v''. \end{aligned} \tag{2.9}$$

$\boldsymbol{X}''(s)$ is the curvature vector of the surface curve. Now if we introduce the surface trihedron $(\boldsymbol{X}', \boldsymbol{S}, \boldsymbol{N})$, where $\boldsymbol{S} = \boldsymbol{N} \times \boldsymbol{X}'$, then the curvature vector can be decomposed as follows:

$$\boldsymbol{X}''(s) = \kappa_n \boldsymbol{N} + \kappa_g \boldsymbol{S}. \tag{2.10}$$

κ_n is called the *normal curvature* of the surface, and κ_g is called the *geodesic curvature*. The normal curvature can be explicitly computed (with s as a parameter) as

$$\kappa_n = \boldsymbol{X}'' \cdot \boldsymbol{N} = \boldsymbol{X}_{uu} \cdot \boldsymbol{N}\, (u')^2 + 2\boldsymbol{X}_{uv} \cdot \boldsymbol{N}\, u'v' + \boldsymbol{X}_{vv} \cdot \boldsymbol{N}\, (v')^2. \tag{2.11a}$$

Now with respect to the parameter t, (2.11a) combined with the equality

$$ds^2 = |\dot{X}|^2\, dt^2 = E\, du^2 + 2F\, du\, dv + G\, dv^2, \tag{2.12}$$

(cf. (2.1) and (2.7a)) implies

$$\kappa_n = \frac{L(\dot{u})^2 + 2M\dot{u}\dot{v} + N(\dot{v})^2}{E(\dot{u})^2 + 2F\dot{u}\dot{v} + G(\dot{v})^2}. \tag{2.11b}$$

Here we have used the coefficients of the *first* and *second fundamental forms* defined by

$$\begin{aligned} E &= \boldsymbol{X}_u \cdot \boldsymbol{X}_u, \quad & F &= \boldsymbol{X}_u \cdot \boldsymbol{X}_v, \quad & G &= \boldsymbol{X}_v \cdot \boldsymbol{X}_v \\ L &= \boldsymbol{X}_{uu} \cdot \boldsymbol{N}, \quad & M &= \boldsymbol{X}_{uv} \cdot \boldsymbol{N}, \quad & N &= \boldsymbol{X}_{vv} \cdot \boldsymbol{N}. \end{aligned} \tag{2.13}$$

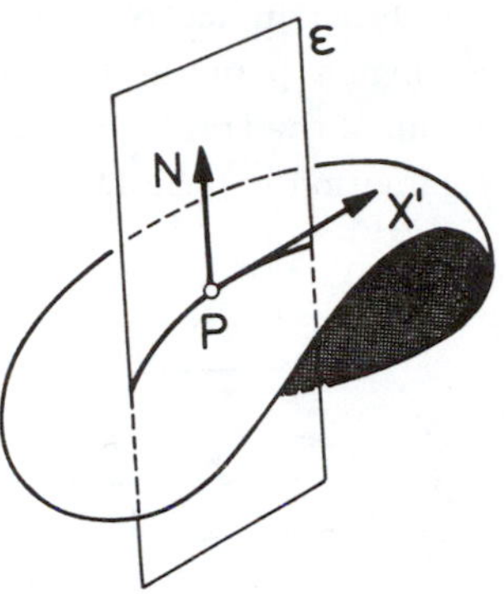

Fig. 2.6. Normal section.

The normal curvature also has a geometric interpretation. Suppose we cut the surface at $\boldsymbol{P}$ with a normal section plane ϵ, *i.e.*, with a plane which contains the normal vector at $\boldsymbol{P}$, cf. Fig. 2.6.

The plane ϵ intersects the surface along a plane curve $\boldsymbol{X}(s)$ whose curvature at $\boldsymbol{P}(u,v)$ is precisely the *normal curvature* κ_n of the surface in the direction $\dot{\boldsymbol{X}}(s)$ at that point. In general, κ_n varies as this section plane is rotated about the normal vector. Clearly, κ_n is periodic, and moreover, it has at most two extreme values, the *principal curvatures* κ_1 and κ_2. The principal curvatures, κ_1 and κ_2, and the normal curvature κ_n at an arbitrary point $\boldsymbol{P}(u,v)$ are related by the *Euler relation*

$$\kappa_n = \kappa_1 \cos^2\phi + \kappa_2 \sin^2\phi, \tag{2.14a}$$

where ϕ is the angle between the section and the first principal direction (which is the direction of the principal curvature κ_1). Now if we introduce the coordinates $\xi = \cos\phi/\sqrt{\kappa_n}$, $\eta = \sin\phi/\sqrt{\kappa_n}$, in the tangent plane at the point $\boldsymbol{P}(u,v)$, (2.14a) then reduces to the equation

$$1 = \kappa_1\xi^2 + \kappa_2\eta^2 \tag{2.14b}$$

of a conic section.

This conic section is called the *Dupin indicatrix*. The Dupin indicatrix characterizes the curvature behavior at a point $\boldsymbol{P}(u,v)$. In particular, if both principal curvatures have the same sign, then (2.14b) is an ellipse, and $\boldsymbol{P}$ is called an *elliptic point*; if one principal curvature is positive and the other negative, then $\boldsymbol{P}$ is called a *hyperbolic point*; if one of the principal curvatures vanishes, then $\boldsymbol{P}$ is called *parabolic*. If both principal curvatures vanish, then we have a *flat point*. The Dupin indicatrix can be obtained by a (small) parallel translation of the tangent plane at the point $\boldsymbol{P}$ in direction of the normal vector: the intersection of the translated tangent plane with the given surface produces (an approximation to) the Dupin indicatrix, see Fig. 2.7.

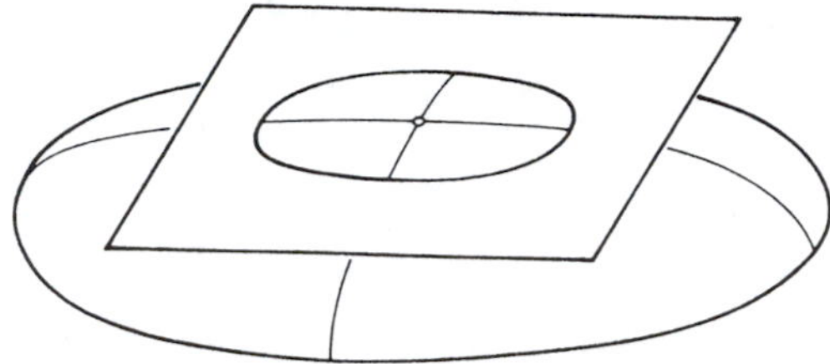

Fig. 2.7a. Surface patch with positive Gaussian curvature (elliptic surface point).

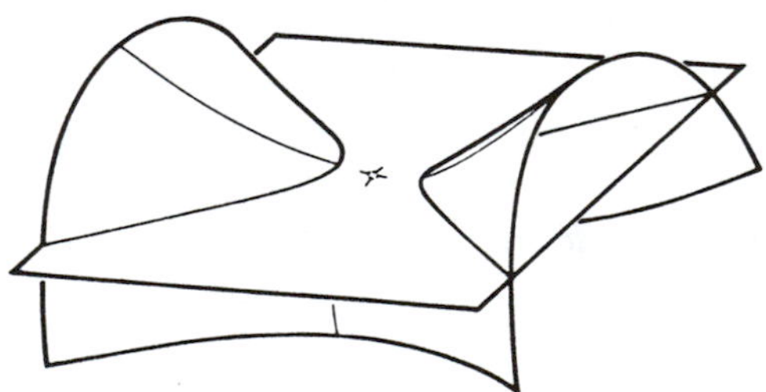

Fig. 2.7b. Surface patch with negative Gaussian curvature (hyperbolic surface point).

The product of the principal curvatures

$$K = \kappa_1 \cdot \kappa_2 \tag{2.15a}$$

is called the *Gaussian curvature*. The arithmetic mean of the principal curvatures

$$H = \tfrac{1}{2}(\kappa_1 + \kappa_2) \tag{2.15b}$$

is called the *mean curvature*. The Gaussian curvature can be used to classify surface patches. If $K > 0$ (elliptic points), then the surface patch has elliptic form, *i.e.*, it is convex (cf. Fig. 2.7a); if $K < 0$ (hyperbolic points), then

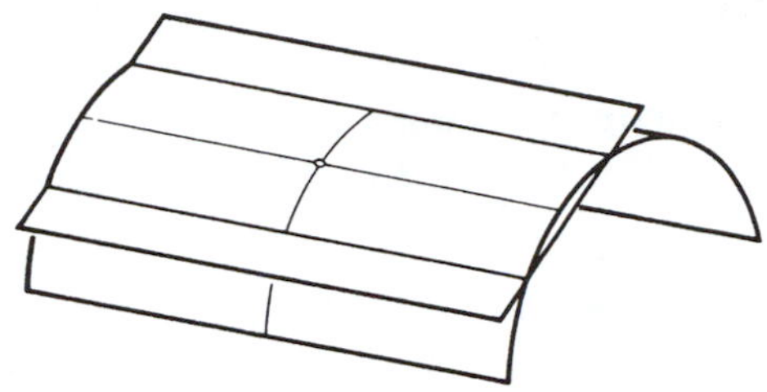

Fig. 2.7c. Surface patch with vanishing Gaussian curvature (parabolic surface point).

the surface patch has "saddle" form (cf. Fig. 2.7b); and if $K = 0$ (parabolic points), then the surface patch has "cylindric" form (cf. Fig. 2.7c).

Gaussian curvature and mean curvature can be calculated directly (see e.g., [LAU 65], [STR 50]):

$$K = \frac{LN - M^2}{EG - F^2} = \frac{(\boldsymbol{N}_u, \boldsymbol{N}_v, \boldsymbol{N})}{(\boldsymbol{X}_u, \boldsymbol{X}_v, \boldsymbol{N})}, \tag{2.16a}$$

$$2H = \frac{EN - 2FM + GL}{EG - F^2}. \tag{2.16b}$$

The directions of the normal sections with extremal normal curvature are perpendicular to each other, and are called *principal directions of curvature*. The corresponding vector field defines the *lines of curvature* which can be computed from the following differential equation (cf. [LAU 65], [STR 50]):

$$\begin{vmatrix} dv^2 & -du\,dv & du^2 \\ E & F & G \\ L & M & N \end{vmatrix} = 0. \tag{2.17}$$

An *example* of lines of curvature are the meridians and circles of latitude of a surface of revolution. This approach fails on the sphere, since there all normal curvatures are equal, *i.e.*, are extremal.

Surfaces with everywhere vanishing Gaussian curvature are called *developable*. They can be rolled out in the plane without distortion. *Examples* of developable surfaces include the *plane* itself, *cylinders*, *cones*, and the *tangent surfaces* generated by the tangents of a space curve, see e.g., [LAU 65], [STR 50].

Surface curves with vanishing normal curvature κ_n are called *asymptotic curves*. An example of asymptotic curves is provided by the straight lines on a ruled surface.

Surface curves with vanishing geodesic curvature κ_g are called *geodesics*. They are the locally shortest surface curves between two points on the surface.

The curvature κ_q *perpendicular* to a prescribed curve is of interest for applications in the geometry of cutting tool motion (see Chap. 16). First we consider the angle γ between two curve directions

$$\dot{\boldsymbol{X}} = \boldsymbol{X}_u \dot{u} + \boldsymbol{X}_v \dot{v}, \qquad \dot{\boldsymbol{X}}^* = \boldsymbol{X}_u \dot{u}^* + \boldsymbol{X}_v \dot{v}^*. \tag{2.18a}$$

Then γ is given by

$$\cos\gamma = \frac{\dot{\boldsymbol{X}} \cdot \dot{\boldsymbol{X}}^*}{|\dot{X}| \cdot |\dot{X}^*|} = \frac{E\dot{u}\dot{u}^* + F(\dot{u}\dot{v}^* + \dot{v}\dot{u}^*) + G\dot{v}\dot{v}^*}{\sqrt{E\dot{u}^2 + 2F\dot{u}\dot{v} + G\dot{v}^2}\sqrt{E(\dot{u}^*)^2 + 2F\dot{u}^*\dot{v}^* + G(\dot{v}^*)^2}}. \tag{2.18b}$$

If $\dot{\boldsymbol{X}}^*$ is orthogonal to $\dot{\boldsymbol{X}}$ ($\gamma = \frac{\pi}{2}$), then it follows that

$$\frac{\dot{u}^*}{\dot{v}^*} = -\frac{F\dot{u} + G\dot{v}}{E\dot{u} + F\dot{v}}. \tag{2.19}$$

Substituting (2.19) in (2.11b), it follows that the cross curvature of the curve $\boldsymbol{X}(t)$ is given by

$$\kappa_q = \frac{L(F\dot{u} + G\dot{v})^2 - 2M(F\dot{u} + G\dot{v})(E\dot{u} + F\dot{v}) + N(E\dot{u} + F\dot{v})^2}{E(F\dot{u} + G\dot{v})^2 - 2F(F\dot{u} + G\dot{v})(E\dot{u} + F\dot{v}) + G(E\dot{u} + F\dot{v})^2}. \tag{2.20a}$$

In particular, if we consider parametric lines, e.g., $u =$ const, we have

$$\kappa_q = \frac{LG^2 - 2MGF + NF^2}{G(EG - F^2)}. \tag{2.20b}$$

2.1.4. Special Types of Surfaces

If a curve $\boldsymbol{K}(u)$ is moved along another curve $\boldsymbol{M}(v)$, then the resulting surface

$$\boldsymbol{X}(u, v) = \boldsymbol{K}(u) + \boldsymbol{M}(v) \tag{2.21a}$$

is called a *translation surface*, see Fig. 2.8.

Translation surfaces have many applications; for example, many parts of automobile bodies, tools, molds for plastic parts, etc. are described by translation surfaces.

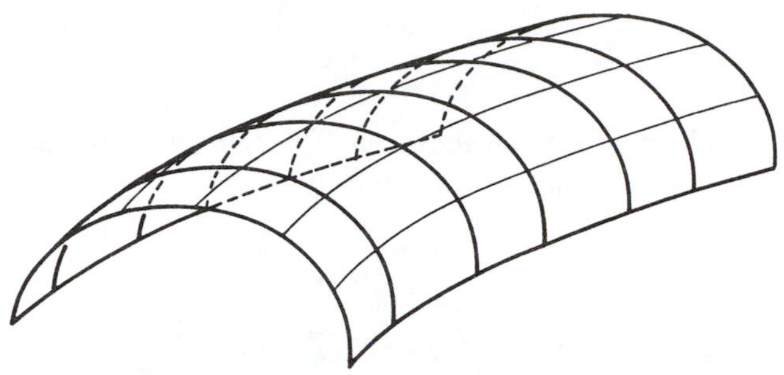

Fig. 2.8. Translation surface.

A variant of the translation surface is the *blending surface*: here two noncongruent (plane) curves $\boldsymbol{K}_1$ and $\boldsymbol{K}_2$ are appropriately connected, see e.g., [Woo 87], [Guj 88], [Fil 89a], [Rog 89], [Cho 90], and [Mar 90].

We now consider an example of a blending surface:

$$\begin{aligned} \boldsymbol{X}_1 &= (0,\, \cos t,\, \sin t)^T, \qquad t \in [0, \frac{\pi}{2}], \\ \boldsymbol{X}_2 &= (3,\, 2\sqrt[3]{\cos t},\, 2\sqrt[3]{\sin t})^T. \end{aligned} \tag{$*$}$$

A *linear* blending surface between the two curves can be obtained (cf. Fig. 2.9a) by taking a convex combination of the parametrizations $\boldsymbol{X}_i$:

$$\boldsymbol{X}(t, \lambda) = \boldsymbol{X}_1(t)(1-\lambda) + \boldsymbol{X}_2(t)\lambda, \qquad \lambda \in [0, 1]. \tag{2.21b}$$

Fig. 2.9a. The blending surface (2.21b) using parametrization (∗).

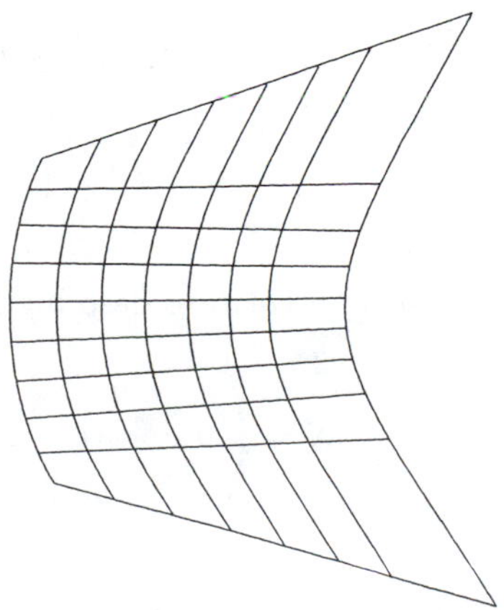

If in $(*)$ the parametrization of $\boldsymbol{X}_2$ is changed to

$$t = \frac{\pi}{2}\sin^2\tau, \qquad \tau \in [0, \frac{\pi}{2}], \qquad (**)$$

then we get the blending surface shown in Fig. 2.9b. The modified parametrization produces a different set of straight lines.

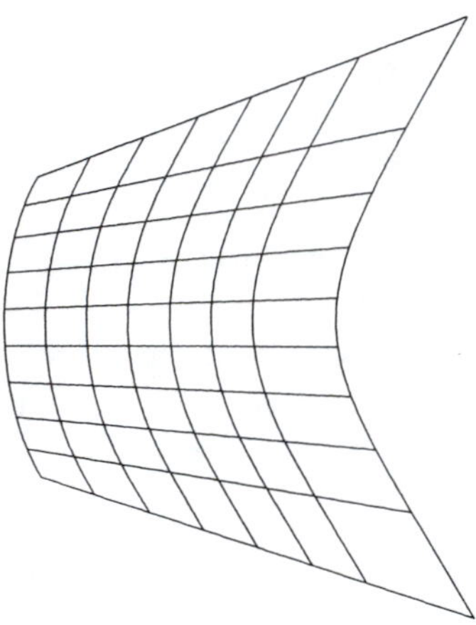

Fig. 2.9b. Blending surface (2.21b) after the parametric transformation $(**)$.

The blending surface described in this way is a piece of a ruled surface, where a *ruled surface* is one which is produced by continuous movement of a straight line. Such surfaces have the parametrization

$$\boldsymbol{X}(u,v) = \boldsymbol{r}(u) + v\boldsymbol{a}(u), \qquad \text{with } v \in \mathbb{R}. \qquad (2.22)$$

Examples of ruled surfaces are *cylindrical surfaces* described in parametric form by

$$\boldsymbol{X} = \begin{pmatrix} x(u) \\ y(u) \\ 0 \end{pmatrix} + v \begin{pmatrix} 0 \\ 0 \\ 1 \end{pmatrix},$$

and *hyperboloids of revolution* described by

$$\boldsymbol{X} = \begin{pmatrix} a\cos u \\ a\sin u \\ 0 \end{pmatrix} + v \begin{pmatrix} -\cos\gamma\sin u \\ \cos\gamma\cos u \\ \sin\gamma \end{pmatrix},$$

where γ is a constant. The hyperboloid of revolution is a special *surface of revolution*. A surface of revolution is produced by rotating a (planar) curve about some axis. If the rotation axis is the z-axis, for example, and the (planar) curve is given by $\boldsymbol{K} = (r(u), 0, f(u))^T$, then the resulting surface of revolution has the parametric form

$$\boldsymbol{X} = \begin{pmatrix} r(u)\cos v \\ r(u)\sin v \\ f(u) \end{pmatrix}, \qquad u \in I,\ v \in [0, 2\pi]. \tag{2.23}$$

Translation surfaces and surfaces of revolution are special cases of *swept surfaces*, which are the result of moving a given (planar) profile, surface, or volume along a prescribed curve $\boldsymbol{X}(s)$. The object which is being moved can be described relative to a coordinate system $(\boldsymbol{v}_1, \boldsymbol{v}_2, \boldsymbol{v}_3)$ associated with the given curve $\boldsymbol{X}(s)$, see [Klo 86]. For example, this coordinate system can be taken to be a parallel translation of a fixed coordinate system, or it could be the corresponding *Frenet frame* of the curve for which we have

$$\begin{aligned} \boldsymbol{v}_1 &= \boldsymbol{X}'(s) && \text{(tangent)}, \\ \boldsymbol{v}_2 &= \frac{\boldsymbol{X}''(s)}{|\boldsymbol{X}''(s)|} && \text{(principal normal)}, \\ \boldsymbol{v}_3 &= \boldsymbol{v}_1 \times \boldsymbol{v}_2 && \text{(binormal)}, \end{aligned} \tag{2.24a}$$

($|\boldsymbol{X}'| = 1$ implies $\boldsymbol{X}' \cdot \boldsymbol{X}'' = 0$, and thus $\boldsymbol{v}_1$ and $\boldsymbol{v}_2$ are orthogonal). If $\boldsymbol{X}$ is a curve on a surface $\boldsymbol{X}(u, v)$ and has the parametrization $u = u(t)$, $v = v(t)$, then the frame can be selected as

$$\begin{aligned} \boldsymbol{v}_1 &= \frac{\dot{\boldsymbol{X}}(t)}{|\dot{\boldsymbol{X}}(t)|} && \text{(tangent)}, \\ \boldsymbol{v}_2 &= \boldsymbol{N} && \text{(normal)}, \\ \boldsymbol{v}_3 &= \boldsymbol{v}_1 \times \boldsymbol{v}_2 && \text{(side vector)}. \end{aligned} \tag{2.24b}$$

If the profile to be moved is described in a Cartesian coordinate system by $\boldsymbol{P}(\tau) = (\xi(\tau), \eta(\tau), \zeta(\tau))^T$, then the parametrization of the swept surface, obtained by moving $\boldsymbol{P}(\tau)$ along the curve $\boldsymbol{X}(s)$, is given by

$$\boldsymbol{X}(s, \tau) = \boldsymbol{X}(s) + \xi(\tau)\boldsymbol{v}_1 + \eta(\tau)\boldsymbol{v}_2 + \zeta(\tau)\boldsymbol{v}_3. \tag{2.25}$$

Fig. 2.10 shows a swept surface obtained from a curve on a surface, and a moving profile which is described in terms of the coordinate system (2.24b).

If a surface or a volume is moved along a curve $\boldsymbol{X}(s)$, then the swept surface is the hull of the moving surface (resp. volume), also called the *envelope*, see e.g., [Mar 90]. A detailed discussion of the envelope conditions is given in Chapter 16 where the geometry of cutting tools is considered.

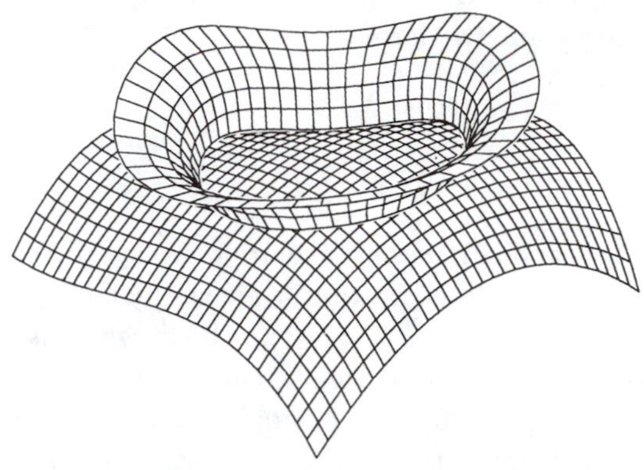

Fig. 2.10. Swept surface along a surface curve.

2.2. Interpolation of Curves and Surfaces

The classical interpolation problem involves replacing a "complicated" function, $y = f(x)$ or $z = f(x, y)$, by a "simpler" function, $y = a(x)$ or $z = a(x, y)$, in such a way that the interpolating function and the given function f have the same values at a prescribed set of points. For our applications, however, we are not only interested in interpolating functions, but also profiles, curves on surfaces, and surface patches of given objects. These might include everyday objects, profiles of tools, surfaces of car bodies or turbine blades, and airplane parts, and also less classical things like flow lines and moving boundaries arising in chemical processes.

Such curves and surfaces are described by points which either come from some mathematical construction, or are obtained as data using some mechanical device such as a digitizer. Thus, in general, we are given a table of values of the unknown function f as follows:

x_0	x_1	$\ldots$	x_n
f_0	f_1	$\ldots$	f_n

	y_0	$\ldots$	y_m
x_0	$f_{0,0}$	$\ldots$	$f_{0,m}$
$\vdots$	$\vdots$		$\vdots$
x_n	$f_{n,0}$	$\ldots$	$f_{n,m}$

In this chapter, we shall discuss interpolation of curves and surfaces given in parametric form (since this is more common in practice), rather than functions of the form $y = f(x)$ or $z = f(x, y)$. Thus, our interpolation problems correspond to tables of values of the form

t_0	t_1	...	t_n
$\boldsymbol{P}_0$	$\boldsymbol{P}_1$	...	$\boldsymbol{P}_n$

	v_0	v_1	...	v_m
u_0	$\boldsymbol{P}_{00}$	$\boldsymbol{P}_{01}$	...	$\boldsymbol{P}_{0m}$
u_1	$\boldsymbol{P}_{10}$	$\boldsymbol{P}_{11}$	...	$\boldsymbol{P}_{1m}$
$\vdots$	$\vdots$		...	$\vdots$
u_n	$\boldsymbol{P}_{n0}$	$\boldsymbol{P}_{n1}$	...	$\boldsymbol{P}_{nm}$

where we denote the parameter values corresponding to points $\boldsymbol{P}_i$ on a curve by t_i, and the parameter values corresponding to points $\boldsymbol{P}_{ik}$ on a surface by (u_i, v_k).

2.2.1. Interpolation of Curves Using Monomials

Given $n+1$ pairwise distinct points $\boldsymbol{P}_i$ in $\mathbb{R}^3$, $i = 0(1)n$, associated with (appropriately selected) parameter values t_i, we are looking for a polynomial

$$\boldsymbol{p}(t) = \sum_{j=0}^{n} \boldsymbol{A}_j t^j, \qquad \boldsymbol{A}_j \in \mathbb{R}^3, \quad t \in [a, b], \tag{2.26}$$

so that

$$\boldsymbol{p}(t_i) = \boldsymbol{P}_i, \qquad i = 0(1)n. \tag{2.27}$$

We refer to the points $\boldsymbol{P}_i$ as *interpolation points*, and to the t_i as *parameter values.* The pairs $(t_i, \boldsymbol{P}_i)$ describe the *nodes* of the interpolation problem.

We have formulated the interpolation problem in *vector-valued* form. The interpolation of functions in $\mathbb{R}^2$ corresponds to choosing the interpolation points as $f(x_i)$, and the associated parameter values to be the x-coordinates x_i, cf. Fig. 2.11.

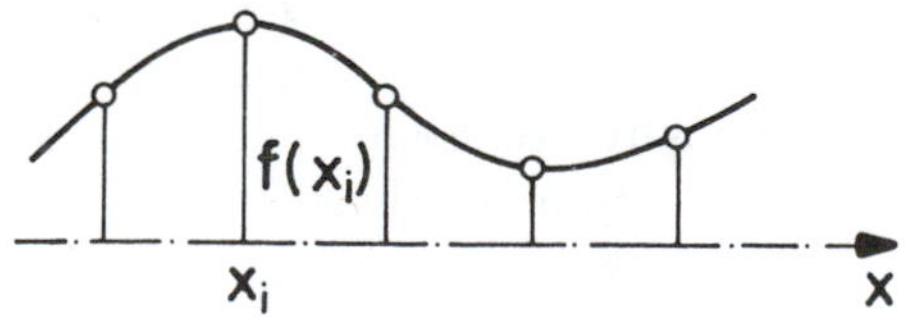

Fig. 2.11. Interpolation.

In (2.26) we have chosen the *monomials* as basis functions. Later we shall introduce others. The monomials m_k form a basis since the functions t^k with different exponents are always linearly independent.

In order to find a solution of the interpolation problem (2.27), we substitute (2.26) in (2.27) to obtain the following linear system of equations:

$$\boldsymbol{P}_i = \sum_{j=0}^{n} \boldsymbol{A}_j m_j(t_i), \qquad i = 0(1)n. \tag{2.28}$$

Concerning the solution of (2.28), we have

Theorem 2.1. *Given $n+1$ nodes $(t_i, \boldsymbol{P}_i)$ with pairwise distinct parameter values t_i, $i = 0(1)n$, there exists exactly one polynomial*

$$\boldsymbol{p}(t) = \sum_{j=0}^{n} \boldsymbol{A}_j m_j(t)$$

satisfying

$$\boldsymbol{P}_i = \boldsymbol{p}(t_i) = \sum_{j=0}^{n} \boldsymbol{A}_j (t_i)^j, \qquad i = 0(1)n.$$

Proof: We write (2.28) in matrix form

$$\begin{pmatrix} 1 & t_0 & \dots & t_0^n \\ \vdots & & & \vdots \\ 1 & t_n & \dots & t_n^n \end{pmatrix} \begin{pmatrix} \boldsymbol{A}_0 \\ \vdots \\ \boldsymbol{A}_n \end{pmatrix} = \begin{pmatrix} \boldsymbol{P}_0 \\ \vdots \\ \boldsymbol{P}_n \end{pmatrix}, \tag{$*$}$$

with unknown vectors $\boldsymbol{A}_j$. The matrix arising here is the well-known Vandermonde matrix, which is nonsingular for any choice of pairwise distinct t_i. This means that the $\boldsymbol{A}_j$ are uniquely determined, cf. e.g., [ENG 85]. ■

Remark. For high values of n, the matrix appearing in $(*)$ is usually not very well conditioned.

Remark. It is computationally expensive to change one of the knots, since then the entire system $(*)$ has to be resolved.

Remark. The same theorem holds, of course, if other basis functions for the space of polynomials are used in place of the monomials in (2.28), and if properly chosen, they may have computational advantages, see Section 4.4.

2.2.2. Interpolation of Curves with Lagrange Polynomials

We can eliminate some of the disadvantages of using monomials for interpolation by choosing other basis functions. In this section we shall work with a different set of polynomial basis functions L_i of degree n defined by the property that

$$L_i(t_k) = \delta_{ik} := \begin{cases} 1, & \text{for } i = k \\ 0, & \text{for } i \neq k, \end{cases} \tag{2.29}$$

where δ_{ik} is the Kronecker delta. This condition assures that L_i has value 1 at $t = t_i$ and vanishes at the other parameter values, see Fig. 2.12. The existence of unique polynomials L_i satisfying (2.29) follows from Theorem 2.1, and we refer to them as *Lagrange polynomials*. They can be constructed explicitly as follows:

$$\begin{aligned} L_k(t) &= \frac{(t-t_0)(t-t_1)\cdots(t-t_{k-1})(t-t_{k+1})\cdots(t-t_n)}{(t_k-t_0)(t_k-t_1)\cdots(t_k-t_{k-1})(t_k-t_{k+1})\cdots(t_k-t_n)} \\ &= \prod_{\substack{j=0 \\ j\neq k}}^{n} \frac{(t-t_j)}{(t_k-t_j)}. \end{aligned} \tag{2.30}$$

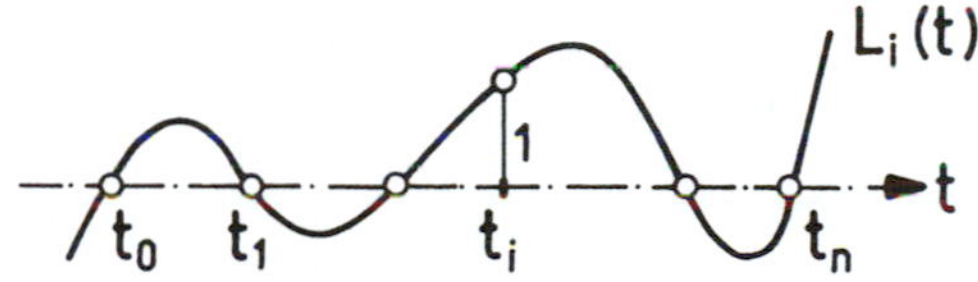

Fig. 2.12. Lagrange polynomials.

In terms of the Lagrange polynomials, the solution of the interpolation problem is given by

$$\boldsymbol{p}(t) = \sum_{j=0}^{n} L_j(t)\boldsymbol{P}_j, \tag{2.31}$$

with $\boldsymbol{P}_j$ the given interpolation points.

Example 1. Suppose we want to interpolate the points $\boldsymbol{P}_i = (x_i, y_i) \in \mathbb{R}^2$ given in the following table:

t_i	0	1	2
x_i	0	1	2
y_i	8	5	3

The associated Lagrange polynomials are then

$$L_0 = \frac{(t-1)(t-2)}{2}, \qquad L_1 = -t(t-2), \qquad L_2 = \frac{t(t-1)}{2},$$

and the Lagrange formula for the solution is given by

$$\boldsymbol{p}(t) = \boldsymbol{P}_0 \frac{(t-1)(t-2)}{2} - \boldsymbol{P}_1 t(t-2) + \boldsymbol{P}_2 \frac{t(t-1)}{2}.$$

There remains, however, one disadvantage in using Lagrange polynomials to solve the interpolation problem: if we add one additional knot, then all of the basis functions have to be recomputed. This disadvantage can be eliminated if we use the Newton interpolation method discussed in the following section.

2.2.3. Interpolation of Curves with Newton Polynomials

The *Newton polynomials* are defined by

$$n_i(t) := (t-t_0)(t-t_1)\cdots(t-t_{i-1}), \tag{2.32}$$

with $n_0(t) = 1$, see Fig. 2.13. Now suppose we look for a solution of the interpolation problem in the form

$$\boldsymbol{p}(t) = \sum_{j=0}^{n} n_j(t)\boldsymbol{A}_j. \tag{2.33}$$

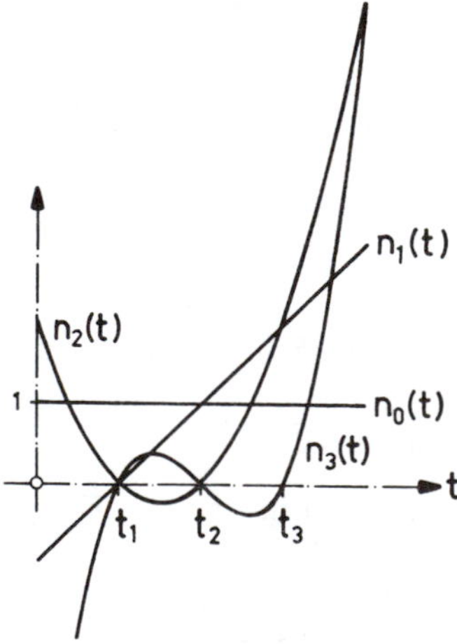

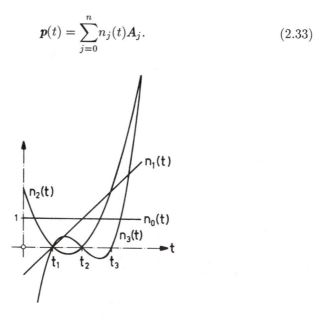

Fig. 2.13. Newton polynomials $n_0, \ldots, n_3$.

The coefficients $\boldsymbol{A}_j$ in (2.33) can be found directly:

$$\begin{aligned} \boldsymbol{p}(t_0) &= \boldsymbol{P}_0 = \boldsymbol{A}_0, \\ \boldsymbol{p}(t_1) &= \boldsymbol{P}_1 = \boldsymbol{A}_0 + \boldsymbol{A}_1(t_1 - t_0), \\ \boldsymbol{p}(t_2) &= \boldsymbol{P}_2 = \boldsymbol{A}_0 + \boldsymbol{A}_1(t_2 - t_0) + \boldsymbol{A}_2(t_2 - t_0)(t_2 - t_1), \end{aligned}$$

which gives

$$\boldsymbol{A}_0 = \boldsymbol{P}_0, \qquad \boldsymbol{A}_1 = \frac{\boldsymbol{P}_1 - \boldsymbol{P}_0}{t_1 - t_0}, \tag{2.34}$$

$$\boldsymbol{A}_2 = \frac{\boldsymbol{P}_2 - \boldsymbol{P}_0 - \dfrac{(\boldsymbol{P}_1 - \boldsymbol{P}_0)(t_2 - t_0)}{(t_1 - t_0)}}{(t_2 - t_0)(t_2 - t_1)} = \frac{\dfrac{\boldsymbol{P}_2 - \boldsymbol{P}_1}{t_2 - t_1} - \dfrac{\boldsymbol{P}_1 - \boldsymbol{P}_0}{t_1 - t_0}}{t_2 - t_0},$$

etc. To make this computation systematic, we introduce the idea of a *divided difference*:

$$[t_i t_k] := \frac{\boldsymbol{P}_i - \boldsymbol{P}_k}{t_i - t_k}. \tag{$*$}$$

Then the coefficients of (2.33) can be written as

$$\begin{aligned} \boldsymbol{A}_0 &= \boldsymbol{P}_0, \\ \boldsymbol{A}_1 &= [t_1 t_0] = \frac{\boldsymbol{P}_1 - \boldsymbol{P}_0}{t_1 - t_0}, \\ \boldsymbol{A}_2 &= [t_2 t_1 t_0] = \frac{[t_2 t_1] - [t_1 t_0]}{t_2 - t_0}, \\ \boldsymbol{A}_3 &= [t_3 t_2 t_1 t_0] = \frac{[t_3 t_2 t_1] - [t_2 t_1 t_0]}{t_3 - t_0}, \\ &\vdots \\ \boldsymbol{A}_n &= [t_n t_{n-1} \ldots t_1 t_0] = \frac{[t_n \ldots t_1] - [t_{n-1} \ldots t_0]}{t_n - t_0}. \end{aligned} \tag{2.35}$$

It is clear that if we add a new point $\boldsymbol{P}_{n+1}$, then we simply have to carry out one more step in this divided difference scheme.

2.2.4. Other Solutions of the Interpolation Problem for Curves

In this section we describe several other interpolation methods based on

- other choices of knots,
- other basis functions.

2.2.4.1. Hermite Interpolation

Suppose again that we are given $n+1$ pairwise distinct parameter values t_k, $k = 0(1)n$, and that, in addition, for each abscissa t_k we are given $m_k + 1$ vectors $\boldsymbol{P}_k, \boldsymbol{P}'_k, \ldots, \boldsymbol{P}_k^{m_k}$ which the curve $\boldsymbol{H}(t)$ is to interpolate in the sense that

$$\boldsymbol{H}(t_k) = \boldsymbol{P}_k, \quad \boldsymbol{H}'(t_k) = \boldsymbol{P}'_k, \quad \ldots, \quad \boldsymbol{H}^{(m_k)}(t_k) = \boldsymbol{P}_k^{m_k}. \tag{2.36}$$

Here we assume $\boldsymbol{H}(t)$ belongs to $C^{\max(m_k)}$. This problem is referred to as the *Hermite interpolation problem.* In this section we are looking for an interpolating polynomial $\boldsymbol{H}(t)$, of lowest possible degree, satisfying the conditions (2.36). It turns out (see e.g., [BER 70] or [FIN 77]) that there is always a unique polynomial $\boldsymbol{H}(t)$ of degree at most

$$N = \sum_{k=0}^{n}(m_k + 1) - 1 \tag{2.36a}$$

satisfying (2.36).

We now give an explicit solution of the Hermite interpolation problem in the case where $m_0 = m_1 = \cdots = m_n = 1$. By (2.36a), in this case we should look for a polynomial of degree $N = 2n + 1$. Suppose we write

$$\boldsymbol{H}(t) = \sum_{j=0}^{n} \boldsymbol{P}_j f_{nj}(t) + \sum_{j=0}^{n} \boldsymbol{P}'_j g_{nj}(t),$$

where $f_{nj}(t)$ and $g_{nj}(t)$ are polynomials of degree $2n+1$. Then with $m_k = 1$, to satisfy (2.36) we need

$$\begin{aligned} f_{nj}(t_k) &= \delta_{jk}, & g_{nj}(t_k) &= 0, \\ f'_{nj}(t_k) &= 0, & g'_{nj}(t_k) &= \delta_{jk}, \end{aligned} \qquad j, k = 0(1)n. \tag{2.37}$$

Polynomials satisfying (2.37) can be constructed in terms of the Lagrange polynomials

$$L_{nk}(t) = \prod_{\substack{j=0 \\ k\neq j}}^{n} \frac{t - t_j}{t_k - t_j}, \qquad j = 0(1)n,$$

as follows:

$$f_{nk}(t) = [1 - 2L'_{nk}(t_k)(t - t_k)]L_{nk}^2(t), \qquad g_{nk}(t) = (t - t_k)L_{nk}^2(t),$$

for $k = 0(1)n$. With a "good" choice of derivatives at the boundary points, the Hermite interpolant can produce excellent results. On the other hand, with a "poor" choice of derivatives there, the results can be disastrous. In Fig. 2.14a we show the effect of changing the length of the tangent vector at the rightmost abscissa, and in Fig.2.14b we show what happens when the direction of that tangent vector is varied.

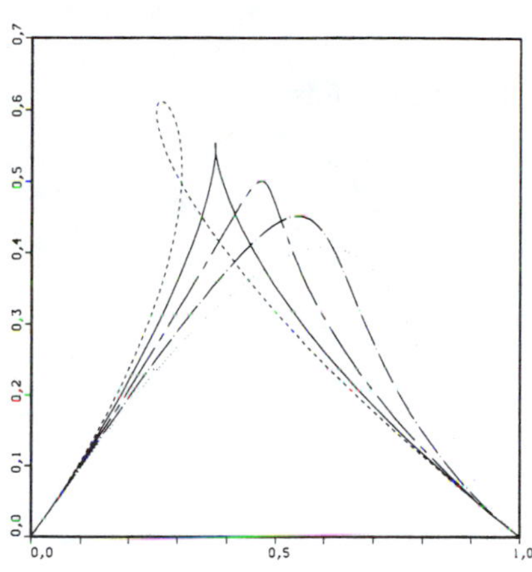

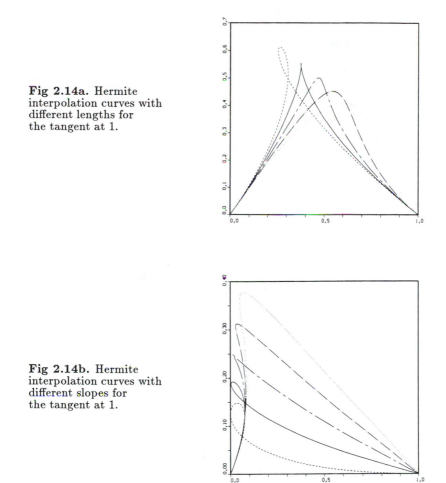

Fig 2.14a. Hermite interpolation curves with different lengths for the tangent at 1.

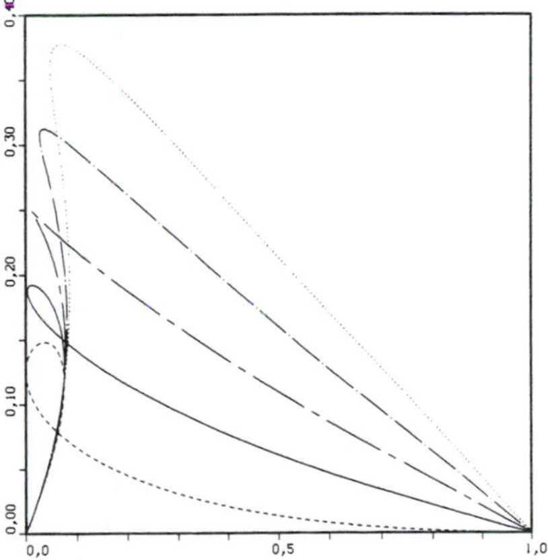

Fig 2.14b. Hermite interpolation curves with different slopes for the tangent at 1.

The problem of interpolating data where not all consecutive derivatives are given at each parameter value (*i.e.*, there are gaps in the derivative information) is called *lacunary interpolation*.

Interpolation problems can, of course, also be solved using other basis functions such as trigonometric polynomials, Chebychev polynomials, etc.

2.2.4.2. Rational Interpolation

Polynomials have the tendency to oscillate between the interpolation points, which is undesirable for applications. Rational functions tend to exhibit less of this problem, and have the additional advantage that we can also interpolate in the neighborhood of poles. This suggests that we look for interpolating functions in the form

$$x(t) = \frac{\sum_{j=0}^{n} a_j t^j}{\sum_{j=0}^{n} d_j t^j}, \quad y(t) = \frac{\sum_{j=0}^{n} b_j t^j}{\sum_{j=0}^{n} d_j t^j}, \quad z(t) = \frac{\sum_{j=0}^{n} c_j t^j}{\sum_{j=0}^{n} d_j t^j}, \tag{2.38}$$

with $a_j, b_j, c_j, d_j \in \mathbb{R}$ and $t \in I$. Here, for simplicity we have chosen the numerator and denominator polynomials to have the same degree, but of course, this is not necessary in general. A general solution of the rational interpolation problem based on continued fractions can be found in [WER 79a].

The conditions (2.38) can also be interpreted as polynomial interpolation in $\mathbb{R}^4$ if we work with *homogeneous coordinates*. Let

$$\begin{aligned} x_0(t) &= \sum_{j=0}^{n} d_j t^j, & x_1(t) &= \sum_{j=0}^{n} a_j t^j, \\ x_2(t) &= \sum_{j=0}^{n} b_j t^j, & x_3(t) &= \sum_{j=0}^{n} c_j t^j. \end{aligned} \tag{2.39a}$$

We must also write the points $\boldsymbol{P}_i$ to be interpolated in homogeneous form, which means that the coordinate of each point must be multiplied by a factor ρ_i. These factors also have to be computed, which in general is a nonlinear problem.

The rational interpolation problem can, however, be solved in terms of a linear system in some special cases, including the cases where four points are given in $\mathbb{R}^2$ and $n = 2$, and where five points are given in $\mathbb{R}^3$ and $n = 3$. To show this, suppose we rewrite (2.39a) in matrix form as

$$\boldsymbol{X}^T = \boldsymbol{F}(t)^T \boldsymbol{K}, \tag{2.39b}$$

where $\boldsymbol{F}(t) = (1, t, t^2, t^3)^T$ is the vector of basis monomials t^k, $k = 0(1)3$, (for the case of $\mathbb{R}^3$), and where $\boldsymbol{K}$ is the matrix of coefficients. Suppose the points

$\boldsymbol{P}_i$ correspond to the parameter values t_i and (unknown) homogenization factors ρ_i. Inserting the points $\boldsymbol{P}_0, \ldots, \boldsymbol{P}_3$ in (2.39b) leads to the system

$$\begin{pmatrix} \rho_0 \boldsymbol{P}_0^T \\ \rho_1 \boldsymbol{P}_1^T \\ \rho_2 \boldsymbol{P}_2^T \\ \rho_3 \boldsymbol{P}_3^T \end{pmatrix} = \begin{pmatrix} \boldsymbol{F}(t_0)^T \\ \boldsymbol{F}(t_1)^T \\ \boldsymbol{F}(t_2)^T \\ \boldsymbol{F}(t_3)^T \end{pmatrix} \boldsymbol{K} =: \boldsymbol{M}\boldsymbol{K},$$

which implies

$$\boldsymbol{K} = \boldsymbol{M}^{-1} \begin{pmatrix} \rho_0 \boldsymbol{P}_0^T \\ \rho_1 \boldsymbol{P}_1^T \\ \rho_2 \boldsymbol{P}_2^T \\ \rho_3 \boldsymbol{P}_3^T \end{pmatrix} =: \boldsymbol{M}^{-1}\boldsymbol{P}(\rho_k).$$

If the fifth point $\boldsymbol{P}_4$ has the normalization factor $\rho_4 = 1$, then substitution leads to

$$\boldsymbol{P}_4 = \boldsymbol{F}(t_4)^T \boldsymbol{M}^{-1} \boldsymbol{P}(\rho_k),$$

which is a linear system for the four unknown factors ρ_i, $i = 1(1)4$.

2.2.5. Interpolation of Surfaces

In this section we consider approximating a surface in $\mathbb{R}^3$ by an interpolating surface $\boldsymbol{X}(u, v)$. Suppose we are given $N + 1$ points $\boldsymbol{P}_i$ in $\mathbb{R}^3$ associated with parameter values (u_i, v_i), $i = 0(1)N$. We seek a polynomial

$$\boldsymbol{X}(u, v) = \sum_{j=0}^{m} \sum_{k=0}^{n} \boldsymbol{A}_{jk} u^j v^k \tag{2.40}$$

of lowest possible degree which interpolates the given points $\boldsymbol{P}_i$ at the parameter values (u_i, v_i).

In general, we cannot be assured of the existence and uniqueness of a solution of this interpolation problem since the corresponding linear system of equations can in some cases be singular; e.g., if the given parameter values lie on certain *forbidden curves* in the (u, v) parameter plane, see [BER 70], [Wun 77].

As an example, consider the case $m = n = 2$, $\max(j+k) = 2$; cf. (2.40). In this case, the determinant of the linear system of equations for the coefficients has the form

$$\det \begin{pmatrix} 1 & u_0 & v_0 & u_0^2 & u_0 v_0 & v_0^2 \\ \vdots & & & & & \vdots \\ 1 & u_N & v_N & u_N^2 & u_N v_N & v_N^2 \end{pmatrix},$$

and this determinant vanishes for any choice of six points lying on a conic section, see [GRO 57].

Because of these difficulties, we need to make some additional assumptions on the nature of the data: We assume either

a) the parameter values of the data points are intersection points of the parameter lines of the surface to be constructed,

   or

b) the parameter values have *generic position*, that is, they do not lie on forbidden curves.

We also assume that the polynomial (2.40) is *complete* in the sense that all terms up to the highest order term $\boldsymbol{A}_{mn}u^m v^n$ are included.

Under hypothesis (a), the parameter values can be expressed in matrix form

$$\begin{matrix} (u_0, v_0) & (u_0, v_1) & \cdots & (u_0, v_n) \\ (u_1, v_0) & (u_1, v_1) & \cdots & (u_1, v_n) \\ \vdots & & & \vdots \\ (u_m, v_0) & (u_m, v_1) & \cdots & (u_m, v_n). \end{matrix}$$

In this case, the interpolation conditions lead to an $(n+1) \times (m+1)$ non-singular linear system of equations for the unknowns $\boldsymbol{A}_{ik}$ in (2.40). Under this hypothesis, we can also solve the interpolation problem using the Lagrange or Newton polynomials, see e.g., [ENG 85]. In terms of Lagrange polynomials $L_i(u)$, $L_k(v)$, the surface (2.40) can be written in *tensor-product form* as

$$\boldsymbol{X}(u, v) = \sum_{i=0}^{m} \sum_{k=0}^{n} L_i(u) L_k(v) \boldsymbol{P}_{ik}. \tag{2.41}$$

In case (b), the interpolation problem can be solved if the interpolation polynomial is complete, and if the $(n+1) \times (m+1)$ points $\boldsymbol{P}_i$, as well as the associated parameter values (u_i, v_i), are all pairwise distinct.

Interpolation with B-splines will be discussed in Sections 4.4 and 6.2.5. The problem of interpolating scattered data is treated in Chapter 9.

2.2.6. Error Bounds for the Approximation of Curves by Interpolation

Suppose we have constructed a polynomial $\boldsymbol{P}_n(t)$ of degree n which interpolates a curve $\boldsymbol{X}(t)$ at a given set of $n+1$ points $\boldsymbol{P}_i$, $i = 0(1)n$. In this section we consider the problem of estimating the size of the deviation between the interpolating polynomial $\boldsymbol{P}_n(t)$ of degree n and the given curve $\boldsymbol{X}(t)$. Let

$$\boldsymbol{X}(t) = \boldsymbol{P}_n(t) + \boldsymbol{R}_{n+1}(t), \tag{2.42}$$

where $\boldsymbol{R}_{n+1}(t)$ is the remainder term, see e.g., [ENG 85]. This remainder term must vanish for the parameter values $t_0, \ldots, t_n$, and so has the general form shown in Fig. 2.15.

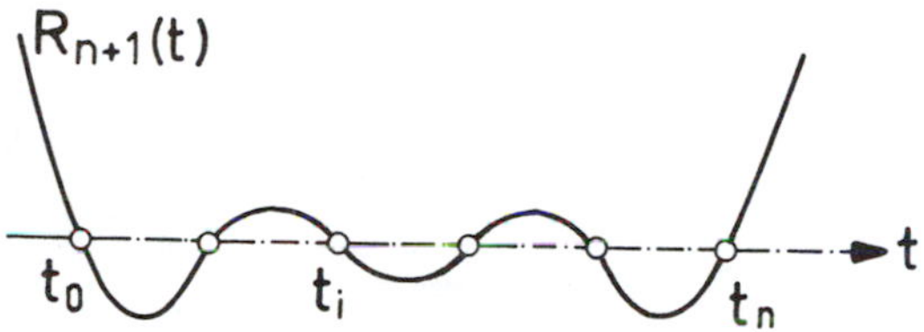

Fig. 2.15. Shape of the remainder term.

It can be shown (see e.g., [GOU 74]) that the remainder term is equal to

$$\boldsymbol{R}_{n+1}(t) = (t - t_0)(t - t_1) \cdots (t - t_n) \frac{\bar{\boldsymbol{X}}^{(n+1)}}{(n+1)!}, \tag{2.43}$$

where $\bar{\boldsymbol{X}}^{(n+1)}$ is a vector whose components have the form $\boldsymbol{X}_i^{(n+1)}(\xi_i)$, with ξ_i a (in general unknown) point in the interval $[t_0, t_n]$. For example, this remainder expression was used in [Hölz 83], [Dan 85] to give error bounds for spline transformations.

2.2.7. Comparison of the Various Interpolation Methods

First, we emphasize that for any given set of interpolation knots, there is one and only one interpolating polynomial. Thus, we have to compare the methods from a practical computational standpoint.

The Lagrange formula is simple, but has the disadvantages that the individual Lagrange polynomials are complicated and depend on the location of the parameter values, and thus, all of them have to be recomputed whenever we modify any one of the parameter values. For practical applications, the Newton formula is more convenient.

We would also like to point out that interpolation using polynomials of degrees $n > 5$ is not recommended, since higher degree polynomials have a strong tendency to oscillate, see [Boni 76]. The example in Fig. 2.16 shows this phenomenon clearly, see [Run 01]. Here we are interpolating the function $y = 1/(1 + x^2)$ on $-5 \leq x \leq 5$ at an increasing number of equally spaced points (with polynomials of increasing degree). The figures show the increasing tendency to oscillate as the degree of the polynomial increases. Interpolation of curves and surfaces using splines is treated in Sections 4.4 and 6.2.5.

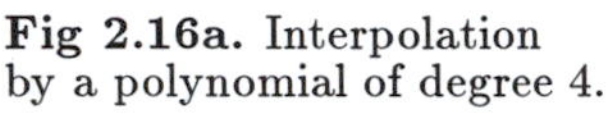

Fig 2.16a. Interpolation by a polynomial of degree 4.

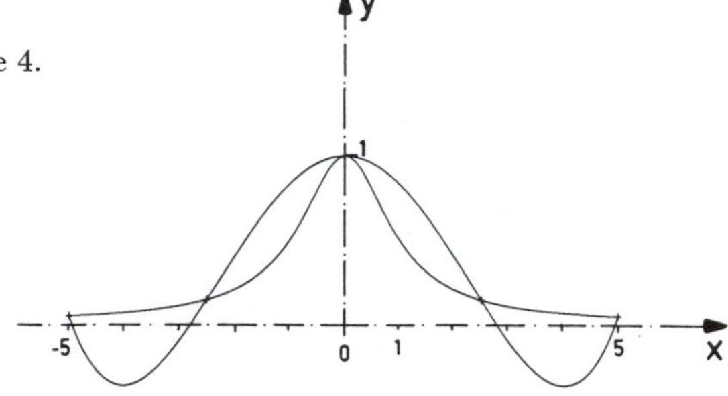

Fig 2.16b. Interpolation by a polynomial of degree 7.

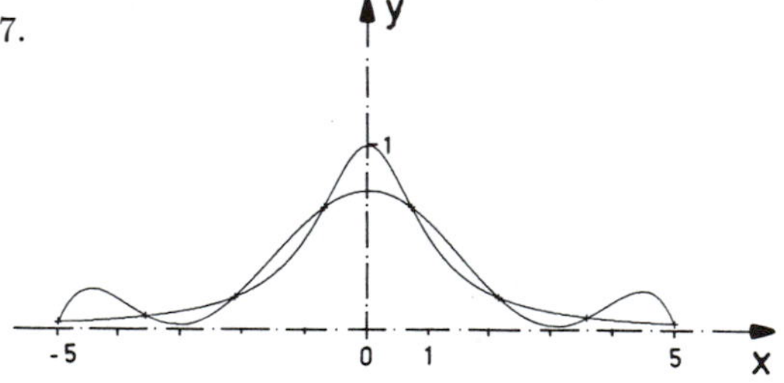

Fig 2.16c. Interpolation by a polynomial of degree 14.

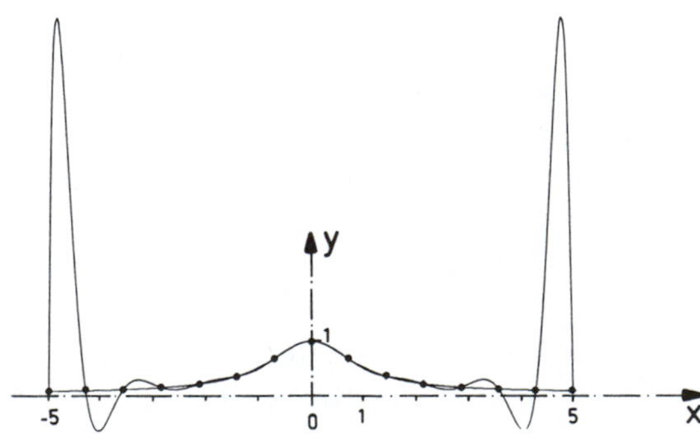

2.3. Approximation of Curves and Surfaces

One way to approximate a given function is to construct an interpolant to it using given knots. In this section we look at an alternative way of approximating functions. We try to make the deviation of the approximating function from the given function as small as possible, without trying to make the function interpolate at any set of knots. This approach is often more desirable than interpolation, particularly for data obtained empirically by measurement, since

- empirical data are usually subject to measurement errors, so it is better not to interpolate exactly; instead, we try to stay within a given tolerance,
- in many applications the data will consist of a very large number of measurements, and so many interpolation methods will have trouble since the number of basis functions is so large, and since adding an additional measurement point will change the structure of the solution.

In view of these observations, our *goal* now is

- to replace a given function f by a *simpler function* Φ in such a way as to minimize a prescribed error measure,
- to describe a given set of empirical data corresponding to measurements on an unknown function f by a function Φ in such a way as to minimize a prescribed error measure.

Fig. 2.17 shows a typical approximating function, and should be compared with Fig. 2.11 which shows a typical interpolating function.

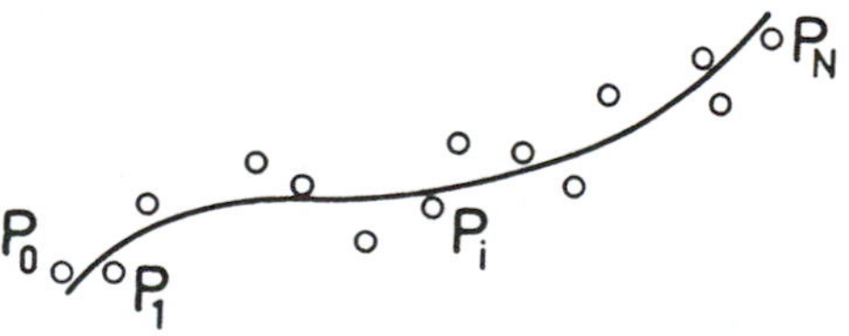

Fig. 2.17. Approximation.

To construct approximations, we look for an *optimal* function Φ as a linear combination of given basis functions, where optimal means that a prescribed *error norm* should be as small as possible.

In order to introduce a norm $\|f\|$, we consider the set $\boldsymbol{M}$ of all continuous functions f on an interval $[a,b]$, and for each function $f \in \boldsymbol{M}$, we associate a real-valued nonnegative number $\|f\|$ (called the norm of f) with the properties

$$\begin{aligned} &\|f\| \geq 0, && \|f\| = 0 \iff f \equiv 0 \text{ in } [a,b], \\ &\|\alpha f\| = |\alpha|\|f\|, && \alpha \text{ arbitrary, } \alpha \in \mathbb{R}, \\ &\|f+g\| \leq \|f\| + \|g\|, && f,g \in \boldsymbol{M}. \end{aligned} \tag{2.44}$$

With respect to a given norm, we can now define the *distance* ρ between two functions $f,g \in \boldsymbol{M}$ by

$$\rho(f,g) := \|f-g\|.$$

It is easy to check that ρ satisfies the axioms of a distance function.

Given a system of $n+1$ basis functions ϕ_k, $k=0(1)n$, we consider the *linear* approximation family consisting of functions of the form

$$\boldsymbol{X}(t) = \sum_{k=0}^{n} \boldsymbol{A}_k \phi_k(t), \qquad \text{with } t \in [a,b], \quad \boldsymbol{A}_k \in \mathbb{R}^3. \tag{2.45}$$

The approximation problem now reduces to finding the best coefficients $\boldsymbol{A}_k$; i.e., to solving the following problem: given $\boldsymbol{Y} \in \boldsymbol{M}$ and a prescribed basis system $\phi_0, \ldots, \phi_n$, find a function

$$\boldsymbol{X}^0(t) = \sum_{k=0}^{n} \boldsymbol{A}_k^{(0)} \phi_k(t)$$

in $\boldsymbol{M}$ such that

$$\rho(\boldsymbol{Y}, \boldsymbol{X}^0) = \|\boldsymbol{Y}(t) - \boldsymbol{X}^0(t)\| = \min_{\boldsymbol{X} \in \boldsymbol{M}} \|\boldsymbol{Y} - \boldsymbol{X}\|. \tag{2.46}$$

Such a function $\boldsymbol{X}^0$ is called a *best approximation* of $\boldsymbol{Y}$ with respect to the given basis system $\{\phi_k\}$ and the selected norm $\|\cdot\|$. We have the following *existence theorem*:

Theorem 2.2. *Suppose ϕ_0, ϕ_1, …, $\phi_n \in \boldsymbol{M}[a,b]$ is a prescribed system of basis functions, and $\|\cdot\|$ is a given norm. Then for any function $\boldsymbol{Y} \in \boldsymbol{M}[a,b]$, there exists at least one best approximation of the form (2.45) satisfying (2.46).*

A proof of this theorem can be found in, e.g., [BER 70]. *Examples* of appropriate basis systems include:

a) *monomials* $\phi_0 = 1$, $\phi_k = t^k$,

b) *trigonometric functions* $\phi_0 = 1$, $\phi_1 = \cos t$, $\phi_2 = \sin t$, $\phi_3 = \cos 2t$, $\phi_4 = \sin 2t$, ...,

c) *Chebychev polynomials* $T_n(t) := \cos n\phi$ for $t \in [-1, 1]$, where $t = \cos\phi$. Since $\cos(\alpha - \beta) + \cos(\alpha + \beta) = 2\cos\alpha\cos\beta$, choosing $\alpha = n\phi$ and $\beta = \phi$ leads to the recurrence relation

$$T_{n-1}(t) + T_{n+1}(t) = 2tT_n(t),$$

which implies the following explicit formulae:

$$\begin{aligned}
T_0(t) &= \cos(0) = 1, \\
T_1(t) &= \cos\phi = t, \\
T_2(t) &= \cos 2\phi = -1 + 2t^2, \\
T_3(t) &= \cos 3\phi = -3t + 4t^3, \\
T_4(t) &= \cos 4\phi = 1 - 8t^2 + 8t^4, \\
T_5(t) &= \cos 5\phi = 5t - 20t^3 + 16t^5.
\end{aligned}$$

Remark. The simplest way to obtain an approximation of a given function is by truncating some series expansion of the function, such as

- the Taylor series (monomials),
- the Fourier series (trigonometric functions),
- the expansion in terms of Chebychev polynomials.

2.3.1. The Discrete Least Squares Method for Curves

Suppose we are given $N+1$ measured points $\boldsymbol{P}_i$, $i = 0(1)N$, in $\mathbb{R}^2$ or $\mathbb{R}^3$, and associated parameter values t_i. Then the deviation $\boldsymbol{D}_i$ of an approximating function $\boldsymbol{X}(t)$ from a point $\boldsymbol{P}_i$ can be written as

$$\boldsymbol{D}_i := \boldsymbol{X}(t_i) - \boldsymbol{P}_i. \tag{2.47}$$

Our aim is to minimize the *overall error*

$$d = \sum_{i=0}^{N} |\boldsymbol{D}_i|^2 = \sum_{i=0}^{N} (\boldsymbol{X}(t_i) - \boldsymbol{P}_i)^2, \tag{2.48}$$

or to minimize each of the components of the error vector. If the measurements are given with different accuracies, then it may be desirable to introduce weights w_i so that less accurate measurements are weighted less than those that are more accurate. In this case the error expression becomes

$$d = \sum_{i=0}^{N} w_i(\boldsymbol{X}(t_i) - \boldsymbol{P}_i)^2. \tag{2.49}$$

We now assume that $\boldsymbol{X}(t)$ has the form

$$\boldsymbol{X}(t) := \sum_{j=0}^{n} \boldsymbol{A}_j \phi_j(t), \qquad \text{with } n < N, \tag{2.50}$$

and try to determine the coefficients so that

$$\sum_{i=0}^{N} w_i(\boldsymbol{X}(t_i) - \boldsymbol{P}_i)^2$$

is minimized.

It is easy to see that a necessary condition for (2.49) to be minimized is that

$$\frac{\partial d}{\partial \boldsymbol{A}_k} = -2\sum_{i=0}^{N} w_i(\boldsymbol{P}_i - \boldsymbol{X}(t_i))\frac{\partial \boldsymbol{X}(t_i)}{\partial \boldsymbol{A}_k} = -2\sum_{i=0}^{N} w_i(\boldsymbol{P}_i - \boldsymbol{X}(t_i))\phi_k(t_i) = 0.$$

Substituting (2.50) in this expression leads to the system of linear equations

$$\sum_{i=0}^{N} w_i \boldsymbol{P}_i \phi_k(t_i) = \sum_{j=0}^{n} \boldsymbol{A}_j \sum_{i=0}^{N} w_i \phi_j(t_i)\phi_k(t_i), \qquad k = 0(1)n, \tag{2.51}$$

for each component of the unknown coefficient vectors $\boldsymbol{A}_\ell$. This system of $n+1$ linear equations (called the *normal equations*) has a unique solution if and only if (see e.g., [BER 70])

1) $N \geq n$,
2) the ϕ_k form a basis, i.e., they are linearly independent and thus form a *Chebychev system.*

If we now introduce the notation

$$[\phi_k, \phi_j] := \sum_{i=0}^{N} w_i \phi_k(t_i)\phi_j(t_i),$$

$$[\boldsymbol{P}, \phi_k] := \sum_{i=0}^{N} w_i \boldsymbol{P}_i \phi_k(t_i),$$

then the system (2.51) can be written as

$$\begin{pmatrix} [\phi_0,\phi_0] & \dots & [\phi_0,\phi_n] \\ [\phi_1,\phi_0] & \dots & [\phi_1,\phi_n] \\ \vdots & & \vdots \\ [\phi_n,\phi_0] & \dots & [\phi_n,\phi_n] \end{pmatrix} \begin{pmatrix} \boldsymbol{A}_0 \\ \boldsymbol{A}_1 \\ \vdots \\ \boldsymbol{A}_n \end{pmatrix} = \begin{pmatrix} [\boldsymbol{P},\phi_0] \\ [\boldsymbol{P},\phi_1] \\ \vdots \\ [\boldsymbol{P},\phi_n] \end{pmatrix}, \tag{2.52}$$

which has to be solved for each component of the coefficient vectors. It can be shown that the coefficient matrix in (2.52) is positive definite for $n \le N$, see e.g., [BOEH 85], but is often not very well-conditioned. The system can be solved by direct QR-decomposition, or by Cholesky decomposition, see e.g., [FIN 77], [GOU 74], and [HIL 56].

If $n = N$, then the least squares approximant reduces to the interpolant. For approximating given functions, the discrete least squares method can be replaced by the *continuous least squares method,* where in place of the sum, we have an *integral.*

2.3.2. The Discrete Least Squares Method for Functions in $\mathbb{R}^3$

In this section we discuss the least squares method for *functions* in $\mathbb{R}^3$. Suppose we are given $N+1$ points $\boldsymbol{P}_i(x_i, y_i, z_i)$ in $\mathbb{R}^3$, $i = 0(1)N$. We seek a function

$$z = f(x, y, b_0, b_1, \dots, b_k, \dots, b_n),$$

where b_k are unknown parameters corresponding to the coefficients $\boldsymbol{A}_k$ in (2.45). The problem now is to minimize the deviations d_i of $\boldsymbol{P}_i$ from f, i.e., to choose the parameters so that

$$\sum_{i=0}^{N} (d_i)^2 = \sum_{i=0}^{N} \Big(z_i - f(x_i, y_i, b_0, b_1, \dots, b_k, \dots, b_n) \Big)^2 \tag{2.53}$$

is minimized. Setting

$$\sum_{i=0}^{N} (d_i)^2 =: [d, d],$$

a necessary condition for a minimum in (2.53) is that

$$\frac{\partial [d,d]}{\partial b_k} = 0, \qquad k = 0(1)N.$$

This leads to the system

$$\Big[z, \frac{\partial f}{\partial b_k} \Big] - \Big[f, \frac{\partial f}{\partial b_k} \Big] = 0, \qquad k = 0(1)n, \tag{2.54}$$

of *normal equations.* For example, if we want to fit a paraboloid

$$f = ax^2 + bxy + cy^2, \qquad (a, b, c \text{ free parameters}),$$

then for $N > 3$, (2.54) leads to the system

$$\begin{array}{lllllll} a[x^4] & + & b[x^3y] & + & c[x^2y^2] & = & [zx^2], \\ a[x^3y] & + & b[x^2y^2] & + & c[xy^3] & = & [zxy], \\ a[x^2y^2] & + & b[xy^3] & + & c[y^4] & = & [zy^2], \end{array}$$

where $[x^2] := \sum_{i=0}^{N}(x_i)^2$, etc.

In general, this least squares method can easily be described in matrix form. Suppose we seek an approximation of the form

$$z = \sum_{k=0}^{n} \phi_k(x, y) b_k, \tag{2.55}$$

where $\phi_k(x, y)$ are linearly independent functions of x and y, and where the b_k are unknown coefficients. Now if we are given $N + 1$ points $\boldsymbol{P}_i(x_i, y_i, z_i)$, then at each point the error is given by

$$d_i = z_i - \sum_{k=0}^{n} \phi_k(x_i, y_i) b_k, \tag{2.56a}$$

or in matrix form as

$$\boldsymbol{D} = \boldsymbol{Z} - \boldsymbol{M}\boldsymbol{B}, \tag{2.56b}$$

where $\boldsymbol{D}$ is the vector with components d_i, $\boldsymbol{Z}$ is the vector with components z_i, $\boldsymbol{B}$ is the vector of the b_k, and $\boldsymbol{M}$ is the coefficient matrix built out of the $\phi_k(x_i, y_i)$ as in (2.56a). Thus $\boldsymbol{D}, \boldsymbol{Z} \in \mathbb{R}^{N+1}$, $\boldsymbol{B} \in \mathbb{R}^{n+1}$, and $\boldsymbol{M}$ is an $(N + 1) \times (n + 1)$ matrix, where we suppose $n < N$. We now have

Lemma 2.1. *The square of the error $\boldsymbol{D}^T\boldsymbol{D}$ will be minimal provided that $\boldsymbol{M}^T\boldsymbol{D} = 0$.*

Proof: We note that in view of (2.56b), the error corresponding to a coefficient vector $\widetilde{\boldsymbol{B}} := \boldsymbol{B} + \boldsymbol{V}$ is given by $\widetilde{\boldsymbol{D}} = \boldsymbol{D} - \boldsymbol{M}\boldsymbol{V}$. Now if $\boldsymbol{M}^T\boldsymbol{D} = 0$, then

$$\begin{aligned} \|\widetilde{\boldsymbol{D}}^2\| &= \widetilde{\boldsymbol{D}}^T\widetilde{\boldsymbol{D}} = \boldsymbol{D}^T\boldsymbol{D} - \boldsymbol{V}^T\boldsymbol{M}^T\boldsymbol{D} - \boldsymbol{D}^T\boldsymbol{M}\boldsymbol{V} + \boldsymbol{V}^T\boldsymbol{M}^T\boldsymbol{M}\boldsymbol{V} \\ &= \boldsymbol{D}^T\boldsymbol{D} + \boldsymbol{V}^T\boldsymbol{M}^T\boldsymbol{M}\boldsymbol{V} = \boldsymbol{D}^T\boldsymbol{D} + (\boldsymbol{V}^T\boldsymbol{M}^T)(\boldsymbol{M}\boldsymbol{V}) \\ &= \|\boldsymbol{D}\|^2 + \|\boldsymbol{M}\boldsymbol{V}\|^2, \end{aligned}$$

and it follows that if $\boldsymbol{V} \neq 0$, then $\|\widetilde{\boldsymbol{D}}\|^2 > \|\boldsymbol{D}\|^2$ for all $\widetilde{\boldsymbol{D}} \neq \boldsymbol{D}$. But this means that $\boldsymbol{D}$ is the desired minimum. ■

Applying this lemma to (2.56b), we get the linear system

$$M^T M B = M^T Z \tag{2.57}$$

of normal equations for the unknown vector $\boldsymbol{B}$.

For simplicity, here we have discussed the case where the deviations were measured in the z-coordinates. This, of course, is an arbitrary choice, and is not invariant with respect to coordinate transformations. Geometrically, it would make more sense to measure the actual Euclidean distances between the data points $\boldsymbol{P}_i$ and the approximating surface. This, however, would have led to a nonlinear system of equations for the best approximation. While these kinds of nonlinear systems can be reduced to the computation of the zeros of polynomials in some special cases, in general they must be approximately solved by some kind of Newton method.

2.3.3. The Discrete Least Squares Method for Parametric Surfaces

We choose our approximating surface to be a bicubic surface of the form

$$\boldsymbol{X}(u,v) = \sum_{k=0}^{3}\sum_{i=0}^{3} \boldsymbol{A}_{ik} u^i v^k, \tag{2.58}$$

or in matrix form

$$\boldsymbol{X}(u,v) = (1,\ u,\ u^2,\ u^3)\boldsymbol{A}\begin{pmatrix}1\\ v\\ v^2\\ v^3\end{pmatrix},$$

where $\boldsymbol{A}$ is a 4×4 matrix with vector-valued elements $\boldsymbol{A}_{ik}$. Componentwise we have

$$\boldsymbol{X}(u,v) = (x(u,v), y(u,v), z(u,v)), \qquad \boldsymbol{A} = \{\boldsymbol{A}_{ik}\} = (a_{ik}, b_{ik}, c_{ik}).$$

Now suppose we are given $N+1$ points $\boldsymbol{P}_\ell$ and their associated parameter values (u_ℓ, v_ℓ), $\ell = 0(1)N$. We assume $N > 15$, and introduce the vectors

$$\begin{aligned} \boldsymbol{W} &:= (u^0v^0, u^0v^1, \ldots, u^0v^3, u^1v^0, u^1v^1, \ldots, u^3v^3)^T, \\ \boldsymbol{a} &:= (a_{00}, a_{01}, \ldots, a_{03}, a_{10}, a_{11}, \ldots, a_{33})^T, \end{aligned} \tag{2.59}$$

in $\mathbb{R}^{16}$, and define the vectors $\boldsymbol{b}, \boldsymbol{c}$ similarly. Then, for example, the x-component of (2.58) can now be written as

$$x = \boldsymbol{W}^T \boldsymbol{a}. \tag{2.60}$$

Now inserting the given points $\boldsymbol{P}_\ell$ and the associated parameter values (u_ℓ, v_ℓ) in (2.60), and introducing

$$\boldsymbol{X} := \begin{pmatrix} x_0 \\ x_1 \\ \vdots \\ x_N \end{pmatrix}, \qquad \boldsymbol{M} := \begin{pmatrix} \boldsymbol{W}_0^T \\ \boldsymbol{W}_1^T \\ \vdots \\ \boldsymbol{W}_N^T \end{pmatrix},$$

we find that the error equation is given by

$$\boldsymbol{d}_x = \boldsymbol{X} - \boldsymbol{M}\boldsymbol{a}. \tag{2.61a}$$

Then in analogy with (2.57), we get the following system of normal equations

$$\boldsymbol{M}^T \boldsymbol{M}\boldsymbol{a} = \boldsymbol{M}^T \boldsymbol{X} \tag{2.61b}$$

for the unknown vector $\boldsymbol{a}$. If we solve (2.61) along with the analogous systems of equations for the other two components y, z, then (2.58) leads explicitly to the parametrization of the approximating surface. As before, the solution depends on the parametrization being used (cf. Sections 4.4 and 6.2.5).

For large numbers N of data points, the multiplications arising in (2.57) and (2.61b) are computationally very expensive. This suggests that (2.56b) or (2.61a) should be solved by direct methods such as the *Householder transformation,* see [BOEH 85]. In this approach, the coefficient matrix $\boldsymbol{M}$ is transformed into an upper triangular matrix using Householder reflections, after which the solution can be found by back substitution.

Interpolation and approximation of parametric curves and surfaces will be discussed in detail in Sections 4.4 and 6.2.5.

3
Spline Curves

3.1. The Idea of a Spline Function

For many applications in modelling, the interpolation and approximation methods discussed in Chapter 2 are not adequate. Usually the user does not want the curvature of his curve or surface to vary too much, i.e., it should appear to be *smooth*. Most of the classical interpolating functions (and in particular polynomials) have a tendency to oscillate, see Sect. 2.2.7. Fig. 3.1 illustrates this phenomenon in the case where we are trying to approximate a curve by a polynomial of degree 8 which interpolates at nine points.

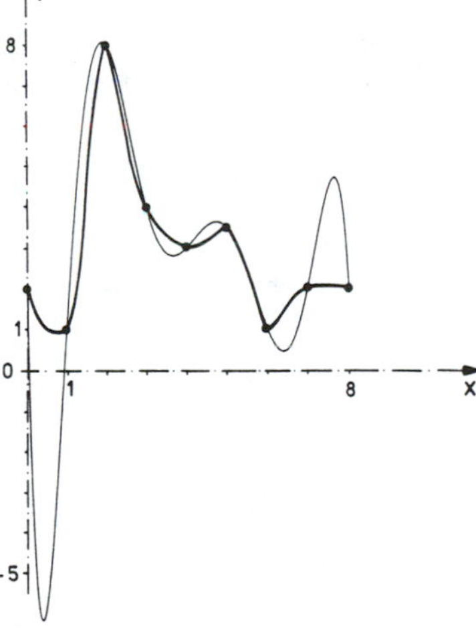

Fig. 3.1. A curve and its polynomial interpolant.

This example shows that at several points, the interpolating polynomial deviates significantly from the desired curve, suggesting that it may not be a good idea to use polynomials to fit functions. We already saw in Fig. 2.16 that using a large number of interpolation points does not help either.

One way to improve the situation is to divide the given interval into smaller subintervals, and to construct an approximating curve consisting of pieces of curves. This idea has been used with considerable success for designing numerical quadrature formulae. For example, the trapezoidal rule uses piecewise linear functions, while Simpson's rule uses piecewise quadratic functions, cf. Fig. 3.2. In this method, the individual parabolas are joined together only continuously. In many applications, however, we would like our curves to be smoother. Thus, it seems appropriate to consider piecewise polynomial functions whose first derivatives (or more generally whose first k derivatives) join continuously.

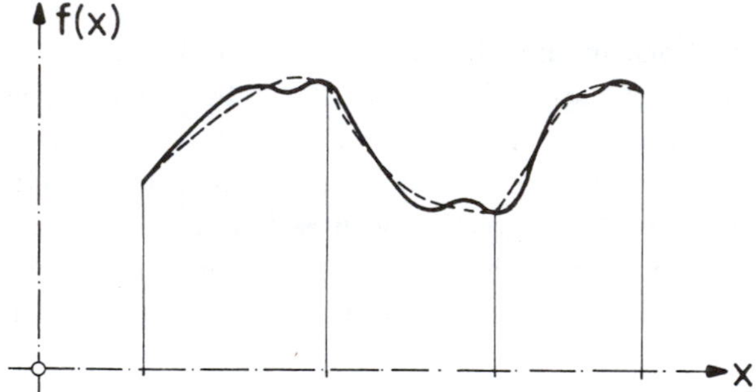

Fig. 3.2. Approximation of a curve by continuously-joined parabolas (dashed line).

To see how this works, suppose that we want to construct a function s such that

- $s \in C^2[a, b]$,
- $s(x_i) = r_i$, for $i = 0(1)N$, i.e., s interpolates the points (x_i, r_i),
- $\frac{d^4 s(x)}{dx^4} = 0$, for $x \in [a, b] \setminus \{x_i,\ i = 0(1)N\}$.

Integrating the last equation, we find that s consists of a cubic polynomial

$$s(x) = a_3 x^3 + a_2 x^2 + a_1 x + a_0$$

in each subinterval $[x_i, x_{i+1}]$.

Remark. The third derivatives of s are piecewise constant with jumps at the break points x_i.

This example suggests the following

Definition 3.1. *A piecewise function s consisting of polynomial pieces of degree n is called*

a) a *spline function* provided that it is $(n-1)$-times continuously differentiable,
b) a *subspline function* provided that it is at least continuous, but is not smooth enough to be a spline.

Remark. The word *spline* comes from the name of a tool (consisting of a thin flexible rod) used by ship builders for drawing smooth curves approximating the cross sections of ship hulls.

Fig. 3.3. A thin bent rod.

A cubic spline function is an approximation to the centerline (elastica) of a thin rod which has been forced to pass through a given set of points, cf. Fig. 3.3. To see this, note that the bending energy E of such a rod is proportional to the integral of the square of its curvature κ; i.e., the elastica is the solution of the problem of minimizing

$$E = c \int_0^\ell \kappa^2 ds, \tag{3.1}$$

where ℓ is the length and ds is the arc length of the rod, and where c is a constant depending on the material from which the rod is made. As shown in Sect. 2.1.1, the curvature of a curve is given by

$$\kappa = \frac{|\boldsymbol{X}' \times \boldsymbol{X}''|}{|\boldsymbol{X}'|^3}.$$

For functions $y = y(x)$, the position vector can be written as $\boldsymbol{X} := (x, y(x))^T$. In this case,

$$|\boldsymbol{X}'| = \sqrt{1 + (y')^2} \qquad \text{and} \qquad ds = \sqrt{1 + (y')^2}\ dx,$$

and thus (3.1) becomes

$$E = c\int_0^\ell \frac{(y'')^2}{(1+(y')^2)^{\frac{5}{2}}}dx \approx c\int_0^\ell (y'')^2\,dx, \tag{3.2a}$$

assuming that $|y'| << 1$; i.e., the slope of the curve is very small. The problem of minimizing (3.1) can be solved using methods of the *calculus of variations* (cf. e.g., [ELS 70]), and it can be shown that a necessary condition for a minimum of

$$\int f(y, y', y''; x)\,dx \tag{3.3}$$

is the so-called *Euler-Lagrange equation*

$$f_y - \frac{d}{dx}(f_{y'}) + \frac{d^2}{dx^2}(f_{y''}) = 0. \tag{3.4}$$

Applying (3.4) to (3.2) yields

$$f_{y''} = 2y'',$$

and so the Euler-Lagrange equation for the variational problem (3.2a) is

$$\frac{d^2}{dx^2}(y'') = y^{(4)} = 0. \tag{3.2b}$$

Integrating this equation shows that the solution must be a cubic polynomial.

Remark. This property of cubic splines is frequently referred to in the literature as the *minimal curvature property*. Among all interpolating curves y, the cubic spline is the curve which minimizes $\int_0^\ell (y'')^2 dx$.

Remark. In general, it can be shown [SPÄ 83] that a spline $s(x)$ of degree $2m+1$ minimizes the integral

$$\int_a^b [f^{(m+1)}(x)]^2 dx \tag{$*$}$$

over all functions $f \in C^m[a,b]$ with square integrable derivatives up to order $m+1$ which satisfy the interpolation conditions

$$f(x_k) = s(x_k) = y_k, \qquad k = 0(1)N,$$

along with one of the following sets of boundary conditions:

a) $s^{(i)}(a) = f_a^i, \quad s^{(i)}(b) = f_b^i, \qquad i = m+1(1)2m,$

b) $s^{(i)}(a) = f^{(i)}(a), \quad s^{(i)}(b) = f^{(i)}(b), \qquad i = 1(1)m.$

Remark. [Mei 87] has studied other measures of energy involving the third and fourth derivatives as well as certain mixed criteria, and has compared the results, see also [Now 90], [Hag 90].

3.2. Conic Sections as Subsplines

In engineering applications (e.g., in the aircraft industry), profiles are frequently modelled by pieces of conic sections joined together to form a C^1 continuous curve. Fig. 3.4 shows a typical such curve consisting of:

$1 \cup 2$ circle ,	$2 \cup 3$ parabola,	$3 \cup 4$ ellipse,
$4 \cup 5$ straight line,	$5 \cup 6$ circle ,	$6 \cup 7$ straight line.

Reflection and translation of this profile produces a cylindrical object; if it is rotated about the z-axis, we get a surface of revolution.

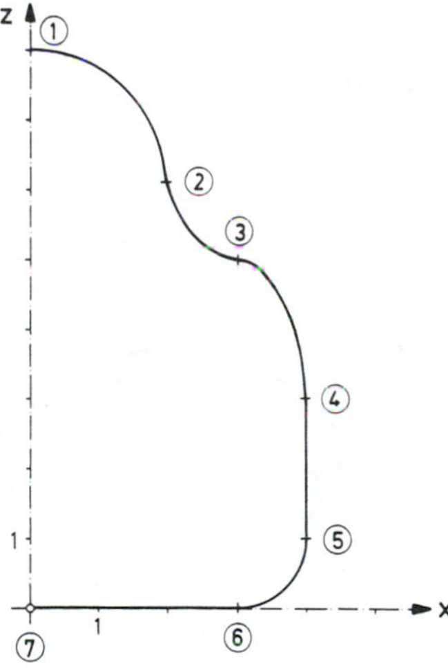

Fig. 3.4. A profile defined by a piecewise conic section subspline.

These *conic section subsplines* seem to have been introduced by [LIM 44], where the segments of the conic sections were constructed to pass through given points with prescribed tangents. This construction leaves one free parameter, since the general form of a conic section in the (x, y) plane is given by

$$y^2 + axy + bx^2 + cx + dy + e = 0. \tag{3.5}$$

To uniquely determine this free parameter, we can require that the curve pass through one additional point, or we can specify the type of the curve (parabola, ellipse, or hyperbola).

The type of a conic section is determined by the eigenvalues of the quadratic form (3.5):

- if it has two real positive (or negative) eigenvalues, then the conic section is an *ellipse* (or a double point);
- if it has one positive and one negative eigenvalue, then the conic section is a *hyperbola* (or a pair of intersecting lines);
- if one eigenvalue is zero and the other is nonzero, then the conic section is a *parabola* (or a double line or a pair of parallel lines).

We now consider an example. Suppose we want to construct a conic section passing through two points $\boldsymbol{P}_1$ and $\boldsymbol{P}_2$ with tangents t_1 and t_2. As an additional condition, we require that the curve be a parabola. To be specific, let $\boldsymbol{P}_1 = (2,0)$, $\boldsymbol{P}_2 = (0,1)$, and suppose t_1 and t_2 are taken to be the directions of the coordinate axes, cf. Fig. 3.5.

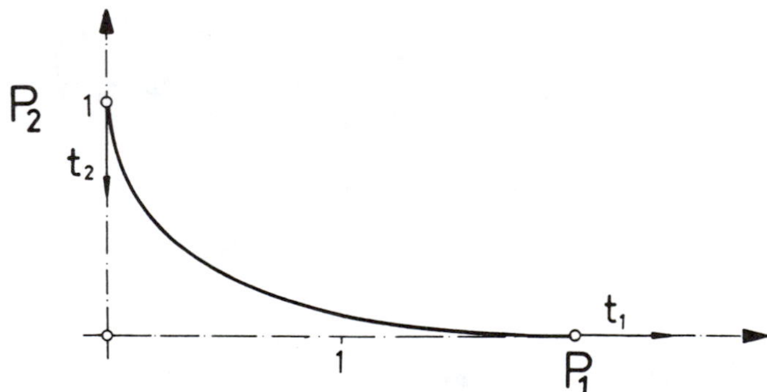

Fig. 3.5. Parabolic piece.

Implicit differentiation of (3.5) yields

$$2yy' + ay + axy' + 2bx + c + dy' = 0,$$

or

$$y' = -\frac{ay + 2bx + c}{2y + ax + d}.$$

By substitution, it follows that

$$\begin{aligned}
&y'(2,0) = 0 && \Rightarrow \quad 4b + c = 0 \quad \text{and} \quad 2a + d \neq 0,\\
&y'(0,1) = \infty && \Rightarrow \quad 2 + d = 0 \quad \text{and} \quad a + c \neq 0,\\
&y = 0 \quad \text{for} \quad x = 2 && \Rightarrow \quad 4b + 2c + e = 0,\\
&y = 1 \quad \text{for} \quad x = 0 && \Rightarrow \quad 1 + d + e = 0,
\end{aligned}$$

which gives

$$b = \frac{1}{4}, \quad c = -1, \quad d = -2, \quad e = 1.$$

In order to force the conic section to be a parabola, we must require that one eigenvalue is zero, i.e., the determinant of the quadratic form (3.5) must vanish. This leads to

$$\begin{vmatrix} 1 & \frac{a}{2} \\ \frac{a}{2} & b \end{vmatrix} = 0 \quad \Rightarrow \quad a = \pm 2\sqrt{b} = \pm 1.$$

Taking account of the condition $a + c \neq 0$, it follows that $a = -1$.

In addition to the analytic approach, a conic section can also be described in *geometric form*. To this end, we need the concept of a pencil of conic sections, see e.g., [GRO 57]. Given two conic sections with the equations

$$S_1(x, y) = 0 \quad \text{and} \quad S_2(x, y) = 0,$$

then the linear combination

$$(1 - \lambda)S_1 + \lambda S_2 = 0, \qquad \lambda \in \mathbb{R}, \tag{3.6}$$

describes a family of conic sections. This family is called a *pencil of conic sections*, and the original two conic sections S_1, S_2 are called the *support of the pencil*. Every conic section in the pencil passes through the points in $S_1 \cap S_2$, since when $S_1 = 0$ and $S_2 = 0$, (3.6) is automatically satisfied, see Fig. 3.6.

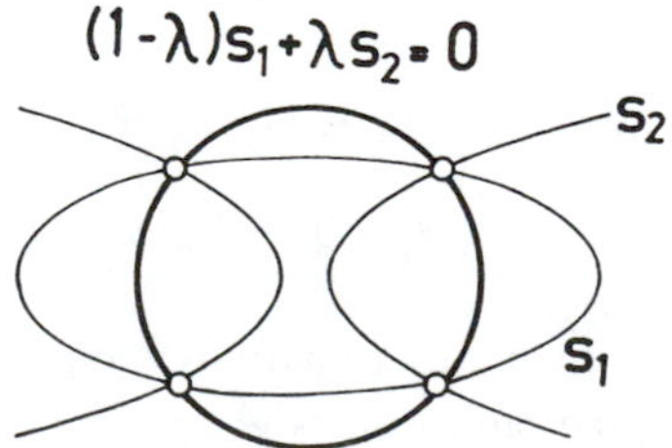

Fig. 3.6. A conic section (heavy line) in the pencil with support S_1 and S_2.

Each choice of the parameter λ corresponds to one conic section of the pencil. For the pencil supported by S_1, S_2, we can solve (3.6) to find the parameter

$$\lambda = \frac{S_1(x_1, y_1)}{S_1(x_1, y_1) - S_2(x_1, y_1)}$$

describing the conic section which passes through the point $\boldsymbol{P}_1 = (x_1, y_1)$, where $\boldsymbol{P}_1 \notin S_1 \cap S_2$. The support of a pencil of conic sections can also be a *singular* conic section. For example, if the quadratic polynomial describing a conic section can be factored into two linear polynomials, then the conic section reduces to two straight lines. Conversely, starting with four lines

$$\ell_i := a_i x + b_i y + c_i = 0,$$

then the product $\ell_1 \cdot \ell_2 = 0$ corresponds to a degenerate conic section, and the combination

$$(1 - \lambda)\ell_1\ell_2 + \lambda\ell_3\ell_4 = 0$$

describes a pencil of conic sections which passes through the four intersection points $(\ell_1, \ell_2) \cap (\ell_3, \ell_4)$, see Fig. 3.7.

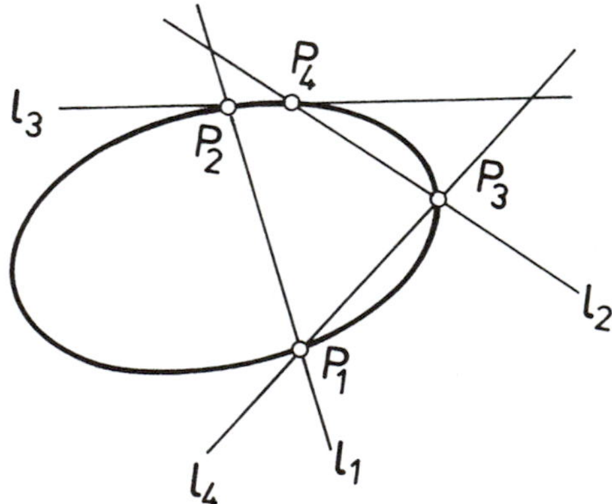

Fig. 3.7. Pencil of conic sections supported by four lines.

Specializing further by setting $\ell_3 = \ell_4$, we get the pencil equation

$$(1 - \lambda)\ell_1\ell_2 + \lambda\ell_3^2 = 0. \tag{3.7}$$

In this case, all conic sections of the pencil are tangent to the lines ℓ_1 and ℓ_2 and pass through the intersection points $\boldsymbol{A} := \ell_1 \cap \ell_3$ and $\boldsymbol{B} := \ell_2 \cap \ell_3$, see Fig. 3.8. The lines ℓ_1 and ℓ_2 are tangents at $\boldsymbol{A}$ and $\boldsymbol{B}$, respectively, since $\boldsymbol{A}$ and $\boldsymbol{B}$ can be regarded as double points. See Figs. 3.7 and 3.8, and in Fig. 3.7 let the points $\boldsymbol{P}_1$ and $\boldsymbol{P}_3$ converge to $\boldsymbol{P}_2$ and $\boldsymbol{P}_4$, respectively so that $\ell_3 \to \ell_4$.

When a conic section is selected from a pencil of conic sections of the form (3.7) by requiring it to pass through an additional point $\boldsymbol{S}$, then this point $\boldsymbol{S}$ is usually referred to as the *shoulder point*, see Fig. 3.8. If $\boldsymbol{S}$ lies in the triangle $(\boldsymbol{A}, \boldsymbol{B}, \boldsymbol{C})$, then the conic section is a continuous curve in the triangle $(\boldsymbol{A}, \boldsymbol{S}, \boldsymbol{B})$, *i.e.*, there are no asymptotes between $\boldsymbol{A}$ and $\boldsymbol{B}$.

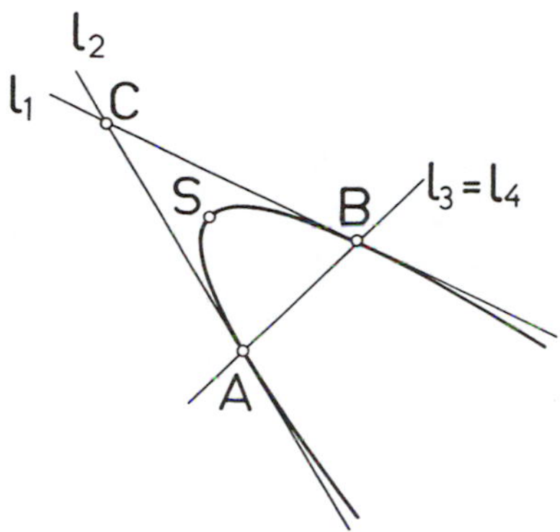

Fig. 3.8. A curve from the pencil with two coincident straight lines as support.

Theorem 3.1. *If the shoulder point* $\boldsymbol{S}$ *is the midpoint of the line joining the midpoints of* $\boldsymbol{AC}$ *and* $\boldsymbol{BC}$*, then the associated conic section is a parabola. If* $\boldsymbol{S}$ *lies inside the parabola (and not on* ℓ_3*), then the conic section is an ellipse. If* $\boldsymbol{S}$ *lies outside of the parabola (and not on* ℓ_1, ℓ_2*), then the conic section is a hyperbola.*

For a proof of this theorem, see e.g., [HOS 87]. We now present an example of the construction of a conic section profile using the *pencil method*. Suppose we are given the break points $\boldsymbol{S}_1$, $\boldsymbol{S}_2$, $\boldsymbol{S}_3$ for the segments, the corresponding tangents ℓ_1, ℓ_2, ℓ_3, and two additional points $\boldsymbol{P}_1, \boldsymbol{P}_2$. To construct the conic section subspline curve, we now introduce support lines ℓ_4 and ℓ_5, see Fig. 3.9.

An analytic description of conic section splines can be found in [Pav 83]. See also [Pratt 85], [Far 89a], and [Hu 91].

The pencil method can also be used to construct *blending patches*. Blending patches are surfaces which join two given curves S_1 and S_2 with a continuous pencil of curves. For example, if $S_1(x, y)$ and $S_2(x, y)$ are conic sections lying in the planes $z = 0$ and $z = z_0$, respectively, then these two curves can be blended by the surface whose contour at height $z = \lambda z_0$ for each $\lambda \in [0, 1]$ is the conic section

$$(1 - \lambda)S_1(x, y) + \lambda S_2(x, y) = 0. \tag{$*$}$$

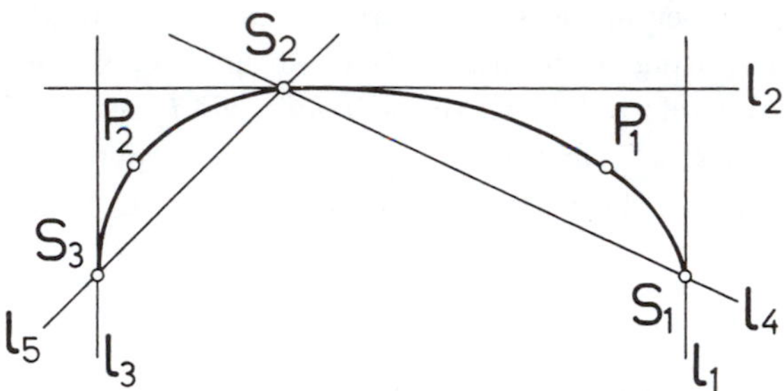

Fig. 3.9. Conic section profile using the pencil method.

As an example, we now construct a blending surface between circular and elliptic cross sections. Suppose the circle is described by

$$z = 0, \qquad x^2 + y^2 = 4,$$

and the ellipse is described by

$$z = 16, \qquad \frac{x^2}{9} + 4y^2 = 1.$$

Then the intermediate cross sections are given by

$$(1 - \lambda)(x^2 + y^2 - 4) + \lambda(\frac{x^2}{9} + 4y^2 - 1) = 0, \qquad \lambda \in [0, 1],$$

which is equivalent to

$$\frac{x^2}{A^2} + \frac{y^2}{B^2} = 1,$$

with

$$A = A(\lambda) = 3\sqrt{\frac{4 - 3\lambda}{9 - 8\lambda}}, \qquad B = B(\lambda) = \sqrt{\frac{4 - 3\lambda}{1 + 3\lambda}}.$$

This leads to the following parametric description of the blending surface (see Fig. 3.10):

$$\begin{aligned}
x(\lambda, \mu) &= A(\lambda) \cos \mu, \\
y(\lambda, \mu) &= B(\lambda) \sin \mu, \qquad \lambda \in [0, 1], \quad \mu \in [0, 2\pi], \\
z(\lambda, \mu) &= 16\lambda.
\end{aligned}$$

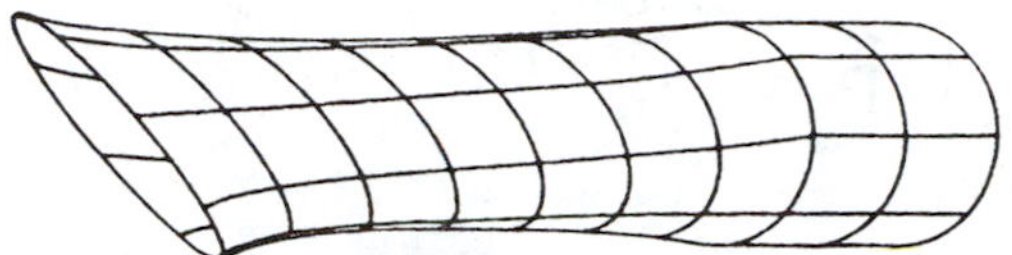

Fig. 3.10. Blending an ellipse and a boundary circle of a cylinder.

3.3. Cubic Spline Curves

In Sect. 3.1 we have shown that a spline function of degree three minimizes bending energy. In this section we discuss how to find the coefficients of the polynomial pieces of such a spline. Throughout this section we shall consider these coefficients to be vectors in $\mathbb{R}^2$ or $\mathbb{R}^3$ so that we are in fact discussing *spline interpolation of parametric curves*. We have already seen the advantages of parametric curves in Chapter 2.

Suppose that we are given $N+1$ points $\boldsymbol{P}_0, \ldots, \boldsymbol{P}_N$ and associated parameter values $t_i \in [a,b]$ satisfying

$$a = t_0 < t_1 < t_2 < \cdots < t_{N-1} < t_N = b.$$

We denote the length of the subintervals $[t_i, t_{i+1}]$ by

$$\Delta t_i := t_{i+1} - t_i.$$

For each neighboring pair of points $\boldsymbol{P}_i$, $\boldsymbol{P}_{i+1}$, we wish to construct an interpolating *cubic curve segment*

$$\begin{aligned} \boldsymbol{X}_i(t) := \boldsymbol{A}_i(t-t_i)^3 + \boldsymbol{B}_i(t-t_i)^2 + \boldsymbol{C}_i(t-t_i) + \boldsymbol{D}_i, \\ t \in [t_i, t_{i+1}], \qquad i = 0(1)N\text{–}1, \end{aligned} \tag{3.8}$$

see Fig. 3.11. The collection of all curve segments will make up the cubic interpolating spline curve s.

We require that s be twice continuously differentiable at each of the break points $\boldsymbol{P}_1, \ldots, \boldsymbol{P}_{N-1}$:

$$\begin{aligned} \boldsymbol{X}_i(t_i) &= \boldsymbol{X}_{i-1}(t_i) \qquad \text{or} \qquad \boldsymbol{X}_i(t_{i+1}) = \boldsymbol{X}_{i+1}(t_{i+1}), \\ \boldsymbol{X}'_i(t_i) &= \boldsymbol{X}'_{i-1}(t_i) \qquad \text{or} \qquad \boldsymbol{X}'_i(t_{i+1}) = \boldsymbol{X}'_{i+1}(t_{i+1}), \\ \boldsymbol{X}''_i(t_i) &= \boldsymbol{X}''_{i-1}(t_i) \qquad \text{or} \qquad \boldsymbol{X}''_i(t_{i+1}) = \boldsymbol{X}''_{i+1}(t_{i+1}). \end{aligned} \tag{3.9}$$

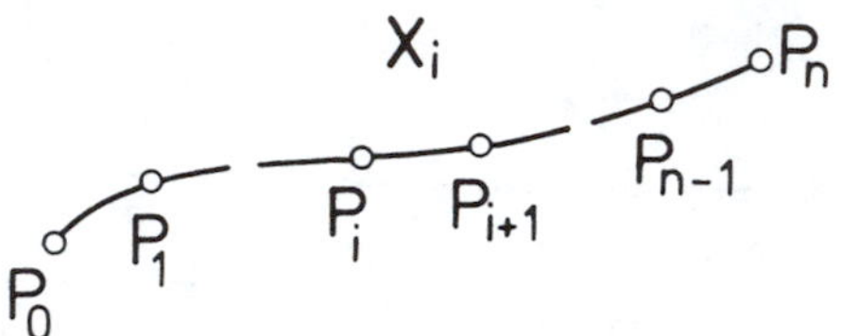

Fig. 3.11. The spline segment $\boldsymbol{X}_i$.

Inserting these conditions into (3.8) gives

$$\begin{aligned}\boldsymbol{X}_i(t_i) &= \boldsymbol{P}_i = \boldsymbol{D}_i, & \boldsymbol{X}_i(t_{i+1}) &= \boldsymbol{P}_{i+1} = \boldsymbol{A}_i\Delta t_i^3 + \boldsymbol{B}_i\Delta t_i^2 + \boldsymbol{C}_i\Delta t_i + \boldsymbol{D}_i,\\ \boldsymbol{X}_i'(t_i) &= \boldsymbol{P}_i' = \boldsymbol{C}_i, & \boldsymbol{X}_i'(t_{i+1}) &= \boldsymbol{P}_{i+1}' = 3\boldsymbol{A}_i\Delta t_i^2 + 2\boldsymbol{B}_i\Delta t_i + \boldsymbol{C}_i,\end{aligned} \tag{3.10}$$

where for convenience we have introduced the (unknown) values $\boldsymbol{P}_i'$ of the derivatives at the break points. Solving (3.10), we get

$$\begin{aligned}\boldsymbol{A}_i &= \frac{1}{(\Delta t_i)^3}[2(\boldsymbol{P}_i - \boldsymbol{P}_{i+1}) + \Delta t_i(\boldsymbol{P}_i' + \boldsymbol{P}_{i+1}')],\\ \boldsymbol{B}_i &= \frac{1}{(\Delta t_i)^2}[3(\boldsymbol{P}_{i+1} - \boldsymbol{P}_i) - \Delta t_i(2\boldsymbol{P}_i' + \boldsymbol{P}_{i+1}')].\end{aligned}$$

Now substituting these formulae and (3.10) in (3.8) leads to the *Hermite* or *Ferguson* formula [Fer 64] for a cubic spline curve:

$$\begin{aligned}&\boldsymbol{X}_i(t) =\\ &\boldsymbol{P}_i\left(2\frac{(t-t_i)^3}{(\Delta t_i)^3} - 3\frac{(t-t_i)^2}{(\Delta t_i)^2} + 1\right) + \boldsymbol{P}_{i+1}\left(-2\frac{(t-t_i)^3}{(\Delta t_i)^3} + 3\frac{(t-t_i)^2}{(\Delta t_i)^2}\right)\\ &+ \boldsymbol{P}_i'\left(\frac{(t-t_i)^3}{(\Delta t_i)^2} - 2\frac{(t-t_i)^2}{\Delta t_i} + (t-t_i)\right) + \boldsymbol{P}_{i+1}'\left(\frac{(t-t_i)^3}{(\Delta t_i)^2} - \frac{(t-t_i)^2}{\Delta t_i}\right).\end{aligned} \tag{3.11}$$

In the special case of equally spaced parameter points ($\Delta t_i = 1$), we can choose $[0, 1]$ as the parameter interval for each curve segment, and the polynomials appearing in (3.11) are the classical *Hermite polynomials*

$$\begin{aligned}H_{00}(t) &= 2t^3 - 3t^2 + 1, & H_{01}(t) &= -2t^3 + 3t^2,\\ H_{10}(t) &= t^3 - 2t^2 + t, & H_{11}(t) &= t^3 - t^2,\end{aligned} \tag{3.12}$$

satisfying

$$\begin{aligned} H_{0i}(k) &= \delta_{ik}, \qquad & H'_{0i}(k) &= 0, \\ H_{1i}(k) &= 0, \qquad & H'_{1i}(k) &= \delta_{ik}, \end{aligned} \tag{3.13}$$

for $i, k = 0, 1$. This shows immediately that the curve defined in (3.11) interpolates the points $\boldsymbol{P}_i$ and the derivatives $\boldsymbol{P}'_i$.

In general, the derivatives $\boldsymbol{P}'_i$ appearing in (3.11) are unknown. They can be *estimated* (see Chapter 4), or alternatively, they can be determined by the requirement that s be C^2 continuous. Taking this approach, we first observe that the second derivatives are given by

$$\begin{aligned} \boldsymbol{X}''_i(t) = 6\boldsymbol{P}_i &\left(\frac{2(t-t_i)}{(\Delta t_i)^3} - \frac{1}{(\Delta t_i)^2} \right) + 6\boldsymbol{P}_{i+1} \left(-2\frac{(t-t_i)}{(\Delta t_i)^3} + \frac{1}{(\Delta t_i)^2} \right) \\ &+ 2\boldsymbol{P}'_i \left(3\frac{(t-t_i)}{(\Delta t_i)^2} - \frac{2}{\Delta t_i} \right) + 2\boldsymbol{P}'_{i+1} \left(\frac{3(t-t_i)}{(\Delta t_i)^2} - \frac{1}{\Delta t_i} \right). \end{aligned} \tag{3.14}$$

Now the C^2 continuity condition

$$\boldsymbol{X}''_{i-1}(t_i) = \boldsymbol{X}''_i(t_i) \tag{3.15}$$

at t_i leads to the *recurrence formula*

$$\begin{aligned} \Delta t_i \boldsymbol{P}'_{i-1} &+ 2(\Delta t_{i-1} + \Delta t_i)\boldsymbol{P}'_i + \Delta t_{i-1}\boldsymbol{P}'_{i+1} \\ &= 3\frac{\Delta t_{i-1}}{\Delta t_i}(\boldsymbol{P}_{i+1} - \boldsymbol{P}_i) + 3\frac{\Delta t_i}{\Delta t_{i-1}}(\boldsymbol{P}_i - \boldsymbol{P}_{i-1}). \end{aligned} \tag{3.16}$$

From (3.16) for $i = 1(1)N{-}1$, we obtain a *tridiagonal* linear system of equations consisting of $N-1$ equations in the $N+1$ unknowns $\boldsymbol{P}'_0, \ldots, \boldsymbol{P}'_N$. Thus, we have

Theorem 3.2. *In interpolating a set of $N+1$ points $\boldsymbol{P}_0, \ldots, \boldsymbol{P}_N$ (with $\boldsymbol{P}_0 \neq \boldsymbol{P}_N$ and $\boldsymbol{P}_i \neq \boldsymbol{P}_k$ for $i \neq k$) by a C^2 cubic spline curve s, there are always two free boundary conditions.*

We now discuss how to choose these two free boundary conditions in a sensible way.

a) We can require that the second derivatives vanish at the boundary points $\boldsymbol{P}_0$ and $\boldsymbol{P}_N$. Splines satisfying this condition are called *natural splines* because they describe a mechanical spline (thin flexible rod) which passes through the points t_0 and t_N, and which remains unbent (and hence is linear) outside of $[t_0, t_N]$. The requirement that

$$\boldsymbol{X}''(t_0) = 0, \qquad \boldsymbol{X}''_{N-1}(t_N) = 0,$$

leads to the equations

$$\begin{aligned} \Delta t_0(2\boldsymbol{P}'_0 + \boldsymbol{P}'_1) &= -3\boldsymbol{P}_0 + 3\boldsymbol{P}_1, \\ \Delta t_{N-1}(\boldsymbol{P}'_{N-1} + 2\boldsymbol{P}'_N) &= -3\boldsymbol{P}_{N-1} + 3\boldsymbol{P}_N. \end{aligned} \tag{3.17}$$

Combining (3.16) and (3.17) leads to the linear system of equations

$$\begin{pmatrix} 2\Delta t_0 & \Delta t_0 & 0 & \cdots & 0 \\ \Delta t_1 & 2(\Delta t_0 + \Delta t_1) & \Delta t_0 & & 0 \\ \vdots & & \ddots & & \vdots \\ 0 & & \Delta t_{N-1} & 2(\Delta t_{N-2} + \Delta t_{N-1}) & \Delta t_{N-2} \\ 0 & \cdots & 0 & \Delta t_{N-1} & 2\Delta t_{N-2} \end{pmatrix} \begin{pmatrix} \boldsymbol{P}'_0 \\ \boldsymbol{P}'_1 \\ \vdots \\ \boldsymbol{P}'_{N-1} \\ \boldsymbol{P}'_N \end{pmatrix} =$$

$$\begin{pmatrix} -3 & 3 & 0 & \cdots & 0 \\ -3\frac{\Delta t_1}{\Delta t_0} & 3(\frac{\Delta t_1}{\Delta t_0} - \frac{\Delta t_0}{\Delta t_1}) & 3\frac{\Delta t_0}{\Delta t_1} & & 0 \\ \vdots & & \ddots & & \vdots \\ 0 & & -3\frac{\Delta t_{N-1}}{\Delta t_{N-2}} & 3(\frac{\Delta t_{N-1}}{\Delta t_{N-2}} - \frac{\Delta t_{N-2}}{\Delta t_{N-1}}) & 3\frac{\Delta t_{N-2}}{\Delta t_{N-1}} \\ 0 & \cdots & 0 & -3 & 3 \end{pmatrix} \begin{pmatrix} \boldsymbol{P}_0 \\ \boldsymbol{P}_1 \\ \vdots \\ \boldsymbol{P}_{N-1} \\ \boldsymbol{P}_N \end{pmatrix} \tag{3.18}$$

b) We can prescribe the slopes at the boundary:

$$\boldsymbol{X}'(a) = \boldsymbol{P}'_0 =: \boldsymbol{T}_0, \qquad \boldsymbol{X}'_{N-1}(b) = \boldsymbol{P}'_N =: \boldsymbol{T}_N.$$

The directions $\boldsymbol{T}_0, \boldsymbol{T}_N$ can be determined from the relative position of the points near the boundaries. For example, we can pass a parabola through $\boldsymbol{P}_0, \boldsymbol{P}_1, \boldsymbol{P}_2$ defined by

$$\boldsymbol{Y} = \boldsymbol{A}_0 + \boldsymbol{A}_1 t + \boldsymbol{A}_2 t^2,$$

with

$$\boldsymbol{Y}(-1) = \boldsymbol{P}_0, \qquad \boldsymbol{Y}(0) = \boldsymbol{P}_1, \qquad \boldsymbol{Y}(1) = \boldsymbol{P}_2.$$

This leads to the equations

$$\boldsymbol{P}_0 = \boldsymbol{A}_0 - \boldsymbol{A}_1 + \boldsymbol{A}_2, \qquad \boldsymbol{P}_1 = \boldsymbol{A}_0, \qquad \boldsymbol{P}_2 = \boldsymbol{A}_0 + \boldsymbol{A}_1 + \boldsymbol{A}_2,$$

which imply

$$\boldsymbol{A}_2 = \tfrac{1}{2}(\boldsymbol{P}_0 - 2\boldsymbol{P}_1 + \boldsymbol{P}_2), \qquad \boldsymbol{A}_1 = \tfrac{1}{2}(\boldsymbol{P}_2 - \boldsymbol{P}_0).$$

This results in the following estimate for the boundary tangent

$$\boldsymbol{T}_0 = \boldsymbol{Y}'(-1) = \boldsymbol{A}_1 - 2\boldsymbol{A}_2, \tag{3.19}$$

also called the *Bessel tangent*, see Sect. 4.1.3 and cf. [Brew 77] and [FAR 90]. For a similar method, see [Mcc 70]. After finding an analogous estimate of $\boldsymbol{T}_N$ and combining it with (3.16), we again get a linear system of $N+1$ equations.

c) Another approach is to use the *not-a-knot condition* of de Boor [BOO 78], [GAN 85]. Here we require that in passing from the first to the second segment and from the next to last to the last segment of the spline, no actual knot is present, that is, there is no jump in the third derivatives at these points. This leads to

$$\boldsymbol{X}_0'''(t_1) = \boldsymbol{X}_1'''(t_1), \qquad \boldsymbol{X}_{N-1}'''(t_{N-1}) = \boldsymbol{X}_{N-2}'''(t_{N-1}). \tag{3.20}$$

In order to expand (3.20), we differentiate (3.14) to obtain

$$\boldsymbol{X}_i'''(t) = \frac{12}{(\Delta t_i)^3}\boldsymbol{P}_i - \frac{12}{(\Delta t_i)^3}\boldsymbol{P}_{i+1} + \frac{6}{(\Delta t_i)^2}\boldsymbol{P}_i' + \frac{6}{(\Delta t_i)^2}\boldsymbol{P}_{i+1}'.$$

Substituting in (3.20) leads to

$$\begin{aligned} -\frac{1}{(\Delta t_0)^2}\boldsymbol{P}_0' &+ \left(\frac{1}{(\Delta t_1)^2} - \frac{1}{(\Delta t_0)^2}\right)\boldsymbol{P}_1' + \frac{1}{(\Delta t_1)^2}\boldsymbol{P}_2' \\ &= \frac{2}{(\Delta t_0)^3}\boldsymbol{P}_0 - \left(\frac{2}{(\Delta t_0)^3} + \frac{2}{(\Delta t_1)^3}\right)\boldsymbol{P}_1 + \frac{2}{(\Delta t_1)^3}\boldsymbol{P}_2. \end{aligned}$$

From the recurrence (3.16), it now follows that

$$\Delta t_1 \boldsymbol{P}_0' + 2(\Delta t_0 + \Delta t_1)\boldsymbol{P}_1' + \Delta t_0 \boldsymbol{P}_2' = 3\frac{\Delta t_0}{\Delta t_1}(\boldsymbol{P}_2 - \boldsymbol{P}_1) + 3\frac{\Delta t_1}{\Delta t_0}(\boldsymbol{P}_1 - \boldsymbol{P}_0).$$

Eliminating $\boldsymbol{P}_2'$, we get the following equation which does not destroy the diagonal structure of the system:

$$\begin{aligned} \frac{1}{\Delta t_0}\left(\frac{1}{\Delta t_0} + \frac{1}{\Delta t_1}\right)\boldsymbol{P}_0' &+ \left(\frac{1}{\Delta t_0} + \frac{1}{\Delta t_1}\right)^2 \boldsymbol{P}_1' = -\left(\frac{2}{(\Delta t_0)^3} + \frac{3}{\Delta t_1(\Delta t_0)^2}\right)\boldsymbol{P}_0 \\ &+ \left(\frac{2}{(\Delta t_0)^3} - \frac{1}{(\Delta t_1)^3} + \frac{3}{\Delta t_1(\Delta t_0)^2}\right)\boldsymbol{P}_1 + \frac{1}{(\Delta t_1)^3}\boldsymbol{P}_2. \end{aligned} \tag{3.21}$$

There is an analogous equation for the other end of the spline curve.

So far we have discussed only open curves. Closed curves can also be constructed using *periodic boundary conditions*. For a closed curve we set $t_0 = t_N$. Since we want

$$\begin{aligned} \boldsymbol{X}_0(t_0) &= \boldsymbol{X}_{N-1}(t_N), \qquad \boldsymbol{X}_0'(t_0) = \boldsymbol{X}_{N-1}'(t_N), \\ \boldsymbol{X}_0''(t_0) &= \boldsymbol{X}_{N-1}''(t_N), \end{aligned} \tag{3.22}$$

we must take $\boldsymbol{P}_0 = \boldsymbol{P}_N$ and $\boldsymbol{P}'_0 = \boldsymbol{P}'_N$. Now using (3.16), we get for $i = N-1$

$$\Delta t_{N-1}\boldsymbol{P}'_{N-2} + 2(\Delta t_{N-1} + \Delta t_{N-2})\boldsymbol{P}'_{N-1} + \Delta t_{N-2}\boldsymbol{P}'_0$$
$$= 3\frac{\Delta t_{N-2}}{\Delta t_{N-1}}(\boldsymbol{P}_0 - \boldsymbol{P}_{N-1}) + 3\frac{\Delta t_{N-1}}{\Delta t_{N-2}}(\boldsymbol{P}_{N-1} - \boldsymbol{P}_{N-2}),$$

and for $i = N$

$$\Delta t_0\boldsymbol{P}'_{N-1} + 2(\Delta t_0 + \Delta t_{N-1})\boldsymbol{P}'_0 + \Delta t_{N-1}\boldsymbol{P}'_1$$
$$= 3\frac{\Delta t_{N-1}}{\Delta t_0}(\boldsymbol{P}_1 - \boldsymbol{P}_0) + 3\frac{\Delta t_0}{\Delta t_{N-1}}(\boldsymbol{P}_0 - \boldsymbol{P}_{N-1}).$$

Substituting these equations in place of the first and the last ones in (3.18), we get a linear system of equations for the computation of the spline coefficients of a periodic spline curve. It is, however, certainly no longer tridiagonal.

As an example, we now construct a cubic spline curve which interpolates a point $\boldsymbol{P}_0$ with tangent $\boldsymbol{T}_0$ at one end, and a point $\boldsymbol{P}_4$ with a natural boundary condition at the other end (see Fig. 3.12), using equally spaced knots with $\Delta t_i = 1$.

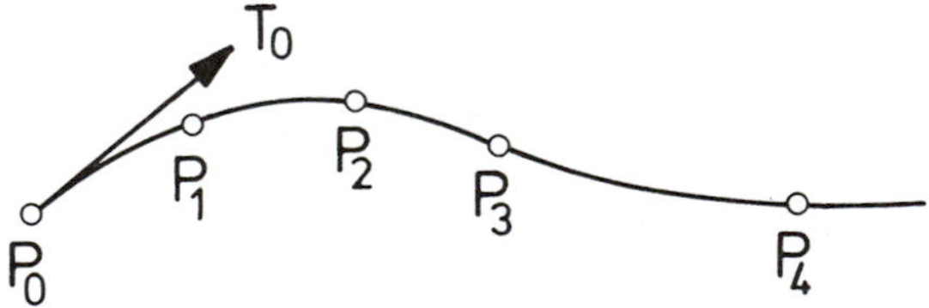

Fig. 3.12. Spline curve with initial slope.

The above discussion leads formally to the following linear system of equations

$$\boldsymbol{A}\{\boldsymbol{P}'\} = \boldsymbol{B}\{\boldsymbol{P}^*\},$$

where $\{\boldsymbol{P}'\} = (\boldsymbol{P}'_0, \ldots, \boldsymbol{P}'_N)^T$ and $\{\boldsymbol{P}^*\} = (\boldsymbol{T}_0, \boldsymbol{P}_0, \ldots, \boldsymbol{P}_N)^T$, and where $\boldsymbol{A}$ and $\boldsymbol{B}$ are the associated coefficient matrices. Writing this system of equations out, we have

$$\begin{pmatrix} 1 & 0 & 0 & 0 & 0 \\ 1 & 4 & 1 & 0 & 0 \\ 0 & 1 & 4 & 1 & 0 \\ 0 & 0 & 1 & 4 & 1 \\ 0 & 0 & 0 & 1 & 2 \end{pmatrix} \begin{pmatrix} \boldsymbol{P}'_0 \\ \boldsymbol{P}'_1 \\ \boldsymbol{P}'_2 \\ \boldsymbol{P}'_3 \\ \boldsymbol{P}'_4 \end{pmatrix} = \begin{pmatrix} 1 & 0 & 0 & 0 & 0 & 0 \\ 0 & -3 & 0 & 3 & 0 & 0 \\ 0 & 0 & -3 & 0 & 3 & 0 \\ 0 & 0 & 0 & -3 & 0 & 3 \\ 0 & 0 & 0 & 0 & -3 & 3 \end{pmatrix} \begin{pmatrix} \boldsymbol{T}_0 \\ \boldsymbol{P}_0 \\ \boldsymbol{P}_1 \\ \boldsymbol{P}_2 \\ \boldsymbol{P}_3 \\ \boldsymbol{P}_4 \end{pmatrix}. \tag{3.23}$$

Inverting the coefficient matrix on the left leads to the solution

$$\{\boldsymbol{P}'\} = \boldsymbol{A}^{-1}\boldsymbol{B}\{\boldsymbol{P}^*\}.$$

Except for the periodic case, the computation of cubic spline curves reduces to solving a system of equations with (often symmetric) tridiagonal coefficient matrix $\boldsymbol{A}$. It can be shown (cf. e.g., [SPÄ 83, 90]) that with equally spaced knots, $\boldsymbol{A}$ is diagonally dominant and positive definite, and hence we have

Theorem 3.3. *For any given interpolation points $\boldsymbol{P}_i$ and appropriate boundary conditions, there is a unique cubic interpolating spline curve using equally spaced knots in the parameter interval.*

In this case, it can be shown that the coefficient matrix $\boldsymbol{A}$ is well conditioned, see e.g., [ENG 85]. The corresponding tridiagonal system can be efficiently solved using specially designed algorithms which make use of the band structure of $\boldsymbol{A}$.

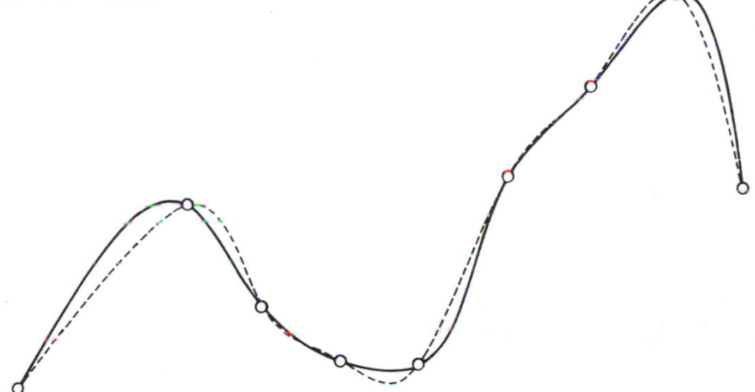

Fig. 3.13a. Interpolation of points $\boldsymbol{P}_i$ using a chord length parametrization (solid line) and a uniform parametrization (dashed line).

The boundary conditions influence the form of the interpolating curve (for examples, see [BART 87], [YAM 88], [FAR 90], [SPÄ 90]) as well as the order of approximation of a given curve $\boldsymbol{X}(t)$ by the interpolating spline (see [Beh 79]). The shape of the curve is also influenced by the choice of parametrization. Figs. 3.13a,b show a natural cubic spline interpolating the points $\boldsymbol{P}_0, \ldots, \boldsymbol{P}_8$. In Fig. 3.13a we illustrate both equally spaced knots (the dashed curve), and the chordal parametrization where the subinterval lengths are proportional to the distances between neighboring points (the solid curve). Fig. 3.13b shows the corresponding results using the chord lengths in all subintervals except for the fourth, which is replaced by 1 (for the dashed curve) and by 100 (for the solid curve). For further examples, see e.g., [BART 87], [FAR 90].

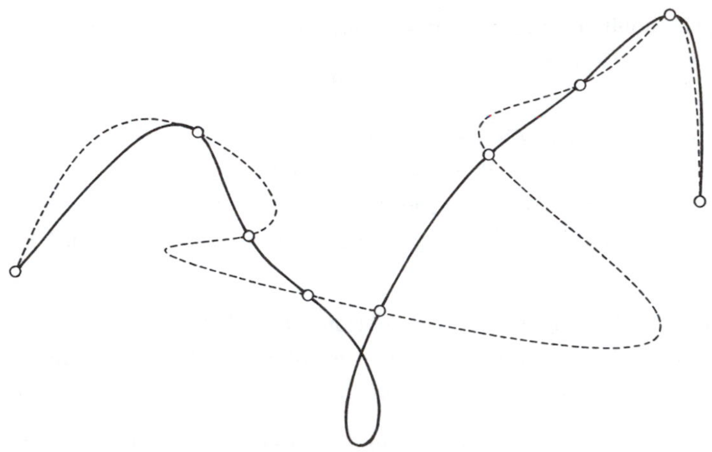

Fig. 3.13b. Interpolation of the same points $\boldsymbol{P}_i$ using chord length parametrization, where the length of the fourth segment is increased to 100 (solid line) and decreased to 1 (dashed line).

3.4. Quintic Spline Curves

In some applications, e.g., in the smoothing of curves, it may be desirable to interpolate second derivative information at the knots. This is not possible with cubic splines, and so we have to use splines of higher degree. While we could use splines of degree 4, in order to be able to enforce symmetric boundary conditions, it is more convenient to work with degree 5, i.e., with *quintic splines*. In this section we show how to construct an interpolating quintic spline. For each $i = 0(1)N{-}1$, we suppose that the i-th segment of the spline is given by

$$\boldsymbol{X}_i(t) = \boldsymbol{A}_i(t-t_i)^5+\boldsymbol{B}_i(t-t_i)^4+\boldsymbol{C}_i(t-t_i)^3+\boldsymbol{D}_i(t-t_i)^2+\boldsymbol{E}_i(t-t_i)+\boldsymbol{F}_i. \quad (3.24)$$

Now suppose we are given $N+1$ data points $\boldsymbol{P}_i$ and corresponding parameter values t_i, $i = 0(1)N$. Then the requirement that all derivatives up to order 4 be continuous across the knots (coupled with the interpolation condition), leads to the following equations:

$$\begin{aligned}
\boldsymbol{X}_i(t_i) &= \boldsymbol{P}_i, \\
\boldsymbol{X}_i(t_{i+1}) &= \boldsymbol{P}_{i+1}, \\
\boldsymbol{X}'_{i-1}(t_i) &= \boldsymbol{P}'_i = \boldsymbol{X}'_i(t_i), \\
\boldsymbol{X}''_{i-1}(t_i) &= \boldsymbol{P}''_i = \boldsymbol{X}''_i(t_i), \\
\boldsymbol{X}'''_{i-1}(t_i) &= \boldsymbol{P}'''_i = \boldsymbol{X}'''_i(t_i), \\
\boldsymbol{X}^{(4)}_{i-1}(t_i) &= \boldsymbol{P}^{(4)}_i = \boldsymbol{X}^{(4)}_i(t_i),
\end{aligned} \quad (3.25)$$

with still to be determined derivatives $\boldsymbol{P}_i', \ldots, \boldsymbol{P}_i^{(4)}$. Taking derivatives of (3.24), we have

$$\begin{aligned} \boldsymbol{X}_i'(t) &= 5\boldsymbol{A}_i(t-t_i)^4 + 4\boldsymbol{B}_i(t-t_i)^3 + 3\boldsymbol{C}_i(t-t_i)^2 + 2\boldsymbol{D}_i(t-t_i) + \boldsymbol{E}_i, \\ \boldsymbol{X}_i''(t) &= 20\boldsymbol{A}_i(t-t_i)^3 + 12\boldsymbol{B}_i(t-t_i)^2 + 6\boldsymbol{C}_i(t-t_i) + 2\boldsymbol{D}_i, \\ \boldsymbol{X}_i'''(t) &= 60\boldsymbol{A}_i(t-t_i)^2 + 24\boldsymbol{B}_i(t-t_i) + 6\boldsymbol{C}_i, \\ \boldsymbol{X}_i^{(4)}(t) &= 120\boldsymbol{A}_i(t-t_i) + 24\boldsymbol{B}_i. \end{aligned} \tag{3.26}$$

The conditions on the zeroth and fourth derivatives lead immediately to

$$\boldsymbol{F}_i = \boldsymbol{P}_i, \qquad \boldsymbol{B}_i = \tfrac{1}{24}\boldsymbol{P}_i^{(4)}.$$

The requirement that

$$\boldsymbol{X}_i^{(4)}(t_{i+1}) = \boldsymbol{X}_{i+1}^{(4)}(t_{i+1}) \tag{3.27a}$$

results in

$$\boldsymbol{A}_i = \frac{1}{120\Delta t_i}(\boldsymbol{P}_{i+1}^{(4)} - \boldsymbol{P}_i^{(4)}), \tag{3.27b}$$

where $\Delta t_i = t_{i+1} - t_i$. Looking at the second derivative at t_i and t_{i+1}, we get

$$\boldsymbol{D}_i = \tfrac{1}{2}\boldsymbol{P}_i'', \qquad \boldsymbol{C}_i = \frac{1}{6\Delta t_i}(\boldsymbol{P}_{i+1}'' - \boldsymbol{P}_i'') - \frac{\Delta t_i}{36}(\boldsymbol{P}_{i+1}^{(4)} + 2\boldsymbol{P}_i^{(4)}). \tag{3.27c}$$

Finally, using $\boldsymbol{X}_i(t_{i+1}) = \boldsymbol{P}_{i+1}$ yields

$$\boldsymbol{E}_i = \frac{1}{\Delta t_i}(\boldsymbol{P}_{i+1} - \boldsymbol{P}_i) - \frac{1}{6}\Delta t_i(\boldsymbol{P}_{i+1}'' + 2\boldsymbol{P}_i'') + \frac{1}{360}(\Delta t_i)^3(7\boldsymbol{P}_{i+1}^{(4)} + 8\boldsymbol{P}_i^{(4)}). \tag{3.27d}$$

Using conditions (3.25) for the first and third derivatives, we can eliminate $\boldsymbol{P}_i'$ and $\boldsymbol{P}_i'''$. Indeed, combining

$$\boldsymbol{X}_{i-1}'(t_i) = \boldsymbol{X}_i'(t_i)$$

with (3.27) leads to

$$\begin{aligned} &\frac{1}{\Delta t_{i-1}}(\boldsymbol{P}_i - \boldsymbol{P}_{i-1}) + \frac{1}{6}\Delta t_{i-1}(\boldsymbol{P}_{i-1}'' + 2\boldsymbol{P}_i'') - \frac{1}{360}(\Delta t_{i-1})^3(7\boldsymbol{P}_{i-1}^{(4)} + 8\boldsymbol{P}_i^{(4)}) \\ &\quad = \frac{1}{\Delta t_i}(\boldsymbol{P}_{i+1} - \boldsymbol{P}_i) - \frac{1}{6}\Delta t_i(\boldsymbol{P}_{i+1}'' + 2\boldsymbol{P}_i'') + \frac{1}{360}(\Delta t_i)^3(7\boldsymbol{P}_{i+1}^{(4)} + 8\boldsymbol{P}_i^{(4)}), \end{aligned} \tag{3.28a}$$

and then requiring

$$\boldsymbol{X}_{i-1}'''(t_i) = \boldsymbol{X}_i'''(t_i)$$

implies the following recurrence formula for the second and fourth derivatives:

$$\begin{aligned} -\frac{1}{\Delta t_{i-1}}\boldsymbol{P}_{i-1}'' + \Big(\frac{1}{\Delta t_i} + \frac{1}{\Delta t_{i-1}}\Big)\boldsymbol{P}_i'' - \frac{1}{\Delta t_i}\boldsymbol{P}_{i+1}'' + \frac{\Delta t_{i-1}}{6}\boldsymbol{P}_{i-1}^{(4)} \\ + \frac{1}{3}(\Delta t_i + \Delta t_{i-1})\boldsymbol{P}_i^{(4)} + \frac{\Delta t_i}{6}\boldsymbol{P}_{i+1}^{(4)} = 0. \end{aligned} \qquad (3.28b)$$

The equations (3.28a), (3.28b) hold for $i = 1(1)N{-}1$, i.e., we have a total of $2(N-1)$ equations for the $2(N+1)$ unknown coefficients $\boldsymbol{P}_0'', \ldots, \boldsymbol{P}_N''$ and $\boldsymbol{P}_0^{(4)}, \ldots, \boldsymbol{P}_N^{(4)}$. In order to get a square system of equations, we again need to provide appropriate boundary conditions. Here we consider the *natural boundary conditions*

$$\boldsymbol{P}_0''' = \boldsymbol{P}_N''' = \boldsymbol{P}_0^{(4)} = \boldsymbol{P}_N^{(4)} = 0. \qquad (3.29)$$

Now $\boldsymbol{P}_0''' = \boldsymbol{X}_0'''(t_0) = 0$ together with (3.26) implies $\boldsymbol{C}_0 = 0$. Using (3.27c) and (3.29) we get

$$\boldsymbol{P}_0'' = \boldsymbol{P}_1'' - \tfrac{1}{6}(\Delta t_0)^2\boldsymbol{P}_1^{(4)}. \qquad (3.29a)$$

Similarly, $\boldsymbol{P}_N''' = \boldsymbol{X}_{N-1}'''(t_N)$ implies

$$\boldsymbol{P}_N'' = \boldsymbol{P}_{N-1}'' - \tfrac{1}{6}(\Delta t_{N-1})^2\boldsymbol{P}_{N-1}^{(4)}, \qquad (3.29b)$$

and using (3.29), we get $\boldsymbol{B}_0 = 0$.

We can now elimininate $\boldsymbol{P}_0''$ and $\boldsymbol{P}_N''$ from (3.28a) and (3.28b) for the indices $i = 1$ and $i = N-1$. Then we get the following "boundary conditions" for $i = 1$:

$$\begin{aligned} (3\Delta t_0 + 2\Delta t_1)\boldsymbol{P}_1'' + \Delta t_1\boldsymbol{P}_2'' - \tfrac{1}{60}(18(\Delta t_0)^3 + 8(\Delta t_1)^3)\boldsymbol{P}_1^{(4)} \\ - \tfrac{7}{60}(\Delta t_1)^3\boldsymbol{P}_2^{(4)} = 6\left[-\frac{(\boldsymbol{P}_1 - \boldsymbol{P}_0)}{\Delta t_0} + \frac{(\boldsymbol{P}_2 - \boldsymbol{P}_1)}{\Delta t_1}\right], \end{aligned} \qquad (3.29c)$$

with

$$\frac{6}{\Delta t_1}\boldsymbol{P}_1'' - \frac{6}{(\Delta t_1)}\boldsymbol{P}_2'' + (3\Delta t_0 + 2\Delta t_1)\boldsymbol{P}_1^{(4)} + \Delta t_1\boldsymbol{P}_2^{(4)} = 0,$$

and similarly for $i = N-1$,

$$\begin{aligned} \Delta t_{N-2}\boldsymbol{P}_{N-2}'' + (2\Delta t_{N-2} + 3\Delta t_{N-1})\boldsymbol{P}_{N-1}'' \\ - \tfrac{7}{60}(\Delta t_{N-2})^3\boldsymbol{P}_{N-2}^{(4)} - \tfrac{1}{60}(8\,(\Delta t_{N-2})^3 + 18(\Delta t_{N-1})^3)\,\boldsymbol{P}_{N-1}^{(4)} \\ = 6\left(\frac{\boldsymbol{P}_N - \boldsymbol{P}_{N-1}}{\Delta t_{N-1}} - \frac{\boldsymbol{P}_{N-1} - \boldsymbol{P}_{N-2}}{\Delta t_{N-2}}\right), \end{aligned} \qquad (3.29d)$$

with

$$-\frac{6}{\Delta t_{N-2}}\boldsymbol{P}''_{N-2}+\frac{6}{\Delta t_{N-2}}\boldsymbol{P}''_{N-1}+\Delta t_{N-2}\boldsymbol{P}^{(4)}_{N-3}+(2\Delta t_{N-2}+3\Delta t_{N-1})\boldsymbol{P}^{(4)}_{N-1}=0.$$

Combining the recurrence formulae (3.28a)–(3.28b) with equations (3.29c)–(3.29d), we get the following system of equations (see e.g., [SPÄ 83]):

$$\begin{pmatrix} \boldsymbol{A} & -\boldsymbol{B} \\ \boldsymbol{C} & \boldsymbol{A} \end{pmatrix}\begin{pmatrix} \boldsymbol{P}'' \\ \boldsymbol{P}^{(4)} \end{pmatrix}=\begin{pmatrix} \boldsymbol{D} \\ \boldsymbol{0} \end{pmatrix}, \tag{3.30}$$

where

$$\boldsymbol{P}'' := (\boldsymbol{P}''_1,\ldots,\boldsymbol{P}''_{N-1})^T,$$
$$\boldsymbol{P}^{(4)} := (\boldsymbol{P}^{(4)}_1,\ldots,\boldsymbol{P}^{(4)}_{N-1})^T,$$

$\boldsymbol{D} :=$

$$\left(6\left(\frac{(\boldsymbol{P}_2-\boldsymbol{P}_1)}{\Delta t_1}-\frac{(\boldsymbol{P}_1-\boldsymbol{P}_0)}{\Delta t_0}\right),\ldots,6\left(\frac{(\boldsymbol{P}_N-\boldsymbol{P}_{N-1})}{\Delta t_{N-1}}-\frac{(\boldsymbol{P}_{N-1}-\boldsymbol{P}_{N-2})}{\Delta t_{N-2}}\right)\right)^T.$$

The matrices $\boldsymbol{A}$, $\boldsymbol{B}$, and $\boldsymbol{C}$ appearing in (3.30) have the following form:

$$\boldsymbol{A}=\begin{pmatrix} 3\Delta t_0+2\Delta t_1 & \Delta t_1 & \cdots & & & 0 \\ \Delta t_1 & 2(\Delta t_1+\Delta t_2) & \Delta t_2 & & & 0 \\ & \Delta t_2 & 2(\Delta t_2+\Delta t_3) & \Delta t_3 & & 0 \\ \vdots & & & \ddots & & \vdots \\ 0 & & & \Delta t_{N-3} & 2(\Delta t_{N-3}+\Delta t_{N-2}) & \Delta t_{N-2} \\ 0 & & & \cdots & \Delta t_{N-2} & 2\Delta t_{N-2}+3\Delta t_{N-1} \end{pmatrix}$$

$$\boldsymbol{B}=\frac{1}{60}\begin{pmatrix} 18(\Delta t_0)^3+8(\Delta t_1)^3 & 7(\Delta t_1)^3 & \cdots & & & 0 \\ 7(\Delta t_1)^3 & 8[(\Delta t_1)^3+(\Delta t_2)^3] & 7(\Delta t_2)^3 & & & 0 \\ & 7(\Delta t_2)^3 & 8[(\Delta t_2)^3+(\Delta t_3)^3] & 7(\Delta t_3)^3 & & 0 \\ \vdots & & & \ddots & & \vdots \\ 0 & & & 7(t_{N-3})^3 & 8[(\Delta t_{N-3})^3+(\Delta t_{N-2})^3] & 7(\Delta t_{N-2})^3 \\ 0 & & & \cdots & 7(\Delta t_{N-2})^3 & 8(\Delta t_{N-2})^3+18(\Delta t_{N-1})^3 \end{pmatrix}$$

$$\boldsymbol{C}=\begin{pmatrix} \frac{6}{\Delta t_1} & -\frac{6}{\Delta t_1} & \cdots & & & 0 \\ -\frac{6}{\Delta t_1} & 6(\frac{1}{\Delta t_1}+\frac{1}{\Delta t_2}) & -\frac{6}{\Delta t_2} & & & 0 \\ & -\frac{6}{\Delta t_2} & 6(\frac{1}{\Delta t_2}+\frac{1}{\Delta t_3}) & -\frac{6}{\Delta t_3} & & 0 \\ \vdots & & & \ddots & & \vdots \\ 0 & & & -\frac{6}{\Delta t_{N-3}} & 6(\frac{1}{\Delta t_{N-3}}+\frac{1}{\Delta t_{N-2}}) & -\frac{6}{\Delta t_{N-2}} \\ 0 & & & \cdots & \frac{6}{t_{N-2}} & \frac{6}{\Delta t_{N-2}} \end{pmatrix}$$

The block submatrices $\boldsymbol{A}$, $\boldsymbol{B}$, and $\boldsymbol{C}$ appearing in the matrix on the right-hand side of the system (3.30) are tridiagonal and symmetric. Since $\boldsymbol{A}$ and $\boldsymbol{B}$ are diagonally dominant while $\boldsymbol{C}$ is nonnegative definite, the coefficient matrix in (3.30) is invertible. Thus (see e.g., [SPÄ 83, 90]) we have

Theorem 3.4. *Given interpolation data $(t_i, \boldsymbol{P}_i)$ and appropriate boundary conditions as described above, there exists a unique quintic interpolating spline.*

In solving the linear system (3.30), we can take advantage of the block structure of the coefficient matrix by using, for example, a block-iteration method, see e.g., [ENG 85].

As in the cubic spline case, using different boundary derivatives has a strong influence on the shape and approximation quality of the interpolating spline, cf. [Beh 81].

Fig. 3.14 shows both a cubic and a quintic spline curve interpolating a given set of data, using the chord length parametrization. The cubic spline curve (dotted line) appears to fit the prescribed data better than the quintic spline curve (solid line), but the quintic spline has fewer strong changes in curvature, and therefore appears "smoother". This is one reason why quintic splines are often used in practice.

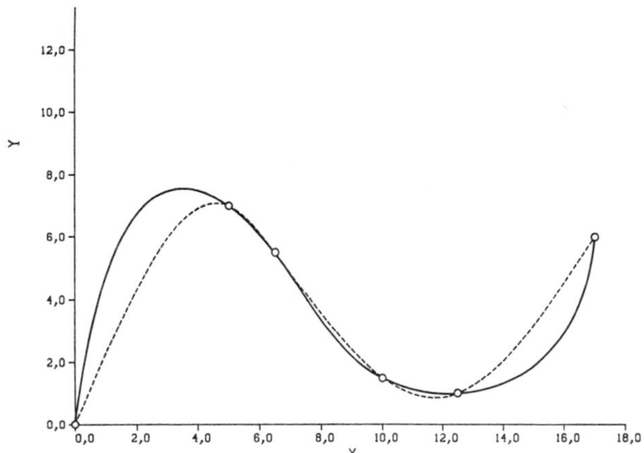

Fig. 3.14. Cubic and quintic spline interpolants.

3.5. Hermite Splines

In anology to classical Hermite interpolation, Hermite splines are functions which interpolate not only values, but also a sequence of derivatives at each knot. Suppose we are given $N+1$ parameter values $t_0 < t_1 < \cdots < t_N$, and data points $\boldsymbol{P}_i$ and derivatives $\boldsymbol{P}_i^{(k)}$ for $k = 1(1)L$ and $i = 0(1)N$. A *Hermite spline curve* is a function of the form

$$\boldsymbol{X}_i(t) = \sum_{\ell=0}^{1}\sum_{k=0}^{L} \boldsymbol{P}_{i+\ell}^{(k)}\, \boldsymbol{H}_{k,\ell}\left(\frac{t-t_i}{\Delta t_i}\right), \qquad i = 0(1)N\text{–}1, \tag{3.32}$$

where $\boldsymbol{H}_{k,\ell}(t)$ are the Hermite basis functions satisfying

$$\boldsymbol{H}_{k,0}^{(j)}(0) = \delta_{jk}, \qquad \boldsymbol{H}_{k,0}^{(j)}(1) = 0, \tag{3.33}$$

$$\boldsymbol{H}_{k,1}^{(j)}(1) = 0, \qquad \boldsymbol{H}_{k,1}^{(j)}(1) = \delta_{jk}, \tag{3.34}$$

for $j, k = 0(1)L$.

In the cubic case we take $L = 1$, and the data consist of the prescribed points $\boldsymbol{P}_i$ and the first derivatives $\boldsymbol{P}_i'$. The Hermite basis functions in this case are given explicitly in (3.12). In the quintic case $L = 2$, and the Hermite functions satisfying (3.33)–(3.34) are given by

$$\begin{aligned}
\boldsymbol{H}_{0,0}(\tau) &= -6\tau^5 + 15\tau^4 - 10\tau^3 + 1, \\
\boldsymbol{H}_{0,1}(\tau) &= \ \ 6\tau^5 - 15\tau^4 + 10\tau^3, \\
\boldsymbol{H}_{1,0}(\tau) &= -3\tau^5 + 8\tau^4 - 6\tau^3 + \tau, \\
\boldsymbol{H}_{1,1}(\tau) &= -3\tau^5 + 7\tau^4 - 4\tau^3, \\
\boldsymbol{H}_{2,0}(\tau) &= -\tfrac{1}{2}\tau^5 + \tfrac{3}{2}\tau^4 - \tfrac{3}{2}\tau^3 + \tfrac{1}{2}\tau^2, \\
\boldsymbol{H}_{2,1}(\tau) &= \ \ \tfrac{1}{2}\tau^5 - \tau^4 + \tfrac{1}{2}\tau^3.
\end{aligned} \tag{3.35}$$

In order to use cubic Hermite spline interpolation when only the points $\boldsymbol{P}_i$ are given, we need some way of estimating the tangents $\boldsymbol{P}_i'$. This must be done carefully, since as we already observed (cf. Fig. 2.19), a poor choice of tangents will cause problems. One way to get a reasonable estimate for the tangent $\boldsymbol{P}_i'$ corresponding to the point $\boldsymbol{P}_i$ is to use the four closest neighboring points to compute

$$\boldsymbol{P}_i' = (1-\alpha)\boldsymbol{S}_i + \alpha\boldsymbol{S}_{i+1}, \tag{3.36}$$

(see [Renn 82]), where

$$\boldsymbol{S}_i := \boldsymbol{P}_i - \boldsymbol{P}_{i-1} \qquad \text{(bi-secant of the points } \boldsymbol{P}_i\text{)},$$

and

$$\alpha = \frac{|\boldsymbol{S}_{i-1} \times \boldsymbol{S}_i|}{|\boldsymbol{S}_{i-1} \times \boldsymbol{S}_i| + |\boldsymbol{S}_{i+1} \times \boldsymbol{S}_{i+2}|} \qquad \text{(average of the areas inside the neighboring bi-secants of } \boldsymbol{P}_i\text{)}.$$

Special formulae have to be used at the end points to compute $\boldsymbol{P}_0'$, $\boldsymbol{P}_1'$, $\boldsymbol{P}_{N-1}'$, and $\boldsymbol{P}_N'$, cf. e.g., [BOO 78], [Renn 82]. An overview of possible ways to compute tangents can be found in [Boeh 84] and [FAR 90], see also Sect. 4.1.3.

Remark. As a generalization of the Hermite splines, it is also possible to define *lacunary splines*, see e.g., [Fau 86].

3.6. Splines in Tension

The usual splines have a tendency to oscillate, causing undesirable inflection points. In order to construct spline functions which are *stiffer*, [Schw 66] made use of a result from mechanics: the solution of the differential equation corresponding to a beam under tension involves hyperbolic sine functions, see e.g., [TIM 56]. Schweikert observed that the tension parameter can be used as a smoothing parameter to dampen the undesired oscillations.

In this section we show how the *classical splines in tension* (also called *exponential splines*) of Schweikert, as well as certain polynomial alternatives introduced later in [Nie 74], can be treated in terms of minimization problems with boundary conditions, see e.g., [Hag 90].

3.6.1. Exponential Splines

We have seen that in analogy to (3.2), the cubic spline minimizes the integral

$$\int_0^\ell \|X''(t)\|^2 dt,$$

and thus produces an approximation to the curve of a bent thin rod. Undesired points of inflection in such a rod can be removed by pulling on its ends to create tension, which in turn reduces the arc length of the rod. The arc length of a curve is given by

$$\int_0^\ell \|X'(t)\|^2 dt,$$

which led Schweikert to consider minimizing the expression

$$\int_0^\ell \|X''(t)\|^2 dt + p^2 \int_0^\ell \|X'(t)\|^2 dt, \tag{3.37}$$

subject to interpolation, continuity, and boundary conditions.

The free parameter $p \geq 0$ influences the entire curve, and is thus a kind of *global tension parameter*. This is a disadvantage of the classical spline in tension. Local control over the smoothing can be accomplished, however, if we modify the expression (3.37) to be minimized slightly:

$$\int_0^\ell \|X''(t)\|^2 dt + \sum_{k=1}^{N} p_k^2 \int_{t_{k-1}}^{t_k} \|X'(t)\|^2 dt. \tag{3.38}$$

Here we have a smoothing parameter $p_k \geq 0$ associated with each subinterval.

An extremal function for this variational problem must satisfy an associated *Euler-Lagrange equation* on each subinterval $[t_{k-1}, t_k]$, $k = 1(1)N$. In analogy to (3.4), these equations are

$$\boldsymbol{X}_k^{(4)}(t) - p_k^2 \boldsymbol{X}_k''(t) = 0 \quad \text{or} \quad \ddddot{\boldsymbol{X}}_k(\tau) - c_k^2 \ddot{\boldsymbol{X}}_k(\tau) = 0, \tag{3.39}$$

with $c_k := \Delta t_{k-1} p_k$ and $\Delta t_k := t_{k+1} - t_k$, where $\tau = (t - t_{k-1})/\Delta t_{k-1} \in [0, 1]$ serves as a *local parameter* on $[t_{k-1}, t_k]$. The solution of (3.39) is

$$\boldsymbol{X}_k(\tau) = \boldsymbol{a}_k + \boldsymbol{b}_k \tau + \boldsymbol{c}_k e^{c_k \tau} + \boldsymbol{d}_k e^{-c_k \tau}, \tag{3.40a}$$

which can be rewritten as

$$\boldsymbol{X}_k(\tau) = \boldsymbol{A}_k(1 - \tau) + \boldsymbol{B}_k \tau + \boldsymbol{C}_k \theta_k(1 - \tau) + \boldsymbol{D}_k \theta_k(\tau), \tag{3.40b}$$

with

$$\theta_k(\tau) := \frac{\sinh c_k \tau - \tau \sinh c_k}{\sinh c_k - c_k}.$$

The formula (3.40b) has better numerical properties than (3.40a), especially in the cases where $p_k \to 0$ or $p_k \to \infty$. The constants $\boldsymbol{A}_k$, $\boldsymbol{B}_k$, $\boldsymbol{C}_k$, $\boldsymbol{D}_k$ and $\boldsymbol{a}_k$, $\boldsymbol{b}_k$, $\boldsymbol{c}_k$, $\boldsymbol{d}_k$ have to be computed from the interpolation, continuity, and boundary conditions. Suppose

$$\boldsymbol{X}(t) = \sum_{k=1}^{N} \boldsymbol{X}_k(\tau),$$

with t the *global* and τ the *local* parameter, is to interpolate the points $\boldsymbol{P}_0, \ldots, \boldsymbol{P}_N$ at $0 = t_0 < t_1 < \cdots < t_N = \ell$. Then it follows from (3.40) that

$$\boldsymbol{X}(t_{k-1}) = \boldsymbol{X}_k(0) = \boldsymbol{P}_{k-1} = \boldsymbol{A}_k, \qquad \boldsymbol{X}(t_k) = \boldsymbol{X}_k(1) = \boldsymbol{P}_k = \boldsymbol{B}_k. \tag{3.41a}$$

Taking the first derivative of (3.40b) and using the chain rule gives

$$\begin{aligned}
\boldsymbol{P}'_{k-1} &:= \boldsymbol{X}'(t_{k-1}^+) = \frac{1}{\Delta t_{k-1}} \dot{\boldsymbol{X}}_k(0) \\
&= \frac{1}{\Delta t_{k-1}}(-\boldsymbol{A}_k + \boldsymbol{B}_k - \boldsymbol{C}_k \dot{\theta}_k(1) + \boldsymbol{D}_k \dot{\theta}_k(0)), \\
\boldsymbol{P}'_k &:= \boldsymbol{X}'(t_k^-) = \frac{1}{\Delta t_{k-1}} \dot{\boldsymbol{X}}_k(1) \\
&= \frac{1}{\Delta t_{k-1}}(-\boldsymbol{A}_k + \boldsymbol{B}_k - \boldsymbol{C}_k \dot{\theta}_k(0) + \boldsymbol{D}_k \dot{\theta}_k(1)),
\end{aligned}$$

where again we write $\boldsymbol{P}'_k$ for the still unknown derivatives. Using (3.41a), we have

$$\boldsymbol{C}_k = -\frac{(\boldsymbol{P}_{k-1}-\boldsymbol{P}_k)(\dot\theta_k(1)+1)+(\boldsymbol{P}'_k+\boldsymbol{P}'_{k-1}\dot\theta_k(1))\Delta t_{k-1}}{N_k},$$
$$\boldsymbol{D}_k = \frac{(\boldsymbol{P}_{k-1}-\boldsymbol{P}_k)(\dot\theta_k(1)+1)+(\boldsymbol{P}'_k\dot\theta_k(1)+\boldsymbol{P}'_{k-1})\Delta t_{k-1}}{N_k}, \qquad (3.41b)$$

with

$$N_k := [\dot\theta_k(1)]^2 - 1.$$

The equation (3.41) still contains the unknown derivatives $\boldsymbol{P}'_k$ and $\boldsymbol{P}'_{k-1}$, but these can be determined from the continuity conditions

$$\boldsymbol{X}''(t_k^-) = \frac{1}{(\Delta t_{k-1})^2}\ddot{\boldsymbol{X}}_k(1) = \frac{1}{(\Delta t_k)^2}\ddot{\boldsymbol{X}}_{k+1}(0) = \boldsymbol{X}''(t_k^+).$$

The equality of the second derivatives leads to

$$\ddot\theta_k(1)(\Delta t_k)^2\boldsymbol{D}_k = \ddot\theta_{k+1}(1)(\Delta t_{k-1})^2\boldsymbol{C}_{k+1}.$$

Combining this with (3.41) gives the following recurrence formula for the $\boldsymbol{P}'_k$:

$$f_k\boldsymbol{P}'_{k-1} + \big(\dot\theta_{k+1}(1)g_k + \dot\theta_k(1)f_k\big)\boldsymbol{P}'_k + g_k\boldsymbol{P}'_{k+1} = g_k\boldsymbol{R}_{k+1} + f_k\boldsymbol{R}_k, \qquad (3.42)$$

for $k = 1(1)N–1$, where

$$f_k := \ddot\theta_k(1)\Delta t_k N_{k+1}, \quad g_k := \ddot\theta_{k+1}(1)\Delta t_{k-1}N_k, \qquad k = 1(1)N–1,$$
$$\boldsymbol{R}_k := \frac{\boldsymbol{P}_k - \boldsymbol{P}_{k-1}}{\Delta t_{k-1}}\big(\dot\theta_k(1)+1\big), \qquad k = 1(1)N.$$

If we introduce appropriate boundary conditions, we get a square system of equations, and the associated matrix is triadiagonal diagonally dominant (cf. e.g., [SPÄ 83]), and thus nonsingular. Solving it and using (3.41b) gives the coefficients of the unique interpolating spline in tension. Efficient and stable algorithms can be found in [Rent 80].

Fig. 3.15a shows an interpolating spline in tension where all of the tension parameters are taken to have the same value p, and so it is in fact an *exponential spline*, i.e., one of Schweikert's classical splines in tension. Setting $p = 0$ results in the cubic interpolating spline, while $p = 100$ leads to a curve which is close to being the piecewise linear interpolant corresponding to the limiting case $p \to \infty$. Fig. 3.15b shows the interpolant associated with the tension vector $\boldsymbol{p} = (5, 100, 10, 3, 3, 3, 3, 100, 4, 100)^T$, where the dashed curve is the cubic spline corresponding to $p_k = p = 0$ for all k.

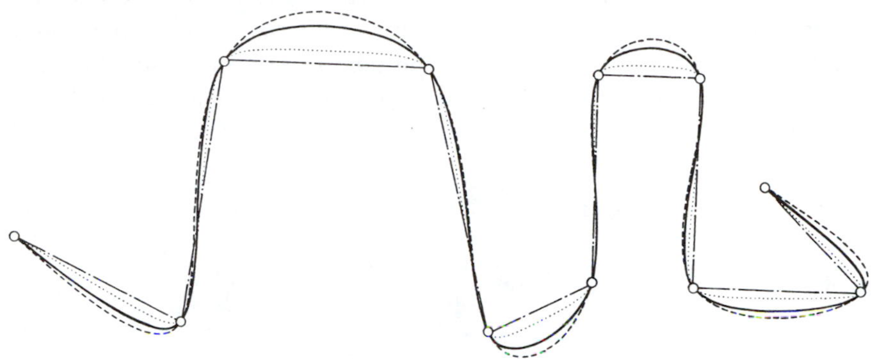

Fig. 3.15a. Splines in tension with different tension parameters and $\Delta t_k = 1$: dashed curve $p = 0$ (cubic spline), solid curve $p = 3$, dotted curve $p = 10$, and dash/dotted curve $p = 100$.

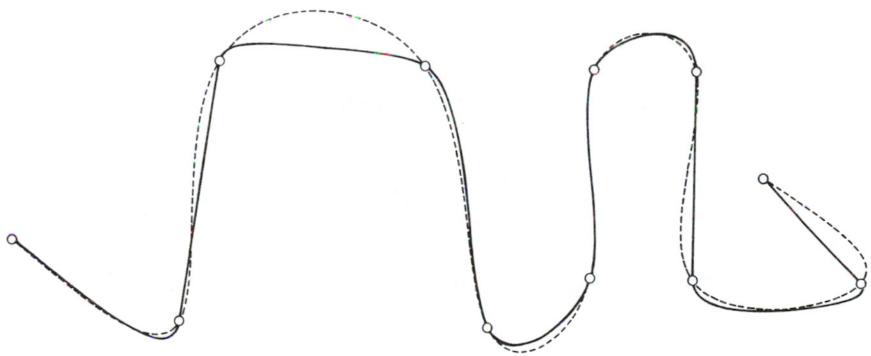

Fig. 3.15b. Splines in tension with different tension parameters and $\Delta t_k = 1$: dashed curve cubic spline; solid curve $\boldsymbol{p} = (5, 100, 10, 3, 3, 3, 3, 100, 4, 100)^T$.

Remark. The integral

$$\int_0^\ell \|\boldsymbol{X}''(t) + p\boldsymbol{X}'(t)\|^2 dt \tag{3.43}$$

has the same physical meaning as (3.37), and similarly leads to (3.39). Thus splines in tension also minimize (3.43), see [Boo 66].

Remark. While above we started with (3.40) and determined the unknown first derivatives from the conditions on the continuity of the second derivatives at the knots, it is also possible to work with the second derivatives as unknowns to be determined from the conditions on the continuity of the first derivatives, see [Cli 74] (for the case $p_k = p$, $k = 1(1)N$) and [Bars 83]. Now we begin with

$$\boldsymbol{X}_k(\tau) = \boldsymbol{A}_k(1-\tau) + \boldsymbol{B}_k\tau + \boldsymbol{C}_k\theta_k(1-\tau) + \boldsymbol{D}_k\theta_k(\tau), \tag{3.44}$$

where

$$\theta_k(\tau) = \frac{\sinh c_k\tau}{c_k^2 \sinh c_k},$$

and c_k are the constants introduced above. Then the coefficients are given by

$$\boldsymbol{A}_k = \boldsymbol{P}_{k-1} - \frac{\boldsymbol{P}''_{k-1}}{p_k^2}, \quad \boldsymbol{B}_k = \boldsymbol{P}_k - \frac{\boldsymbol{P}''_k}{p_k^2},$$

$$\boldsymbol{C}_k = (\Delta t_{k-1})^2 \boldsymbol{P}''_{k-1}, \quad \boldsymbol{D}_k = (\Delta t_{k-1})^2 \boldsymbol{P}''_k,$$

and the continuity condition on the first derivatives

$$\boldsymbol{X}'(t_k^-) = \frac{1}{\Delta t_{k-1}}\dot{\boldsymbol{X}}_k(1) = \frac{1}{\Delta t_k}\dot{\boldsymbol{X}}_{k+1}(0) = \boldsymbol{X}'(t_k^+)$$

leads to the following recurrence formula for computing the $\boldsymbol{P}''_k$:

$$F_k\boldsymbol{P}''_{k-1} + (G_k + G_{k+1})\boldsymbol{P}''_k + F_{k+1}\boldsymbol{P}''_{k+1} = \boldsymbol{r}_{k+1} - \boldsymbol{r}_k, \tag{3.45}$$

where

$$F_k := \left(\frac{1}{c_k} - \frac{1}{\sinh c_k}\right)\frac{1}{p_k}, \qquad G_k := \left(\coth c_k - \frac{1}{c_k}\right)\frac{1}{p_k},$$

$$\boldsymbol{r}_k := (\boldsymbol{P}_k - \boldsymbol{P}_{k-1})\frac{1}{\Delta t_{k-1}}.$$

For arbitrary tension values p_k, equations (3.45) result in a tridiagonal, diagonally dominant, symmetric linear system of equations [Bars 83].

Remark. In working with splines in tension, we are immediately faced with the question of how to choose the tension parameters in order to remove undesirable oscillations in the curve while at the same time preserving various properties of the data such as positivity, monotonicity, and convexity (see also Sect. 3.8 and Chap. 13). The positivity and convexity of the data can be preserved using exponential splines provided that the tension factors are chosen to be sufficiently large [Pru 76]. A method for determining the p_i in order to preserve convexity was given in [Spä 69]. [Pru 76] also presents a method for choosing the tension factors so that convexity is preserved, and a second method which preserves monotonicity. [Hess 86] has obtained necessary and sufficient conditions for convexity preservation. For methods which preserve positivity, monotonicity and convexity, see [Lyn 82] (where a global tension parameter p is computed), and [Ren 87] and [Sap 88] (where local tension parameters p_i are computed iteratively). For automatic apriori determination of the p_i, see [Fen 88].

Remark. Exponential splines can also be used for least squares fitting, see Sect. 2.4. For example, [Hei 86] has developed a method where the user inputs only the data and the number of desired pieces, i.e., the number of knots. The knots themselves are determined by a cluster analysis. First a cubic spline (with all $p_k = 0$) is constructed. If undesirable inflection points are present, then a method of [Rent 80] is applied iteratively to find tension parameters which provide a satisfactory solution. The final fitting spline consists of both cubic and exponential spline pieces.

Remark. There are, of course, natural bivariate analogs of the exponential spline; see [Spä 71].

3.6.2. Polynomial Splines in Tension

Compared to polynomial splines, exponential splines have the disadvantage that they are more expensive to compute. In this section we discuss an interpolating weighted cubic spline which involves certain tension parameters. It is defined to be the minimizer of the expression

$$\sum_{k=1}^{N} p_k \int_{t_{k-1}}^{t_k} \|X''(t)\|^2 dt + \sum_{k=0}^{N} \nu_k \|X'(t_k)\|^2, \tag{3.46}$$

where $p_k \geq 0$ are prescribed *interval weights* and $\nu_k \geq 0$ are prescribed *point weights*. If $p_k = 1$ and $\nu_k = 0$ for all k, then (3.46) reduces to the case of C^2 continuous cubic splines treated in detail above. When $\nu_k = 0$ for all k, we have the *interval weighted* cubic spline (see [Salk 74], [Fol 86a]), while if

$p_k = 1$ for all k, we have Nielson's ν-spline (point weighted cubic splines), see [Nie 74, 86].

To compute these splines, suppose the spline $\boldsymbol{X}$ restricted to the k-th interval is given by

$$\boldsymbol{X}_k(\tau) = \boldsymbol{P}_{k-1}H_{00}(\tau) + \boldsymbol{P}'_{k-1}H_{10}(\tau) + \boldsymbol{P}_kH_{01}(\tau) + \boldsymbol{P}'_kH_{11}(\tau), \qquad (3.47)$$

where $H_{ik}(\tau)$ are the cubic Hermite functions given in (3.12). Then it follows from (3.46) that the unknown derivatives $\boldsymbol{P}'_k$ must satisfy the recurrence formula

$$2c_k\boldsymbol{P}'_{k-1} + (\nu_k + 4c_k + 4c_{k+1})\boldsymbol{P}'_k + 2c_{k+1}\boldsymbol{P}'_{k+1} = 6c_{k+1}\boldsymbol{r}_{k+1} + 6c_k\boldsymbol{r}_k, \quad (3.48)$$

where $c_k := p_k/\Delta t_{k-1}$, $\Delta t_k := t_{k+1} - t_k$, and $\boldsymbol{r}_k$ is as in (3.45), see [Fol 87a].

In contrast to Sect. 3.3, the natural boundary conditions are now

$$p_1\boldsymbol{X}''_1(0) - \nu_0\boldsymbol{X}'_1(0) = 0, \qquad p_N\boldsymbol{X}''_N(1) + \nu_N\boldsymbol{X}'_N(1) = 0,$$

while the boundary conditions for a periodic spline are

$$\boldsymbol{X}_1(0) = \boldsymbol{X}_N(1), \quad \boldsymbol{X}'_1(0) = \boldsymbol{X}'_N(1), \quad p_1\boldsymbol{X}''_1(0) - p_N\boldsymbol{X}''_N(1) = (\nu_0 + \nu_N)\boldsymbol{X}'_1(0).$$

For non-negative tension values p_k, ν_k, the equations (3.48) lead to a tridiagonal, diagonally dominant (in general, nonsymmetric) system of equations, see e.g., [Salk 74], [Fol 86a, 87]. For the case of ν-splines (where $p_k = 1$ for all k), [Bars 83] gave intervals of values $\underline{\nu}_k < \nu_k < \overline{\nu}_k$ for the tension parameters ν_k which lead to nonsingular systems of equations, even for negative tension values, see also Sect. 5.5.4.

It follows from (3.46) that the first derivatives with respect to the parameters are in general only C^1. However, since the second derivative satisfies

$$p_{k+1}\boldsymbol{X}''(t_k^+) - p_k\boldsymbol{X}''(t_k^-) = \nu_k\boldsymbol{X}'(t_k), \qquad (3.49)$$

(as can be seen from (3.47) and (3.48)), we conclude that if $p_k = p_{k+1}$, then $\boldsymbol{X}(t)$ is curvature continuous at $\boldsymbol{X}(t_k)$ as long as $\|\boldsymbol{X}'(t_k)\| \neq 0$. Thus, in particular, the ν-splines are curvature continuous, cf. [Nie 74, 86].

As $p_k \to \infty$, $\boldsymbol{X}(t)$ becomes linear on $[t_{k-1}, t_k]$, provided that p_{k-1}, p_{k+1} and ν_{k-1} and ν_k remain bounded (cf. the fourth interval in Fig. 3.16b and the sixth interval in Fig. 3.16d).

As $\nu_k \to \infty$, $\boldsymbol{X}(t)$ develops a corner at $\boldsymbol{X}(t_k)$ (since then $\|\boldsymbol{X}'(t_k)\| \to 0$) provided that p_k and p_{k+1} remain bounded (cf. the fourth and fifth interpolation points in Figs. 3.16c and 3.16d).

In the following figures we illustrate the behavior of the various tension splines as compared to the cubic spline.

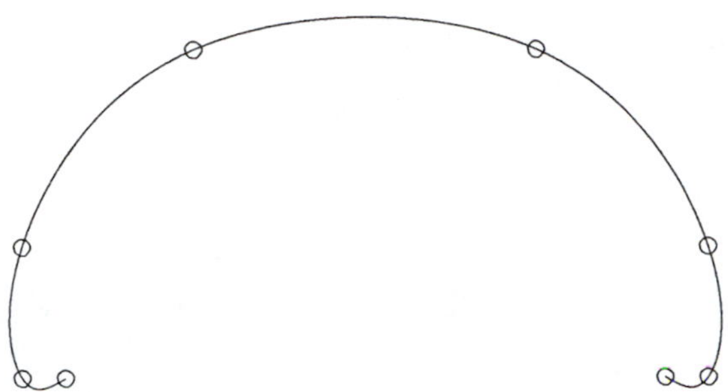

Fig. 3.16a. Cubic spline with $p_k = 1$, $\nu_k = 0$, $\Delta t_k = 1$ for all k.

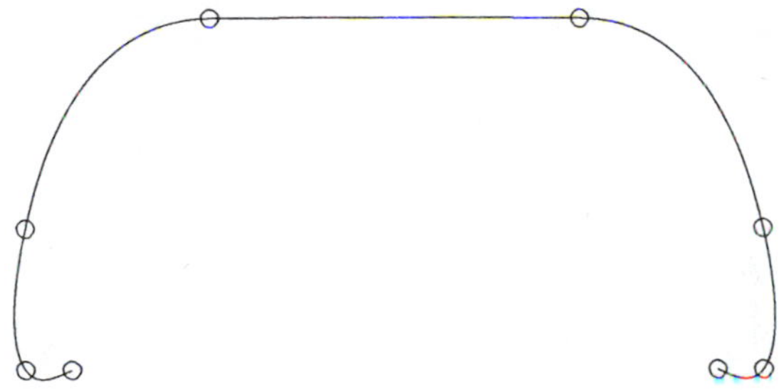

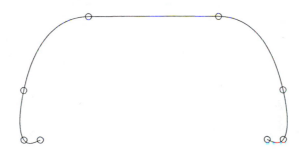

Fig. 3.16b. Interval weighted cubic spline with $\Delta t_k = 1$, $\nu_k = 0$ for all k and $p_k = 1, 1, 1, 100, 1, 1, 1$.

Remark. For interpolating function values, [Salk 74] has suggested choosing

$$p_k = \left[1 + \left(\frac{y_{k+1}(x) - y_k(x)}{\Delta t_k}\right)^2\right]^{-m},$$

with $m = 3$, since then $\int (y'')^2 dx$ approximates the L_2 norm of the curvature of $y(x)$. Choosing $m = 0$ gives $p_k = 1$ for all k, and we get the ν-splines, see [Fol 86a].

Remark. The same question arises for polynomial splines in tension as for exponential splines: how can one automatically choose the tension parameters so that shape properties such as convexity and/or monotonicity of the data are

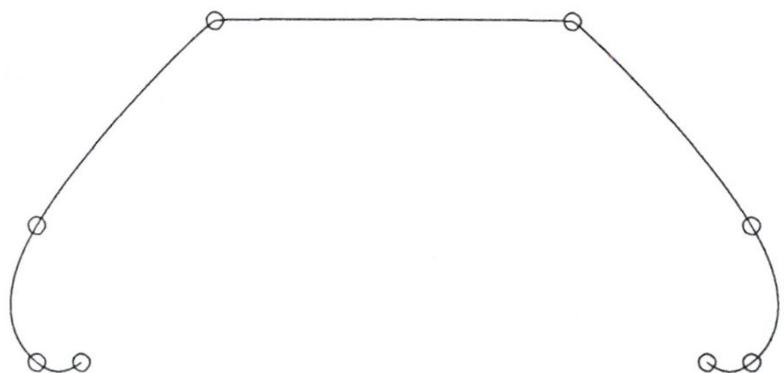

Fig. 3.16c. Point weighted cubic spline (ν-spline) with $p_k = 1$ and $\Delta t_k = 1$ for all k and $\nu_k = 0, 0, 0, 100, 100, 0, 0, 0$.

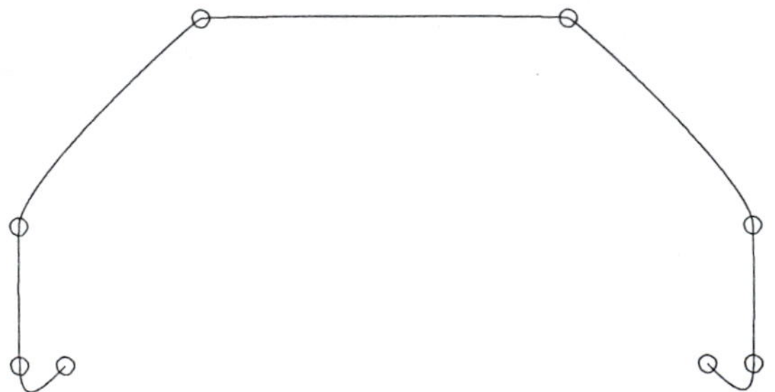

Fig. 3.16d. Interval and point weighted spline with $\Delta t_k = 1$ for all k and $p_k = 1, 100, 1, 1, 1, 100, 1$, $\nu_k = 0, 0, 0, 100, 100, 0, 0, 0$.

preserved. By (3.49), a polynomial spline solving (3.46) is a tangent continous cubic subspline satisfying the interpolation conditions $f(t_i) = y_i$, and hence is uniquely determined by its first derivatives $y'_i = f'(t_i)$ at the points $f(t_i)$. This observation was used in [Fol 88] to develop an algorithm for polynomial splines in tension which preserves both monotonicity and convexity. The interval and point weights are not used, and remain as free design parameters which can be chosen by the user.

Since $f(t)$ is piecewise cubic, integrating by parts in (3.46) and noting that $f'''(t)$ is piecewise constant, we get the following equivalent quadratic form:

$$Q(\boldsymbol{Y}') = \boldsymbol{A}^T \boldsymbol{Y}' + \tfrac{1}{2} \boldsymbol{Y}'^T \boldsymbol{H} \boldsymbol{Y}', \qquad (*)$$

with

$$Y' = (y'_0, \ldots, y'_N)^T, \qquad A = (a_0, \ldots, a_N)^T,$$

where

$$a_i = -3S_i c_i - 3S_{i+1}c_{i+1}, \quad S_i = \frac{\Delta y_i}{\Delta t_i}, \qquad \Delta y_i = y_{i+1} - y_i,$$

and H is the symmetric, tridiagonal $(n+1) \times (n+1)$ matrix with

$$H_{i,i+1} = H_{i+1,i} = c_{i+1}, \qquad H_{i,i} = \tfrac{1}{2}\nu_i + 2a_i + 2a_{i+1}.$$

Now we can enforce convexity and monotonicity conditions by requiring that y' remains between given upper and lower bounds, see e.g., [Fri 86], [Fol 88]. Then the unknown first derivatives can be found by minimizing $(*)$ subject to the monotonicity and convexity conditions on y'.

Remark. The expression (3.46) can be generalized to deal with higher order polynomial splines:

$$\sum_{k=1}^{N} p_k \int_{t_{k-1}}^{t_k} \|X^{(L)}(t)\|^2 dt + \sum_{k=0}^{N} \sum_{\ell=1}^{L-1} \nu_{k,\ell} \|X^{(\ell)}(t_k)\|^2. \tag{3.50}$$

Here we choose $L \geq 2$, and prescribe the first $L-1$ derivatives with shape parameters $\nu_{k,\ell} \geq 0$ at each knot, see [Las 92].

At each knot, $X(t)$ satisfies the smoothness conditions

$$\begin{aligned} X^{(\epsilon)}(t_k^+) &= X^{(\epsilon)}(t_k^-), \qquad \epsilon = 0(1)L\text{–}1, \\ p_{k+1}X^{(\epsilon)}(t_k^+) - p_k X^{(\epsilon)}(t_k^-) &= \nu_{k,2L-\epsilon-1}(-1)^{L-\epsilon} X^{(2L-\epsilon-1)}(t_k), \\ \epsilon &= L(1)2\ell\text{–}2. \end{aligned} \tag{3.51}$$

For $p_k = p$, $k = 1(1)N$, equations (3.50) and (3.51) cover Hagen's geometric spline curves [Hag 85], and, in particular, for $L = 3$ include the torsion and curvature-continuous quintic splines (τ-splines), see [Las 90] and Sect. 5.6.4.

Remark. The methods discussed here for computing polynomial splines in tension have been extended to bicubic surfaces by [Nie 86] and [Fol 87a] for the cases of ν-splines and interval weighted ν-splines. Biquintic surfaces based on interval weighted τ-splines have been treated by [Neu 92].

Remark. Polynomial splines in tension can, of course, also be written in Bézier or B-spline form (see Chap. 4). This has been done for interval weighted cubic splines in [Fol 87], for ν- and τ-splines in [Las 90], for interval weighted ν- and τ-splines in [Las 90, 92], for interval weighted ν-splines in [Gre 91], and for Q-splines in [Kul 91].

Remark. In [Pot 90] the expression (3.46) was extended to include the integral

$$\int_0^\ell \|\boldsymbol{X}'''(t)\|^2 dt,$$

which leads to solutions of the form

$$\boldsymbol{X}_k(\tau) = \boldsymbol{a}_k + \boldsymbol{b}_k\tau + \boldsymbol{c}_k\tau^2 + \boldsymbol{d}_k\tau^3 + \boldsymbol{\lambda}_k e^{p_k \Delta t_{k-1}\tau} + \boldsymbol{\mu}_k e^{-p_k \Delta t_{k-1}\tau},$$

i.e., to a mixture of exponential and cubic polynomial splines in tension. Because of this extra term, the pieces $\boldsymbol{X}_k(\tau)$ join together with both κ- and τ-continuity, as well as both κ'- and τ'-continuity. In the case $p_k = p$, for all k, also κ'' is continuous. Thus, these splines possess a considerably higher degree of geometric smoothness than the τ-splines which come from (3.50) with $L = 3$, where a similar integral involving $\|\boldsymbol{X}'''(t)\|^2$ appears.

3.7. Nonlinear Splines

We have seen above that cubic spline functions do a very good job of describing the shape of flexible rods which are not bent too much. More strongly bent rods can be modelled using parametric cubic splines or higher order spline functions. Since the continuity conditions between the pieces of these splines are linear, they form linear spaces of functions.

It is also of interest to try to approximate as closely as possible the true shape of a bent rod. Since this involves nonlinear continuity conditions, we are no longer dealing with a linear space of functions. We refer to these kinds of splines as *nonlinear splines*, see e.g., [Wer 79], [Hag 90]. We shall discuss two models:

- *wooden splines* which satisfy

$$\frac{d^2\kappa}{ds^2} = 0, \tag{3.52}$$

  where κ is the curvature and s is arc length,
- *mechanical splines* which minimize the bending energy

$$U = \int \kappa^2 ds. \tag{3.53}$$

Wooden splines can be regarded as a direct generalization of the cubic splines, since $\kappa \approx \frac{d^2y}{dx^2}$ whenever $s \approx x$. Thus (3.52) implies (3.2b). Now integrating (3.52) directly, between the knots we get the *natural curve equation*

$$\kappa = as + b, \qquad a, b \in \mathbb{R},$$

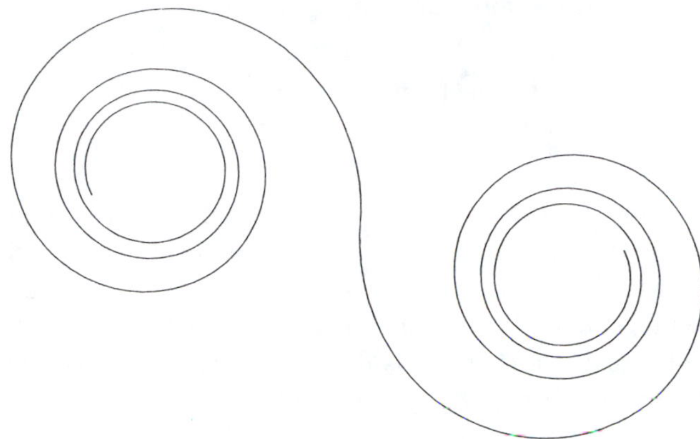

Fig. 3.17. Cornu spiral.

cf. [STR 50], [LAU 65]. This describes a plane curve called a *Cornu spiral*, *Euler spiral*, or *clothoid*, which looks like the spring in a clock, cf. Fig. 3.17. Such curves and their offsets are often used in designing roads [Baa 84], [Mee 90].

Definition 3.2. *Given points* $\{\boldsymbol{P}_i\} \in \mathbb{R}^2$, *an interpolating spline curve is called a wooden spline provided that*

1) *between any two neighboring knots, the spline reduces to a Cornu spiral,*
2) *the curvatures of the pieces of the spline match up at the points* $\boldsymbol{P}_i$.

These splines are also often referred to as *geometric* or *intrinsic* splines since they can be described by purely geometric or so-called *intrinsic quantities*, such as arc length and curvature.

Integrating the natural equation (3.52) in the xy-plane leads to an integral formula for the associated curve, see [STR 50]. In terms of the so-called *angle of contingence* θ of the curve, we have

$$\kappa = \frac{d\theta}{ds},$$

where s denotes the arc length. Integrating this equation yields

$$\theta(s) = \int_0^s \kappa(\sigma)d\sigma + \theta_s = \tfrac{1}{2}as^2 + bs + \theta_s. \tag{3.54a}$$

Since from the triangle of elevation of a curve we also have

$$\frac{dx}{ds} = \cos\theta, \qquad \frac{dy}{ds} = \sin\theta,$$

we are led to the integral formulae

$$x = \int_0^s \cos\theta(t)dt, \qquad y = \int_0^s \sin\theta(t)dt \tag{3.54b}$$

for the associated curve.

Our aim now is to develop an "approximate" analytic formula for the wooden spline curve. As shown in Fig. 3.18, we associate $s = 0$ with the point $\boldsymbol{P}_0$, and the point $\boldsymbol{P}(s)$ is described by the chord $\boldsymbol{P}_0\boldsymbol{P}$ of length L. The angle between the tangents at $\boldsymbol{P}_0$ and the line $\boldsymbol{L}$ is denoted by θ_1, while the angle between the tangents at $\boldsymbol{P}$ and $\boldsymbol{L}$ is denoted by θ_2. The angle between the two end tangents is $\psi = \theta_2 - \theta_1$, see [SU 89].

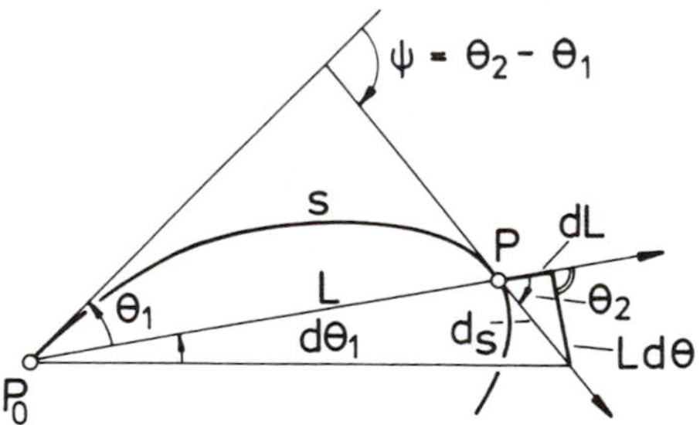

Fig. 3.18. Wooden spline curve.

From the supporting triangle at $\boldsymbol{P}$, we get dL, ds and $Ld\theta_1$ as an approximation to the circular arc length $(L + dL)d\theta_1$, neglecting terms of order two. Since $\kappa = \frac{d\psi}{ds}$ (see [STR 50]), we get the following system of differential equations:

$$\frac{dL}{ds} = \cos\theta_2, \quad \frac{d\theta_1}{ds} = -\frac{\sin\theta_2}{L}, \quad \frac{d\theta_2}{ds} = \kappa - \frac{\sin\theta_2}{L}, \tag{3.55}$$

with initial conditions

$$L(0) = 0, \qquad \theta_1(0) = \theta_2(0) = 0.$$

From the first equation in (3.55) it follows that

$$\left.\frac{dL}{ds}\right|_{s=0} = \cos\theta_2\Big|_{s=0} = 1, \qquad \left.\frac{d^2L}{ds^2}\right|_{s=0} = -\frac{d\theta_2}{ds}\sin\theta_2\Big|_{s=0} = 0,$$

that is, the chord length L of the spline curve is given by

$$L = s + \mathcal{O}(s^3).$$

Since the solutions (3.55) should be Cornu spirals, (3.54a) implies

$$\theta_1 := A_1 s + \tfrac{1}{2} B_1 s^2 + \mathcal{O}(s^3), \qquad \theta_2 := A_2 s + \tfrac{1}{2} B_2 s^2 + \mathcal{O}(s^3). \tag{3.56}$$

Thus, (3.55) leads to the system

$$\frac{d\theta_1}{ds} \approx A_1 + B_1 s = -\frac{\sin\theta_2}{L} \approx -A_2 - \tfrac{1}{2} B_2 s,$$

and comparing coefficients yields

$$A_1 = -A_2, \qquad B_1 = -\tfrac{1}{2} B_2.$$

Coupling the third equation in (3.55) with (3.52), we get

$$A_2 = \frac{b}{2}, \qquad B_2 = \frac{2}{3} a.$$

Now inserting in (3.52) for $s = 0$ and $s = L_0$ gives

$$\kappa(0) =: \kappa_1 = b, \qquad \kappa(L_0) =: \kappa_2 = aL_0 + b,$$

after which (3.56) leads to

$$\theta_1 = -\frac{L_0}{6}(2\kappa_1 + \kappa_2) + \mathcal{O}(L_0^3), \qquad \theta_2 = \frac{L_0}{6}(\kappa_1 + 2\kappa_2) + \mathcal{O}(L_0^3). \tag{3.57}$$

We can now construct an interpolating wooden spline curve. Given $N+1$ points $\boldsymbol{P}_i$, then by (3.57) each curve segment satisfies

$$\begin{aligned} \theta_{1,i+1} &= -\frac{L_{i+1}}{6}(2\kappa_i + \kappa_{i+1}) + \mathcal{O}(L_{i+1}^3), \\ \theta_{2,i} &= \frac{L_i}{6}(\kappa_{i-1} + 2\kappa_i) + \mathcal{O}(L_i^3). \end{aligned}$$

Writing

$$\phi_i := \theta_{2,i} - \theta_{1,i+1}$$

for the angle between two neighboring spline pieces, we have the following recurrence formula for the curvatures κ_i:

$$\mu_i \kappa_{i-1} + 2\kappa_i + \lambda_i \kappa_{i+1} = 3K_i + \mathcal{O}(\phi^2), \tag{3.58}$$

with

$$\lambda_i := \frac{L_{i+1}}{L_i + L_{i+1}}, \qquad \mu_i := \frac{L_i}{L_i + L_{i+1}}, \qquad K_i := \frac{2\phi_i}{L_i + L_{i+1}}.$$

In analogy with the construction of cubic spline curves, under appropriate boundary conditions, (3.58) leads to a linear system of equations for the values of the curvature at knots of the spline. Equation (3.54) together with (3.52) describes the individual curve segments of the wooden spline curve.

As a supplement to the method developed here, we mention [Mee 89, 90] where drawing, editing, and offsetting of Cornu spirals is treated, and [Nut 72] where Cornu spiral splines are developed using a special local coordinate system. See also [Ada 75], [Pal 77, 78, 78a], and [Sche 78, 78a], where in addition to point, tangent and curvature values, torsion values are also given at the knots. In addition, these papers also construct blending surfaces with prescribed curvature or torsion profiles, using linear blending functions in the sense of the pencil method of Sect. 3.2.

We turn now to *mechanical splines*. We call a curve $\boldsymbol{X}(s)$ which minimizes the bending energy $u = \int \kappa^2 ds$ an M-curve. We define an interpolating mechanical spline to be a piecewise GC^2 continuous M-curve which passes through a prescribed set of points $\{\boldsymbol{P}_i\} \in \mathbb{R}^3$, cf. [Glass 66], [Lee 73], [Meh 74], and [Mal 77].

Our analysis of mechanical splines will be based on the *Frenet equations* associated with a space curve. If a curve $\boldsymbol{X} = \boldsymbol{X}(s)$ is parametrized by arc length s, then the tangent vector $\boldsymbol{v}_1$ (cf. (2.24a)) is given by

$$\boldsymbol{v}_1 := \boldsymbol{X}', \qquad |\boldsymbol{v}_1| = 1.$$

The second derivative $\boldsymbol{X}''(s)$ points in the principal normal direction $\boldsymbol{v}_2$, and its value gives the curvature κ of the curve (cf. [STR 50], [LAU 65]), i.e., we have

$$\boldsymbol{X}'' = \boldsymbol{v}_1' = \kappa \boldsymbol{v}_2, \tag{3.59a}$$

with $\boldsymbol{v}_2$ the *principal normal vector*. Then introducing the *binormal vector* $\boldsymbol{v}_3 = \boldsymbol{v}_1 \times \boldsymbol{v}_2$ leads to the *Frenet frame* $(\boldsymbol{v}_1, \boldsymbol{v}_2, \boldsymbol{v}_3)$, whose derivatives are

$$\begin{aligned} \boldsymbol{v}_1' &= \kappa \boldsymbol{v}_2, \\ \boldsymbol{v}_2' &= -\kappa \boldsymbol{v}_1 + \tau \boldsymbol{v}_3, \\ \boldsymbol{v}_3' &= -\tau \boldsymbol{v}_2, \end{aligned} \tag{3.59b}$$

where τ is the *torsion* of the curve ($\tau = 0$ corresponds to a planar curve).

In terms of these quantities, we have

$$\kappa^2 = (\boldsymbol{v}_1')^2 \qquad \text{with side conditions } \boldsymbol{v}_1^2 = 1,\ \boldsymbol{X}' = \boldsymbol{v}_1,$$

and thus the minimization problem (3.53) reduces to the following variational problem with side conditions:

$$U = \int F ds,$$

with

$$F = (v_1')^2 + \lambda(s)(v_1^2 - 1) + 2\Phi(s) \cdot (X' - v_1), \tag{3.60}$$

where λ, Φ are *Lagrange parameters*. A necessary condition for a solution of this problem is that the *Euler-Lagrange equations*

$$\frac{d}{ds}(F_{X'}) - \frac{\partial F}{\partial X} = 0, \qquad \frac{d}{ds}(F_{v_1'}) - \frac{\partial F}{\partial v_1} = 0$$

be satisfied, and it follows that

$$-v_1'' + \lambda v_1 - \Phi = 0, \qquad \Phi' = 0.$$

Substituting the derivatives of the Frenet formulae gives

$$(\lambda + \kappa^2)v_1 - \kappa' v_2 - \kappa\tau v_3 = \Phi. \tag{3.61}$$

Differentiating again, comparing coefficients, and noting that $\Phi' = 0$, we have

$$\begin{aligned} [v_1]: &\qquad \lambda' + 3\kappa'\kappa = 0 \to \lambda + \frac{3}{2}\kappa^2 =: D \quad (= \text{constant}), \\ [v_3]: &\qquad -2\kappa'\tau - \kappa\tau' = 0 \to \kappa^2\tau =: C \quad (= \text{constant}). \end{aligned} \tag{3.62}$$

Taking the absolute value in (3.61) leads to the following differential equation for the unknown curvature κ:

$$\kappa'^2 + (\tfrac{1}{2}\kappa^2 - D)^2 + \frac{C^2}{\kappa^2} = |\Phi|^2. \tag{3.63}$$

After substituting

$$\omega = \frac{1}{3}D - \frac{1}{4}\kappa^2,$$

(3.63) becomes

$$\omega'^2 = 4\omega^3 - C_1\omega - C_3,$$

where C_i are constants. The solution of this differential equation is a *Weierstrass elliptic p-function*

$$\omega = p(s + A; C_1, C_2),$$

where A is a constant of integration. It follows that the curvature and torsion of the curve are given by

$$\kappa^2 = \frac{4}{3}D - 4p(s + A; C_1, C_2), \qquad \tau = \frac{C}{\kappa^2}.$$

Now the pieces of the spline can be computed by integrating (3.59), see [LAU 65], [SU 89].

Solutions of (3.63) subject to the additional constraint that the competing curves be of constant length were obtained by [Jou 92]. Modelling algorithms based on parametric energy minimizing curves and approximations to them by cubic splines have been developed by [Schul 90], see also [Hag 90].

3.8. Shape Preserving Splines

The methods developed so far have only taken account of the *local* behavior of the spline function. *Global* properties such as *shape preservation* are also important. Shape preservation means that a spline function interpolating or approximating monotone or convex data should also be monotone or convex, respectively. Most of the results in this area are for the case of nonparametric splines.

As preparation for studying shape preserving splines, we begin by deriving a criterion for the nonnegativity of a polynomial on a given interval; i.e., for when

$$x(t) = \sum_{i=0}^{n} a_i t^i \geq 0 \qquad \text{for all } t \in [0, 1]. \tag{3.64}$$

Clearly, a sufficient condition is that $a_i \geq 0$. The substitution $t = \frac{\tau}{1-\tau}$ maps $[0, 1]$ to $\mathbb{R}^+$, and (3.64) becomes

$$\sum_{i=0}^{n} \alpha_i \tau^i \geq 0 \qquad \text{for all } \tau \geq 0,$$

with

$$\alpha_i = \sum_{k=0}^{i} \binom{n-k}{i-k} a_k. \tag{3.65}$$

This leads to the (sharper) sufficient condition that $\alpha_i \geq 0$ for $i = 0(1)n$. Necessary and sufficient conditions for (3.64) are known only for small values of n. For example, for $n = 2$, it is known (see [Schm 89]) that (3.64) is satisfied if and only if

$$\alpha_0 \geq 0, \qquad \alpha_2 \geq 0, \qquad \alpha_1 \geq -2\sqrt{\alpha_0 \alpha_2}, \tag{3.66a}$$

while for $n = 3$, it holds if and only if

$$(\alpha_0, \alpha_1, \alpha_2, \alpha_3) \in A \cup B,$$

with

$$\begin{aligned} A &= \{(\alpha_0, \alpha_1, \alpha_2, \alpha_3) : \alpha_i \geq 0\}, \\ B &= \{(\alpha_0, \alpha_1, \alpha_2, \alpha_3) : \alpha_0 \geq 0, \alpha_3 \geq 0, \\ &\qquad 4\alpha_0\alpha_2^3 + 4\alpha_3\alpha_1^3 + 27\alpha_0^2\alpha_3^2 - 18\alpha_0\alpha_2\alpha_3 - \alpha_1^2\alpha_2^2 \geq 0\}. \end{aligned} \tag{3.66b}$$

Little is known about the case $n \geq 4$ ([Schm 89]).

To discuss criteria for shape preserving splines interpolating data (x_i, z_i), $i = 0(1)n$, we assume now that the interpolation nodes lie in the interval $[0, 1]$, i.e., $0 = x_0 < x_1 < \cdots < x_n = 1$.

We begin by discussing quadratic C^1 interpolating splines. Suppose the interpolating spline curve defined on the interval $I = [x_i, x_{i+1}]$ is given by

$$y_i(x) = a_1 + b_1(x - x_i) + c_1(x - x_i)^2. \tag{3.67}$$

Assume that at the endpoints of this interval we have

$$\begin{aligned} y_i(x_i) &= z_i, \qquad & y_i(x_{i+1}) &= y_{i+1}(x_i) = z_{i+1}, \\ y_i'(x_i) &= s_i, \qquad & y_i'(x_{i+1}) &= y_{i+1}'(x_i) = s_{i+1}, \end{aligned} \tag{3.68}$$

and let $h = x_{i+1} - x_i$. Inserting these equations in (3.67), we get

$$\begin{aligned} a_1 &= z_i, \qquad & a_1 + hb_1 + h^2c_1 &= z_{i+1}, \\ b_1 &= s_i, \qquad & b_1 + 2hc_1 &= s_{i+1}, \end{aligned} \tag{3.67a}$$

from which it follows that

$$\begin{aligned} a_1 &= \frac{1}{h^2}(z_{i+1} - z_i - hs_i), \\ c_1 &= \frac{1}{2h}(s_{i+1} - s_i), \end{aligned} \tag{3.67b}$$

which means that our interpolation problem can only have a solution provided

$$\frac{z_{i+1} - z_i}{h} = \frac{s_i + s_{i+1}}{2}. \tag{3.69}$$

If $s_i s_{i+1} > 0$ and (3.69) holds, then by (3.67), the quadratic polynomial piece y_i of the spline is monotone in the interval I. If in addition, $s_i < s_{i+1}$, then y_i is convex, while if $s_i > s_{i+1}$, it is concave.

In general, the condition (3.69) will not be satisfied, and hence we now insert one additional knot $x_i < \xi < x_{i+1}$ in the interval, and interpolate with two quadratic pieces

$$y_i(x) = \begin{cases} a_1 + b_1(x - x_i) + c_1(x - x_i)^2, & x_i < x < \xi \\ a_2 + b_2(x - \xi) + c_2(x - \xi)^2, & \xi < x < x_{i+1}, \end{cases} \tag{3.70}$$

which are joined together with C^1 continuity. We introduce the notation $y_i'(\xi) = \bar{s}$ for the first derivative at ξ. Inserting this in (3.68) leads to

$$a_1 = z_i, \qquad b_1 = s_i, \qquad c_1 = \frac{\bar{s} - s_i}{2\alpha}, \tag{3.70a}$$

$$a_2 = a_1 + b_1\alpha + c_1\alpha^2, \qquad b_2 = \bar{s}, \qquad c_2 = \frac{s_{i+1} - \bar{s}}{2\beta}, \tag{3.70b}$$

with

$$\bar{s} = y_i'(\xi) = \frac{2(z_{i+1} - z_i) - (\alpha s_i + \beta s_{i+1})}{h},$$
$$\alpha = \xi - x_i, \qquad \beta = x_{i+1} - \xi.$$

It was shown in [Schu 83] that if $s_i s_{i+1} \geq 0$, then the interpolating spline in (3.70) is monotone on the interval I if and only if $s_i\bar{s} \geq 0$. Moreover, if $s_i < s_{i+1}$, then the spline is convex on I if and only if $s_i \leq \bar{s} \leq s_{i+1}$. Analogously, if $s_i > s_{i+1}$, then it is concave on I if and only if $s_{i+1} \leq \bar{s} \leq s_i$. The proof follows directly from a computation of the second derivative of (3.70).

Let $\delta = (z_{i+1} - z_i)/h$. Then the spline y_i in (3.70) has a point of inflection in I whenever $(s_{i+1} - \delta)(s_i - \delta) \geq 0$. If $(s_{i+1} - \delta)(s_i - \delta) < 0$ and $|z_{i+1} - \delta| < |z_i - \delta|$, then for all ξ satisfying

$$x_i < \xi \leq \bar{\xi} \qquad \text{with} \qquad \bar{\xi} = x_i + \frac{2h(s_{i+1} - \delta)}{s_{i+1} - s_i},$$

the interpolating spline y_i of (3.70) is convex on I when $s_i < s_{i+1}$, and is concave on I for $s_i > s_{i+1}$. If $s_i s_{i+1} \geq 0$, then y_i is also monotone.

Similarly, if $|s_{i+1} - \delta| > |s_i - \delta|$, then for all ξ satisfying

$$\bar{\xi} \leq \xi < x_{i+1} \qquad \text{with} \qquad \bar{\xi} = x_{i+1} + \frac{2h(s_i - \delta)}{s_{i+1} - s_i},$$

the spline (3.70) is convex on I if $s_i < s_{i+1}$, and is concave on I if $s_i > s_{i+1}$. If $s_i s_{i+1} \geq 0$, then y_i is monotone.

In order to discuss shape preserving properties of cubic curves, we transform the expression (3.11) to local coordinates $t = \Delta t_i \tau$, $\tau \in [0,1]$, see [Fri 80]. Then we have

$$y_i(t) = z_i + m_i \Delta t_i(\tau^3 - 2\tau^2 + \tau) + m_{i+1}\Delta t_i(\tau^3 - \tau^2) + \Gamma_i \Delta t_i(3\tau^2 - 2\tau^3), \quad (3.71)$$

where $m_i = s_i'(t)$ and $\Gamma_i = (z_{i+1} - z_i)/\Delta t_i$. In this case, s_i is monotone increasing on I if and only if

$$m_i \geq 0, \qquad m_{i-1} - \sqrt{m_{i-1} m_i} + m_i \leq 3\Gamma_{i-1}, \qquad \text{for } i = 1(1)n. \quad (3.72a)$$

By (3.65) a sufficient condition for this to happen is that

$$m_i \geq 0, \qquad m_{i-1} + m_i \leq 3\Gamma_{i-1}, \qquad \text{for } i = 1(1)n. \quad (3.72b)$$

The cubic spline is convex precisely when

$$2m_{i-1} + m_i \leq 3\Gamma_{i-1} \leq m_{i-1} + 2m_i, \quad \text{for } i = 1(1)n. \quad (3.73)$$

For an overview of other criteria, see [Schm 89] and [Bro 91] (also [Fle 86], [Mont 87]). An algorithm for shape preserving interpolating splines of arbitrary degree was described by [Cos 88]. For shape preserving interpolating splines in tension, see [Wev 88a], and for approximating splines in tension, see [Koz 86]. An alternative to shape preserving splines in tension involving rationals was developed by [Gre 86]. Shape preserving interpolating rational splines can be found in [Gre 86] and [Sak 88]. For a survey of available methods, see [SPÄ 90].

Shape preserving interpolating curvature continuous parametric splines are discussed in [Goo 88, 89]. Further references can be found in Sect. 3.6.1, Sect. 3.6.2, and in [Schab 92]. Extensions to surfaces are treated in [Fon 87], [Fle 89], and [Cos 90].

4
Bézier and B-spline Curves

In our study of cubic spline curves (using monomials as basis functions), we were not able to give a geometric interpretation to the coefficients of the spline. There are, however, other polynomial basis functions for which the coefficient vectors $\boldsymbol{b}_i$ of a spline have *geometric significance*. By this we mean that the positions of the $\boldsymbol{b}_i$ more or less determine the shape of the curve (or surface), and that we can determine various geometric properties of it from the position of these coefficients. Clearly, such basis functions are of great importance for interactive work, since they *geometrize* all of our formulae. We shall examine essentially two types of spline functions:

- *Bézier spline curves*,
- *B-spline curves*.

From a mathematical point of view, the spline functions considered here are simply obtained by appropriately transforming the monomial basis to a new basis. Thus, formally, the new spline coefficients $\boldsymbol{b}_i$ can be computed from the coefficients of (ordinary) splines using the basis transformation equations.

In this chapter we shall distinguish two cases of interest:

- integral curves,
- rational curves,

depending on whether the basis functions are polynomials or rational functions. The integral curves are often referred to as *ordinary Bézier/B-spline curves* or as *polynomial Bézier/B-spline curves*.

4.1. Integral Bézier Curves

The geometric properties of (integral) Bézier curves were developed independently by P. de Casteljau starting in 1959, and by P. Bézier starting in 1962, where they were used in the CAD systems of Renault and Citroën, respectively. In 1970, R. Forrest discovered the connection between the work of Bézier and the classical *Bernstein polynomials*: the Bernstein polynomials are in fact the basis functions used for Bézier curves, see e.g., [BEZ 72], [Cas 59], [For 72].

Bernstein polynomials can be derived from the binomial formula

$$1 = [(1-t)+t]^n = \sum_{r=0}^{n} \binom{n}{r} (1-t)^{n-r} t^r. \tag{4.1}$$

The polynomials

$$B_r^n(t) := \binom{n}{r} (1-t)^{n-r} t^r, \qquad r = 0(1)n, \tag{4.2}$$

of degree n are called *Bernstein polynomials*, see e.g., [DAV 75], [Gon 83], [Sta 81], [LORE 53].

It follows immediately from the definition of the Bernstein polynomials that they satisfy the recurrence relation

$$B_i^n(t) = (1-t) B_i^{n-1}(t) + t B_{i-1}^{n-1}(t). \tag{4.2a}$$

The derivatives are given by

$$\frac{d^p}{dt^p} B_i^n(t) = \frac{n!}{(n-p)!} \Delta^p B_i^n(t), \tag{4.2b}$$

where the difference operators are defined by

$$\Delta^0 B_i^n(t) = B_i^n(t), \qquad \Delta^p B_i^n(t) = \Delta^{p-1}[B_{i-1}^{n-1}(t) - B_i^{n-1}(t)].$$

We shall also make use of the degree raising formulae

$$(1-t)^p B_i^{n-p}(t) = \frac{\binom{n-p}{i}}{\binom{n}{i}} B_i^n(t), \qquad t^p B_i^{n-p}(t) = \frac{\binom{n-p}{i}}{\binom{n}{i+p}} B_{i+p}^n(t), \tag{4.2c}$$

and the product formula [Las 92a]

$$\prod_{k=1}^{\alpha} B_{i_k}^{n_k}(t) = \frac{\prod_{k=1}^{\alpha} \binom{n_k}{i_k}}{\binom{|\boldsymbol{n}|}{|\boldsymbol{i}|}} B_{|\boldsymbol{i}|}^{|\boldsymbol{n}|}(t), \tag{4.2d}$$

where $\boldsymbol{i} = (i_1, \ldots, i_\alpha)$, $|\boldsymbol{i}| = i_1 + \cdots + i_\alpha$, $\boldsymbol{n} = (n_1, \ldots, n_\alpha)$, and $|\boldsymbol{n}| = n_1 + \cdots + n_\alpha$.

For ease of exposition, we now focus on the parameter interval $I = [0,1]$. On this interval, the Bernstein polynomials have the following properties:

$$\begin{aligned} &B_r^n(0) = B_r^n(1) = 0, && r \neq 0, r \neq n, \\ &B_0^n(0) = B_n^n(1) = 1, && B_0^n(1) = B_n^n(0) = 0, \\ &B_r^n(t) \geq 0, \quad t \in I, && \max_I B_r^n(t) = B_r^n(r/n), \\ &B_r^n(t) = B_{n-r}^n(1-t), && \sum_{r=0}^n B_r^n(t) = 1. \end{aligned} \tag{4.3}$$

Fig. 4.1 shows the graphs of the quintic Bernstein polynomials $B_i^5(t)$ for $t \in [0,1]$.

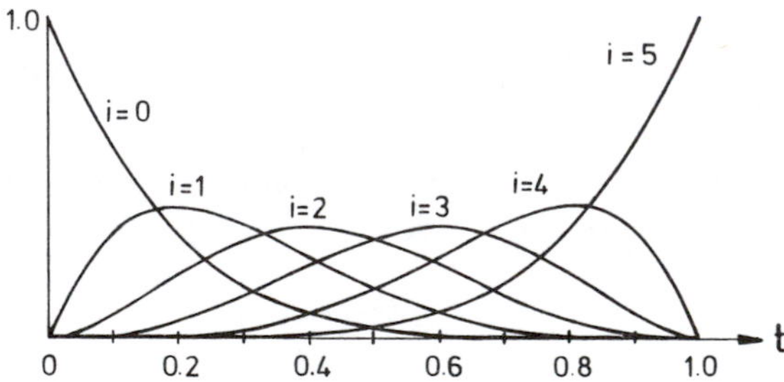

Fig. 4.1. Bernstein polynomials of degree five.

Using the Bernstein polynomials as basis functions, we can define a *Bézier curve* or *Bézier polynomial* of degree n to be a curve in the parametric form

$$\boldsymbol{X}(t) = \sum_{i=0}^n \boldsymbol{b}_i B_i^n(t), \tag{4.4}$$

with coefficients $\boldsymbol{b}_i \in \mathbb{R}^d$, $d = 1,2,3$, cf. [BEZ 72], [Boeh 84], [For 72]. In general, we take the $\boldsymbol{b}_i$ to be vectors in $\mathbb{R}^2$ or $\mathbb{R}^3$, and refer to them as *Bézier points*. In the case where the $\boldsymbol{b}_i$ are real numbers, we call them *Bézier ordinates*. The polygon formed by connecting the Bézier points is called the *Bézier polygon*. Fig. 4.2 shows several Bézier polygons $\{\boldsymbol{b}_i\}$ and their associated Bézier curves $\boldsymbol{X}(t)$.

Remark. While it is usual to define the Bézier curve in terms of the vertices $\boldsymbol{b}_i$ of the polygon, it is also possible to define it in terms of the edges $\boldsymbol{a}_i$ of the polygon (cf. e.g., [BEZ 72, 86], [ENG 85], [MÜL 80]) as

$$\boldsymbol{X}(t) = \sum_{i=0}^n \boldsymbol{a}_i f_i^n(t), \tag{4.4a}$$

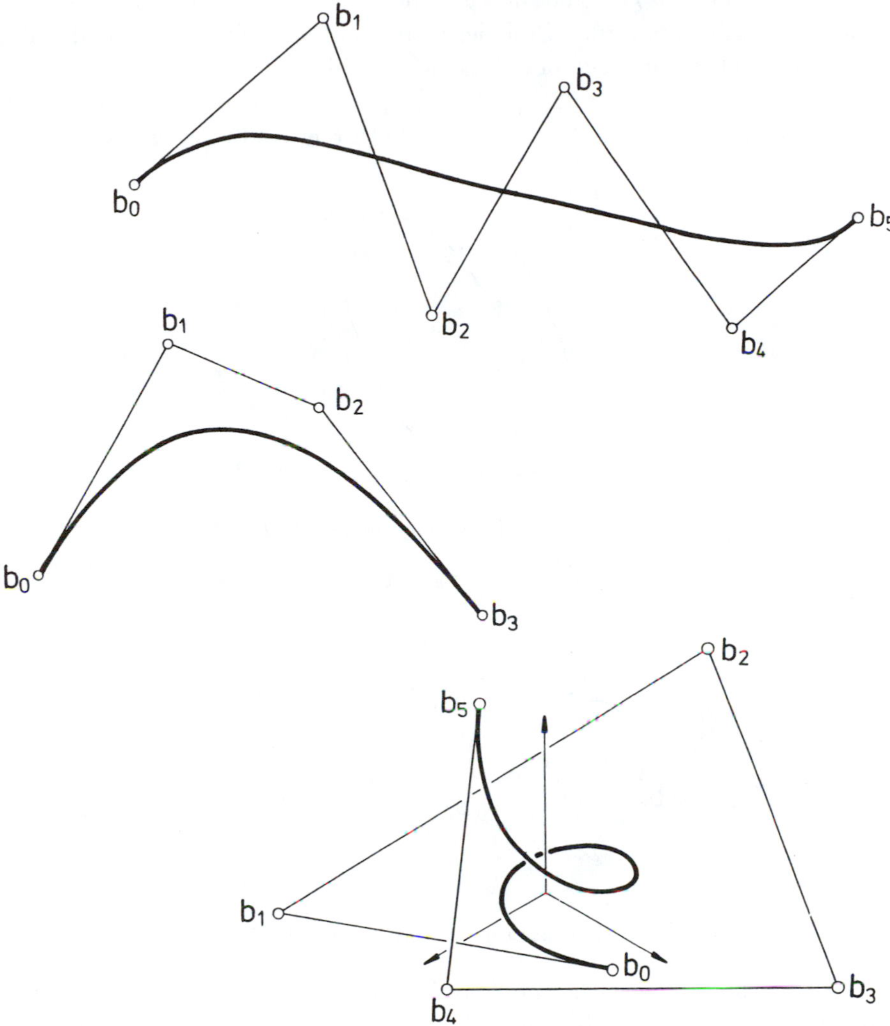

Fig. 4.2. Bézier curves and their associated Bézier polygons.

using the basis functions

$$f_i^n(t) = \sum_{k=i}^{n} (-1)^{k+i} \binom{k-1}{k-i} \binom{n}{k} t^k.$$

Here $\boldsymbol{a}_0 = \boldsymbol{b}_0$ and $\boldsymbol{a}_i = \boldsymbol{b}_i - \boldsymbol{b}_{i-1}$ for $i = 1(1)n$, i.e.,

$$\boldsymbol{b}_i = \sum_{j=0}^{i} \boldsymbol{a}_j \qquad \text{and} \qquad B_i^n(t) = f_i^n(t) - f_{i+1}^n(t).$$

Definition (4.4a) is sometimes used in connection with *hodograph curves*, which are defined in terms of the derivatives of the original curve, and which are important in the study of points of inflection, singularities, and intersections, see [BEZ 86], [For 72], [Sed 87a]. Fig. 4.1a depicts the polynomials $f_i^n(t)$ on $[0,1]$, while Fig. 4.2a illustrates both representation methods.

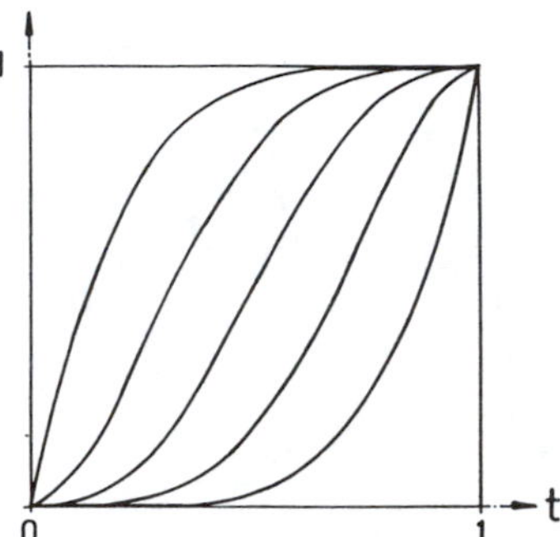

Fig. 4.1a. Basis polynomials $f_i^5(t)$.

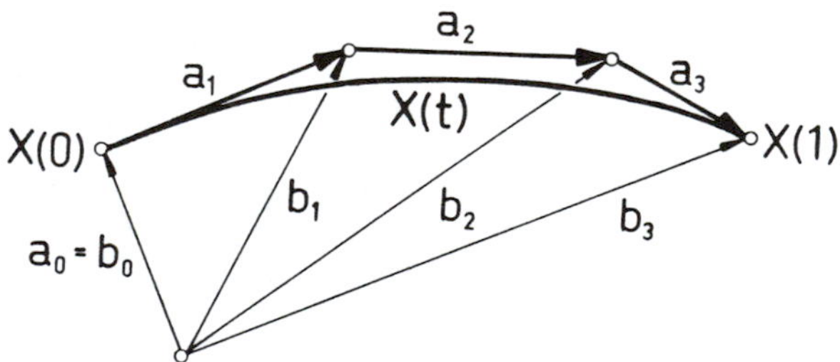

Fig. 4.2a. Alternate representations of a Bézier polygon.

If we choose the coefficients $\boldsymbol{b}_i$ to be real numbers and define the Bézier points to be $(i/n, \boldsymbol{b}_i) \in \mathbb{R}^2$, then the parametric formula (4.4) for a Bézier curve reduces to a *Bézier function* of the variable t. This interpretation makes sense since the parameter t can also be represented formally as a Bézier function:

$$t = \sum_{r=0}^{n-1} \binom{n-1}{r} (1-t)^{n-(1+r)}\, t^{1+r} = \sum_{i=0}^{n} \frac{i}{n} B_i^n(t). \qquad (*)$$

Indeed, by the binomial formula,

$$t = 1 \cdot t = [(1-t)+t]^{n-1} t = \left(\sum_{r=0}^{n-1} \binom{n-1}{r} (1-t)^{n-1-r} t^r \right) t.$$

Now changing the index of summation by $i := r + 1$ and taking account of the fact that

$$\binom{n-1}{i-1} = \frac{i}{n}\binom{n}{i},$$

equation $(*)$ follows immediately from (4.2). Comparing $(*)$ and (4.4) for $\boldsymbol{b}_i \in \mathbb{R}$ shows that in this case, the Bézier points are given by $(i/n, \boldsymbol{b}_i)$, see Fig. 4.3.

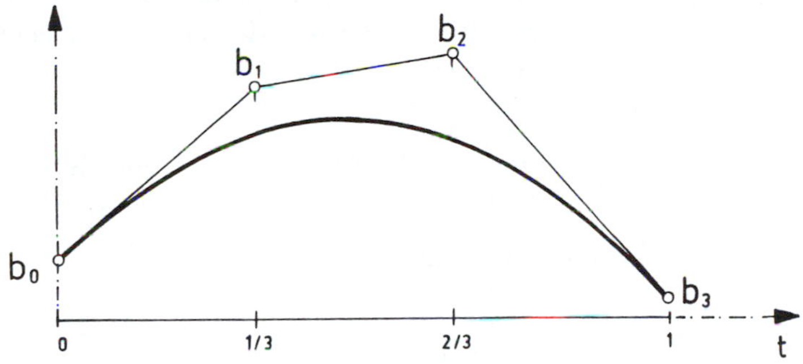

Fig. 4.3. A Bézier function.

In looking at Figs. 4.2 and 4.3, we notice that the Bézier curves are always tangent to the Bézier polygons at the endpoints. To establish this property in general, we introduce the alternate formula for a Bézier curve

$$\boldsymbol{X}(t) = (1 - t + tE)^n \boldsymbol{b}_0, \tag{4.5}$$

using *operator* notation, cf. [Hosa 78]. Here the operator E is defined by

$$E\boldsymbol{b}_i := \boldsymbol{b}_{i+1}, \quad E^m E^n = E^{m+n}, \quad Ex = xE, \quad E^0 = I, \quad EE^{-1} = 1.$$

The equivalence of (4.4) and (4.5) can easily be seen by multiplying out the factors in (4.5). Differentiation of (4.5) leads to the formulae

$$\begin{aligned} \boldsymbol{X}'(t) &= n(1 - t + tE)^{n-1}(E - 1)\, \boldsymbol{b}_0, \qquad (4.6)\\ \boldsymbol{X}''(t) &= n(n-1)(1 - t + tE)^{n-2}(E - 1)^2\, \boldsymbol{b}_0. \end{aligned}$$

Inserting the endpoint parameter values $t = 0$ and $t = 1$, we get

$$\begin{aligned} \boldsymbol{X}(0) &= \boldsymbol{b}_0, & \boldsymbol{X}(1) &= \boldsymbol{b}_n, \\ \boldsymbol{X}'(0) &= n(\boldsymbol{b}_1 - \boldsymbol{b}_0), & \boldsymbol{X}'(1) &= n(\boldsymbol{b}_n - \boldsymbol{b}_{n-1}), \\ \boldsymbol{X}''(0) &= n(n-1)(\boldsymbol{b}_2 - 2\boldsymbol{b}_1 + \boldsymbol{b}_0), & \boldsymbol{X}''(1) &= n(n-1)(\boldsymbol{b}_n - 2\boldsymbol{b}_{n-1} + \boldsymbol{b}_{n-2}), \end{aligned} \tag{4.6a}$$

which shows that

- the Bézier curve starts at $\boldsymbol{b}_0$ and ends at $\boldsymbol{b}_n$,
- the edges $\overline{\boldsymbol{b}_0\boldsymbol{b}_1}$ and $\overline{\boldsymbol{b}_{n-1}\boldsymbol{b}_n}$ of the Bézier polygon are tangent to the curve (if one or more neighboring Bézier points all fall at $\boldsymbol{b}_0$, then the tangent at $\boldsymbol{b}_0$ corresponds to the edge of the polygon passing through $\boldsymbol{b}_0$ and the first distinct neighbor of $\boldsymbol{b}_0$),
- the second derivative of the curve at the endpoints depends only on the Bézier point at the end and its two distinct neighbors.

The k-th derivative of a Bézier curve is given by

$$\boldsymbol{X}^{(k)}(t) = \frac{n!}{(n-k)!}(1 - t + tE)^{n-k}(E-1)^k \boldsymbol{b}_0. \tag{4.7a}$$

Using the forward difference operators

$$\begin{aligned} \Delta \boldsymbol{b}_i &:= \boldsymbol{b}_{i+1} - \boldsymbol{b}_i, \\ \Delta^2 \boldsymbol{b}_i &:= \Delta(\Delta \boldsymbol{b}_i) = \Delta(\boldsymbol{b}_{i+1} - \boldsymbol{b}_i) = \boldsymbol{b}_{i+2} - 2\boldsymbol{b}_{i+1} + \boldsymbol{b}_i, \\ &\vdots \\ \Delta^k \boldsymbol{b}_i &:= \boldsymbol{b}_{i+k} - \binom{k}{1}\boldsymbol{b}_{i+k-1} + \binom{k}{2}\boldsymbol{b}_{i+k-2} - \cdots + \boldsymbol{b}_i = \sum_{\ell=0}^{k}(-1)^\ell \binom{k}{\ell}\boldsymbol{b}_{i+k-\ell}, \end{aligned} \tag{4.8}$$

we can rewrite (4.7a) as

$$\boldsymbol{X}^{(k)}(t) = \frac{n!}{(n-k)!}\sum_{i=0}^{n-k} \Delta^k \boldsymbol{b}_i B_i^{n-k}(t). \tag{4.7b}$$

By (4.7a,b), we now have (see e.g., [BEZ 72]) the following lemma.

Lemma 4.1. *The derivatives of order k of a Bézier curve at the endpoints $t = 0$ and $t = 1$ depend only on the boundary point and its k neighbors.*

Proof: Evaluating the formula (4.7b) for the k-th derivative at 0 and 1 gives

$$\boldsymbol{X}^{(k)}(0) = \frac{n!}{(n-k)!}\Delta^k \boldsymbol{b}_0, \qquad \boldsymbol{X}^{(k)}(1) = \frac{n!}{(n-k)!}\Delta^k \boldsymbol{b}_{n-k}, \tag{4.7c}$$

and the assertion follows from (4.8). ∎

To evaluate a polynomial written in terms of the monomial basis at a given point $t = t_0$, we can use the classical (numerically stable) Horner scheme, see e.g., [Gol 84]. The analogous method for evaluating polynomials written in terms of the Bernstein basis is the *de Casteljau algorithm*. To explain this algorithm, it is convenient to introduce the following notation for Bézier curves of different degrees:

$$\begin{aligned}
\boldsymbol{X}_n(t) &:= (1-t+tE)^n \boldsymbol{b}_0 =: \boldsymbol{b}_{0,\dots,n}, \\
\boldsymbol{X}_{n-1}(t) &:= (1-t+tE)^{n-1} \boldsymbol{b}_0 =: \boldsymbol{b}_{0,\dots,n-1}, \\
&\vdots \\
\boldsymbol{X}_{n-k}(t) &:= (1-t+tE)^{n-k} \boldsymbol{b}_0 =: \boldsymbol{b}_{0,\dots,n-k}, \\
E^m \boldsymbol{X}_{n-k}(t) &=: \boldsymbol{b}_{m,\dots,n-k+m}.
\end{aligned} \tag{4.9}$$

Then

$$\boldsymbol{X}_n(t) = (1-t+tE)\boldsymbol{X}_{n-1}(t) = (1-t)\boldsymbol{X}_{n-1}(t) + tE\boldsymbol{X}_{n-1}(t)$$

becomes

$$\boldsymbol{b}_{0,\dots,n} = (1-t)\boldsymbol{b}_{0,\dots,n-1} + t\boldsymbol{b}_{1,\dots,n},$$

while

$$\boldsymbol{X}_{n-1}(t) = (1-t+tE)\boldsymbol{X}_{n-2}(t) = (1-t)\boldsymbol{X}_{n-2}(t) + tE\boldsymbol{X}_{n-2}(t)$$

can be written as

$$\boldsymbol{b}_{0,\dots,n-1} = (1-t)\boldsymbol{b}_{0,\dots,n-2} + t\boldsymbol{b}_{1,\dots,n-1},$$

and

$$E\boldsymbol{X}_{n-1}(t) = E(1-t)\boldsymbol{X}_{n-2}(t) + tE^2\boldsymbol{X}_{n-2}(t)$$

gives

$$\boldsymbol{b}_{1,\dots,n} = (1-t)\boldsymbol{b}_{1,\dots,n-1} + t\boldsymbol{b}_{2,\dots,n}.$$

Continuing this process, for $s = k + r$ we have

$$\boldsymbol{b}_{r,\dots,s} = (1-t)\boldsymbol{b}_{r,\dots,s-1} + t\boldsymbol{b}_{r+1,\dots,s}, \tag{$*$}$$

where the $\boldsymbol{b}$ on the left has a subscript with $k+1$ indices, while those on the right have subscripts with k indices. In the last step (where $s = r+1$)) we get

$$\boldsymbol{b}_{rs} = (1-t)\boldsymbol{b}_r + t\boldsymbol{b}_s. \qquad (**)$$

Now if we work backwards through this process of repeated linear interpolation, starting with the given Bézier points $\boldsymbol{b}_i$ and using the formula $(**)$ first, then the formula $(*)$ repeatedly, it follows by (4.9) that after the last step, we have the value of the Bézier curve at the point $t = t_0$. To illustrate the algorithm, we consider the following (de Casteljau scheme) which shows the individual linear interpolation steps:

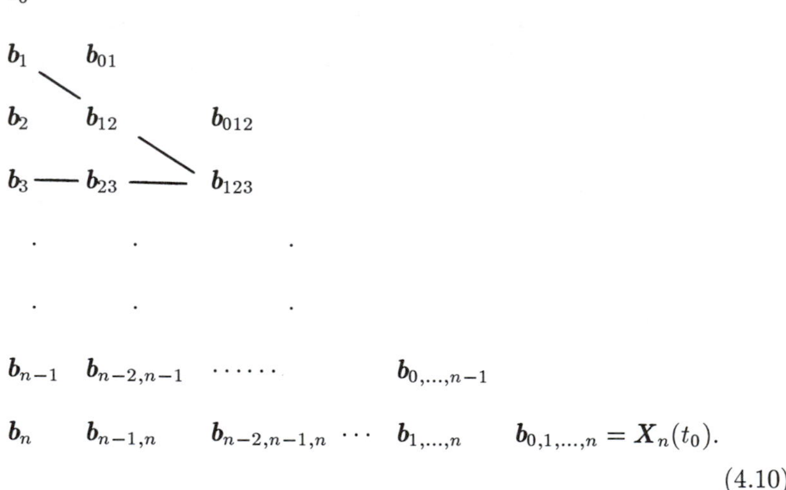

(4.10)

The computation proceeds as follows: for each step in the horizontal direction, we multiply by t_0, while in the diagonal direction we multiply by $(1-t_0)$. The lines in the table show the calculation of $\boldsymbol{b}_{123}$. The table also provides the values of the derivatives of $\boldsymbol{X}_n(t)$ at t_0, since by (4.6) and (4.9)

$$\boldsymbol{X}'_n(t) = n(1-t+tE)^{n-1}(\boldsymbol{b}_1 - \boldsymbol{b}_0) = n(\boldsymbol{b}_{1,\dots,n} - \boldsymbol{b}_{0,\dots,n-1}),$$

i.e., the first derivative can be computed from the entries in the next to last column. Similarly, the second and higher derivatives can be computed from the earlier columns of the table.

The de Casteljau algorithm can also be derived geometrically, see Fig. 4.4. If we divide the edges of the Bézier polygon in the ratio $(1-t_0)$ to t_0, connect the resulting points by straight lines, divide these new edges again in the

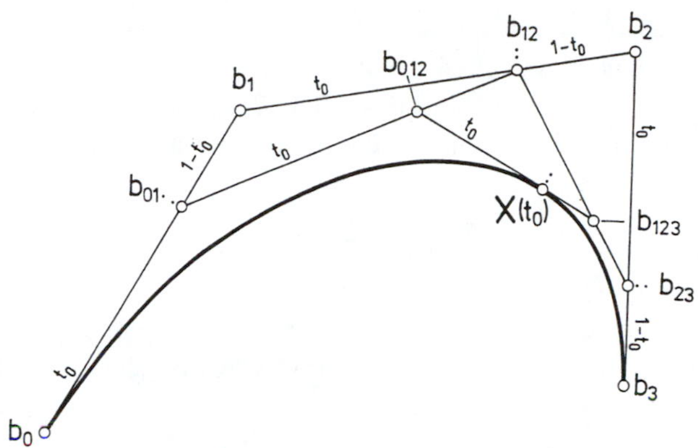

Fig. 4.4. The de Casteljau algorithm – a geometric interpretation.

same ratios, and repeat this process a total of n times, then the dividing point obtained in the last step is $\boldsymbol{X}_n(t_0)$, and the last edge is tangent to the Bézier curve at the point $\boldsymbol{X}_n(t_0)$.

We now consider an example. Suppose we are given the planar Bézier curve of degree four

$$\boldsymbol{X}_4(t) = \begin{pmatrix}1\\0\end{pmatrix}(1-t)^4 + \begin{pmatrix}0\\2\end{pmatrix}4(1-t)^3t + \begin{pmatrix}1\\5.5\end{pmatrix}6(1-t)^2t^2 + \begin{pmatrix}6\\5.5\end{pmatrix}4(1-t)t^3 + \begin{pmatrix}7.5\\0.5\end{pmatrix}t^4,$$

and we want to find the point $\boldsymbol{X}_4(0.6)$ on the curve. Applying the de Casteljau algorithm to both the x and y components using the parameter value $t = 0.6$ gives

x-component:

1				
0	0.4			
1	0.6	0.52		
6	4.0	2.64	1.792	
7.5	6.9	5.74	4.5	3.4168,

y-component:

0				
2	1.2			
5.5	4.1	2.94		
5.5	5.5	4.94	4.14	
0.5	2.5	3.7	4.196	4.1736.

Another approach to the de Casteljau algorithm (and thus also to the properties of Bézier curves and also B-spline curves, see Sect. 4.3) is via the *theory of polar forms*, also called the *blossoming principle*, see [CAS 86], [Ram 87, 88, 89], [Sei 88, 88a, 89, 89a], [Boeh 90]. The basic idea is the concept of a *symmetric multiaffine mapping*. A mapping $f : \mathbb{R} \to \mathbb{R}^d$ is called *affine*, provided it satisfies

$$f(\sum_{i=1}^{m} a_i t_i) = \sum_{i=1}^{m} a_i f(t_i)$$

for all $a_1, \ldots, a_m, t_1, \ldots, t_m \in \mathbb{R}$ with $\sum_i a_i = 1$. A mapping $f : \mathbb{R}^p \to \mathbb{R}^d$ with p arguments is called *p-affine*, or simply *multiaffine*, provided that for arbitrary $j \in \mathbb{N}$ and arbitrary $a_1, \ldots, a_{j-1}, a_{j+1}, \ldots, a_p \in \mathbb{R}$, each of the mappings

$$f_{a_1,\ldots,\bar{a}_j,\ldots,a_p} : \mathbb{R} \to \mathbb{R}^d, \qquad t \mapsto f(a_1, \ldots, a_{j-1}, t, a_{j+1}, \ldots, a_p)$$

is affine. The mapping $f : \mathbb{R}^p \to \mathbb{R}^d$ is called *symmetric* provided that f has the same value for all permutations of its variables.

We have the following *blossoming principle*: given any polynomial $F : \mathbb{R} \to \mathbb{R}^d$ of degree n, there exists a uniquely defined symmetric n-affine mapping $f : \mathbb{R}^n \to \mathbb{R}^d$ with

$$f(\underbrace{t, \ldots, t}_{n \text{ fold}}) = F(t).$$

The function f is called the *blossom* or *polar form* of F.

To give an example, the polynomial

$$F(t) = a_0 + a_1 t + a_2 t^2 + a_3 t^3$$

corresponds to the symmetric 3-affine function

$$f(t_1, t_2, t_3) = a_0 + \frac{a_1}{3}(t_1 + t_2 + t_3) + \frac{a_2}{3}(t_1 t_2 + t_2 t_3 + t_3 t_1) + a_3 t_1 t_2 t_3.$$

To see how the blossoming principle leads to the de Casteljau algorithm, we introduce the substitution

$$t = (1 - t) \cdot 0 + t \cdot 1.$$

Then it follows from the affine invariance that

$$f(\underbrace{t, \ldots, t}_{n \text{ fold}}) = (1 - t) f(0, \underbrace{t, \ldots, t}_{n-1 \text{ fold}}) + t f(1, \underbrace{t, \ldots, t}_{n-1 \text{ fold}}).$$

Recursively substituting and using the symmetry of f, it follows that

$$F(t) = f(\underbrace{t, \dots, t}_{n \text{ fold}}) = \sum_{i=0}^{n} \binom{n}{i} (1-t)^{n-i} t^i f(\underbrace{0, \dots, 0}_{n-i \text{ fold}}, \underbrace{1, \dots, 1}_{i \text{ fold}}) = \sum_{i=0}^{n} \boldsymbol{b}_i \; B_i^n(t).$$

A geometric interpretation of this recurrence can be found in Fig. 4.4a.

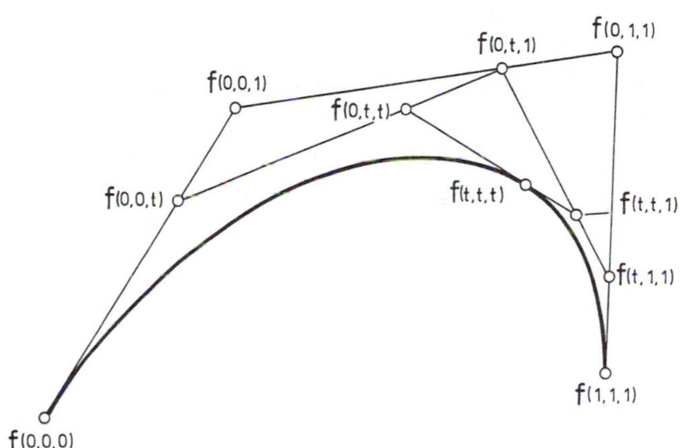

Fig. 4.4a. The de Casteljau algorithm from the blossoming principle.

Regarding the de Casteljau algorithm, we should also note that

1) it is *numerically stable* since only additions and multiplications are required [Faro 87],
2) it is *affinely invariant* since the subdivision process is affinely invariant,
3) it provides a way to *subdivide* Bézier curves. Indeed, the given Bézier curve is subdivided at the point $t = t_0$ into two Bézier curves whose Bézier points can be found in the de Casteljau table.

Regarding this last point [Stä 76], we have

Lemma 4.2. (Subdivision) *Using the de Casteljau algorithm, a Bézier curve can be subdivided into two Bézier curve segments which join together (up to the n-th derivative) at the point corresponding to $t = t_0$. The Bézier points of the two new Bézier curve segments are given by the entries on the boundary of the de Casteljau table.*

Fig. 4.4 illustrates Lemma 4.2 for (polynomial) degree $n = 3$. The boundary points in the associated de Casteljau table are

$$\boldsymbol{b}_0, \boldsymbol{b}_{01}, \boldsymbol{b}_{012}, \boldsymbol{b}_{0123} = \boldsymbol{X}(t_0) \qquad \text{and} \qquad \boldsymbol{X}(t_0) = \boldsymbol{b}_{0123}, \boldsymbol{b}_{123}, \boldsymbol{b}_{23}, \boldsymbol{b}_3.$$

These points form the Bézier points of the two new Bézier curve segments lying between $\boldsymbol{b}_0$ and $\boldsymbol{X}(t_0)$ and $\boldsymbol{X}(t_0)$ and $\boldsymbol{b}_3$, respectively, see Fig. 4.5.

Proof of Lemma 4.2. We show that the subdivision splits the Bézier curve $\boldsymbol{X}(t)$ into two curve segments: $\boldsymbol{X}^I(t)$ with the parameter interval $[0, t_0]$, and $\boldsymbol{X}^{II}(t)$ with the parameter interval $[t_0, 1]$. Let

$$\boldsymbol{X}^I(t) = (1 - t + tE)^n \,\tilde{\boldsymbol{b}}_0,$$

where the Bézier points (cf. (4.9)) are given by

$$\begin{aligned}
\tilde{\boldsymbol{b}}_0 &= \boldsymbol{b}_0 \\
\tilde{\boldsymbol{b}}_1 &= (1 - t_0 + t_0E)\,\boldsymbol{b}_0 \\
\tilde{\boldsymbol{b}}_2 &= (1 - t_0 + t_0E)^2\,\boldsymbol{b}_0 \\
&\vdots \\
\tilde{\boldsymbol{b}}_{n-1} &= (1 - t_0 + t_0E)^{n-1}\,\boldsymbol{b}_0 \\
\tilde{\boldsymbol{b}}_n &= (1 - t_0 + t_0E)^n\,\boldsymbol{b}_0.
\end{aligned} \tag{$*$}$$

Using ($*$) and the binomial formula, it follows that

$$\boldsymbol{X}^I(t) = (1 - t + t(1 - t_0 + t_0E))^n \,\boldsymbol{b}_0,$$

or

$$\boldsymbol{X}^I(t) = (1 - tt_0 + tt_0E)^n \,\boldsymbol{b}_0. \tag{$**$}$$

But now if we transform the original curve $\boldsymbol{X}(t)$ to the new parameter interval $[0, t_0]$, we get $\boldsymbol{X}(\tilde{t})$ which, in the operator form (4.5), is precisely ($**$) if we identify $\tilde{t} = tt_0$. This shows that the first half of $\boldsymbol{X}$ coincides with $\boldsymbol{X}^I$. The fact that the second half of $\boldsymbol{X}$ coincides with $\boldsymbol{X}^{II}$ is established similarly, where the parameter transformation is now given by $\tilde{t} = (1 - t_0)t + t_0$. ■

Given a Bézier curve $\boldsymbol{X}(t)$ defined on $[0, 1]$, the method developed here for subdividing a Bézier curve can also be used to compute a segment $\boldsymbol{X}_m(t)$, defined for $t \in [c, d]$ with $0 < c < d < 1$. To accomplish this, we first subdivide $\boldsymbol{X}(t)$ at the point $t = d$. Then we find the parameter value $\tilde{t}_c$ of the left-hand piece $\boldsymbol{X}_\ell(\tilde{t})$ of $\boldsymbol{X}(t)$ corresponding to the parameter value c. We then subdivide $\boldsymbol{X}_\ell(\tilde{t})$ at $\tilde{t}_c$, and the right-hand piece of $\boldsymbol{X}_\ell$ is the desired middle segment of $\boldsymbol{X}(t)$. This method of computing a middle segment can also be interpreted as a "generalized de Casteljau algorithm" which can be obtained directly from the original de Casteljau algorithm, see e.g., [Gol 82], [FAR 90].

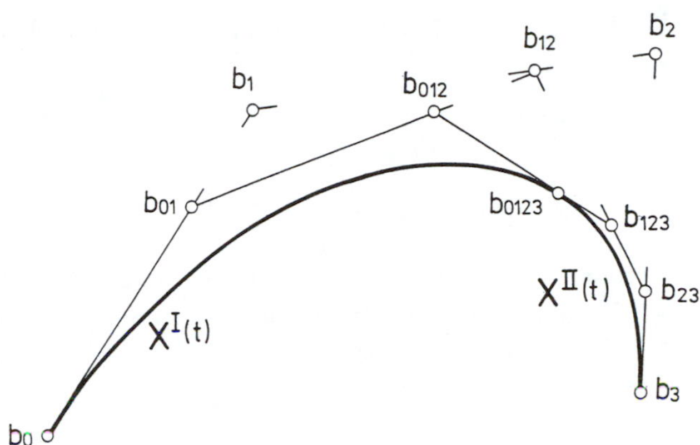

Fig. 4.5. Decomposition of a Bézier curve into two C^3 continuous curve segments (cf. Fig. 4.4).

If the subdivision described above is iterated (i.e., each segment is again subdivided, etc.), then the corresponding sequence of Bézier polygons converges to the original Bézier curve whenever the set of partition points created are dense in $[0,1]$ (for the case of repeated bisection where $\Delta t_0 = 0.5$, see [Lane 80]). The convergence is very fast [Coh 85], [Dah 86], and thus repeated subdivision can be used to render the curve graphically by finding a sufficiently close refined polygon (see [BART 87]).

In the above discussion of the Bernstein polynomials, we have assumed that we are working on the parameter interval $[0,1]$. As we have just seen, it is often not sufficient or reasonable to make this assumption, and so we now discuss Bernstein polynomials defined on a *general parameter interval* $[a,b]$. We shall see that all of the previous properties carry over after reparametrization.

The Bernstein polynomials corresponding to the parameter interval $[a,b]$ with $a < b$ are defined by

$$B_k^n(t) = \binom{n}{k} \frac{(t-a)^k (b-t)^{n-k}}{(b-a)^n}. \tag{4.11}$$

The identity $[(t-a)+(b-t)] = (b-a)$ leads to this definition in the same way as in the special case of (4.2). Now inserting these Bernstein polynomials as basis functions in the parametric formula (4.4) leads to the following formula (generalizing (4.7b)) for the derivatives

$$\boldsymbol{X}^{(k)}(t) = \frac{1}{(b-a)^k} \frac{n!}{(n-k)!} \sum_{i=0}^{n-k} \Delta^k \boldsymbol{b}_i \, B_i^{n-k}(t), \tag{4.12a}$$

where here again we are using the Bernstein polynomials defined in (4.11). In analogy with (4.7c), the values of these derivatives at the end points of the parameter interval are now

$$\begin{aligned} \boldsymbol{X}^{(k)}(a) &= \frac{1}{(b-a)^k}\frac{n!}{(n-k)!}\Delta^k\, \boldsymbol{b}_0, \\ \boldsymbol{X}^{(k)}(b) &= \frac{1}{(b-a)^k}\frac{n!}{(n-k)!}\Delta^k\, \boldsymbol{b}_{n-k}. \end{aligned} \tag{4.12b}$$

All of the properties of Bernstein polynomials which do not involve derivatives can be established for the general parameter interval $[a, b]$ by reduction to the interval $[0, 1]$ via the substitution

$$\tilde{t} = \frac{t-a}{b-a}, \tag{4.13a}$$

which gives

$$t = a(1-\tilde{t}) + b\,\tilde{t} \qquad \text{for} \qquad \tilde{t} \in [0, 1]. \tag{4.13b}$$

This means that the Bernstein polynomials can be considered as functions of the local parameter $\tilde{t}$ on the unit interval $[0, 1]$, while t is the global parameter.

On the other hand, we emphasize that the parameter interval does play a role when dealing with derivatives of Bézier spline curves consisting of Bézier curves pieced together, since changing the parameter intervals changes the values of the derivatives.

4.1.1. Geometric Properties of Bézier Curves

After this extensive introduction, we now proceed to establish the essential geometric properties of Bézier curves which account for their practical importance. We have already noted above that the Bézier curve interpolates the end points of its Bézier polygon, and that the first and last edges of this polygon are tangent to the curve. In this connection, we can establish the following convex hull property (the convex hull of a polygon $\{\boldsymbol{b}_i\}$, see Fig. 4.6, is the intersection of all convex sets containing $\boldsymbol{b}_i$, where a set B is *convex* provided that whenever $x, y \in B$, then all points on the line joining x and y are also in B):

Theorem 4.1. (Convex hull property). *A planar Bézier curve lies entirely inside the convex hull of its associated Bézier polygon.*

Proof of Theorem 4.1. Every point lying on a Bézier curve can be computed using the de Casteljau algorithm. But the de Casteljau algorithm is based on constructing lines joining points lying on the edges of the Bézier polygon. Clearly, at each step these lines certainly must lie in the convex hull of the Bézier points, and the theorem is established. ■

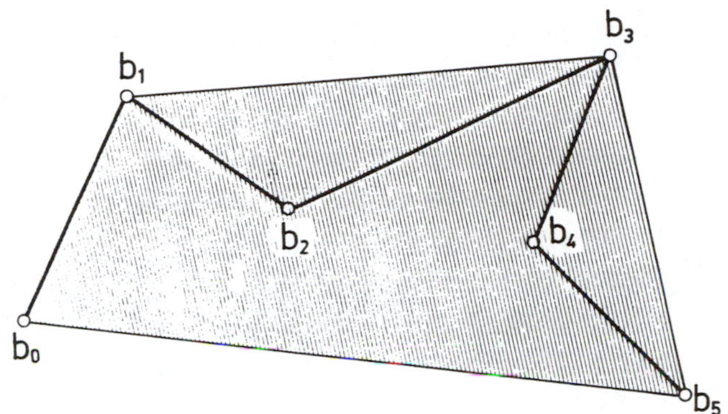

Fig. 4.6. Convex hull of a set of points $\boldsymbol{b}_0, \ldots, \boldsymbol{b}_5$.

For planar Bézier curves we also have

Theorem 4.2. (Variation-diminishing property). *If a straight line intersects the Bézier polygon of a planar Bézier curve k times, then it can intersect the Bézier curve at most k times.*

Proof: See e.g., [Scho 67]. ■

Remark. For $k = 2$, this means that the Bézier curve is always convex whenever the Bézier polygon is convex (cf. Fig. 4.7a). The curve can, however, be convex even though its Bézier polygon is not (cf. Fig. 4.7b). If the Bézier curve has a point of inflection (see Fig. 4.7c), then the Bézier polygon must have at least one point of inflection.

Remark. If a Bézier point $\boldsymbol{b}_i$ is moved to a new position $\boldsymbol{b}_i^*$, then all points on the Bézier curve move towards $\boldsymbol{b}_i^*$ in a direction parallel to $\boldsymbol{b}_i^* - \boldsymbol{b}_i$. Thus, changes in the Bézier curve can be controlled visually as shown in Fig. 4.7d, where the Bézier point $\boldsymbol{b}$ is moved to $\boldsymbol{b}^*$, and points on the Bézier curve (shown as a solid curve) move parallel to $\boldsymbol{b}^* - \boldsymbol{b}$.

Remark. Singularities of Bézier curves are discussed in [SU 89], and Bézier curves whose curvatures are always positive are described in [Rou 88].

It is known from linear algebra that a vector in $\mathbb{R}^2$ can be expressed in terms of vectors in $\mathbb{R}^3$ using an appropriate basis transformation. We now want to discuss the analogous question for *basis functions*. Given a Bézier curve of degree n, we want to express the same curve in terms of a new basis of degree $n+1$. This is known as *degree raising*. We have (cf. e.g., [For 72])

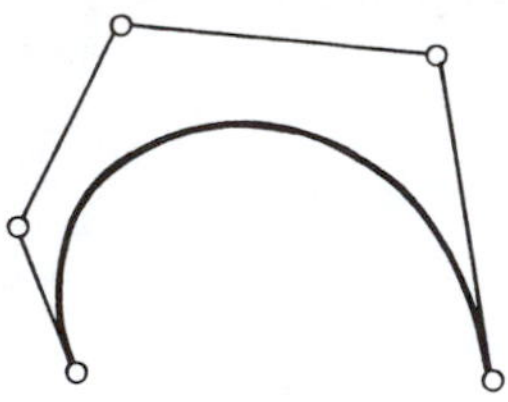

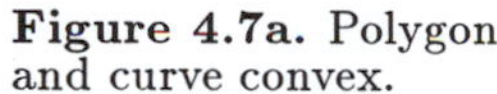

Figure 4.7a. Polygon and curve convex.

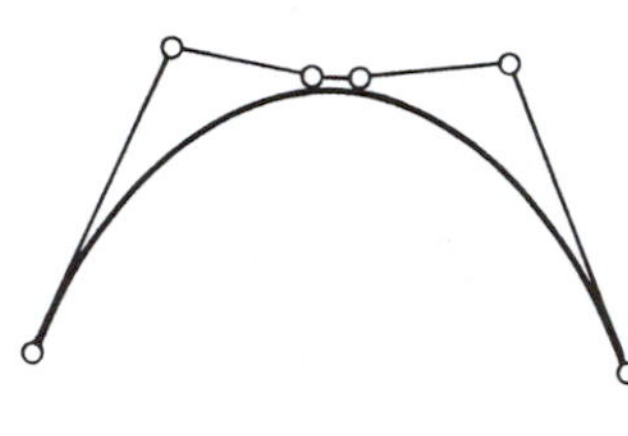

Figure 4.7b. Polygon not convex and curve convex.

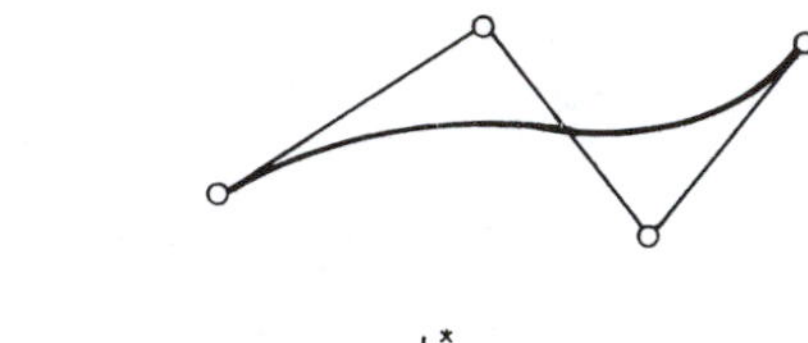

Figure 4.7c. Curve and polygon with a point of inflection.

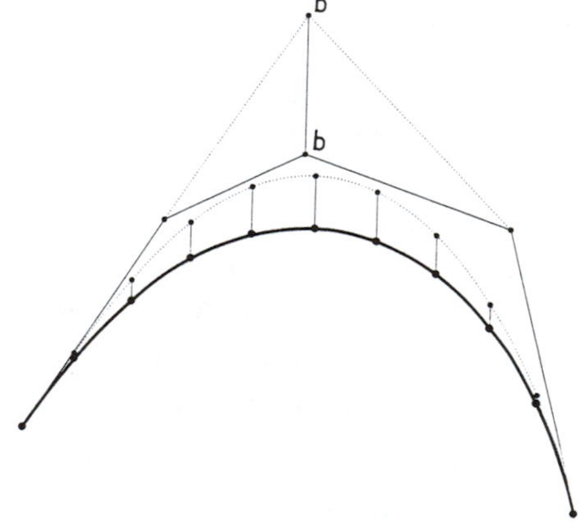

Figure 4.7d. Effect of moving one of the Bézier points.

Lemma 4.3. (Degree raising) *Suppose $\boldsymbol{X}(t)$ is a Bézier curve of degree n corresponding to Bézier points $\boldsymbol{b}_i$. Then $\boldsymbol{X}$ can be written as a Bézier curve of degree $n+1$ with new Bézier points*

$$\overline{\boldsymbol{b}}_k = \frac{k}{n+1}\boldsymbol{b}_{k-1} + (1 - \frac{k}{n+1})\boldsymbol{b}_k, \quad k = 0(1)n{+}1, \qquad \boldsymbol{b}_{-1} = \boldsymbol{b}_{n+1} = 0. \quad (4.14)$$

Proof: Since we are trying to represent the same curve with respect to two different bases, in terms of the operator notation of (4.5), we must have

$$(1 - t + tE)^n \, \boldsymbol{b}_0 = (1 - t + tE)^{n+1} \, \overline{\boldsymbol{b}}_0.$$

If we multiply the terms in the expansion of the left-hand side

$$\cdots + \binom{n}{k}(1-t)^{n-k}\, t^k E^k\, \boldsymbol{b}_0 + \binom{n}{k-1}(1-t)^{n-k+1}\, t^{k-1} E^{k-1}\, \boldsymbol{b}_0 + \cdots$$

by $[(1-t)+t]$, we get

$$\cdots + \binom{n}{k}(1-t)^{n-k+1}\, t^k E^k\, \boldsymbol{b}_0 + \binom{n}{k}(1-t)^{n-k}\, t^{k+1} E^k\, \boldsymbol{b}_0$$
$$+ \binom{n}{k-1}(1-t)^{n-k+2}\, t^{k-1} E^{k-1}\, \boldsymbol{b}_0 + \binom{n}{k-1}(1-t)^{n-k+1}\, t^k E^{k-1}\, \boldsymbol{b}_0 + \cdots.$$

Now expanding out the right-hand side gives

$$\cdots + \binom{n+1}{k}(1-t)^{n-k+1}\, t^k E^k\, \bar{\boldsymbol{b}}_0 + \cdots,$$

and comparing the coefficients of the factor $(1-t)^{n-k+1}\, t^k$, we get

$$\binom{n+1}{k}\bar{\boldsymbol{b}}_k = \binom{n}{k}\boldsymbol{b}_k + \binom{n}{k-1}\boldsymbol{b}_{k-1}.$$

Inserting the definition of the binomial coefficients and simplifying leads to (4.14). ■

It will be convenient to describe the process of degree elevation of a Bézier curve by an operator B. Thus, if Φ is the Bézier polygon corresponding to a Bézier curve $B_n\Phi$ of degree n, then we can express the transformation described in (4.14) by $B_n\Phi = B_{n+1}(B\Phi)$. Writing $\Phi, B\Phi, B^2\Phi, \ldots$ for the sequence of degree raised curves, we can now state

Theorem 4.3. (Convergence). *The sequence $B^p\Phi$ of Bézier polygons corresponding to degree raising of a given Bézier polygon Φ converges to the corresponding Bézier curve; i.e.,*

$$\lim_{p\to\infty} B^p\Phi = B_n\Phi.$$

Theorem 4.3 follows from the *Weierstrass approximation theorem*, see e.g., [DAV 75] (p. 108), [FAR 87, 90]. The rate of convergence is too slow, however, for practical applications [Coh 85].

Fig. 4.8 illustrates the assertion of Theorem 4.3. In Fig. 4.8a, degree raising of a Bézier curve of degree $n = 5$ to degree $n = 6$ is shown. Fig. 4.8b presents the result of degree raising of a Bézier curve of degree $n = 3$ to one of degree $n = 10$, and clearly shows the convergence of the Bézier polygons to the Bézier curve.

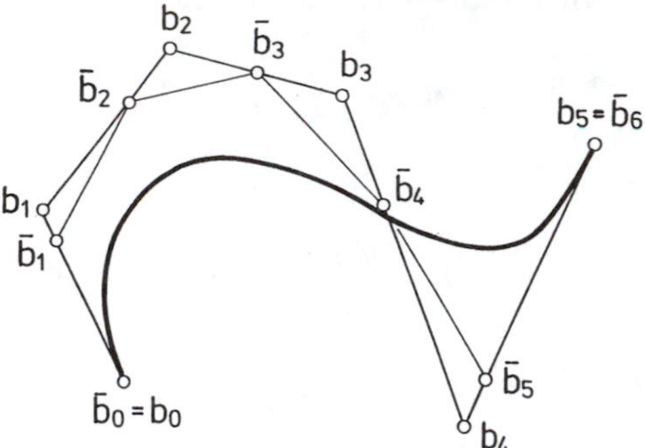

Fig. 4.8a. Bézier polygons for degree raising from $n = 5$ to $n = 6$.

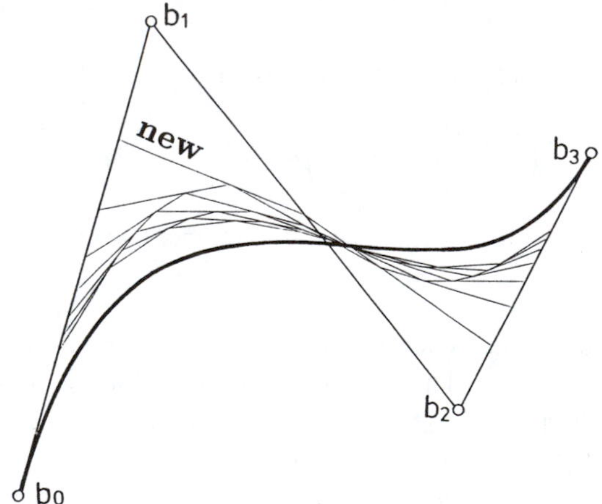

Fig. 4.8b. Degree raised Bézier polygons approximate the Bezier curve.

Remark. Formula (4.14) can be generalized to the case of degree raising from degree n to degree $n + \mu$:

$$\bar{\boldsymbol{b}}_k = \sum_{j=k-\mu}^{k} \frac{\binom{n}{j}\binom{\mu}{k-j}}{\binom{n+\mu}{k}} \boldsymbol{b}_j, \qquad k = 0(1)n{+}\mu, \quad j \geq 0. \tag{4.14a}$$

Remark. Degree raising can be of use in interactively working with a curve, since it leads to a Bézier polygon with more degrees of freedom to manipulate, see e.g., [For 72], [HOS 87].

Remark. Degree raising is often needed in order to compare coefficients in the construction of smoothness conditions for joining spline surfaces together.

Remark. In general, *exact degree reduction* is not possible. In particular, if we solve for $\boldsymbol{b}_k$ in (4.14), we obtain a recurrence formula for $\boldsymbol{b}_k$ in terms of the $\overline{\boldsymbol{b}}_k$. But in general, this doesn't work, since the end points of the Bézier polygons do not coincide. Equations for *approximate* degree reduction can be found in [For 72], [FAR 87, 90]. Various strategies for degree reduction have also been discussed in [Watk 88]. We present a general approach to approximate degree reduction in Chap. 10.

In view of Lemma 4.3, it makes only limited sense to say that a given Bézier curve has a particular degree. The representation at hand may involve a higher degree than necessary, which suggests defining the polynomial degree to be that of the representation with minimal degree. This suggests the question of how to recognize when the degree can be reduced, since in dealing with polynomials numerically, generally, lower degree polynomials are simpler and more stable.

If a given Bézier curve is represented with too high a degree N, then according to (4.14), the associated Bézier points must be dependent. These dependencies can be found by taking differences:

Lemma 4.4. *Given a Bézier curve of degree N with lowest possible polynomial degree $n < N$. Then taking repeated differences of the Bézier points must terminate after the n-th step; i.e., all higher differences must vanish.*

Proof: Consider the formula (4.7b) for the derivatives. Since $\boldsymbol{X}$ is of degree n, all derivatives of order $k > n$ must vanish; i.e.,

$$\boldsymbol{X}^{(k)} = \frac{N!}{(N-k)!} \sum_{i=0}^{N-k} \Delta^k \boldsymbol{b}_i \; B_i^{N-k}(t) = 0$$

for $k > n$. In view of the linear independence of the basis functions, this is only possible if

$$\Delta^{n+k}\, \boldsymbol{b}_\ell = 0, \qquad \text{for} \qquad k = 1(1)N-n. \quad \blacksquare \tag{4.15}$$

4.1.2. Bézier Spline Curves

From the standpoint of analysis, the Bézier spline curves are nothing more than polynomial splines represented in terms of a different basis than the usual monomials. Formally, the spline coefficients in terms of the new basis functions can be found from the basis transformation equations, see Chap. 10. In contrast to working with bases built from monomials, the advantage of writing splines in Bézier form is that the Bézier points have geometric significance. Moreover, the smoothness conditions between neighboring spline

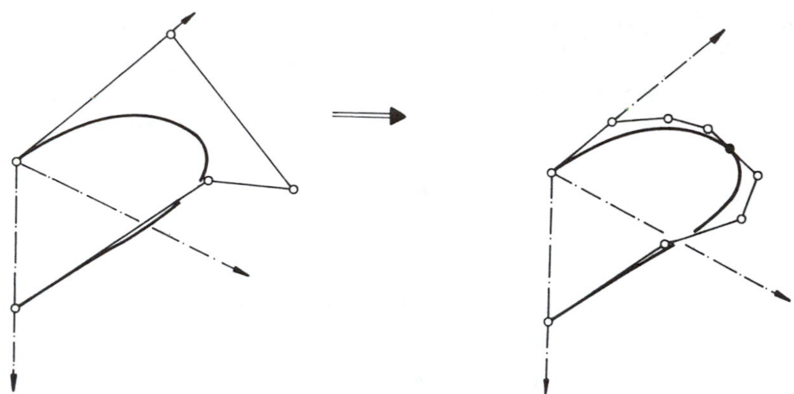

Fig. 4.9. A curve represented in Bézier form and as a Bézier spline curve.

segments also can be interpreted geometrically, see e.g., [Boeh 76], [Stä 76]. Fig. 4.9 shows a Bézier curve with associated Bézier polygon, and the same curve as a Bézier spline curve with its associated Bézier polygon.

In order to discuss the geometric meaning of smoothness conditions between neighboring spline segments, we assume

- that neighboring spline curve segments are of the same order and that the segments are connected with C^p continuity,
- that the parametrization of the individual segments can be chosen freely, *i.e.*, we transform the parameter $t \in [0,1]$ via

$$t = \frac{u}{\mu_j}, \qquad \mu_j \in \mathbb{R} \tag{4.16}$$

to the parameter interval $u \in [0, \mu_j]$, so that the segment with index j corresponds to the parameter interval $[0, \mu_j]$.

Remark. Changing the parameter interval for an individual curve does not affect the shape of the curve. But by the chain rule, the length of the parameter interval does play a role in computing derivatives, and so the choice of the parameter interval does affect the continuity conditions for a spline curve.

Suppose for each $i = 0(1)m$, the i-th piece of the spline curve is given in Bézier form by

$$\boldsymbol{X}_i(u) = \sum_{k=0}^{n} \boldsymbol{b}_{ki} B_k^n\left(\frac{u}{\mu_i}\right), \qquad u \in [0, \mu_i]. \tag{4.17}$$

We also could have used Bernstein polynomials written in the more general form (4.11), in which case the parameter t associated with the i-th piece would run over the interval $[t_i, t_{i+1}]$. Then the spline curve would correspond to a global parameter interval $t_0 < t_1 < \cdots < t_m < t_{m+1}$, and t would be a global parameter. The u in (4.17) is a local parameter. In terms of these join points, we have

$$\mu_i = t_{i+1} - t_i.$$

We prefer to work with the local parametrization of (4.17).

First we require that neighboring pieces of the spline should join with C^0 *continuity*. This requires that

$$\boldsymbol{X}_i(\mu_i) = \boldsymbol{X}_{i+1}(0) \qquad \text{in local parameters,}$$

or

$$\boldsymbol{X}_i(t_{i+1}^-) = \boldsymbol{X}_{i+1}(t_{i+1}^+) \qquad \text{in global parameters.}$$

In terms of Bézier points, this condition can be written as

$$\boldsymbol{b}_{ni} = \boldsymbol{b}_{0,i+1}. \tag{4.18}$$

If the first derivatives are also to agree at the points where the segments join together to give C^1 *continuity*, then we also need

$$\boldsymbol{X}_i'(\mu_i) = \boldsymbol{X}_{i+1}'(0)$$

using local parameters, with an analogous condition in the case of global parameters. From (4.6a) and the chain rule, using (4.16) we get the condition

$$\frac{n}{\mu_i}(\boldsymbol{b}_{ni} - \boldsymbol{b}_{n-1,i}) = \frac{n}{\mu_{i+1}}(\boldsymbol{b}_{1,i+1} - \boldsymbol{b}_{0,i+1}), \tag{4.18a}$$

or, since $\boldsymbol{b}_{ni} = \boldsymbol{b}_{0,i+1}$, equivalently

$$\boldsymbol{b}_{ni}(\mu_{i+1} + \mu_i) = \mu_i \boldsymbol{b}_{1,i+1} + \mu_{i+1}\boldsymbol{b}_{n-1,i}. \tag{4.18b}$$

Thus, C^1 continuity requires that the points $\boldsymbol{b}_{ni} = \boldsymbol{b}_{0,i+1}$ along with the two neighboring Bézier points must be collinear, *i.e.*, these Bézier points lie on the common tangent at the join point to the two segments of the Bézier spline curve. The μ_i can be considered as weights, see Fig. 4.10a. If in particular, $\mu_i = \mu_{i+1}$ (the two neighboring segments have the same parameter interval), then $\boldsymbol{b}_{ni}$ is at the midpoint of the line from $\boldsymbol{b}_{n-1,i}$ to $\boldsymbol{b}_{1,i+1}$, see Fig. 4.10b.

The effect of using different parameter intervals for neighboring Bézier spline segments is shown in Fig. 4.11 (see also Fig. 3.13a,b). Here we are working with cubic Bézier curves where the segment on the left corresponds to the Bézier points $\boldsymbol{b}_0, \boldsymbol{b}_1, \boldsymbol{b}_2, \boldsymbol{b}_3$, and where the curve on the right is to pass

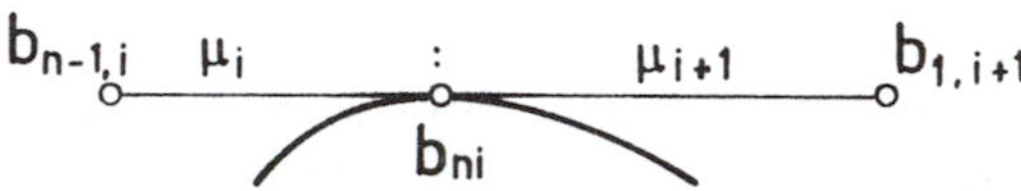

Fig. 4.10a. C^1 continuity in the case of distinct parameter intervals.

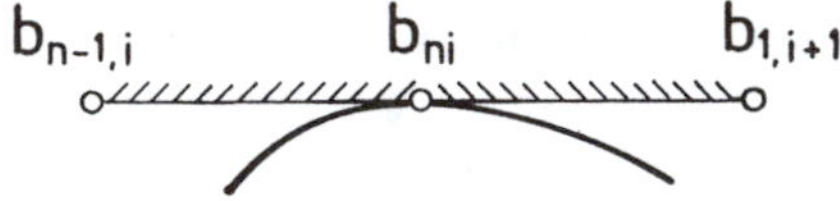

Fig. 4.10b. C^1 continuity in the case of a common parameter interval.

through the point $\boldsymbol{b}_6$ with tangent determined by $\boldsymbol{b}_5$. Then the C^1 continuity condition (4.18b) requires that the Bézier point $\boldsymbol{b}_4$ lie on the tangent line to the curve at $\boldsymbol{b}_3$. Figs. 4.11a,b show the effect of the choice of $\boldsymbol{b}_5$ in the case where the two parameter intervals are of the same length. Fig. 4.11c shows what happens when one interval is four times as large as the other. The result in this case is probably closer to what we "expect" the curve to look like. This phenomenon was discussed already in [Stä 76].

If the neighboring Bézier spline segments are to join with C^2 *continuity*, then from (4.6a), we require

$$\frac{n(n-1)}{{\mu_i}^2}(\boldsymbol{b}_{ni} - 2\boldsymbol{b}_{n-1,i} + \boldsymbol{b}_{n-2,i}) = \frac{n(n-1)}{\mu_{i+1}^2}(\boldsymbol{b}_{2,i+1} - 2\boldsymbol{b}_{1,i+1} + \boldsymbol{b}_{0,i+1}),$$

or in view of $\boldsymbol{b}_{ni} = \boldsymbol{b}_{0,i+1}$ and (4.18b),

$$\boldsymbol{b}_{n-1,i} + \frac{\mu_{i+1}}{\mu_i}(\boldsymbol{b}_{n-1,i} - \boldsymbol{b}_{n-2,i}) = \boldsymbol{b}_{1,i+1} + \frac{\mu_i}{\mu_{i+1}}(\boldsymbol{b}_{1,i+1} - \boldsymbol{b}_{2,i+1}) = \boldsymbol{D}_{i+1}. \quad (4.19)$$

The points $\boldsymbol{D}_i$ introduced in (4.19) are auxiliary points, and are the intersections of the straight lines which according to (4.19) pass through the points $(\boldsymbol{b}_{n-1,i}, \boldsymbol{b}_{n-2,i})$ and $(\boldsymbol{b}_{1,i+1}, \boldsymbol{b}_{2,i+1})$. Fig. 4.12a illustrates the geometric meaning of the C^2 continuity conditions (4.19). The case of C^3 continuity is shown in Fig. 4.12b.

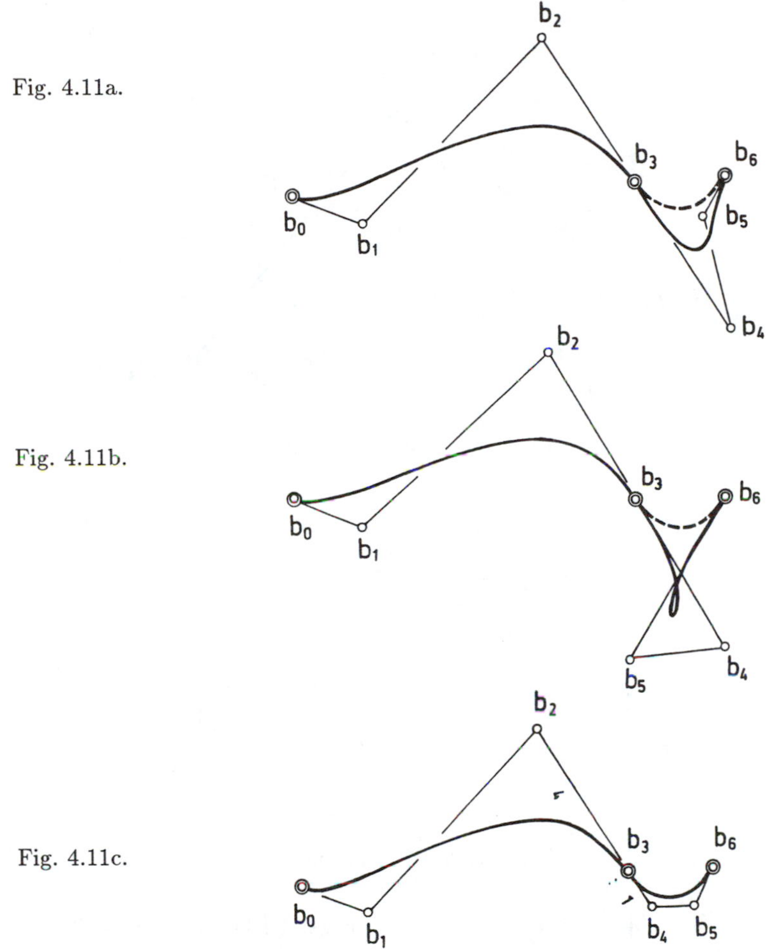

Fig. 4.11. Cubic Bézier spline curves: a) and b) show poor choices for the second segments, while c) shows a better one.

Remark. If a given Bézier curve of degree n is subdivided into several segments using the de Casteljau algorithm, the resulting segments always join with C^n smoothness. This means that the Bézier points of the segment $\boldsymbol{X}_2$ which are determined by the smoothness conditions arising from requiring that it join with $\boldsymbol{X}_1$ at $\boldsymbol{X}_1(1)$ can be found by a *de Casteljau extrapolation*. The construction for the case of C^n smoothness differs from the de Casteljau construction for the determination of a point on the curve corresponding to the parameter λ with $\lambda > 1$ only in the order in which the operations are

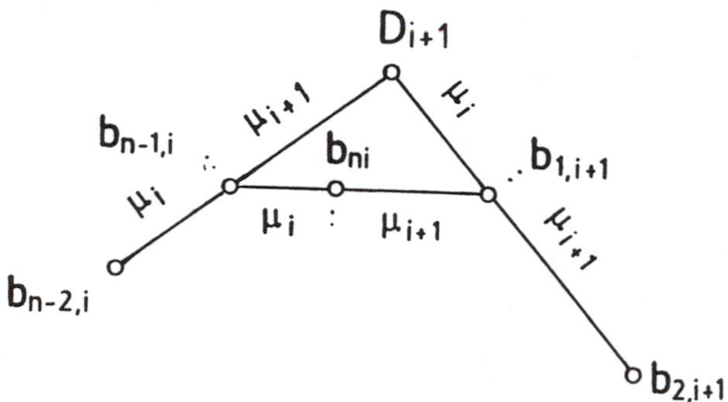

Fig. 4.12a. C^2 continuity.

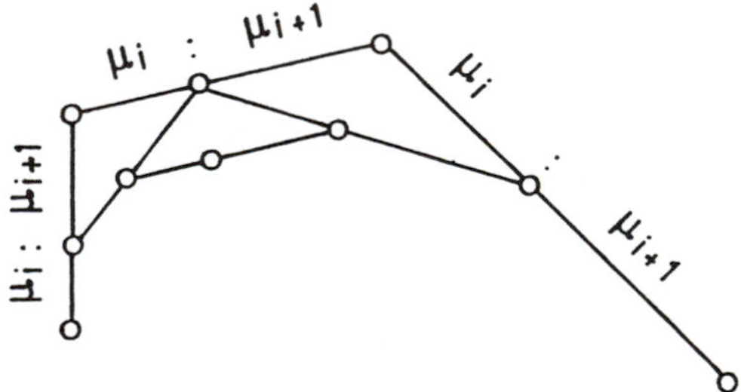

Fig. 4.12b. C^3 continuity.

carried out [Stä 76]. To see this, compare Figs. 4.13a and Fig. 4.13b, where the order of the operations is shown.

4.1.3. Cubic Bézier Splines

The auxiliary points $\boldsymbol{D}_i$ introduced in (4.19) permit a unified treatment of interpolation using Bézier spline curves. In this section we restrict our attention to the cubic case. The need to choose two additional boundary conditions (see Chap. 3) translates here into the free choice of the second and next to last Bézier points of the Bézier spline curve. Suppose now that for $i = 0(1)n–1$, the i-th segment of the desired spline is given by

$$\boldsymbol{X}_i(t) = \boldsymbol{b}_{0,i}B_0^3(t) + \boldsymbol{b}_{1,i}B_1^3(t) + \boldsymbol{b}_{2,i}B_2^3(t) + \boldsymbol{b}_{3,i}B_3^3(t), \qquad (4.20)$$

and that we want to interpolate points $\boldsymbol{P}_i$ for $i = 0(1)n$. Then clearly we

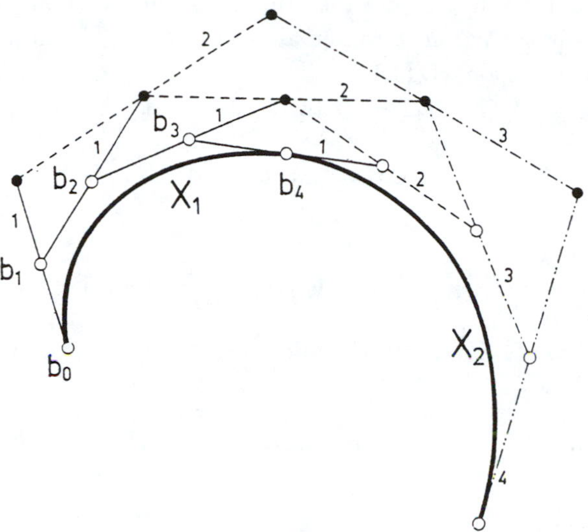

Fig. 4.13a. de Casteljau for $t_0 = 2$.

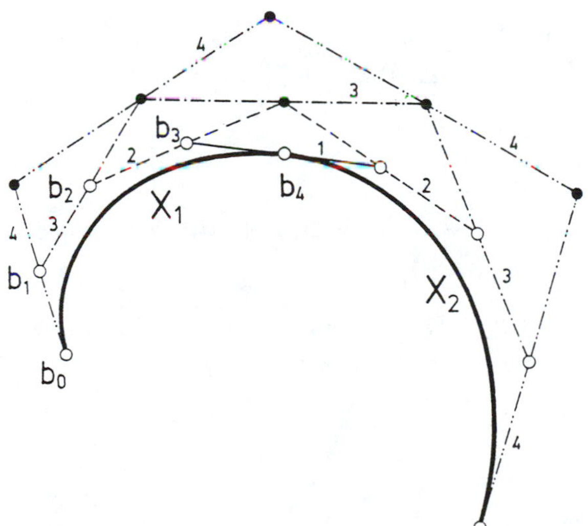

Fig. 4.13b. C^4 continuity construction.

must have

$$P_i = b_{0,i} = b_{3,i-1}.$$

We now assume that for each $i = 0(1)n$–1, the point P_i is to be associated with the parameter value t_i, and that the Bernstein polynomial in (4.20) is parametrized over the parameter interval $[a, b] = [t_i, t_{i+1}]$. Thus, we are using a global parametrization, cf. (4.11).

Setting $\mu_i = \Delta t_i = t_{i+1} - t_i$, (4.18) and (4.19) imply the *connecting conditions*

$$\boldsymbol{b}_{3,i}(\Delta t_{i+1} + \Delta t_i) = \Delta t_i\, \boldsymbol{b}_{1,i+1} + \Delta t_{i+1}\, \boldsymbol{b}_{2,i}, \tag{4.21a}$$

$$\boldsymbol{D}_{i+1} = \boldsymbol{b}_{2,i} + \frac{\Delta t_{i+1}}{\Delta t_i}(\boldsymbol{b}_{2,i} - \boldsymbol{b}_{1,i}) = \boldsymbol{b}_{1,i+1} + \frac{\Delta t_i}{\Delta t_{i+1}}(\boldsymbol{b}_{1,i+1} - \boldsymbol{b}_{2,i+1}), \tag{4.21b}$$

with auxiliary points $\boldsymbol{D}_i$. Shifting indices on the right-hand side and eliminating $\boldsymbol{b}_{1,i}$ leads to

$$\boldsymbol{D}_i \Delta t_{i+1} + \boldsymbol{D}_{i+1}(\Delta t_{i-1} + \Delta t_i) = \boldsymbol{b}_{2,i}(\Delta t_{i-1} + \Delta t_i + \Delta t_{i+1}) =: \boldsymbol{b}_{2,i}\Delta_i, \tag{4.22a}$$

with $\Delta_i = \Delta t_{i-1} + \Delta t_i + \Delta t_{i+1}$. Now eliminating $\boldsymbol{b}_{2,i}$

$$\boldsymbol{D}_{i+1}\Delta t_{i-1} + \boldsymbol{D}_i(\Delta t_i + \Delta t_{i+1}) = \boldsymbol{b}_{1,i}(\Delta t_{i-1} + \Delta t_i + \Delta t_{i+1}) = \boldsymbol{b}_{1,i}\Delta_i, \tag{4.22b}$$

and shifting indices again, adding (4.22a), and using (4.21a) leads to the recurrence relation

$$\begin{aligned} \boldsymbol{D}_i(\Delta t_{i+1})^2\Delta_{i+1} + \boldsymbol{D}_{i+1}(\Delta t_{i+1}(\Delta ti + \Delta t_{i-1})\Delta_{i+1} + \Delta t_i(\Delta t_{i+1} + \Delta t_{i+2})\Delta_i) \\ + \boldsymbol{D}_{i+2}(\Delta t_i)^2\Delta_i = \boldsymbol{b}_{3,i}(\Delta t_{i+1} + \Delta t_i)\Delta_i\Delta_{i+1}. \end{aligned} \tag{4.23}$$

Introducing the abbreviations

$$\begin{aligned} \alpha_i &= (\Delta t_i)^2\Delta_i, \\ \beta_i &= \Delta t_{i+1}(\Delta t_i + \Delta t_{i-1})\Delta_{i+1} + \Delta t_i(\Delta t_{i+1} + \Delta t_{i+2})\Delta_i, \\ \gamma_i &= (\Delta t_{i+1} + \Delta t_i)\Delta_i\Delta_{i+1}, \end{aligned} \tag{4.24a}$$

we can combine (4.22a), (4.22b) into the following linear system of equations for the computation of the auxiliary points $\boldsymbol{D}_i$:

$$\begin{pmatrix} (\Delta t_1 + \Delta t_0) & \Delta t_{-1} & 0 & \cdots & & & 0 \\ \alpha_1 & \beta_0 & \alpha_0 & & & & \\ 0 & \alpha_2 & \beta_1 & & & & \\ \vdots & & & \ddots & & & \vdots \\ & & & & \beta_{n-3} & \alpha_{n-3} & 0 \\ & & & & \alpha_{n-1} & \beta_{n-2} & \alpha_{n-2} \\ 0 & & & \cdots & 0 & \Delta t_n & (\Delta t_{n-1} + \Delta t_{n-2}) \end{pmatrix} \begin{pmatrix} \boldsymbol{D}_0 \\ \boldsymbol{D}_1 \\ \boldsymbol{D}_2 \\ \vdots \\ \\ \boldsymbol{D}_{n-1} \\ \boldsymbol{D}_n \end{pmatrix} = \begin{pmatrix} \boldsymbol{b}_{1,0}\Delta_0 \\ \boldsymbol{b}_{3,0}\gamma_0 \\ \\ \vdots \\ \\ \boldsymbol{b}_{3,n-2}\gamma_{n-2} \\ \boldsymbol{b}_{2,n-1}\Delta_{n-1} \end{pmatrix} \tag{4.24b}$$

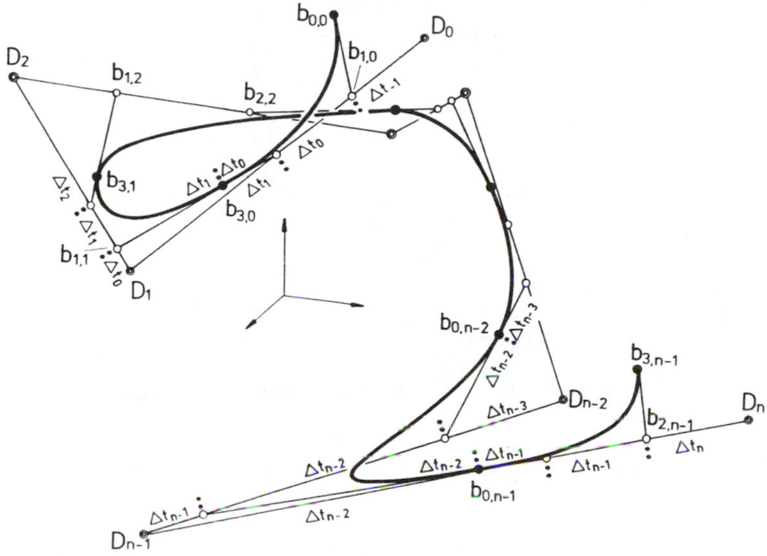

Fig. 4.14. Cubic Bézier spline curve and control polygon.

Here Δt_{-1}, Δt_n can be selected arbitrarily for an open Bézier spline curve, but for the closed case, the knot vector should be extended periodically, so that these two differences are determined, see [Her 83].

For open Bézier spline curves, the points $\boldsymbol{b}_{1,0}$ and $\boldsymbol{b}_{2,n-1}$ can be chosen arbitrarily. For an equally spaced parametrization, the coefficient matrix in (4.24b) is of the same form as the coefficient matrix in (3.18). Additional results on interpolation with Bézier splines can be found e.g., in [Har 82, 84].

Fig. 4.14 shows a cubic Bézier spline curve with associated Bézier polygon. The ratios of the lengths of the edges of the polygon induced by the parametrization are indicated in the figure.

The construction above produces a C^2 continuous cubic Bézier spline. It is also possible to construct cubic Bézier subsplines which are only C^1 continuous. In view of the form (4.20) we have chosen for the pieces, for each knot point $\boldsymbol{P}_i = \boldsymbol{b}_{0,j}$, we need a way to determine the neighboring Bézier points $\boldsymbol{b}_{2,j-1}$ and $\boldsymbol{b}_{1,j}$. These Bézier points lie on the tangent to the curve at $\boldsymbol{b}_{0,j}$ with direction $\boldsymbol{T}_j$, where $\boldsymbol{T}_j$ must be determined from the position of the neighboring data points $\boldsymbol{P}_{j-1}$ and $\boldsymbol{P}_{j+1}$. For example, setting

$$\Delta \boldsymbol{P}_j = \boldsymbol{P}_{j+1} - \boldsymbol{P}_j, \qquad \boldsymbol{c}_j = \frac{\Delta \boldsymbol{P}_j}{\Delta t_j},$$

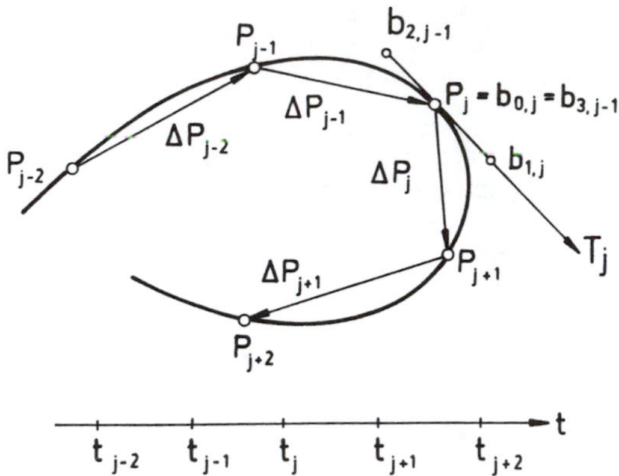

Fig. 4.15. Tangent $\boldsymbol{T}_j$ and neighboring Bézier points.

we can choose one of the following linear combinations of neighboring points (cf. Fig. 4.15):

$$\boldsymbol{T}_j = (1-\alpha_j)\boldsymbol{c}_{j-1} + \alpha_j \boldsymbol{c}_j \qquad \text{or} \qquad \boldsymbol{T}_j = (1-\beta_j)\Delta \boldsymbol{P}_{j-1} + \beta_j \Delta \boldsymbol{P}_j. \quad (4.25)$$

Some of the suggested ways for selecting the factors α_j and β_j (see e.g., [Boeh 84], [BOO 78], [FAR 90]) include:

- the *Bessel scheme*, where

$$\alpha_j = \frac{\Delta t_{j-1}}{\Delta t_{j-1} + \Delta t_j}, \tag{4.26a}$$

- the *Akima scheme*, where

$$\alpha_j = \frac{|\Delta \boldsymbol{c}_{j-2}|}{|\Delta \boldsymbol{c}_{j-2}| + |\Delta \boldsymbol{c}_j|}, \tag{4.26b}$$

- the *Renner/Pochop scheme*, where

$$\beta_j = \frac{|\Delta \boldsymbol{q}_{j-2}|}{|\Delta \boldsymbol{q}_{j-2}| + |\Delta \boldsymbol{q}_j|} \qquad \text{with} \qquad \boldsymbol{q}_j = \frac{\Delta \boldsymbol{P}_j}{|\Delta \boldsymbol{P}_j|}. \tag{4.26c}$$

Now for each knot point $\boldsymbol{b}_{0,j}$, the desired neighboring Bézier points can be taken, for example, to be

$$\boldsymbol{b}_{2,j-1} = \boldsymbol{P}_j - \tfrac{1}{3}\Delta t_{j-1}\boldsymbol{T}_j, \qquad \boldsymbol{b}_{1,j} = \boldsymbol{P}_j + \tfrac{1}{3}\Delta t_j \boldsymbol{T}_j.$$

The factor $\frac{1}{3}$ appears because, by (4.20), the tangent at the endpoint of a cubic Bézier curve $\boldsymbol{X}(t)$ is given by

$$\dot{\boldsymbol{X}}_j(0) = \frac{3}{\Delta t_j}(\boldsymbol{b}_{1,j} - \boldsymbol{b}_{0,j}),$$

which implies

$$\boldsymbol{b}_{1,j} = \boldsymbol{b}_{0,j} + \frac{\Delta t_j}{3}\dot{\boldsymbol{X}}_j(0).$$

4.1.4. Rational Bézier Curves

Bézier curves and Bézier spline curves can be used to model a wide variety of curves. However, in practice it is often desirable to use conic sections (which cannot be represented in Bézier form) for blending. In order to be able to include conic sections in the set of representable curves, we turn to *rational Bézier curves*. This larger class of Bézier curves provides considerably more flexibility in curve design than the usual (nonrational) Bézier curves, which, of course, are included as a special case. See e.g., [Far 83], [Boeh 84], [Pie 86, 87a, 88].

To give a unified treatment of rational curves and surfaces, it is convenient to work with homogeneous coordinates. In order to give a geometric interpretation of homogeneous coordinates in the Euclidean plane E^2 (or in E^3), we think of the plane E^2 as being imbedded in the Euclidean space E^3. Let $\bar{O}$ be a point in E^3 with $\bar{O} \notin E^2$. Then, every point $\boldsymbol{P}$ in E^2 can be uniquely identified with the straight line $g_{\bar{O}P}$. This line can be regarded as a one dimensional subspace of the linear space $V^3_{\bar{O}}$ consisting of all position vectors in E^3 relative to $\bar{O}$. This allows us to introduce (in a natural way) an invertible one-to-one mapping Φ such that

$$\boldsymbol{P} \in E^2 \iff g_{\bar{O}P} \in E^3 \iff \text{one dimensional subspace of } V^3_{\bar{O}}.$$

Suppose now that $(\bar{O}; x_0, x_1, x_2)$ is a Cartesian coordinate system for E^3, and that E^2 is the plane $x_0 = 1$. Then under the mapping Φ, the point $\boldsymbol{P} = (1, x, y)^T$ is associated with the set of vectors $\mathcal{X} = \lambda(1, x, y)^T = (x_0, x_1, x_2)^T$ with nonzero $\lambda \in \mathbb{R}^+$ (see Fig. 4.16), *i.e.*, in E^2 we have

$$x = \frac{x_1}{x_0}, \quad y = \frac{x_2}{x_0}, \qquad x_0 \neq 0.$$

Under the mapping Φ, all straight lines g_∞ in E^3 which are parallel to E^2 are excluded. Each such line can be associated with a point $\boldsymbol{P}_\infty$ at infinity in the plane E^2. In E^3, the line g_∞ has direction

$$\boldsymbol{P}_\infty = \begin{pmatrix} 0 \\ x \\ y \end{pmatrix}.$$

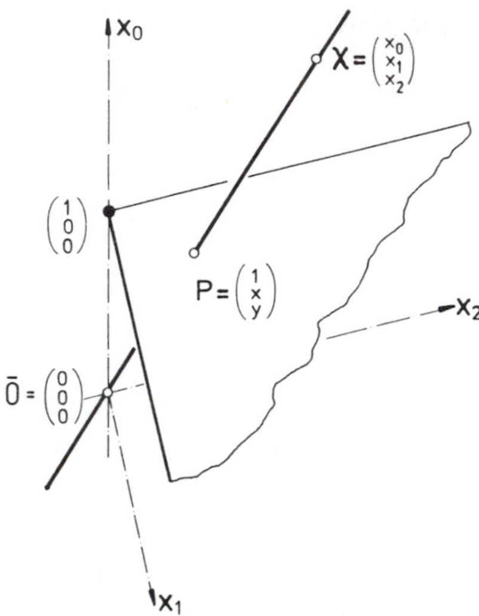

Fig. 4.16. Introducing homogeneous coordinates via projection of points of E^3 into the hyperplane $x_0 = 1$.

We call $\boldsymbol{P}_\infty$ a *point at infinity*. In E^2 the points at infinity are described by direction vectors $\vec{\boldsymbol{p}} = (x, y)^T$.

To introduce homogeneous coordinates in E^3, we imbed E^3 in the Euclidean space E^4. As before, we choose a point $\bar{O} \in E^4$ with $\bar{O} \notin E^3$. Each point $\boldsymbol{P} \in E^3$ is associated with a line $g_{\bar{O}P}$ which can be considered as an element of the vector space $V^4_{\bar{O}}$.

As above, we now assume that $(\bar{O}; x_0, x_1, x_2, x_3)$ is a Cartesian coordinate system for E^4, and that E^3 is the hyperplane defined by $x_0 = 1$ in E^4. We now define the mapping Φ which associates a point $\boldsymbol{P} = (1, x, y, z)^T$ with the set of vectors of the form $\mathcal{X} = \lambda(1, x, y, z)^T = (x_0, x_1, x_2, x_3)^T$ with nonzero $\lambda \in \mathbb{R}^+$. Thus in E^3,

$$x = \frac{x_1}{x_0}, \quad y = \frac{x_2}{x_0}, \quad z = \frac{x_3}{x_0}, \qquad x_0 \neq 0.$$

Again the lines g_∞ parallel to E^3 are excluded. Each of these lines is associated with a point at infinity $\boldsymbol{P}_\infty$ of E^3, and in E^4 has the direction

$$\boldsymbol{P}_\infty = (0, x, y, z)^T.$$

In E^3 they are described by the direction vector

$$\vec{\boldsymbol{p}} = (x, y, z)^T.$$

Analogous to the geometric interpretation of homogeneous coordinates, we understand rational curves in the plane E^2 as projections of the ordinary (integral) Bézier curves in E^3, and similarly, rational curves in E^3 as projections of ordinary Bézier curves in E^4. The control points of an integral Bézier curve in E^3 can be written as

$$\boldsymbol{B}_j = \begin{cases} (\beta_j, \beta_j u_j, \beta_j v_j)^T = (\beta_j, \beta_j \boldsymbol{b}_j)^T, & \beta_j \neq 0, \quad \boldsymbol{b}_j \in E^2 \\ (0, u_j, v_j)^T = (0, \vec{\boldsymbol{b}}_j)^T, & \beta_j = 0, \quad \vec{\boldsymbol{b}}_j \in E^2. \end{cases} \tag{4.27a}$$

The corresponding parametric formula for a planar Bézier curve in the space E^3 is then

$$\mathcal{X}(t) = \sum_{j=0}^{n} \boldsymbol{B}_j B_j^n(t), \tag{4.28a}$$

and after projecting into E^2, we get the following parametric formula for a planar rational Bézier curve:

$$\boldsymbol{X}(t) = \frac{\sum_{j=0}^{n} \beta_j \boldsymbol{b}_j B_j^n(t)}{\sum_{j=0}^{n} \beta_j B_j^n(t)}, \qquad \text{for } \beta_j \neq 0. \tag{4.29a}$$

If one of the control points $\boldsymbol{B}_k$ is a point at infinity, then we set $\beta_k = 0$ in the denominator of (4.29a) and replace $\beta_k \boldsymbol{b}_k$ by $\vec{\boldsymbol{b}}_k$ in the numerator. If we project an integral Bézier curve in E^4, we get a rational Bézier curve in E^3. The control points in E^4 have the form

$$\boldsymbol{B}_j = \begin{cases} (\beta_j, \beta_j u_j, \beta_j v_j, \beta_j w_j)^T = (\beta_j, \beta_j \boldsymbol{b}_j)^T, & \beta_j \neq 0, \quad \boldsymbol{b}_j \in E^3 \\ (0, u_j, v_j, w_j)^T = (0, \vec{\boldsymbol{b}}_j)^T, & \beta_j = 0, \quad \vec{\boldsymbol{b}}_j \in E^3. \end{cases} \tag{4.27b}$$

Correspondingly, the parametric formula for a rational Bézier curve in E^3 in the inverse image space E^4 is given by

$$\mathcal{X}(t) = \sum_{j=0}^{n} \boldsymbol{B}_j B_j^n(t), \tag{4.28b}$$

or as a projection into E^3 (parametric formula for a rational Bézier curve in space)

$$\boldsymbol{X}(t) = \frac{\sum_{j=0}^{n} \beta_j \boldsymbol{b}_j B_j^n(t)}{\sum_{j=0}^{n} \beta_j B_j^n(t)}, \qquad \text{for} \quad \beta_j \neq 0. \tag{4.29b}$$

If a control point $\boldsymbol{B}_k$ is a point at infinity, then we set $\beta_k = 0$ in the denominator of (4.29b), and replace $\beta_k \boldsymbol{b}_k$ by $\vec{\boldsymbol{b}}_k$ in the numerator. As in the

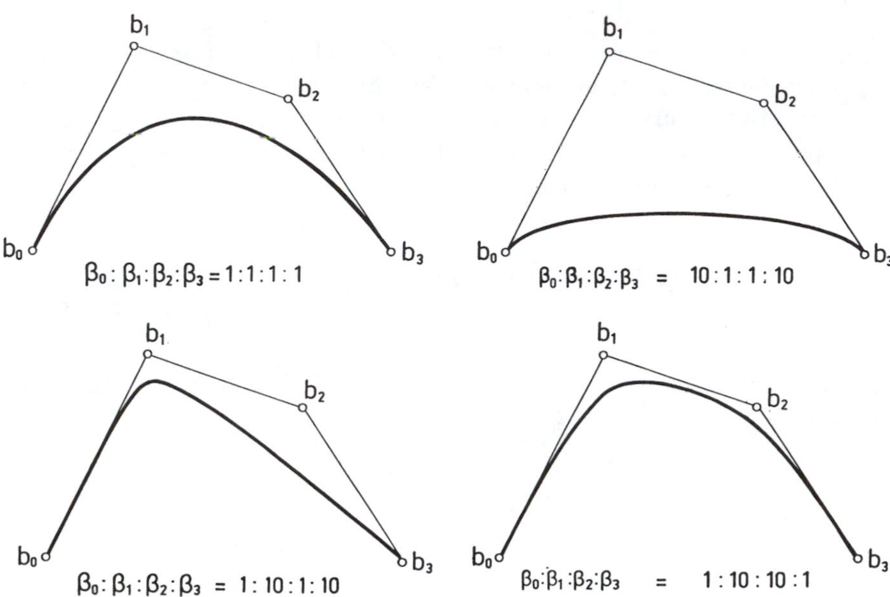

Fig. 4.17a. Rational Bézier curves with different weights and the same Bézier polygon.

integral case, the $\{\boldsymbol{b}_j\}$ form the control polygon of the rational Bézier curve, where the β_j are the associated *weights*.

Now as before, we choose $[0, \mu]$ as the parameter interval. Then at the ends of this interval, the curve passes through the control points as long as they are not points at infinity. From now on, we assume that this is the case, which means that neither β_0 nor β_n can be zero.

If the weights are chosen to be $\beta_j = 1$ for all j, then since $\sum_{j=0}^{n} B_j^n(t) = 1$, (4.29a,b) becomes the parametric formula (4.4) for an ordinary Bézier curve.

The weights can be used as additional design elements, as is illustrated in the figures. Fig 4.17a shows rational Bézier curves with different weights corresponding to a given Bézier polygon. The first curve is an ordinary Bézier curve. Since the Bézier polygon is convex while all of the weights were selected to be positive, all of the curves are convex.

Fig. 4.17b shows that increasing a weight causes the curve to move towards the associated Bezier point. This is true in general:

Increasing a weight β_j causes all points on the curve to move towards the Bézier point $\boldsymbol{b}_j$, while decreasing the weight causes all points to move away from $\boldsymbol{b}_j$.

We can establish this as follows. Given a rational Bézier curve

$$\boldsymbol{X}(t) = \frac{\sum_{j=0}^{n} \beta_j \boldsymbol{b}_j B_j^n(t)}{\sum_{j=0}^{n} \beta_j B_j^n(t)}, \tag{4.28c}$$

suppose that the weight β_k is changed to $\bar{\beta}_k = \beta_k + \Delta\beta_k$. We denote the resulting Bézier curve by $\bar{\boldsymbol{X}}(t)$. Writing both curves in the rational form (4.28c) leads to

$$(\boldsymbol{X}(t) - \bar{\boldsymbol{X}}(t))(\sum \bar{\beta}_j B_j^n(t)) = -\Delta\beta_k B_k^n(t)(\boldsymbol{b}_k - \boldsymbol{X}(t)),$$

which immediately implies the assertion.

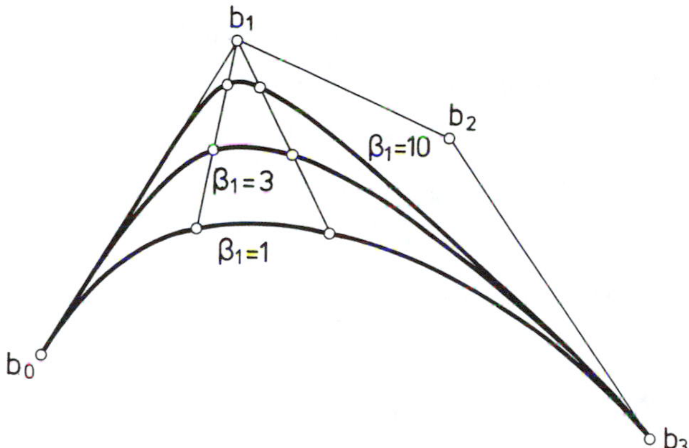

Fig. 4.17b. Effect of varying a weight.

Varying a weight also affects the parametric structure of the Bézier curve. For equally spaced parameter values, increasing the weights causes the points to move closer together, as can be seen in Fig. 4.17b. In [Pie 86] it is shown how weights can be computed to bring about desired changes in the Bézier curve.

In general, *two weights* β_i, β_j can always be chosen to be 1 without affecting the shape of the curve, see e.g, [Pat 85]. Indeed, since $\mathcal{X}(t)$ and $\beta\mathcal{X}(t)$ with $\beta \neq 0$ describe the same rational Bézier curve, it is always possible to assign an arbitrary weight to one of the Bézier points as long as it is not a point at infinity. Thus, for example, we can take $\beta_0 = 1$. Now if we care only about the form of the curve and not about its parametric structure, then by an additional rational parameter transformation, we can transform another β_i to 1. For example, if $\beta_0 = 1$, then β_n can also be transformed to 1 by choosing

$$t = \frac{\tau}{a - c\tau} \qquad \text{so} \qquad 1 - t = \frac{a(1-\tau)}{a - c\tau}, \tag{$*$}$$

with $a - c = 1$ and $a^n = \beta_n$. The effect of the changes on the parametric structure is shown in Fig. 4.17c.

Utilizing the interpretation of rational curves as projections of integral curves from associated higher dimensional inverse image spaces, with appropriate assumptions we can generalize the known properties of integral Bézier curves to the case of rational Bézier curves. Assuming that neither boundary point of the Bézier curve is a point at infinity, we have

a) the Bézier polygon approximately describes the shape of the curve,
b) the first and last points of the curve coincide with the Bézier points $\boldsymbol{b}_0$ and $\boldsymbol{b}_n$, respectively,
c) the first and last edges of the Bézier polygon are tangent to the curve,
d) if all $\beta_k > 0$, then the convex hull property holds,
e) if all $\beta_k > 0$, then the variation diminishing property holds.

It was shown above that integral Bézier curves are *affinely invariant*. Because of the way in which they are defined, rational Bézier curves are even *projectively invariant*.

If negative weights are permitted, then the rational Bézier curve no longer lies entirely in the convex hull of the Bézier polygon. Goldman and DeRose [Gol 86] have given a bound on the deviation of a rational Bézier curve from the convex hull of its Bézier polygon. Suppose we denote the rational basis functions [Pie 86a] by

$$B_i^{n*}(t) := \frac{\beta_i B_i^n(t)}{\sum_{k=0}^n \beta_k B_k^n(t)}.$$

Let

$$M_k = |\text{max negative value of } B_k^{n*}(t)|, \qquad M = \max\{M_k\},$$

and

$$p_t = \{\text{number of basis functions which are negative at } t\}, \quad p = \sup_{t\in[0,1]} \{p_t\}.$$

Then $\delta = 1 + 2Mp$ is a bound on the deviation of the curve from the convex hull of its Bézier polygon.

We now show that the *de Casteljau algorithm* can be extended to rational Bézier curves. The simplest way to do this is to apply the de Casteljau scheme (4.10) to homogeneous coordinates, *i.e.*, to vectors in E^4 as in (4.27). This means that the de Casteljau scheme should be applied both to the weights and to the weighted Bézier points, see e.g, [Far 83]. Thus, points $\mathcal{X}(t_0)$ on

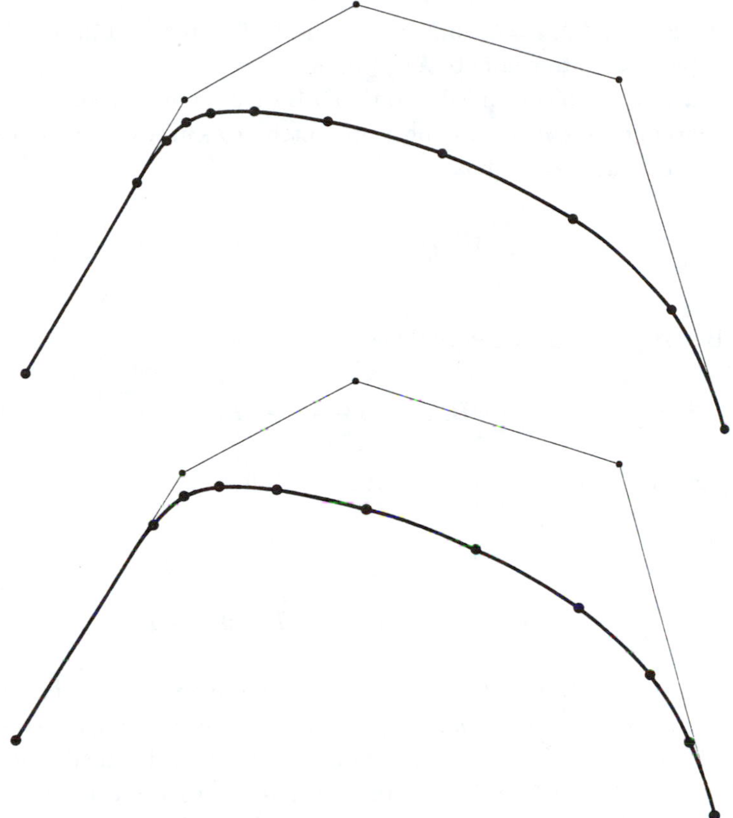

Fig. 4.17c. Changes in the parametric structure by reparametrization using equi-length parametric intervals. $\beta_i = (1, 10, 1, 5, 10)$ (top) and $\beta_4 \to 1$ ($a = \sqrt[4]{10}$) (bottom) .

the curve with $t_0 \in [0, \mu]$ can be computed by repeated linear interpolation in E^4 using the recurrence formula in homogeneous form

$$\boldsymbol{B}_k^i(t_0) = (1 - t_0)\boldsymbol{B}_k^{i-1}(t_0) + t_0\boldsymbol{B}_{k+1}^i(t_0), \tag{4.30a}$$

with $\boldsymbol{B}_k^k = \boldsymbol{B}_k$ and $\mathcal{X}(t_0) = \boldsymbol{B}_0^n$, or in E^3 as

$$\begin{aligned} \beta_k^i(t_0)\boldsymbol{b}_k^i(t_0) &= (1 - t_0)\beta_k^{i-1}(t_0)\boldsymbol{b}_k^{i-1}(t_0) + t_0\beta_{k+1}^i(t_0)\boldsymbol{b}_{k+1}^i(t_0) \\ \beta_k^i(t_0) &= (1 - t_0)\beta_k^{i-1}(t_0) + t_0\beta_{k+1}^i(t_0), \end{aligned} \tag{4.30b}$$

with $\boldsymbol{b}_k^k = \boldsymbol{b}_k$, $\beta_k^k = \beta_k$ and $\boldsymbol{X}(t_0) = \boldsymbol{b}_0^n$. The point $\boldsymbol{X}(t_0)$ subdivides the rational Bézier curve into two segments $\boldsymbol{X}^1(t)$ and $\boldsymbol{X}^2(t)$, which can again be represented as rational Bézier curves of the same degree. The associated

Bézier points $\boldsymbol{B}_0^i$ and $\boldsymbol{B}_i^n$, $i = 0(1)n$, are automatically produced in using the de Casteljau recurrence to compute $\boldsymbol{X}(t_0)$ (compare with Sect. 4.1).

Degree raising can be extended to rational Bézier curves in a similar way. A curve (4.26a) of degree n can also be represented as a curve of degree $n+1$. In terms of homogeneous coordinates, we have

$$\mathcal{X}(t) = \sum_{k=0}^{n+1} \bar{\boldsymbol{B}}_k B_k^{n+1}(t), \qquad t \in [0,1], \tag{4.30c}$$

where the Bézier points $\bar{\boldsymbol{B}}_i$ are given by

$$\bar{\boldsymbol{B}}_k = \frac{k}{n+1}\boldsymbol{B}_{k-1} + (1 - \frac{k}{n+1})\boldsymbol{B}_k. \tag{4.30d}$$

In terms of inhomogeneous coordinates, it follows that

$$\begin{aligned} \bar{\beta}_k \bar{\boldsymbol{b}}_k &= \frac{k}{n+1}\beta_{k-1}\boldsymbol{b}_{k-1} + (1 - \frac{k}{n+1})\beta_k \boldsymbol{b}_k, \\ \bar{\beta}_k &= \frac{k}{n+1}\beta_{k-1} + (1 - \frac{k}{n+1})\beta_k, \qquad k = 0(1)n{+}1. \end{aligned} \tag{4.30e}$$

Here we should note that this is not the most general representation of a rational Bézier curve of degree n as a rational Bézier curve of degree $n+1$. All curves of the form $\lambda(t)\mathcal{X}(t) \in E^4$, with an arbitrary real-valued function $\lambda(t)$ which does not vanish on $[0, \mu]$, are associated with the same rational Bézier curve (the factor $\lambda(t)$ drops out when we form the quotients x_i/x_0, $i = 1, 2, 3$).

Now if, for example, we choose $\lambda(t)$ to be linear, $\lambda(t) = \alpha(1-t) + \gamma t$, then $\lambda(t)\mathcal{X}(t)$ is of degree $n+1$, and we get the following formula of degree $n+1$ for the Bézier points (in homogeneous coordinates):

$$\bar{\boldsymbol{B}}_k = \gamma \frac{k}{n+1}\boldsymbol{B}_{k-1} + \alpha\left(1 - \frac{k}{n+1}\right)\boldsymbol{B}_k, \qquad \alpha, \gamma \in \mathbb{R}.$$

This means that every rational Bézier curve $\boldsymbol{X}(t)$ of degree n can be represented as a rational Bézier curve of degree $n+1$ in an infinite number of ways!

Combining formal degree raising (4.30) in $\mathbb{R}^4$ with reparametrization shows that the *same rational Bézier curve can be described by different Bézier polygons*. The curves differ in their parametrization [Far 89].

Fig. 4.18 illustrates this fact for the second cubic rational Bézier curve of Fig. 4.17a. First we use (4.30) to raise the degree of this curve from three to four (the new Bézier polygon is shown). Now we return to the second

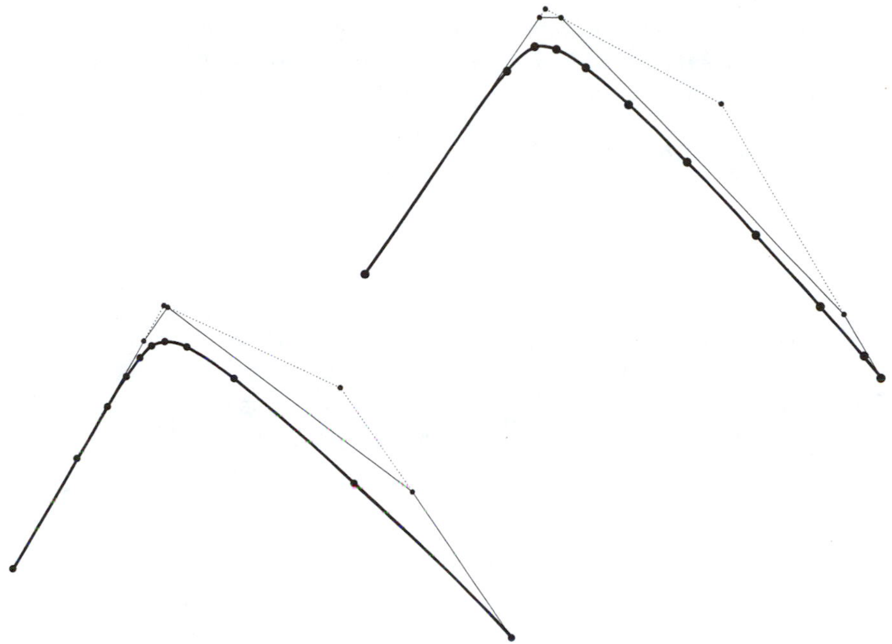

Fig. 4.18. Two control polygons describing the same rational Bézier curve.

curve in Fig. 4.17a, and perform a parameter transformation according to (4.29) with $a = \sqrt[3]{100}$ so that the last weight is mapped to 0.1. Once again raising the degree from three to four, we get a second control polygon whose associated curve is exactly the same as the one we first constructed. The difference between the two curves is in the parametrization, as can be seen by looking at the sequence of indicated points on the curve: they correspond to the same equally spaced points in the parameter domain $[0, 1]$ (at a distance 0.125 apart), but are distributed differently on the curves.

We now show how *quadratic rational Bézier curves* can be used to represent *conic sections*, see e.g., [Lee 87]. As in (4.27), we take

$$\boldsymbol{X}(t) = \frac{\beta_0 \boldsymbol{b}_0 B_0^2(t) + \beta_1 \boldsymbol{b}_1 B_1^2(t) + \beta_2 \boldsymbol{b}_2 B_2^2(t)}{\beta_0 B_0^2(t) + \beta_1 B_1^2(t) + \beta_2 B_2^2(t)}. \tag{4.31}$$

To simplify the geometric interpretation, we specialize to $\beta_0 = \beta_2 = 1$, and introduce the *middle point* $\boldsymbol{M} = \frac{1}{2}(\boldsymbol{b}_0 + \boldsymbol{b}_2)$ and the *shoulder point* $\boldsymbol{S} = (1 - s)\boldsymbol{M} + s\boldsymbol{b}_1$, where

$$s = \frac{\beta_1}{1 + \beta_1}. \tag{$*$}$$

Then (see Fig. 4.19a and [HOS 87] for a proof) we have the following lemma:

Lemma 4.5. *The set of rational Bézier curves of order two in parametric form (4.31) describes the set of (nondegenerate) conic sections. If the parameter s is introduced as in $(*)$, then*

$s = \frac{1}{2}$ *gives a parabolic arc,*

$s < \frac{1}{2}$ *gives an elliptic arc,*

$s > \frac{1}{2}$ *gives a hyperbolic arc.*

For $\beta_1 > 0$ and for t in the parameter interval $[0, \mu]$, the conic section curve lies entirely inside of the Bézier polygon, while for $\beta_1 < 0$ it lies entirely outside. We can describe the complete conic section without this exchange of weights if we do not restrict ourselves to the parameter interval $[0, \mu]$, but instead use all of $\mathbb{R}$, see Fig. 4.19b and [Dero 91].

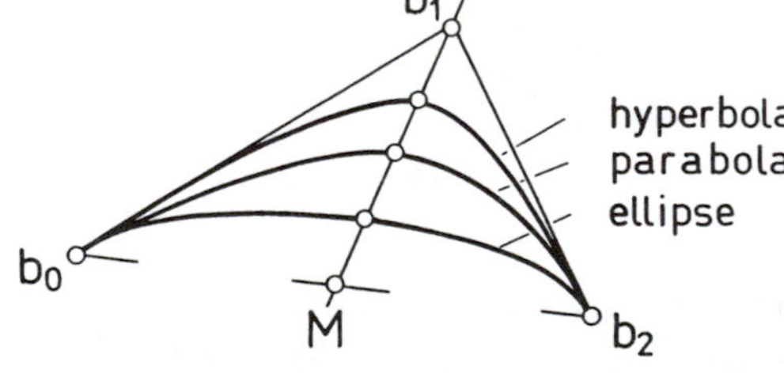

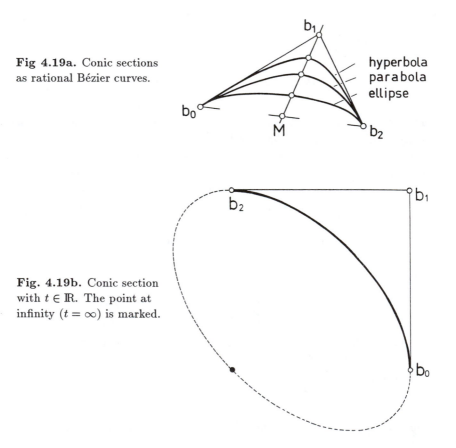

Fig 4.19a. Conic sections as rational Bézier curves.

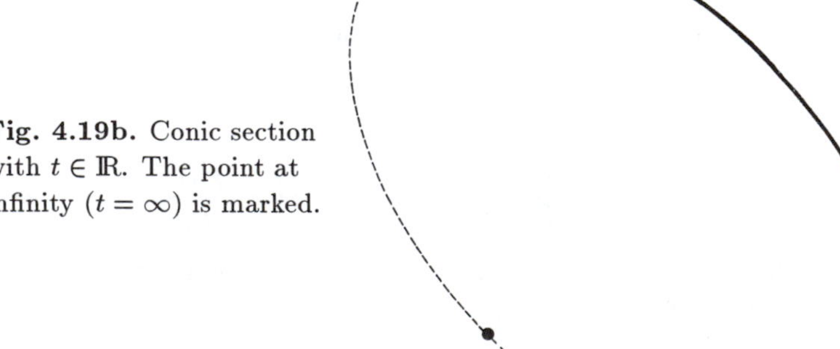

Fig. 4.19b. Conic section with $t \in \mathbb{R}$. The point at infinity ($t = \infty$) is marked.

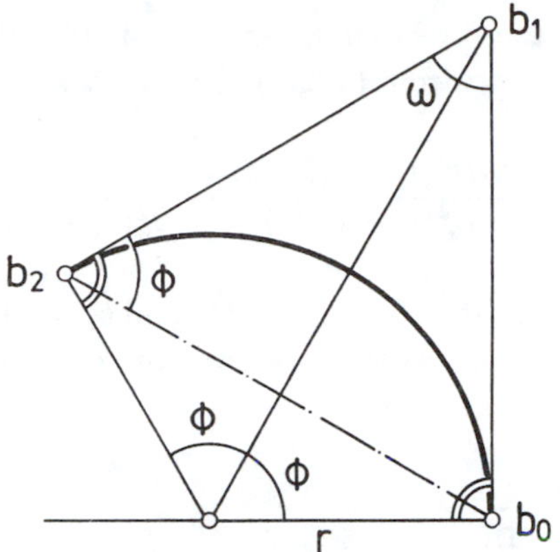

Fig. 4.20a. Bézier points for a circular arc.

If the two edges of a Bézier polygon $\boldsymbol{b}_0\boldsymbol{b}_1$ and $\boldsymbol{b}_1\boldsymbol{b}_2$ are of equal length, then we can use (4.31) to represent a *circle*. To show this, we make the following special selection of the Bézier points (cf. Fig. 4.20a):

$$\boldsymbol{b}_0 = (r, 0)^T, \qquad \boldsymbol{b}_1 = (r, r\tan\phi)^T, \qquad \boldsymbol{b}_2 = (r\cos 2\phi, r\sin 2\phi)^T. \quad (4.32a)$$

Setting $\beta_1 := \cos\phi$, it follows from (4.31) that the components of the Bézier curve are

$$\begin{aligned} x &= \frac{r(1-t)^2 + 2rt(1-t)\cos\phi + rt^2\cos 2\phi}{(1-t)^2 + 2t(1-t)\cos\phi + t^2}, \\ y &= \frac{2rt(1-t)\sin\phi + rt^2\sin 2\phi}{(1-t)^2 + 2t(1-t)\cos\phi + t^2}. \end{aligned} \quad (4.32b)$$

Squaring and adding, we see that $x^2 + y^2 = r^2$.

We can extend this special case to the general one by making a linear transformation. For $\beta_1 = \cos\phi > 0$, the circular arc lies entirely inside of the Bézier polygon (see Fig. 4.20a), while for $\beta_1 < 0$, it lies outside.

Remark. The well-known rational parametrization of the circle

$$x = r\frac{1-t^2}{1+t^2}, \qquad y = r\frac{2t}{1+t^2}$$

follows immediately from (4.31) by choosing $\beta_0 = \beta_1 = 1$, $\beta_2 = 2$ and $\boldsymbol{b}_0 = (r, 0)^T$, $\boldsymbol{b}_1 = (r, r)^T$, $\boldsymbol{b}_2 = (0, r)^T$.

Another approach to describing circles is to use Bézier points at infinity, see e.g., [Pie 87]. If we choose $\beta_1 = 0$, then the associated Bézier point is a point at infinity with the direction $\vec{\boldsymbol{b}}_1$, see Fig. 4.20b. If in addition, we set $\beta_0 = \beta_2 = 1$, then it follows from (4.29b) and (4.32a) (cf. [Pie 87]) that

$$\boldsymbol{X}(t) = \frac{\boldsymbol{b}_0 B_0^2(t) + \boldsymbol{b}_2 B_2^2(t)}{B_0^2(t) + B_2^2(t)} + \frac{\vec{\boldsymbol{b}}_1 B_1^2(t)}{B_0^2(t) + B_2^2(t)}. \tag{$*$}$$

Now choosing $\boldsymbol{b}_0 = (r,0)^T$, $\boldsymbol{b}_2 = (-r,0)^T$, and $\vec{\boldsymbol{b}}_1 = (0,r)^T$ (see Fig. 4.20b), it follows immediately from $(*)$ that we have the equation of a half circle of radius r.

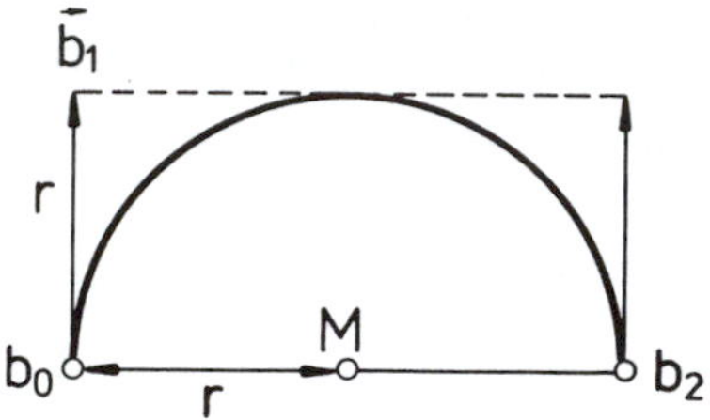

Fig. 4.20b. Bézier points for a half circle, where $\vec{\boldsymbol{b}}_1$ is a point at infinity.

Equations (4.32a,b) correspond to the special case where the circle has center at O. If we rotate the Bézier points (4.32a) by an angle α, then

$$\begin{aligned} \boldsymbol{b}_0 &= r(\cos\alpha, \sin\alpha)^T, & \boldsymbol{b}_1 &= \frac{r}{\cos\phi}\left(\cos(\phi+\alpha), \sin(\phi+\alpha)\right)^T, \\ \boldsymbol{b}_2 &= r\left(\cos(2\phi+\alpha), \sin(2\phi+\alpha)\right)^T, & \beta_1 &= \cos\phi. \end{aligned} \tag{4.33a}$$

Now if in addition, we translate the center to $\boldsymbol{M}$, it follows that the general Bézier formula for a circle is

$$\boldsymbol{X}(t) = \frac{\tilde{\boldsymbol{b}}_0 B_0^2(t) + \beta_1 \tilde{\boldsymbol{b}}_1 B_1^2(t) + \tilde{\boldsymbol{b}}_2 B_2^2(t)}{B_0^2(t) + \beta_1 B_1^2(t) + B_2^2(t)}, \tag{4.33b}$$

where $\tilde{\boldsymbol{b}}_i = \boldsymbol{M} + \boldsymbol{b}_i$, and $\boldsymbol{b}_i$ are given by (4.33a). Other circle representations and techniques for dealing with circles are developed in [Bli 87] and [Pie 89c].

It is also possible to use cubic rational Bézier curves (and not just degree-raised versions of quadratic ones) to represent circular arcs. Necessary and sufficient conditions for which this can be done can be found in [Wan 91].

For the numerical control of cutting tools, it is common to utilize processors which are capable of describing circles and straight lines exactly. For this purpose we can use interpolating or approximating *circular splines*. Such splines are usually constructed for the case of interpolation using elementary methods, see e.g., [SU 89], [Pie 86a]. These methods are usually not effective for finding approximations. Here we shall discuss *circular spline curves* as an application of rational Bézier curves. It will be convenient to introduce the following notation for basis functions:

$$B_0^{2*}(t) := \frac{(1-t)^2}{N}, \qquad B_1^{2*}(t) := \frac{2t(1-t)}{N}, \qquad B_2^{2*}(t) := \frac{t^2}{N},$$

where $N := (1-t)^2 + 2\beta_1 t(1-t) + t^2$ and $\beta_1 := \cos\phi$.

Since the auxiliary points $\boldsymbol{d}_i$ play an important role in the Bézier spline formula, we shall use them as the primary tool for defining circular splines. In view of Figs. 4.20a and 4.20c, we can write

$$\begin{aligned}\boldsymbol{X}(t) &= \boldsymbol{d}_0 B_0^{2*}(t) + \boldsymbol{d}_1\beta_1 B_1^{2*}(t) + [\boldsymbol{d}_1(1-\lambda_1) + \boldsymbol{d}_2\lambda_1]B_2^{2*}(t) \qquad (4.34)\\ &+ \sum_{i=2}^{n-1}\Big([\boldsymbol{d}_{i-1}(1-\lambda_{i-1}) + \boldsymbol{d}_i\lambda_{i-1}]B_0^{2*}(t) + \boldsymbol{d}_i\beta_i B_1^{2*}(t)\\ &+ [\boldsymbol{d}_i(1-\lambda_i) + \boldsymbol{d}_{i+1}\lambda_i]B_2^{2*}(t)\Big) + [\boldsymbol{d}_{n-1}(1-\lambda_{n-1}) + \boldsymbol{d}_n\lambda_{n-1}]B_0^{2*}(t)\\ &+ \boldsymbol{d}_n\beta_n B_1^{2*}(t) + \boldsymbol{d}_{n+1}B_2^{2*}(t),\end{aligned}$$

where the λ_i are chosen so that the line segments $(\boldsymbol{d}_{i-1}, \boldsymbol{d}_i)$ are divided in such a way that the corresponding Bézier polygon of each spline segment has sides of equal length, *i.e.*, (see Fig. 4.20c)

$$|\boldsymbol{S}_{i-1}\boldsymbol{d}_i| = |\boldsymbol{S}_i\boldsymbol{d}_i| \qquad \text{with} \qquad \boldsymbol{S}_i = (1-\lambda_{i+1})\boldsymbol{d}_i + \lambda_{i+1}\boldsymbol{d}_{i+1}.$$

An elementary calculation gives

$$\lambda_1 = \frac{|\boldsymbol{d}_0\boldsymbol{d}_1|}{|\boldsymbol{d}_1\boldsymbol{d}_2|}, \qquad \lambda_i = (1-\lambda_{i-1})\frac{|\boldsymbol{d}_i\boldsymbol{d}_{i-1}|}{|\boldsymbol{d}_i\boldsymbol{d}_{i+1}|}, \qquad i = 2(1)n\text{–}1.$$

If the control points $\boldsymbol{d}_i$ are given, then these conditions, together with (4.34), lead to an explicit formula for the circular spline curve. However, the last segment is a circle only if the following additional condition holds:

$$|\boldsymbol{S}_{n-1}\boldsymbol{d}_n| = |\boldsymbol{d}_n\boldsymbol{d}_{n+1}|.$$

We can use circular splines to uniquely *interpolate* prescribed points $\boldsymbol{P}_i$ at parameter values t_i, $i = 0(1)n$. In particular, we can use (4.34) provided that the $\boldsymbol{P}_i$ are chosen as the join points of the curve, and the direction of the tangent at the initial point $\boldsymbol{P}_0 = \boldsymbol{d}_0$ is given. The remaining control points $\boldsymbol{d}_i$ can be found by elementary geometry from the symmetry properties of a circular segment.

To attack the problem of *approximating* a given set of points $\boldsymbol{P}_i$ (with associated parameter values t_i, $i = 0(1)n$) by circular splines (using $m < n$ segments), we rewrite (4.34) by representing the control points $\boldsymbol{d}_i$ by the middle points $\boldsymbol{M}_i$, the radii r_i, and the (half) angles ϕ_i of the circular segments. By Fig. 4.20c, we have

$$\begin{aligned} \boldsymbol{d}_0 &= \boldsymbol{M}_1 + r_1\tilde{\boldsymbol{d}}_0, \qquad \boldsymbol{d}_i = \boldsymbol{M}_i + r_i\tilde{\boldsymbol{d}}_i, \qquad i = 1(1)m \\ \boldsymbol{d}_{m+1} &= \boldsymbol{M}_m + r_m\tilde{\boldsymbol{d}}_{m+1}, \end{aligned} \tag{4.34a}$$

where according to Fig. 4.20a,

$$\begin{aligned} \tilde{\boldsymbol{d}}_0 &= \begin{pmatrix} \cos\alpha_1 \\ \sin\alpha_1 \end{pmatrix}, \qquad \tilde{\boldsymbol{d}}_i = \frac{1}{\cos\phi_i}\begin{pmatrix} \cos(\phi_i + \alpha_i) \\ \sin(\phi_i + \alpha_i) \end{pmatrix}, \\ \tilde{\boldsymbol{d}}_{m+1} &= \begin{pmatrix} \cos\alpha_{m+1} \\ \sin\alpha_{m+1} \end{pmatrix}. \end{aligned} \tag{4.34b}$$

By Fig. 4.20a, the angles α_i should be chosen to be

$$\alpha_i = \alpha_1 + 2\sum_{j=1}^{i-1}\phi_j, \qquad i = 2(1)m{+}1,$$

while the centers of the circles $\boldsymbol{M}_i$ can be found from

$$\boldsymbol{M}_i = \boldsymbol{M}_{i-1} + (r_{i-1} - r_i)\begin{pmatrix} \cos\alpha_i \\ \sin\alpha_i \end{pmatrix}, \qquad i = 2(1)m, \tag{4.34c}$$

assuming $\boldsymbol{M}_1$ is prescribed. We can find the λ_i from the recurrence formula

$$\lambda_i = \frac{r_i \tan\phi_i}{r_i\tan\phi_i + r_{i+1}\tan\phi_{i+1}}, \qquad i = 1(1)m{-}1. \tag{4.34d}$$

Here again the last segment must be constructed from symmetry considerations [Hos 92c].

This setup leaves $2m - 1$ nonlinear unknowns which can be used in an approximation process: α_1, and the ϕ_i, r_i for $i = 1(1)m{-}1$, see [Hos 92c]. We use these to minimize the distance between the given points $\boldsymbol{P}_i$ and the circular spline curve. The lengths of the parameter intervals could also be varied to improve the order of approximation.

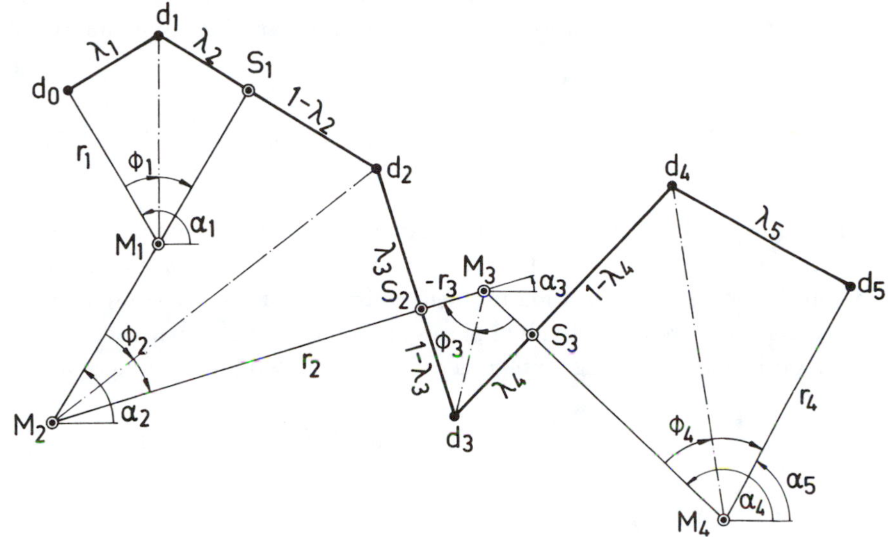

Fig. 4.20c. Bézier polygon for a circular spline curve.

The Bézier formula for circular spline curves developed here can be interpreted as a rational B-spline curve. We discuss this in Section 4.3.6.

Recently, conic sections (in Bézier form) have been used to describe fonts for printer and plotter programs, see e.g., [Pit 67], [Pav 83], [Pla 83], [Pratt 85], [Wils 87], [Hus 89], [Hu 91].

Generalizations of the rational Bézier curves based on the *duality principle* can be found in [Hos 83]. There the curves are not considered to be point sets, but rather sets of lines (envelopes of tangents).

4.2. Application of the Bernstein-Bézier Technique to Finite Elements

Many problems in the application of mathematics to chemistry, physics, engineering, architecture, etc., can be formulated as minimization problems involving some kind of energy functional. In most cases, no closed solution of the minimization problem can be found, and we have to find an approximate solution. One way to do this is to look for an approximation made up of a collection of small pieces (*finite elements*), each of which has some simple form.

Here we describe this approach only for the univariate case, where the desired function is defined on an interval $[a, b]$. The analogous use of tensor-product Bézier surfaces and triangular Bézier surfaces can be found, e.g., in

[Lus 87]; see also any of the numerous textbooks on finite elements. The energy functionals which arise in practice generally are linear combinations of the following integrals:

$$\int_a^b \boldsymbol{X}(u)du, \quad \tfrac{1}{2}\int_a^b (\boldsymbol{X}(u))^2 du, \quad \tfrac{1}{2}\int_a^b (\boldsymbol{X}'(u))^2 du, \quad \tfrac{1}{2}\int_a^b (\boldsymbol{X}''(u))^2 du, \tag{4.35a}$$

where $\boldsymbol{X}(u)$ is the desired function. We now decompose the interval $[a, b]$ into m subintervals $[u_\ell, u_{\ell+1}]$, $\ell = 0(1)m–1$, where $a = u_0 < \cdots < u_m = b$. We seek a solution of the minimization problem consisting of a piecewise polynomial of degree n, where the individual pieces are joined together with some degree of smoothness. The energy integrals can be computed piece by piece as

$$\begin{gathered}\int_{u_\ell}^{u_{\ell+1}} \boldsymbol{X}(u)du, \quad \tfrac{1}{2}\int_{u_\ell}^{u_{\ell+1}} (\boldsymbol{X}(u))^2 du, \quad \tfrac{1}{2}\int_{u_\ell}^{u_{\ell+1}} (\boldsymbol{X}'(u))^2 du, \\ \tfrac{1}{2}\int_{u_\ell}^{u_{\ell+1}} (\boldsymbol{X}''(u))^2 du, \qquad \ell = 0(1)m–1.\end{gathered} \tag{4.35b}$$

For our purposes it will be convenient to express the approximate solution as a Bézier (sub)spline curve of order n consisting of m pieces. Suppose the Bézier points of the ℓ-th piece are

$$\boldsymbol{b}_\ell = (\boldsymbol{b}_{\ell n}, \boldsymbol{b}_{\ell n+1}, \ldots, \boldsymbol{b}_{(\ell+1)n})^T, \qquad l = 0(1)m–1. \tag{4.36a}$$

Then in terms of the Bézier vector

$$\boldsymbol{B}^n(t) = (B_0^n(t), B_1^n(t), \ldots, B_n^n(t))^T, \tag{4.36b}$$

the ℓ-th piece of the Bézier curve can be written as the scalar product

$$\boldsymbol{X}(t) = (\boldsymbol{b}_\ell)^T \boldsymbol{B}^n(t). \tag{4.37}$$

Then, for example, the first integral in (4.35b) becomes

$$\int_{u_\ell}^{u_{\ell+1}} \boldsymbol{X}(u)du = \Delta u_\ell (\boldsymbol{b}_\ell)^T \int_0^1 \boldsymbol{B}^n(t)dt = \frac{\Delta u_\ell}{n+1}(1, 1, \ldots, 1)\boldsymbol{b}_\ell. \tag{4.38}$$

The other integrals in (4.35b) involve integrating products of the type $\boldsymbol{B}(\boldsymbol{B})^T$. But since (cf. (4.2d))

$$B_i^r(t)B_j^r(t) = \binom{r}{i}\binom{r}{j}(1-t)^{2r-(i+j)}\, t^{i+j} = \frac{\binom{r}{i}\binom{r}{j}}{\binom{2r}{i+j}} B_{i+j}^{2r}(t), \tag{4.39a}$$

we get

$$(2r+1)\int_0^1 B_i^r(t)B_j^r(t)dt = \frac{\binom{r}{i}\binom{r}{j}}{\binom{2r}{i+j}} =: \binom{r}{i,j}, \qquad (4.39b)$$

and thus

$$\boldsymbol{B}_r := \int_0^1 \boldsymbol{B}^r(\boldsymbol{B}^r)^T dt = \frac{1}{2r+1}\begin{pmatrix} \binom{r}{0,0} & \binom{r}{0,1} & \cdots & \binom{r}{0,r} \\ \binom{r}{1,0} & \binom{r}{1,1} & \cdots & . \\ \vdots & \vdots & & \vdots \\ \binom{r}{r,0} & \binom{r}{r,1} & \cdots & \binom{r}{r,r} \end{pmatrix}. \qquad (4.39c)$$

As an example, we take cubic Bézier curves as approximating functions. In this case

$$\boldsymbol{B}_0 = \int_0^1 \boldsymbol{B}^0(\boldsymbol{B}^0)^T dt = 1, \quad \boldsymbol{B}_1 = \int_0^1 \boldsymbol{B}^1(\boldsymbol{B}^1)^T dt = \tfrac{1}{3}\begin{vmatrix} 1 & \frac{1}{2} \\ \frac{1}{2} & 1 \end{vmatrix} = \tfrac{1}{6}\begin{vmatrix} 2 & 1 \\ 1 & 2 \end{vmatrix},$$

$$\boldsymbol{B}_2 = \int_0^1 \boldsymbol{B}^2(\boldsymbol{B}^2)^T dt = \tfrac{1}{5}\begin{vmatrix} 1 & \frac{1}{2} & \frac{1}{6} \\ \frac{1}{2} & \frac{2}{3} & \frac{1}{2} \\ \frac{1}{6} & \frac{1}{2} & 1 \end{vmatrix} = \tfrac{1}{30}\begin{vmatrix} 6 & 3 & 1 \\ 3 & 4 & 3 \\ 1 & 3 & 6 \end{vmatrix},$$

$$\boldsymbol{B}_3 = \int_0^1 \boldsymbol{B}^3(\boldsymbol{B}^3)^T dt = \tfrac{1}{140}\begin{vmatrix} 20 & 10 & 4 & 1 \\ 10 & 12 & 9 & 4 \\ 4 & 9 & 12 & 10 \\ 1 & 4 & 10 & 20 \end{vmatrix}.$$

Similarly, taking account of the formula for the derivative of a Bézier curve (cf. (4.7)), the other integrals in (4.35b) become

$$\int_{u_\ell}^{u_{\ell+1}} (\boldsymbol{X}(u))^2 du = \Delta u_\ell (\boldsymbol{b}_\ell)^T \left[\int_0^1 \boldsymbol{B}^n(\boldsymbol{B}^n)^T dt\right] \boldsymbol{b}_\ell,$$

$$\int_{u_\ell}^{u_{\ell+1}} (\boldsymbol{X}'(u))^2 du = \frac{n^2}{\Delta u_\ell}\Delta^1(\boldsymbol{b}_\ell)^T \left[\int_0^1 \boldsymbol{B}^{n-1}(\boldsymbol{B}^{n-1})^T dt\right] \Delta^1 \boldsymbol{b}_\ell, \qquad (4.40)$$

$$\int_{u_\ell}^{u_{1+1}} (\boldsymbol{X}''(u))^2 du = \frac{n^2(n-1)^2}{(\Delta u_\ell)^3}\Delta^2(\boldsymbol{b}_\ell)^T \left[\int_0^1 \boldsymbol{B}^{n-2}(\boldsymbol{B}^{n-2})^T dt\right] \Delta^2 \boldsymbol{b}_\ell.$$

In order to write integrals involving derivatives in a more convenient form, we introduce the notation

$$\begin{aligned} \boldsymbol{b}_\ell^\alpha &:= (\boldsymbol{b}_{\ell n}, \boldsymbol{b}_{\ell n+1}, \ldots, \boldsymbol{b}_{\ell n+\alpha-1}, \boldsymbol{b}_{\ell n+\alpha+1}, \ldots, \boldsymbol{b}_{\ell n+n}), && (4.40a)\\ \boldsymbol{b}_\ell^{\alpha,\beta} &:= (\boldsymbol{b}_{\ell n}, \boldsymbol{b}_{\ell n+1}, \ldots, \boldsymbol{b}_{\ell n+\alpha-1}, \boldsymbol{b}_{\ell n+\alpha+1}, \ldots, \boldsymbol{b}_{\ell n+\beta-1}, \boldsymbol{b}_{\ell n+\beta+1}, \ldots, \boldsymbol{b}_{\ell n+n}). \end{aligned}$$

Then

$$\int_{u_\ell}^{u_{\ell+1}} (\boldsymbol{X}'(u))^2 du = \frac{n^2}{\Delta u_\ell}(\boldsymbol{b}_\ell^0 - \boldsymbol{b}_\ell^n)^T \boldsymbol{B}_{n-1}(\boldsymbol{b}_\ell^0 - \boldsymbol{b}_\ell^n), \tag{4.40b}$$

$$\begin{aligned} &\int_{u_\ell}^{u_{\ell+1}} (\boldsymbol{X}''(u))^2 du && (4.40c)\\ &= \frac{n^2(n-1)^2}{(\Delta u_\ell)^3}(\boldsymbol{b}_\ell^{01} - 2\boldsymbol{b}_\ell^{0n} + \boldsymbol{b}_\ell^{n-1,n})^T \boldsymbol{B}_{n-2}(\boldsymbol{b}_\ell^{01} - 2\boldsymbol{b}_\ell^{0n} + \boldsymbol{b}_\ell^{n-1,n}). \end{aligned}$$

Combining the above, we see that the total energy corresponding to a linear combination of the integrals in (4.35) can always be written in the form

$$I(\boldsymbol{b}_0, \ldots, \boldsymbol{b}_{m-1}) = \tfrac{1}{2}\sum_{i=0}^{m-1} \boldsymbol{b}_i^T \boldsymbol{M}_i \boldsymbol{b}_i + \sum_{i=0}^{m-1} \boldsymbol{m}_i^T \boldsymbol{b}_i, \tag{4.41}$$

with $(n+1) \times (n+1)$ coefficient matrices $\boldsymbol{M}_i$ and coefficient vectors $\boldsymbol{m}_i$.

Now suppose that we require that the pieces meet at the join points with C^0 continuity. This means that since the elements have common values at the join points, we do not need to repeat the corresponding Bézier points in our vector of unknowns. Thus, the matrices $\boldsymbol{M}_i$ and vectors $\boldsymbol{m}_i$ can be built by assembling pieces which *overlap*. We illustrate this process in the case of two pieces ($m = 2$). In this case the energy is given by

$$I(\boldsymbol{b}_0, \boldsymbol{b}_1) := \tfrac{1}{2}\boldsymbol{b}^T \boldsymbol{M} \boldsymbol{b} + \boldsymbol{m}^T \boldsymbol{b}, \tag{4.42}$$

where $\boldsymbol{b} = (\boldsymbol{b}_0, \boldsymbol{b}_1, \ldots, \boldsymbol{b}_n, \ldots, \boldsymbol{b}_{2n})^T$, $\boldsymbol{m}$ is the $2n+1$ vector obtained by appending the last n entries in the vector $\boldsymbol{m}_1^T$ to the vector $\boldsymbol{m}_0^T$, and $\boldsymbol{M}$ is the $(2n+1) \times (2n+1)$ matrix

$$\boldsymbol{M} = \left(\begin{array}{cc|c} \boldsymbol{M}_0 & & 0 \\ \hline & * & \\ \hline 0 & & \boldsymbol{M}_1 \end{array}\right),$$

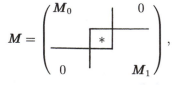

where the $*$ indicates that the two matrices overlap by one entry.

If we want C^1 smoothness at the first joint, for example, then in view of (4.18b), we must enforce the condition

$$\boldsymbol{b}_n = \frac{\Delta u_0}{\Delta u_0 + \Delta u_1}\boldsymbol{b}_{n+1} + \frac{\Delta u_1}{\Delta u_0 + \Delta u_1}\boldsymbol{b}_{n-1} =: \alpha \boldsymbol{b}_{n+1} + (1-\alpha)\boldsymbol{b}_{n-1}.$$

This equation can be solved for one of the points $\boldsymbol{b}_n$, $\boldsymbol{b}_{n-1}$, or $\boldsymbol{b}_{n+1}$, which can then be eliminated from (4.41). For example, if $\boldsymbol{b}_n$ is eliminated, we get

$$\boldsymbol{b} = \begin{pmatrix} \boldsymbol{b}_0 \\ \boldsymbol{b}_1 \\ \vdots \\ \boldsymbol{b}_{n-1} \\ \boldsymbol{b}_n \\ \boldsymbol{b}_{n+1} \\ \vdots \\ \boldsymbol{b}_{2n} \end{pmatrix} = \begin{pmatrix} 1 & 0 & & & & & & 0 \\ 0 & 1 & & & & & & \\ & & \ddots & & & & & \\ & & & 1 & & & & \\ & & & 1-\alpha & \alpha & & & \\ & & & & & 1 & & \\ & & & & & & \ddots & \\ & & & & & & 1 & 0 \\ 0 & & & & & & 0 & 1 \end{pmatrix} \begin{pmatrix} \boldsymbol{b}_0 \\ \boldsymbol{b}_1 \\ \vdots \\ \boldsymbol{b}_{n-1} \\ \boldsymbol{b}_{n+1} \\ \vdots \\ \boldsymbol{b}_{2n} \end{pmatrix} =: \boldsymbol{S}\hat{\boldsymbol{b}}. \tag{4.43}$$

If smoothness conditions are enforced at the other join points, then other (end) Bézier points can be eliminated in a similar way. It follows that in the C^1 case, the functional to be minimized has the form

$$I(\hat{\boldsymbol{b}}) = \tfrac{1}{2}\hat{\boldsymbol{b}}^T \hat{\boldsymbol{M}}\hat{\boldsymbol{b}} + \hat{\boldsymbol{m}}^T\hat{\boldsymbol{b}}, \qquad \boldsymbol{M} := \boldsymbol{S}^T\boldsymbol{M}\boldsymbol{S}, \quad \hat{\boldsymbol{m}} = \boldsymbol{S}^T\boldsymbol{m}. \tag{4.44a}$$

To illustrate the effectiveness of the method, we consider the following simple example, cf. [SCHW 80], [Lus 87]. Suppose we have a beam of constant cross-section and length L, which is clamped on one end and is supported at the points $2L/3$ and L. In addition, suppose that the beam is subject to a downward force Q uniformly distributed along the beam between $2L/3$ and L, together with a downward point force F at the point $L/2$ (cf. Fig. 4.21). Then in order to determine the centerline $\boldsymbol{X}(u)$ of the beam (the elastica), we need to minimize the functional

$$I = \tfrac{1}{2}c\int_0^L (\boldsymbol{X}''(u))^2 du - Q\int_{\frac{2}{3}L}^L \boldsymbol{X}(u)du - F\boldsymbol{X}(\tfrac{1}{2}L), \tag{4.44b}$$

subject to the boundary conditions $\boldsymbol{X}(0) = 0, \boldsymbol{X}'(0) = 0, \boldsymbol{X}(L/2) = 0, \boldsymbol{X}(L) = 0$.

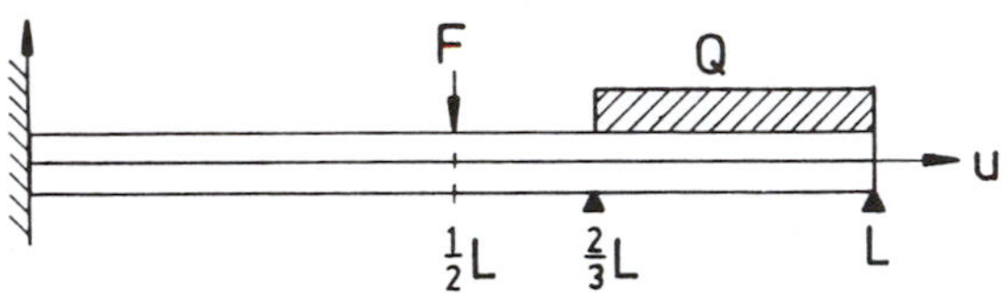

Fig. 4.21. A loaded beam.

We divide the beam into three elements with join points at $L/2$ and $2L/3$. As elements we use cubic Bézier curves joined together with C^1 continuity. The desired Bézier curve $\boldsymbol{X}(u)$ then can be written as

$$\boldsymbol{X}(u) = \begin{cases} \boldsymbol{b}_0 B_0^3(t_0) + \cdots + \boldsymbol{b}_3 B_3^3(t_0), & u \in [0, \tfrac{1}{2}L] \\ \boldsymbol{b}_3 B_0^3(t_1) + \cdots + \boldsymbol{b}_6 B_3^3(t_1), & u \in [\tfrac{1}{2}L, \tfrac{2}{3}L] \\ \boldsymbol{b}_6 B_0^3(t_2) + \cdots + \boldsymbol{b}_9 B_3^3(t_2), & u \in [\tfrac{2}{3}L, L], \end{cases} \tag{4.45}$$

with local parameters t_i. Here the C^0 continuity condition is already incorporated, and the C^1 smoothness conditions are

$$\boldsymbol{b}_3 = \tfrac{1}{4}\,\boldsymbol{b}_2 + \tfrac{3}{4}\,\boldsymbol{b}_4, \qquad \boldsymbol{b}_6 = \tfrac{2}{3}\,\boldsymbol{b}_5 + \tfrac{1}{3}\,\boldsymbol{b}_7, \tag{4.46}$$

where $\Delta u_0 = L/2$, $\Delta u_1 = L/6$, and $\Delta u_2 = L/3$. The boundary condition implies

$$\boldsymbol{b}_0 = 0, \quad \boldsymbol{b}_1 = 0, \quad \boldsymbol{b}_6 = 0, \quad \boldsymbol{b}_9 = 0. \tag{4.47}$$

Substituting (4.45) in the energy functional and simplifying, we get

$$I(\bar{\boldsymbol{b}}) = \tfrac{1}{2}\bar{\boldsymbol{b}}^T \bar{\boldsymbol{M}}\,\bar{\boldsymbol{b}} + \bar{\boldsymbol{m}}^T\,\bar{\boldsymbol{b}},$$

where $\bar{\boldsymbol{b}} = (\boldsymbol{b}_0, \boldsymbol{b}_1, \boldsymbol{b}_3, \boldsymbol{b}_4, \boldsymbol{b}_6, \boldsymbol{b}_7, \boldsymbol{b}_8, \boldsymbol{b}_9)^T$,

$$\bar{\boldsymbol{M}} = \frac{6c}{L^3}\begin{pmatrix} 16 & -24 & 8 & 0 & \cdots & & \cdots & 0 \\ -24 & 48 & -96 & 72 & 0 & & & \\ 8 & -96 & 1024 & -1152 & 216 & 0 & 0 & \vdots \\ 0 & 72 & -1152 & 1728 & -972 & 324 & 0 & 0 \\ & 0 & 216 & -972 & 1456 & -729 & 0 & 27 \\ \vdots & & 0 & 324 & -729 & 486 & -81 & 0 \\ & & 0 & 0 & 0 & -81 & 162 & -81 \\ 0 & \cdots & \cdots & 0 & 27 & 0 & -81 & 54 \end{pmatrix},$$

and

$$\bar{\boldsymbol{m}} = -(0, 0, F, 0, QL/12, QL/12, QL/12, QL/12).$$

This functional can now be minimized using a standard optimization method (cf. Chap. 2).

4.3. Integral B-spline Curves

We have seen above that Bézier curves have a number of major advantages, such as the property that the Bézier polygon gives a good indication of the shape of the corresponding curve, and the fact that we can modify the shape of the Bézier curve by making appropriate changes in the Bézier polygon. On the other hand, one disadvantage is the fact that the number of vertices of the Bézier polygon is directly coupled to the degree of the Bézier curve, *i.e.*, increasing the number of vertices of the Bézier polygon (in order to get more control over the shape of the curve) also requires increasing the degree of the polynomial. One way to avoid this is to work with Bézier spline curves, but this involves having to solve a linear system of equations. Another disadvantage of Bézier curves is the global nature of the Bézier points: each point has an effect on the entire parameter interval so that changes in it induce a change in the entire Bézier curve.

These disadvantages can be eliminated by working with *B-spline curves (basis-spline curves)*. B-spline functions are defined locally, *i.e.*, changes in one of the points of the associated control polygon affect the corresponding curve locally, and indeed, the region of influence of each control point can be determined exactly. In addition, it is possible to introduce new control points without increasing the polynomial degree, and even to change the order of continuity between neighboring spline segments by choosing the control points appropriately. Moreover, it is also easy to obtain analogs of the geometric properties of Bézier curves such as the convex-hull property, the variation-diminishing property, etc.

From a mathematical point of view, there is no *formal* difference between (integral) B-spline curves and Bézier spline curves; we are simply talking about different ways of representing the same curve with respect to different basis systems, see e.g., Chap. 10. But the B-spline formulation has the advantage that, in comparison with Bézier spline curves, fewer control points are needed, *i.e.*, B-spline curves require less storage in the computer. As we shall see, Bézier curves are actually just a special case of B-spline curves.

4.3.1. B-splines

A B-spline of order k is made up of pieces of polynomials of degree $k-1$, joined together with C^{k-2} continuity at the break points, see e.g., [Boo 72], [BOO 78], [Cox 71].

Corresponding to the set of break points $t_0 < t_1 < t_2 < \cdots < t_{m-1} < t_m$ of a B-spline function, we define the associated *knot vector*

$$\boldsymbol{T} = (t_0, t_1, t_2, \ldots, t_{m-1}, t_m). \tag{4.48a}$$

We refer to the individual points t_k of $\boldsymbol{T}$ as *knots*.

Definition 4.1. *Given a knot vector* $\boldsymbol{T} = (t_0, t_1, \ldots, t_{n-1}, t_n, t_{n+1}, \ldots, t_{n+k})$, *the associated normalized B-spline* N_{ik} *of order* k *(of degree* $k-1$*) is defined to be the following function:*

$$N_{i1}(t) = \begin{cases} 1, & \text{for } t_i \le t < t_{i+1} \\ 0, & \text{otherwise,} \end{cases} \tag{4.48b}$$

for $k = 1$, *and*

$$N_{ik}(t) = \frac{(t-t_i)}{(t_{i+k-1}-t_i)} N_{i,k-1}(t) + \frac{(t_{i+k}-t)}{(t_{i+k}-t_{i+1})} N_{i+1,k-1}(t), \tag{4.48c}$$

for $k > 1$ *and* $i = 0(1)n$.

These basis functions have the following properties:

a) $N_{ik}(t) > 0$, for $t_i < t < t_{i+k}$,

b) $N_{ik}(t) = 0$, for $t_0 \le t \le t_i$, $t_{i+k} \le t \le t_{n+k}$,

c) $\sum_{i=0}^{n} N_{ik}(t) = 1$, $t \in [t_{k-1}, t_{n+1}]$,

d) $N_{ik}(t)$ has continuity C^{k-2} at each of the knots t_ℓ.

Remark. In view of properties a)–c), we say that the B-splines form a *partition of unity*.

Remark. The function N_{ik} has support on the interval $[t_i, t_{i+k}]$.

Remark. For *equally spaced knots*, say $t_i = i$, the recurrence relation (4.48c) becomes

$$N_{ik}(t) = \frac{(t-i)}{(k-1)} N_{i,k-1}(t) + \frac{(i+k-t)}{(k-1)} N_{i+1,k-1}(t), \tag{4.49}$$

and we refer to the N_{ik} as *uniform* B-splines. B-splines defined over non-equally-spaced knots are called *non-uniform* B-splines.

Remark. We shall also make use of *non-normalized* B-splines M_{ik} defined by

$$M_{i1} = \begin{cases} (t_{i+1} - t_i)^{-1}, & \text{for } t_i \le t < t_{i+1} \\ 0, & \text{otherwise,} \end{cases}$$

and

$$M_{ik} = \frac{(t - t_i)}{(t_{i+k} - t_i)} M_{i,k-1}(t) + \frac{(t_{i+k} - t)}{(t_{i+k} - t_i)} M_{i+1,k-1}(t).$$

The normalized and unnormalized B-splines are related by the equation

$$N_{ik}(t) = (t_{i+k} - t_i) M_{ik}(t).$$

Remark. In the literature, the normalized B-splines $N_{ik}(t)$ of degree $k-1$ are sometimes denoted by $N_i^{k-1}(t)$.

Remark. Another approach to defining B-splines is via integrals. Starting with the linear B-spline N_{02} and taking the moving average (see Fig. 4.22a), we get

$$N_{03}(t) = \int_{t-1}^{t} N_{02}(t)dt,$$

and, more generally,

$$N_{ik}(t) = \int_{t-1}^{t} N_{i,k-1}(t)dt,$$

see [Sab 76a] and [Boeh 87b].

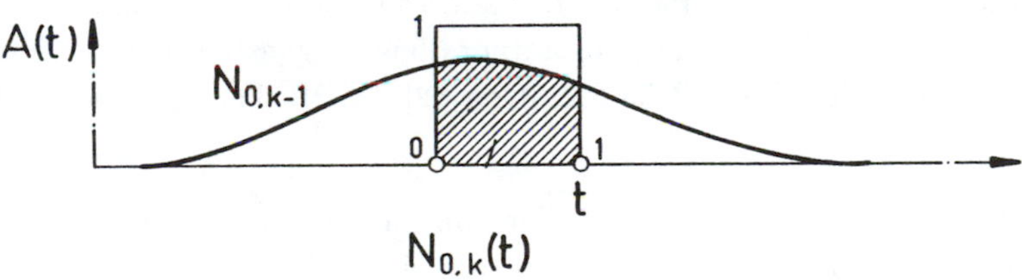

Fig. 4.22a. A B-spline as a moving average of a lower degree B-spline.

Remark. B-splines can also be obtained as *projections of k-dimensional simplices* σ into the one dimensional space $\mathbb{R}$. Indeed, if we think of σ as casting a shadow in $\mathbb{R}$ whose intensity is proportional to the thickness of the simplex in the direction of projection, then the value of the B-spline at each point in $\mathbb{R}$ is precisely the value of the associated projected volume (see Fig. 4.22b and, e.g., [Pra 84]).

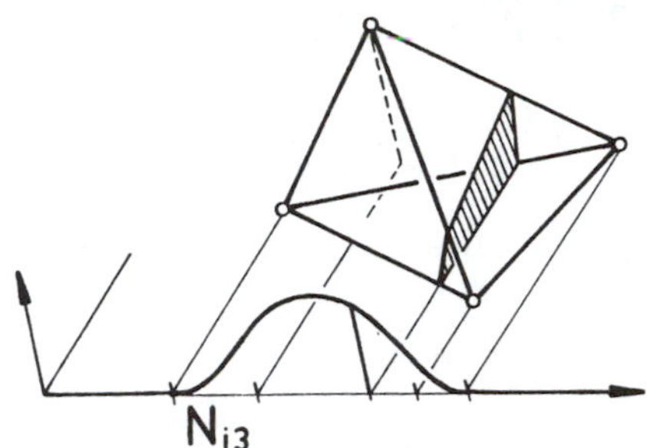

Fig. 4.22b. N_{i3} as the shadow of a three dimensional simplex.

Remark. B-splines can also be introduced by taking differences of truncated power functions (cf. e.g., [BOO 78], [SCHU 81]), by certain statistical methods (cf. e.g., [Gol 88, 88a]), and with the help of blossoming (polar forms), see Sect. 4.1 and [CAS 86], [Ram 87], [Sei 88, 89]. In Sects. 4.3.3, 4.3.4 we discuss the use of the blossoming principle to construct an algorithm for the computation of B-spline curves and for degree raising.

Remark. A historical discussion of the development of B-splines can be found in [But 88].

Remark. Generally, the recursions (4.48) or (4.49) are not used for the computation of B-splines in practice. Instead, the de Boor algorithm discussed in Sect. 4.3.3 is used. A number of authors have suggested matrix representations, see e.g., [FAU 81], [Cha 82], [Coh 82], [MOR 85], [QUI 87], [YAM 88] and [Cho 90a].

Before discussing algorithms for computing with B-splines, we take a closer look at some uniform B-splines. Fig. 4.22c shows the simplest B-splines N_{01} and N_{21}, which according to Definition 4.1 are piecewise constants. For $k = 2$ and $i = 0$, $i = 1$, it follows from (4.49) that

$$\begin{aligned} N_{02}(t) &= \frac{t}{1} N_{01}(t) + \frac{(2-t)}{1} N_{11}(t), \\ N_{12}(t) &= \frac{(t-1)}{1} N_{11}(t) + \frac{(3-t)}{1} N_{21}(t), \end{aligned} \tag{4.50a}$$

see Fig. 4.22d. It is clear from the figure that these two B-splines are simply translates of each other, a fact which is not obvious from (4.50a).

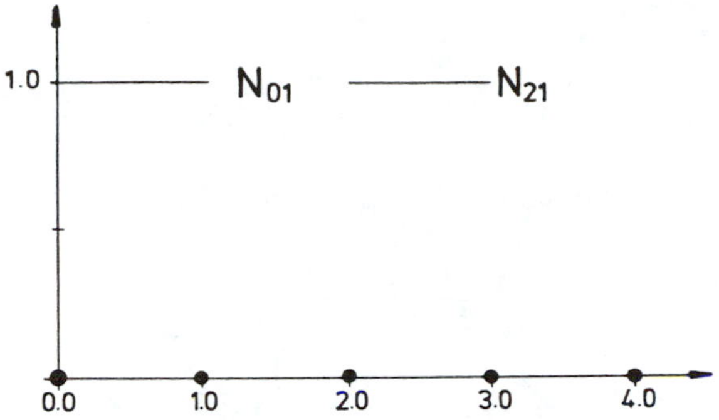

Fig. 4.22c. The B-splines N_{01}, N_{21}.

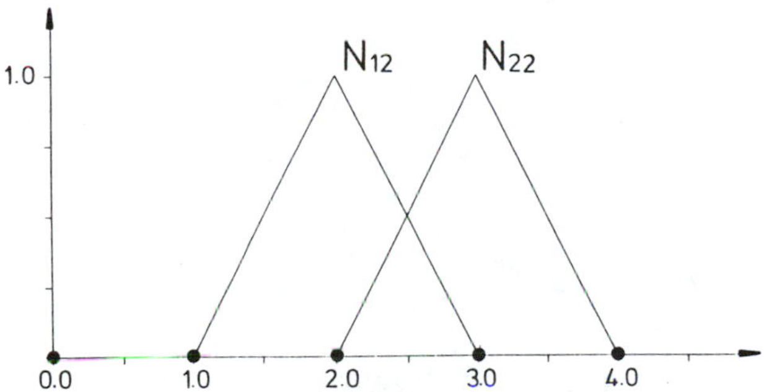

Fig. 4.22d. The B-splines N_{12}, N_{22}.

The translation invariance of uniform B-splines can be established by an appropriate parametric transformation: setting

$$t \equiv w \; mod \; 1, \tag{4.51}$$

i.e.,

$$t = \begin{cases} w, & \text{on } [0,1] \\ w+1, & \text{on } [1,2] \\ w+2, & \text{on } [2,3], \end{cases}$$

etc. (where $w \in [0,1]$), it follows from (4.50a) that

$$N_{02} = wN_{01} + (1-w)N_{11}, \qquad N_{12} = wN_{11} + (1-w)N_{21}. \tag{4.50b}$$

Consider now the case $k = 3$. From (4.49) we have

$$N_{03}(t) = \frac{t}{2}N_{02}(t) + \frac{(3-t)}{2}N_{12}(t),$$

and by (4.50b) we get

$$N_{03}(t) = \frac{t}{2}\left[wN_{01} + (1-w)N_{11}\right] + \frac{(3-t)}{2}\left[wN_{11} + (1-w)N_{21}\right].$$

Now making the transformation (4.51), it follows that

$$\begin{aligned} N_{03}(w) &= \frac{w^2}{2}N_{01} + \left[\frac{(1-w)(1+w)}{2} + \frac{(2-w)}{2}w\right]N_{11} + \frac{(1-w)^2}{2}N_{21} \\ &= \frac{w^2}{2}N_{01} + (-w^2 + w + \tfrac{1}{2})N_{11} + \frac{(1-w)^2}{2}N_{21}. \end{aligned} \tag{4.50c}$$

Fig. 4.22e shows the graph of this B-spline.

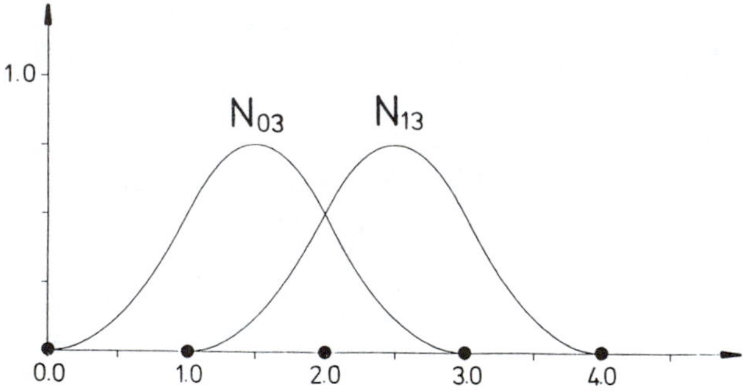

Fig. 4.22e. The B-splines N_{03}, N_{13}.

Finally, we look at $N_{04}(t)$. From (4.49) and (4.51) it follows that

$$\begin{aligned} N_{04} = \frac{w^3}{6}N_{01} &+ (-\tfrac{1}{2}w^3 + \tfrac{1}{2}w^2 + \tfrac{1}{2}w + \tfrac{1}{6})N_{11} \\ &+ (\tfrac{1}{2}w^3 - w^2 + \tfrac{2}{3})N_{21} + \frac{(1-w)^3}{6}N_{31}, \end{aligned} \tag{4.50d}$$

see Fig. 4.22f.

So far in our discussion of B-splines, we have assumed that each point in the knot vector is *simple*. On the other hand, the recursive definition (4.48) of B-splines also makes sense when knots are considered with *multiplicity* $r \leq k$. Typical examples of such knot vectors include

$\boldsymbol{T} = (t_0, t_0, t_0, t_1, t_2, t_3, \ldots, t_{n+k})$, t_0 triple,

$\boldsymbol{T} = (t_0, t_1, t_1, t_2, t_3, t_3, t_3, \ldots, t_{n+k})$, t_1 double, t_3 triple,

$\boldsymbol{T} = (t_0, t_0, t_0, t_0, t_1, t_1, t_1, t_1)$, first and last knots quadruple.

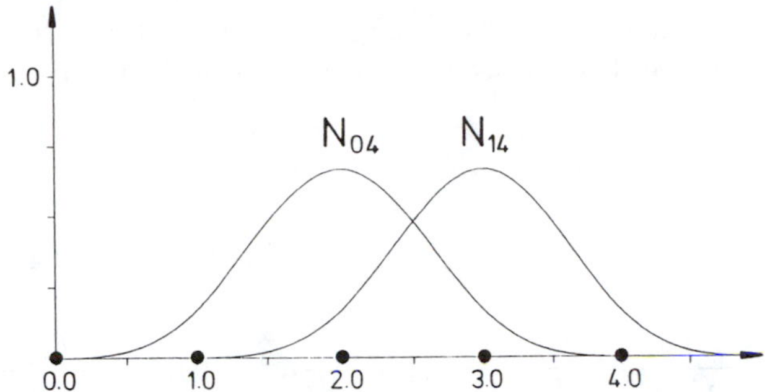

Fig. 4.22f. The B-splines N_{04}, N_{14}.

To see what happens for multiple knots, let $k = 3$ and let $\boldsymbol{T} = (0, 0, 0, 1, 1, 1)$ be the knot vector. Then by (4.48)

$$N_{03} = \frac{(t - t_0)}{(t_2 - t_0)} N_{02} + \frac{(t_3 - t)}{(t_3 - t_1)} N_{12}. \tag{$*$}$$

We now insert the desired knot values $t_0 = t_1 = t_2 = 0$ and $t_3 = 1$ in $(*)$. Using (4.50b) and the fact that for this knot vector, $N_{01} = N_{11} = 0$, and so $N_{02} = 0$ (and setting $0/0 = 0$), we get

$$N''_{03} = (1 - w)^2 N_{21}, \tag{4.52a}$$

where here the double prime *does not denote a derivative*, but rather indicates two additional coinciding parameter values.

It is of interest to note that N''_{03} coincides with the Bernstein polynomial $B_0^2(t)$. Analogously,

$$N'_{13} = \frac{(t - t_1)}{(t_3 - t_1)} N_{12} + \frac{(t_4 - t)}{(t_4 - t_2)} N_{22} = (-2w^2 + 2w) N_{21} = B_1^2(w).$$

In general, (cf. e.g., [Gor 74a], [Rie 73]) we have

Lemma 4.6. *The B-splines $N_{ik}(w)$ of order k are identical to the Bernstein polynomials $B_i^{k-1}(w)$ of degree $k-1$ provided that the knot vector $\boldsymbol{T}$ consists of $2k$ knots of which k are located at the parameter value 0, and k at the parameter value 1, i.e., the knot vector is given by*

$$\boldsymbol{T} = (\underbrace{0, \ldots, 0}_{k \text{ fold}}, \underbrace{1, \ldots, 1}_{k \text{ fold}}).$$

For later use, we also consider the case where there are k knots at each of the ends of the parameter interval, along with additional interior knots. For $k = 2$ this means we should consider knot vectors of the form

$$\boldsymbol{T} = (t_0, t_0, t_1, \ldots, t_{m-1}, t_m, t_m).$$

Then, for example, for N'_{03} with $t_0 = t_1 = 0$, $t_2 = 1$, $t_3 = 2$, (4.49) leads to

$$N'_{03} = \frac{(t - t_0)}{(t_2 - t_0)} N_{02} + \frac{(t_3 - t)}{(t_3 - t_1)} N_{12} = (-\frac{3}{2} w^2 + 2w) N_{11} + \frac{(1 - w)^2}{2} N_{21}. \tag{4.52b}$$

So far we have discussed multiple knots in a rather formal way. The following theorem deals with the geometric significance of multiple knots:

Theorem 4.4. *Suppose that a B-spline has a knot of multiplicity ℓ at a parameter value t. Then at this point the continuity of the B-spline N_{ik} is reduced from C^{k-2} to $C^{k-\ell-1}$. Moreover, the support interval of the B-spline is reduced from length k to length $k - (\ell - 1)$.*

Proof: See [Gor 74a]. ■

For applications, it is important to note that by Theorem 4.4, if a B-spline has a k-tuple knot at a parameter value $t = t_j$, then it may be discontinuous at this point. Fig. 4.22g shows a B-spline of order 3 with a double knot (compare with Fig. 4.22e).

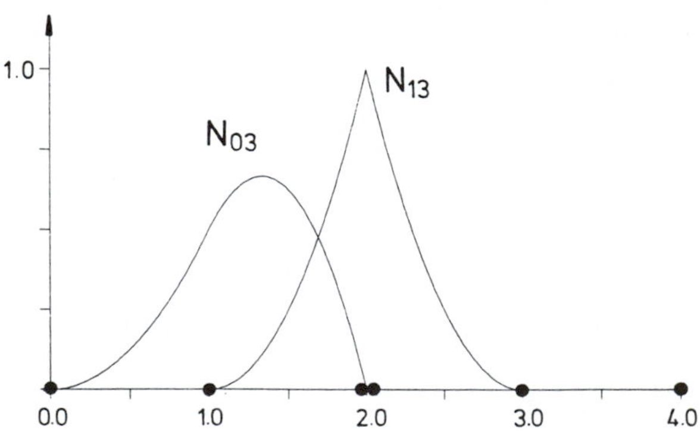

Fig. 4.22g. A quadratic B-spline with a double knot.

4.3.2. B-spline Curves

Analogous to Bézier curves, in this section we introduce B-spline curves. The basis for our development is a theorem of Schoenberg (see e.g., [Scho 67], [Boo 72]) which asserts that the B-splines corresponding to a given knot vector form a linear *basis*.

While for Bézier curves the order of the curve and the number of Bézier points are related, here the situation is different: since the B-splines have *local support* (in the parameter domain), we can construct a B-spline curve using arbitrarily many B-splines. This (see [Boeh 77, 77a], [Gor 74a], [Rie 73]) leads us to

Definition 4.2. *Suppose we are given points $\boldsymbol{d}_i$ in $\mathbb{R}^2$ or $\mathbb{R}^3$ for $i = 0(1)n$, and that $\boldsymbol{T}$ is a knot vector as in Definition 4.1. Then*

$$\boldsymbol{X}(t) = \sum_{i=0}^{n} \boldsymbol{d}_i N_{ik}(t), \qquad n \geq k-1, \qquad t \in [t_{k-1}, t_{n+1}],$$

is called a B-spline curve of order k with knot vector $\boldsymbol{T}$. The points $\boldsymbol{d}_i$ are called the control points or de Boor points and form the de Boor polygon.

Remark. This definition of a B-spline curve takes account of the fact that we have a full function space (*i.e.* all basis functions are present) only in the parameter interval from t_{k-1} to t_{n+1}.

Remark. The choice of the knot vector automatically determines the continuity of the curve at each of the knots t_j, i.e.,

- if t_j has multiplicity $1 \leq \ell < k$, then the curve has $C^{k-\ell-1}$ continuity at t_j.
- a knot of multiplicity k at t_j means C^{-1} continuity (*i.e.*, no continuity at all).

Remark. If $t_i = t_{i+k}$, then $N_{ik} \equiv 0$.

Remark. The choice of the first and last $k-1$ knots is free, and can be adjusted to give the desired boundary behavior of the B-spline curve, see the following sections or [Bars 82], [BART 87].

Remark. For a direct approach to B-spline curves starting with the control polygon, see e.g., [Chai 74], [Rie 75] and Sect. 14.3.

The local influence of the individual de Boor points follows from the definition of the B-splines. We have the following lemma of [Gor 74]:

Lemma 4.7. *Suppose* $\boldsymbol{X}(t) = \sum_{i=0}^{n} \boldsymbol{d}_i N_{ik}(t)$ *is a B-spline curve associated with the knot vector* $\boldsymbol{T} = (t_0, t_1, \ldots, t_n, t_{n+1}, \ldots, t_{n+k})$. *Then the de Boor point* $\boldsymbol{d}_j$ *influences* $\boldsymbol{X}(t)$ *only for* $t_j < t < t_{j+k}$. *Moreover, the curve at the point associated with a given parameter value* $t_r < t^* < t_{r+1}$ *is determined completely by the de Boor points* $\boldsymbol{d}_{r-(k-1)}, \ldots, \boldsymbol{d}_r$.

The *derivatives* of a B-spline curve can be computed recursively [Boo 72] by

$$\boldsymbol{X}^{(j)}(t) = (k-1)(k-2)\cdots(k-j)\sum_i \boldsymbol{d}_i^{[j]} \boldsymbol{N}_{i,k-j}(t),$$

where

$$\boldsymbol{d}_i^{[j]} = \begin{cases} \boldsymbol{d}_i, & j = 0 \\ \dfrac{\boldsymbol{d}_i^{[j-1]} - \boldsymbol{d}_{i-1}^{[j-1]}}{t_{i+k-j} - t_i}, & j > 0. \end{cases} \tag{4.53}$$

4.3.2.1. Open B-spline Curves

For a general choice of the de Boor points $\boldsymbol{d}_i$ and the knot vector $\boldsymbol{T}$, the B-spline curve does not have any obvious geometric relationship to its de Boor polygon, see Fig. 4.23.

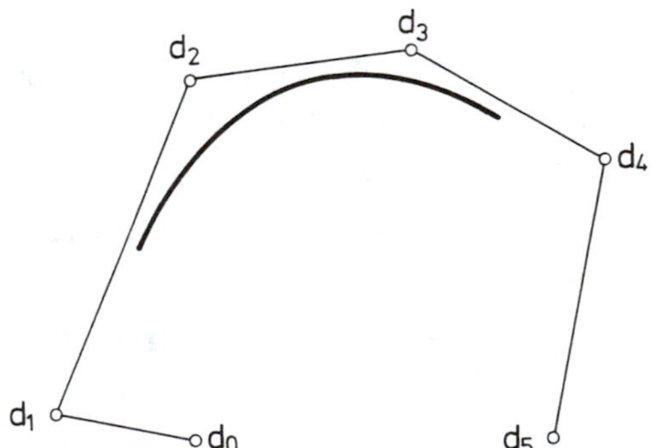

Fig. 4.23. A B-spline curve of order $k = 5$ and its de Boor polygon.

As in the case of open Bézier curves, we can get convenient geometric properties if we select the first and last k knots to be

$$t_0 = t_1 = \cdots = t_{k-1}$$
$$t_{n+1} = t_{n+2} = \cdots = t_{n+k}.$$

This assures that the points $\boldsymbol{d}_0$ and $\boldsymbol{d}_n$ are points on the curves. In addition, since the first knot has multiplicity k, the first B-spline reduces to the Bernstein polynomial $B_0^{k-1}(t)$, and thus the tangent property of the Bézier curve holds for the spline.

Thus, for open B-spline curves of order k, we choose the *knot vector*

$$\boldsymbol{T} = (t_0 = t_1 = t_2 = \cdots = t_{k-1}, t_k, t_{k+1}, \ldots, t_n, t_{n+1} = t_{n+2} = \cdots = t_{n+k}). \tag{4.54}$$

Here the interior knots can be multiple knots, of course. From (4.54) it follows (cf. [Boo 72], Lemma 4.6) that

Lemma 4.8. *For an open B-spline curve of order k,*

- *for simple knots, the number of intervals is $n - k + 2$,*
- *if $n = k - 1$ and the knot vector is given by (4.54), then the B-spline curve reduces to a Bézier curve.*

The choice of the knot vector in (4.54) influences the shape of the associated B-spline as shown in Figs. 4.22a,b. In Fig. 4.24 we graph the B-splines N_{i3} for a knot vector with $t_0 = 0$ and $t_{n+1} = t_5 = 5$. In particular, Fig. 4.24a corresponds to the uniform knot vector $\boldsymbol{T} = (0,0,0,1,2,3,4,5,5,5)$. The non-uniform knot vector $\boldsymbol{T} = (0,0,0,1,2.75,3.25,4,5,5,5)$ is used in Fig. 4.24b.

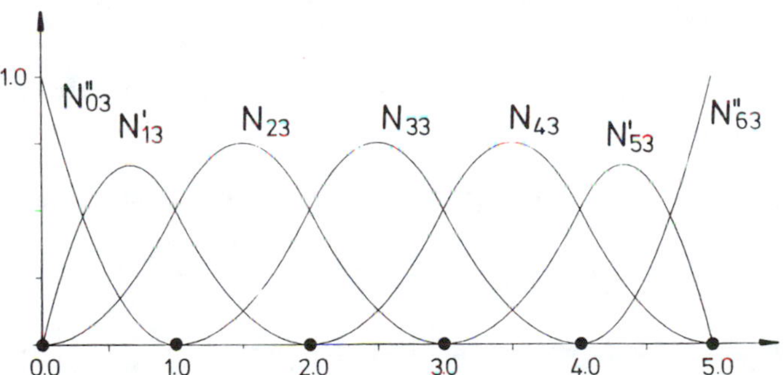

Fig. 4.24a. B-splines for an open B-spline curve with uniform knot vector.

As an example, we now explicitly compute the parametric formula of a B-spline curve of order $k = 3$ with a set of six prescribed de Boor points $\boldsymbol{d}_j$ ($n = 5$), using normalized B-splines with equally spaced parametrization. Then by (4.54),

$$\boldsymbol{T} = (0,0,0,1,2,3,4,4,4).$$

In this case, the B-splines correspond to the following knot vectors:

$$\begin{array}{ll} N''_{03} \text{ to } [0,0,0,1], & N'_{13} \text{ to } [0,0,1,2], \\ N_{23} \text{ to } [0,1,2,3], & N_{33} \text{ to } [1,2,3,4], \\ N'_{43} \text{ to } [2,3,4,4], & N''_{53} \text{ to } [3,4,4,4]. \end{array}$$

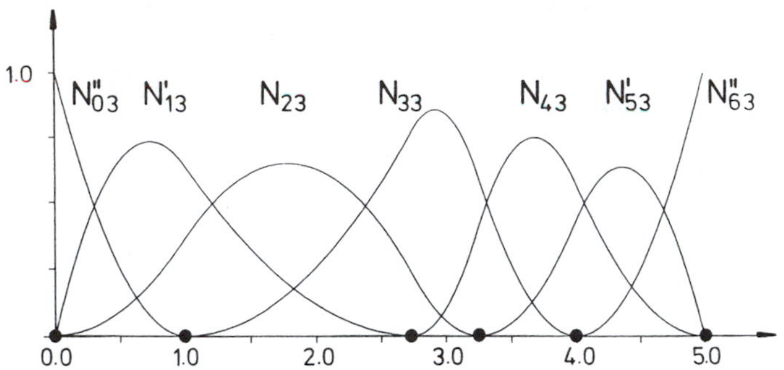

Fig. 4.24b. B-splines for an open B-spline curve with non-uniform knot vector.

This shows that we have symmetry at the boundaries. Thus, in terms of the parametrization with parameter $w \in [0,1]$, we have

$$N''_{03}(w) = N''_{53}(1-w), \qquad N'_{13}(w) = N'_{43}(1-w).$$

It follows that the parametric formula for the B-spline curve corresponding to basis functions (4.50c), (4.52a), (4.52b) is

$$\begin{aligned}
\boldsymbol{X}(t) &= \boldsymbol{d}_0 N''_{03}(t) + \boldsymbol{d}_1 N'_{13}(t) + \boldsymbol{d}_2 N_{23}(t) + \boldsymbol{d}_3 N_{33}(t) + \boldsymbol{d}_4 N'_{43}(t) + \boldsymbol{d}_5 N''_{53}(t) \\
&= \boldsymbol{d}_0[w^2 - 2w + 1]N_{21} + \boldsymbol{d}_1\Big[(-\frac{3}{2}w^2 + 2w)N_{21} + (\frac{w^2}{2} - w + \frac{1}{2})N_{31}\Big] \\
&+ \boldsymbol{d}_2\Big[\frac{w^2}{2}N_{21} + (-w^2 + w + \frac{1}{2})N_{31} + (\frac{1}{2}w^2 - w + \frac{1}{2})N_{41}\Big] \\
&+ \boldsymbol{d}_3\Big[\frac{w^2}{2}N_{31} + (-w^2 + w + \frac{1}{2})N_{41} + (\frac{1}{2}w^2 - w + \frac{1}{2})N_{51}\Big] \\
&+ \boldsymbol{d}_4\Big[\frac{w^2}{2}N_{41} + (-\frac{3}{2}w^2 + w + \frac{1}{2})N_{51}\Big] + \boldsymbol{d}_5 w^2 N_{51},
\end{aligned}$$

for $w \in [0,1]$, which in terms of the N_{i1} becomes

$$\begin{aligned}
\boldsymbol{X}(t) = &\Big[\boldsymbol{d}_0(w^2 - 2w + 1) + \boldsymbol{d}_1(-\frac{3}{2}w^2 + 2w) + \boldsymbol{d}_2\frac{w^2}{2}\Big]N_{21} \\
&+ \Big[\boldsymbol{d}_1(\frac{w^2}{2} - w + \frac{1}{2}) + \boldsymbol{d}_2(-w^2 + w + \frac{1}{2}) + \boldsymbol{d}_3\frac{w^2}{2}\Big]N_{31} \\
&+ \Big[\boldsymbol{d}_2(\frac{w^2}{2} - w + \frac{1}{2}) + \boldsymbol{d}_3(-w^2 + w + \frac{1}{2}) + \boldsymbol{d}_4\frac{w^2}{2}\Big]N_{41} \\
&+ \Big[\boldsymbol{d}_3(\frac{w^2}{2} - w + \frac{1}{2}) + \boldsymbol{d}_4(-\frac{3}{2}w^2 + w + \frac{1}{2}) + \boldsymbol{d}_5 w^2\Big]N_{51}.
\end{aligned} \tag{4.54a}$$

From (4.54a) we note that a de Boor point has an effect on at most $k = 3$ parameter intervals, and moreover in view of the k equal knots at the beginning and at the end of the interval, this B-spline curve consists of a total of $\ell = (n+1) - (k-1) = n - k + 2 = 4$ segments.

Fig. 4.25a shows a B-spline curve as in (4.54a). Fig. 4.25b shows a B-spline curve with $n = 7$ and $k = 4$. In Fig. 4.25c we choose $n = 9$, $k = 3$, and show the Bézier curve which arises if we choose the de Boor points to be the Bézier points.

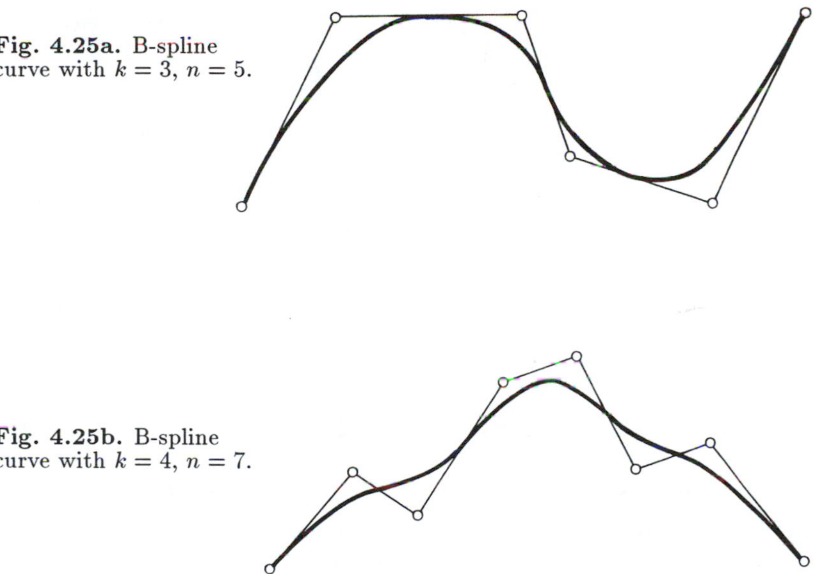

Fig. 4.25a. B-spline curve with $k = 3$, $n = 5$.

Fig. 4.25b. B-spline curve with $k = 4$, $n = 7$.

The computation of B-spline curves for $k = 3$ from (4.54a) or from the analogous formulae for $k > 3$ is relatively expensive. Here we have chosen to work with this representation to illustrate the use of B-splines, and to develop some familiarity with B-splines. However, in the practical use of B-splines, one usually works with the algorithm of de Boor to be introduced below.

4.3.2.2. Closed B-spline Curves

In order to construct closed B-spline curves, we periodically extend the given de Boor points $\boldsymbol{d}_0, \ldots, \boldsymbol{d}_n$ by setting

$$\boldsymbol{d}_0 = \boldsymbol{d}_{n+1},$$

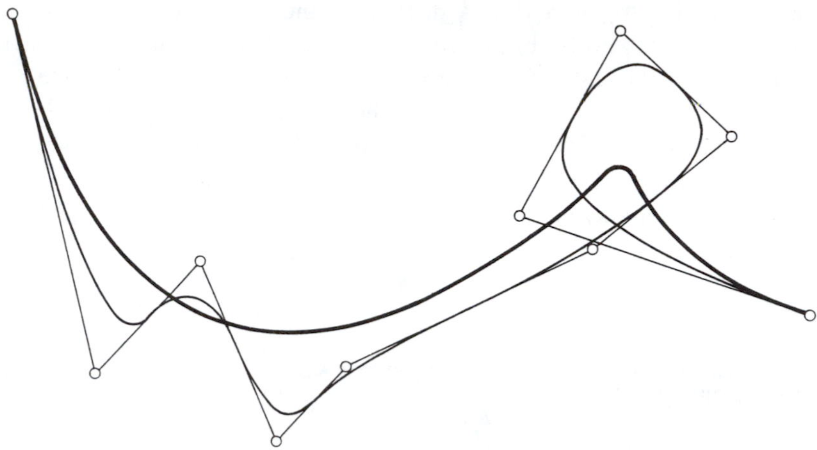

Fig. 4.25c. B-spline curve with $k=3, n=9$ and the Bézier curve of degree 9 with the same control polygon.

and expanding the knot vector periodically by

$$t_{n+1} \hat{=} t_0, \quad t_{n+2} = t_{n+1} + (t_1 - t_0) \hat{=} t_1, \quad t_{n+3} = t_{n+2} + (t_2 - t_1), \ldots \quad (4.55)$$

This leads to the knot vector

$$\boldsymbol{T} = (t_0, t_1, \ldots, t_n, t_{n+1} \hat{=} t_0, t_{n+2} \hat{=} t_1, \ldots, t_{n+k} \hat{=} t_{k-1}).$$

Figs. 4.26a,b show the B-splines N_{i3} for uniform knots as in (4.26a), and for non-uniform knots as in (4.26b), where the knots are marked with dots.

The parametric formula for a closed B-spline curve is thus

$$\boldsymbol{X}(t) = \sum_{i=0}^{n} \boldsymbol{d}_i N_{ik}(t), \qquad \text{with} \qquad t \in [t_0, t_{n+1}].$$

Here we have

$$\begin{array}{ll}
N_{0k} & \text{has support on } [t_0, t_k], \\
N_{1k} & \text{has support on } [t_1, t_{k+1}], \\
\vdots & \\
N_{n-2,k} & \text{has support on } [t_{n-2}, t_{k-3}], \\
N_{n-1,k} & \text{has support on } [t_{n-1}, t_{k-2}], \\
N_{nk} & \text{has support on } [t_n, t_{k-1}].
\end{array}$$

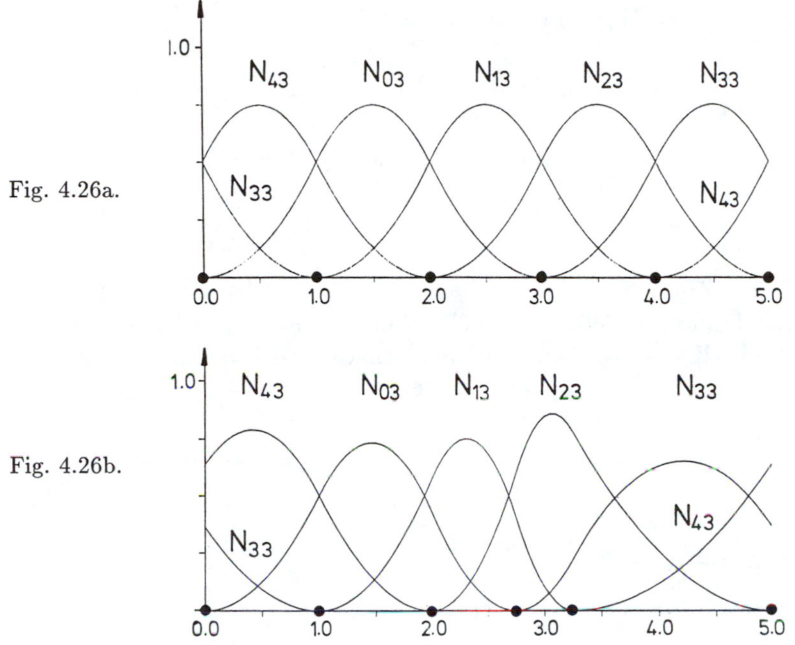

Fig. 4.26. B-splines with uniform and non-uniform knot vectors for a closed B-spline curve.

As an example, we consider the case $n = 3$, $k = 3$. Then the parametric formula for the B-spline curve is given by

$$\boldsymbol{X}(t) = \sum_{i=0}^{3} \boldsymbol{d}_i N_{i3} = \boldsymbol{d}_0 N_{03} + \boldsymbol{d}_1 N_{13} + \boldsymbol{d}_2 N_{23} + \boldsymbol{d}_3 N_{33},$$

or by (4.50c),

$$\begin{aligned}\boldsymbol{X}(t) = \boldsymbol{d}_0 &\Big[\frac{w^2}{2} N_{01} + (-w^2 + w + \frac{1}{2}) N_{11} + \frac{1}{2}(1-w)^2 N_{21}\Big] \\ + \boldsymbol{d}_1 &\Big[\frac{w^2}{2} N_{11} + (-w^2 + w + \frac{1}{2}) N_{21} + \frac{1}{2}(1-w)^2 N_{31}\Big] \\ + \boldsymbol{d}_2 &\Big[\frac{w^2}{2} N_{21} + (-w^2 + w + \frac{1}{2}) N_{31} + \frac{1}{2}(1-w)^2 N_{01}\Big] \\ + \boldsymbol{d}_3 &\Big[\frac{w^2}{2} N_{31} + (-w^2 + w + \frac{1}{2}) N_{01} + \frac{1}{2}(1-w)^2 N_{11}\Big]\end{aligned}$$

$$
\begin{aligned}
&= \left[\frac{1}{2}(1-w)^2\boldsymbol{d}_2 + (-w^2+w+\frac{1}{2})\boldsymbol{d}_3 + \frac{w^2}{2}\boldsymbol{d}_0\right]N_{01} \\
&+ \left[\frac{1}{2}(1-w)^2\boldsymbol{d}_3 + (-w^2+w+\frac{1}{2})\boldsymbol{d}_0 + \frac{w^2}{2}\boldsymbol{d}_1\right]N_{11} \\
&+ \left[\frac{1}{2}(1-w)^2\boldsymbol{d}_0 + (-w^2+w+\frac{1}{2})\boldsymbol{d}_1 + \frac{w^2}{2}\boldsymbol{d}_2\right]N_{21} \\
&+ \left[\frac{1}{2}(1-w)^2\boldsymbol{d}_1 + (-w^2+w+\frac{1}{2})\boldsymbol{d}_2 + \frac{w^2}{2}\boldsymbol{d}_3\right]N_{31}.
\end{aligned}
\tag{4.56}
$$

Fig. 4.27a shows a closed B-spline curve of order $k = 3$ as given by the explicit formula (4.56). Fig. 4.27b shows a closed B-spline curve of order $k = 4$, and Fig. 4.27c shows a B-spline curve with $k = 3$ and $n = 8$.

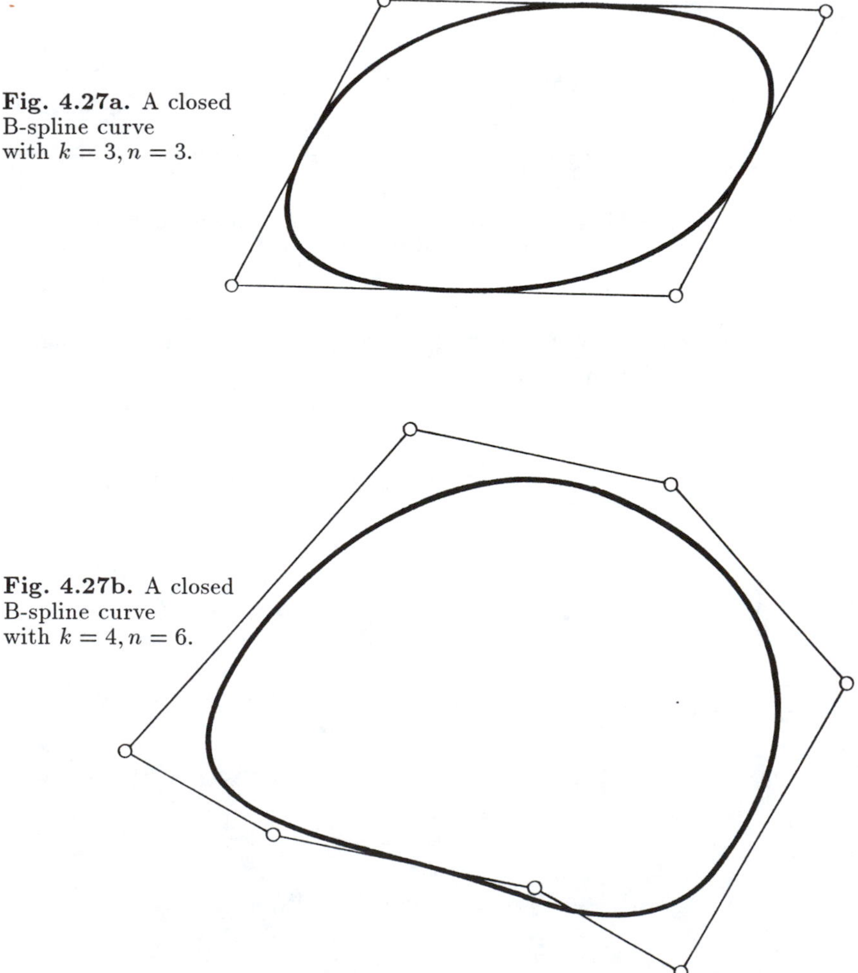

Fig. 4.27a. A closed B-spline curve with $k = 3, n = 3$.

Fig. 4.27b. A closed B-spline curve with $k = 4, n = 6$.

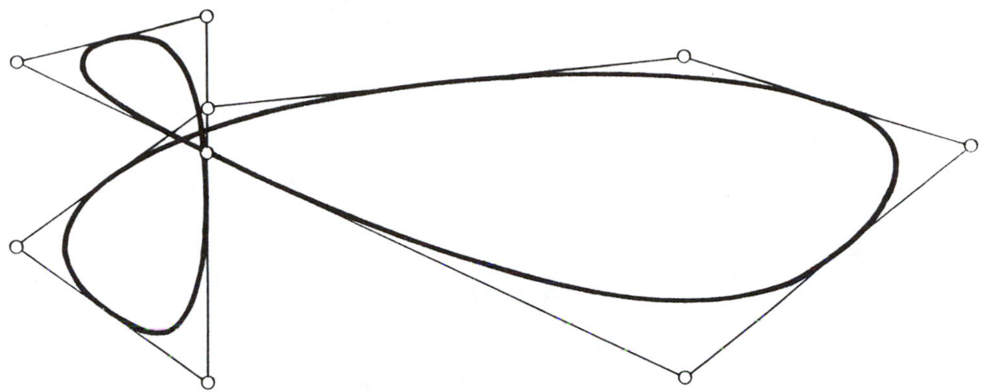

Fig. 4.27c. A closed B-spline curve with $k = 3, n = 8$.

Figs. 4.27d–f show B-spline curves of order $k = 3$ (solid lines), $k = 4$ (dotted) and $k = 5$ (dot/dashed) corresponding to the same de Boor polygon. In Fig. 4.27d we are assuming a uniform knot vector, while in Fig. 4.27e a non-uniform knot vector (with simple knots) is being used. Fig. 4.27f is based on a knot vector with a double knot.

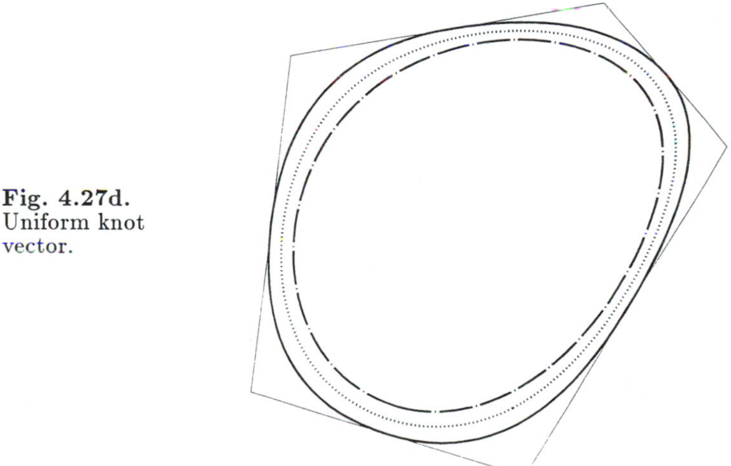

Fig. 4.27d. Uniform knot vector.

4.3.3. The de Boor Algorithm

The de Boor algorithm allows the computation of points on a B-spline curve without explicit knowledge of the B-spline basis functions. It is a *general-*

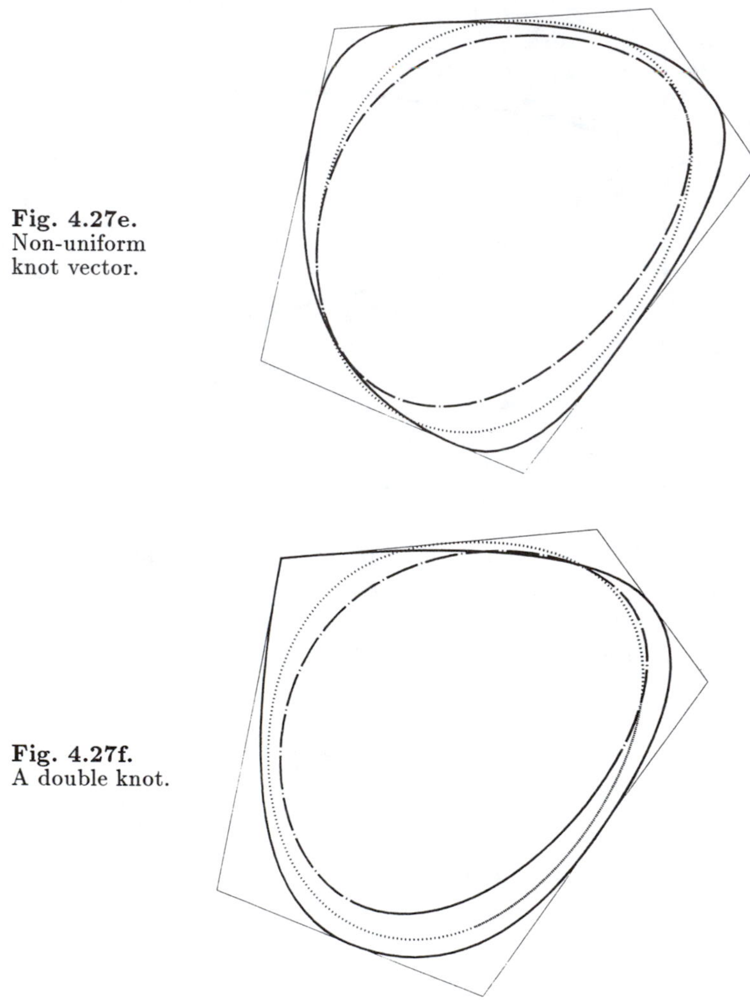

Fig. 4.27e. Non-uniform knot vector.

Fig. 4.27f. A double knot.

ization of the de Casteljau algorithm, and also computes the desired function value via a linear subdivision, see e.g., [Boeh 84], [Boo 72]. In [Boeh 88] several algorithms of this type are discussed.

We begin with the recursive definition (4.48c) of the normalized B-splines

$$N_{ik}(t) = \frac{(t - t_i)}{(t_{i+k-1} - t_i)} N_{i,k-1}(t) + \frac{(t_{i+k} - t)}{(t_{i+k} - t_{i+1})} N_{i+1,k-1}(t)$$

corresponding to an (arbitrary) knot vector $\boldsymbol{T}$, and insert these basis functions in the parametric formula of Definition 4.2 for the B-spline curve:

$$\begin{aligned}\boldsymbol{X}(t) &= \sum_{i=0}^{n} \boldsymbol{d}_i N_{ik}(t) \\ &= \sum_{i=0}^{n} \boldsymbol{d}_i \frac{(t-t_i)}{(t_{i+k-1}-t_i)} N_{i,k-1}(t) + \sum_{i=0}^{n} \boldsymbol{d}_i \frac{(t_{i+k}-t)}{(t_{i+k}-t_{i+1})} \boldsymbol{N}_{i+1,k-1}(t).\end{aligned}$$

Now shifting the indices in the second term by $i = i-1$, it follows that

$$\boldsymbol{X}(t) = \sum_{i=0}^{n+1} \frac{\boldsymbol{d}_i(t-t_i) + \boldsymbol{d}_{i-1}(t_{i+k-1}-t)}{(t_{i+k-1}-t_i)} N_{i,k-1}(t) =: \sum_{i=0}^{n+1} \boldsymbol{d}_i^1 N_{i,k-1}(t),$$

where we take $\boldsymbol{d}_{-1} = 0$ and $\boldsymbol{d}_{n+1} = 0$.

This re-indexing can be continued by repeated insertion of the appropriate recurrence formula for the basis functions, and finally leads to

$$\boldsymbol{X}(t) = \sum_{i=0}^{n+j} \boldsymbol{d}_i^j(t) N_{i,k-j}(t), \qquad j = 0(1)k\text{–}1. \tag{4.57a}$$

Here

$$\boldsymbol{d}_i^j = (1-\alpha_i^j)\boldsymbol{d}_{i-1}^{j-1} + \alpha_i^j \boldsymbol{d}_i^{j-1}, \qquad j > 0 \tag{4.57b}$$

with

$$\alpha_i^j = \frac{t-t_i}{t_{i+k-j}-t_i} \qquad \text{and} \qquad \boldsymbol{d}_j^0 = \boldsymbol{d}_j. \tag{4.57c}$$

Now if the algorithm is carried out according to (4.57), then for $j = k-1$ we get the basis functions N_{r1}, i.e., for $t \in [t_r, t_{r+1}]$ we get the function value

$$\boldsymbol{X}(t) = \boldsymbol{d}_r^{k-1}. \tag{4.57d}$$

The computational steps given in (4.57) can be systematically organized into a *table* analogous to the one for the de Casteljau algorithm. Here we must take into consideration the fact that for a given parameter value $t \in [t_r, t_{r+1}]$, all $N_{ik}(t)$ vanish except for the basis functions N_{ik} corresponding to indices $r-(k-1) \leq i \leq r$. Thus, the algorithm of de Boor can be described by the following table:

$$
\begin{array}{llllllll}
\boldsymbol{d}_{r-k+1} & = \boldsymbol{d}^0_{r-k+1} & & & & & & \\
\boldsymbol{d}_{r-k+2} & = \boldsymbol{d}^0_{r-k+2} & \boldsymbol{d}^1_{r-k+2} & & & & & \\
\cdot & & & \boldsymbol{d}^2_{r-k+3} & & & & \\
\cdot & & & & \cdot & & & \\
\cdot & & & & & \cdot & & \\
\cdot & & & & & & \cdot & \\
\boldsymbol{d}_{r-1} & = \boldsymbol{d}^0_{r-1} & \boldsymbol{d}^1_{r-1} & \boldsymbol{d}^2_{r-1} & \cdots & \boldsymbol{d}^{k-2}_{r-1} & & \\
\boldsymbol{d}_{r} & = \boldsymbol{d}^0_{r} & \boldsymbol{d}^1_{r} & \boldsymbol{d}^2_{r} & \cdots & \boldsymbol{d}^{k-2}_{r} & \boldsymbol{d}^{k-1}_{r} = \boldsymbol{X}(t).
\end{array}
\tag{4.58}
$$

In the horizontal direction we always multiply by α_i^j, *and in the diagonal direction by* $1 - \alpha_i^j$. *The factor* α_i^j *is defined in* (4.57c).

Fig. 4.28 shows the geometric interpretation of the de Boor algorithm as repeated linear interpolation. Here we chose $t^* = 3.5$ and an equally spaced knot vector. The figure shows clearly that we are dealing with a generalization of the de Casteljau algorithm (cf. Fig. 4.4). As was the case for the de Casteljau algorithm, the de Boor algorithm can also be used to decompose a B-spline curve into two segments. The upper boundary elements (and the previous de Boor points) and the lower boundary elements (and the de Boor points following it) in the table are the new control points of the two B-spline segments.

The computation of the coefficients α_i^j can be clarified with the help of ratios as shown in Fig. 4.28. The intermediate points on the edges of the de Boor polygon are in the same ratio as the associated knots with respect to the parameter t^*.

We consider a simple example. Suppose we are given a closed B-spline curve of order $k = 3$ with $n + 1$ de Boor points (e.g., $n = 12$), and we want to compute the curve point $\boldsymbol{X}(t^*)$ corresponding to $t^* = 7.75$, assuming an equally spaced parametrization. Because of our assumption of an equidistant parametrization, we must work with the knot vector

$$\boldsymbol{T} = (0, 1, 2, ..., 12, 13 \hat{=} 0, 14 \hat{=} 1, 2).$$

Then t^* lies in the interval [7,8] and $r = 7$. We now proceed in two different ways (which yield the same result):

- analogous to (4.56), we find the desired curve point directly,
- we find the curve point with the help of the de Boor algorithm.

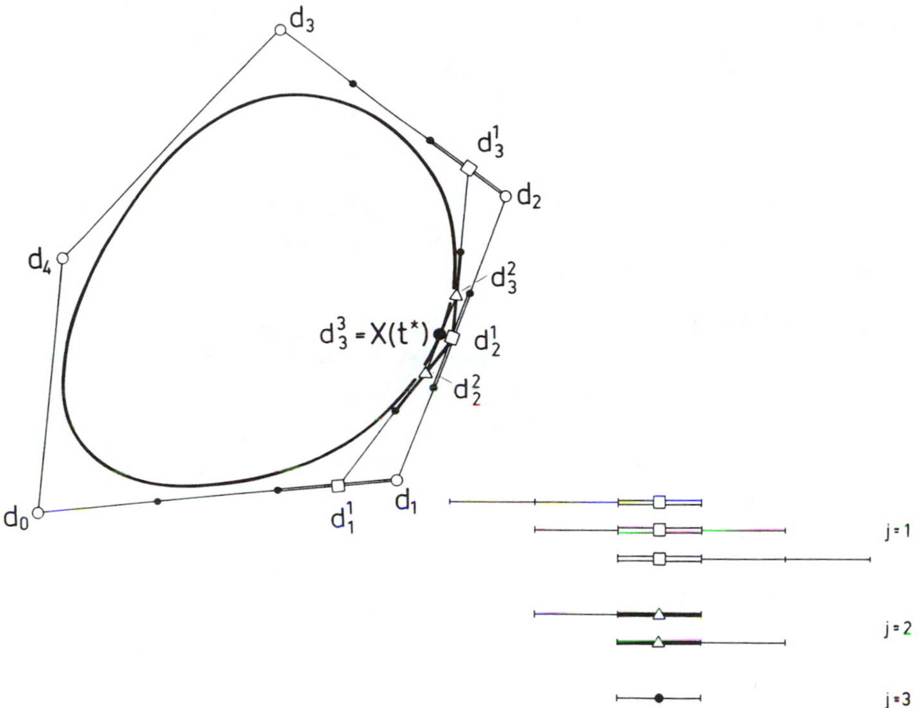

Fig. 4.28. Geometric interpretation of the de Boor algorithm ($k = 4, r = 3$).

a) *Direct computation.* If we work with the analog of (4.56) for $n = 12$ and set $w^* = 0.75$, we get

$$\boldsymbol{X}(t^*) = \boldsymbol{X}(w^*) = \tfrac{9}{32}\boldsymbol{d}_7 + \tfrac{11}{16}\boldsymbol{d}_6 + \tfrac{1}{32}\boldsymbol{d}_5.$$

b) *Using the de Boor algorithm.* Since $r = 7$ and $k = 3$, it follows from (4.58) that in the first column of the de Boor algorithm we have the points $(\boldsymbol{d}_5, \boldsymbol{d}_6, \boldsymbol{d}_7)$, and we are to find the values $\boldsymbol{d}_6^1$ and $\boldsymbol{d}_7^1$. This requires α_6^1 and α_7^1. For the above knot vector and $t^* = 7.75$, (4.57c) implies that

$$\alpha_6^1 = \tfrac{7}{8}, \qquad \alpha_7^1 = \tfrac{3}{8}.$$

Now for the second column of the de Boor algorithm, we need α_7^2, and by (4.57c) we have $\alpha_7^2 = \frac{3}{4}$. Thus, carrying out the de Boor algorithm leads to the following table of values:

$$\begin{array}{lll}
\boldsymbol{d}_5 & & \\
\boldsymbol{d}_6 & \frac{7}{8}\boldsymbol{d}_6 + \frac{1}{8}\boldsymbol{d}_5 & \\
\boldsymbol{d}_7 & \frac{3}{8}\boldsymbol{d}_7 + \frac{5}{8}\boldsymbol{d}_6 & \frac{3}{4}(\frac{3}{8}\boldsymbol{d}_7 + \frac{5}{8}\boldsymbol{d}_6) + \frac{1}{4}(\frac{7}{8}\boldsymbol{d}_6 + \frac{1}{8}\boldsymbol{d}_5) \\
& & = \frac{9}{32}\boldsymbol{d}_7 + \frac{11}{16}\boldsymbol{d}_6 + \frac{1}{32}\boldsymbol{d}_5 = \boldsymbol{X}(t^*).
\end{array}$$

Because of the importance of the de Boor algorithm as an effective means of computing the points on a B-spline curve, we now describe the individual steps of the de Boor algorithm again:

Given: • $n+1$ de Boor points $\boldsymbol{d}_0, \boldsymbol{d}_1, \ldots, \boldsymbol{d}_n$ with $n \geq k-1$, k the order of the B-spline curve

- For a closed B-spline curve, the knot vector $\boldsymbol{T} = (t_0, t_1, \ldots, t_n, t_{n+1})$
  By (4.55) it is recommended to extend the knot vector to the left and to the right, *i.e.*, to introduce
  $t_{-1} = t_0 - (t_{n+1} - t_n), \quad t_{-2} = t_{-1} - (t_n - t_{n-1}), \ldots,$
  $t_{n+2} = t_{n+1} + (t_1 - t_0), \quad t_{n+3} = t_{n+2} + (t_2 - t_1), \ldots,$
- For open B-spline curves, the knot vector
  $\boldsymbol{T} = (t_0 = t_1 = t_2 = \cdots = t_{k-1}, t_k, \ldots, t_n, t_{n+1} = \cdots = t_{n+k}).$

Compute: The point on the curve corresponding to the parameter value t^*:

1) Find the index r with $t_r \leq t^* < t_{r+1}$
2) Compute $\alpha_j^1 = (t^* - t_j)/(t_{j+k-1} - t_j)$ for $j = (r-k+2)(1)r$
   and find $\boldsymbol{d}_j^1 = (1 - \alpha_j^1)\boldsymbol{d}_{j-1} + \alpha_j^1 \boldsymbol{d}_j$
3) For $\ell = 2(1)k$–1, and for $j = (r-k+l+1)(1)r$
   find $\alpha_j^\ell = (t^* - t_j)/(t_{j+k-\ell} - t_j)$
   set $\boldsymbol{d}_j^\ell = (1 - \alpha_j^\ell)\boldsymbol{d}_{j-1}^{\ell-1} + \alpha_j^\ell \boldsymbol{d}_j^{\ell-1}$,
4) Compute the function value $\boldsymbol{X}(t^*) = \boldsymbol{d}_r^{k-1}$.

The de Boor algorithm can also be used to compute *derivatives*. The first derivative of a B-spline curve

$$\boldsymbol{X}(t) = \sum_{i=0}^{n} \boldsymbol{d}_i N_{ik}(t)$$

is given by

$$\frac{d\boldsymbol{X}}{dt} = \sum_{i=1}^{n} \boldsymbol{d}_i^{(1)} N_{i,k-1}(t),$$

with

$$\boldsymbol{d}_i^{(1)} = (k-1)\frac{(\boldsymbol{d}_i - \boldsymbol{d}_{i-1})}{(t_{i+k-1} - t_i)}. \tag{4.59}$$

Repeating this computation leads to higher derivatives, see e.g., (4.53). Another approach to computing the derivatives can be found in [Lee 82] and [Boeh 84a]. [Lee 86] discusses the numerical stability of the various algorithms.

The de Boor algorithm can be generalized [Boeh 90] to construct the de Boor points, points on the curve, and the derivatives. The method is based on the blossoming principle of [Sei 89, 89a]. Starting with a B-spline curve

$$\boldsymbol{X}_r(t) = \sum_{i=r-(k-1)}^{r} \boldsymbol{d}_i N_{ik}(t)$$

on the interval $I = [t_r, t_{r+1}]$, following the suggestion of [Boeh 90], we apply the following linear recurrence relation

$$\boldsymbol{d}_j^\ell = (1 - \alpha_j^\ell)\boldsymbol{d}_{j-1}^{\ell-1} + \alpha_j^\ell \boldsymbol{d}_j^{\ell-1}, \tag{$*$}$$

with

$$\alpha_j^\ell = \frac{t - a_j^\ell}{b_j^\ell - a_j^\ell}.$$

Putting

$$a_j^\ell = t_j, \qquad b_j^\ell = t_{j+k-\ell},$$

in this recurrence, $(*)$ leads to the de Boor algorithm, while choosing $a_j^\ell = a$ and $b_j^\ell = b$ leads to the de Casteljau algorithm for a Bézier curve defined on $[a, b]$. Let $x_1, \ldots, x_{k-1}$ be an ordered set of points in I, and let

$$\alpha_j^\ell(x_\ell) = \frac{x_\ell - a_j^\ell}{b_j^\ell - a_j^\ell}.$$

Then by (4.57d), the *blossom* of $\boldsymbol{X}_r$ is given by

$$\boldsymbol{X}_r(x_1, \ldots, x_{k-1}) = \boldsymbol{d}_r^{k-1}(x_1, \ldots, x_{k-1}).$$

For the special choice

$$x_1 = \cdots = x_{k-1} = t^*,$$

we immediately see that the de Boor algorithm produces (see Fig. 4.29)

$$\boldsymbol{d}_r^{k-1}(x_1, \ldots, x_1) = \boldsymbol{X}_r(t^*).$$

On the other hand,

$$\boldsymbol{X}(t_{r+1}, \ldots, t_{\ell+k-1}) = \boldsymbol{d}_r.$$

Thus, for the uniform knot vector with $t_j = j$ we get the control polygon shown in Fig. 4.29.

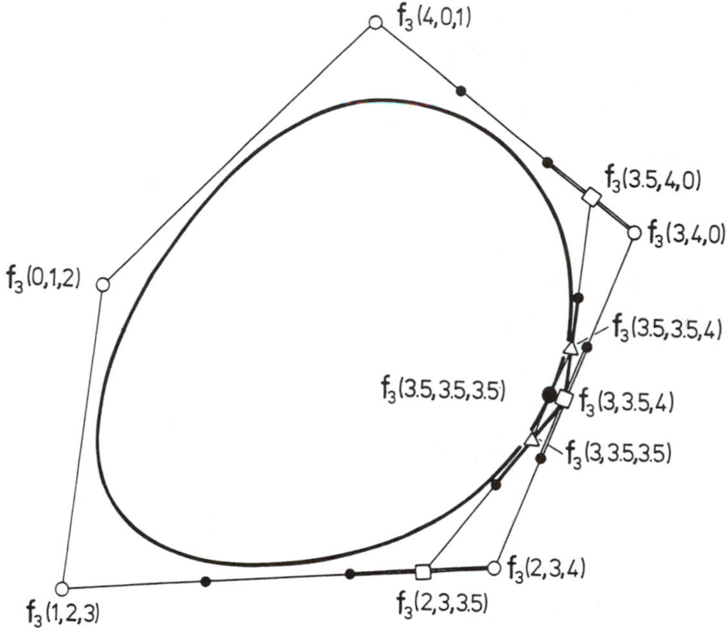

Fig. 4.29. Control points of a B-spline curve using the blossoming principle.

Remark. As for the de Casteljau algorithm, the de Boor polygon converges to the B-spline curve if we repeatedly refine at a set of dense parameter values in $[t_0, t_n]$, see [Coh 85].

4.3.4. Inserting Additional de Boor Points

In interactively working with a Bézier curve, it is very useful to be able to insert additional Bézier points (thus raising the degree of the curve) without changing the shape of the curve. We now show that the same kind of thing can be done for B-spline curves by inserting an additional point $\boldsymbol{d}_r^*$ into a given sequence of de Boor points $\boldsymbol{d}_0, \ldots, \boldsymbol{d}_n$. Here, however, in contrast to Bézier curves, it is not necessary to increase the degree of the curve. The position of the new de Boor point $\boldsymbol{d}_r^*$ will be determined by the position of its associated knot t^* in the knot vector $\boldsymbol{T}$. We recall that for a B-spline curve of order k, precisely k of its de Boor points affect any given segment of the curve.

Suppose we are given the B-spline curve

$$\boldsymbol{X}(t) = \sum_{i=0}^{n} \boldsymbol{d}_i N_{ik}(t)$$

of order k, and that we insert a new parameter value t^* between t_r and t_{r+1}. This gives us a new knot vector $\boldsymbol{T}^*$ with knots

$$\begin{aligned} t_i^* &= t_i, & 0 \le i \le r, \\ t_{r+1}^* &= t^*, & t^* \in [t_r, t_{r+1}], \\ t_{i+1}^* &= t_i, & r+1 \le i \le n. \end{aligned} \tag{4.60}$$

Now we want to write the given B-spline curve in terms of the new basis:

$$\boldsymbol{X}^*(t) = \sum_{i=0}^{n+1} \boldsymbol{d}_i^* N_{ik}^*(t) = \boldsymbol{X}(t).$$

By the local support properties of the B-splines, it is clear that the basis functions

$$N_{0k}, \ldots, N_{r-k,k} \qquad \text{and} \qquad N_{r+1,k}, \ldots, N_{nk}$$

(in the old numbering system) are not affected by the insertion of the new knot t^*, and so

$$\begin{aligned} \boldsymbol{d}_i^* &= \boldsymbol{d}_i, & \text{for} \quad & 0 \le i \le r-k+1, \\ \boldsymbol{d}_{i+1}^* &= \boldsymbol{d}_i, & \text{for} \quad & r \le i \le n. \end{aligned} \tag{4.61a}$$

For $r-k+2 \le i \le r$, we can compute the new de Boor points using the blossoming formula. If we set

$$t^* = t_{r+1}^* = \alpha_i t_{i+k-1} + (1-\alpha_i) t_i,$$

then by the affine invariance, it follows [Boeh 80] that the new de Boor points are given by

$$\boldsymbol{d}_i^* = (1-\alpha_i)\boldsymbol{d}_{i-1} + \alpha_i \boldsymbol{d}_i, \tag{4.61b}$$

with

$$\alpha_i = \frac{t_{r+1}^* - t_i}{t_{i+k-1} - t_i}.$$

The proof of this assertion follows from the de Boor algorithm and a comparison with (4.57b,c), see e.g., [Boeh 80]. Other approaches to the insertion of new knots can be found e.g., in [Coh 80] or [Lee 82].

Fig. 4.30a illustrates the insertion of a new de Boor point for $k=4$, and also shows where the new points lie in the de Boor scheme.

In [Pie 89a] a kind of *inverse knot insertion* is studied. We choose a new de Boor point $\boldsymbol{d}^*$ on the control polygon, and try to find the position of a new knot in the knot vector to correspond to $\boldsymbol{d}^*$. We recall that if we want to

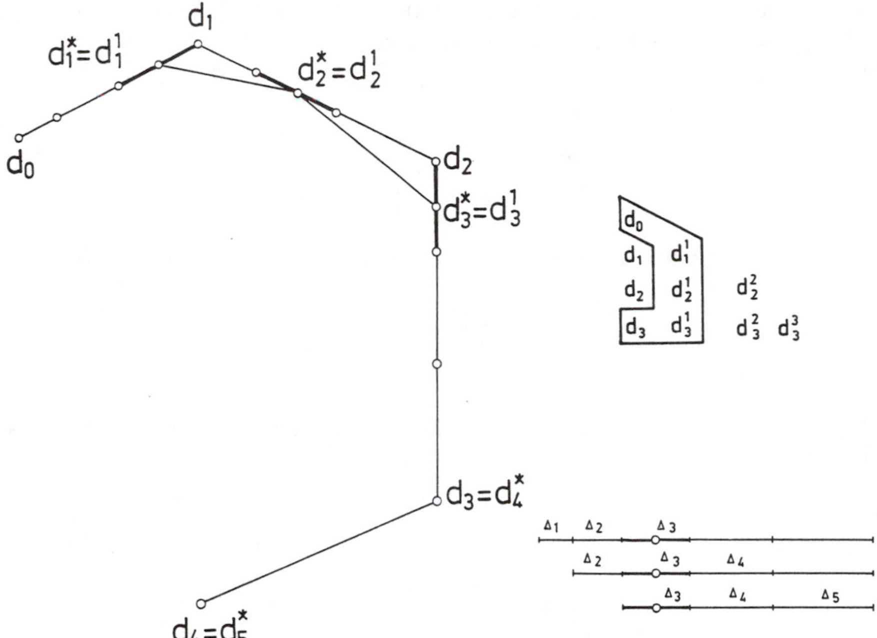

Fig. 4.30a. Insertion of a new de Boor point using the de Boor algorithm.

modify the shape by moving the control point $\boldsymbol{d^*}$ in the direction of a point $\boldsymbol{Q}$, then all points on the curve which lie in the supports of the associated basis functions move in a direction parallel to $\boldsymbol{d^*} - \boldsymbol{Q}$. If we assume that $\boldsymbol{d^*}$ lies on the edge of the control polygon with endpoints $\boldsymbol{d}_{i-1}$, $\boldsymbol{d}_i$, then it follows that $\boldsymbol{d}^*$ is given by the convex combination $\boldsymbol{d}^* = (1-s)\boldsymbol{d}_{i-1} + s\boldsymbol{d}_i$. Solving this equation for s gives

$$s = \frac{|\boldsymbol{d}^* - \boldsymbol{d}_{i-1}|}{|\boldsymbol{d}_i - \boldsymbol{d}_{i-1}|}.$$

In view of (4.61b), it follows that the new knot must be

$$t_i^* = t_i + s(t_{i+k-1} - t_i).$$

We now present an example of knot insertion, cf. Fig 4.30b. Suppose we are given a closed B-spline curve of order k associated with an equally spaced parametrization $t_i = i$, *i.e.*, its knot vector is

$$\boldsymbol{T} = (0, 1, 2, 3, 4, ...).$$

We want to insert the knot $t^* = (t_r + t_{r+1})/2 = (2r+1)/2$. By (4.61b) we have

$$\alpha_i = \frac{2r + 1 - 2i}{2(k-1)}. \tag{4.62}$$

Assuming $k = 4$, the new de Boor point can be found from the de Boor algorithm

$$
\begin{array}{lll}
\boldsymbol{d}_{r-3} = \boldsymbol{d}^*_{r-3} & & \\
 & \frac{5}{6}\boldsymbol{d}_{r-2} + \frac{1}{6}\boldsymbol{d}_{r-3} = & \boldsymbol{d}^*_{r-2} \\
\boldsymbol{d}_{r-2} & & \\
 & \frac{1}{2}\boldsymbol{d}_{r-1} + \frac{1}{2}\boldsymbol{d}_{r-2} = & \boldsymbol{d}^*_{r-1} \\
\boldsymbol{d}_{r-1} & & \\
 & \frac{1}{6}\boldsymbol{d}_{r} + \frac{5}{6}\boldsymbol{d}_{r-1} = & \boldsymbol{d}^*_{r} \\
\boldsymbol{d}_{r} = \boldsymbol{d}^*_{r+1} & &
\end{array}
\tag{4.63a}
$$

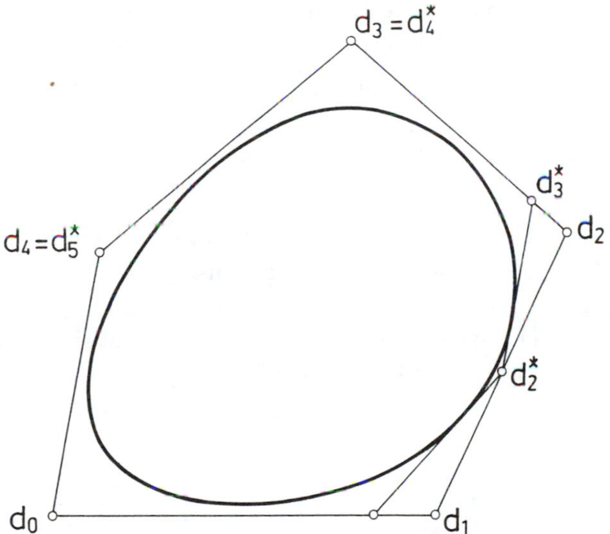

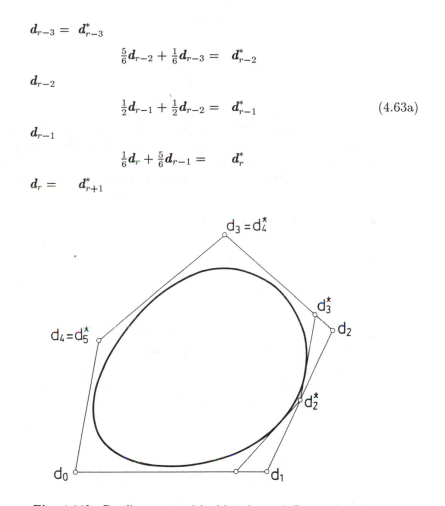

Fig. 4.30b. B-spline curve with old and new de Boor polygons.

The de Boor algorithm can also be used to find new de Boor points, even when the parameter value t^* is inserted with multiplicity greater than one. The following portion of the de Boor scheme for $k = 5$ associated with insertion of the parameter t^* with multiplicity two produces the new de Boor points enclosed in the box.

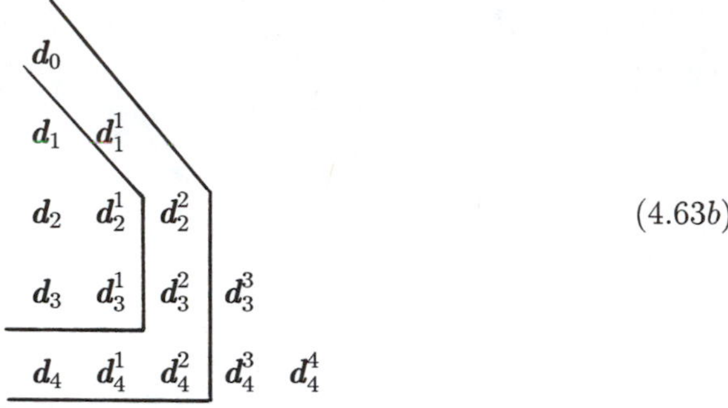

(4.63b)

If a parameter value is to be inserted with multiplicity three, then instead of the second new column, we include the third new column; for a parameter value of multiplicity four, all the points on the boundary of the de Boor scheme would be de Boor points for the new representation. The curve touches the new de Boor polygon.

Remark. The insertion of multiple knots can also be easily derived using the blossoming principle [Sei 88a, 89a].

Remark. The inverse problem of *knot removal* is useful for data reduction. However, knot removal cannot be done exactly, in general. In Sect. 10.2 we discuss approximate methods for knot removal, see e.g., [Lyc 87, 88], [Han 87], [Wev 88, 91].

In some applications it is also useful to *raise the polynomial degree* of a B-spline. This process can also be studied using the blossoming principle. Indeed, suppose we increase the multiplicity of each knot by 1, including the boundary knots. This introduces $n-k+3$ new elements in the knot vector. Now associated with the new knot vector

$$\boldsymbol{T}^* = (\underbrace{t_0^* = t_1^* = \cdots t_k^*}_{k+1 \text{ fold}}, t_{k+1}^* = t_{k+2}^*,\ t_{k+3}^* = t_{k+4}^*, \ldots\ldots, t_{2n-k-1}^* = t_{2n-k}^*,$$
$$t_{2n-k+1}^* = t_{2n-k+2}^*,\ \underbrace{t_{2n-k+3}^* = t_{2n-k+4}^* = \cdots = t_{2n+3}^*}_{k+1 \text{ fold}}),$$

we represent the curve in the parametric form

$$\boldsymbol{X}^*(t) = \sum_{i=0}^{n^*} \boldsymbol{d}_i^* N_{i,k^*}(t), \qquad n^* = n-2+k,$$

with increased polynomial degree $k^* = k+1$. It is shown in [Sei 89a] that the new de Boor points are given by

$$\boldsymbol{d}_l^* = f^*(t_{l+1}^*, \ldots, t_{l+k}^*) = \frac{1}{k}\sum_{i=1}^{k} f(t_{l+1}^*, \ldots, \hat{t}_{l+i}^*, \ldots, t_{l+k}^*),$$

where the hat on the parameter value $\hat{t}^*_{\ell+i}$ means that it is not to be included.

As an application of this result, we consider the given knot vector

$$\boldsymbol{T} = (0,0,0,0,1,2,3,3,3,3), \tag{4.64a}$$

with $k = 4$ and $n = 5$. Raising the degree to $k^* = 5$, ($n^* = 8$), leads to the knot vector

$$\boldsymbol{T}^* = (0,0,0,0,0,1,1,2,2,3,3,3,3,3). \tag{4.64b}$$

By the above result, the new de Boor points are then

$$\begin{aligned}
\boldsymbol{d}_0^* &= f_5^*(0,0,0,0) = \tfrac{1}{4}\cdot 4f_4(0,0,0) = \boldsymbol{d}_0,\\
\boldsymbol{d}_1^* &= f_5^*(0,0,0,1) = \tfrac{1}{4}[3f_4(0,0,1) + f_4(0,0,0)] = \tfrac{1}{4}(3\boldsymbol{d}_1 + \boldsymbol{d}_0),\\
\boldsymbol{d}_2^* &= f_5^*(0,0,1,1) = \tfrac{1}{4}[2f_4(0,1,1) + 2f_4(0,0,1)]\\
&= \tfrac{1}{4}[f_4(0,0,1) + f_4(0,1,2) + 2f_4(0,0,1)] = \tfrac{1}{4}(3\boldsymbol{d}_1 + \boldsymbol{d}_2),\\
\boldsymbol{d}_3^* &= f_5^*(0,1,1,2) = \tfrac{1}{4}[f_4(0,1,1) + f_4(1,1,2) + 2f_4(0,1,2)]\\
&= \tfrac{1}{4}[\tfrac{1}{2}f_4(0,0,1) + \tfrac{1}{2}f_4(0,1,2) + \tfrac{2}{3}f_4(0,1,2) + \tfrac{1}{3}f_4(1,2,3) + 2f_4(0,1,2)]\\
&= \tfrac{1}{24}[3f_4(0,0,1) + 19f_4(0,1,2) + 2f_4(1,2,3)] = \tfrac{1}{24}(2\boldsymbol{d}_3 + 19\boldsymbol{d}_2 + 3\boldsymbol{d}_1),\\
\boldsymbol{d}_4^* &= f_5^*(1,1,2,2) = \tfrac{1}{4}[2f_4(1,1,2) + 2f_4(1,2,2)] = \tfrac{1}{4}[2(\tfrac{2}{3}f_4(0,1,2)+\\
&\qquad + \tfrac{1}{3}f_4(1,2,3)) + 2(\tfrac{1}{3}f_4(0,1,2) + \tfrac{2}{3}f_4(1,2,3))] = \tfrac{1}{2}(\boldsymbol{d}_3 + \boldsymbol{d}_2),\\
\boldsymbol{d}_5^* &= f_5^*(1,2,2,3) = \tfrac{1}{24}(3\boldsymbol{d}_4 + 19\boldsymbol{d}_3 + 2\boldsymbol{d}_2),\\
\boldsymbol{d}_6^* &= f_5(2,2,3,3) = \tfrac{1}{4}(3\boldsymbol{d}_4 + \boldsymbol{d}_3),\\
\boldsymbol{d}_7^* &= f_5(2,3,3,3) = \tfrac{1}{4}(\boldsymbol{d}_5 + 3\boldsymbol{d}_4),\\
\boldsymbol{d}_8^* &= f_5(3,3,3,3) = \boldsymbol{d}_5.
\end{aligned} \tag{4.64c}$$

Other algorithms for degree raising can be found in [Pra 91], [Coh 85a]. Fig. 4.31 shows the control polygons for B-spline curves of order 4 and order 5 representing the same curve.

4.3.5. Properties of B-spline Curves

We have seen in the examples above that a quadratic B-spline curve touches the edges of its de Boor polygon. The following theorem (cf. e.g., [Gor 74], [Rie 73]) describes the general situation.

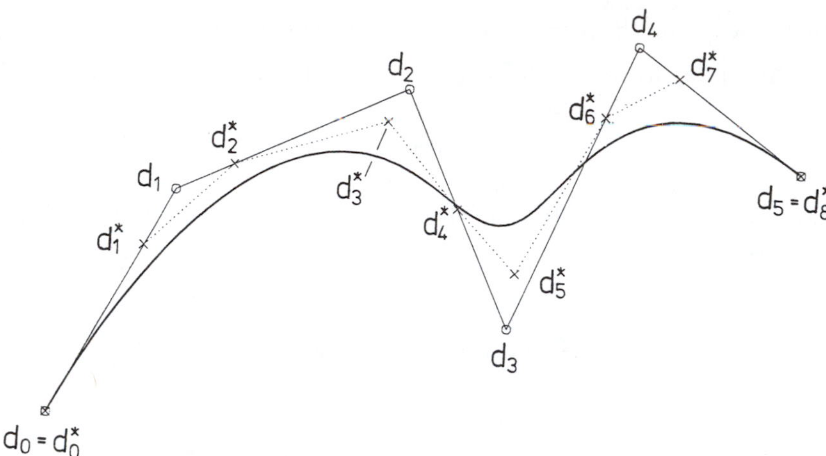

Fig. 4.31. Degree raising of a B-spline curve using (4.64) de Boor points $\boldsymbol{d}_i$ for $k=4$, $\boldsymbol{d}_j^*$ for $k=5$.

Theorem 4.5. *Let $\boldsymbol{X}$ be a B-spline curve of order k. Then*

- *If ℓ knots coincide, then the differentiability of $\boldsymbol{X}$ is reduced to $C^{(k-1-\ell)}$ at the corresponding point;*
- *if $(k-1)$ points of the de Boor polygon are collinear, then $\boldsymbol{X}$ touches the polygon (see Fig. 4.27a,c);*
- *if k points of the de Boor polygon are collinear, then $\boldsymbol{X}$ and the de Boor polygon have a common segment (see Fig. 4.33a);*
- *if $(k-1)$ de Boor points coincide, then $\boldsymbol{X}$ interpolates the common point and the two adjacent sides of the polygon are tangent to $\boldsymbol{X}$. The point may be a cusp (see Fig. 4.33a);*
- *the convex-hull property holds for each set of k neighboring de Boor points, i.e., each segment of the B-spline curve lies in the convex hull of the associated k de Boor points. Globally, the curve $\boldsymbol{X}$ lies in the union of the convex hulls of its curve segments (see Fig. 4.32);*
- *if $\boldsymbol{X}$ lies in a plane, then it satisfies the variation-diminishing property for each set of k neighboring de Boor points.*

Proof: The proof of the variation-diminishing property is analogous to that for Bézier curves, see e.g., [Lane 83]. The convex-hull-property follows from the de Boor algorithm in the same way as for Bézier curves where the convex-hull property follows from the de Casteljau algorithm. If $k-1$ de Boor points coincide with the de Boor point $\boldsymbol{d}_i$, then the convex hull of these points reduces to the point $\boldsymbol{d}_i$. Since the curve lies in the convex hull of their segments, it follows that $\boldsymbol{d}_i$ must lie on the curve, and that the sides of the de Boor polygon are tangents at this point. Properties 2 and 3 follow analogously. Finally, the first assertion can be established by inserting an ℓ-tuple knot as in Sect. 4.3.4. ■

Fig. 4.32 shows the convex hulls of the de Boor polygons for several B-spline curves of different orders.

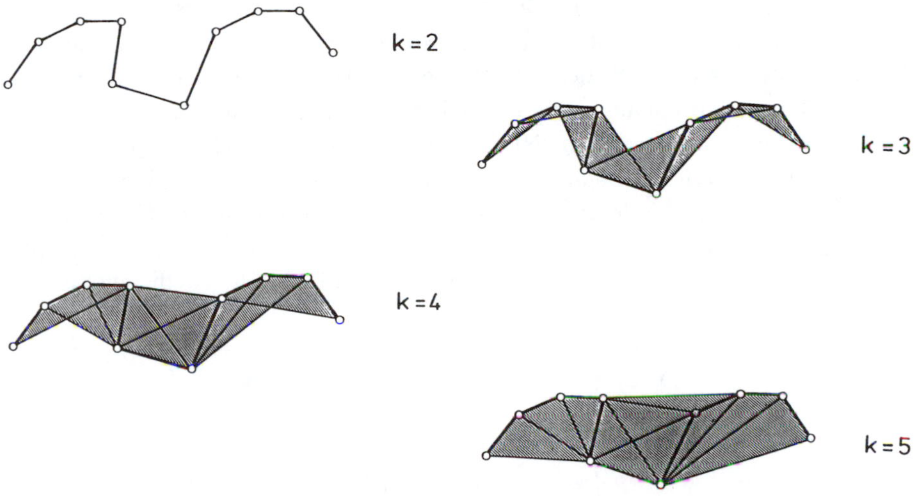

Fig. 4.32. Convex hulls of B-spline segments of various orders.

4.3.6. Rational B-spline Curves

In this section we develop the analog of the rational Bézier curves for B-spline curves. Given a knot vector $\boldsymbol{T}$ and a set of de Boor points in homogeneous coordinates

$$\boldsymbol{D}_j = \begin{cases} (\beta_j, u_j\beta_j, v_j\beta_j, w_j\beta_j)^T = (\beta_j, \beta_j\boldsymbol{d}_j)^T, & \beta \neq 0,\ \boldsymbol{d}_j \in E^3 \\ (0, u_j, v_j, w_j)^T = (0, \vec{\boldsymbol{d}}_j)^T, & \beta_j = 0,\ \vec{\boldsymbol{d}}_j \in E^3, \end{cases} \tag{4.65a}$$

we define a *rational B-spline curve* of order k in parametric form (with homogeneous coordinates) by

$$\mathcal{X}(t) = \sum_{j=0}^{n} \boldsymbol{D}_j N_{jk}(t), \qquad n \geq k-1,$$

or with inhomogeneous coordinates (x, y, z) of $\mathbb{R}^3$ as

$$x = \frac{\sum_{j=0}^{n} u_j \beta_j N_{jk}(t)}{\sum_{j=0}^{n} \beta_j N_{jk}(t)}, \quad y = \frac{\sum_{j=0}^{n} v_j \beta_j N_{jk}(t)}{\sum_{j=0}^{n} \beta_j N_{jk}(t)}, \quad z = \frac{\sum_{j=0}^{n} w_j \beta_j N_{jk}(t)}{\sum_{j=0}^{n} \beta_j N_{jk}(t)}. \tag{4.65b}$$

Matrix representations of rational B-spline curves can be found in [Cho 90a], [Gra 91]. For $\beta_j > 0$, all properties of the ordinary B-spline curves carry over ([Til 83], [Vers 75], [Pie 87b, 91]), and the weight factors β_j provide us with additional degrees of freedom for design. Since B-spline segments can also be interpreted as Bézier curves, the assertions of Sect. 4.1.5 also carry over. Rational B-spline curves defined with respect to a non-uniform knot vector are often referred to as NURBS (non-uniform rational B-splines).

For the practical evaluation of rational B-spline curves with given weights, we can again use the de Boor algorithm. As in the rational de Casteljau algorithm, in addition to the weighted de Boor points, we also have to apply the algorithm (4.57) to the weights. By (4.57)–(4.58), the coefficients in the de Boor scheme are

$$\boldsymbol{D}_i^j = (1 - \alpha_i^j)\boldsymbol{D}_{i-1}^{j-1} + \alpha_i \boldsymbol{D}_i^{j-1}, \qquad j > 0$$

with

$$\alpha_i^j = \frac{t - t_i}{t_{i+k-j} - t_i}, \quad \boldsymbol{D}_j^0 = \boldsymbol{D}_j,$$

and

$$\mathcal{X}(t) = \boldsymbol{D}_r^{k-1}, \qquad \text{for } t \in [t_r, t_{r+1}],$$

or in E^3

$$\begin{aligned} \beta_i^j \boldsymbol{d}_i^j &= (1 - \alpha_i^j)\beta_{i-1}^{j-1}\boldsymbol{d}_{i-1}^{j-1} + \alpha_i^j \beta_i^{j-1}\boldsymbol{d}_i^{j-1} \\ \beta_i^j &= (1 - \alpha_i^j)\beta_{i-1}^{j-1} + \alpha_i^j \beta_i^{j-1}, \qquad j > 0 \end{aligned} \tag{4.66}$$

with $\boldsymbol{d}_j^0 = \boldsymbol{d}_j$, $\beta_j^0 = \beta_j$ and $\boldsymbol{X}(t) = \boldsymbol{d}_r^{k-1}$ for $t \in [t_r, t_{r+1}]$.

Algorithms for knot insertion and degree raising can be formulated in a similar way. A detailed treatment of methods for dealing with rational B-spline curves can be found in [Pie 89a, 91].

Fig. 4.33a,b shows some examples of rational B-spline curves. Fig. 4.33a shows a closed rational B-spline curve of order $k = 4$, where the de Boor points at each end have multiplicity three. The solid curve is again the ordinary B-spline curve. The dotted curve corresponds to the weights (10:10:10:1:1:1:1:10), and therefore is closer to the edges of the left-hand side of the de Boor polygon. The dot–dashed curve corresponds to weights (1:1:1:1:1:1:10:10), and is thus closer to the upper de Boor points.

Fig. 4.33a.

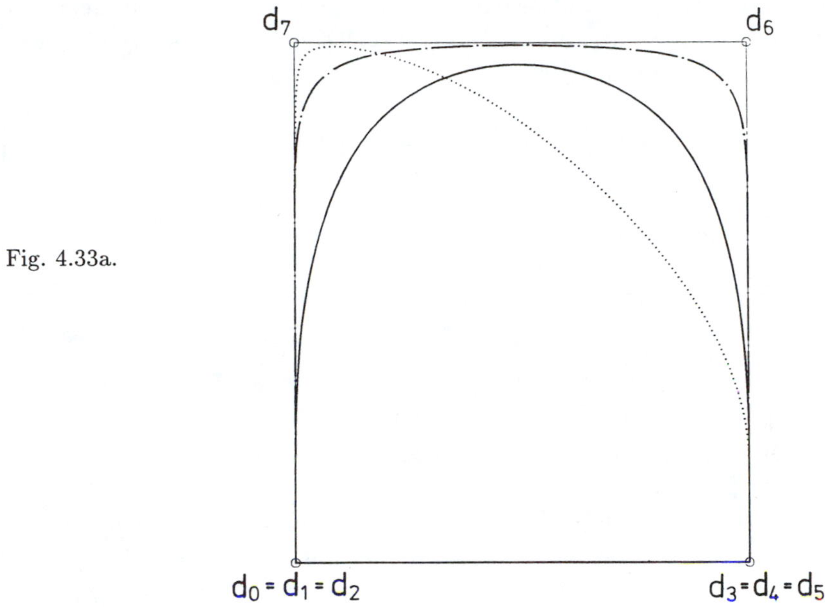

Fig. 4.33b.

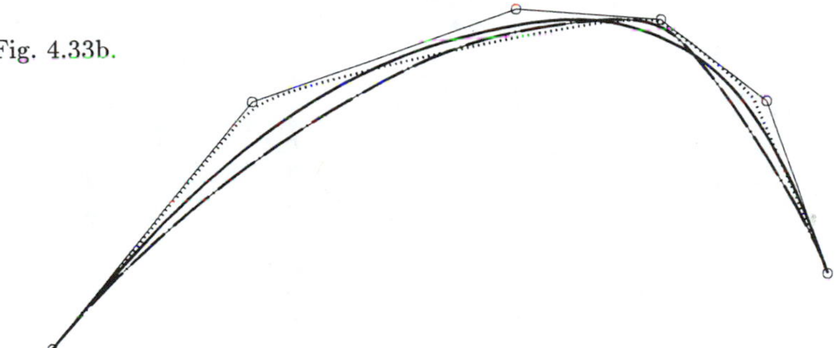

Fig. 4.33a,b. Rational B-spline curves corresponding to the same de Boor polygon with various weights.

Fig. 4.33b shows open rational B-spline curves of order $k = 3$. The solid curve is the associated ordinary B-spline curve. The dotted curve corresponds to weights (1:100:100:1000:100:1), and hence is closer to the middle part of the de Boor polygon. The dot-dashed curve corresponds to the weights $(1 : 0.3 : 0.3 : 1 : 0.3 : 1\ )$.

Rational quadratic B-spline curves can be used to represent circular splines, see e.g., the rational Bézier circular splines in (4.33). Indeed, (see [Til 83, 84]), we can write

$$\boldsymbol{X}(t) = \frac{\sum_{i=0}^{2n} \beta_i \boldsymbol{d}_i N_{i3}(t)}{\sum_{i=0}^{2n} \beta_i N_{i3}(t)}, \tag{4.67a}$$

with the knot vector

$$\boldsymbol{T} = (0,0,0,t_1,t_1,t_2,t_2,\ldots,t_{n-2},t_{n-2},t_{n-1},t_{n-1},t_n,t_n,t_n), \tag{4.67b}$$

and the weights

$$\begin{aligned} \beta_i &= 1, && \text{for } i = 2\ell \text{ with } \ell = 0(1)n, \\ \beta_i &= \cos\phi_k, && \text{for } i = 2k-1 \text{ with } k = 1(1)n. \end{aligned} \tag{4.67c}$$

The control points in (4.67) correspond to the auxiliary points $\boldsymbol{d}_i$ in the Bézier formula (4.34) for a circle. Using (4.67), we clearly can also represent a complete circle. Fig. 4.34 corresponds to $\phi_i = 60°$ and $t_i = i$ for $i = 0(1)3$.

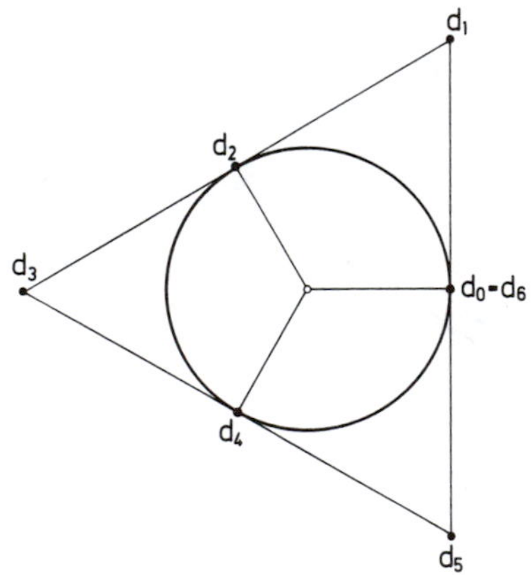

Fig. 4.34. The de Boor polygon for a circle.

The relationship between (4.67) and the Bézier formula (4.33) for circular segments can be found by comparing basis functions:

$$\begin{aligned} &\text{B-spline over } [0,0,0,t_1] \mathrel{\hat{=}} B_0^{2*}(t_1), \\ &\text{B-spline over } [0,0,t_1,t_1] \mathrel{\hat{=}} B_1^{2*}(t_1), \\ &\text{B-spline over } [0,t_1,t_1,t_2] \mathrel{\hat{=}} B_2^{2*}(t_1) \oplus B_0^{2*}(t_2), \end{aligned} \tag{4.68}$$

where the $\oplus$ denotes formal addition since $B_2^{2*}(t_1)$ and $B_2^{2*}(t_2)$ are defined on $[0, t_1]$ and $[t_1, t_2]$, respectively.

4.4. Interpolation and Approximation with Spline Curves

In this section we study parametric interpolation and approximation of point sets using spline curves. Given points $\boldsymbol{P}_i$ with associated parameter values t_i, $i = 0(1)n$, we solve the following problems:

- find an interpolating spline curve passing through the points $\boldsymbol{P}_i$,
- find an approximating spline curve passing near the given points $\boldsymbol{P}_i$ which minimizes some prescribed error norm.

The choice of *parametrization* is critical for both interpolation and approximation, see e.g., [AHL 67], [Eps 76], [BOO 78], [Hartl 80], [Töp 82], [Mas 86], [Fol 87a, 89] and [Nie 89]. Thus, we first discuss methods for parametrizing point sets. Since none of our methods lead to an optimal parametrization corresponding to a given point set, in the case of approximation we shall also employ some kind of *parameter correction*, see e.g., [Pla 83], [Hos 88, 89], [Rog 89] and [Sark 91, 91a].

4.4.1. Parametrization of Curves

We assume that $[0, a]$ is our parameter interval. The following parametrization strategies can be used ([Fol 89], [Nie 89], [Lee 89]):

a) *equally spaced* parametrization, *i.e.*, the parameter interval is divided into equally spaced points $t_i = ia/n$ corresponding to the $\boldsymbol{P}_i$, $i = 0(1)n$,

b) *chordal* parametrization, *i.e.*, we compute the distances $\Delta_i = |\boldsymbol{P}_i - \boldsymbol{P}_{i-1}|$ (with $\Delta_0 = 0$) and the total length $s = \sum_{i=0}^{n} \Delta_i$. Then we set

$$t_i = \sum_{j=0}^{i} \frac{a\Delta_j}{s}, \qquad i = 0(1)n. \tag{4.69a}$$

Geometrically, the chordal parametrization can be interpreted as an approximation to the arc length,

c) *centripetal* parametrization [Lee 89], which in analogy to the chordal parametrization works with $\Delta_i = \sqrt{|\boldsymbol{P}_i - \boldsymbol{P}_{i-1}|}$, or more generally with

$$\Delta_i = |\boldsymbol{P}_i - \boldsymbol{P}_{i-1}|^{\alpha}. \tag{4.69b}$$

[Lee 89] motivates the choice of $\alpha = \frac{1}{2}$ as follows: if a car is driven through a slalom course, then the driver must be careful that the normal acceleration does not become too large. To accomplish this, we may require that the normal force along the path being travelled is proportional to the change in angle. The centripetal parametrization is an approximation to this model,

d) *geometric* parametrization, *i.e.*, we form the geometric mean of a) and b),

e) *affinely invariant* parametrization of [Fol 89], [Nie 89]: First we construct a least squares fit to the given point set using a conic section. The associated coefficient matrix $Q = \{q_{ij}\}$, $i, j = 1, 2$, is formed from the standard deviation

$$q_{11} = \frac{\sigma_y}{d}, \qquad q_{22} = \frac{\sigma_x}{d}, \qquad q_{12} = q_{21} = \frac{-\sigma_{xy}}{d},$$

where $d = \sigma_x \sigma_y - (\sigma_{xy})^2$. Here

$$\sigma_x = \frac{\sum_{i=0}^{n}(x_i - \bar{x})^2}{n+1}, \quad \sigma_y = \frac{\sum_{i=0}^{n}(y_i - \bar{y})^2}{n+1},$$
$$\sigma_{xy} = \frac{\sum_{i=0}^{n}(x_i - \bar{x})(y_i - \bar{y})}{n+1},$$

with

$$\bar{x} = \frac{\sum_{i=0}^{n} x_i}{n+1}, \qquad \bar{y} = \frac{\sum_{i=0}^{n} y_i}{n+1}.$$

The distance between two points $\boldsymbol{U}$ and $\boldsymbol{V}$ is then defined by

$$M^2[\boldsymbol{P}](\boldsymbol{U}, \boldsymbol{V}) = (\boldsymbol{U} - \boldsymbol{V})Q(\boldsymbol{U} - \boldsymbol{V})^T. \tag{4.69c}$$

This gives an affinely invariant chordal parametrization which generalizes b).

f) [Fol 89] suggested a parametrization which not only takes account of the distances between the interpolation points, but also how the angles vary. He takes

$$\Delta t_i = d_i \left[1 + \frac{3\widehat{\theta}_i d_{i-1}}{2(d_{i-1} + d_i)} + \frac{3\widehat{\theta}_{i+1} d_{i+1}}{2(d_i + d_{i+1})} \right], \tag{4.69d}$$

where $d_i = M[\boldsymbol{P}](\boldsymbol{P}_i, \boldsymbol{P}_{i+1})$ is the Nielson metric of (4.69c), $\widehat{\theta}_i = \min(\theta_i, \pi/2)$, and θ_i is the angle between $\overline{\boldsymbol{P}_{i-1}\boldsymbol{P}_i}$ and $\overline{\boldsymbol{P}_i\boldsymbol{P}_{i+1}}$, see Fig. 4.35.

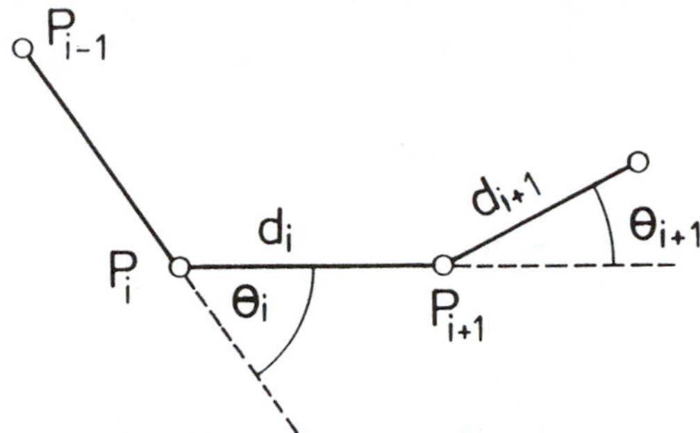

Fig. 4.35. Parametrization according to (4.69d).

The expression (4.69d) can be simplified using the chordal or centripetal metric. It should be noted that methods e) and f) can only be used for plane curves.

4.4.2. Interpolation with B-spline Curves

To interpolate or approximate a given point set, we can choose between monomial splines, Bézier splines, and B-splines. B-splines require the least amount of storage; for example, for the cubic case, monomial splines and Bézier splines require two (vector valued) coefficients per spline segment, while B-spline curves use only one control point per spline segment. This means that for storing monomial spline curves and Bézier spline curves, we need a total of $3n + 1$ coefficients or control points, while for B-spline curves only $n + 1$ are required. Thus, in the following we work only with B-spline curves as our interpolants.

Suppose we want to interpolate with a B-spline curve of the form

$$\boldsymbol{X}(t) = \sum_{i=0}^{\ell} \boldsymbol{d}_i N_{ik}(t), \tag{4.70}$$

where $\ell = n$ and the $\boldsymbol{d}_i$ are unknown de Boor points. For closed curves (without multiple knots) we choose the corresponding knot vector to be

$$\boldsymbol{T} = (\bar{t}_0, \bar{t}_1, \ldots, \bar{t}_n),$$

while for open curves (without multiple interior knots) we use

$$\boldsymbol{T} = (\bar{t}_0 = \bar{t}_1 = \cdots = \bar{t}_{k-1}, \bar{t}_k, \bar{t}_{k+1}, \ldots, \bar{t}_n, \bar{t}_{n+1} = \bar{t}_{n+2} = \cdots = \bar{t}_{n+k}).$$

The entries in the knot vector can be chosen to be uniformly spaced, although arbitrarily spaced knots can also be used. The length of the knot vector is, however, restricted by Lemma 4.8. Tiller (see [Pie 91]) suggested choosing the interior knots as the mean values

$$t_j = \frac{1}{k} \sum_{i=j}^{j+k-1} u_i$$

of the parameter values u_i of the points to be interpolated.

Thus, for example, if we choose a uniform knot vector with $t_0 = 0$, then the parameter interval becomes $[0, n-k+2]$. In particular, if $n = k-1$, then the B-spline curve reduces to a Bézier curve on $[0,1]$.

If we choose to work with a non-uniform knot vector, then we must take care that the parameter values are more or less "uniformly" distributed. The interpolation problem will not be solvable if the parameter values of the knots accumulate in certain intervals, leaving gaps elsewhere. Indeed, if this happens, then because of the local support of the B-spline basis functions, we get gaps in the corresponding linear system of equations (see [Schm 79] and Fig. 4.39). The effects of different parametrizations and different choices of the knot vector are illustrated in Figs. 4.36a,b,c,d. Fig. 4.36a shows the interpolants of a given set of points using a uniform knot vector corresponding to different choices of parametrization. Fig. 4.36b shows the interpolants of a point set with strong variations in the distance between points, using various parametrizations. Fig. 4.36c displays the interpolant of a point set using chordal parametrization and a non-uniform knot vector. Finally, Fig. 4.36d shows the interpolant of the point set in 4.36c using a knot vector with knots which are distributed as uniformly as possible.

Fig. 4.36 suggests that centripetal parametrization is the most efficient with regard to computational effort.

4.4.3. Approximation with B-spline Curves

If the value of ℓ in (4.70) is smaller than n, then the interpolation problem discussed in Sect. 4.4.2 becomes an *approximation problem*. We discuss first the case of closed curves, and assume that the knot vector is given by

$$\boldsymbol{T} = (\bar{t}_0, \bar{t}_1, \ldots, \bar{t}_\ell), \qquad \ell < n,$$

where the parameter values t_i corresponding to the approximating points $\boldsymbol{P}_i$ are projected onto the parameter interval $[\bar{t}_0, \bar{t}_\ell]$, with $\bar{t}_0 = t_0$ and $\bar{t}_\ell = t_n$. Because of the local support properties of the B-spline basis functions, we must choose ℓ to be greater than or equal $k-1$, where k is the order of the B-splines. The case $\ell = k-1$ corresponds to approximation by a Bézier curve.

We denote the approximation errors by

$$\boldsymbol{D}_i = \boldsymbol{X}(t_i) - \boldsymbol{P}_i, \qquad i = 0(1)n. \tag{4.71a}$$

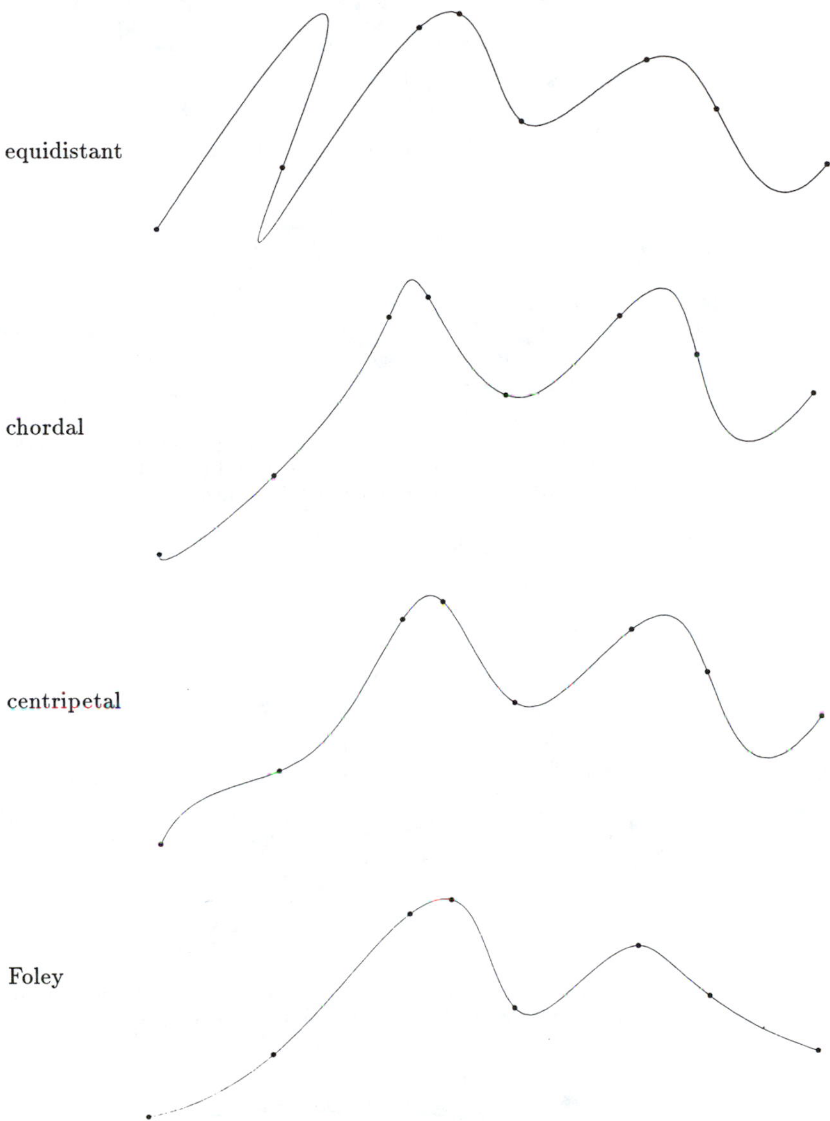

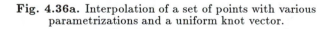

Fig. 4.36a. Interpolation of a set of points with various parametrizations and a uniform knot vector.

Our goal is to minimize the total error

$$D^2 := \sum_{i=0}^{n} w_i \boldsymbol{D}_i^2 = \sum_{i=0}^{n} w_i (\boldsymbol{X}(t_i) - \boldsymbol{P}_i)^2, \tag{4.71b}$$

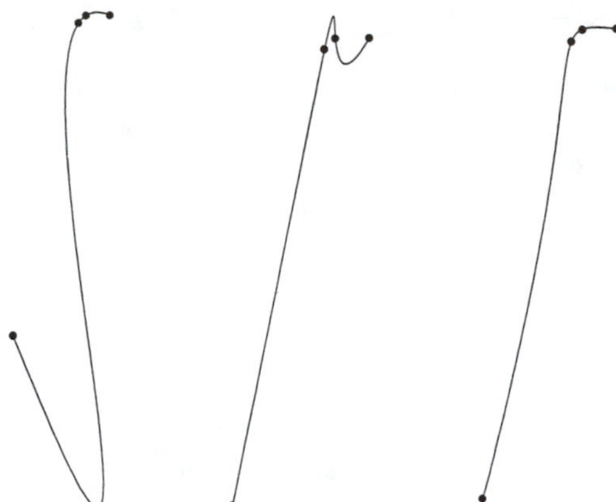

Fig. 4.36b. Interpolation of a point set with strongly varying distances between points using equidistant (left), chordal (center), and centripetal (right) parametrizations.

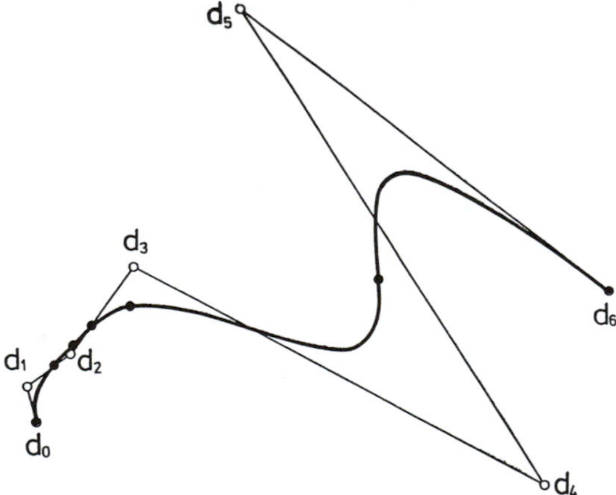

Fig. 4.36c. Interpolation of a point set using the chordal parametrization and knot vector $\boldsymbol{T} = (0, 0, 0, 0.5, 1.2, 4, 4, 4, 4)$.

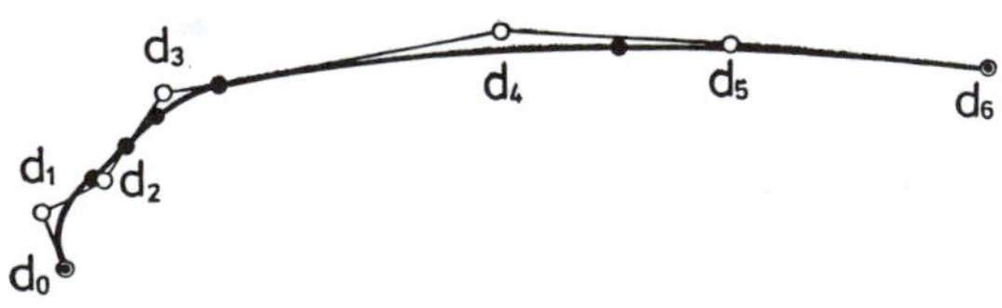

Fig. 4.36d. Interpolation of the point set in Fig. 4.36c using the knot vector $\boldsymbol{T} = (0, 0, 0, 0, 0.4, 0.8, 1.2, 4, 4, 4, 4)$.

where the $w_i > 0$ are weight functions. Inserting (4.70) in (4.71b) and differentiating with respect to the unknown de Boor points, we get the following necessary conditions for a minimum:

$$\sum_{i=0}^{n} w_i \boldsymbol{P}_i N_{m,k}(t_i) = \sum_{j=0}^{\ell} \boldsymbol{d}_j \sum_{i=0}^{n} w_i N_{jk}(t_i) N_{mk}(t_i), \qquad m = 0(1)\ell, \qquad (4.72)$$

where $\ell < n$. These $\ell+1$ linear equations for the $\ell+1$ unknown de Boor points are called the *normal equations*. The de Boor points can also be computed directly from the error equations (4.71a) using the Householder transformation, see Chap. 2. For large sets of data (where n is large in comparison with ℓ), it is recommended that the Householder transformation be used since this method is faster and numerically more stable than solving (4.72).

The situation is similar for open curves. We begin by finding the parameter values t_i in the interval I corresponding to the points $\boldsymbol{P}_i$, where $t_0 = 0$ and $t_n = n - k + 2$.

The solution of (4.71b), or equivalently (4.72), depends on the parametrization used. If we insert the parameter values t_i in the solution

$$\boldsymbol{X}(t) = \sum_{i=0}^{\ell} \boldsymbol{d}_i N_{i,k}(t),$$

then as shown in Fig. 4.37, the errors $\boldsymbol{D}_i$ do not necessarily give the *shortest distance* between the given points $\boldsymbol{P}_i$ and the approximating curve $\boldsymbol{X}(t_i)$. To minimize these distances, we need $\boldsymbol{D}_i$ to be orthogonal to the approximating curve $\boldsymbol{X}(t)$ at the points $\boldsymbol{X}(t_i)$. We can accomplish this in two ways:

- by looking for the approximating curve with the shortest distances to the given points $\boldsymbol{P}_i$ instead of minimizing the error (4.71a). This requires solving a nonlinear optimization problem;

- by iteratively *adjusting the parameter values* t_i in order to force the error vectors $\boldsymbol{D}_i$ to be perpendicular to the approximating curve $\boldsymbol{X}(t)$ at the points $\boldsymbol{X}(t_i)$.

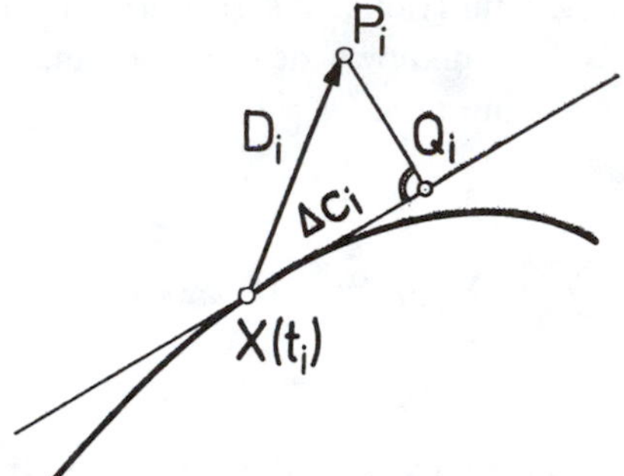

Fig. 4.37. Parameter correction.

In order to avoid nonlinear methods, we focus on the second approach. First we construct the tangent (a local linear approximation) to the (desired) curve $\boldsymbol{X}(t)$ at the point $\boldsymbol{X}(t_i)$, and project $\boldsymbol{P}_i$ onto this tangent, see Fig. 4.37. Let $\boldsymbol{Q}_i$ be the foot of the line through $\boldsymbol{P}_i$ perpendicular to the tangent. Then the distance Δc_i from the point $\boldsymbol{X}(t_i)$ to $\boldsymbol{Q}_i$ is given by (see [Hos 88])

$$\Delta c_i = (\boldsymbol{P}_i - \boldsymbol{X}(t_i)) \cdot \frac{\boldsymbol{X}'(t_i)}{|\boldsymbol{X}'(t_i)|}, \tag{4.73a}$$

and is an approximation of the amount by which the parameter must be changed in order to make the error vector $\boldsymbol{D}_i = \boldsymbol{P}_i - \boldsymbol{X}(t_i)$ be orthogonal to the curve. Assuming that the overall parameter interval is given by $\mu = \mu_n - \mu_0$ and that ℓ is the approximate length of the curve (as computed from a polygonal approximation to the curve), the size of the parameter correction at the point $\boldsymbol{X}(t_i)$ is given by

$$t_i^* = t_i + \frac{\Delta c_i}{\ell}\mu. \tag{4.73b}$$

In [Rog 89] it is proposed to locally minimize the error vector using a series expansion instead of (4.73a) and (4.73b):

$$\boldsymbol{D}_i = \boldsymbol{P}_i - \boldsymbol{X}(t_i) \approx \boldsymbol{P}_i - \boldsymbol{X}(t_i) - \dot{\boldsymbol{X}}(t_i)\Delta c_i.$$

Taking the absolute value and differentiating, we get

$$\Delta c_i = (\boldsymbol{P}_i - \boldsymbol{X}(t_i)) \cdot \frac{\boldsymbol{X}'(t_i)}{|\boldsymbol{X}'(t_i)|^2}, \tag{4.73c}$$

which leads to the parameter correction formula

$$t_i^* = t_i + \Delta c_i. \tag{4.73d}$$

Another method for minimizing the local error vector was suggested by [Pla 83], [Hos 89]. The idea is to minimize $\boldsymbol{D}^2 = (\boldsymbol{P}_i - \boldsymbol{X}(t))^2$, which after differentiation leads to

$$f := (\boldsymbol{P}_i - \boldsymbol{X}(t)) \cdot \dot{\boldsymbol{X}}(t) = 0.$$

Now if we use the well-known Newton iteration formula to compute a zero of f, we are led to the correction formula

$$\Delta c_i = -\frac{f}{\dot{f}} = \frac{-(\boldsymbol{P}_i - \boldsymbol{X}(t_i)) \cdot \dot{\boldsymbol{X}}(t_i)}{(\boldsymbol{P}_i - \boldsymbol{X}(t_i)) \cdot \ddot{\boldsymbol{X}}(t_i) - \dot{\boldsymbol{X}}(t_i)^2}. \tag{4.73e}$$

We now solve the normal equations (4.72) again, using the new parameter values. If not all error vectors are (approximately) orthogonal to the corresponding approximating curve, then we repeat the parameter transformation described by (4.73), and again solve the corresponding normal equations, etc. The iteration should be continued until all error vectors are (approximately) orthogonal to the approximating curve. This curve is then optimal in the sense that it minimizes the actual distances between the given points $\boldsymbol{P}_i$ and the approximating curve. In practice, we stop the iteration when we are within some tolerance (e.g., $\epsilon = \pm 0.1°$) of orthogonality. In general, three or four iteration steps suffice to get a good approximation to the optimal approximating curve. Continuing the iteration beyond this point can sometimes lead to a deterioration in the approximation.

Some other approaches to improving the parametrization are based on intrinsic measures of the curve such as arc length and curvature, see e.g., [Hartl 80], [Sha 82], [THO 85], [Mas 86].

A further reduction in the error can be achieved by *adjusting the knots* in the knot vector, but this requires nonlinear optimization methods, see e.g., [Hos 90, 92c].

Fig 4.38 shows an example of the approximation of a set of data using a cubic B-spline curve with nine segments. Fig. 4.38a shows the approximation with centripetal parametrization, while Fig. 4.38b illustrates the effect of parameter correction. After making the correction, the maximal error is reduced from 0.37 to 0.10, while the sum of errors is reduced from 0.6594 to 0.0409.

As pointed out in [Coh 89], the final result obtained using parameter optimization depends on the starting parametrization since, in general, only local extrema can be found.

To facilitate the solution of the normal equations, we should distribute the points to be approximated in as uniform a way as possible over the parameter intervals. In particular, if altogether we have $\ell + 1$ subintervals in

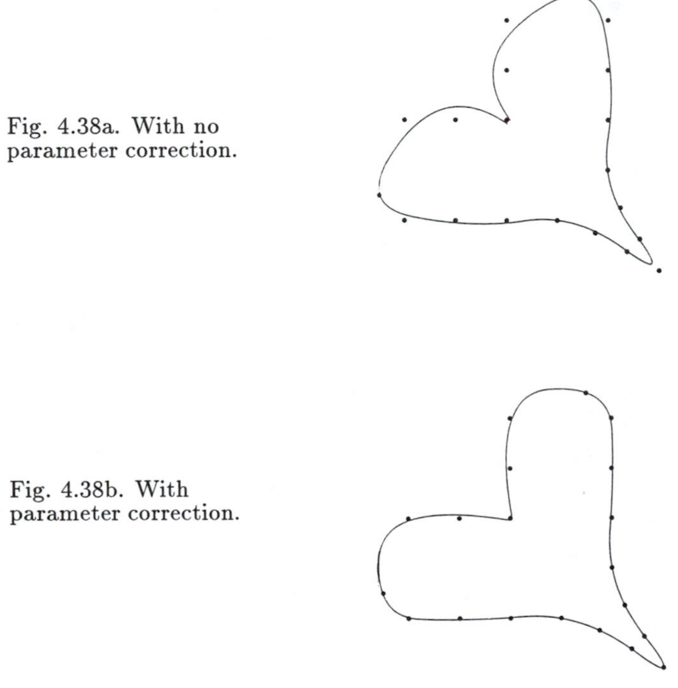

Fig. 4.38a. With no parameter correction.

Fig. 4.38b. With parameter correction.

Fig. 4.38. Approximation of data using a cubic B-spline.

the knot vector, and $n+1$ points are to be approximated, then an average of $(n+1)/(\ell+1)$ points should lie in each parameter subinterval. If some parameter subinterval contains "essentially fewer" points, then the neighboring subintervals of the knot vector should have correspondingly "more" points, since otherwise we get gaps in the system of normal equations.

We now give an example to show that, in trying to find the de Boor points, it can happen that we get an approximation problem which cannot be solved. Suppose we want to approximate with a cubic spline curve with $k=4$ and some given knot vector. Then for each subinterval, there are four de Boor points which have an effect there, so that the parametric representation of the B-spline curve has the following structure:

$$\begin{aligned}\boldsymbol{X}(u) = N_{01}(\boldsymbol{d}_0, \boldsymbol{d}_1, \boldsymbol{d}_2, \boldsymbol{d}_3) + N_{11}(\boldsymbol{d}_1, \boldsymbol{d}_2, \boldsymbol{d}_3, \boldsymbol{d}_4) + N_{21}(\boldsymbol{d}_2, \boldsymbol{d}_3, \boldsymbol{d}_4, \boldsymbol{d}_5) \\ + N_{31}(\boldsymbol{d}_3, \boldsymbol{d}_4, \boldsymbol{d}_5, \boldsymbol{d}_6) + \cdots .\end{aligned}$$

If the first and third subintervals contain exactly one point to be approximated while the second does not, then there are only two equations for the computation of the three de Boor points $\boldsymbol{d}_0, \boldsymbol{d}_1, \boldsymbol{d}_2$, and the problem cannot be solved. Such gaps can be avoided, however, by adjusting the parameter values $\bar{t}_j$ of the knot vector so that "sufficiently" many points to be approximated fall in each parameter subinterval, see Fig. 4.39. The overall length of the parameter interval can remain unchanged, see [Schm 79]. [Ferg 86] suggested modifying the norm appropriately.

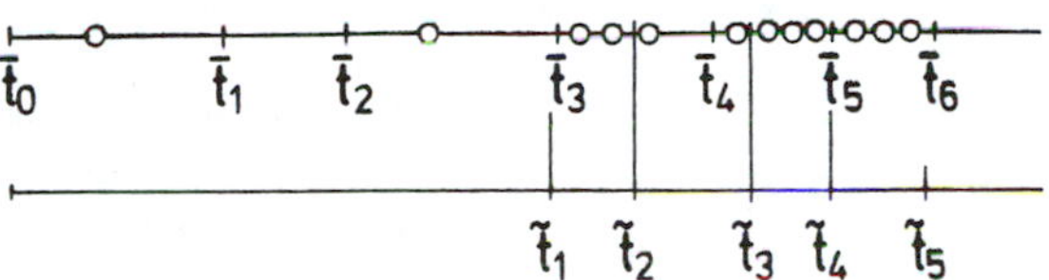

Fig. 4.39. Modifying the knot vector $\{\bar{t}_j\} \Rightarrow \{\tilde{t}_j\}$.

Another criterion for an optimal approximation is the amount of data needed to describe the approximating curve. [Lyc 87, 88] developed strategies for removing relatively unimportant knots from the knot vector. This is accomplished by assigning weights to knots, and then deciding how many knots can be removed (in order according to their weights), while staying within a prescribed tolerance. In [Wev 88, 91] an algorithm is described for removing as many knots as possible using cubic B-splines and an appropriate norm.

Recently, effective methods for interpolation and approximation with rational B-spline curves have also been developed [Schn 92]. In this case, the weight factors (which can be kept positive) and the de Boor points are used as free parameters to help reduce the error. The goal of the approximation problem is to minimize the absolute value of the error

$$\sum_{i=0}^{m} \left| \boldsymbol{P}_i - \frac{\sum_{j=0}^{n} \beta_j \boldsymbol{d}_j N_{jk}(t_i)}{\sum_{j=0}^{n} \beta_j N_{jk}(t_i)} \right|^2 =: \sum_{\ell=1}^{3} \left| \boldsymbol{P}^\ell - \boldsymbol{\Phi}(\boldsymbol{\beta}) \boldsymbol{D}^\ell \right|^2 , \tag{4.74}$$

given $m+1$ points, where here $\boldsymbol{\beta}$ stands for the vector of weight factors and

$\boldsymbol{P}^\ell = (P_0^\ell, P_1^\ell, \ldots, P_m^\ell)^T$ for the components of the given points, $\ell = 1, 2, 3$,

$\boldsymbol{D}^\ell = (d_0^\ell, d_1^\ell, \ldots, d_n^\ell)^T$ for the components of the de Boor points, $\ell = 1, 2, 3$.

Here Φ is the matrix of basis functions with entries

$$\Phi_{ij}(\beta) = \frac{\beta_j N_{jk}(t_i)}{\sum_{q=0}^{n} \beta_q N_{qk}(t_i)}, \qquad i = 0(1)m, \quad j = 0(1)n.$$

Introducing

$$\boldsymbol{D}^\ell := \Phi^+(\beta)\boldsymbol{P}^\ell, \tag{4.75}$$

where $\Phi^+(\beta)$ is the Moore-Penrose or pseudo inverse of Φ (see [CAM 79]), we can minimize the reduced error functional

$$\boldsymbol{G}(\beta) = \sum_{\ell=1}^{3} |\boldsymbol{P}^\ell - \Phi(\beta)\Phi^+(\beta)\boldsymbol{P}^\ell|^2 \tag{4.76}$$

instead of (4.74). While (4.76) is actually nonlinear in the weight factors, with the help of appropriate projectors, it can be linearized [Schn 92]. In order to get positive weight factors, we make the further substitution

$$\beta_i = \epsilon + \frac{\pi}{2} + \arctan(\alpha_i), \qquad \epsilon > 0,$$

where α_i are the new unknowns. This restricts the admissible weights to lie in the interval $[\epsilon, \pi + \epsilon]$. Then the linearized system coming from (4.76) can be solved using the iterative Gauss-Newton relaxation method. We can take $\beta_i = \epsilon + \pi/2$ as starting values. The Gauss-Newton method leads to an optimal weight vector $\boldsymbol{\beta}^*$, after which the associated de Boor points can be found from (4.75). This (initial) solution is then used to adjust the parameter values according to (4.73e), and then the whole process is repeated. It turns out that approximation with rational curves using parameter correction is even more effective than with integral curves.

4.5. Final Remarks

In this chapter we have presented the Bézier curves, the Bézier spline curves, and the B-spline curves side by side. We do not want to leave the impression, however, that these two kinds of spline curves are different: indeed geometrically, the Bézier spline curves and the B-spline curves are exactly the same curves, but expressed in terms of different bases. In Chap. 10 we show how to find the Bézier points corresponding to a B-spline curve. The advantage of the B-spline method is that fewer control points are needed in order to describe a curve than are needed for representing the same curve as a Bézier spline curve. On the other hand, the advantages of the Bézier method are that there is a "closer" relationship between the curve and its control polygon, and that the basis functions can be easily computed.

5
Geometric Spline Curves

The spline curves treated in earlier chapters were based on the requirement that neighboring segments of the splines be joined together with C^r continuity so that the first r derivatives of the segments match up at the joint. This very formal approach is natural from the viewpoint of analysis, but in some cases, leads to a very unsatisfactory model of practical smoothness, as the example in Fig. 5.1 shows. In addition, for many applications, C^r continuity proves to be *too stiff*. Interpolation of variably spaced data can turn out to be very problematic, even when using a non-uniform parametrization, for example, if the curvature varies greatly from one segment to the next. We also note that certain (surface) segment configurations cannot be constructed with C^r continuity at all (see Chap. 7). Finally, C^r continuity is not invariant with respect to reparametrization, and so it will be *destroyed* under a reparametrization. On the other hand, changing the parametrization can be very useful, for example

- in constructing an *optimal approximation* by means of iterative improvement of parameter values [Hos 88],
- in *smoothing out curvature* in a spline curve or spline surface [Schel 84],
- in dealing with the extremely important problem of *conversion* from one geometric modelling system to another [Dan 85], [Hos 87, 88b],
- in constructing an *optimal offset curve or surface approximation* using Bézier splines [Hos 88a, 88b].

These considerations suggest extending the concept of C^r continuity. In Sect. 5.1 we first consider an extension which is based on the idea of continuity of the curvatures κ_i of a curve. In Sections 5.2 and 5.3 we discuss smoothness conditions which do not depend on the space dimension. One approach, coming from differential geometry, is based on order of contact, while the second approach uses methods from the variational calculus.

Since this is still an active research area, and so far there is no universally accepted terminology, in this book we use the following

Convention: The general terminology *geometric spline curve* will be applied to all curves which satisfy some kind of smoothness conditions which are more general than C^r continuity, and can be interpreted geometrically.

This will be the case for all spline curves to be discussed in this chapter. For reasons to be explained in Sect. 5.2, we reserve the term GC^r *continuity* for the case of spline curves defined in terms of order of contact.

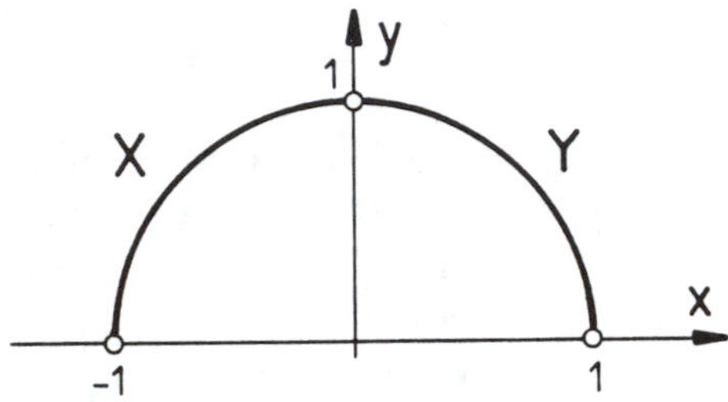

Fig. 5.1. $\boldsymbol{X} = (-\cos(\pi u^2/2), \sin(\pi u^2/2))^T$, $\boldsymbol{Y} = (\sin(\pi t^2/2), \cos(\pi t^2/2))^T$, $u, t \in [0,1]$. $\boldsymbol{X}$ and $\boldsymbol{Y}$ are curvature continuous at $\boldsymbol{P} = (0,1)$ but not C^2.

5.1. FC^r Continuous Spline Curves

Suppose two parametric curves $\boldsymbol{X}(u)$, $u \in [u_0, u_1]$, and $\boldsymbol{Y}(t)$, $t \in [t_0, t_1]$, lying in $\mathbb{R}^d$ join continuously at a regular point $\boldsymbol{P} = \boldsymbol{X}(u_1) = \boldsymbol{Y}(t_0)$ (*i.e.*, $|\boldsymbol{X}'| \neq 0$, $|\boldsymbol{Y}'| \neq 0$ at $\boldsymbol{P}$). Then we say that they join with C^1 continuity provided that at $\boldsymbol{P}$ they have the same derivatives, and thus the same tangents. On the other hand, the *weaker*, but more geometric condition that the tangents to the curves have the same direction (but possibly different lengths) is already satisfied when the derivative vectors point in the same direction. Thus, using "+" for evaluation from the right and "−" for evaluation from the left, the condition for *tangent continuity* can be written as

$$\boldsymbol{Y}'(t_0^+) = \omega_{11} \boldsymbol{X}'(u_1^-), \tag{5.1}$$

where $\omega_{11} > 0$ in view of the regularity assumption. The more geometric requirement of common tangents thus leads to an additional degree of freedom, *i.e.*, to an extra *design parameter*. Since the tangents are independent of the actual parametrization used, this concept of smoothness is also parametrically invariant.

For the usual C^2 and C^3 continuous curves, we also require that the second and third derivative vectors agree at $\boldsymbol{P}$, respectively. By (2.2) and

(2.3), the second and third derivatives involve the geometric invariants κ and τ [STR 50]. Thus, in the case of a continuous Frenet frame, the additional requirement that the curvature or curvature and torsion match up at join points is clearly a geometrically reasonable modification of the C^2 and C^3 continuity conditions, respectively.

From the equations for computing the curvature and the torsion of a curve, we see that provided (5.1) holds, *curvature continuity* is equivalent to

$$\boldsymbol{Y}''(t_0^+) = \omega_{11}^2 \boldsymbol{X}''(u_1^-) + \omega_{12} \boldsymbol{X}'(u_1^-), \tag{5.2}$$

while if both (5.1) and (5.2) hold, then *torsion continuity* is equivalent to

$$\boldsymbol{Y}'''(t_0^+) = \omega_{11}^3 \boldsymbol{X}'''(u_1^-) + \omega_{13} \boldsymbol{X}''(u_1^-) + \omega_{23} \boldsymbol{X}'(u_1^-), \tag{5.3}$$

see e.g., [Dyn 85], [Hag 86]. In order to avoid division by zero, we also have to assume that the derivative vectors are linearly independent! Equation (5.2) can be interpreted physically as follows:

> *The addition of a multiple $\omega_{12}\boldsymbol{X}'$ of the velocity vector (pointing in the direction of the tangent) does not cause a change in a direction perpendicular to the curve, i.e., it does not change the curvature of the curve.*

Equation (5.3) has an analogous interpretation.

This concept of continuity can be extended to higher order for curves in $\mathbb{R}^d$ only for $d \geq r > 3$. For a curve in $\mathbb{R}^d$, the *Frenet-Serret formulae* [SPI 79] are

$$\begin{array}{lcll}
\dot{\boldsymbol{e}}_1 & = & & \kappa_1 \boldsymbol{e}_2 \\
\dot{\boldsymbol{e}}_2 & = & -\kappa_1 \boldsymbol{e}_1 & + \; \kappa_2 \boldsymbol{e}_3 \\
\dot{\boldsymbol{e}}_3 & = & -\kappa_2 \boldsymbol{e}_2 & + \; \kappa_3 \boldsymbol{e}_4 \\
 & \cdot & & \\
 & \cdot & & \\
 & \cdot & & \\
\dot{\boldsymbol{e}}_{d-1} & = & -\kappa_{d-2}\, \boldsymbol{e}_{d-2} & + \; \kappa_{d-1} \boldsymbol{e}_d \\
\dot{\boldsymbol{e}}_d & = & -\kappa_{d-1}\, \boldsymbol{e}_{d-1}, &
\end{array}$$

where the $\boldsymbol{e}_i$ denote the *Frenet basis vectors* of the curve in terms of the arc length parametrization, and the *dot* denotes the derivative with respect to arc length. The coefficients κ_i appearing in the Frenet-Serret formulae are independent of the parametrization, and are called *curvatures*. They uniquely determine the curve up to its orientation and position, see [KLI 78], [SPI 79], or [Hag 86]; for the case of $\mathbb{R}^3$ see [STR 50].

In analogy with (5.2) and (5.3), it can be shown (cf. e.g. [Dyn 85], [Las 88a] and [Gre 89]) that two curves $\boldsymbol{X}(u)$, $u \in [u_0, u_1]$, and $\boldsymbol{Y}(t)$, $t \in [t_0, t_1]$, in $\mathbb{R}^d$ meeting at a regular point $\boldsymbol{P} = \boldsymbol{X}(u_1) = \boldsymbol{Y}(t_0)$ with linearly independent derivatives $\boldsymbol{X}^{(\rho)}$, $\boldsymbol{Y}^{(\rho)}$, $\rho = 1(1)r\text{–}1$, exhibit continuity of the first $r-1$ curvatures κ_i (with $r \leq d$) and the Frenet basis if and only if

$$\begin{aligned}
\boldsymbol{Y}(t_0^+) &= \boldsymbol{X}(u_1^-) \\
\boldsymbol{Y}'(t_0^+) &= \omega_{11}\boldsymbol{X}'(u_1^-) \\
\boldsymbol{Y}''(t_0^+) &= \omega_{11}^2\boldsymbol{X}''(u_1^-) + \omega_{12}\boldsymbol{X}'(u_1^-) \\
\boldsymbol{Y}'''(t_0^+) &= \omega_{11}^3\boldsymbol{X}'''(u_1^-) + \omega_{13}\boldsymbol{X}''(u_1^-) + \omega_{23}\boldsymbol{X}'(u_1^-) \\
\boldsymbol{Y}^{(4)}(t_0^+) &= \omega_{11}^4\boldsymbol{X}^{(iv)}(u_1^-) + \omega_{14}\boldsymbol{X}'''(u_1^-) + \omega_{24}\boldsymbol{X}''(u_1^-) + \omega_{34}\boldsymbol{X}'(u_1^-) \\
&\;\vdots \\
\boldsymbol{Y}^{(r)}(t_0^+) &= \omega_{11}^r\boldsymbol{X}^{(r)}(u_1^-) + \omega_{1r}\boldsymbol{X}^{(r-1)}(u_1^-) + \cdots + \omega_{r-1,r}\boldsymbol{X}'(u_1^-).
\end{aligned} \tag{5.4}$$

We refer to this as *Frenet C^r continuity* (FC^r or F^r continuity for short). In $\mathbb{R}^2$ and $\mathbb{R}^3$, κ_1 reduces to ordinary curvature κ, and κ_2 reduces to the torsion τ. The continuity conditions (5.4) can be written in matrix form as

$$Y_+ = AX_-,$$

where A is a lower triangular matrix which is referred to in the literature as the *connection matrix*.

Dyn, Edelman and Micchelli [Dyn 85, 85a, 87] have shown that results on the existence of local basis functions for geometric spline curves can be obtained by studying the connection matrix. In particular, they showed:

> *If the connection matrix A is totally positive, then there exists a basis similar to the B-splines consisting of locally supported non-negative elements adding up to one.*

In view of the geometric interpretation of the curvatures κ_i, and their invariance under parametric transformation, in the literature spline curves satisfying (5.4) are sometimes called *geometric C^r continuous* (GC^r or G^r for short), contrary to the convention we have adopted, see e.g. [Dyn 85, 85a], [Hag 86], [Boeh 87, 87a], [Las 88a], [Pot 89].

The generalization of C^r continuity given above depends on the space dimension in the sense that for a plane curve, all curvatures κ_i with $i > 1$ are identically zero, for a curve in $\mathbb{R}^3$, all curvatures κ_i with $i > 2$ are identically zero, etc. This means that the complete set of design parameters in (5.4) are in fact only available for curves in $\mathbb{R}^d$ with $d \geq r$. Thus, in $\mathbb{R}^2$ and $\mathbb{R}^3$, only curvature continuity and torsion continuity, respectively, can really be fully exploited.

Referring to the smoothness conditions (5.4), we see that the set of FC^r continuous curves contains two subsets of geometric spline curves of special practical interest:

- geometric spline curves whose smoothness is defined in terms of order of contact; we shall refer to them as GC^r continuous,
- geometric spline curves which satisfy a minimization condition similar to the one for C^r continuous polynomials (see Sect. 3.6.2).

The interest in these curves, which will be discussed in more detail below, is in part due to the fact that both types of continuity are independent of the space dimension. Thus, for example, in $\mathbb{R}^2$ we can easily construct GC^r continuous curves with $r > 3$.

Remark. The curves discussed in Sects. 5.1, 5.2 and 5.3 have been used in [Pot 89] along with generalized tangent surfaces to define the even more abstract so-called TC^r *continuous curves*. These curves have continuous and continuously differentiable elliptic (conic) curvatures, which can again be expressed as quotients of Euclidean curvatures as κ_2/κ_1, κ_3/κ_1, etc.

5.2. $\mathrm{GC^r}$ Continuous Spline Curves

Suppose two curves $\boldsymbol{X}(u)$, $u \in [u_0, u_1]$, and $\boldsymbol{Y}(t)$, $t \in [t_0, t_1]$, in $\mathbb{R}^d$ meet at a regular point $\boldsymbol{P} = \boldsymbol{X}(u_1) = \boldsymbol{Y}(t_0)$. We say the two curves meet with *geometric C^r continuity* (GC^r or G^r for short), if there exists an algebraic curve which meets both curves $\boldsymbol{X}(u)$ and $\boldsymbol{Y}(t)$ at $\boldsymbol{P}$ with *contact of order* r in the sense that the first r terms in the Taylor series expansions about the point $\boldsymbol{P}$ of the two given curves and the algebraic curve all agree at $\boldsymbol{P}$.

In order to be able to compare the Taylor series of $\boldsymbol{X}(u)$ and $\boldsymbol{Y}(t)$ with each other, we first have to make sure they are developed with respect to the same parameter, so that we can find the derivatives with respect to the same parameter. For this, it suffices to reparametrize one of the curves using an admissible, orientation-preserving transformation. For example, we reparametrize $\boldsymbol{X}(u)$ via $u \mapsto u(t)$ with $\omega_0 \equiv u(t_0^-) = u_1^-$. Thus, $\boldsymbol{X}(u)$ and $\boldsymbol{Y}(t)$ join with GC^r continuity means that

$$\frac{d^\rho}{dt^\rho}\boldsymbol{Y}(t)\Big|_{t_0^+} = \frac{d^\rho}{dt^\rho}\boldsymbol{X}(u(t))\Big|_{u_1^- = u(t_0^-)}, \qquad \rho = 0(1)r, \tag{5.5}$$

i.e., $\boldsymbol{X}(u)$ and $\boldsymbol{Y}(t)$ join with C^r continuity after a reparametrization.

In particular, a GC^1 continuous join can be converted to a C^1 join by a linear transformation, and a GC^2 continuous join can be converted to a C^2 continuous join by a quadratic parameter transformation, see [Vero 76], [Herr 87], [Boeh 88b], and [Fri 86], [Nie 86] for $r = 1$. For the purposes of CAGD, this is important since a linear parameter transformation does not change the polynomial degree of a non-rational curve.

If $\boldsymbol{X}$ and $\boldsymbol{Y}$ join with C^r continuity with respect to a global parameter t, then they also join with C^r continuity with respect to the natural arc length parametrization s (see [SCHEF 01], [Dero 85], [Bars 84a, 88a, 89]), *i.e.*,

$$\frac{d^\rho}{ds^\rho}\boldsymbol{Y}(s)\Big|_{s_0^+=s(t_0^+)} = \frac{d^\rho}{ds^\rho}\boldsymbol{X}(s)\Big|_{s_0^-=s(t_0^-)} \qquad \rho = 0(1)r,$$

and conversely, as can easily be seen with the help of the chain rule. Thus GC^r continuity is sometimes referred to as *arc length continuity* in the literature.

In carrying out the differentiation on the right-hand side of (5.5), we need to use the chain rule. For $r = 1, 2$, for example, we get

$$\frac{d}{dt}\boldsymbol{Y}(t)\Big|_{t_0^+} = \left(\frac{d}{dt}u(t)\Big|_{t_0^-}\right)\left(\frac{d}{du}\boldsymbol{X}(u)\Big|_{u_1^-}\right),$$

and

$$\frac{d^2}{dt^2}\boldsymbol{Y}(t)\Big|_{t_0^+} = \left(\frac{d}{dt}u(t)\Big|_{t_0^-}\right)^2\left(\frac{d^2}{du^2}\boldsymbol{X}(u)\Big|_{u_1^-}\right) + \left(\frac{d^2}{dt^2}u(t)\Big|_{t_0^-}\right)\left(\frac{d}{du}\boldsymbol{X}(u)\Big|_{u_1^-}\right),$$

which we can write as

$$\boldsymbol{Y}'(t_0^+) = \omega_1 \boldsymbol{X}'(u_1^-) \tag{5.6}$$

$$\boldsymbol{Y}''(t_0^+) = \omega_1^2 \boldsymbol{X}''(u_1^-) + \omega_2 \boldsymbol{X}'(u_1^-) \tag{5.7}$$

in terms of the abbreviations

$$\omega_\rho = \frac{d^\rho}{dt^\rho}u(t)\Big|_{t=t_0^-}, \tag{5.8}$$

where because of the regularity condition, $\omega_1 > 0$.

This means that if the two curves join with GC^2 continuity, then the equations (5.6) and (5.7) must be satisfied, where the variables ω_ρ are given by (5.8).

On the other hand, given two curves $\boldsymbol{X}(u)$ and $\boldsymbol{Y}(t)$ meeting at a point, if there exist real numbers ω_0, ω_1, ω_2 such that (5.6) and (5.7) hold, then we can reparametrize to a common parameter in such a way that $\boldsymbol{X}(u)$ and $\boldsymbol{Y}(t)$

join with C^2 continuity with respect to this global parameter, and we have GC^2 continuity.

In this case, the parameter transformation would have to be chosen so that the boundary conditions in (5.8) are satisfied, e.g. by using a Taylor series expansion.

Equations for contact of higher order follow analogously, and were first presented explicitly by Geise [Gei 62]. For example, for $r = 3$ we have

$$\boldsymbol{Y}'''(t_0^+) = \omega_1^3 \boldsymbol{X}'''(u_1^-) + 3\omega_1\omega_2 \boldsymbol{X}''(u_1^-) + \omega_3 \boldsymbol{X}'(u_1^-), \tag{5.9}$$

and for $r = 4$,

$$\begin{aligned} \boldsymbol{Y}^{(4)}(t_0^+) = \omega_1^4 \boldsymbol{X}^{(4)}(u_1^-) &+ 6\omega_1^2\omega_2 \boldsymbol{X}'''(u_1^-) \\ &+ (3\omega_2^2 + 4\omega_1\omega_3)\boldsymbol{X}''(u_1^-) + \omega_4 \boldsymbol{X}'(u_1^-), \end{aligned} \tag{5.10}$$

see [Cohe 82], [Goo 85a], [Bars 84a], [Dero 85].

The following recurrence formula was given in [Hos 88c]:

$$\boldsymbol{Y}^{(r)}(t_0^+) = \sum_{i=1}^{r} \alpha_{ri} \boldsymbol{X}^{(i)}(u_1^-),$$

with $\alpha_{11} = \omega_1$, $\alpha_{j0} = 0$, $\alpha_{jk} = 0$ for $j < k$, and

$$\alpha_{jk} = \omega_1 \alpha_{j-1,k-1} + E\alpha_{j-1,k},$$

where the operator E is defined by

$$E\omega_j = \omega_{j+1}, \qquad E(\omega_j\omega_k) = \omega_{j+1}\omega_k + \omega_j\omega_{k+1},$$

cf. e.g., [Goo 85a].

Here we remark that since the *visual smoothness* of a GC^r continuous curve corresponds to that of a C^r continuous curve, GC^r continuous curves are also frequently referred to as *visually C^r continuous* (VC^r or V^r continuous for short), see e.g., [Far 82, 82a, 85], [Boeh 88b], [Herr 87], [Las 88a, 90a], [Pot 89].

If we compare the continuity conditions of GC^r continuity (which in the form (5.6) – (5.7) are parametrically invariant) with the continuity conditions given in (5.4), then it is obvious that they are equivalent for $r \leq 2$. Equation (5.6) stands for tangent continuity, while (5.7) stands for curvature continuity. On the other hand, GC^r continuity for $r > 2$ is more restrictive in the sense of matching conditions: GC^r continuity implies the continuity of the curvatures

κ_i, but the converse is not true. For example, torsion continuous curves $\boldsymbol{X}$ and $\boldsymbol{Y}$ must satisfy

$$\omega_{13} = 3\omega_{11}\omega_{12}$$

at $\boldsymbol{P} = \boldsymbol{X}(u_1) = \boldsymbol{Y}(t_0)$ in order to meet with third order contact.

Although in a direct comparison, GC^r continuity is more restrictive, it is nevertheless more useful since is does not depend on the space dimension (the definition of visual continuity does not depend on space dimension, *i.e.*, r does not depend on d). Thus, for example, we can construct planar quintic spline curves with GC^4 continuity at the knots. This means that in addition to geometric spline curves of higher polynomial degree, we also can work with spline curves with a higher degree of geometric smoothness at their joins, since in $\mathbb{R}^2$, (5.4) permits at most curvature continuous (cubic) spline curves.

The concept of order of contact can be given a differential geometric interpretation, which in turn leads to a further distinction from the curves defined by (5.4). Using the Frenet-Serret formulae, the derivatives of $\boldsymbol{X}$ and $\boldsymbol{Y}$ can be easily computed. It follows by an application of the product rule [STR 50] that

$$\begin{aligned}
\dot{\boldsymbol{X}} &= \boldsymbol{e}_1 \\
\ddot{\boldsymbol{X}} &= \kappa_1 \boldsymbol{e}_2 \\
\dddot{\boldsymbol{X}} &= -\kappa_1^2 \boldsymbol{e}_1 + \dot{\kappa}_1 \boldsymbol{e}_2 + \kappa_1 \kappa_2 \boldsymbol{e}_3 \\
\ddddot{\boldsymbol{X}} &= -3\kappa_1 \dot{\kappa}_1 \boldsymbol{e}_1 + (\ddot{\kappa}_1 - \kappa_1^3 - \kappa_1 \kappa_2^2)\boldsymbol{e}_2 + (2\dot{\kappa}_1 \kappa_2 + \kappa_1 \dot{\kappa}_2)\boldsymbol{e}_3 + \kappa_1 \kappa_2 \kappa_3 \boldsymbol{e}_4,
\end{aligned}$$

etc., and correspondingly for $\boldsymbol{Y}$. Thus, the ρ-th derivatives of $\boldsymbol{X}$ and $\boldsymbol{Y}$ (with $\rho > 2$) with respect to s can be found from the curvatures κ_k, $1 \le k < \rho$, and the derivatives $\kappa_k^{(\sigma)}$, $1 \le \sigma \le \rho - 1 - k$, of the curvatures κ_k, $1 \le k < \rho - 2$. Since GC^r continuity means C^r continuity with respect to the arc length parametrization, this leads to the following geometric interpretation of GC^r continuity:

- for plane curves, GC^r continuity is equivalent to the continuity of $\boldsymbol{e}_1$ and $\boldsymbol{e}_2$, the tangent and normal vectors, as well as κ and its derivatives $\kappa^{(\rho)}$, $1 \le \rho < r - 2$;
- for curves in $\mathbb{R}^3$, GC^r continuity is equivalent to the continuity of $\boldsymbol{e}_1$, $\boldsymbol{e}_2$ and $\boldsymbol{e}_3$, the Frenet frame, and the κ, $\kappa^{(\rho)}$, $1 \le \rho < r - 2$, τ and $\tau^{(\rho)}$, $1 \le \rho < r - 3$ [Pot 88].

On the other hand, the smoothness conditions (5.4) assure that in the case $r = 3$, for example, since $\omega_{13} \neq 3\omega_{11}\omega_{12}$ in general, only the curvatures κ_ρ, $1 \le \rho < r$ are continuous, but not their derivatives.

Thus, the differential geometric invariants κ_i of FC^r continuous curves are only C^0 continuous. This means that in terms of differential geometric variables, the concept of FC^r continuity is more restrictive than the concept of GC^r continuity, which leaves more geometric variables invariant [Pot 88], [Deg 88].

Finally, we remark that we have made no assumptions about the representation of the curves $\boldsymbol{X}(u)$, $\boldsymbol{Y}(t)$, *i.e.*, $\boldsymbol{X}(u)$ and $\boldsymbol{Y}(t)$ can just as well be written in terms of polynomial or non-polynomial basis systems. In the following, we shall focus our discussion on the polynomial representation, since non-polynomial geometric spline curves were already discussed in Sect. 3.6.1, Remark 6 in Sect. 3.6.2, and in Sect. 3.7.

5.3. Geometric Spline Curves with Minimum Norm Property

In Sect. 3.6.2 we considered spline curves (3.50) which minimized certain functionals. These splines satisfy continuity conditions (3.51) which can be given a very abstract geometric interpretation [Pot 89]. For $p_k = p$, the splines in (3.50)–(3.51) reduce to Hagen's geometric spline curves [Hag 85] with *tension parameters* $\nu_{k,\ell}/p$. For $L = 3$ these become torsion continuous quintic C^2 splines (τ-splines), and for $L = 2$, curvature continuous cubic C^1 splines (ν-splines) [Nie 74, 86]. It should be noted that τ-splines are very special torsion continuous curves which, except for the C^3 continuous quintic subsplines, in general are different from the GC^3 continuous quintic subsplines. In general, GC^3 continuous curves do not satisfy a minimization property as do the τ-splines.

Hagen's geometric spline curves for $L > 2$ ($r > 2$) with nonvanishing $\nu_{k,\ell}$ are a subset of the geometric splines defined in (5.4). The intersection of this set with the set of GC^r continuous splines contains only the C^r continuous splines, as can be seen immediately by comparing the continuity conditions. A *general* GC^r continuous spline does not minimize the functional (3.50).

Generalizations of ν-splines [Nie 74] and τ-splines [Hag 85] in the form of interval-weighted ν- and τ-splines can be found in [Fol 87], [Las 92], [Neu 92], and [Gre 91]. A generalization of the interval-weighted cubic splines of [Salk 74] is provided by the Q-splines of [Kul 91]. GC^3 and GC^4 continuous splines in tension are discussed in [Pot 90], see also Sect. 3.6.2.

Tangent, curvature, and torsion continuous smoothness conditions will be discussed below in several examples, where the curves are given in Bézier form.

5.4. Tangent Continuous Spline Curves

In this section we derive conditions on the coefficients of a spline curve written in Bézier form which assure that it satisfies the continuity conditions (5.1). Suppose

$$\boldsymbol{X}(u) = \sum_{k=0}^{n} \boldsymbol{b}_k B_k^n(\mu), \qquad u = (1-\mu)u_0 + \mu u_1, \quad \mu \in [0,1], \tag{5.11}$$

$$\boldsymbol{Y}(t) = \sum_{k=0}^{\bar{n}} \bar{\boldsymbol{b}}_k B_k^{\bar{n}}(\bar{\mu}), \qquad t = (1-\bar{\mu})t_0 + \bar{\mu} t_1, \quad \bar{\mu} \in [0,1]. \tag{5.12}$$

Then differentiating and comparing coefficients leads to the conditions

$$\bar{\boldsymbol{b}}_0 = \boldsymbol{b}_n, \tag{5.13}$$

$$(1 + qN_1\omega_{11})\boldsymbol{b}_n = qN_1\omega_{11}\boldsymbol{b}_{n-1} + \bar{\boldsymbol{b}}_1, \tag{5.14}$$

with

$$q = \frac{\Delta t}{\Delta u}, \qquad \Delta u = u_1 - u_0, \quad \Delta t = t_1 - t_0, \qquad N_1 = \frac{n}{\bar{n}}.$$

This means that $\bar{\boldsymbol{b}}_0 = \boldsymbol{b}_n$ divides the line segment $\boldsymbol{b}_{n-1}\bar{\boldsymbol{b}}_1$ in the ratio $1 : qN_1\omega_{11}$. Then in terms of the distance $a_1 = |\Delta \boldsymbol{b}_{n-1}| = |\boldsymbol{b}_n - \boldsymbol{b}_{n-1}|$ between $\boldsymbol{b}_{n-1}$ and $\boldsymbol{b}_n$, and the distance $\bar{a}_1 = |\Delta\bar{\boldsymbol{b}}_0| = |\bar{\boldsymbol{b}}_1 - \bar{\boldsymbol{b}}_0|$ between $\bar{\boldsymbol{b}}_0$ and $\bar{\boldsymbol{b}}_1$, we have (Fig. 5.2)

$$a_1 : \bar{a}_1 = 1 : qN_1\omega_{11}.$$

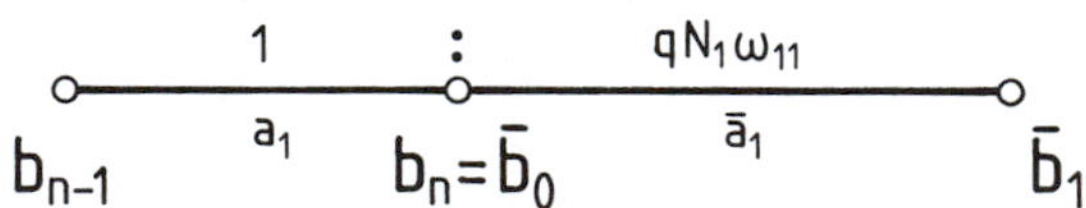

Fig. 5.2. Bézier polygon for a tangent continuous join.

Assuming that $\boldsymbol{b}_n$ and $\boldsymbol{b}_{n-1}$ are given, the "next" Bézier point $\bar{\boldsymbol{b}}_1$ is then

$$\bar{\boldsymbol{b}}_1 = \boldsymbol{b}_n + qN_1\omega_{11}\Delta\boldsymbol{b}_{n-1}, \tag{5.15}$$

where ω_{11} is a free *design parameter*. Thus, the Bézier point $\bar{\boldsymbol{b}}_1$ lies on the tangent line defined by $\boldsymbol{b}_n$ and $\boldsymbol{b}_{n-1}$ as required by (5.1). Moreover, since

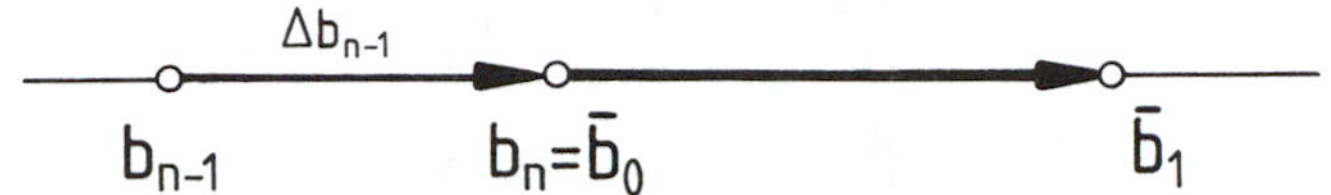

Fig. 5.3. Geometric interpretation of (5.15).

$q > 0$, $N_1 > 0$, and $\omega_{11} \geq 0$, $\boldsymbol{b}_{n-1}$ and $\bar{\boldsymbol{b}}_1$ lie on opposite sides of $\bar{\boldsymbol{b}}_0 = \boldsymbol{b}_n$, see Fig. 5.3.

Tangent continuous quadratic spline curves are treated in [Bars 84a] and [Dero 88]; see also [Bars 90]. Tangent continuous cubic subsplines can be found in [Bars 84a], [Dero 88], and [Hos 87, 88b].

5.5. Curvature Continuous Spline Curves

5.5.1. Bézier Representation of Curvature Continuous Splines

If we insert the Bézier representations (5.11) and (5.12) in (5.4) with $r = 2$, differentiate and compare coefficients, we find that the Bézier points $\boldsymbol{b}_{n-2}$ and $\bar{\boldsymbol{b}}_2$ of the two neighboring Bézier curves must satisfy (5.13), (5.14), and

$$\begin{aligned} (1 + qN_2\omega_{11}^2\gamma)\boldsymbol{b}_{n-1} &= qN_2\omega_{11}^2\gamma\boldsymbol{b}_{n-2} + \boldsymbol{s} \\ (\gamma + q)\bar{\boldsymbol{b}}_1 &= q\boldsymbol{s} + \gamma\bar{\boldsymbol{b}}_2, \end{aligned} \tag{5.16}$$

see Fig. 5.4 and compare with Fig. 4.12. Here $\boldsymbol{s}$ denotes an auxiliary point,

$$\gamma = \frac{1 + qN_1\omega_{11}}{N_1\omega_{11}(1 + q\frac{N_2}{N_1}\omega_{11}) + N_1\frac{\Delta t}{\bar{n}-1}\omega_{12}} \tag{5.17}$$

is a *design parameter*, and $N_2 = n(n-1)/\bar{n}(\bar{n}-1)$. This shows that $\boldsymbol{b}_{n-1}$ divides the line segment $\boldsymbol{b}_{n-2}\boldsymbol{s}$ in the ratio $1 : qN_2\omega_{11}^2\gamma$, while $\bar{\boldsymbol{b}}_1$ divides the segment $\boldsymbol{s}\bar{\boldsymbol{b}}_2$ in the ratio $\gamma : q$, (see Fig. 5.4). Writing a_2 and $\bar{a}_2$ for the distances of $\boldsymbol{b}_{n-2}$ and $\bar{\boldsymbol{b}}_2$ from the tangent $\boldsymbol{t}$ at $\bar{\boldsymbol{b}}_0 = \boldsymbol{b}_n$, respectively, it follows from the similarity of the two triangles shown in Fig. 5.4 that

$$a_2 : \bar{a}_2 = 1 : q^2N_2\omega_{11}^2.$$

We assume that the Bézier points $\boldsymbol{b}_n$, $\boldsymbol{b}_{n-1}$ and $\boldsymbol{b}_{n-2}$ are given, and that $\bar{\boldsymbol{b}}_0$ and $\bar{\boldsymbol{b}}_1$ have been determined from (5.13) and (5.14). Then $\bar{\boldsymbol{b}}_2$ can be found with the help of the construction point $\boldsymbol{s}$ from (5.16), or also directly from

$$\bar{\boldsymbol{b}}_2 = \boldsymbol{b}_n - q^2N_2\omega_{11}^2\Delta\boldsymbol{b}_{n-2} + \mu\Delta\boldsymbol{b}_{n-1}, \tag{5.18}$$

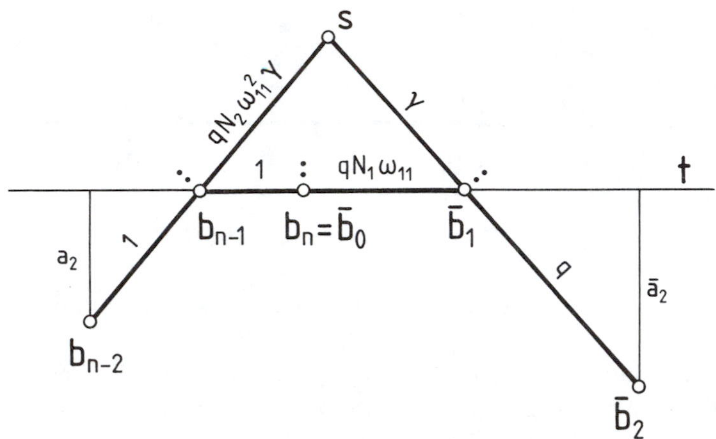

Fig. 5.4. Bézier polygon for a curvature continuous join.

with $\mu = \mu(\omega_{12})$. Since ω_{11} and ω_{12} are free *design parameters*, at a curvature continuous join, the Bézier point $\bar{\boldsymbol{b}}_2$ lies on a straight line $\boldsymbol{g}$ parallel to the tangent to the curve at $\boldsymbol{b}_n = \bar{\boldsymbol{b}}_0$; see [Kah 82], [Hag 86], and Fig. 5.5. The distance between $\boldsymbol{g}$ and the tangent is determined by the (previously chosen) parameter ω_{11}, while the actual location of $\bar{\boldsymbol{b}}_2$ is determined by ω_{12}.

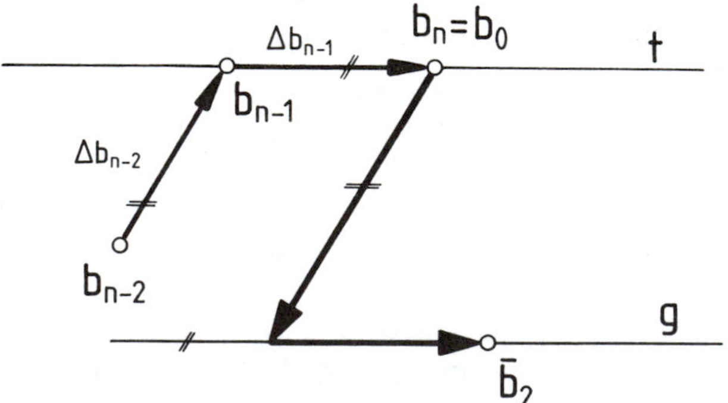

Fig. 5.5. Geometric interpretation of (5.18).

It follows from (5.14), (5.16) and (5.18) that the entire construction (and thus also $\bar{\boldsymbol{b}}_2$ and the straight line $\boldsymbol{g}$) lies in the osculating plane π associated with $\boldsymbol{X}$ and $\boldsymbol{Y}$ at $\boldsymbol{P} = \boldsymbol{X}(u_1) = \boldsymbol{Y}(t_0)$ which is determined by $\Delta\boldsymbol{b}_{n-1}$ and

$\Delta \boldsymbol{b}_{n-2}$. Moreover, because of the equivalence of (5.1)–(5.2) with

$$\frac{|\boldsymbol{Y}'' \times \boldsymbol{Y}'|}{\|\boldsymbol{Y}'\|^3} = \frac{|\boldsymbol{X}'' \times \boldsymbol{X}'|}{\|\boldsymbol{X}'\|^3}$$

for a continuous Frenet frame, inserting the Bézier representations of $\boldsymbol{X}$ and $\boldsymbol{Y}$, we get

$$\frac{|\Delta^2 \bar{\boldsymbol{b}}_0 \times \Delta \bar{\boldsymbol{b}}_0|}{\|\Delta \bar{\boldsymbol{b}}_0\|^3} = \frac{\bar{n}(n-1)}{n(\bar{n}-1)} \frac{|\Delta^2 \boldsymbol{b}_{n-2} \times \Delta \boldsymbol{b}_{n-1}|}{\|\Delta \boldsymbol{b}_{n-1}\|^3},$$

or equivalently

$$\frac{|\Delta \bar{\boldsymbol{b}}_0 \times \Delta \bar{\boldsymbol{b}}_1|}{|\bar{a}_1|^3} = \frac{\bar{n}(n-1)}{n(\bar{n}-1)} \frac{|\Delta \boldsymbol{b}_{n-2} \times \Delta \boldsymbol{b}_{n-1}|}{|a_1|^3},$$

which implies that $\boldsymbol{b}_{n-2}, \ldots, \bar{\boldsymbol{b}}_2$ not only lie in a plane, but in fact are the vertices of a convex polygon. This means that $\boldsymbol{b}_{n-2}$ and $\bar{\boldsymbol{b}}_2$ lie on the same side of the tangent $\boldsymbol{t}$, as can also be seen from (5.18). The surface areas of the triangles $\bar{\boldsymbol{b}}_0, \bar{\boldsymbol{b}}_1, \bar{\boldsymbol{b}}_2$ and $\boldsymbol{b}_{n-2}, \boldsymbol{b}_{n-1}, \boldsymbol{b}_n$ are given by $|\Delta \bar{\boldsymbol{b}}_0 \times \Delta \bar{\boldsymbol{b}}_1|$ and $|\Delta \boldsymbol{b}_{n-2} \times \Delta \boldsymbol{b}_{n-1}|$ (up to a constant factor). It follows that the relation

$$\frac{\text{area}\{\bar{\boldsymbol{b}}_0, \bar{\boldsymbol{b}}_1, \bar{\boldsymbol{b}}_2\}}{\bar{a}_1^3} = \frac{\bar{n}(n-1)}{n(\bar{n}-1)} \frac{\text{area}\{\boldsymbol{b}_n, \boldsymbol{b}_{n-1}, \boldsymbol{b}_{n-2}\}}{a_1^3},$$

between the two triangular areas, area$\{\bar{\boldsymbol{b}}_0, \bar{\boldsymbol{b}}_1, \bar{\boldsymbol{b}}_2\}$ and area$\{\boldsymbol{b}_n, \boldsymbol{b}_{n-1}, \boldsymbol{b}_{n-2}\}$, is equivalent to the curvature continuity of $\boldsymbol{X}$ and $\boldsymbol{Y}$, see [Far 82a] and Fig. 5.6.

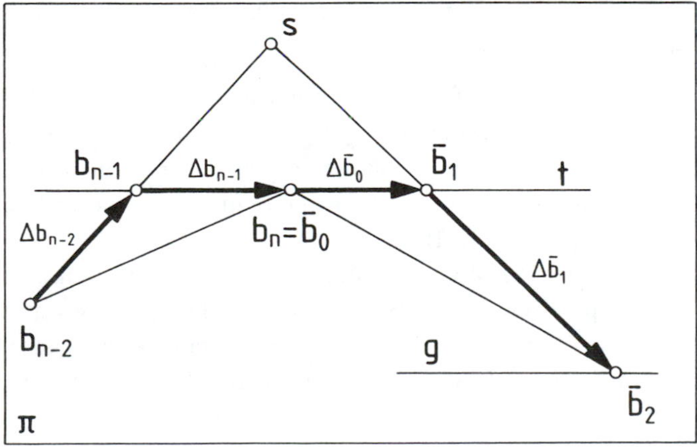

Fig. 5.6. Geometry of a curvature continuous join.

Curvature continuous cubic spline curves are of particular practical interest, see e.g. [Boo 87], [Hos 87, 88a], [Goo 88], [Schab 89], [Bars 90], [Eck 90] as are the curvature continuous quintic subsplines, especially for interpolation.

Here we treat cubic spline curves $\boldsymbol{X}(t)$, whose segments $\boldsymbol{X}_i : [t_i, t_{i+1}] \rightarrow \mathbb{R}^d$, with $d = 2, 3$ are given by $\boldsymbol{X}_i(\mu) = \sum_{k=0}^{3} \boldsymbol{b}_{3i+k} B_k^3(\mu)$, $\mu = (t - t_i)/\Delta t_i$, $\mu \in [0, 1]$, $\Delta t_i = t_{i+1} - t_i$. For a discussion of curvature continuous quintic subsplines, see [Dero 88], [Hos 87, 88b].

5.5.2. B-spline-Bézier Representation of Curvature Continuous Spline Curves

A B-spline-Bézier representation of curvature continuous C^1 splines, *i.e.*, curvature continuous cubic splines with $\omega_{i,11} = 1$ (γ-splines), was given in [Boeh 85a]. For γ-splines, the continuity conditions simplify to

$$\begin{aligned}(1 + q_i)\boldsymbol{b}_{3i} &= q_i \boldsymbol{b}_{3i-1} + \boldsymbol{b}_{3i+1} \\ (1 + q_i\gamma_i)\boldsymbol{b}_{3i-1} &= q_i\gamma_i \boldsymbol{b}_{3i-2} + \boldsymbol{d}_i \\ (\gamma_i + q_i)\boldsymbol{b}_{3i+1} &= q_i \boldsymbol{d}_i + \gamma_i \boldsymbol{b}_{3i+2},\end{aligned} \tag{5.19}$$

with

$$\gamma_i = \frac{1}{1 + \frac{1}{1+q_i}\frac{\Delta t_i}{2}\omega_{i,12}}. \tag{5.20}$$

As was the case for C^2 continuous cubic splines (see Sect. 4.1.4), for γ-splines we get

$$\begin{aligned}\Delta_{i,1}\boldsymbol{b}_{3i+1} &= (\Delta t_i + \Delta t_{i+1}\gamma_{i+1})\boldsymbol{d}_i + \Delta t_{i-1}\gamma_i \boldsymbol{d}_{i+1} \\ \Delta_{i,1}\boldsymbol{b}_{3i+2} &= \Delta t_{i+1}\gamma_{i+1}\boldsymbol{d}_i + (\Delta t_{i-1}\gamma_i + \Delta t_i)\boldsymbol{d}_{i+1},\end{aligned} \tag{5.21}$$

with

$$\Delta_{i,1} = \Delta t_{i-1}\gamma_i + \Delta t_i + \Delta t_{i+1}\gamma_{i+1}, \tag{5.22}$$

see Fig. 5.7. Equation (5.21) together with a modified version of equation (5.14) leads to an algoritm for computing the Bézier points of a γ-spline associated with a prescribed control polygon $\{\boldsymbol{d}_i\}$, knot vector $\{t_i\}$, and design parameters $\{\gamma_i\}$; see [Boeh 85a, 87] and also [Die 89]. On the other hand, the $\boldsymbol{d}_i$ can be found from the prescribed Bézier points, knot vector, and design parameters using equations (5.19).

The above discussion of C^2 continuous cubic Bézier splines shows that the points $\boldsymbol{d}_i$ are precisely the de Boor points of the B-spline representation of C^2 splines. For GC^2 continuous Bézier spline curves, we can also find a representation in terms of local basis functions in which the $\boldsymbol{d}_i$ act as control points. To this end, we write

$$\boldsymbol{X}(t) = \sum_i \boldsymbol{d}_i G_i^3(t)$$

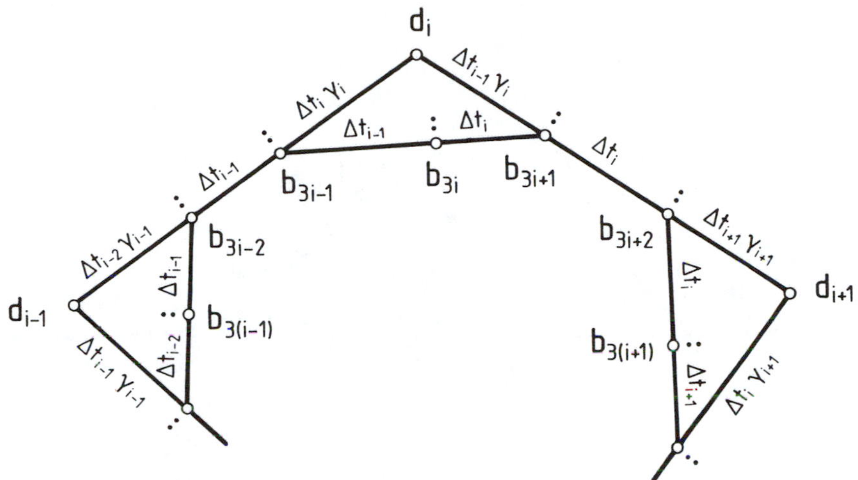

Fig. 5.7. de Boor and Bézier polygon of a γ-spline.

in terms of the *geometric B-splines* $G_i^3(t)$, which depend not only on the knot vector, as is the case for the ordinary B-splines $N_i^3(t)$, but also on the γ_i. The geometric B-splines $G_i^3(t)$ can be found from the identity

$$G_i^3(t) = \sum_j \delta_{ij} G_j^3(t). \tag{5.23}$$

Using (5.21), it follows from (5.23) that the nonzero interior Bézier ordinates of $G_i^3(t)$ are given by

$$b_{3i-2,i} = \frac{1}{\Delta_{i-1,1}} \gamma_{i-1} \Delta t_{i-2}, \qquad b_{3i-1,i} = \frac{1}{\Delta_{i-1,1}} (\gamma_{i-1} \Delta t_{i-2} + \Delta t_{i-1}),$$
$$b_{3i+2,i} = \frac{1}{\Delta_{i,1}} \gamma_{i+1} \Delta t_{i+1}, \qquad b_{3i+1,i} = \frac{1}{\Delta_{i,1}} (\gamma_{i+1} \Delta t_{i+1} + \Delta t_i),$$

while the nonzero junction points are given by

$$\begin{aligned} b_{3(i-1),i} &= \frac{1}{\Delta_{i-1,2}} \Delta t_{i-2} b_{3i-2,i}, \\ b_{3i,i} &= \frac{1}{\Delta_{i,2}} (\Delta t_{i-1} b_{3i+1,i} + \Delta t_i b_{3i-1,1}), \\ b_{3(i+1),i} &= \frac{1}{\Delta_{i+1,2}} \Delta t_{i+1} b_{3i+2,i}, \end{aligned}$$

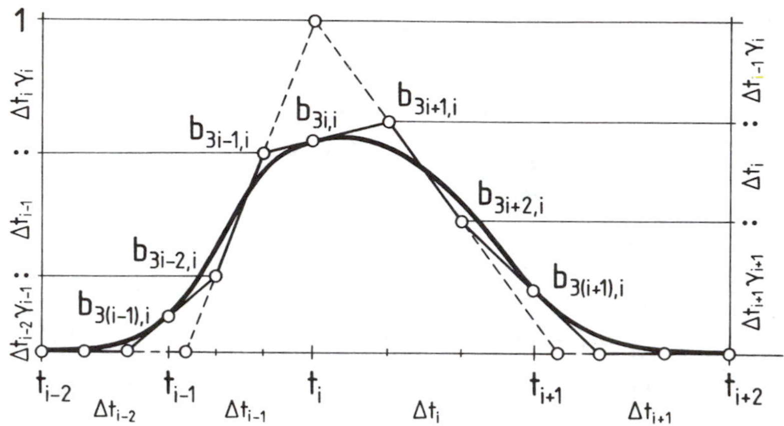

Fig. 5.8. Bézier points of a geometric B-spline $G_i^3(t)$.

where $\Delta_{i,1}$ is as in (5.22), and $\Delta_{i,2} = \Delta t_{i-1} + \Delta t_i$, see [Boeh 85a]. Fig. 5.8 shows the Bézier ordinates $b_{3i+j,i}$ of a geometric B-spline associated with the abscissae $t_i + j\Delta t/3$.

Formulae for the $G_i^k(t)$ as linear combinations of the ordinary B-splines $N_i^k(t)$ of Sect. 4.3.1 can be found in [Coh 87], and a recurrence equation for the basis functions is given in [Eck 90]. A de Boor algorithm was developed in [Boeh 85a], while in [Eck 90] degree raising, knot insertion (see also [Joe 89, 90], [Die 89a]), and the interesting problem of knot elimination are all treated. The system of equations corresponding to interpolating γ-splines given in [Eck 87] was studied further by [Far 90a], where the singular cases are treated.

In using γ-splines for design, we recommend first setting all γ_i to 1, which corresponds to ordinary C^2 splines, and then adjusting the individual γ_i values to achieve the desired curve. Fig. 5.9 shows the effect of choosing different values for the γ_i. The solid line corresponds to the cubic C^2 spline (*i.e.*, $\gamma_i = 1$ for all i).

Because of the importance of cubic splines in practice, various representations of curvature continuous cubic splines have been developed in the literature, based on a wide variety of applications. As examples, we mention the Bézier and γ-splines discussed above, as well as ν-splines, β-splines, Wilson-Fowler-splines, Manning's splines, and Bär's splines [Bär 77]. For a historical overview, see e.g. [Boeh 88b], [Las 88a]. An extensive list of references can be found in [Gre 89], [Las 88a]. The various representations in terms of monomial, Hermite, Bézier, B-spline, and other basis functions can

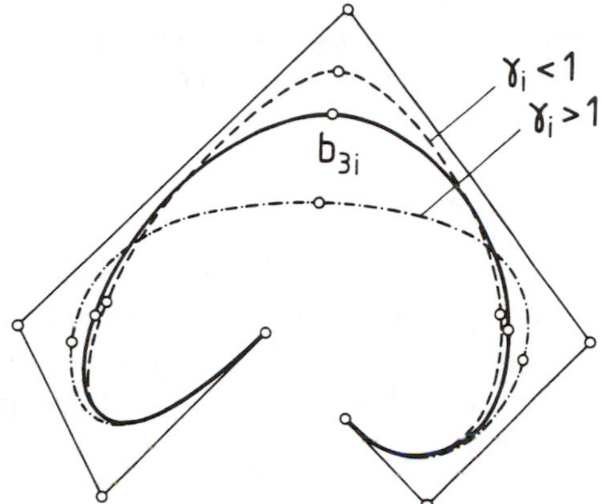

Fig. 5.9. Influence of the design parameters γ_i.

of course be transformed into each other (see e.g. Chap. 10). In this sense we already know all about these representations from the results of Sects. 5.5.1 and 5.5.2, and thus there is no real need to discuss them in further detail. But since the examples play an important role in practice, we now discuss them briefly.

5.5.3. Manning's Spline Curves

In 1974, Manning [Man 74] found a formula in terms of monomials for the curvature continuous cubic spline interpolating given points $\boldsymbol{P}_i$, $i = 0(1)N$. In this case, each segment of the spline can be written in the form

$$\boldsymbol{X}_i(u) = \boldsymbol{R}_{i,0} + u\boldsymbol{R}_{i,1} + u^2\boldsymbol{R}_{i,2} + u^3\boldsymbol{R}_{i,3}, \qquad u \in [0,1].$$

Manning's approach to determining the unknown coefficients $\boldsymbol{R}_{i,k}$ is in essence the same as the method used in Sect. 3.3 to determine the coefficients of a C^2 continuous interpolating cubic spline. In particular, for $\boldsymbol{X}_i(u)$ we have

$$\begin{aligned}\boldsymbol{X}_i(u) = \boldsymbol{P}_i + u\boldsymbol{X}'_{i+} + u^2 \left[3(\boldsymbol{P}_{i+1} - \boldsymbol{P}_i) - 2\boldsymbol{X}'_{i+} - \boldsymbol{X}'_{i+1-}\right] \\ + u^3 \left[\boldsymbol{X}'_{i+} + \boldsymbol{X}'_{i+1-} - 2(\boldsymbol{P}_{i+1} - \boldsymbol{P}_i)\right],\end{aligned}$$

where as in Sect. 3.3, we write

$$\begin{aligned}\boldsymbol{P}_i &\equiv \boldsymbol{X}_i(0) = \boldsymbol{R}_{i,0} \\ \boldsymbol{P}_{i+1} &\equiv \boldsymbol{X}_i(1) = \boldsymbol{R}_{i,0} + \boldsymbol{R}_{i,1} + \boldsymbol{R}_{i,2} + \boldsymbol{R}_{i,3} \\ \boldsymbol{X}'_{i+} &\equiv \boldsymbol{X}'_i(0^+) = \boldsymbol{R}_{i,1} \\ \boldsymbol{X}'_{i+1-} &\equiv \boldsymbol{X}'_i(1^-) = \boldsymbol{R}_{i,1} + 2\boldsymbol{R}_{i,2} + 3\boldsymbol{R}_{i,3}.\end{aligned}$$

The unknown derivatives can now be computed from the tridiagonal system of equations which arises by requiring that we have continuity of the tangent vectors and curvature at the knots, *i.e.*, the equations (5.1)–(5.2) are satisfied, see [Man 74].

5.5.4. ν-splines

Nielson's ν-splines [Nie 74, 86] (see also Sect. 3.6.2, as well as [Bars 81a, 84], [Fri 86], [Fol 87]) are defined similarly as the solution of an interpolation problem, where now $\boldsymbol{X}(t)$ must minimize (3.50) with $p_k = p = 1$ over the space H^L with $L = 2$. By (3.51), the ν-splines must satisfy the continuity conditions

$$\begin{aligned} \boldsymbol{X}(t_i^+) &= \boldsymbol{X}(t_i^-) \\ \boldsymbol{X}'(t_i^+) &= \boldsymbol{X}'(t_i^-) \\ \boldsymbol{X}''(t_i^+) &= \boldsymbol{X}''(t_i^-) + \nu_i \boldsymbol{X}'(t_i^-) \end{aligned}$$

at the $(t_i, \boldsymbol{P}_i)$, $i = 1(1)N\text{–}1$.

Thus, a ν-spline is a curvature continuous C^1 spline as well as a GC^2 continuous spline with $\omega_{i,11} = 1$ and $\nu_i \equiv \omega_{i,12}$. Hence, *$\nu$-splines are interpolating γ-splines*. A description of the different approaches to ν-splines and interpolating γ-splines along with a discussion of their similarities can be found in [Far 90a]. In addition to the Hermite-formula for ν-splines given in Sect. 3.6.2, we also have the B-spline Bézier representation for them. Converting between the two representations can be accomplished using (5.20) or

$$\nu_i = \frac{2}{\Delta t_i}(1 + q_i)\Big(\frac{1}{\gamma_i} - 1\Big).$$

In view of this relation, the influence of the so-called *tension parameters* ν_i on the shape of the curve is reciprocally proportional to the influence of the γ_i, see Figs. 5.9 and 3.16.

It should be remarked that the restriction $\nu_i \geq 0$ arising from the minimization characterization of ν-splines can be weakened to allow certain negative values. This follows from the discussion of the system of equations (3.48) [Bars 84], and also from the positivity requirement on the γ_i of the Bézier formula [Las 90]. For negative ν_i, a ν-spline no longer necessarily minimizes the integral (3.46) (or (3.50) with $L = 2$ and $p_k = p = 1$). A comparison of the *true energies* as given by (3.1) shows, however, that ν-splines with "mildly negative" ν-values better describe an actual elastic spline than the ν-splines minimizing (3.46), see [Fri 87]. Interval-weighted ν-splines are treated in [Fol 87] (Hermite representation) and in [Las 92], [Gre 91] (B-spline and Bézier representations).

5.5.5. β-splines

In 1981, Barsky [Bars 81] (see also [BARS 88]) presented a class of (special) curvature continuous cubic splines with local support. He began with the continuity conditions

$$
\begin{aligned}
\boldsymbol{X}_i(0) &= \boldsymbol{X}_{i-1}(1)\\
\boldsymbol{X}_i'(0) &= \beta_1 \boldsymbol{X}_{i-1}'(1)\\
\boldsymbol{X}_i''(0) &= \beta_1^2 \boldsymbol{X}_{i-1}''(1) + \beta_2 \boldsymbol{X}_i'(1),
\end{aligned}
$$

in terms of global design parameters β_1 and β_2 which are to be used for all segments, and used equally spaced knots, *i.e.*, $t_i = i$, $i \in \mathbb{N}$. Each segment of the β-spline curve $\boldsymbol{X}(t)$ is written in the form

$$
\boldsymbol{X}_i(u) = \sum_{k=-3}^{0} \boldsymbol{V}_{i+k} B_k(u), \qquad u \in [0,1],
$$

where the $\boldsymbol{V}_0, \boldsymbol{V}_1, \ldots, \boldsymbol{V}_N$ are called the control points. Here the basis functions $B_k(u) = B_k(\beta_1, \beta_2, u)$ are chosen to be piecewise cubic polynomials satisfying the continuity conditions, and having local support in the sense that $B_k \equiv 0$ outside of (t_k, t_{k+4}). This means that the derivatives must also be identically zero outside of (t_k, t_{k+4}), and it follows that the four cubic pieces of a β-B-spline can be written in terms of monomials as

$$
B_k(u) = \begin{cases}
\frac{1}{\beta}[2u^3], & t \in [t_i, t_{i+1}]\\
\frac{1}{\beta}[2 + 6\beta_1 u + (3\beta_2 + 6\beta_1^2)u^2 \\
\qquad -(2\beta_2 + 2\beta_1^2 + 2\beta_1 + 2)u^3], & t \in [t_{i+1}, t_{i+2}]\\
\frac{1}{\beta}[(\beta_2 + 4\beta_1^2 + 4\beta_1) + (6\beta_1^3 - 6\beta_1)u - (3\beta_2 \\
\qquad +6\beta_1^3 + 6\beta_1^2)u^2 + (2\beta_2 + 2\beta_1^3 + 2\beta_1^2 + 2\beta_1)u^3], & t \in [t_{i+2}, t_{i+3}]\\
\frac{1}{\beta}[(2\beta_1^3 - 6\beta_1^3 u + 6\beta_1^3 u^2 - 2\beta_1^3 u^3], & t \in [t_{i+3}, t_{i+4}]
\end{cases}
$$

with

$$
\beta = 2\beta_1^3 + 4\beta_1^2 + 4\beta_1 + \beta_2 + 2,
$$

for $u \in [0,1]$, see [Bars 81] or [BART 87].

Because the knots are equally spaced and the design parameters β_1 and β_2 are chosen uniformly and globally, we refer to the $B_k(u)$ as *uniformly-shaped* β-B-splines. A comparison with Sect. 5.5.4 shows that uniformly-shaped β-splines are very special ν-splines, and thus γ-splines. A comparison [Boeh 85a], [Fri 86] of the continuity conditions leads to

$$
\beta_1 = q_i = \frac{\Delta t_i}{\Delta t_{i-1}} = \text{constant}, \qquad \text{for all } i.
$$

Thus, β_1 is *not a true design parameter*, and

$$\beta_2 = q_i \Delta t_i \nu_i = \text{constant}, \qquad \text{for all } i.$$

From (5.20) and (5.24) it follows immediately that

$$\beta_2 = 2(1+q_i)\Big(\frac{1}{\gamma_i} - 1\Big) = \text{constant}, \qquad \text{for all } i,$$

connecting uniformly-shaped β-splines with γ-splines, see [Dero 88].

While as in the case of ordinary B-splines, a control point only influences four (successive) curve segments, an obvious weakness of uniformly-shaped β-splines is the fact that both β_1 and β_2 have global influence, and so individual pieces of the curve cannot be separately manipulated. Uniformly-shaped β-splines can be localized as follows. For each knot, we prescribe $\beta_{i,1}$ and $\beta_{i,2}$, which are then interpolated using Hermite polynomials $\beta_{i,1}(u)$ and $\beta_{i,2}(u)$ in such a way that the geometric continuity conditions are not affected. This can be done with Hermite polynomials of fifth degree. As before, for these *continuously-shaped β-splines*, a control point $\boldsymbol{V}_i$ influences only four successive segments, but now the effect of the β's is completely local, [Bars 83] (see also [Bart 84], [BART 87]). By the construction of the *blending functions* $\beta_{i,1}(u)$ and $\beta_{i,2}(u)$, $\beta_{i,1}$ and $\beta_{i,2}$ have an effect only on the two curve segments $\boldsymbol{X}_i$ and $\boldsymbol{X}_{i+1}$. The price we have to pay for this strong localness is that we have to use rational functions formed from polynomials of higher degree, *i.e.*, continuously-shaped β-splines are no longer piecewise cubic polynomials. In particular, the β-B-splines $B_k(u)$ and hence the associated spline curves are now piecewise rational functions involving quotients of polynomials of degree 18 and 15.

These disadvantages of the continuously-shaped β-splines are not shared by the *generalized cubic β-splines* corresponding to given $\beta_{i,1}$ and $\beta_{i,2}$. For a treatment of generalized β-splines in terms of divided differences and truncated power functions which is analogous to the classical treatment of ordinary B-splines, see e.g. [SCHU 81], [Bart 84] or [BART 87]. An explicit description of the *discretely-shaped β-splines* [Joe 87], analogous to the above treatment of uniformly-shaped β-splines, was given by [Goo 85a, 86], where, however, the β values are no longer uniform, and can be different in each interval.

General β-splines can be expressed in terms of B-splines [Höll 86], and thus are equivalent to γ- and ν-splines. The connection between the β's, γ's and ν's is given above, where now β_1 and β_2 have to be replaced by $\beta_{i,1}$ and $\beta_{i,2}$.

5.5.6. Wilson-Fowler Splines

Wilson-Fowler splines were introduced by Wilson and Fowler [Fow 66] (see also [Fri 86a]) as planar, curvature continuous spline curves which interpolate the points $\boldsymbol{P}_i = (x_i, y_i)$, for $i = 0(1)N$. For each segment of $\boldsymbol{X}$ we introduce a local (t, ν) coordinate system, where $t \in [0, L_i]$ is defined by the line segment connecting $\boldsymbol{P}_i$ and $\boldsymbol{P}_{i+1}$ (cf. Fig. 3.18), and $\nu_i = \nu_i(t)$ is taken to be the cubic Hermite polynomial

$$\nu_i(t) = \frac{t(t-L_i)^2}{L_i^2}\tan\alpha_i + \frac{t^2(t-L_i)}{L_i^2}\tan\delta_i, \qquad i = 0(1)N\text{–}1. \tag{5.25}$$

Thus, we have

$$\nu_i(0) = 0, \quad \nu_i'(0) = \tan\alpha_i, \quad \nu_i(L_i) = 0, \quad \nu_i'(L_i) = \tan\delta_i,$$

where $\tan\alpha_i$ and $\tan\delta_i$ are determined by the GC^2 continuity conditions, and where L_i denotes the Euclidean distance between $\boldsymbol{P}_i$ and $\boldsymbol{P}_{i+1}$.

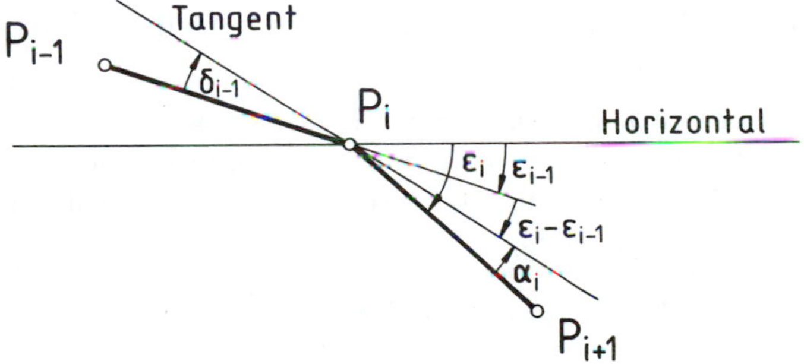

Fig. 5.10. Definition of the angles.

If we find the connection between the natural local coordinate systems of a *Wilson-Fowler spline* (*WF-spline* for short) and the global (x, y) coordinate system, we obtain parametric formulae for the individual segments $\boldsymbol{X}_i = (x_i(t), y_i(t))$. This can be accomplished by means of a translation and rotation via

$$\begin{aligned} x_i(t) &= x_i + (t-t_i)\cos\delta_i - \nu_i(t-t_i)\ \sin\delta_i \\ y_i(t) &= y_i + (t-t_i)\sin\epsilon_i + \nu_i(t-t_i)\cos\epsilon_i, \end{aligned} \tag{5.26}$$

where $t \in [t_i, t_{i+1}]$, and we are using the chordal parametrization

$$t_0 = 0, \qquad t_i = t_{i-1} + L_{i-1}.$$

Since $\nu_i(t - t_i)$ is given by the cubic Hermite polynomial (5.25), we now see that a WF-spline consists of cubic segments. Substituting the WF formula (5.26) for a curvature continuous spline into the GC^2 continuity conditions, we can find the relationship between the GC^2 parameters $\omega_{i,11}$, $\omega_{i,12}$ and the WF parameters $\tan \alpha_i$ and $\tan \delta_i$ [Fri 86]:

$$\omega_{i,11} = \cos(\epsilon_i - \epsilon_{i-1}) - \tan \alpha_i \sin(\epsilon_i - \epsilon_{i-1}),$$

and

$$\omega_{i,12} = \frac{2}{L_i}(2 \tan \alpha_i + \tan \delta_i) \sin(\epsilon_i - \epsilon_{i-1}).$$

From (5.17) we can find the connection of the WF parameters $\tan \alpha_i$, $\tan \delta_i$ to the design parameters γ_i of the B-spline-Bézier representation of Sect. 5.5.2, and to the ν_i's in the ν-spline representation [Fri 86]. [Fri 87] has shown that the WF-splines which had been expected to provide a better model of the *true energy* (3.1) of a spline than the parametric cubic splines, do not differ much visually or in energy from the latter, which are *much simpler* to compute.

5.6. Torsion Continuous Spline Curves

5.6.1. Bézier Representation of Torsion Continuous Spline Curves

Given two curves written in Bézier form (5.11), (5.12), suppose we want them to join together with continuous torsion, *i.e.*, (5.4) should hold with $r = 3$. Then substituting the Bézier representations in (5.4), differentiating, and using (5.13), (5.14) and (5.16), we get the following additional conditions on the Bézier points $\boldsymbol{b}_{n-3}$ and $\bar{\boldsymbol{b}}_3$:

$$\begin{aligned}(1 + qN_3\omega_{11}^3\delta)\boldsymbol{b}_{n-2} &= qN_3\omega_{11}^3\delta\boldsymbol{b}_{n-3} + \boldsymbol{e}^- \\ (\delta + q\epsilon)\boldsymbol{s} &= q\epsilon\boldsymbol{e}^- + \delta\boldsymbol{e}^+ \\ (\epsilon + q)\bar{\boldsymbol{b}}_2 &= q\boldsymbol{e}^+ + \epsilon\bar{\boldsymbol{b}}_3,\end{aligned} \tag{5.27}$$

see Fig. 5.11, and compare with Fig. 4.14b. Here $\boldsymbol{e}^-$ and $\boldsymbol{e}^+$ are intermediate construction points, and

$$\begin{aligned}\delta &= \frac{q}{2N_3\omega_{11}^3 + qJ - qN_2\omega_{11}^2 G} \\ \epsilon &= \frac{q}{2 - H}\end{aligned} \tag{5.28}$$

are *design parameters*, where

$$G = \frac{N_1\omega_{11}(2qN_3\omega_{11}^4 + 3q\frac{N_3}{N_1}\omega_{11}^3 - 1) + qJ\,(2 + qN_1\omega_{11}) + K}{qN_2\omega_{11}^2(1 + qN_1\omega_{11}) + N_1\omega_{11}\left(1 + q\frac{N_2}{N_1}\omega_{11}\right) + N_1\frac{\Delta t}{(\bar{n}-1)}\omega_{12}},$$

$$H = \frac{2 + 3qN_1\omega_{11} - q^3 N_3 \omega_{11}^3 - q^2 J - qK}{1 + 2qN_1\omega_{11} + q^2 N_2 \omega_{11}^2 + qN_1 \frac{\Delta t}{(\bar{n}-1)}\omega_{12}},$$

$$J = N_2 \frac{\Delta t}{(\bar{n}-2)}\omega_{13}, \qquad K = N_1 \frac{(\Delta t)^2}{(\bar{n}-1)(\bar{n}-2)}\omega_{23},$$

and

$$N_3 = \frac{n(n-1)(n-2)}{\bar{n}(\bar{n}-1)(\bar{n}-2)},$$

cf. [Las 88a]. Equation (5.27) can be interpreted as follows: $\boldsymbol{b}_{n-2}$ divides the line segment $\boldsymbol{b}_{n-3}\boldsymbol{e}^+$ in the ratio $1 : qN_3\omega_{11}^3\delta$, $\boldsymbol{s}$ divides the line segment $\boldsymbol{e}^-\boldsymbol{e}^+$ in the ratio $\delta : q\epsilon$, and $\bar{\boldsymbol{b}}_2$ divides the line segment $\boldsymbol{e}^+\bar{\boldsymbol{b}}_3$ in the ratio $\epsilon : q$ (Fig. 5.11). Writing a_3 and $\bar{a}_3$ for the distances of $\boldsymbol{b}_{n-3}$ and $\bar{\boldsymbol{b}}_3$, respectively, from the osculating plane π passing through $\bar{\boldsymbol{b}}_0 = \boldsymbol{b}_n$ and defined by $\Delta\boldsymbol{b}_{n-1}$ and $\Delta\boldsymbol{b}_{n-2}$, it follows as in Sect. 5.5.1 that

$$a_3 : \bar{a}_3 = 1 : q^3 N_3 \omega_{11}^3.$$

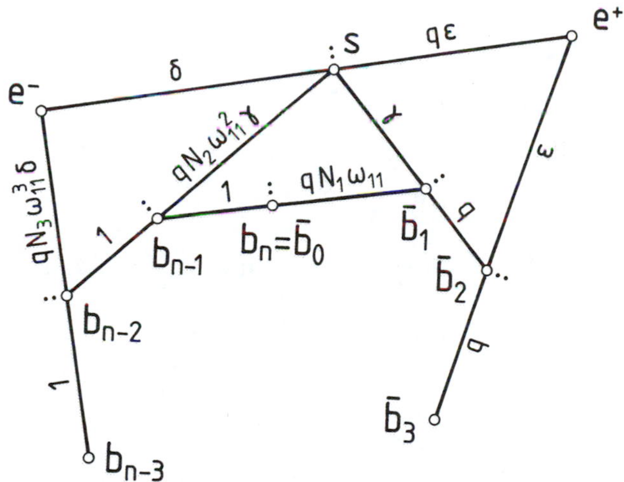

Fig. 5.11. Bézier polygon for a torsion continuous join.

If the Bézier points $\boldsymbol{b}_n, \ldots, \boldsymbol{b}_{n-3}$ are given while $\bar{\boldsymbol{b}}_0, \ldots, \bar{\boldsymbol{b}}_2$ are determined from the tangent and curvature continuity conditions in Sects. 5.4 and 5.5, then $\bar{\boldsymbol{b}}_3$ can be computed in terms of the intermediate construction points $\boldsymbol{e}^+$ and $\boldsymbol{e}^-$ from (5.27), or directly from

$$\bar{\boldsymbol{b}}_3 = \boldsymbol{b}_n + q^3 N_3 \omega_{11}^3 \Delta\boldsymbol{b}_{n-3} - \lambda\Delta\boldsymbol{b}_{n-2} + \mu\Delta\boldsymbol{b}_{n-1}, \tag{5.29}$$

where $\lambda = \lambda(\omega_{13})$ and $\mu = \mu(\omega_{23})$. Thus, for a torsion continuous join, the Bézier points $\bar{\boldsymbol{b}}_3$ lie in a plane $\boldsymbol{E}$ parallel to the osculating plane π at $\bar{\boldsymbol{b}}_0 = \boldsymbol{b}_n$. The distance of $\boldsymbol{E}$ to the osculatory plane π is determined by the (previously chosen) parameter ω_{11}, while the parameters ω_{13} and ω_{23} determine $\bar{\boldsymbol{b}}_3$, cf. [Hag 86], and see Fig. 5.12.

In addition, it follows from (5.29) that $\boldsymbol{b}_{n-3}$ and $\bar{\boldsymbol{b}}_3$ lie on opposite sides of the osculating plane.

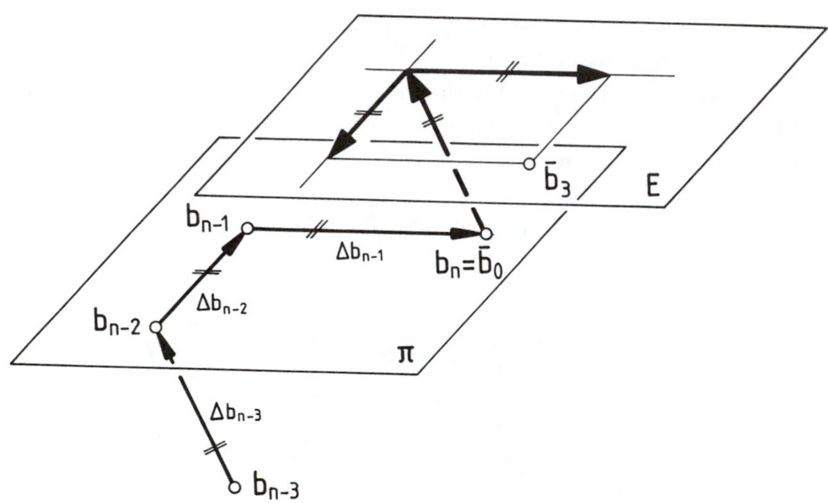

Fig. 5.12. Geometric interpretation of (5.29).

From the equivalence of (5.1)–(5.3) with

$$\frac{(\boldsymbol{Y}', \boldsymbol{Y}'', \boldsymbol{Y}''')}{\|\boldsymbol{Y}' \times \boldsymbol{Y}''\|^2} = \frac{(\boldsymbol{X}', \boldsymbol{X}'', \boldsymbol{X}''')}{\|\boldsymbol{X}' \times \boldsymbol{X}''\|^2}$$

for a continuous Frenet frame (as in Sect. 5.5), inserting the Bézier formulae for $\boldsymbol{X}$ and $\boldsymbol{Y}$, we get

$$\frac{(\Delta\bar{\boldsymbol{b}}_0, \Delta\bar{\boldsymbol{b}}_1, \Delta\bar{\boldsymbol{b}}_2)}{\bar{a}_1^6} = \frac{\bar{n}(n-2)}{n(\bar{n}-2)} \frac{(\Delta\boldsymbol{b}_n, \Delta\boldsymbol{b}_{n-1}, \Delta\boldsymbol{b}_{n-2})}{a_1^6}.$$

Since $(\Delta\bar{\boldsymbol{b}}_0, \Delta\bar{\boldsymbol{b}}_1, \Delta\bar{\boldsymbol{b}}_2)$ and $(\Delta\boldsymbol{b}_n, \Delta\boldsymbol{b}_{n-1}, \Delta\boldsymbol{b}_{n-2})$ give the volumes of the simplices defined by $\boldsymbol{b}_{n-3}, \ldots, \boldsymbol{b}_n$ and $\bar{\boldsymbol{b}}_3, \ldots, \bar{\boldsymbol{b}}_0$, resp., it follows that the relation

$$\frac{\text{volume}\{\bar{\boldsymbol{b}}_0, \bar{\boldsymbol{b}}_1, \bar{\boldsymbol{b}}_2, \bar{\boldsymbol{b}}_3\}}{\bar{a}_1^6} = \frac{\bar{n}(n-2)}{n(\bar{n}-2)} \frac{\text{volume}\{\boldsymbol{b}_n, \boldsymbol{b}_{n-1}, \boldsymbol{b}_{n-2}, \boldsymbol{b}_{n-3}\}}{a_1^6} \tag{5.30}$$

between the volumes of the two simplices is equivalent to the torsion continuity of $\boldsymbol{X}$ and $\boldsymbol{Y}$ [Far 82a].

Torsion continuous quartic splines are of practical interest, but involve unsymmetric boundary conditions, which makes quintic subsplines even more useful. Symmetric boundary conditions can be generated "artificially", of course, e.g., by requiring that the segments of the spline curve pass through given points $\boldsymbol{X}(t^*)$ at given parameter values $t^* \in (t_i, t_{i+1})$. For t^* it is best to choose the center of the intervals. This corresponds to determining shoulder points, see Chap. 3, [Long 87], and also [Boeh 82].

Examples of B-spline and Bézier representations of quartic and quintic (sub)splines will be given below.

5.6.2. B-spline-Bézier Representation of Torsion Continuous Spline Curves

B-spline representations of torsion continuous splines of fourth degree have been discussed in [Dyn 85, 87], [Boeh 87], [Eck 89, 90] as well as [Joe 90a]. In addition to the above equations with $n = \bar{n} = 4$, i.e., $N_1 = N_2 = N_3 = 1$, we now get (cf. Fig. 5.13)

$$\begin{aligned}
\Delta_{i,1}\boldsymbol{e}_i^- &= \Delta t_{i-2}\phi_i\, \boldsymbol{d}_{i-2} + (\Delta t_{i-1}\delta_i + \Delta t_i\, \epsilon_i + \Delta t_{i+1}\psi_i)\, \boldsymbol{d}_{i-3} \\
\Delta_{i,1}\boldsymbol{s}_i &= (\Delta t_{i-2}\phi_i + \Delta t_{i-1}\, \delta_i)\, \boldsymbol{d}_{i-2} + (\Delta t_i\, \epsilon_i + \Delta t_{i+1}\psi_i)\, \boldsymbol{d}_{i-3} \qquad (5.31)\\
\Delta_{i,1}\boldsymbol{e}_i^+ &= (\Delta t_{i-2}\phi_i + \Delta t_{i-1}\, \delta_i + \Delta t_i\, \epsilon_i)\, \boldsymbol{d}_{i-2} + \Delta t_{i+1}\psi_i\, \boldsymbol{d}_{i-3},
\end{aligned}$$

with

$$\Delta_{i,1} = \Delta t_{i-2}\phi_i + \Delta t_{i-1}\delta_i + \Delta t_i\epsilon_i + \Delta t_{i+1}\psi_i.$$

Here

$$\begin{aligned}
\Delta_i\psi_i &= (\Delta t_{i-1}\epsilon_i + \Delta t_i + \Delta t_{i+1}\gamma_{i+1})\Delta t_i\, \epsilon_i\, \delta_{i+1} \\
\Delta_i\phi_i &= (\Delta t_{i-1}\gamma_i + \Delta t_i + \Delta t_{i+1}\delta_{i+1})\Delta t_i\, \epsilon_i\, \delta_{i+1},
\end{aligned}$$

with

$$\Delta_i = (\Delta t_{i-1}\gamma_i + \Delta t_i)(\Delta t_{i+1}\gamma_{i+1} + \Delta t_i) - \Delta t_{i-1}\, \epsilon_i\, \Delta t_{i+1}\, \delta_{i+1}.$$

For the inner Bézier points, we have

$$\begin{aligned}
\Delta_{i,2}\boldsymbol{b}_{4i-2} &= \Delta t_{i-2}\epsilon_{i-1}\boldsymbol{s}_i + (\Delta t_{i-1} + \Delta t_i\gamma_i)\, \boldsymbol{e}_{i-1}^+ \\
\Delta_{i,2}\boldsymbol{b}_{4i-1} &= (\Delta t_{i-2}\epsilon_{i-1} + \Delta t_{i-1})\, \boldsymbol{s}_i + \Delta t_i\gamma_i\, \boldsymbol{e}_{i-1}^+ \qquad (5.32)\\
\Delta_{i,3}\boldsymbol{b}_{4i+1} &= \Delta t_{i-1}\gamma_i\, \boldsymbol{e}_{i+1}^- + (\Delta t_i + \Delta t_{i+1}\delta_{i+1})\, \boldsymbol{s}_i \\
\Delta_{i,3}\boldsymbol{b}_{4i+2} &= (\Delta t_{i-1}\gamma_i + \Delta t_i)\, \boldsymbol{e}_{i+1}^- + \Delta t_{i+1}\delta_{i+1}\, \boldsymbol{s}_i,
\end{aligned}$$

with

$$\Delta_{i,2} = \Delta t_{i-2}\epsilon_{i-1} + \Delta t_{i-1} + \Delta t_i \gamma_i$$
$$\Delta_{i,3} = \Delta t_{i-1}\gamma_i + \Delta t_i + \Delta t_{i+1}\delta_{i+1},$$

while for the Bézier knot points we have

$$(\Delta t_{i-1} + \Delta t_i)\, \boldsymbol{b}_{4i} = \Delta t_{i-1}\boldsymbol{b}_{4i+1} + \Delta t_i\, \boldsymbol{b}_{4i-1}. \tag{5.33}$$

If the construction points $\boldsymbol{d}_i$ are used as the de Boor control points of a B-spline type representation

$$\boldsymbol{X}(t) = \sum_i \boldsymbol{d}_i G_i^4(t),$$

then (5.31)–(5.33) lead to an algorithm for determining the Bézier points of a torsion continuous quartic spline associated with a prescribed control polygon $\{\boldsymbol{d}_i\}$, knot vector $\{t_i\}$, and design parameters $\{\gamma_i, \delta_i, \epsilon_i\}$. On the other hand (see Fig. 5.13), the $\boldsymbol{d}_i$ can also be determined from (5.27) along with

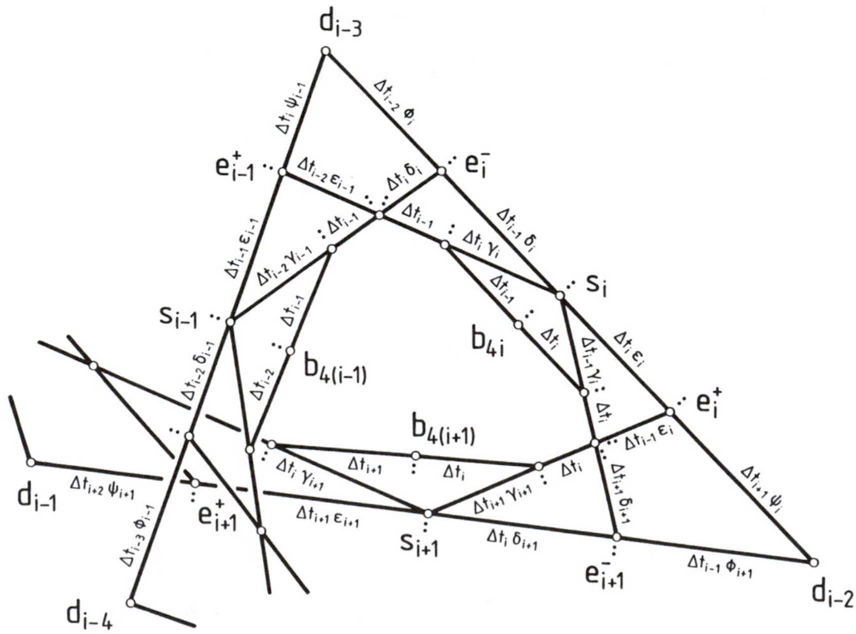

Fig. 5.13. de Boor and Bézier polygon of a torsion continuous quartic spline.

the equations

$$(\Delta t_i \epsilon_i + \Delta t_{i+1} \psi_i)\, \boldsymbol{e}_i^+ = \Delta t_i \epsilon_i \, \boldsymbol{d}_{i-2} + \Delta t_{i+1} \psi_i \, \boldsymbol{s}_i$$
$$(\Delta t_{i-2} \phi_i + \Delta t_{i-1} \delta_i)\, \boldsymbol{e}_i^- = \Delta t_{i-2} \phi_i \, \boldsymbol{s}_i + \Delta t_{i-1} \delta_i \, \boldsymbol{d}_{i-3}.$$

In working with torsion continuous splines, we recommend proceeding as with curvature continuous cubic splines. Fig. 5.14 shows the influence of the various δ_i's on the shape of the curve. Because of (5.27), ϵ_i has an opposite effect from that of δ_i, while the influence of the γ_i's is as in the cubic case, see Fig. 5.9. The solid line shows the quartic C^3 spline which corresponds to choosing all weights equal to one.

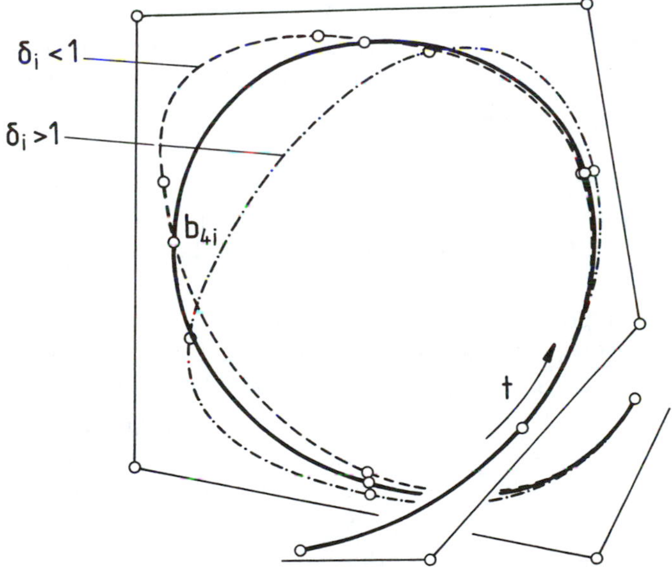

Fig. 5.14. Influence of the design parameters δ_i.

The system of equations associated with interpolating FC^3 splines can be found in [Eck 87]. The local bases $G_i^4(u)$ can be found as for the case of γ-splines treated in Sect. 5.5.2 by construction of the associated Bézier polygon, see [Eck 87, 89].

5.6.3. GC^3 Continuous Spline Curves

The GC^3 continuous spline curves are special torsion continuous spline curves with

$$\omega_{i,13} = 3\omega_{i,11}\,\omega_{i,12}, \tag{5.34}$$

cf. (5.3) with (5.9).

In terms of a B-spline-Bézier representation, we get continuity conditions of the type (5.27). Now, however, we have to take (5.34) into account. As a consequence, for GC^3 continuity, (5.30) is necessary but not sufficient [Far 82a]. Moreover, the design parameters δ_i and ϵ_i for GC^3 continuous splines are no longer independent. Indeed, we have

$$\delta_i(\epsilon_i) = \frac{2\gamma_i(q_i\gamma_i + 1)\epsilon_i}{(1+q_i)(9-5\gamma_i)\epsilon_i - 2\gamma_i(\gamma_i + q_i)}, \tag{5.35}$$

where $n = \bar{n} = 4$ (and so $N_1 = N_2 = N_3 = 1$) for the case of quartic splines [Pot 88], [Las 90a]. This means that in order for a torsion continuous spline to be a GC^3 continuous spline, the δ_i and ϵ_i must satisfy equation (5.35). Thus, at each knot, in addition to the parameter γ_i, we have an additional design parameter, either δ_i or ϵ_i at our disposal. For example, if ϵ_i is chosen as the independent design parameter, then δ_i is determined by (5.35). While torsion continuous curves which use a full set of design parameters can only be constructed in $\mathbb{R}^3$, GC^3 continuous curves can also be realized in the plane, see Sect. 5.2.

A B-spline representation of a GC^3 continuous quartic spline was also given in [Bars 84a], where the local bases were given in monomial form. B-spline-Bézier representations, as well as the convex hull and variation diminishing properties of GC^3 continuous quartic and also GC^4 continuous quintic curves, can be found in [Las 90a] and [Eck 87, 89]. [Joe 90a] gave a detailed treatment of multiple knots and the influence of the design parameters on the shape of the curve.

GC^3 continuous curves of seventh degree and GC^4 continuous curves of ninth degree in Bézier form were used in [Hos 88b] for conversion and approximation of spline curves. [Dero 88a] gave a recurrence relation for the determination of the control points of the Bézier representations of GC^r continuous curves (r arbitrary) $\boldsymbol{X}(u)$ and $\boldsymbol{Y}(t)$. In the first step we calculate the Bézier points determined by the GC^r continuity conditions on the curve $\boldsymbol{X}(u(t))$ reparametrized in terms of t. Then the Bézier points of $\boldsymbol{Y}(t)$ determined by the GC^r continuity conditions are recursively computed from the C^r continuity condition (5.5).

5.6.4. τ-splines

A comparison of the continuity conditions of a τ-spline, where $p_k = p = 1$ in (3.51), with the continuity conditions (5.4) shows that for a τ-spline,

$$\begin{array}{llll} \omega_{i,11} = 1 & \omega_{i,12} = 0 & \omega_{i,13} = \nu_{i,2} & \omega_{i,14} = 0 \\ & & \omega_{i,23} = 0 & \omega_{i,24} = 0 \\ & & & \omega_{i,34} = -\nu_{i,1}. \end{array}$$

Thus, τ-splines are a subset of torsion continuous quintic subsplines. A Bézier representation for τ-splines was given in [Las 90]. This leads to the continuity conditions shown in Fig. 5.15, where

$$\delta_i = \frac{1}{1 + \frac{\nu_{i,2}\Delta_i}{3(1+q_i)^2}}, \qquad \epsilon_i = \frac{1}{1 + \frac{q_i\nu_{i,2}\Delta_i}{3(1+q_i)^2}}, \tag{5.36}$$

and

$$\rho_i = \frac{1}{3q_i - R_i}, \qquad \sigma_i = q_i^2 \frac{1}{T_i - 3 + (T_i - S_i)\frac{q_i}{\epsilon_i}}, \qquad \tau_i = q_i \frac{1}{3 - T_i}, \tag{5.37}$$

with

$$R_i = \frac{3q_i - 1 + \frac{\Delta_i^3}{24(1+q_i)^3}\nu_{i,1}}{1 + \frac{\Delta_i}{3(1+q_i)^3}\nu_{i,2}}, \qquad S_i = \frac{2(1 - q_i) + \frac{(1-q_i)\Delta_i^3}{24(1+q_i)^3}\nu_{i,1}}{1 + \frac{2q_i\Delta_i}{3(1+q_i)^3}\nu_{i,2}},$$

$$T_i = \frac{3 - q_i + \frac{q_i\Delta_i^3}{24(1+q_i)^3}\nu_{i,1}}{1 + \frac{q_i^2\Delta_i}{3(1+q_i)^3}\nu_{i,2}}.$$

Equations (5.36) and (5.37) can be used to find the design parameters for the Bézier representation of a τ-spline. The design parameters can, of course, no longer be independent, since the τ-spline definition involves two form parameters while the construction in Fig. 5.15 involves five. The situation is similar to that for the Bézier representation of GC^3 continuous spline curves. The dependencies inherent in (5.36) and (5.37) are

$$\delta_i = \frac{q_i\epsilon_i}{1 - (1 - q_i)\epsilon_i}, \tag{5.38}$$

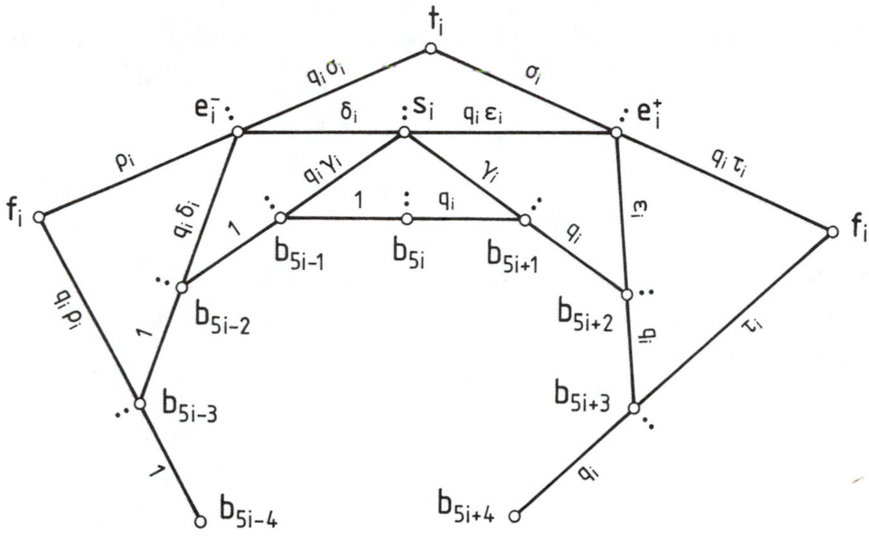

Fig. 5.15. Construction of the Bézier polygon for a τ-spline.

and

$$\rho_i = \frac{(1+q_i\delta_i)(\epsilon_i+\delta_i)\sigma_i}{\delta_i[3(1+q_i)\sigma_i-(\delta_i+q_i\epsilon_i)]}, \qquad \tau_i = \frac{(\epsilon_i+q_i)(\epsilon_i+\delta_i)\sigma_i}{\epsilon_i[3(1+q_i)\sigma_i-(\delta_i+q_i\epsilon_i)]}. \tag{5.39}$$

This means that in order for a torsion continuous quintic Bézier subspline curve to be a τ-spline, the design parameters of the Bézier representation must satisfy (5.38) and (5.39). Thus at each knot, either ϵ_i or δ_i can be selected as an *independent design parameter*. In the first case, δ_i is determined by (5.38) to be $\delta_i = \delta_i(\epsilon_i)$, and ρ_i, or σ_i, or τ_i can be chosen as an *independent design parameter*. If σ_i is selected as the free parameter, then ρ_i and τ_i are forced by (5.39) to be $\rho_i = \rho_i(\sigma_i)$ and $\tau_i = \tau_i(\sigma_i)$. The B-spline-Bézier representation of τ-splines is, of course, identical with the B-spline-Bézier representation of torsion continuous quintic (sub)splines, as shown in [Eck 87, 89]. It results by embedding the Bézier construction of Fig. 5.15 into a B-spline construction, as was done in the quintic B-spline-Bézier construction (see e.g. [Sabl 78]), but where the design parameters for τ-splines cannot be selected independently. Indeed, we must set $\gamma_i = 1$, and $\delta_i, \epsilon_i, \rho_i, \sigma_i$ and τ_i are to be selected subject to (5.38) and (5.39), see [Las 90]. Interval-weighted τ-splines are treated in [Las 92] (for B-spline and Bézier representations), and in [Neu 92] (in terms of a Hermite representation).

5.7. Rational Geometric Spline Curves

As we saw in Sect. 4.1.4, rational spline curves have the advantage that they can model conic sections exactly. In Sect. 4.1.4 we discussed rational splines with C^r continuity. Here we discuss rational geometric spline curves which combine the free-form and design properties of rational curves with the design and invariance properties (with respect to parameter transformations) of geometric spline curves. The following theorem (see [Boeh 88a], [Deg 88], [Pot 89]) is of special importance.

Theorem 5.1. *The GC^r continuity conditions of Sect. 5.2 and the continuity conditions of Sect. 5.1 are invariant under projective transformations, and thus in particular, under projections.*

Invariance properties have been discussed in detail in [Gol 89, 89a]. In addition to opening up the possibility of treating a rational curve in $\mathbb{R}^d$ as the projection of a nonrational curve in $d+1$ dimensional homogeneous coordinate space, this theorem also allows the derivation of continuity conditions in two essentially different ways: by starting with the representation in terms of homogeneous coordinates in $\mathbb{R}^4$, or by using inhomogeneous coordinates in $\mathbb{R}^3$. Since in practice we frequently want to work with inhomogeneous representations of the "object space", *i.e.*, in $\mathbb{R}^3$, here we shall use inhomogeneous coordinates to construct rational curvature and torsion continuous curves. However, in order to illustrate the advantages of working with homogeneous coordinates, we will use them to construct rational GC^r continuous curves.

5.7.1. Rational FC^r Continuous Spline Curves

In order to write out the continuity conditions (5.4) which two adjoining curves $\boldsymbol{X}(u)$, $u \in [u_0, u_1]$, and $\boldsymbol{Y}(t)$, $t \in [t_0, t_1]$, must satisfy, we need to compute their derivatives at the point $\boldsymbol{P} = \boldsymbol{X}(u_1) = \boldsymbol{Y}(t_0)$. In view of the rational nature of $\boldsymbol{X}(u)$ and $\boldsymbol{Y}(t)$, this means that we need to use the quotient rule. Thus, for example, if

$$\boldsymbol{X}(u) = \frac{\boldsymbol{Z}(u)}{N(u)},$$

then

$$\boldsymbol{X}'(u) = \frac{\boldsymbol{Z}'N - \boldsymbol{Z}N'}{N^2},$$
$$\boldsymbol{X}''(u) = \frac{\boldsymbol{Z}''N - \boldsymbol{Z}N''}{N^2} - 2\frac{N'}{N}\boldsymbol{X}'(u).$$

Thus if the curves are given in Bézier form,

$$\boldsymbol{X}(u) = \frac{\sum_{k=0}^{n} \beta_k \boldsymbol{b}_k B_k^n(\mu)}{\sum_{k=0}^{n} \beta_k B_k^n(\mu)}, \qquad u = (1-\mu)u_0 + \mu u_1, \quad \mu \in [0,1],$$

with a similar formula for $\boldsymbol{Y}(t)$, then we have

$$\boldsymbol{Z}^{(p)}(u) = \frac{\partial^p}{\partial u^p}\boldsymbol{Z}(\mu) = \frac{n!}{(\Delta u)^p(n-p)!}\sum_{k=0}^{n-p}\Delta^p\beta_k\,\boldsymbol{b}_k\,B_k^{n-p}(\mu),$$

$$N^{(p)}(u) = \frac{\partial^p}{\partial u^p}N(\mu) = \frac{n!}{(\Delta u)^p(n-p)!}\sum_{k=0}^{n-p}\Delta^p\beta_k B_k^{n-p}(\mu),$$

with $\Delta u = u_1 - u_0$, and the forward differences

$$\Delta^p\beta_k\boldsymbol{b}_k = \sum_{j=0}^{p}(-1)^j\binom{p}{j}\beta_{k+p-j}\boldsymbol{b}_{k+p-j},$$

$$\Delta^p\beta_k = \sum_{j=0}^{p}(-1)^j\binom{p}{j}\beta_{k+p-j}.$$

Thus, assuming a continuous Frenet frame, the condition for *tangent continuity* becomes

$$\omega_{11}\frac{n\beta_{n-1}}{\Delta u\beta_n}\Delta\boldsymbol{b}_{n-1} = \frac{\bar{n}\bar{\beta}_1}{\Delta t\bar{\beta}_0}\Delta\bar{\boldsymbol{b}}_0.$$

For *curvature continuity*, we must also have (cf. [Boeh 82])

$$\left(\frac{\bar{n}-1}{\bar{n}}\right)\frac{\bar{\beta}_0\bar{\beta}_2}{\bar{\beta}_1^2}\frac{\Delta\bar{\boldsymbol{b}}_0\times\Delta\bar{\boldsymbol{b}}_1}{|\Delta\bar{\boldsymbol{b}}_0|^3} = \left(\frac{n-1}{n}\right)\frac{\beta_n\beta_{n-2}}{\beta_{n-1}^2}\frac{\Delta\boldsymbol{b}_{n-1}\times\Delta\boldsymbol{b}_{n-2}}{|\Delta\boldsymbol{b}_{n-1}|^3}.$$

Finally, for *torsion continuity* we also need

$$\left(\frac{\bar{n}-2}{\bar{n}}\right)\frac{\bar{\beta}_0\bar{\beta}_3}{\bar{\beta}_1\bar{\beta}_2}\frac{(\Delta\bar{\boldsymbol{b}}_0,\Delta\bar{\boldsymbol{b}}_1,\Delta\bar{\boldsymbol{b}}_2)}{|\Delta\bar{\boldsymbol{b}}_0\times\Delta\bar{\boldsymbol{b}}_1|^2} = \left(\frac{n-2}{n}\right)\frac{\beta_n\beta_{n-3}}{\beta_{n-1}\beta_{n-2}}\frac{(\Delta\boldsymbol{b}_{n-1},\Delta\boldsymbol{b}_{n-2},\Delta\boldsymbol{b}_{n-3})}{|\Delta\boldsymbol{b}_{n-1}\times\Delta\boldsymbol{b}_{n-2}|^2},$$

see also [Boeh 87a]. Rewriting these continuity conditions in terms of the Bézier polygon, then analogous to the non-rational case discussed in the previous sections, for $r=1$ (tangent continuity), we require

$$(\bar{\beta}_1 + qN_1\omega_{11}\beta_{n-1})\boldsymbol{b}_n = \bar{\beta}_1\bar{\boldsymbol{b}}_1 + qN_1\omega_{11}\beta_{n-1}\boldsymbol{b}_{n-1},$$

with q, N_1 as in Sect. 5.4, assuming that the denominators of $\boldsymbol{X}(u)$ and $\boldsymbol{Y}(t)$ (the homogenizing weight functions) join together continuously (so that $\bar{\beta}_0 = \beta_n$). Fig. 5.16 shows the continuity conditions for a curvature continuous join ($r=2$), and also shows the embedding of the curvature continuity

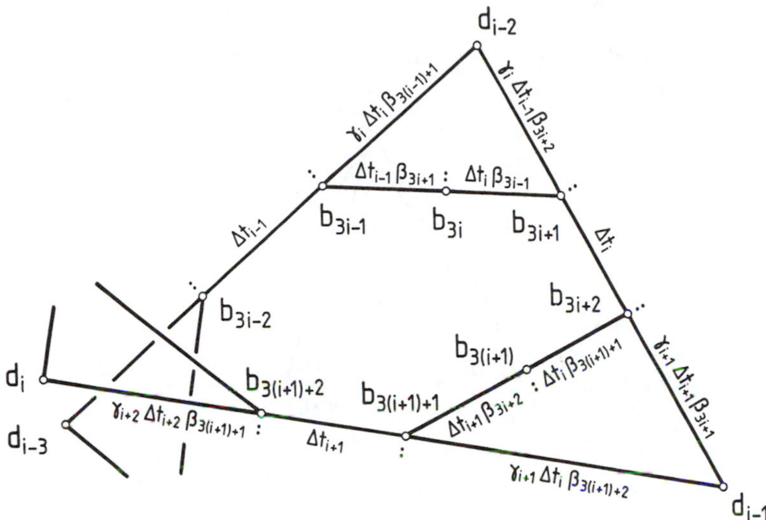

Fig. 5.16. de Boor and Bézier Polygon for a curvature continuous rational cubic spline.

conditions into a cubic B-spline representation [Boeh 87a]. Corresponding β-spline representations (using the monomial basis) can be found in [Bars 88] for a uniform knot vector, and in [Joe 89] for a general non-uniform knot vector. The de Boor points are denoted by $\boldsymbol{d}_i$ as before. Curvature continuous quadratic rational curves were discussed in [Far 89a]; torsion continuous quartic rational curves in [Boeh 87a], [Las 91], and [Joe 90a]; and FC^4 continuous quintic rational curves in [Las 91].

The Bézier ordinates $b_{ni+j,i}$ of the rational geometric B-splines $G_i^n(t)$ appearing in the B-spline representation

$$\boldsymbol{X}(t) = \sum_i \boldsymbol{d}_i G_i^n(t)$$

can be found in an analogous way to the approach in Sect. 5.5.2. But now the local basis functions $G_i^n(t)$ are, in general, only C^0 continuous. This is due to the unequal choice of the weights β_{ni-1} and β_{ni+1} which results in a bend in the polygon at the point $\boldsymbol{b}_{ni}$. Fig. 5.17 illustrates this behavior for the case $n = 3$.

We also want to mention papers of Nielson [Nie 84] and Jordan and Schindler [Jor 84] which deal with curves of the form

$$\boldsymbol{X}(t) = \sum_{j=0}^{2k} \boldsymbol{P}_j R_j(t), \tag{5.40}$$

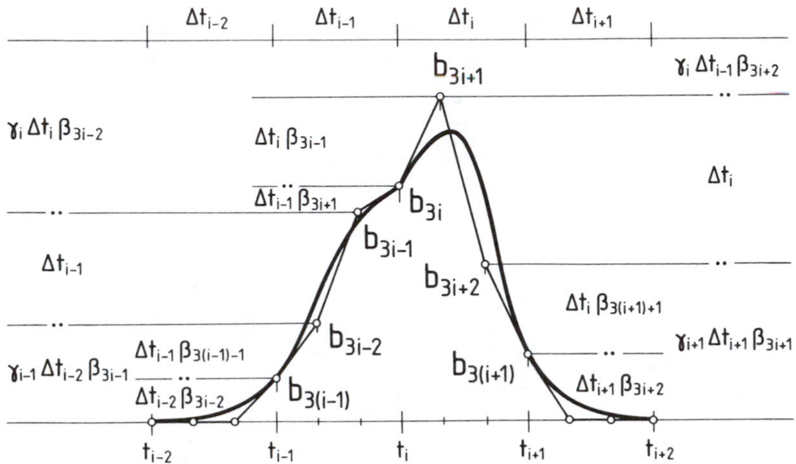

Fig. 5.17. Local basis function $G_i^3(t)$ for a curvature continuous rational cubic B-spline curve.

with special rational basis functions $R_j(t)$ defined by

$$\begin{aligned} R_0(t) &= \frac{1}{N(t)}(1-t)^{2k+1}, \\ R_j(t) &= \frac{1}{N(t)}[\alpha_j - (-1)^j](1-t)^{2k-j}t^j, \\ R_{2k}(t) &= \frac{1}{N(t)}t^{2k+1}, \end{aligned}$$

with

$$N(t) = (1-t)^{2k+1} + \sum_{i=1}^{2k-1}[\alpha_i - (-1)^i](1-t)^{2k-i}t^i + t^{2k+1}.$$

The $R_j(t)$ possess the properties of the ordinary Bernstein polynomials $B_j^n(t)$, and so the associated control points $\boldsymbol{P}_j$ have a kind of *Bézier point character*, $\boldsymbol{X}(t)$ has the convex hull property, etc. The *design parameters* α_j have an effect similar to that of the weights β_j in the rational Bézier form: an increase in α_j "pulls" the curve in the direction $\boldsymbol{P}_j$. Since

$$\begin{aligned} \boldsymbol{X}'(0) &= (\alpha_1 + 1)(\boldsymbol{P}_1 - \boldsymbol{P}_0), \\ \boldsymbol{X}'(1) &= (\alpha_{2k-1} + 1)(\boldsymbol{P}_{2k} - \boldsymbol{P}_{2k-1}), \end{aligned}$$

and

$$\kappa(0) = \frac{2(\alpha_2 - 1)}{(\alpha_1 + 1)^2} \frac{|\boldsymbol{P}_2 - \boldsymbol{P}_1| \cdot |\sin \gamma_1|}{|\boldsymbol{P}_1 - \boldsymbol{P}_0|^2},$$

$$\kappa(1) = \frac{2(\alpha_{2k-2} - 1)}{(\alpha_{2k-1} + 1)^2} \frac{|\boldsymbol{P}_{2k-2} - \boldsymbol{P}_{2k-1}| \cdot |\sin \gamma_{2k}|}{|\boldsymbol{P}_{2k} - \boldsymbol{P}_{2k-1}|^2},$$

with γ_i the angle formed by the points $\boldsymbol{P}_{i-1}, \boldsymbol{P}_i, \boldsymbol{P}_{i+1}$, using (5.40) to define the segments, it is, for example, possible to construct tangent continuous and curvature continuous cubic ($k = 1$) (sub) splines [Nie 84], [Jor 84].

Because of the special form of the $R_j(t)$, the derivatives of $\boldsymbol{X}(t)$, at the vertices in particular, are easier to find than in the general rational case. Thus, for the torsion of $\boldsymbol{X}(t)$ at $t = 0$ and $t = 1$, we have

$$\tau(0) = \frac{3(\alpha_3 + 1)}{(\alpha_1 + 1)(\alpha_2 - 1)} \frac{|\boldsymbol{P}_3 - \boldsymbol{P}_1|}{|\boldsymbol{P}_1 - \boldsymbol{P}_0| \cdot |\boldsymbol{P}_2 - \boldsymbol{P}_1|} \frac{|\cos \delta_1|}{|\sin \gamma_1|},$$

$$\tau(1) = \frac{3(\alpha_{2k-3} + 1)}{(\alpha_{2k-1} + 1)(\alpha_{2k-2} - 1)} \frac{|\boldsymbol{P}_{2k-3} - \boldsymbol{P}_{2k-1}|}{|\boldsymbol{P}_{2k} - \boldsymbol{P}_{2k-1}| \cdot |\boldsymbol{P}_{2k-2} - \boldsymbol{P}_{2k-1}|} \frac{|\cos \delta_{2k}|}{|\sin \gamma_{2k}|},$$

where δ_1 is the angle between $(\boldsymbol{P}_1 - \boldsymbol{P}_0) \times (\boldsymbol{P}_2 - \boldsymbol{P}_1)$ and $\boldsymbol{P}_3 - \boldsymbol{P}_1$ (which thus measures the deviation from the osculatory plane), and where δ_{2k} is defined analogously. For $k = 3$ this permits the construction of torsion continuous or GC^3 continuous rational quintic (sub)spline curves which are useful for applications.

Generalizations to FC^r continuity of higher order based on simplices can be found in [Las 91], and in [Goo 90] (where the approach follows a similar C^r continuity construction in [Ram 87]).

5.7.2. Rational GC^r Continuous Spline Curves

To construct GC^r continuous rational curves using homogeneous coordinates, we must first describe the equations corresponding to the continuity conditions in (5.6) ff., see [Deg 88]. Suppose we are given two parametrized curves $\boldsymbol{X}(u)$ and $\boldsymbol{Y}(t)$ represented in terms of homogeneous coordinates, *i.e.*, as curves $\mathcal{X}(u) : [u_0, u_1] \to \mathbb{R}^4$ and $\mathcal{Y}(t) : [t_0, t_1] \to \mathbb{R}^4$ in a four dimensional homogeneous space. Then for any nonvanishing function $\eta(u)$, $\bar{\mathcal{X}}(u) = \eta(u)\mathcal{X}(u)$ and $\mathcal{X}(u)$ represent the same curve in $\mathbb{R}^3$, and an analogous assertion holds for $\mathcal{Y}(t)$. In view of this, the two curves meet continuously at a regular point provided that

$$\mathcal{Y}(t_0^+) = \alpha_0 \mathcal{X}(u_1^-),$$

where $\alpha_0 = \alpha(u_1) \neq 0$. Then analogous to Sect. 5.2, applying the chain rule, we get the following GC^r continuity conditions:

$$\frac{d^\rho}{dt^\rho}\mathcal{Y}(t)\Big|_{t_0^+} = \frac{d^\rho}{dt^\rho}\Big(\alpha(t)\mathcal{X}(u(t))\Big)\Big|_{u_1^-=u(t_0^-)} \qquad \rho = 0(1)r, \tag{5.41}$$

where $\mathcal{X}(u)$ is reparametrized by $u \mapsto u(t)$ with $\omega_0 = u(t_0^-) = u_1^-$. Carrying out the differentiation on the right-hand side of (5.41) yields

$$\begin{aligned}
\mathcal{Y}'(t_0^+) &= \alpha_0\omega_1\mathcal{X}'(u_1^-) + \alpha_1\mathcal{X}(u_1^-)\\
\mathcal{Y}''(t_0^+) &= \alpha_0\omega_1^2\mathcal{X}''(u_1^-) + (\alpha_0\omega_2 + 2\alpha_1\omega_1)\mathcal{X}'(u_1^-) + \alpha_2\mathcal{X}(u_1^-)\\
\mathcal{Y}'''(t_0^+) &= \alpha_0\omega_1^3\mathcal{X}'''(u_1^-) + 3(\alpha_0\omega_1\omega_2 + \alpha_1\omega_1^2)\mathcal{X}''(u_1^-)\\
&\quad + (3\alpha_2\omega_1 + 3\alpha_1\omega_2 + \alpha_0\omega_3)\mathcal{X}'(u_1^-) + \alpha_3\mathcal{X}(u_1^-),
\end{aligned} \tag{5.42}$$

etc. (a recurrence formula is given in [Deg 88]). Here

$$\alpha_\rho = \frac{d^\rho}{dt^\rho}\alpha(t)|_{t=t_0^-}, \qquad \omega_\rho = \frac{d^\rho}{dt^\rho}u(t)|_{t=t_0^-}, \tag{5.43}$$

where as a consequence of the regularity assumption, we have $\omega_1 > 0$.

This means that if the two curves join with GC^r continuity, then the equations (5.42) must be satisfied, where the variables α_ρ and ω_ρ are given by (5.43).

On the other hand, given two curves which meet at a point, if we can find real numbers $\omega_0, \ldots, \omega_r$ and $\alpha_0, \ldots, \alpha_r$ so that the equations (5.42) are satisfied, then there is a function $\alpha(t)$ and a reparametrization to a global common parameter such that with respect to this global parametrization, $\mathcal{X}(u)$ and $\mathcal{Y}(t)$ meet with C^r continuity, and thus the curves are GC^r continuous. The parameter transformation $u(t)$ and the function $\alpha(t)$ have to be chosen so that the boundary conditions in (5.43) are satisfied, for example by using a Taylor series expansion.

A GC^1 continuous join can be transformed via a linear rational parameter transformation to a C^1 continuous join [Deg 88]. This is of special interest in CAGD since a linear rational parameter transformation does not change the degree of a rational curve, cf. e.g., [Boeh 88b].

To evaluate the quantities in (5.42) for given curves $\mathcal{X}(u)$ and $\mathcal{Y}(t)$ in Bézier form with homogeneous coordinates (4.27)–(4.28), we may use the homogeneous representations $\mathcal{X}(u)$ and $\mathcal{Y}(t)$. In this case, and now we see the big advantage of homogeneous coordinates: the derivatives can be found without using the quotient rule. Inserting in (5.42) and comparing coefficients leads immediately to

$$\Delta^\rho \bar{\boldsymbol{b}}_0 = \sum_{k=0}^{\rho} A_{\rho k} \frac{(n-\rho)!}{(n-k)!} \Delta^k \boldsymbol{b}_{n-k}, \qquad \rho = 1(1)r, \tag{5.44}$$

where $A_{\rho k}$ denotes the elements of the connection matrix A defined by (5.42), and $A_{00} = \alpha_0 = 1$, by the continuity of the homogenizing weight functions at the point $\boldsymbol{b}_n = \bar{\boldsymbol{b}}_0$. Equation (5.44) allows the recursive computation of the $\bar{\boldsymbol{b}}_i$ from the $\boldsymbol{b}_i$, see [Deg 88].

The affine case of Sect. 5.2 is also contained in (5.41), of course. Writing ξ_i, $i = 0(1)3$, for the homogeneous components of $\mathcal{X}(u)$, then the affine coordinates x_i, $i = 1, 2, 3$, of $\mathbb{R}^3$ are given by $x_i = \xi_i/\xi_0$ (cf. with Sect. 4.1.4). For $\alpha_0 = 1, \alpha_1 = \cdots = \alpha_r = 0$, the equations (5.41) formally reduce to (5.6).

The necessary and sufficient conditions (5.42) of [Deg 88] (see also [Hoh 89]) lead to another important observation. Applying the (unmodified) GC^r conditions of Sect. 5.2 to the homogeneous representations is too restrictive, since (5.42) implies that GC^r continuous rational curves do not have to come from projection into $\mathbb{R}^3$ of GC^r continuous (in the sense of Sect. 5.2) polynomial curves in $\mathbb{R}^4$. Polynomial curves in $\mathbb{R}^4$ which are not GC^r continuous (in the sense of Sect. 5.2), but which satisfy (5.42) with $\alpha_\rho \neq 1$, also produce GC^r continuous rational curves in $\mathbb{R}^3$! The special case of C^r continuous curves (*i.e.*, $\omega_\rho = 1$) was treated in detail in [Hoh 89]. In [Mano 90], basis functions were developed for the homogeneous representation of C^2 continuous cubic rational curves using monomials and one-sided power functions, and the influence of the α_ρ on the shape of the curve was investigated. The influence of α_0 and α_1 is comparable with that of ω_1 and ω_2, while varying α_2 leaves the point where the two segments join invariant.

[Joe 90a] generalized the results of [Joe 87, 89, 90] to rational quartic GC^3 continuous spline curves in both B-spline and Bézier form. In addition to allowing multiple knots and discussing knot insertion, his paper gives an especially detailed treatment of curve design.

6
Spline Surfaces

6.1. Introduction

In earlier chapters we have dealt with (classical) spline curves, Bézier-spline curves, B-splines, and geometric spline curves. We turn now to spline surfaces. We distinguish between

- tensor-product surfaces on a rectangular parameter domain,
- surfaces with triangular or $(2n+1)$-sided parameter domain,
- transfinite methods (Coons patches).

Here we deal mainly with classical continuity conditions. Spline surfaces with geometric smoothness are dealt with in Chap. 7, and transfinite methods will be discussed in Chap. 8.

In constructing interpolating and approximating surfaces, we have to differentiate between two kinds of data:

- regularly spaced data,
- scattered data.

Scattered data methods will be discussed in Chap. 9. The methods developed in this chapter can also be extended to $\mathbb{R}^3$, in which case we can work with parameter domains which have the form of parallelepipeds or tetrahedra, see Chap. 11.

6.2. Tensor-product Surfaces

Tensor-product surfaces can be defined using a variety of basis functions, but we will be interested primarily in Bernstein polynomials and B-splines as basis functions since they have convenient *geometric properties*.

Tensor-product surfaces can be obtained easily from the corresponding representations of curves. We begin with a curve

$$\boldsymbol{X}(u) = \sum_{i=0}^{n} \boldsymbol{C}_i F_i(u)$$

in $\mathbb{R}^2$ or $\mathbb{R}^3$, where $F_i(u)$ are basis functions. Next we move this curve through space, allowing at the same time deformations in $\boldsymbol{X}(u)$. This motion can be described in terms of a parameter v via

$$\boldsymbol{C}_i(v) = \sum_{k=0}^{m} \boldsymbol{A}_{ik} G_k(v).$$

The resulting surface

$$\boldsymbol{X}(u, v) = \sum_{i=0}^{n} \sum_{k=0}^{m} \boldsymbol{A}_{ik}\ F_i(u) G_k(v),$$

is called a *tensor-product surface*, where $a \leq u \leq b$, $c \leq v \leq d$, i.e., the parameters (u, v) run over a rectangular domain.

If the basis functions $F_i(u)$ and $G_k(v)$ are chosen to be monomials, then the tensor-product surface has the form

$$\boldsymbol{X}(u, v) = \sum_{i=0}^{n} \sum_{k=0}^{m} \boldsymbol{A}_{ik} u^i v^k.$$

The above equations for a tensor-product surface can also be written in matrix form as

$$\boldsymbol{X}(u, v) = (F_0(u), \cdots, F_n(u)) \begin{pmatrix} \boldsymbol{A}_{00}, & \cdots, & \boldsymbol{A}_{0m} \\ \vdots & & \vdots \\ \boldsymbol{A}_{n0} & \cdots & \boldsymbol{A}_{nm} \end{pmatrix} \begin{pmatrix} G_0(v) \\ \vdots \\ G_m(v) \end{pmatrix}.$$

The interpolation problem

$$\boldsymbol{X}(u_j, v_\ell) = P_{j\ell}, \qquad j = 0(1)n, \quad \ell = 0(1)m,$$

(assuming that the conditions of Sect. 2.2.5 guaranteeing the existence of a solution are satisfied) leads to the system of equations

$$\boldsymbol{X} = \boldsymbol{FAG}, \tag{$*$}$$

with

$$X = \begin{pmatrix} P_{00} & \cdots & P_{0m} \\ \vdots & & \vdots \\ P_{n0} & \cdots & P_{nm} \end{pmatrix} \qquad A = \begin{pmatrix} A_{00} & \cdots & A_{0m} \\ \vdots & & \vdots \\ A_{n0} & \cdots & A_{nm} \end{pmatrix}$$

$$F = \begin{pmatrix} F_0(u_0) & \cdots & F_n(u_0) \\ \vdots & & \vdots \\ F_0(u_n) & \cdots & F_n(u_n) \end{pmatrix} \qquad G = \begin{pmatrix} G_0(v_0) & \cdots & G_n(v_m) \\ \vdots & & \vdots \\ G_m(v_0) & \cdots & G_m(v_m) \end{pmatrix}.$$

The unknown coefficients are then given by

$$A = F^{-1} X G^{-1}.$$

Under assumption b) of Sect. 2.2.5, the interpolation problem for tensor-product surfaces can be solved in an extremely efficient way. To see this, we write $(*)$ as

$$X = FD \qquad \text{where} \qquad D = AG.$$

This allows us to use the tensor-product structure in order to solve the interpolation problem by repeated application of a curve interpolation scheme. For example, we can perform curve interpolation

- in the u-direction for all k, and then
- in the v-direction for all i.

Starting with the prescribed points $P_{j\ell}$ and associated parameter values (u_j, v_ℓ), the first step leads to intermediate values $d_i(v_\ell)$ satisfying the matrix equation

$$FD = X,$$

i.e.,

$$\begin{pmatrix} F_0(u_0) & \cdots & F_n(u_0) \\ \vdots & & \vdots \\ F_0(u_n) & \cdots & F_n(u_n) \end{pmatrix} \begin{pmatrix} d_0(v_0) & \cdots & d_0(v_m) \\ \vdots & & \vdots \\ d_n(v_0) & \cdots & d_n(v_m) \end{pmatrix} = \begin{pmatrix} P_{00} & \cdots & P_{0m} \\ \vdots & & \vdots \\ P_{n0} & \cdots & P_{nm} \end{pmatrix}.$$

The second step computes the coefficients A_{ik} in terms of the intermediate values $d_i(v_\ell)$ via the matrix equation

$$G^T A^T = D^T,$$

i.e.,

$$\begin{pmatrix} G_0(v_0) & \cdots & G_n(v_m) \\ \vdots & & \vdots \\ G_m(v_0) & \cdots & G_m(v_m) \end{pmatrix} \begin{pmatrix} \boldsymbol{A}_{00} & \cdots & \boldsymbol{A}_{0m} \\ \vdots & & \vdots \\ \boldsymbol{A}_{n0} & \cdots & \boldsymbol{A}_{nm} \end{pmatrix} = \begin{pmatrix} \boldsymbol{d}_0(v_0) & \cdots & \boldsymbol{d}_n(v_0) \\ \vdots & & \vdots \\ \boldsymbol{d}_0(v_m) & \cdots & \boldsymbol{d}_n(v_m) \end{pmatrix}.$$

If we do not make use of the tensor-product structure, we would have to solve a system of $(n+1)(m+1) \times (n+1)(m+1)$ equations. By exploiting the tensor-product structure, we have reduced the problem to solving $m+1$ systems of equations with a fixed coefficient matrix $\boldsymbol{F}$ of size $(n+1) \times (n+1)$, followed by solving $n+1$ systems of equations with a fixed coefficient matrix $\boldsymbol{G}$ of size $(m+1) \times (m+1)$.

The interpolation problem for multivariate tensor-products can be handled in the same way, see [Boo 77], [BOO 78], [Las 85, 87], and [Alf 89].

6.2.1. Bicubic Polynomial Splines

In analogy with cubic spline curves, in this section we develop *bicubic interpolating splines*, see e.g., [Boo 62]. We start with $(n+1)(m+1)$ prescribed points $\boldsymbol{P}_{ij}$ and associated parameter values (u_i, v_j), which form a *rectangular grid* in the (u, v)-parameter plane, see Fig. 6.1.

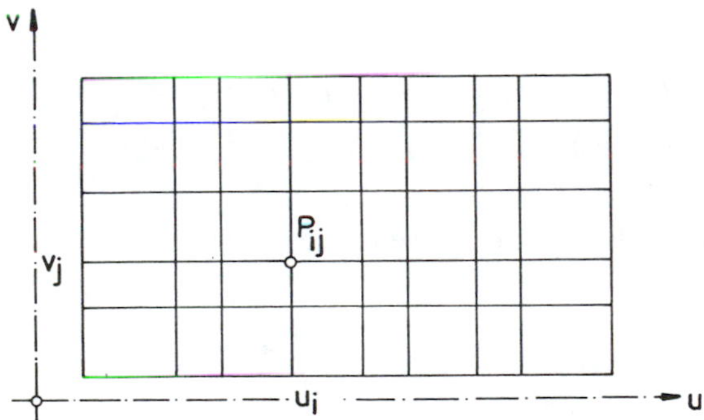

Fig. 6.1. Grid of points in the parameter plane.

Our aim is to find an *interpolating bicubic spline surface* whose polynomial pieces are given by

$$\boldsymbol{X}_{ij}(u, v) = \sum_{k=0}^{3} \sum_{\ell=0}^{3} \boldsymbol{A}_{ijk\ell}(u - u_i)^k (v - v_j)^\ell, \tag{6.1}$$

for $i = 0(1)n\text{–}1$, $j = 0(1)m\text{–}1$.

For fixed (i,j), (6.1) represents a bicubic *patch* which we want to use to interpolate the points $\{\boldsymbol{P}_{ij}, \boldsymbol{P}_{i+1,j}, \boldsymbol{P}_{i+1,j+1}, \boldsymbol{P}_{i,j+1}\}$. We have

$$\begin{array}{ll} \text{interpolation:} & \boldsymbol{X}_{ij}(u_i, v_j) = \boldsymbol{P}_{ij} \\ \text{continuity:} & \dfrac{\partial \boldsymbol{X}_{ij}}{\partial u}, \quad \dfrac{\partial \boldsymbol{X}_{ij}}{\partial v}, \quad \dfrac{\partial^2 \boldsymbol{X}_{ij}}{\partial u \partial v} \quad \text{continuous.} \end{array} \tag{6.2}$$

The expression (6.1) can be written in the following matrix form:

$$\boldsymbol{X}_{ij}(u,v) = \boldsymbol{U}^T(u,u_i)\boldsymbol{A}_{ij}\boldsymbol{V}(v,v_j), \tag{6.3}$$

where

$$\boldsymbol{A}_{ij} = \begin{pmatrix} \boldsymbol{A}_{ij00} & \boldsymbol{A}_{ij01} & \boldsymbol{A}_{ij02} & \boldsymbol{A}_{ij03} \\ \boldsymbol{A}_{ij10} & \boldsymbol{A}_{ij11} & \boldsymbol{A}_{ij12} & \boldsymbol{A}_{ij13} \\ \boldsymbol{A}_{ij20} & \boldsymbol{A}_{ij21} & \boldsymbol{A}_{ij22} & \boldsymbol{A}_{ij23} \\ \boldsymbol{A}_{ij30} & \boldsymbol{A}_{ij31} & \boldsymbol{A}_{ij32} & \boldsymbol{A}_{ij33} \end{pmatrix},$$

and

$$\begin{aligned} \boldsymbol{U}(u,u_i) &= (1, (u-u_i), (u-u_i)^2, (u-u_i)^3)^T \\ \boldsymbol{V}(v,v_j) &= (1, (v-v_j), (v-v_j)^2, (v-v_j)^3)^T. \end{aligned}$$

To compute the coefficients $\boldsymbol{A}_{ijk\ell}$, we proceed as in Chap. 3 and introduce the notation

$$\begin{aligned} &\frac{\partial}{\partial u}\boldsymbol{X}_{ij}(u_i,v_j) =: \boldsymbol{p}_{ij}, \qquad \frac{\partial}{\partial v}\boldsymbol{X}_{ij}(u_i,v_j) =: \boldsymbol{q}_{ij}, \\ &\frac{\partial^2}{\partial u \partial v}\boldsymbol{X}_{ij}(u_i,v_j) =: \boldsymbol{r}_{ij}. \end{aligned} \tag{6.4}$$

Now the coefficients in (6.1) resp. (6.3) can be written as

$$\begin{aligned} \boldsymbol{X}_{ij}(u_i,v_j) &= \boldsymbol{P}_{ij} = \boldsymbol{A}_{ij00}, \\ \frac{\partial}{\partial u}\boldsymbol{X}_{ij}(u_i,v_j) &= \boldsymbol{p}_{ij} = \boldsymbol{A}_{ij10}, \qquad \frac{\partial}{\partial v}\boldsymbol{X}_{ij}(u_i,v_j) = \boldsymbol{q}_{ij} = \boldsymbol{A}_{ij01}, \\ \frac{\partial^2}{\partial u \partial v}\boldsymbol{X}_{ij}(u_i,v_j) &= \boldsymbol{r}_{ij} = \boldsymbol{A}_{ij11}. \end{aligned} \tag{6.5a}$$

In addition, setting $\Delta u_i := u_{i+1} - u_i$ and $\Delta v_j := v_{j+1} - v_j$, we also have

$$\boldsymbol{X}_{ij}(u_{i+1}, v_j) = \boldsymbol{P}_{i+1,j} = (1,\ \Delta u_i,\ (\Delta u_i)^2,\ (\Delta u_i)^3)\ \boldsymbol{A}_{ij} \begin{pmatrix} 1 \\ 0 \\ 0 \\ 0 \end{pmatrix},$$

$$\frac{\partial}{\partial u}\boldsymbol{X}_{ij}(u_{i+1}, v_j) = \boldsymbol{p}_{i+1,j} = (0,\ 1,\ 2\Delta u_i,\ 3(\Delta u_i)^2)\boldsymbol{A}_{ij} \begin{pmatrix} 1 \\ 0 \\ 0 \\ 0 \end{pmatrix}, \tag{6.5b}$$

$$\frac{\partial}{\partial v}\boldsymbol{X}_{ij}(u_i, v_{j+1}) = \boldsymbol{q}_{i,j+1} = (1,0,0,0)\boldsymbol{A}_{ij} \begin{pmatrix} 0 \\ 1 \\ 2\Delta v_j \\ 3(\Delta v_j)^2 \end{pmatrix}.$$

Let

$$\boldsymbol{W}_{ij} := \begin{pmatrix} \boldsymbol{P}_{ij} & \boldsymbol{q}_{ij} & \boldsymbol{P}_{i,j+1} & \boldsymbol{q}_{i,j+1} \\ \boldsymbol{p}_{ij} & \boldsymbol{r}_{ij} & \boldsymbol{p}_{i,j+1} & \boldsymbol{r}_{i,j+1} \\ \boldsymbol{P}_{i+1,j} & \boldsymbol{q}_{i+1,j} & \boldsymbol{P}_{i+1,j+1} & \boldsymbol{q}_{i+1,j+1} \\ \boldsymbol{p}_{i+1,j} & \boldsymbol{r}_{i+1,j} & \boldsymbol{p}_{i+1,j+1} & \boldsymbol{r}_{i+1,j+1} \end{pmatrix}, \tag{6.6}$$

$$\boldsymbol{G}(t_i) := \begin{pmatrix} 1 & 0 & 0 & 0 \\ 0 & 1 & 0 & 0 \\ 1 & \Delta t_i & (\Delta t_i)^2 & (\Delta t_i)^3 \\ 0 & 1 & 2\Delta t_i & 3(\Delta t_i)^2 \end{pmatrix},$$

$$\boldsymbol{G}^{-1}(t_i) = \begin{pmatrix} 1 & 0 & 0 & 0 \\ 0 & 1 & 0 & 0 \\ -\frac{3}{(\Delta t_i)^2} & -\frac{2}{\Delta t_i} & \frac{3}{(\Delta t_i)^2} & -\frac{1}{\Delta t_i} \\ \frac{2}{(\Delta t_i)^3} & \frac{1}{(\Delta t_i)^2} & -\frac{2}{(\Delta t_i)^3} & \frac{1}{(\Delta t_i)^2} \end{pmatrix}.$$

Then

$$\boldsymbol{W}_{ij} = \boldsymbol{G}(u_i)\boldsymbol{A}_{ij}\boldsymbol{G}^T(v_j), \tag{6.7a}$$

and the unknown matrix $\boldsymbol{A}_{ij}$ must satisfy

$$\boldsymbol{A}_{ij} = \boldsymbol{G}^{-1}(u_i)\boldsymbol{W}_{ij}[\boldsymbol{G}^T(v_j)]^{-1}. \tag{6.7b}$$

Thus, the spline coefficients $\boldsymbol{A}_{ijk\ell}$ of the (i,j)-th patch can be computed, once we have the coefficient matrix $\boldsymbol{W}_{ij}$. Our aim now is to compute the derivatives $\boldsymbol{p}_{ij}, \boldsymbol{q}_{ij}, \boldsymbol{r}_{ij}$ appearing in $\boldsymbol{W}_{ij}$.

Because of the assumption that the parameter values lie on a rectangular grid, the lines defined by $u = u_i$ and $v = v_j$ are *parametric lines* of the surface, which by (6.1) can be interpreted as *cubic spline curves*. This means that the derivatives $\boldsymbol{p}_{ij}, \boldsymbol{q}_{ij}, \boldsymbol{r}_{ij}$ can be found from the recurrence relations which lead to the coefficients of the cubic spline curves, see (3.22). For cubic spline interpolation, we enforced boundary conditions on the derivatives. Correspondingly, for bicubic spline surfaces, we shall assume that on the *boundary*, we can freely choose *derivatives*

$$\begin{array}{lll} \boldsymbol{p}_{ij} & \text{for } i = 0, n, & j = 0(1)m, \\ \boldsymbol{q}_{ij} & \text{for } i = 0(1)n, & j = 0, m, \\ \boldsymbol{r}_{ij} & \text{for } i = 0, n, & j = 0, m. \end{array} \tag{$*$}$$

One way to find the derivatives in $(*)$ is to estimate them using nearby data points. For example, we can

1) compute the *boundary derivatives*:

- for each $j = 0(1)m$, to estimate $\boldsymbol{p}_{0j}$, take the derivative with respect to u of the cubic polynomial which interpolates the boundary points $\boldsymbol{P}_{ij}$ with indices $i = 0, 1, 2, 3$. We estimate $\boldsymbol{p}_{nj}$ using the points with indices $i = n{-}3(1)n$,
- for each $i = 0(1)n$, to get $\boldsymbol{q}_{i0}$, take the derivative with respect to v of the cubic polynomial which interpolates the boundary points $\boldsymbol{P}_{ij}$ with indices $j = 0, 1, 2, 3$. We estimate $\boldsymbol{q}_{im}$ using the points with indices $j = m{-}3(1)m$,

2) for each $j = 0(1)m$, compute $\boldsymbol{p}_{ij}$ for $i = 1(1)n{-}1$ by solving

$$\begin{aligned} &\Delta u_i \boldsymbol{p}_{i-1,j} + 2\boldsymbol{p}_{ij}(\Delta u_i + \Delta u_{i-1}) + \Delta u_{i-1}\boldsymbol{p}_{i+1,j} \\ &\qquad = 3\frac{\Delta u_{i-1}}{\Delta u_i}(\boldsymbol{P}_{i+1,j} - \boldsymbol{P}_{ij}) - 3\frac{\Delta u_i}{\Delta u_{i-1}}(\boldsymbol{P}_{i-1,j} - \boldsymbol{P}_{ij}). \end{aligned} \tag{6.8a}$$

By the results of Chap. 3, this linear system has a unique solution once we are given the boundary values $\boldsymbol{p}_{ij}$ for $i = 0, n$ and $j = 0(1)m$,

3) for each $i = 0(1)n$, compute $\boldsymbol{q}_{ij}$ for $j = 1(1)m$–1 by solving

$$\begin{aligned}\Delta v_j \boldsymbol{q}_{i,j-1} + 2\boldsymbol{q}_{ij}(\Delta v_j + \Delta v_{j-1}) + \Delta v_{j-1}\boldsymbol{q}_{i,j+1} \\ = 3\frac{\Delta v_{j-1}}{\Delta v_j}(\boldsymbol{P}_{i,j+1} - \boldsymbol{P}_{ij}) - 3\frac{\Delta v_j}{\Delta v_{j-1}}(\boldsymbol{P}_{i,j-1} - \boldsymbol{P}_{ij}).\end{aligned} \tag{6.8b}$$

By the results of Chap. 3, this linear system has a unique solution once we are given the boundary values $\boldsymbol{q}_{ij}$ for $j = 0, m$ and $i = 0(1)n$,

4) compute the second derivatives $\boldsymbol{r}_{ij}$ for $i = 0, n$ and $j = 0, m$ using cubic polynomials. For example, to find $\boldsymbol{r}_{0m}$ we can find a polynomial $\boldsymbol{Y}_{q_{im}}(u)$ interpolating the data $\boldsymbol{q}_{im}$, $i = 0, 1, 2, 3$. Differentiating this polynomial with respect to u and evaluating at u_0 leads to the desired mixed derivatives $\boldsymbol{r}_{0m}$, since the $\boldsymbol{q}_{ij}$ are already derivatives with respect to v. The mixed derivatives at the corners could also be estimated by interpolating the $\boldsymbol{p}_{ij}$ with a polynomial $\boldsymbol{Y}_{p_{ij}}(v)$, and then differentiating with respect to v. This will, in general, lead to different values, since

$$\frac{\partial}{\partial u}\boldsymbol{Y}_{q_{im}}(u) \neq \frac{\partial}{\partial v}\boldsymbol{Y}_{p_{ij}}(v),$$

5) for $j = 0, m$, compute $\boldsymbol{r}_{ij}$ for $i = 1(1)n$–1 by solving the systems

$$\begin{aligned}\Delta u_i \boldsymbol{r}_{i-1,j} + 2\boldsymbol{r}_{ij}(\Delta u_i + \Delta u_{i-1}) + \Delta u_{i-1}\boldsymbol{r}_{i+1,j} \\ = 3\frac{\Delta u_{i-1}}{\Delta u_i}(\boldsymbol{q}_{i+1,j} - \boldsymbol{q}_{ij}) - 3\frac{\Delta u_i}{\Delta u_{i-1}}(\boldsymbol{q}_{i-1,j} - \boldsymbol{q}_{ij}).\end{aligned} \tag{6.8c}$$

For $j = 0$, given the boundary values $\boldsymbol{r}_{i0}$ for $i = 0, n$, this system is uniquely solvable for $\boldsymbol{r}_{i0}$, $i = 1(1)n$–1. A similar assertion holds for $j = m$,

6) for $i = 0(1)n$ compute $\boldsymbol{r}_{ij}$ for $j = 1(1)m$–1 by solving the systems

$$\begin{aligned}\Delta v_j \boldsymbol{r}_{i,j-1} + 2\boldsymbol{r}_{ij}(\Delta v_j + \Delta v_{j-1}) + \Delta v_{j-1}\boldsymbol{r}_{i,j+1} \\ = 3\frac{\Delta v_{j-1}}{\Delta v_j}(\boldsymbol{p}_{i,j+1} - \boldsymbol{p}_{ij}) - 3\frac{\Delta v_j}{\Delta v_{j-1}}(\boldsymbol{p}_{i,j-1} - \boldsymbol{p}_{ij}).\end{aligned} \tag{6.8d}$$

For each i, given the boundary values $\boldsymbol{r}_{ij}$ for $j = 0, m$, by (4)–(5), this system is uniquely solvable for $\boldsymbol{r}_{ij}$, $j = 1(1)m$–1.

7) insert the values $\boldsymbol{p}_{ij}, \boldsymbol{q}_{ij}, \boldsymbol{r}_{ij}$ into (6.6) resp. (6.7b), and compute the coefficients of the interpolant (6.1).

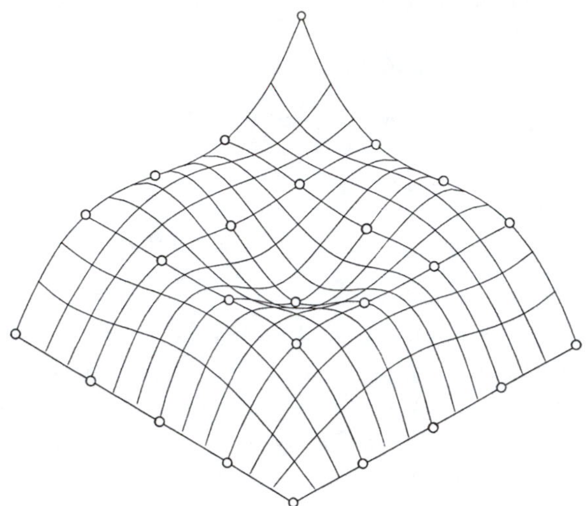

Fig. 6.2. Bicubic spline surface interpolating 5 × 5 points (circles).

Fig. 6.2 shows an example of a bicubic interpolating spline surface. The prescribed points are marked with circles.

6.2.2. Tensor-product Bézier Surfaces

In this section we use the *Bernstein polynomials* $B_i^n(u)$ and $B_k^m(v)$ as basis functions. This leads to the *parametric formula*

$$\boldsymbol{X}(u,v) = \sum_{i=0}^{n} \sum_{k=0}^{m} \boldsymbol{b}_{ik}\, B_i^n(u) B_k^m(v) \tag{6.9}$$

for the *tensor-product Bézier surfaces* of degree (n, m), where the points $u, v \in [0,1] \times [0,1]$. The coefficients $\boldsymbol{b}_{ik}$ are called the *Bézier points*, and the set of Bézier points is referred to as the *Bézier net*. The sequences of Bézier points $\{\boldsymbol{b}_{ik}\}$ in a Bézier net corresponding to a fixed i or k will be called *threads* of the Bézier net. The points $\boldsymbol{b}_{00}, \boldsymbol{b}_{n0}, \boldsymbol{b}_{0m}, \boldsymbol{b}_{nm}$ are corner points of the Bézier surface. The point sets $\{\boldsymbol{b}_{0k}\}$, $\{\boldsymbol{b}_{i0}\}$, $\{\boldsymbol{b}_{im}\}$, $\{\boldsymbol{b}_{nk}\}$ are the Bézier points of the boundary curves lying on the Bézier surface. The points

$\{\boldsymbol{b}_{00}, \boldsymbol{b}_{10}, \boldsymbol{b}_{01}\}$	determine the *tangent plane* at $\boldsymbol{b}_{00}$,
$\{\boldsymbol{b}_{n0}, \boldsymbol{b}_{n-1,0}, \boldsymbol{b}_{n1}\}$	determine the *tangent plane* at $\boldsymbol{b}_{n0}$,
$\{\boldsymbol{b}_{0m}, \boldsymbol{b}_{1m}, \boldsymbol{b}_{0,m-1}\}$	determine the *tangent plane* at $\boldsymbol{b}_{0m}$,
$\{\boldsymbol{b}_{nm}, \boldsymbol{b}_{n-1,m}, \boldsymbol{b}_{n,m-1}\}$	determine the *tangent plane* at $\boldsymbol{b}_{nm}$.

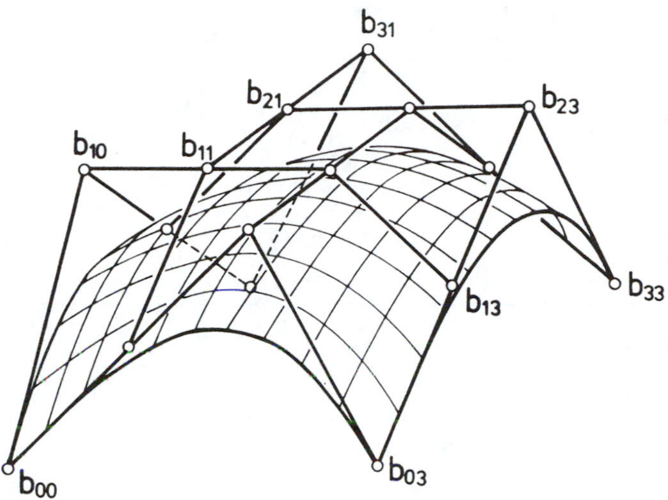

Fig. 6.3. Bézier surface of degree (3,3) and its Bézier net.

Fig. 6.3 shows a Bézier surface of degree (3,3) and its associated Bézier net.

The coefficients in (6.9) can be either vector-valued or real-valued. In the latter case we refer to them as *Bézier ordinates* $b_{ik} \in \mathbb{R}$. Inserting Bézier ordinates in (6.9) leads to a surface in $\mathbb{R}^3$ described by a function defined over the unit square. Then, as is the case for curves, the Bézier ordinates b_{ik} are to be associated with points (abscissae) with the parameter values $(i/n, k/m)$. Fig. 6.4 illustrates schematically the correspondence between Bézier ordinates and their abscissae for a Bézier function of degree (n, m). If in this case we define vector-valued Bézier points $\boldsymbol{b}_{ik}$ by $\boldsymbol{b}_{ik} = (i/n, k/m, b_{ik})$, then all of the following algorithms for the vector-valued case apply to real-valued Bézier surfaces by working with the vector-valued representation $\boldsymbol{X}(u,v) = (x, y, X(u,v))$ with $x = u$, $y = v$, cf. Sect. 4.1.

Along the parametric line $u = u_0$, the surface reduces to a Bézier curve with Bézier points

$$\boldsymbol{b}_k = \sum_{i=0}^{n} \boldsymbol{b}_{ik} B_i^n(u_0). \tag{6.10a}$$

Then with respect to the variable v, (6.9) reduces to

$$\boldsymbol{X}(u_0, v) = \sum_{k=0}^{m} \boldsymbol{b}_k B_k^m(v). \tag{6.10b}$$

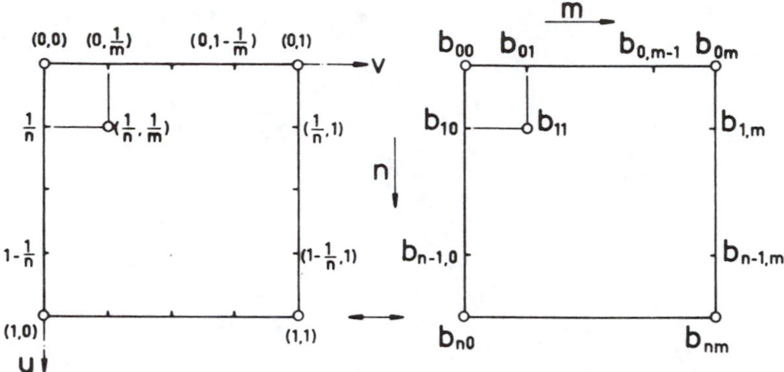

Fig. 6.4. Schematic correspondence between parameter values and Bézier ordinates.

The lines on the surface shown in Fig. 6.3 are in fact a set of parametric lines.

As in Sect. 4.1, the derivatives of $\boldsymbol{X}$ are given by

$$\frac{\partial^r}{\partial u^r}\boldsymbol{X}(u,v) = \frac{n!}{(n-r)!}\sum_{i=0}^{n-r}\sum_{k=0}^{m}\Delta^{r0}\boldsymbol{b}_{ik}\;B_i^{n-r}(u)B_k^m(v),$$

$$\frac{\partial^s}{\partial v^s}\boldsymbol{X}(u,v) = \frac{m!}{(m-s)!}\sum_{i=0}^{n}\sum_{k=0}^{m-s}\Delta^{0s}\boldsymbol{b}_{ik}\;B_i^{n}(u)B_k^{m-s}(v),$$

where the forward differences are given by

$$\Delta^{r0}\boldsymbol{b}_{ik} = \Delta^{r-1,0}\boldsymbol{b}_{i+1,k} - \Delta^{r-1,0}\boldsymbol{b}_{ik}, \qquad \Delta^{0s}\boldsymbol{b}_{ik} = \Delta^{0,s-1}\boldsymbol{b}_{i,k+1} - \Delta^{0,s-1}\boldsymbol{b}_{ik}.$$

The mixed derivatives are given by

$$\frac{\partial^{r+s}}{\partial u^r\partial v^s}\boldsymbol{X}(u,v) = \frac{n!}{(n-r)!}\frac{m!}{(m-s)!}\sum_{i=0}^{n-r}\sum_{k=0}^{m-s}\Delta^{rs}\boldsymbol{b}_{ik}\;B_i^{n-r}(u)B_k^{m-s}(v),$$

with

$$\Delta^{rs}\boldsymbol{b}_{ik} = \sum_{j=0}^{r}\sum_{\ell=0}^{s}(-1)^j(-1)^\ell\binom{r}{j}\binom{s}{\ell}\boldsymbol{b}_{i+r-j,k+s-\ell}.$$

The first partial derivative perpendicular to the boundary $u=0$ is

$$\boldsymbol{X}_u(0,v) = \frac{n}{\Delta u}\sum_{k=0}^{m}(\boldsymbol{b}_{1k} - \boldsymbol{b}_{0k})B_k^m(v), \tag{6.11}$$

where Δu is the length of the parameter interval. It depends only on the boundary thread of the Bézier net and the thread immediately next to it. In this sense, Lemma 4.1 can also be carried over to Bézier surfaces, see [Stä 76]. Fig. 6.5 depicts the vectors $(\boldsymbol{b}_{1k} - \boldsymbol{b}_{0k})$ which determine the first cross boundary derivative of a Bézier surface along the boundary $u = 0$.

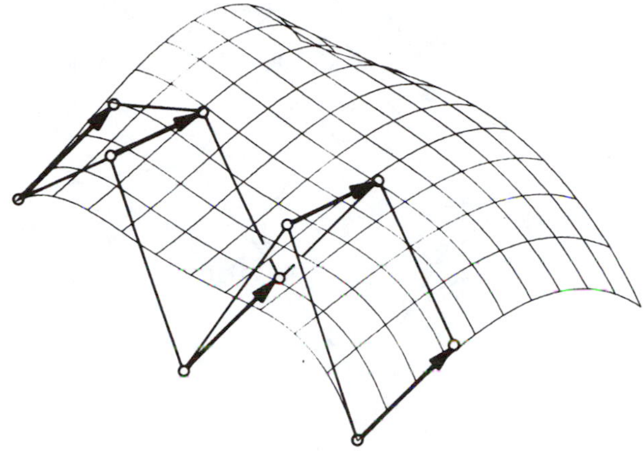

Fig. 6.5. Bézier points for the first cross derivative along a boundary.

As for Bézier curves, it is also possible to *raise the polynomial degree* of a Bézier surface. Analogous to the case of Bézier curves, the Bézier net converges to the surface (Weierstrass approximation theorem).

Raising the degree from n to $n+1$ in the u-direction leads to the (new) Bézier points (cf. Chap. 4)

$$\begin{aligned} \boldsymbol{b}^*_{0j} &= \boldsymbol{b}_{0j}, \\ \boldsymbol{b}^*_{ij} &= \boldsymbol{b}_{ij} + \left(\frac{i}{n+1}\right)(\boldsymbol{b}_{i-1,j} - \boldsymbol{b}_{ij}), \qquad i = 1(1)n \\ \boldsymbol{b}^*_{n+1,j} &= \boldsymbol{b}_{nj}, \end{aligned}$$

for $j = 0(1)m$.

An analogous assertion holds for raising the degree from m to $m+1$ in the v-direction. Fig. 6.6 illustrates the result of raising the degree from 2 to 3 in both the u-direction and the v-direction. The *de Casteljau algorithm* can also be applied to Bézier surfaces by first using the algorithm for $v = v_0$ to compute the Bézier points

$$\boldsymbol{b}_i = \sum_{k=0}^{m} \boldsymbol{b}_{ik} B_k^m(v_0), \qquad i = 0(1)n,$$

and then using the algorithm again for $u = u_0$ with these Bézier points to

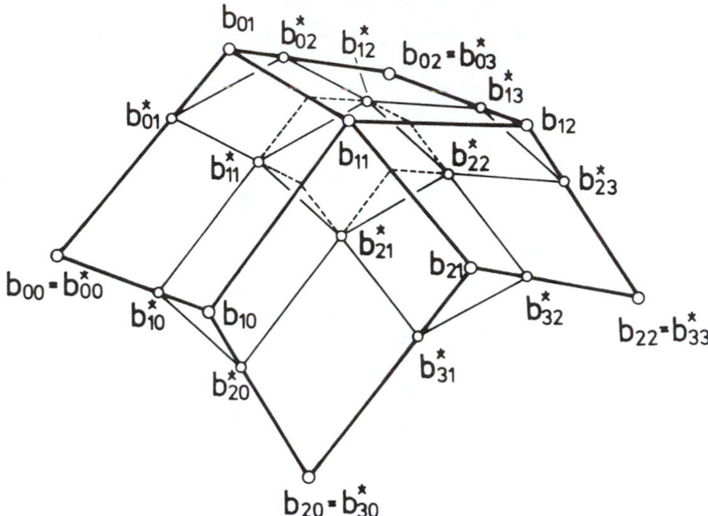

Fig. 6.6. Bézier points $\boldsymbol{b}_{ik}$ (before) and $\boldsymbol{b}^*_{ik}$ (after) degree raising from degree 2 to degree 3.

produce the surface value

$$\boldsymbol{X}(u_0, v_0) = \sum_{i=0}^{n} \boldsymbol{b}_i B_i^n(u_0) = \sum_{i=0}^{n} \sum_{k=0}^{m} \boldsymbol{b}_{ik} B_i^n(u_0) B_k^m(v_0).$$

Fig. 6.7 illustrates the de Casteljau algorithm for a Bézier surface of degree (4,2). Of course, the de Casteljau algorithm can also be applied first with $u = u_0$, and then in the second step with $v = v_0$. As for Bézier curves, the next-to-last elements in the de Casteljau scheme give the directions of the *partial derivatives* $\boldsymbol{X}_u$ resp. $\boldsymbol{X}_v$.

In addition to carrying out the de Casteljau algorithm by performing interpolation in the u and then in the v-direction, it is also possible to work alternately in the two directions. Fig. 6.7b shows how this bilinear interpolation works for $n = m = 3$, see [Kah 82]. In total, there are $(m+n)!/(m!n!)$ essentially different ways to calculate the values of a surface point. This is illustrated in Fig. 6.7c for the bicubic case, where the points at the top are the Bézier points, those in the middle are intermediate points arising in carrying out the de Casteljau scheme, and the point at the bottom is the desired surface value. In terms of computational complexity, it is most efficient to choose one of the schemes in Fig. 6.7c which corresponds to moving along the boundary of the diagram. For $m > n$, it is best to first proceed in the u-direction [Las 84]. The de Casteljau algorithm implies

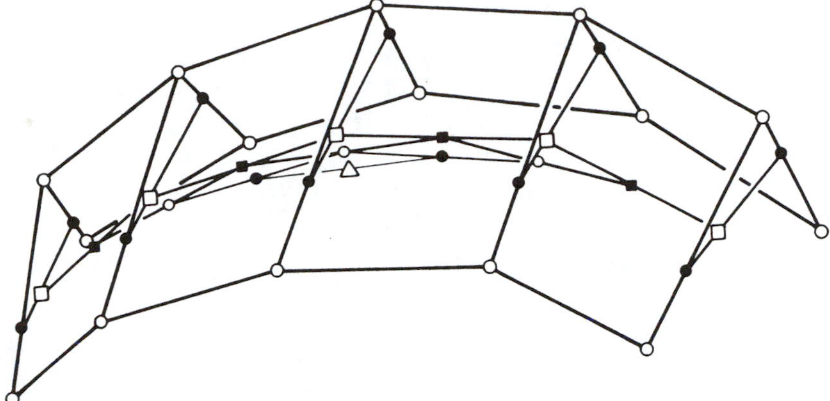

Fig. 6.7a. de Casteljau algorithm for surfaces: darker lines indicate interpolation in the v-direction.

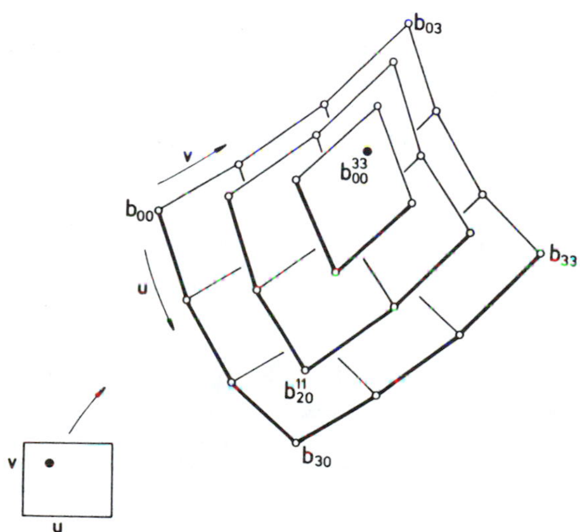

Fig. 6.7b. The de Casteljau algorithm viewed as bilinear interpolation.

Lemma 6.1. *A Bézier surface lies entirely in the convex hull of its Bézier points.*

The *convexity* of Bézier surfaces is not so easy to determine as for curves. This is a consequence of the fact that a surface with all convex parametric lines is not necessarily convex. For example, the Bézier surface shown in Fig. 6.3 has all convex parametric lines, but is itself not convex. Indeed, near the corners it has negative Gausssian curvature. A *sufficient condition* for convexity [Schel 84] is e.g.,

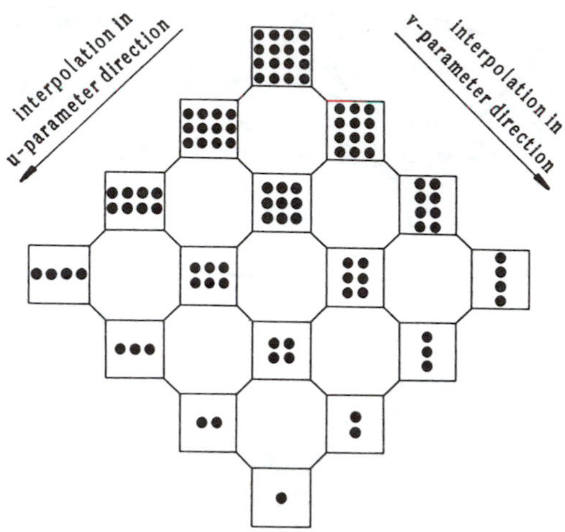

Fig. 6.7c. Possible versions of the de Casteljau algorithm for a bicubic surface.

Lemma 6.2. *A tensor-product Bézier surface of degree (n, m) is convex provided*

- *all edges $(\boldsymbol{b}_{ij}, \boldsymbol{b}_{i,j+1})$ and $(\boldsymbol{b}_{ij}, \boldsymbol{b}_{i+1,j})$ of the Bézier net are edges of the convex hull of the Bézier net,*
- *any set of four Bézier points of the form $\boldsymbol{b}_{ij}, \boldsymbol{b}_{i+1,j}, \boldsymbol{b}_{i,j+1}, \boldsymbol{b}_{i+1,j+1}$ must be the corners of a parallelogram.*

A special case of Lemma 6.2 is provided by a Bézier net with planar facets and planar convex threads.

Remark. The Bézier points $\boldsymbol{b}_{ik}$ completely determine the Bézier surface, and are even related affinely invariantly to the surface.

Remark. The de Casteljau algorithm can be used to subdivide a Bézier surface into Bézier spline patches, see Fig. 6.8.

Remark. The Bézier nets associated with repeated subdivision converge to the Bézier surface. The Bézier net approximates the surface, and can be used to compute intersection curves of Bézier surfaces, see [Las 86]. Fig. 6.9 shows the approximation of a Bézier surface by repeated refining of the Bézier net.

6.2.2.1. Continuity Conditions

If a single Bézier surface is not able to approximate a given surface sufficiently well, then we may use several *Bézier surface patches* which are joined together with some prescribed *continuity conditions*, see [Boeh 76, 76a]. In the simplest

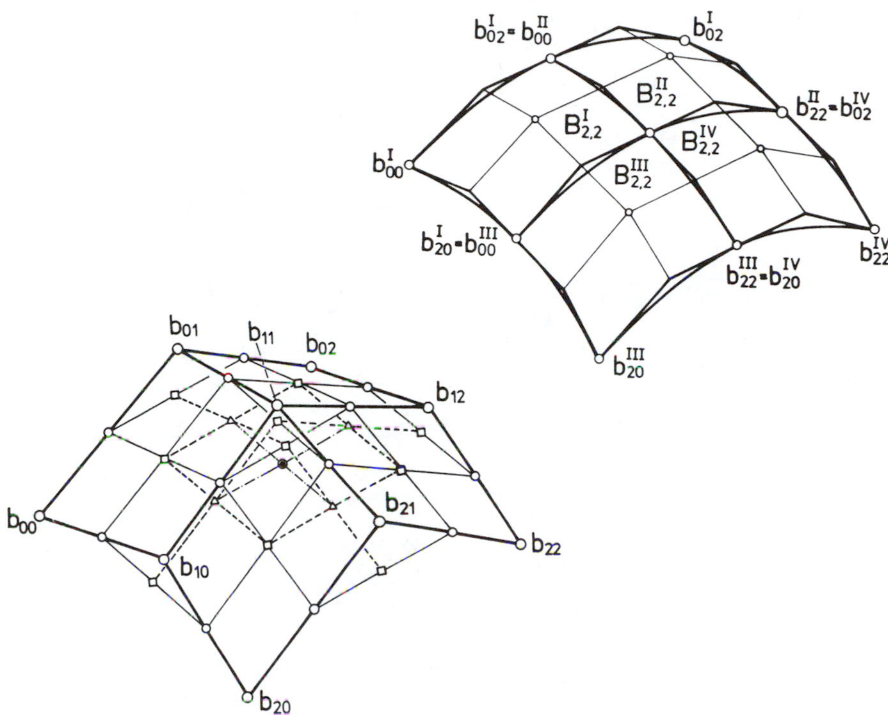

Fig. 6.8. Subdivision of a Bézier surface using the de Casteljau algorithm.

cases, the continuity conditions can be found in the same way as for curves. For example, we can require that

- the first derivatives agree along and across the common boundary curve between two Bézier patches (C^1 *continuity*),
- the first derivatives agree along the common boundary curve, and the cross derivatives along the boundary curve have the same direction (*visual C^1 continuity*),
- the two neighboring Bézier patches have the same tangent planes along the common boundary curve (*geometric C^1 continuity* or *GC^1 continuity*).

Here we should note that visual C^1 continuity implies geometric C^1 continuity. In the following we restrict ourselves to the classical case of C^1 continuity conditions. Other continuity conditions will be discussed in Chap. 7.

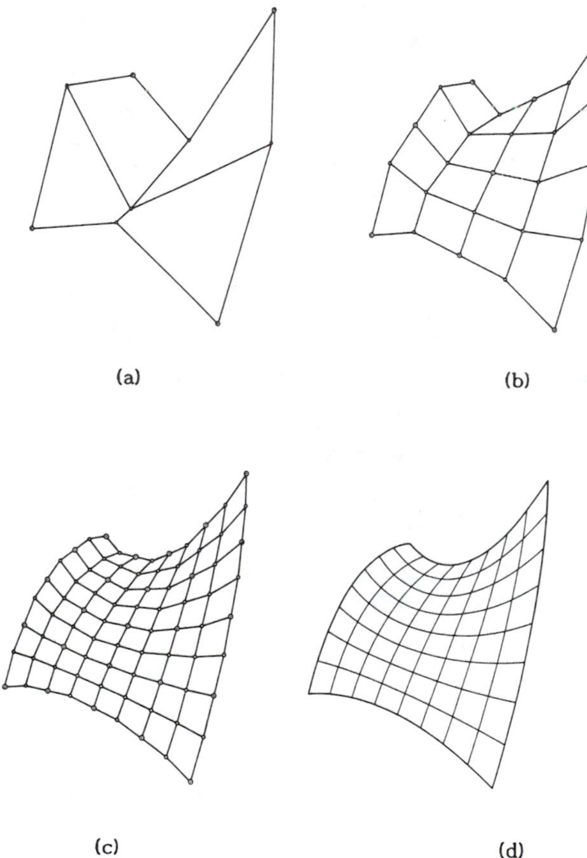

Fig. 6.9. Repeated refinement of a Bézier net (a) leads to approximations (b), (c) of the Bézier surface (d).

In order to formalize geometric continuity criteria in this case, we consider a typical patch $\boldsymbol{X}_{pq}$ of degree (n, m) with parametric representation

$$\boldsymbol{X}_{pq}(u,v) = \sum_{i=0}^{n} \sum_{k=0}^{m} \boldsymbol{b}_{ikpq} \, B_i^n(u) B_k^m(v). \tag{6.12}$$

We assume we are working with a global parametrization, *i.e.*, this patch is defined for $u \in [u_p, u_{p+1}]$, $v \in [v_q, v_{q+1}]$. This yields the situation in the parameter plane for neighboring Bézier patches shown in Fig. 6.10.

Along the common boundary between two patches, the two boundary curves trivially agree. In addition, for C^1 continuity along the parametric line

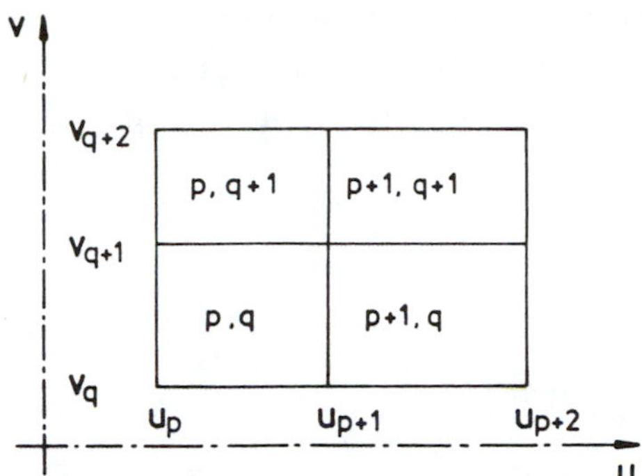

Fig. 6.10. Neighboring patches.

$u = u_{p+1}$, we have

$$\frac{\partial \boldsymbol{X}_{pq}}{\partial u}(u_{p+1}, v) = \frac{\partial \boldsymbol{X}_{p+1,q}}{\partial u}(u_{p+1}, v), \tag{6.13}$$

while along the parametric line $v = v_{q+1}$, we have

$$\frac{\partial \boldsymbol{X}_{pq}}{\partial v}(u, v_{q+1}) = \frac{\partial \boldsymbol{X}_{p,q+1}}{\partial v}(u, v_{q+1}). \tag{6.14}$$

Substituting (6.12) in (6.13)–(6.14), and writing

$$\Delta u_p := u_{p+1} - u_p, \qquad \Delta v_q := v_{q+1} - v_q,$$

it follows that along the line $u = u_{p+1}$

$$\frac{n}{\Delta u_p}(\boldsymbol{b}_{nk} - \boldsymbol{b}_{n-1,k})_{pq} = \frac{n}{\Delta u_{p+1}}(\boldsymbol{b}_{1k} - \boldsymbol{b}_{0k})_{p+1,q}, \qquad k = 0(1)m, \tag{6.15}$$

while along the line $v = v_{q+1}$

$$\frac{m}{\Delta v_q}(\boldsymbol{b}_{im} - \boldsymbol{b}_{i,m-1})_{pq} = \frac{m}{\Delta v_{q+1}}(\boldsymbol{b}_{i1} - \boldsymbol{b}_{i0})_{p,q+1}, \qquad i = 0(1)n.$$

Because we have a common boundary curve,

$$\boldsymbol{b}_{nkpq} = \boldsymbol{b}_{0k,p+1,q} \qquad \text{or} \qquad \boldsymbol{b}_{impq} = \boldsymbol{b}_{i0p,q+1},$$

and it follows from (6.15) that the C^1 *continuity conditions* (cf. (4.18b)) become

$$\begin{aligned} \boldsymbol{b}_{nkpq}(\Delta u_{p+1} + \Delta u_p) &= \Delta u_{p+1}\, \boldsymbol{b}_{n-1,kpq} + \Delta u_p\, \boldsymbol{b}_{1k,p+1,q}, \\ \boldsymbol{b}_{impq}(\Delta v_{q+1} + \Delta v_q) &= \Delta v_{q+1}\, \boldsymbol{b}_{i,m-1,pq} + \Delta v_q\, \boldsymbol{b}_{i1p,q+1}, \end{aligned} \tag{6.16}$$

i.e., for every fixed index k, the Bézier points in (6.16) associated with the u-direction are collinear and divide the connecting edges in the same ratio as the lengths of the parameter intervals $\Delta u_p : \Delta u_{p+1}$. The same holds for the Bézier points associated with the v-direction, cf. Fig. 6.11.

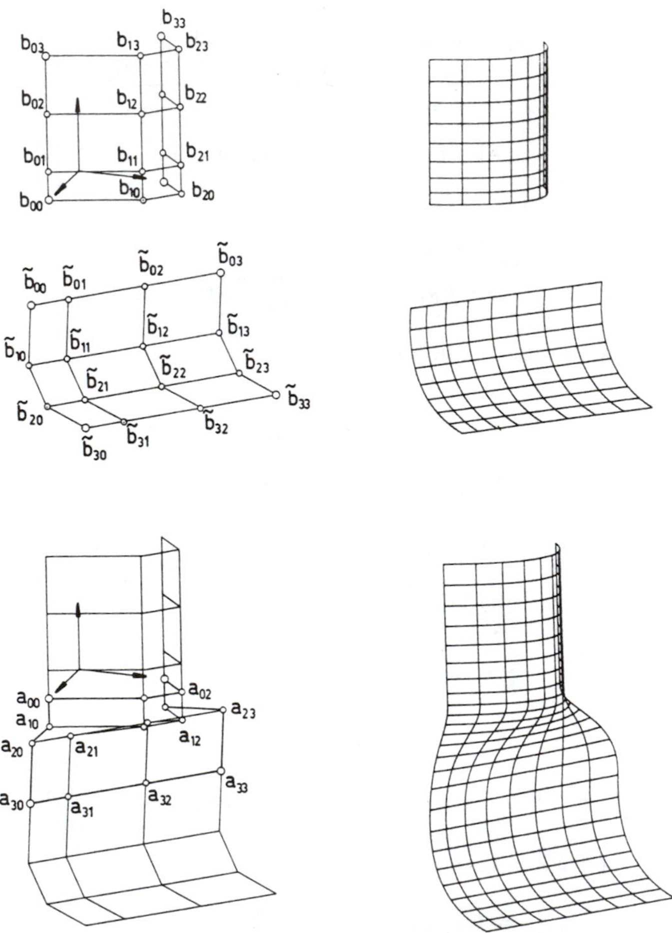

Fig. 6.11. C^1 continuous connection of two Bézier surfaces.

We also note that Bézier patches which join with C^1 continuity along a common boundary curve also have mixed second derivatives (*twist vectors*) which agree along this curve. For example, since

$$
\begin{aligned}
&\frac{\partial^2 \boldsymbol{X}_{pq}}{\partial u \partial v}(u_{p+1}, v) = \\
&\qquad \frac{nm}{\Delta u_p \Delta v_q} \sum_{k=0}^{m-1} [(\boldsymbol{b}_{n,k+1} - \boldsymbol{b}_{n-1,k+1}) - (\boldsymbol{b}_{n,k} - \boldsymbol{b}_{n-1,k})]_{pq} B_k^{m-1}(v), \\
&\frac{\partial^2 \boldsymbol{X}_{p+1,q}}{\partial u \partial v}(u_{p+1}, v) = \\
&\qquad \frac{nm}{\Delta u_{p+1} \Delta v_q} \sum_{k=0}^{m-1} [(\boldsymbol{b}_{1,k+1} - \boldsymbol{b}_{0,k+1}) - (\boldsymbol{b}_{1,k} - \boldsymbol{b}_{0,k})]_{p+1,q} B_k^{m-1}(v),
\end{aligned}
\tag{6.17}
$$

it immediately follows from (6.15) that these derivatives coincide. We will investigate the use of the twist vector later as an important *design element.*

The continuity conditions can also be used to construct C^1 continuous blending surfaces (sweeping), see e.g., [Woo 87], [Fil 89a], [Rog 89], and [Cho 90]. Suppose we are given two Bézier surface patches $\boldsymbol{F}_0$ and $\boldsymbol{F}_2$ which are to be connected using a Bézier surface $\boldsymbol{F}_1$ with C^1 continuity. To illustrate this problem with an example, we consider the bicubic case (where the degree is (3,3)), and assume that $\boldsymbol{F}_0$ is described by the Bézier points $\{\boldsymbol{b}_{ik}\}$, while $\boldsymbol{F}_2$ is described by the Bézier points $\{\tilde{\boldsymbol{b}}_{ik}\}$. We want the surface $\boldsymbol{F}_1$ to join $\boldsymbol{F}_0$ along the boundary curve associated with the Bézier points $\{\boldsymbol{b}_{i3}\}$, and to join $\boldsymbol{F}_2$ along the boundary curve associated with the Bézier points $\{\tilde{\boldsymbol{b}}_{i0}\}$. Then the Bézier points $\{\boldsymbol{a}_{ik}\}$ of $\boldsymbol{F}_1$ must satisfy

$$
\{\boldsymbol{a}_{0i}\} = \{\boldsymbol{b}_{3i}\} \qquad \text{and} \qquad \{\boldsymbol{a}_{3i}\} = \{\tilde{\boldsymbol{b}}_{0i}\}.
$$

From (6.14b) it follows that

$$
\begin{aligned}
\Delta v_0 \, \boldsymbol{a}_{i1} &= \boldsymbol{b}_{i3}(\Delta v_0 + \Delta v_1) - \Delta v_1 \, \boldsymbol{b}_{i2}, \\
\Delta v_2 \, \boldsymbol{a}_{i2} &= \tilde{\boldsymbol{b}}_{i0}(\Delta v_1 + \Delta v_2) - \Delta v_1 \, \tilde{\boldsymbol{b}}_{i1},
\end{aligned}
\tag{6.18}
$$

where Δv_i is the length of the v-parameter intervals for the surfaces $\boldsymbol{F}_i$. The ratios of the lengths of the parameter intervals can serve as a design parameter for the blending surface, see Fig. 6.11.

Continuity conditions of higher order can be derived as in the case of curves, see Chap. 4. For C^2 continuity, in addition to (6.16), we must require the existence of *auxiliary points* $\boldsymbol{d}_k$ satisfying

$$
\begin{aligned}
\boldsymbol{d}_k &= \boldsymbol{b}_{n-1,kpq}\left(1 + \frac{\Delta u_{p+1}}{\Delta u_p}\right) - \boldsymbol{b}_{n-2,kpq}\frac{\Delta u_{p+1}}{\Delta u_p} \\
&= \boldsymbol{b}_{1,k,p+1,q}\left(1 + \frac{\Delta u_p}{\Delta u_{p+1}}\right) - \boldsymbol{b}_{2k,p+1,q}\frac{\Delta u_p}{\Delta u_{p+1}}
\end{aligned}
\tag{6.19}
$$

6.2.3. Bézier Spline Surfaces

Suppose we are given $(L+1)(M+1)$ points $\boldsymbol{P}_{ik}$ and associated parameter values (u_i, v_k) for $i = 0(1)L$ and $k = 0(1)M$. In this section we consider interpolating this data using

- *biquadratic* spline surfaces,
- *bicubic* spline surfaces.

We begin with the biquadratic case. If we express the (i,k)-th piece of this surface in the Bézier form

$$\boldsymbol{X}_{ik}(u,v) = \sum_{j=0}^{2}\sum_{\ell=0}^{2} \boldsymbol{b}_{2i+j,2k+\ell}\, B_j^2(u)B_\ell^2(v), \tag{6.20}$$

then it is convenient to identify the given points with the corners of the patch. This means we should choose

$$\boldsymbol{b}_{2i,2k} = \boldsymbol{P}_{ik}. \tag{6.21}$$

Now we require tangential continuity between patches, *i.e.*, the conditions (6.15) resp. (6.16) should hold. This leaves one free Bézier point (e.g., $\boldsymbol{b}_{2i,1}$ or $\boldsymbol{b}_{1,2k}$) on each parametric line $u = u_i$ and $v = v_k$. After selecting these points, all other Bézier points can be computed recursively from (6.16).

We turn now to *bicubic Bézier spline surfaces*. Here we have two choices for interpolating the prescribed data:

a) by a C^1 continuous *bicubic Bézier subspline surface*,

b) by a C^2 continuous *bicubic Bézier spline surface*.

Suppose the desired Bézier spline surface has the parametric representation

$$\boldsymbol{X}_{ik}(u,v) = \sum_{j=0}^{3}\sum_{\ell=0}^{3} \boldsymbol{b}_{3i+j,3k+\ell}\, B_j^3(u)B_\ell^3(v). \tag{6.22}$$

Then we may again identify the given values with the corners of the patches, *i.e.*, we may take

$$\boldsymbol{b}_{3i,3k} = \boldsymbol{P}_{ik}.$$

Consider case a). We want to construct a C^1 continuous bicubic Bézier subspline surface. The computation of the corresponding Bézier points can be accomplished in two steps (see Fig. 6.12) as follows:

1) Interpolate the $L+1$ rows and $M+1$ columns of the matrix of given points $\boldsymbol{P}_{ik}$ by cubic Bézier spline curves or cubic Bézier subspline curves by the methods of Chap. 4. This produces Bézier points

$$\boldsymbol{b}_{3i,3k+\ell} \qquad \text{and} \qquad \boldsymbol{b}_{3i+\ell,3k}, \qquad \ell = 1, 2,$$

where the boundary points $\boldsymbol{b}_{1,3k}$ and $\boldsymbol{b}_{3i,1}$ and the points $\boldsymbol{b}_{3L-1,3k}$ and $\boldsymbol{b}_{3i,3M-1}$ can all be chosen arbitrarily.

2) For each patch corner $\boldsymbol{P}_{ik} = \boldsymbol{b}_{3i,3k}$, consider the nine Bézier points shown in Fig. 6.12a.

Fig 6.12a. Computing Bézier points surrounding a vertex.

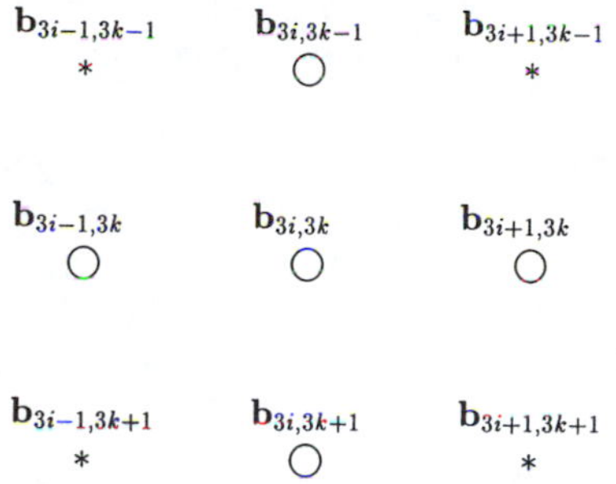

Those marked with circles are already determined. By the C^1 continuity conditions, these points must satisfy

$$\begin{aligned}
\Delta u_i\, \boldsymbol{b}_{3i+1,3k-1} + \Delta u_{i+1}\, \boldsymbol{b}_{3i-1,3k-1} &= \boldsymbol{b}_{3i,3k-1}(\Delta u_i + \Delta u_{i+1}),\\
\Delta u_i\, \boldsymbol{b}_{3i+1,3k+1} + \Delta u_{i+1}\, \boldsymbol{b}_{3i-1,3k+1} &= \boldsymbol{b}_{3i,3k+1}(\Delta u_i + \Delta u_{i+1}),\\
\Delta v_k\, \boldsymbol{b}_{3i-1,3k+1} + \Delta v_{k+1}\, \boldsymbol{b}_{3i-1,3k-1} &= \boldsymbol{b}_{3i-1,3k}(\Delta v_k + \Delta v_{k+1}),\\
\Delta v_k\, \boldsymbol{b}_{3i+1,3k+1} + \Delta v_{k+1}\, \boldsymbol{b}_{3i+1,3k-1} &= \boldsymbol{b}_{3i+1,3k}(\Delta v_k + \Delta v_{k+1}).
\end{aligned} \tag{6.23}$$

If we prescribe one of the four Bézier points marked with a star, we can compute the other three recursively from (6.23). The choice of one of the Bézier points marked with a star corresponds geometrically to prescribing the twist vector. For example, if we choose the Bézier point $\boldsymbol{b}_{3i+1,3k+1}$ to lie in the plane passing through the Bézier points $\boldsymbol{b}_{3i,3k}$, $\boldsymbol{b}_{3i,3k+1}$, $\boldsymbol{b}_{3i-1,3k}$, $\boldsymbol{b}_{3i+1,3k}$ determined in Step 1, then

$$\boldsymbol{b}_{3i+1,3k+1} = \boldsymbol{b}_{3i,3k+1} + (\boldsymbol{b}_{3i+1,3k} - \boldsymbol{b}_{3i,3k}),$$

and by (6.17) the second mixed derivative, and thus also the twist vector, must vanish.

We now consider C^2 *continuous* bicubic Bézier spline surfaces. By (6.15) resp. (6.16), the C^1 continuity conditions are

$$\begin{aligned}(\Delta u_i + \Delta u_{i-1})\, \boldsymbol{b}_{3i,s} &= \Delta u_{i-1}\, \boldsymbol{b}_{3i-1,s} + \Delta u_i\, \boldsymbol{b}_{3i+1,s},\\ (\Delta v_k + \Delta v_{k-1})\, \boldsymbol{b}_{r,3k} &= \Delta v_{k-1}\, \boldsymbol{b}_{r,3k-1} + \Delta v_k\, \boldsymbol{b}_{r,3k+1}.\end{aligned} \tag{6.24}$$

For C^2 continuity in the u-direction, we require

$$\begin{aligned}(\Delta u_i + \Delta u_{i-1})\, \boldsymbol{b}_{3i-1,s} - \Delta u_i\, \boldsymbol{b}_{3i-2,s} &=: \Delta u_{i-1}\, \boldsymbol{f}_{i,s}, \quad i = 1(1)L,\\ (\Delta u_i + \Delta u_{i-1})\, \boldsymbol{b}_{3i+1,s} - \Delta u_{i-1}\, \boldsymbol{b}_{3i+2,s} &=: \Delta u_i\, \boldsymbol{f}_{i,s}, \quad i = 0(1)L\text{–}1,\end{aligned} \tag{6.25}$$

while for C^2 continuity in the v-direction, we require

$$\begin{aligned}(\Delta v_k + \Delta v_{k-1})\, \boldsymbol{b}_{r,3k-1} - \Delta v_k\, \boldsymbol{b}_{r,3k-2} &=: \Delta v_{k-1}\, \boldsymbol{g}_{r,k}, \quad k = 1(1)M,\\ (\Delta v_k + \Delta v_{k-1})\, \boldsymbol{b}_{r,3k+1} - \Delta v_{k-1}\, \boldsymbol{b}_{r,3k+2} &=: \Delta v_k\, \boldsymbol{g}_{r,k}, \quad k = 0(1)M\text{–}1,\end{aligned} \tag{6.26}$$

where $\boldsymbol{f}_{is}$, $s = 0(1)3M$ and $\boldsymbol{g}_{rk}$, $r = 0(1)3L$, are auxiliary points. Using the surface weights $\boldsymbol{d}_{ik}$ defined via

$$\begin{aligned}(\Delta u_i + \Delta u_{i-1})\, \boldsymbol{g}_{3i-1,k} - \Delta u_i\, \boldsymbol{g}_{3i-2,k} &=: \Delta u_{i-1}\, \boldsymbol{d}_{i,k},\\ (\Delta u_i + \Delta u_{i-1})\, \boldsymbol{g}_{3i+1,k} - \Delta u_{i-1}\, \boldsymbol{g}_{3i+2,k} &=: \Delta u_i\, \boldsymbol{d}_{i,k},\end{aligned} \tag{6.27}$$

$$\begin{aligned}(\Delta v_k + \Delta v_{k-1})\, \boldsymbol{f}_{i,3k-1} - \Delta v_k\, \boldsymbol{f}_{i,3k-2} &=: \Delta v_{k-1}\, \boldsymbol{d}_{i,k},\\ (\Delta v_k + \Delta v_{k-1})\, \boldsymbol{f}_{i,3k+1} - \Delta v_{k-1}\, \boldsymbol{f}_{i,3k+2} &=: \Delta v_k\, \boldsymbol{d}_{i,k},\end{aligned} \tag{6.28}$$

it follows (see e.g., [Boeh 76a], [Las 85]) that the following spline conditions must be satisfied:

$$\begin{aligned}\beta_k^2 \alpha_i^2\, \boldsymbol{b}_{3i-2,3k-2} = \beta_k^1 (\alpha_i^1 \boldsymbol{d}_{i-1,k-1} + \alpha_{i-2}^0 \boldsymbol{d}_{i,k-1})\\ + \beta_{k-2}^0 (\alpha_i^1 \boldsymbol{d}_{i-1,k} + \alpha_{i-2}^0 \boldsymbol{d}_{i,k}),\end{aligned} \tag{6.29}$$

$$\beta_k^2 \alpha_i^5\, \boldsymbol{b}_{3i,3k-2} = \beta_k^1 (\alpha_i^3 \boldsymbol{d}_{i-1,k-1} + \alpha_i^4 \boldsymbol{d}_{i,k-1} + \alpha_{i-1}^3 \boldsymbol{d}_{i+1,k-1}) \tag{6.30}$$

$$+ \beta_{k-2}^0 (\alpha_i^3 \boldsymbol{d}_{i-1,k} + \alpha_1^4 \boldsymbol{d}_{i,k} + \alpha_{i-1}^3 \boldsymbol{d}_{i+1,k}),$$

$$\beta_k^2 \alpha_{i+1}^2\, \boldsymbol{b}_{3i+2,3k-2} = \beta_k^1 (\alpha_{i+1}^0 \boldsymbol{d}_{i,k-1} + \alpha_i^1 \boldsymbol{d}_{i+1,k-1})$$

$$+ \beta_{k-2}^0 (\alpha_{i+1}^0 \boldsymbol{d}_{i,k} + \alpha_i^1 \boldsymbol{d}_{i+1,k}), \tag{6.31}$$

with corresponding equations for the $\boldsymbol{b}_{3i+j,3k}$ and $\boldsymbol{b}_{3i+j,3k+2}$, $j = -2, 0, 2$.

Here the $\alpha_i^0, \ldots, \alpha_i^5$ are defined by

$$\alpha_i^0 \equiv \Delta u_i, \qquad \alpha_i^1 \equiv \Delta u_i + \Delta u_{i-1}, \qquad \alpha_i^2 \equiv \Delta u_i + \Delta u_{i-1} + \Delta u_{i-2},$$

$$\alpha_i^3 \equiv (\alpha_i^0)^2 \alpha_{i+1}^2, \qquad \alpha_i^4 \equiv \alpha_i^0 \alpha_{i-1}^1 \alpha_{i+1}^2 + \alpha_{i+1}^0 \alpha_{i+1}^1 \alpha_i^2, \qquad \alpha_i^5 \equiv \alpha_i^1 \alpha_i^2 \alpha_{i+1}^2,$$

while the $\beta_k^0, \ldots, \beta_k^5$ are defined in an analogous way in terms of the parameter interval length Δv_k.

Thus, once we have the $\boldsymbol{d}_{ik}$, the intermediate points $\boldsymbol{f}_{is}$, $s \neq 0, 3M$ and $\boldsymbol{g}_{rk}$, $r \neq 0, 3L$, along with the Bézier points which do not lie on the boundary, are uniquely defined. This means that if an appropriate set of $(L+1)(M+1)$ Bézier points, e.g., those in (6.32), are prescribed, then the remaining interior Bézier points are uniquely determined by the spline conditions. To get these starting points $\boldsymbol{d}_{ik}$, we solve the linear system of equations

$$\boldsymbol{L} \cdot \boldsymbol{D} \cdot \boldsymbol{M} = \boldsymbol{B}, \tag{6.32}$$

where

$$\boldsymbol{L} = \begin{pmatrix} \alpha_1^1 & \alpha_{-1}^0 & 0 & \cdots & 0 & 0 & 0 \\ \alpha_1^3 & \alpha_1^4 & \alpha_0^3 & & 0 & 0 & 0 \\ \vdots & & & & & & \vdots \\ 0 & 0 & 0 & & \alpha_{L-1}^3 & \alpha_{L-1}^4 & \alpha_{L-2}^3 \\ 0 & 0 & 0 & \cdots & 0 & \alpha_L^0 & \alpha_{L-2}^1 \end{pmatrix}$$

$$\boldsymbol{D} = \begin{pmatrix} \boldsymbol{d}_{0,0} & \cdots & \boldsymbol{d}_{0,M} \\ \cdot & & \cdot \\ \cdot & & \cdot \\ \cdot & & \cdot \\ \cdot & & \cdot \\ \cdot & & \cdot \\ \boldsymbol{d}_{L,0} & \cdots & \boldsymbol{d}_{L,M} \end{pmatrix}$$

$$\boldsymbol{M} = \begin{pmatrix} \beta_1^1 & \beta_1^3 & \cdots & 0 & 0 \\ \beta_{-1}^0 & \beta_1^4 & & 0 & 0 \\ 0 & \beta_0^3 & 0 & 0 & \\ \vdots & & & & \vdots \\ 0 & 0 & & \beta_{M-1}^3 & 0 \\ 0 & 0 & & \beta_{M-1}^4 & \beta_M^0 \\ 0 & 0 & \cdots & \beta_{M-2}^3 & \beta_{M-1}^1 \end{pmatrix}$$

and

$$B = \begin{pmatrix} \alpha_1^2\beta_1^2\boldsymbol{b}_{1,1} & \alpha_1^2\beta_1^5\boldsymbol{b}_{1,3} & \cdots\cdot & \alpha_1^2\beta_{M-1}^5\boldsymbol{b}_{1,3M-3} & \alpha_1^2\beta_M^2\boldsymbol{b}_{1,3M-1} \\ \alpha_1^5\beta_1^2\boldsymbol{b}_{3,1} & \alpha_1^5\beta_1^5\boldsymbol{b}_{3,3} & \cdots\cdot & \alpha_1^5\beta_{M-1}^5\boldsymbol{b}_{3,3M-3} & \alpha_1^5\beta_M^2\boldsymbol{b}_{3,3M-1} \\ \cdot & \cdot & \cdots\cdot & \cdot & \cdot \\ \cdot & \cdot & \cdots\cdot & \cdot & \cdot \\ \alpha_{L-1}^5\beta_1^2\boldsymbol{b}_{3L-3,1} & \alpha_{L-1}^5\beta_1^5\boldsymbol{b}_{3L-3,3} & \cdots\cdot & \alpha_{L-1}^5\beta_{M-1}^5\boldsymbol{b}_{3L-3,3M-3} & \alpha_{L-1}^5\beta_M^2\boldsymbol{b}_{3L-3,3M-1} \\ \alpha_L^2\beta_1^2\boldsymbol{b}_{3L-1,1} & \alpha_L^2\beta_1^5\boldsymbol{b}_{3L-1,3} & \cdots\cdot & \alpha_L^2\beta_{M-1}^5\boldsymbol{b}_{3L-1,3M-3} & \alpha_L^2\beta_M^2\boldsymbol{b}_{3L-1,3M-1} \end{pmatrix}$$

For $\boldsymbol{d}_{00}, \boldsymbol{d}_{L0}, \boldsymbol{d}_{0M}, \boldsymbol{d}_{LM}$ we also need Δu_i, $i = -1, L$ and Δv_k, $k = -1, M$. The Bézier points $\boldsymbol{b}_{00}, \boldsymbol{b}_{3L,0}, \boldsymbol{b}_{0,3M}$, and $\boldsymbol{b}_{3L,3M}$ do not depend on the auxiliary points $\boldsymbol{f}_{is}, \boldsymbol{g}_{rk}$, and $\boldsymbol{d}_{ik}$. Since the boundary curves of the bicubic spline surface are cubic splines associated with the auxiliary points $\boldsymbol{f}_{is}$, $s = 0, 3M$, resp. $\boldsymbol{g}_{rk}$, $r = 0, 3M$, the interior boundary curve Bézier points can be computed by cubic spline interpolation.

A cubic spline is defined by a total of $(3L+1)(3M+1)$ Bézier points. The number of Bézier points which completely determine the C^2 Bézier spline is $(L+3)(M+3)$. The Bézier points which are determined by the original data are marked with circles in Fig. 6.12b.

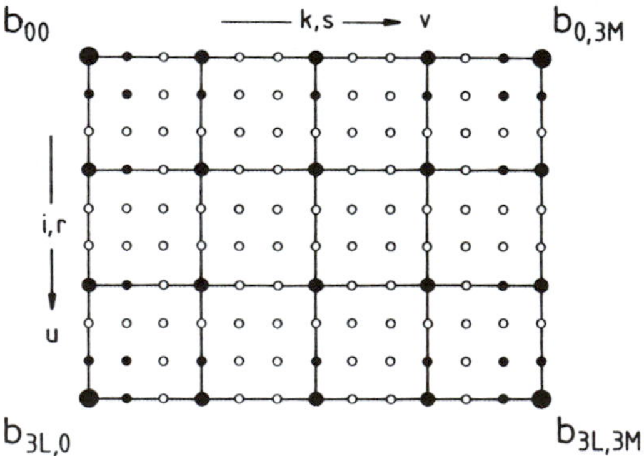

Fig. 6.12b. Prescribed data points (dots) for a Bézier spline surface.

We need $(L+1)(M+1)$ Bézier points to determine the $\boldsymbol{d}_{ik}$. For the $L+1$ auxiliary points $\boldsymbol{f}_{i0}$, we need to use $L+1$ interior Bézier points on the edge $v = 0$. The same holds for $s = 3M$ and $v = 1$. Once we have them, we can compute $\boldsymbol{b}_{r0}$ and $\boldsymbol{b}_{r,3M}$ for $r \neq 0, 3L$. For the $M+1$ auxiliary points $\boldsymbol{g}_{0k}$, we need to use $M+1$ interior Bézier points on the edge $u = 0$. The same holds for $r = 3L$ and $u = 1$. From them, we can compute $\boldsymbol{b}_{0s}$ and $\boldsymbol{b}_{3L,s}$ for $s \neq 0, 3M$. The total number of Bézier points which have to be determined from the given $(L+3)(M+3)$ Bézier points by the spline conditions is $8(LM-1)$, [Las 85].

6.2.4. Tensor-product B-spline Surfaces

Integral tensor-product B-spline surfaces can be defined analogously to Bézier surfaces as

$$\boldsymbol{X}(u,v) = \sum_{i=0}^{m}\sum_{j=0}^{n} \boldsymbol{d}_{ij} N_{ik}(u) N_{j\ell}(v), \tag{6.33}$$

using B-spline basis functions of order k and ℓ and *de Boor points* $\boldsymbol{d}_{ij}$. The set of points $\{\boldsymbol{d}_{ij}\}$ forms the *de Boor net*. These surfaces have properties analogous to those for B-spline curves, such as the *convex-hull property*, and a criterion for *convexity* of the B-spline surface [Loh 81]. As for curves,

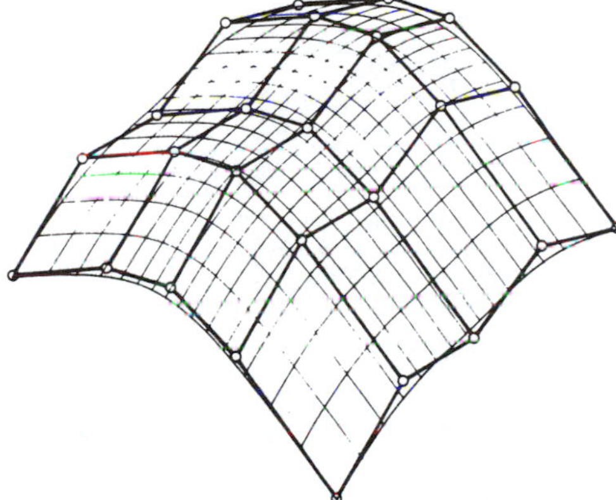

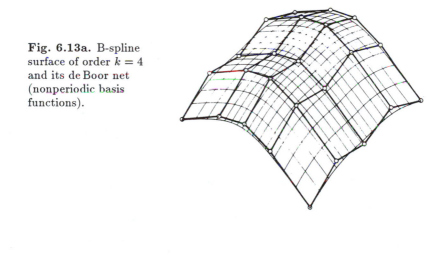

Fig. 6.13a. B-spline surface of order $k = 4$ and its de Boor net (nonperiodic basis functions).

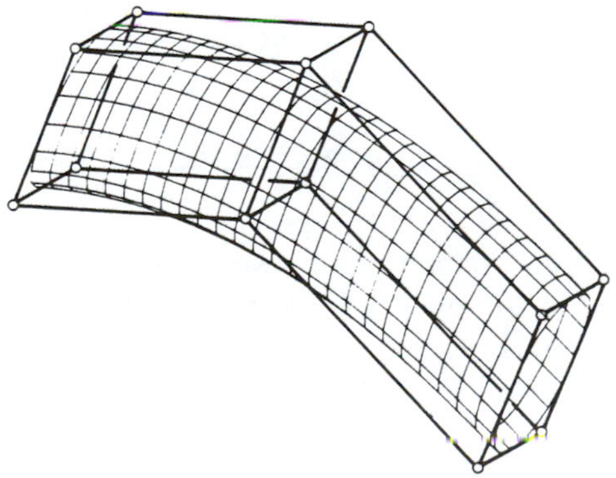

Fig. 6.13b. B-spline surface of order $k = 3$ with basis functions periodic in the u-direction.

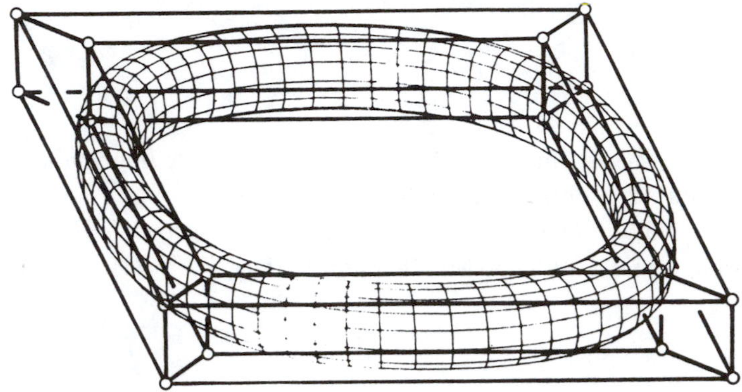

Fig. 6.13c. B-spline surface of order $k = 3$ with periodic basis functions in both the u- and v-directions.

the de Boor points have *local effect*; for example, the surface patch corresponding to $u \in [u_p, u_{p+1}]$, $v \in [v_q, v_{q+1}]$ is only influenced by the points $\boldsymbol{d}_{p-(k-1),q-(\ell-1)}, \ldots, \boldsymbol{d}_{pq}$. Thus, the de Boor point $\boldsymbol{d}_{rs}$ only influences the parameter domain $u \in [u_r, u_{r+k}]$, $v \in [v_s, v_{s+\ell}]$. To compute a point on the surface (6.33), we can again use the *de Boor algorithm*. As for Bézier surfaces, we first apply the algorithm for $u = u_0$ in order to compute the de Boor points

$$\boldsymbol{d}_j(u_0) = \sum_{i=0}^{m} \boldsymbol{d}_{ij} N_{ik}(u_0).$$

Then the algorithm is applied again with $v = v_0$ using the points $\boldsymbol{d}_j(u_0)$. Fig. 6.13 shows some tensor-product B-spline surfaces with nonperiodic, simple periodic, and doubly periodic basis functions and their associated de Boor nets.

6.2.5. Interpolation and Approximation with Integral B-spline Surfaces

We turn now to interpolation and approximation of two–dimensional point sets by spline surfaces. Our aim is to solve the following problem: given not necessarily uniformly spaced points $\boldsymbol{P}_i$ and associated parameter values (u_i, v_i), $i = 0(1)N$, find an interpolating spline surface which passes through the points, or an approximating spline surface with respect to some prescribed error norm.

As in the case of curves, the parametrization plays an essential role for both interpolation and approximation. While for curves there are some simple *ad hoc* methods available, the situation is considerably more complicated for surfaces, particularly in the case of scattered data.

6.2.5.1. Parametrization of Surface Points

The parametrization of surfaces can be reduced to the parametrization of parametric lines, provided that the points to be interpolated lie along possible parametric lines to begin with. If the points are scattered, however, then additional work has to be done to find a parametrization. For example, sometimes the given points can be projected onto an appropriate auxiliary surface whose parameter net is known. However, in doing so, we must be careful to insure that the mapping between the parameter plane and the interpolating or approximating surface is *bijective*, see Fig. 6.14.

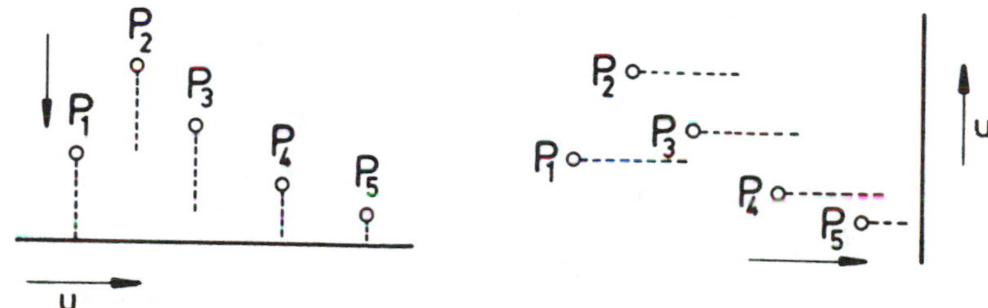

Fig. 6.14. Projection of points $\boldsymbol{P}_i$ onto a parameter plane (a) bijective, (b) not bijective.

If the given set of points is structured in such a way that they (and the expected interpolating or approximating surface) can be mapped bijectively onto a *plane*, then with an appropriate normalization, the desired parameter values can be read off directly, see Fig. 6.15.

Often, a convenient plane π for parametrization is some plane which approximates the point set, such as the one which minimizes the shortest distance between the points $\boldsymbol{P}_i$ and the plane, see [HOS 87]. The plane which arises from least squares fitting of the points (regression plane) is not as well suited as a parameter plane, since it involves minimizing the z-distances of the points from the fitting plane [BOEH 85].

Another possible approach is the following: choose an appropriate surface corner point $\boldsymbol{P}_{e1}$ and another corner point $\boldsymbol{P}_{e2}$ which is at a maximum distance from it. Then find a third corner point $\boldsymbol{P}_{e3}$ which is at a maximum distance

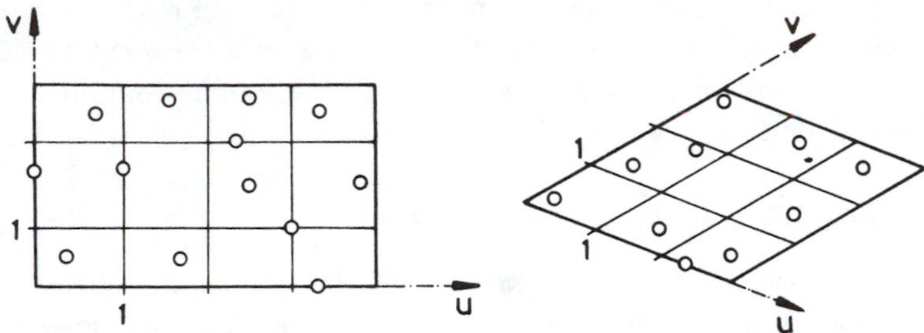

Fig. 6.15. Possible images of points in the parameter plane.

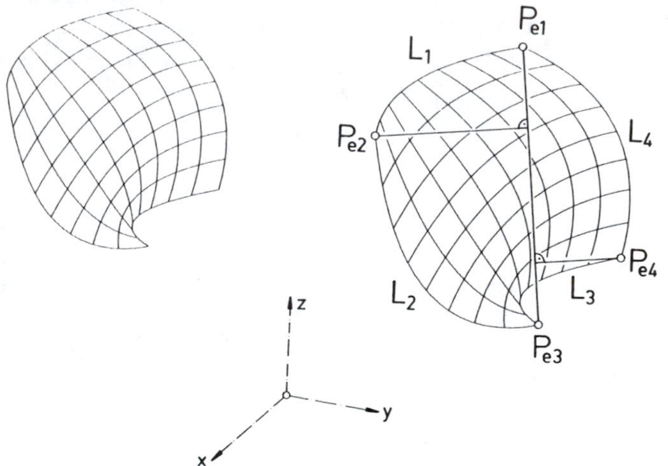

Fig. 6.16a. Auxiliary projection plane.

from the line $\boldsymbol{P}_{e1}\boldsymbol{P}_{e2}$. Then the three points $(\boldsymbol{P}_{e1}, \boldsymbol{P}_{e2}, \boldsymbol{P}_{e3})$ determine an auxiliary projection plane π, see Fig. 6.16a.

For the following, we now assume that the corner points of our point set can be connected by four boundary curves $\boldsymbol{L}_1, \ldots, \boldsymbol{L}_4$ to produce a kind of tensor-product domain. Parametric representations for these boundary curves $\boldsymbol{L}_i$ can be obtained by B-spline approximation and parametric correction. We now project the $\boldsymbol{L}_i$ via orthogonal projection onto a working plane π. Suppose that the images of $\boldsymbol{L}_i$ are denoted by $\bar{\boldsymbol{L}}_i$, and enclose a planar quadrilateral with curved edges. This should be done in such a way that the quadrilateral is (nearly) convex, since for nonconvex domains, the interpolant constructed in (6.34) below can lead to domains which do not lie entirely inside the prescribed boundaries [Vri 91].

We interpolate the boundary curves $\bar{\boldsymbol{L}}_1, \ldots, \bar{\boldsymbol{L}}_4$ by a linear *Gordon-Coons interpolant*, see Chap. 8. This gives the domain bounded by the $\bar{\boldsymbol{L}}_1, \ldots, \bar{\boldsymbol{L}}_4$ a parametric structure, which according to (8.13) can be described by

$$\begin{aligned}\boldsymbol{Q}(\mu,\nu) = {} & (1-\nu)\bar{\boldsymbol{L}}_1(\mu) + \nu\bar{\boldsymbol{L}}_3(\mu) + (1-\mu)\bar{\boldsymbol{L}}_2(\nu) + \mu\bar{\boldsymbol{L}}_4(\nu) \\ & - \Big((1-\mu)[(1-\nu)\bar{\boldsymbol{L}}_1(0) + \nu\bar{\boldsymbol{L}}_3(0)] + \mu[(1-\nu)\bar{\boldsymbol{L}}_2(1) + \nu\bar{\boldsymbol{L}}_4(1)]\Big), \\ & \mu,\nu \in [0,1]. \end{aligned} \tag{6.34}$$

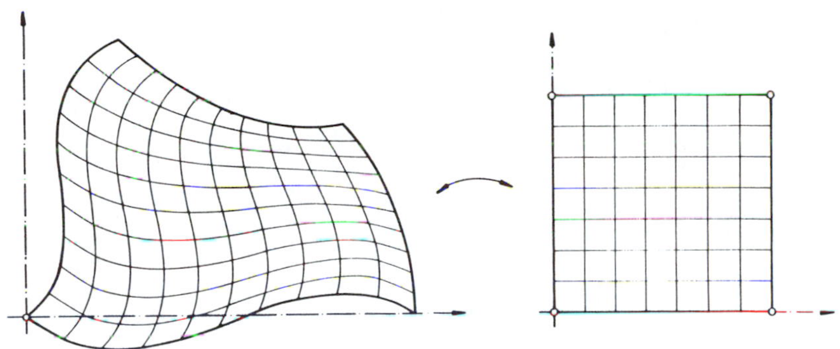

Fig. 6.16b. Mapping of a grid.

Then taking $p+1$ lines with constant μ, and $q+1$ lines with constant ν, leads to a planar grid consisting of (in general) curved lines, but which can be mapped onto a standard $(p+1)\times(q+1)$ grid on the unit square, see Fig. 6.16b.

To get the parameter values associated with interior points $\boldsymbol{P}_i$, we project these points orthogonally onto π. The images $\bar{\boldsymbol{P}}_i$ of the points $\boldsymbol{P}_i$ will in general lie in subrectangles of the curved grid. Since the subsequent fitting uses parameter correction, it suffices to find an approximate location for the parameter values u_i, v_i to be associated with the point $\boldsymbol{P}_i$, for example by replacing each subrectangle of the grid in Fig. 6.16b by a quadrilateral with linear sides, see Fig. 6.16c.

Now every subrectangle can again be fit with a linear Coons interpolant. The parametric formulae for the interpolants associated with these approximate subrectangles can be obtained from (6.34), where the $\bar{\boldsymbol{L}}_i$ are replaced by the lines $\boldsymbol{G}_i$ given by

$$\begin{aligned} &\boldsymbol{G}_1(\bar{\mu}) = (1-\bar{\mu})\boldsymbol{R}_{jk} + \bar{\mu}\boldsymbol{R}_{j+1,k}, \qquad \boldsymbol{G}_2(\bar{\nu}) = (1-\bar{\nu})\boldsymbol{R}_{j+1,k} + \bar{\nu}\boldsymbol{R}_{j+1,k+1}, \\ &\boldsymbol{G}_3(\bar{\mu}) = (1-\bar{\mu})\boldsymbol{R}_{j,k+1} + \bar{\mu}\boldsymbol{R}_{j+1,k+1}, \qquad \boldsymbol{G}_4(\bar{\nu}) = (1-\bar{\nu})\boldsymbol{R}_{jk} + \bar{\nu}\boldsymbol{R}_{j,k+1}, \end{aligned} \tag{6.35}$$

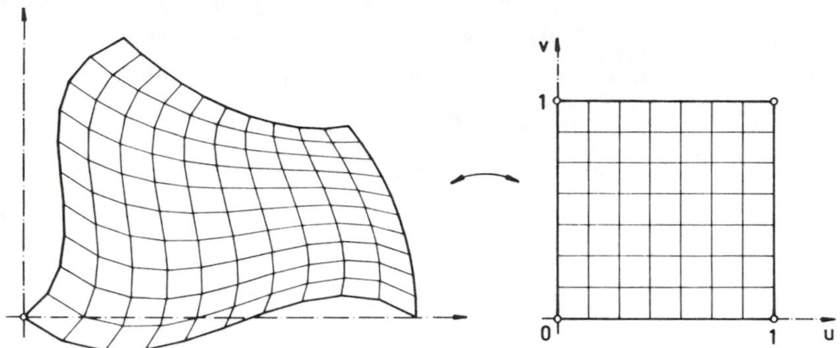

Fig. 6.16c. Linear approximation of a grid.

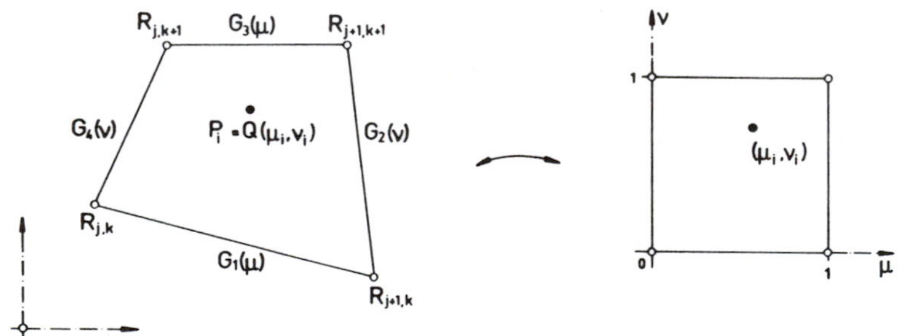

Fig. 6.16d. Coons interpolant for a subrectangle.

where $\bar{\mu}, \bar{\nu} \in [0, 1]$, see Fig. 6.16d.

If $\bar{\boldsymbol{P}}_i$ lies in a subrectangle with vertex $\boldsymbol{R}_{jk}$, then

$$\bar{\boldsymbol{P}}_i = \boldsymbol{Q}(\bar{\mu}_i, \bar{\nu}_i).$$

Solving this nonlinear system of equations leads to the parameter value $\bar{\mu}_i, \bar{\nu}_i$ for $\bar{\boldsymbol{P}}_i$ in the corresponding subrectangle. If the vertex $\boldsymbol{R}_{jk}$ of the subrectangle corresponds to the parameter value (u_j, v_k), and if the grid in the parameter plane is constructed from $p + 1$ lines in the μ direction and $q + 1$ lines in the ν direction, then the (global) parameter value corresponding to $\bar{\boldsymbol{P}}_i$ (assuming linear structure) is

$$u_i = u_j + \frac{\bar{\mu}_i}{p}, \qquad v_i = v_k + \frac{\bar{\nu}_i}{q}.$$

If the point set is near a part of a *cylinder*, then we can project the points onto a cylinder. If this piece of a cylinder is rolled out onto a plane, then possible parameter values can be obtained as in the case of a plane, see Fig. 6.17. If the given point set is *saddle shaped*, then it is recommended to use a projection onto a hyperbolic paraboloid. The two families of ruled lines on the hyperbolic paraboloid can then be directly interpreted as images of the parametric lines $u =$ constant and $v =$ constant in the parameter plane. The hyperbolic paraboloid can be represented as a bilinear tensor-product Bézier surface, see Fig. 6.17. The position of such a comparison surface then can be found with the help of an approximation method.

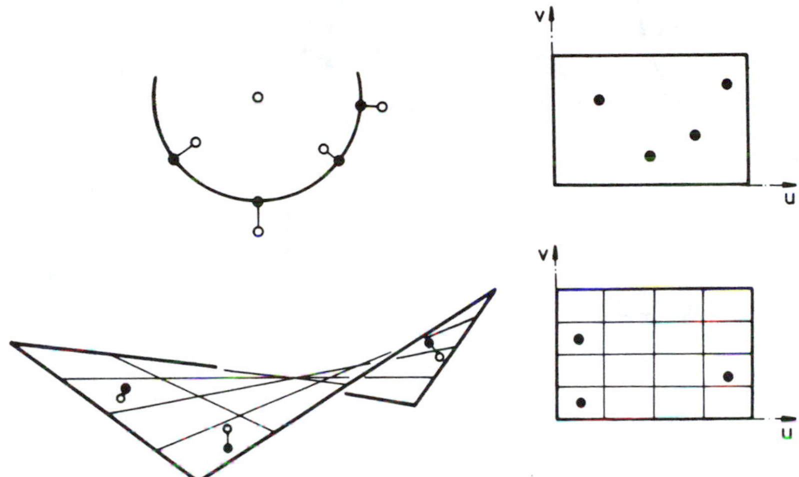

Fig. 6.17. Possible comparison surfaces for parameter location.

If the boundary curves of the point set to be approximated are known, and can be used as the boundaries of a tensor-product surface patch, then the following strategy has been shown to work well:

- first approximate the four boundary curves by B-spline curves;
- then construct a linear Coons surface (8.13) as a comparison surface;
- project the interior points of the given point set perpendicularly onto this comparison surface. The feet of these perpendiculars on the comparison surface yield a first approximation to the parameter values, see Fig. 6.18. These points can be found by using an optimization algorithm to find a surface point with the shortest distance from the prescribed point.

This method is not too reliable when the boundary curves do not provide enough information about the interior of the surface, for example, if all boundary curves are planar, but the interior involves peaks and valleys. In this case, the Coons interpolant of the boundary curves is a plane onto which all points $\boldsymbol{P}_i$ will be projected. Then in the case of especially steep surfaces, we will not get adequate separation of the individual parameter values. In this case one should segment the surface to get boundary curves which better model the behavior of the desired surface in the interior.

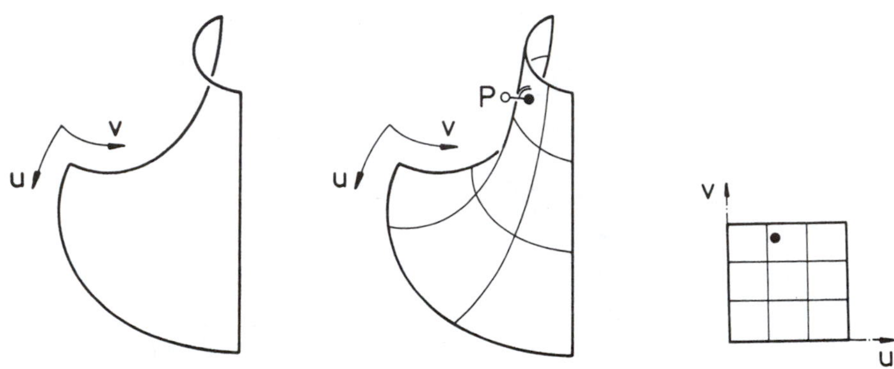

Fig. 6.18. Parametrization using a Coons surface.

In such problematic cases, it is also possible to use a stepwise method. First we parametrize using a Coons surface, then fit an approximating B-spline surface with a small number of pieces. This first approximation surface leads to improved parameter values for the points $\boldsymbol{P}_i$. We then repeat the process by approximating with a B-spline surface with a larger number of pieces. The set of "auxiliary B-spline surfaces" produced serve solely to improve the parametrization.

Similar problems also arise in dealing with data without smooth boundaries. In this case we should construct smooth boundaries by, for example, looking for smooth curves in the interior of the data set. We then parametrize over these points using an auxiliary Coons surface, extrapolate parameter values for points outside of the interior, and then once again parametrize the interior of the entire parameter domain over an appropriately chosen new parameter plane.

Clearly, all of these methods are to a certain extent ad hoc.

6.2.5.2. Interpolation and Approximation with B-spline Surfaces

In this section we discuss the use of B-spline surfaces for interpolation and approximation. For simplicity, we assume throughout that the surface to be constructed has the same order in both parameters. In addition, we treat only (open) surface patches, and assume that the points $\boldsymbol{P}_s$ to be interpolated or approximated are associated with the parameter values (μ_s, ν_s) for $s = 0(1)N$.

We assume that the tensor-product B-spline surface is written in parametric form as

$$\boldsymbol{X}(u,v) = \sum_{i=0}^{p} \sum_{j=0}^{q} \boldsymbol{d}_{ij} N_{ik}(u) N_{jk}(v), \tag{6.36}$$

with associated *knots*

$$\begin{aligned}\boldsymbol{T} =&(u_0 = u_1 = \cdots = u_{k-1}, u_k, \cdots, u_p, u_{p+1} = u_{p+2} = \cdots = u_{p+k}) \\ &\times (v_0 = v_1 = \cdots = v_{k-1}, v_k, \cdots, v_q, v_{q+1} = v_{q+2} = \cdots = v_{q+k}).\end{aligned}$$

In view of this choice of knots, the basis functions in (6.36) are not periodic, and so again $N_{0k} = N_{0k}^{(k-2)}, N_{1k} = N_{1k}^{(k-1)}$, etc. If we further assume that $u_0 = v_0 = 0$ and that the parametrization is equally spaced (with interval length 1), then the maximal parameter values are given by

$$u_{p+1} = p - k + 2, \qquad v_{q+1} = q - k + 2. \tag{$*$}$$

We have an *interpolation problem* when the number of prescribed points is $N + 1 = (p + 1)(q + 1)$. Associating the parameter values (μ_s, ν_s) with the points $\boldsymbol{P}_s$ to be interpolated, (6.36) then leads to the following linear system of equations

$$\boldsymbol{P}_s = \sum_{i=0}^{p} \sum_{j=0}^{q} \boldsymbol{d}_{ij} N_{ik}(\mu_s) N_{jk}(\nu_s), \qquad s = 0(1)N, \tag{6.37}$$

with $p, q \geq k - 1$. The coefficient matrix in (6.37) is *sparse* since the basis functions are *local*. Ordering the equations appropriately, we can arrange that this matrix is banded or block-banded, see [Schm 79], [Bre 82]. The system (6.36) is solvable if the μ's and ν's are chosen so that $N_{ik}(\mu_i) > 0$ for $i = 0(1)p$ and $N_{jk}(\nu_j) > 0$ for $j = 0(1)q$.

If $N + 1 > (p + 1)(q + 1)$, then we have an *approximation problem*. Here we discuss solving it by the *discrete least squares method* of Gauss. It follows from (6.36) that the errors are given by

$$\boldsymbol{D}_s = \boldsymbol{P}_s - \sum_{i=0}^{p} \sum_{j=0}^{q} \boldsymbol{d}_{ij} N_{ik}(\mu_s) N_{jk}(\nu_s), \tag{6.38a}$$

for $s = 0(1)N$. As in (4.72), differentiating the sum of squares of the errors in (6.38a) with respect to the coefficients leads to the system of *normal equations*

$$\begin{aligned}&\sum_{s=0}^{N} \boldsymbol{P}_s N_{\ell k}(\mu_s) N_{mk}(\nu_s) \\ &\quad = \sum_{s=0}^{N} \sum_{i=0}^{p} \sum_{j=0}^{q} \boldsymbol{d}_{ij} N_{ik}(\mu_s) N_{jk}(\nu_s) N_{\ell k}(\mu_s) N_{mk}(\nu_s),\end{aligned} \tag{6.38b}$$

for $\ell = 0(1)p$, $m = 0(1)q$. The system (6.38b) can be solved by matrix inversion taking account of the band structure. Alternatively, the minimization problem can be solved directly using Householder transformations, see e.g., [BOEH 85], [Bre 82], and Sect. 4.4.3.

In approximating a set of points by a surface, it is recommended that parameter correction also be carried out. As was the case for curve approximation, in general the error vectors $\boldsymbol{D}_s = \boldsymbol{P}_s - \boldsymbol{X}(\mu_s, \nu_s)$ are not orthogonal to the approximation surface $\boldsymbol{X}(u, v)$. As was done for parameter correction in approximating a curve, we now project the point $\boldsymbol{P}_s$ onto the tangent plane to the approximating surface $\boldsymbol{X}(u, v)$ at the point $\boldsymbol{X}(\mu_s, \nu_s)$, see Fig. 6.19.

The partial derivatives $\boldsymbol{X}_u(\mu_s, \nu_s)$ and $\boldsymbol{X}_v(\mu_s, \nu_s)$ give the tangent directions along the parametric lines $u = \mu_s$ or $v = \nu_s$. The projection of the image $\bar{\boldsymbol{D}}_s$ of the error vector $\boldsymbol{D}_s$ onto the tangent direction leads to corrections in the parameter values, *i.e.*, we have (see Fig. 6.19)

$$\mu_s^* = \mu_s + \frac{\Delta c_s}{L(\mu_s)} \omega_u, \qquad \nu_s^* = \nu_s + \frac{\Delta d_s}{L(\nu_s)} \omega_v, \tag{6.39a}$$

where

$$\begin{aligned}\Delta c_s &= (\boldsymbol{P}_s - \boldsymbol{X}(\mu_s, \nu_s)) \cdot \frac{\boldsymbol{X}_u(\mu_s, \nu_s)}{|\boldsymbol{X}_u(\mu_s, \nu_s)|}, \\ \Delta d_s &= (\boldsymbol{P}_s - \boldsymbol{X}(\mu_s, \nu_s)) \cdot \frac{\boldsymbol{X}_v(\mu_s, \nu_s)}{|\boldsymbol{X}_v(\mu_s, \nu_s)|},\end{aligned} \tag{6.39b}$$

ω_u, ω_v are the lengths of the u and v parameter intervals, and $L(\mu_s)$, $L(\nu_s)$ are approximations to the lengths of the parametric lines $u = \mu_s$ and $v = \nu_s$, respectively.

Using the first terms in the Taylor expansions of the error vectors defined by $\boldsymbol{D}_s = \boldsymbol{P}_s - \boldsymbol{X}(\mu_s, \nu_s)$, and minimizing the absolute error, the following correction formula was given in [Rog 89]:

$$\widetilde{\boldsymbol{D}} = \boldsymbol{P}_s - \boldsymbol{X}(u, v) - \boldsymbol{X}_u(\mu_s, \nu_s)\Delta\mu_s - \boldsymbol{X}_v(\mu_s, \nu_s)\Delta\nu_s. \tag{6.40a}$$

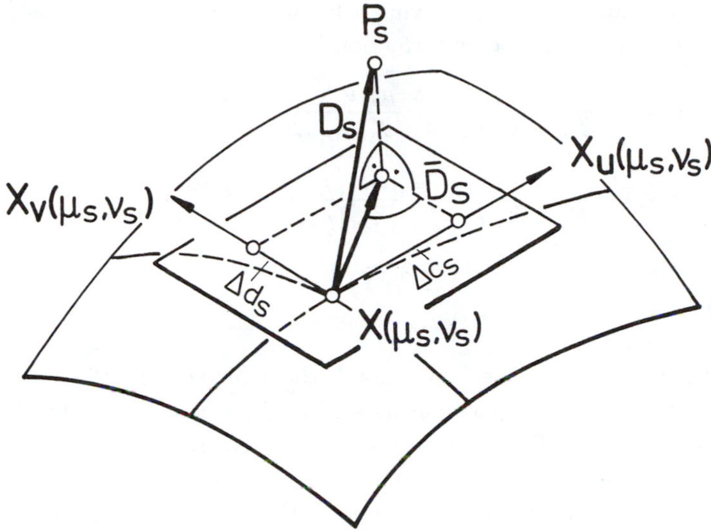

Fig. 6.19. Projection of the error vector onto the tangent plane for parameter correction.

This approximation to the error vector $\boldsymbol{D}_s$ assumes a minimum when

$$\frac{\partial \widetilde{\boldsymbol{D}}^2}{\partial u} = 2\widetilde{\boldsymbol{D}} \cdot \boldsymbol{X}_u = 0, \qquad \frac{\partial \widetilde{\boldsymbol{D}}^2}{\partial v} = 2\widetilde{\boldsymbol{D}} \cdot \boldsymbol{X}_v = 0. \tag{6.40b}$$

Inserting the derivatives of the error components of (6.40a) into (6.40b), and solving the resulting system of equations in $\Delta\mu$ and $\Delta\nu$, we get the following parameter corrections for (μ_s, ν_s):

$$\begin{aligned}
\Delta\mu_s = \Big[&(\boldsymbol{X}_v(\mu_s,\nu_s))^2[\boldsymbol{D}_s \cdot \boldsymbol{X}_u(\mu_s,\nu_s)] \\
&- [\boldsymbol{X}_u(\mu_s,\nu_s) \cdot \boldsymbol{X}_v(\mu_s,\nu_s)](\boldsymbol{D}_s \cdot \boldsymbol{X}_v(\mu_s,\nu_s))\Big]/F, \\
\Delta\nu_s = \Big[&(\boldsymbol{X}_u(\mu_s,\nu_s))^2[\boldsymbol{D}_s \cdot \boldsymbol{X}_v(\mu_s,\nu_s)] \\
&- [\boldsymbol{X}_u(\mu_s,\nu_s) \cdot \boldsymbol{X}_v(\mu_s,\nu_s)](\boldsymbol{D}_s \cdot \boldsymbol{X}_u(\mu_s,\nu_s))\Big]/F, \\
F = (&\boldsymbol{X}_u(\mu_s,\nu_s))^2(\boldsymbol{X}_v(\mu_s,\nu_s))^2 - [\boldsymbol{X}_u(\mu_s,\nu_s) \cdot \boldsymbol{X}_v(\mu_s,\nu_s)]^2.
\end{aligned} \tag{6.41}$$

The convergence of this process can be accelerated by taking account of the second derivatives. To do this, at each point we minimize the error vector

$$(\boldsymbol{P}_s - \boldsymbol{X}(\mu_s + \Delta\mu_s, \nu_s + \Delta\nu_s))^2,$$

using the Taylor expansion and solving a fitting problem as in (4.73e). This leads to the following parameter correction formulae:

$$\begin{aligned}\Delta\mu_s &= \frac{-(\boldsymbol{P}_s - \boldsymbol{X}(\mu_s,\nu_s))\cdot\boldsymbol{X}_u(\mu_s,\nu_s)}{(\boldsymbol{P}_s - \boldsymbol{X}(\mu_s,\nu_s))\cdot\boldsymbol{X}_{uu}(\mu_s,\nu_s) - \boldsymbol{X}_u(\mu_s,\nu_s)^2},\\ \Delta\nu_s &= \frac{-(\boldsymbol{P}_s - \boldsymbol{X}(\mu_s,\nu_s))\cdot\boldsymbol{X}_v(\mu_s,\nu_s)}{(\boldsymbol{P}_s - \boldsymbol{X}(\mu_s,\nu_s))\cdot\boldsymbol{X}_{vv}(\mu_s,\nu_s) - \boldsymbol{X}_v(\mu_s,\nu_s)^2}.\end{aligned} \tag{6.42}$$

The new parameter values are then given by

$$\mu_s^* = \mu_s + \Delta\mu_s, \qquad \nu_s^* = \nu_s + \Delta\nu_s.$$

If this method is iterated, then, in general, the error vector will converge to the perpendicular to the approximating surface $\boldsymbol{X}(u,v)$ at the point $\boldsymbol{P}_s$.

Since in (6.42) the denominators are the derivatives of the numerators (see (4.73a)), the local convergence of this parameter correction method is determined by the Newton-Raphson method [FAU 81], [ENG 85].

Figs. 6.20a,b,c illustrate the approximation of some point sets using B-spline surfaces. Fig. 6.20a shows a biquadratic approximating surface with uniform knot vector of length (2×2) without parameter correction. Fig. 6.20b shows the effect of the parameter correction: the first image shows the result after a parameter correction with 10 steps; and the second image shows the result after 30 steps. We clearly see the smoothing effect of the parameter correction. The maximal approximation error is reduced from 0.68 to 0.0016 in the process. We should note, however, that because of roundoff errors, too many iterations can often lead to numerical instability and result in a poorer fit. A comparable approximation can be obtained using rational approximation after only three iteration steps, see Sect. 6.5.1.

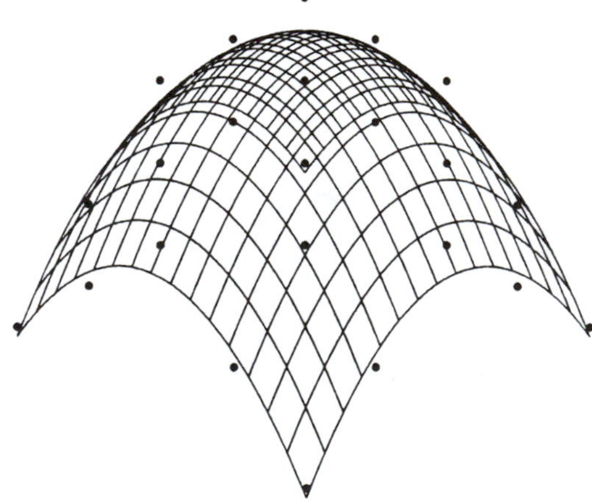

Fig. 6.20a. First approximation step without parameter correction.

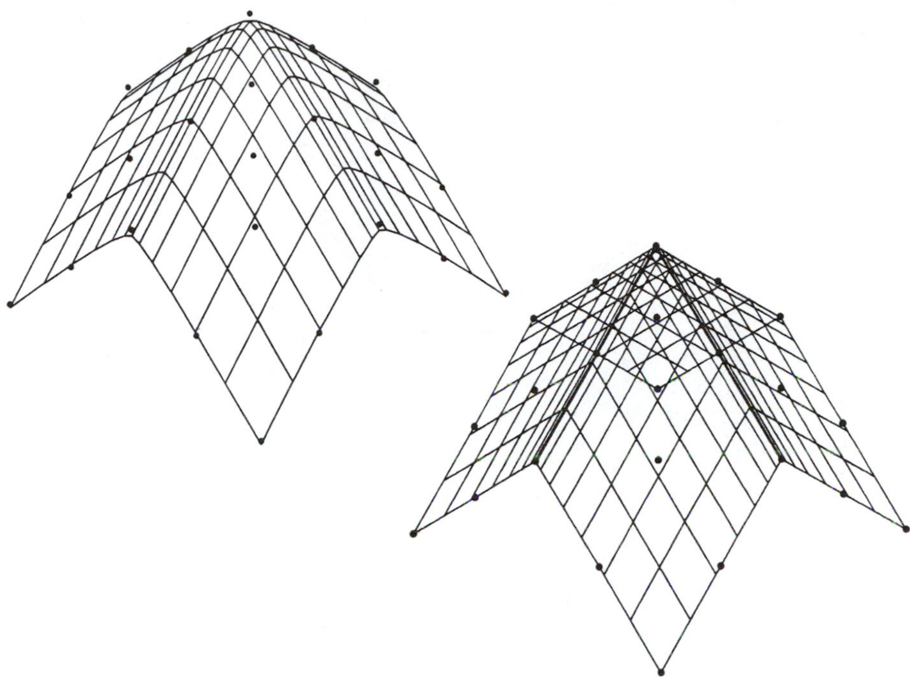

Fig. 6.20b. Approximation by a quadratic B-spline surface and the effect of parameter correction.

Fig. 6.20c shows a set of data obtained using a digitizer, and the corresponding bicubic B-spline approximating surface using a set of 10×8 uniformly spaced knots. The approximating surface is depicted in terms of the cross-section curves produced by the intersections with a set of parallel planes (these can serve as tool cutting paths, see Chap. 16).

6.3. Triangular Bézier Surfaces

In approximating the surface of a prescribed object using Bézier spline or B-spline surfaces, it can happen that it is not sufficient to work with the four-sided patches of a tensor-product surface. For example, it may be necessary to include *three-sided* or *five-sided* (etc.) patches, see Fig. 6.21.

There are two simple ways to obtain a three-sided domain from a four-sided domain:

- shrink one side to a point, see Fig. 6.22a,
- adjust two adjoining edges so that they become collinear. Thus if the two edges share the point $\boldsymbol{P}_{00}$, this means that we force the angle between the tangents to the two edges to be 180°, see Fig. 6.22b.

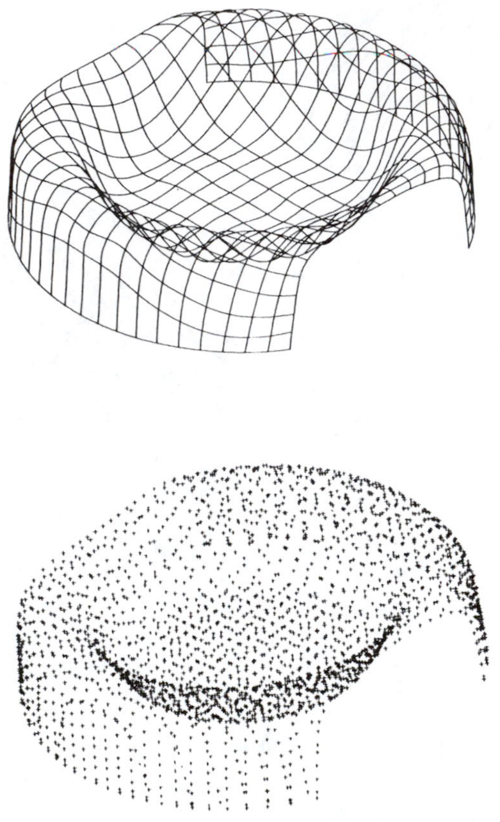

Fig. 6.20c. Approximation of a digitized set of points.

Both approaches are used in practice (see e.g., [FAU 81]), but generally produce unsatisfactory results. If a side of a four-sided patch is shrunk to a point, then all Bézier points on it fall on top of the point $\boldsymbol{b}_{00}$, which means that the resulting triangular patch is not differentiable at $\boldsymbol{b}_{00}$, and that the *normal vector* at $\boldsymbol{b}_{00}$ is not defined, and can only be estimated as the limit of the normal vectors at points near $\boldsymbol{b}_{00}$.

In the second case, the tangents to the two edges joining at the (degenerate) point $\boldsymbol{b}_{00}$ point in opposite directions, see Fig. 6.22b. Since these two tangents must span the *tangent plane* at $\boldsymbol{b}_{00}$, it follows that the tangent plane is *undefinined* at $\boldsymbol{b}_{00}$.

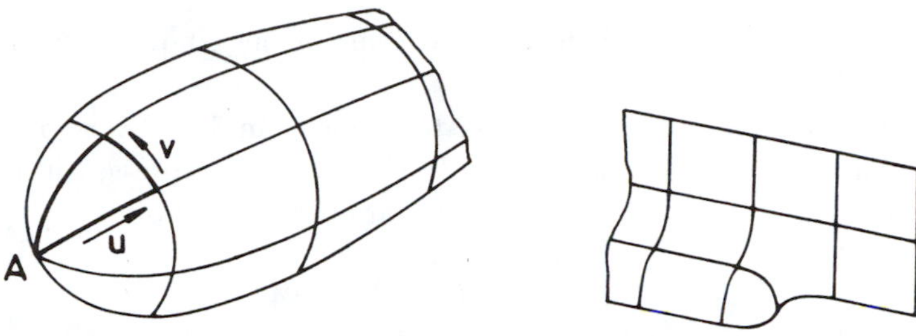

Fig. 6.21. Three-sided and five-sided patches.

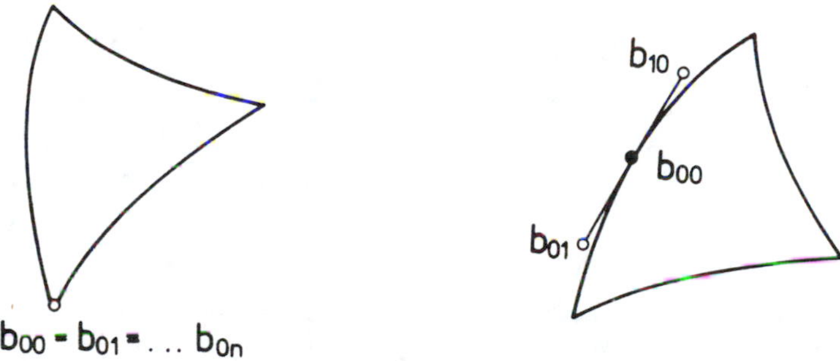

Fig. 6.22. Degenerate tensor-product surfaces.

We now show that these difficulties can be avoided by the use of surfaces constructed over nondegenerate triangular parameter domains.

6.3.1. Barycentric Coordinates

Barycentric coordinates are a useful tool for describing the location of a point in a triangle. To introduce them, we first consider the line segment $I = [0, 1]$. Every point $\boldsymbol{P}$ in I can be described by coordinates (u, v) with

$$u = t, \quad v = 1 - t, \quad t \in I, \qquad \text{subject to } u + v = 1.$$

The coordinates (u, v) are called the *barycentric coordinates* of the line segment I. With respect to these coordinates, the Bernstein polynomials can be written as

$$B_i^n(t) =: B_{ij}^n(u, v) = \frac{n!}{i!j!} u^i v^j,$$

with

$$i + j = n; \qquad i, j \geq 0, \qquad u + v = 1; \qquad u, v \geq 0.$$

The $B_{ij}^n(u, v)$ can thus be thought of as terms in the binomial expansion of $(u + v)^n = 1$.

This principle can be carried over to triangles in $\mathbb{R}^2$. Every point $\boldsymbol{P}$ in the interior of a triangle with vertices $\boldsymbol{U}, \boldsymbol{V}, \boldsymbol{W}$ can be expressed in terms of the *barycentric coordinates* (u, v, w) which are determined by

$$\boldsymbol{P}(u, v, w) = u\boldsymbol{U} + v\boldsymbol{V} + w\boldsymbol{W}, \tag{6.43}$$

subject to the constraints

$$u + v + w = 1, \qquad u, v, w, \geq 0.$$

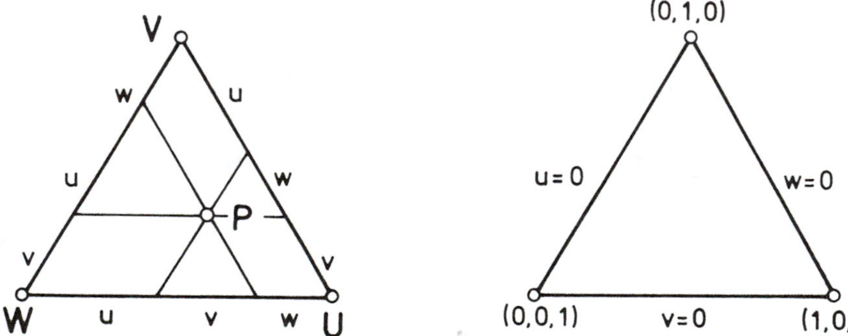

Fig. 6.23a. Barycentric coordinates.

Fig. 6.23a provides a geometric interpretation of barycentric coordinates. The vertex $\boldsymbol{V}$ in the figure has coordinates $(0, 1, 0)$, and the edge $\boldsymbol{UV}$ corresponds to the equation $w = 0$. An arbitrary point $\boldsymbol{P}(u, v, w)$ can be found by (proportional) parallel translation.

We call the triangle $(\boldsymbol{U}, \boldsymbol{V}, \boldsymbol{W})$ the *base triangle* of the barycentric coordinate system. Points inside the base triangle have positive coordinates, while points outside of the base triangle have at least one negative coordinate. If the point is inside the triangle, then the coordinates themselves can be interpreted as areas of subtriangles of the base triangle as follows:

$$u = \frac{|\Delta \boldsymbol{PVW}|}{|\Delta \boldsymbol{UVW}|}, \qquad v = \frac{|\Delta \boldsymbol{PWU}|}{|\Delta \boldsymbol{UVW}|}, \qquad w = \frac{|\Delta \boldsymbol{PUV}|}{|\Delta \boldsymbol{UVW}|},$$

where

$$|\Delta \boldsymbol{UVW}| = \text{area of the triangle } (\boldsymbol{U}, \boldsymbol{V}, \boldsymbol{W}),$$

and

$$|\Delta \boldsymbol{PVW}| = \text{area of the triangle } (\boldsymbol{P}, \boldsymbol{V}, \boldsymbol{W}),$$

etc. This follows immediately from (6.43) and the equality $u + v + w = 1$, using Cramer's rule. Fig. 6.23b shows another important geometric property of the barycentric coordinates which will be useful later for the construction of transfinite triangular interpolants via the "side-vertex method", see Sect. 8.2.

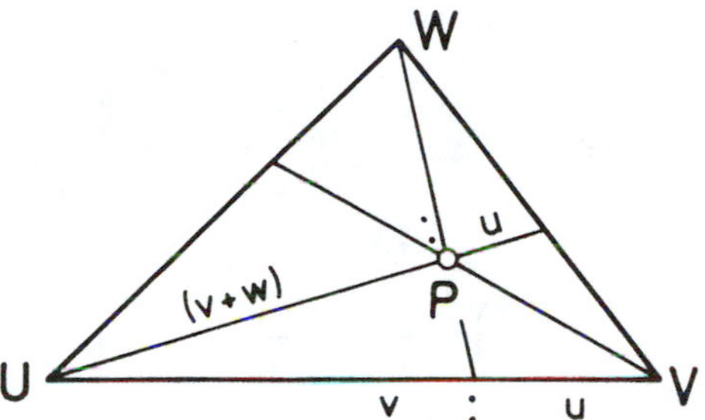

Fig. 6.23b. Barycentric coordinates as ratios.

Barycentric coordinates are invariant under affine transformations, *i.e.*, if $P \to \widehat{P} = LP + T$ is an affine transformation with linear part L and translation part T, then $\widehat{P}$ has the same barycentric coordinates with respect to the transformed triangle as P has with respect to the original one. In view of this, when working with individual triangles, we can assume that they are equilateral. This is useful, e.g., if we want to visualize the symmetry of u, v and w. Fig. 6.23c illustrates affine invariance.

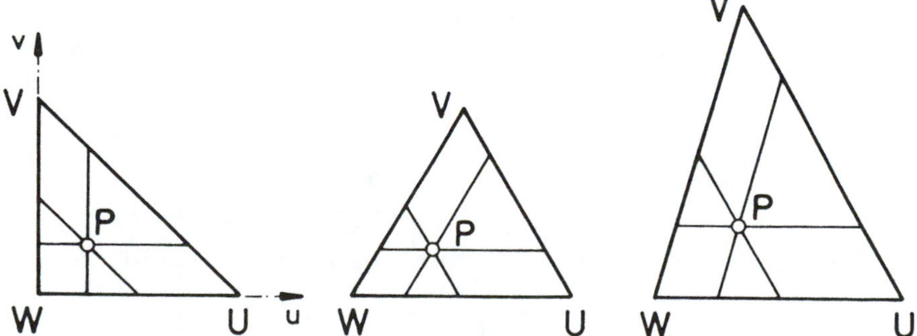

Fig. 6.23c. Affine invariance of barycentric coordinates.

6.3.2. Generalized Bernstein Polynomials and Triangular Bézier Surfaces

With the help of barycentric coordinates, it is now easy to define the Bernstein polynomials associated with a base triangle. As in Chap. 4, we start with the identity

$$(u+v+w)^n = \sum_{\substack{i,j,k\geq 0\\ i+j+k=n}} \frac{n!}{i!j!k!} u^i v^j w^k,$$

and make use of the elementary relations

$$(u+[v+w])^n = \sum_{i=0}^{n} \binom{n}{i} u^i (v+w)^{n-i}, \qquad (v+w)^{n-i} = \sum_{j=0}^{n-1} \binom{n-i}{j} v^j w^{n-i-j},$$

where

$$\binom{n}{i} = \frac{n!}{i!(n-i)!}.$$

Then the *generalized Bernstein polynomials of degree* n are defined as

$$B^n_{ijk}(u,v,w) := \frac{n!}{i!j!k!} u^i v^j w^k, \tag{6.44}$$

with

$$i+j+k=n, \qquad i,j,k\geq 0, \qquad u+v+w=1, \qquad u,v,w\geq 0.$$

For ease of notation, we define

$$\boldsymbol{I} := (i,j,k)^T, \qquad \boldsymbol{u} := (u,v,w)^T, \tag{6.45a}$$

and

$$|\boldsymbol{I}| = i+j+k, \qquad |\boldsymbol{u}| = u+v+w. \tag{6.45b}$$

Then in place of (6.44), we have

$$B^n_{\boldsymbol{I}}(\boldsymbol{u}) = \frac{n!}{i!j!k!} u^i v^j w^k, \qquad \text{with } |\boldsymbol{u}| = 1. \tag{6.46}$$

By their definition, it is clear that the Bernstein polynomials satisfy the normalization condition

$$\sum_{|\boldsymbol{I}|=n} B^n_{\boldsymbol{I}}(\boldsymbol{u}) = 1, \tag{6.47}$$

where the sum is taken over all possible vectors $\boldsymbol{I}$ which satisfy the conditions $|\boldsymbol{I}| = n$ and $\boldsymbol{I} \geq 0$. This involves a total of $\binom{n+2}{2} = (n+1)(n+2)/2$ terms.

On the edges of the base triangle, the generalized Bernstein polynomials reduce to the ordinary Bernstein polynomials. For example,

$$B^n_{0jk}(0, v, w) = B^n_j(v) = B^n_k(w). \tag{6.48}$$

Clearly, every generalized Bernstein polynomial $B^n_{\boldsymbol{I}}$ can be intepreted as a (polynomial) parametric formula for a surface patch (function) over the base triangle Δ. The maximum of this function occurs at $\boldsymbol{u} = \boldsymbol{I}/n$, see Fig. 6.24.

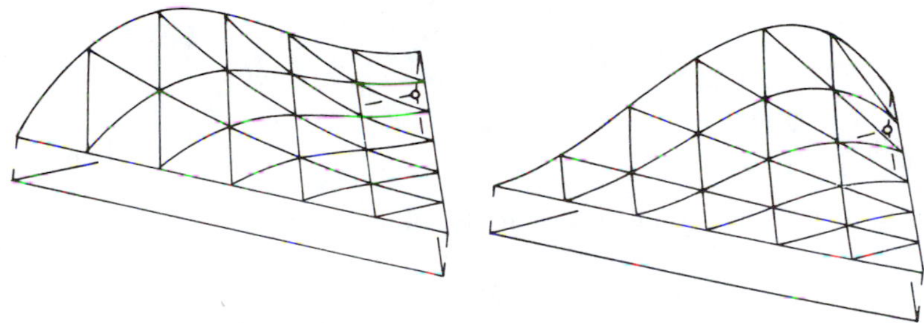

Fig. 6.24. Basis functions B^3_{012} and B^3_{021} on a base triangle.

Since the generalized Bernstein polynomials are linearly independent, they form a basis for an $(n+1)(n+2)/2$ dimensional linear subspace of the space of polynomials of degree (n, n). Every element $\boldsymbol{X}$ in the linear subspace spanned by the $B^n_{\boldsymbol{I}}$ has a unique expansion

$$\boldsymbol{X}(\boldsymbol{u}) = \sum_{|\boldsymbol{I}|=n} \boldsymbol{b}_{\boldsymbol{I}} B^n_{\boldsymbol{I}}(\boldsymbol{u}), \tag{6.49}$$

so that $\boldsymbol{X}(\boldsymbol{u})$ is the *parametric representation of a triangular Bézier surface of degree n*. The coefficients $\boldsymbol{b}_{\boldsymbol{I}}$ in (6.49) are called *Bézier points*. They form the *Bézier net* or Bézier polyhedron associated with the surface, see [Boeh 84], [Far 79, 86]. Applying the de Casteljau algorithm discussed below (see (6.52)), it follows immediately that triangular Bézier surfaces have the *convex hull property*. The corner Bézier points $\boldsymbol{b}_{00n}, \boldsymbol{b}_{0n0}, \boldsymbol{b}_{n00}$ lie on the surface. For each corner point, the point at that vertex and its neighbor on a given edge determine the tangent to the boundary curve at the corner. The corner point and its two neighbors determine the tangent plane at the corner.

The coefficients in (6.49) can be vector-valued (in $\mathbb{R}^2$ or $\mathbb{R}^3$), or they can be real numbers $b_{\boldsymbol{I}} \in \mathbb{R}$, which we then refer to as *Bézier ordinates*. In the latter case, the triangular Bézier surface is actually a function defined on the base triangle. As in the case of Bézier curves, we associate Bézier ordinates with points $\boldsymbol{I}/n$ on the base triangle according to the following scheme: with the Bézier ordinates

$$\begin{array}{ccccccc} & & & b_{0n0} & & & \\ & & b_{0,n-1,1} & & b_{1,n-1,0} & & \\ & \cdots & & \cdots & & \cdots & \\ b_{00n} & & \cdots & & \cdots & & b_{n00}, \end{array}$$

we associate the parameter values

$$\begin{array}{ccccccc} & & & (0,1,0) & & & \\ & & (0,1-\frac{1}{n},\frac{1}{n}) & & (\frac{1}{n},1-\frac{1}{n},0) & & \\ & \cdots & & & & \cdots & \\ (0,0,1) & & & & & & (1,0,0). \end{array}$$

For $n = 3$ the corresponding schemes become

$$\begin{array}{ccccccc} & & & b_{030} & & & \\ & & b_{021} & & b_{120} & & \\ & b_{012} & & b_{111} & & b_{210} & \\ b_{003} & & b_{102} & & b_{201} & & b_{300}, \end{array}$$

with

$$\begin{array}{ccccccc} & & & (0,1,0) & & & \\ & & (0,\frac{2}{3},\frac{1}{3}) & & (\frac{1}{3},\frac{2}{3},0) & & \\ & (0,\frac{1}{3},\frac{2}{3}) & & (\frac{1}{3},\frac{1}{3},\frac{1}{3}) & & (\frac{2}{3},\frac{1}{3},0) & \\ (0,0,1) & & (\frac{1}{3},0,\frac{2}{3}) & & (\frac{2}{3},0,\frac{1}{3}) & & (1,0,0). \end{array}$$

Fig. 6.25a shows a scalar-valued triangular Bézier patch of degree $n = 3$, and Fig. 6.25b shows a triangular Bézier patch in parametric form.

Remark. If we introduce four-component vector-valued Bézier points by $\boldsymbol{b}_{\boldsymbol{I}} = (\boldsymbol{I}/n, b_{\boldsymbol{I}})$ with $b_{\boldsymbol{I}} \in \mathbb{R}$, then all of the following algorithms for vector-valued patches can be applied to the vector-valued representation $\boldsymbol{X}(\boldsymbol{u}) = (u, v, w, X(\boldsymbol{u}))$ of the function case, cf. Sects. 4.1, 6.2. The abscissae $\boldsymbol{I}/n$ and the coordinates u, v, w of $\boldsymbol{X}(\boldsymbol{u})$ are connected via (6.43) with the Cartesian coordinates x, y of $\mathbb{R}^2$.

As for Bézier curves, we can carry out *degree raising* (basis transformation from degree n to degree $n + 1$) for triangular Bézier surfaces.

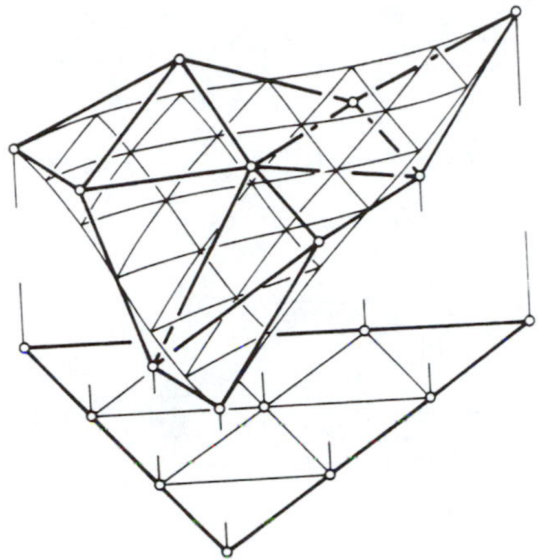

Fig. 6.25a. Scalar-valued triangular Bézier patch of degree $n = 3$ with its Bézier net.

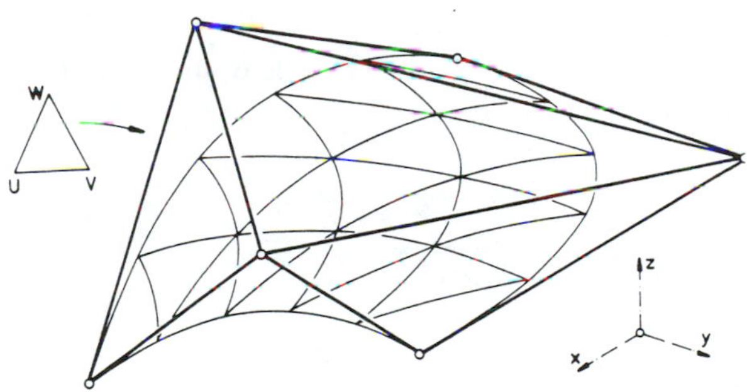

Fig. 6.25b. Quadratic triangular Bézier patch in parametric form with its Bézier net.

Lemma 6.3. *Suppose $\boldsymbol{b}_{ijk}$ are the Bézier points of a triangular Bézier surface of degree n, and $\boldsymbol{b}^*_{ijk}$ are the Bézier points of the same surface after transforming to a basis of degree $n+1$. Then*

$$\boldsymbol{b}^*_{ijk} = \frac{i}{n+1}\boldsymbol{b}_{i-1,j,k} + \frac{j}{n+1}\boldsymbol{b}_{i,j-1,k} + \frac{k}{n+1}\boldsymbol{b}_{i,j,k-1}, \tag{6.50}$$

with $\boldsymbol{b}_{ijk} = 0$ whenever any of the i, j, k is smaller than 0 or greater than n.

Equivalently, we can write

$$\boldsymbol{b}_I^* = \frac{i}{n+1}\boldsymbol{b}_{I-\boldsymbol{e}_1} + \frac{j}{n+1}\boldsymbol{b}_{I-\boldsymbol{e}_2} + \frac{k}{n+1}\boldsymbol{b}_{I-\boldsymbol{e}_3},$$

where

$$\boldsymbol{e}_1 = (1,0,0)^T, \qquad \boldsymbol{e}_2 = (0,1,0)^T, \qquad \boldsymbol{e}_3 = (0,0,1)^T.$$

Proof: Since $u+v+w=1$, we must have

$$\sum_{|I|=n+1} \boldsymbol{b}_I^* \frac{(n+1)!}{i!j!k!} u^i v^j w^k = \sum_{|I|=n} \boldsymbol{b}_I \frac{n!}{i!j!k!} u^i v^j w^k$$

$$= \sum_{|I|=n} \boldsymbol{b}_I \frac{n!}{i!j!k!} (u^{i+1} v^j w^k + u^i v^{j+1} w^k + u^i v^j w^{k+1}).$$

Comparing coefficients leads to

$$\boldsymbol{b}_{ijk}^* \frac{(n+1)!}{i!j!k!} = \boldsymbol{b}_{i-1,jk} \frac{n!}{(i-1)!j!k!} + \boldsymbol{b}_{i,j-1,k} \frac{n!}{i!(j-1)!k!} + \boldsymbol{b}_{ij,k-1} \frac{n!}{i!j!(k-1)!},$$

and dividing by the factor on the left leads to (6.50). ■

This says that $\boldsymbol{b}_{ijk}^*$ can be obtained by computing the values at $(i/(n+1), j/(n+1), k/(n+1))$ of the linear function interpolating the values $\boldsymbol{b}_{I-\boldsymbol{e}_1}$, $\boldsymbol{b}_{I-\boldsymbol{e}_2}$, $\boldsymbol{b}_{I-\boldsymbol{e}_3}$. On the boundary of the net, the degree raising formula (6.50) reduces to the one for curves in (4.14). Fig. 6.26 illustrates the construction of new Bézier points by degree raising from $n=2$ to $n=3$.

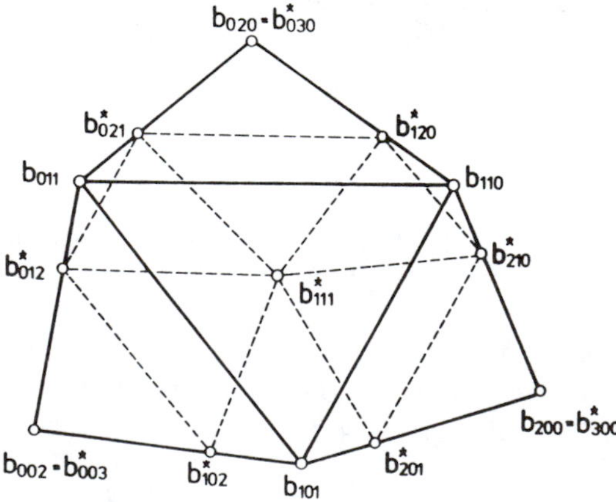

Fig. 6.26. Degree raising from $n=2$ to $n=3$ for a triangular Bézier patch.

The *de Casteljau algorithm* can also be extended to triangular Bézier surfaces. The classical de Casteljau algorithm (see Chap. 4) is based on the following decomposition of the Bernstein polynomials:

$$B_i^n(t) = (1-t)B_i^{n-1}(t) + tB_{i-1}^{n-1}(t). \tag{$*$}$$

Now since $i + j + k = n$, we have

$$\frac{n!}{i!j!k!} = \frac{(n-1)!}{(i-1)!j!k!} + \frac{(n-1)!}{i!(j-1)!k!} + \frac{(n-1)!}{i!j!(k-1)!},$$

and it follows analogous to $(*)$ that

$$B_{ijk}^n(\boldsymbol{u}) = uB_{i-1,j,k}^{n-1}(\boldsymbol{u}) + vB_{i,j-1,k}^{n-1}(\boldsymbol{u}) + wB_{i,j,k-1}^{n-1}(\boldsymbol{u}),$$

or equivalently

$$B_{\boldsymbol{I}}^n(\boldsymbol{u}) = uB_{\boldsymbol{I}-\boldsymbol{e}_1}^{n-1}(\boldsymbol{u}) + vB_{\boldsymbol{I}-\boldsymbol{e}_2}^{n-1}(\boldsymbol{u}) + wB_{\boldsymbol{I}-\boldsymbol{e}_3}^{n-1}(\boldsymbol{u}). \tag{6.51}$$

In view of the basic idea of the algorithm that the desired surface point $\boldsymbol{X}(\boldsymbol{u})$ can be found stepwise by computing with Bernstein polynomials of lower degree, we can write

$$\boldsymbol{X}(\boldsymbol{u}) = \sum_{|\boldsymbol{I}|=n-r} \boldsymbol{b}_{\boldsymbol{I}}^r(\boldsymbol{u}) B_{\boldsymbol{I}}^{n-r}(\boldsymbol{u}), \qquad r = 0(1)n,$$

where $\boldsymbol{b}_{\boldsymbol{I}}^r(\boldsymbol{u})$ are triangular Bézier patches of degree r, and $B_{\boldsymbol{I}}^{n-r}(\boldsymbol{u})$ are generalized Bernstein polynomials of degree $n-r$.

On the other hand, from (6.51) we have

$$\boldsymbol{X}(\boldsymbol{u}) = \sum_{|\boldsymbol{I}|=n-r-1} (u\boldsymbol{b}_{\boldsymbol{I}+\boldsymbol{e}_1}^r(\boldsymbol{u}) + v\boldsymbol{b}_{\boldsymbol{I}+\boldsymbol{e}_2}^r(\boldsymbol{u}) + w\boldsymbol{b}_{\boldsymbol{I}+\boldsymbol{e}_3}^r(\boldsymbol{u}))B_{\boldsymbol{I}}^{n-r-1}(\boldsymbol{u}).$$

Comparing coefficients leads to the de Casteljau *recursion relation*

$$\boldsymbol{b}_{\boldsymbol{I}}^{r+1}(\boldsymbol{u}) = u\boldsymbol{b}_{\boldsymbol{I}+\boldsymbol{e}_1}^r(\boldsymbol{u}) + v\boldsymbol{b}_{\boldsymbol{I}+\boldsymbol{e}_2}^r(\boldsymbol{u}) + w\boldsymbol{b}_{\boldsymbol{I}+\boldsymbol{e}_3}^r(\boldsymbol{u}), \qquad |I| = n-r-1. \tag{6.52}$$

Here we set $\boldsymbol{b}_{\boldsymbol{I}}^0 = \boldsymbol{b}_{\boldsymbol{I}}$ for $r = 0$, and the step $r = n-1$ produces

$$\boldsymbol{b}_{\boldsymbol{I}}^n(\boldsymbol{u}) = \boldsymbol{X}(\boldsymbol{u}).$$

The recurrence formula (6.52) asserts that geometrically, the new coefficient $\boldsymbol{b}_{\boldsymbol{I}}^{r+1}$ is computed as an average of the three coefficients $\boldsymbol{b}_{\boldsymbol{I}+\boldsymbol{e}_i}^r$, $i = 1, 2, 3$, using the barycentric coordinates (u, v, w) with respect to the upright triangle $\boldsymbol{b}^r_{I+e_1}$, $\boldsymbol{b}^r_{I+e_2}$, $\boldsymbol{b}^r_{I+e_3}$, see Fig. 6.27.

Summarizing, we can formulate the *de Casteljau algorithm* as follows:

1) Given: a triangular Bézier patch $\boldsymbol{X}(\boldsymbol{u})$ of degree n and a point $\boldsymbol{u}_0 = (u_0, v_0, w_0)^T$,

2) For $r = 0(1)n\text{–}1$, $|\boldsymbol{I}| = n - r$ compute

$$\boldsymbol{b}^{r+1}_{I} = u_0 \boldsymbol{b}^r_{I+e_1} + v_0 \boldsymbol{b}^r_{I+e_2} + w_0 \boldsymbol{b}^r_{I+e_3},$$

3) $\boldsymbol{X}(\boldsymbol{u}_0) = \boldsymbol{b}^n_0$.

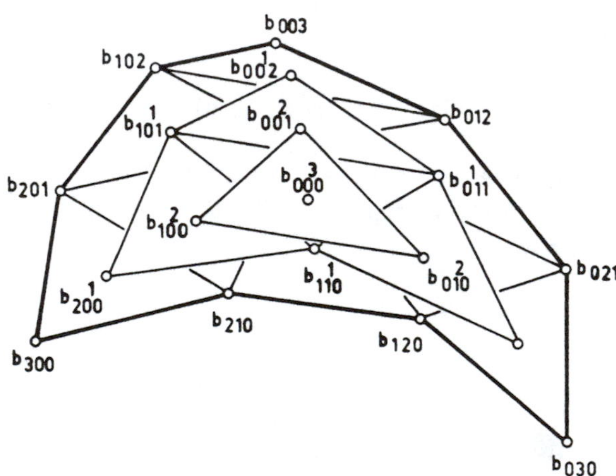

Fig. 6.27. de Casteljau algorithm for triangular Bézier patches.

In view of the de Casteljau algorithm, every triangular Bézier patch must lie in the convex hull of its Bézier points. For assertions about the convexity of triangular patches, see [Cha 84, 84a, 85], [Far 86], [Zho 90], [Cao 91], [Goo 91a], [Gre 91a], and [Dah 91].

As for tensor-product surfaces, the de Casteljau algorithm can also be used to subdivide a triangular Bézier surface into three spline patches touching at a given point (u_0, v_0, w_0), see [Far 79, 86]. Fig. 6.28 shows the boundaries of the three new nets in the case $n = 3$, where the intermediate points of the de Casteljau algorithm are taken from Fig. 6.27.

The Bézier points of the three new surface patches are again boundary points of the de Casteljau scheme (which we now organize as a *tetrahedron*).

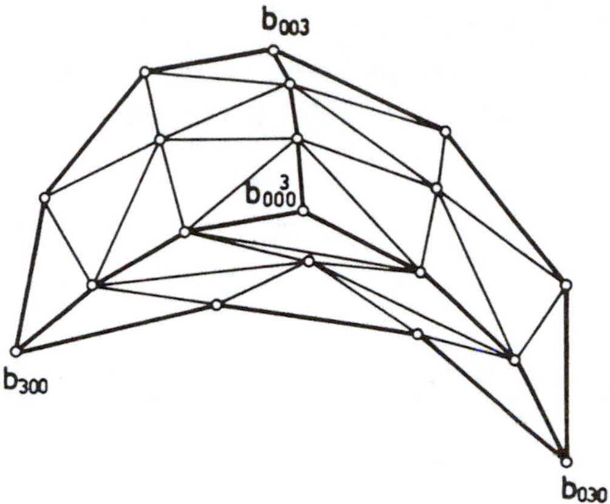

Fig. 6.28. Subdividing a triangular Bézier patch using the de Casteljau algorithm.

Patch I	has Bézier points $\boldsymbol{b}^r_{0jk}$	(vertices $\boldsymbol{b}^n_{000}, \boldsymbol{b}^0_{0n0}, \boldsymbol{b}^0_{00n}$),
Patch II	has Bézier points $\boldsymbol{b}^r_{i0k}$	(vertices $\boldsymbol{b}^n_{000}, \boldsymbol{b}^0_{n00}, \boldsymbol{b}^0_{00n}$),
Patch III	has Bézier points $\boldsymbol{b}^r_{ij0}$	(vertices $\boldsymbol{b}^n_{000}, \boldsymbol{b}^0_{n00}, \boldsymbol{b}^0_{0n0}$),

where $r+j+k=n$. By combining several steps of the de Casteljau algorithm, we can subdivide a triangular Bézier patch into an arbitrary number of subtriangles of degree n, see [Gol 83] and [Boeh 84]. The sequence of piecewise linear surfaces interpolating the Bézier nets converges to the Bézier surface [Fil 86].

6.3.3. Continuity Conditions for Triangular Bézier Patches

In order to build a smooth spline surface using triangular Bézier patches, we need to impose *continuity conditions* between the patches. Here we discuss conditions for joining

- a triangular Bézier patch with another triangular Bézier patch,
- a tensor-product Bézier patch with a triangular Bézier patch,
- a triangular Bézier patch with a tensor-product Bezier patch.

Because of the conditions $u+v+w=1$ and $0 \leq u, v, w \leq 1$, partial derivatives of triangular patches *do not agree with the directional derivatives along parametric lines*. In the following, D will always denote a directional derivative. Directional derivatives can be avoided by inserting $w=1-u-v$ into the function being differentiated, in which case only the partial derivatives with respect to u and v appear, and the parametric lines corresponding to constant values of w are "lost".

To discuss continuity conditions, we need the first and second directional derivatives of a triangular Bézier patch. As in Chap. 4, it will be convenient to work with the *operator notation*

$$\boldsymbol{X}(u,v,w) = (E_1 u + E_2 v + E_3 w)^n \boldsymbol{b}_{000} = \sum_{|\boldsymbol{I}|=n} \boldsymbol{b}_{\boldsymbol{I}} B_{\boldsymbol{I}}^n(\boldsymbol{u}),$$

with

$$\boldsymbol{I} = (i,j,k)^T, \quad \boldsymbol{u} = (u,v,w)^T, \quad u+v+w=1,$$

and

$$E_1 \boldsymbol{b}_{ijk} = \boldsymbol{b}_{i+1,jk}, \qquad E_2 \boldsymbol{b}_{ijk} = \boldsymbol{b}_{i,j+1,k}, \qquad E_3 \boldsymbol{b}_{ijk} = \boldsymbol{b}_{ij,k+1}.$$

Since $u+v+w=1$, the first directional derivative is given by

$$\begin{aligned}\frac{D\boldsymbol{X}}{du} &= n(E_1 u + E_2 v + E_3 w)^{n-1}(E_1 - E_3)\boldsymbol{b}_{000} \\ &= n \sum_{i+j+k=n-1} (\boldsymbol{b}_{i+1,jk} - \boldsymbol{b}_{ij,k+1}) B_{ijk}^{n-1}(u,v,w),\end{aligned} \tag{6.53a}$$

where we have used $w = 1-u-v$. Alternately, we could have eliminated $v = 1-u-w$, in which case the first directional derivative becomes

$$\begin{aligned}\frac{D\boldsymbol{X}}{du} &= n(E_1 u + E_2 v + E_3 w)^{n-1}(E_1 - E_2)\boldsymbol{b}_{000} \\ &= n \sum_{i+j+k=n-1} (\boldsymbol{b}_{i+1,jk} - \boldsymbol{b}_{i,j+1,k}) B_{ijk}^{n-1}(u,v,w).\end{aligned} \tag{6.53b}$$

Similarly, we also have

$$\begin{aligned}\frac{D\boldsymbol{X}}{dv} &= n(E_1 u + E_2 v + E_3 w)^{n-1}(E_2 - E_3)\boldsymbol{b}_{000} \\ &= n \sum_{i+j+k=n-1} (\boldsymbol{b}_{i,j+1,k} - \boldsymbol{b}_{ij,k+1}) B_{ijk}^{n-1}(u,v,w).\end{aligned} \tag{6.53c}$$

Fig. 6.29 gives a geometric interpretation of the directional derivatives along the boundary curve corresponding to constant u. At the corner $(0,0,1)$ of the triangular patch, the directional derivatives are

$$\frac{D\boldsymbol{X}}{du}(0,0,1) = n(\boldsymbol{b}_{10,n-1} - \boldsymbol{b}_{00n}), \qquad \frac{D\boldsymbol{X}}{dv}(0,0,1) = n(\boldsymbol{b}_{01,n-1} - \boldsymbol{b}_{00n}),$$

i.e., the *tangent plane* at a vertex is determined by the Bézier point at the corner and its two neighbors. The *second directional derivatives* are given (e.g.) by

$$\begin{aligned}
\frac{D^2\boldsymbol{X}}{du^2} &= n(n-1)(E_1u+E_2v+E_3w)^{n-2}(E_1-E_3)^2\boldsymbol{b}_{000} \\
&= n(n-1)\sum_{i+j+k=n-2}(\boldsymbol{b}_{i+2,jk}-2\boldsymbol{b}_{i+1,j,k+1}+\boldsymbol{b}_{ij,k+2})B_{ijk}^{n-2}, \\
\frac{D^2\boldsymbol{X}}{dv^2} &= n(n-1)(E_1u+E_2v+E_3w)^{n-2}(E_2-E_1)^2\boldsymbol{b}_{000} \qquad (6.54)\\
&= n(n-1)\sum_{i+j+k=n-2}(\boldsymbol{b}_{i,j+2,k}-2\boldsymbol{b}_{i+1,j+1,k}+\boldsymbol{b}_{i+2,jk})B_{ijk}^{n-2}, \\
\frac{D^2\boldsymbol{X}}{dudv} &= n(n-1)(E_1u+E_2v+E_3w)^{n-2}(E_1-E_3)(E_2-E_3)\boldsymbol{b}_{000} \\
&= n(n-1)\sum_{i+j+k=n-2}(\boldsymbol{b}_{i+1,j+1,k}-\boldsymbol{b}_{i+1,j,k+1} \\
&\qquad -\boldsymbol{b}_{i,j+1,k+1}+\boldsymbol{b}_{ij,k+2})B_{ijk}^{n-2}.
\end{aligned}$$

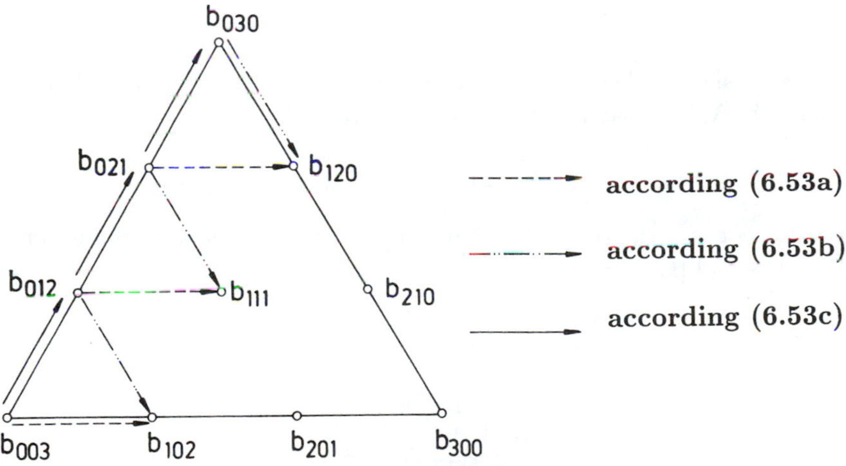

Fig. 6.29. Geometric interpretation of directional derivatives.

We now develop the continuity conditions along the boundary curve corresponding to $u = 0$. Suppose we have two triangular Bézier patches

$$\boldsymbol{X}_1(u,v,w)=\sum_{|\boldsymbol{I}|=n}\boldsymbol{b}_{\boldsymbol{I}}^1B_{\boldsymbol{I}}^n(\boldsymbol{u}), \qquad \boldsymbol{X}_2(u,v,w)=\sum_{|\boldsymbol{I}|=n}\boldsymbol{b}_{\boldsymbol{I}}^2B_{\boldsymbol{I}}^n(\boldsymbol{u}).$$

In order for these two surfaces to meet with C^0 *continuity*, we need

$$\boldsymbol{X}_1(0,v,w) = \boldsymbol{X}_2(0,v,w),$$

or

$$\boldsymbol{b}^1_{0,n-k,k} = \boldsymbol{b}^2_{0,n-k,k}, \qquad k = 0(1)n. \tag{6.55}$$

For GC^1 *continuity*, the tangent planes of the two surfaces must coincide at each point along the boundary. This gives

$$\frac{D\boldsymbol{X}_1}{du}(0,v,w) = \lambda_1 \frac{D\boldsymbol{X}_2}{du}(0,v,w) + \lambda_2 \frac{D\boldsymbol{X}_2}{dw}(0,v,w),$$

where $\lambda_1, \lambda_2 \in \mathbb{R}$ are free "design parameters". Using (6.53) and comparing coefficients leads to

$$\begin{aligned}(\boldsymbol{b}^1_{1,n-1-k,k} &- \boldsymbol{b}^1_{0,n-1-k,k+1}) \\ &= \lambda_1(\boldsymbol{b}^2_{1,n-1-k,k} - \boldsymbol{b}^2_{0,n-k-1,k+1}) + \lambda_2(\boldsymbol{b}^2_{0,n-1-k,k+1} - \boldsymbol{b}^2_{0,n-k,k}).\end{aligned}$$

Then setting $\lambda := 1 - \lambda_1 + \lambda_2$ and using (6.55), we get the GC^1 continuity condition

$$\boldsymbol{b}^1_{1,n-1-k,k} = \lambda \boldsymbol{b}^2_{0,n-1-k,k+1} + \lambda_1 \boldsymbol{b}^2_{1,n-1-k,k} - \lambda_2 \boldsymbol{b}^2_{0,n-k,k}. \tag{6.56a}$$

The special case where the tangents coincide along the parametric line corresponds to $\lambda_1 = -1$ and $\lambda_2 = -1$. In that case

$$\boldsymbol{b}^1_{1,n-1-k,k} - \boldsymbol{b}^1_{0,n-k-1,k+1} = \boldsymbol{b}^2_{0,n-k,k} - \boldsymbol{b}^2_{1,n-1-k,k} \tag{6.56b}$$

i.e., the edges of the Bézier nets are pairwise parallel (*rhombus condition*), see Fig. 6.30 and [Far 79].

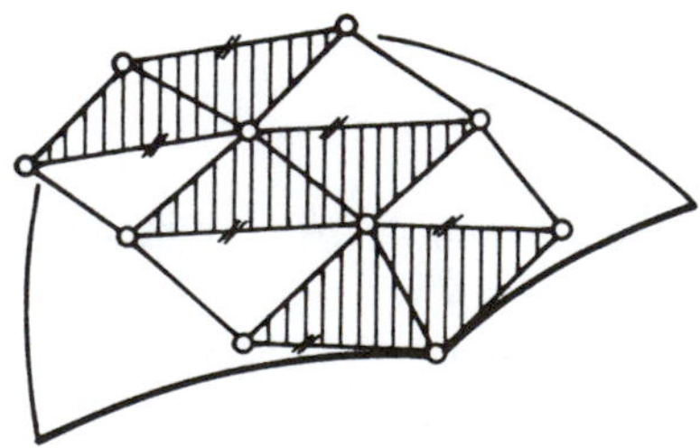

Fig. 6.30. Rhombus condition for C^1 continuity of triangular Bézier patches.

The condition (6.56b) holds for patches defined on two equilateral triangles in parameter space, cf. [Far 79], [Kah 82]. If the two parameter triangles have more general shapes, and in particular have different shapes, as happens for example in scattered data fitting (see Chap. 9), then the analog of (6.56b) for C^1 continuity is an equation of the type (6.56a), but with λ_1 and λ_2 determined by the geometry of the triangles [Alf 84a] (see [Las 87] for the trivariate case). In the general case, the continuity condition requires that the four Bézier points involved in the condition must lie on two coplanar triangles, which because of the factors λ_1 and λ_2 must be the affine images of the two parameter space triangles. This condition must hold for every k, *i.e.*, for all triangular pairs along the common boundary, see Fig. 6.30a.

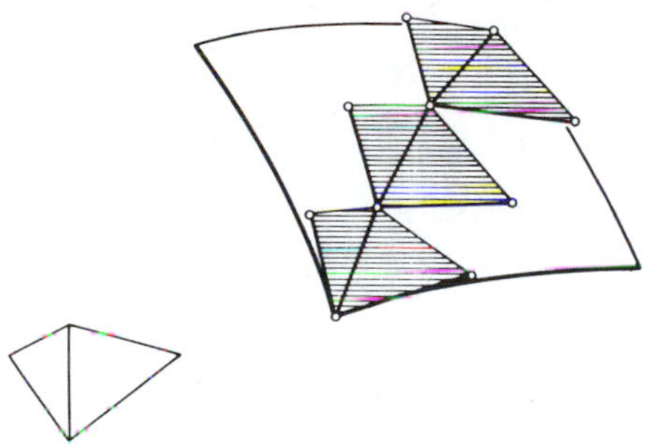

Fig. 6.30a. C^1 continuity of two triangular Bézier patches in parametric form. Every pair of triangles along the common boundary of the Bézier nets must be coplanar, and all cross-hatched triangular pairs must be images of the parameter space triangle under the same affine mapping.

For scalar-valued triangular Bézier patches, only the condition of coplanarity is needed. Since the ordinates b_{ijk} are associated with the abscissae $(i/n, j/n, k/n)$, the second condition is automatically satisfied whenever the first one holds, see Fig. 6.30b. Coplanarity alone is not sufficient for C^1 continuity in the parametric case, as shown by an example in [Far 86].

For C^2 *continuity*, we discuss only the simplest case where the domain triangles are equilateral, and where we also have

$$\frac{D^2\boldsymbol{X}_1}{du^2}(0, v, w) = \frac{D^2\boldsymbol{X}_2}{du^2}(0, v, w).$$

By (6.54)

$$\begin{aligned}\boldsymbol{b}^1_{2,n-2-k,k} - 2\boldsymbol{b}^1_{1,n-2-k,k+1} + \boldsymbol{b}^1_{0,n-2-k,k+2} \\ = \boldsymbol{b}^2_{0,n-k,k} - 2\boldsymbol{b}^2_{1,n-1-k,k} + \boldsymbol{b}^2_{2,n-2-k,k},\end{aligned}$$

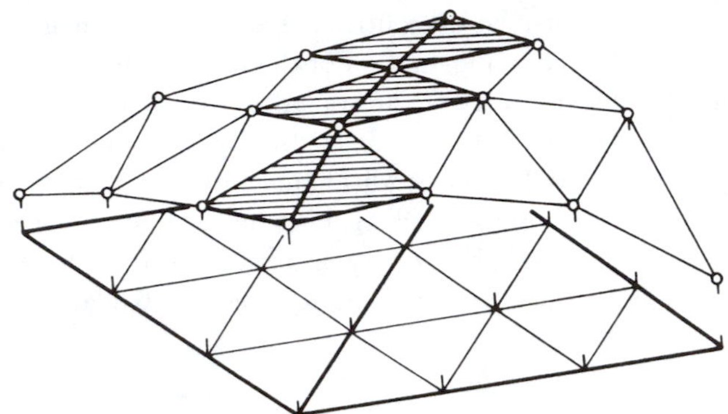

Fig. 6.30b. C^1 continuity for scalar-valued triangular Bézier patches. Every pair of triangles along the common boundary of the Bézier nets must be coplanar. If they are, then all cross-hatched triangular pairs are automatically images of the parameter space triangles w.r.t. the same affine mapping.

and now using (6.56b), we get the following C^2 continuity condition

$$\boldsymbol{b}^2_{2,n-2-k,k} = \boldsymbol{b}^2_{1,n-1-k,k} + \boldsymbol{b}^2_{1,n-2-k,k+1} - \boldsymbol{D}_k, \tag{6.57}$$

where

$$\boldsymbol{D}_k = \boldsymbol{b}^1_{1,n-1-k,k} + \boldsymbol{b}^1_{1,n-2-k,k+1} - \boldsymbol{b}^1_{2,n-2-k,k}$$

is an auxiliary point. This condition can also be interpreted geometrically, cf. (4.19) and Fig. 4.12.

If a patch is joined to another with C^r continuity, then those Bézier points determined by the continuity conditions can be computed using r extrapolation steps of the de Casteljau algorithm, see [Far 83a, 86] and the remark on page 141 of Sect. 4.1.

So far, we have discussed continuity conditions only for triangular patches. We now turn to the case of attaching a *tensor-product Bézier patch* of degree (n, n) to a *triangular Bézier patch* of degree n, see Fig. 6.32. Suppose the triangular Bézier patch is given by

$$\boldsymbol{X}_1(u, v, w) = \sum_{i+j+k=n} \boldsymbol{b}_{ijk} B^n_{ijk}(u, v, w), \tag{6.58a}$$

with $0 \le u, v, w \le 1$, $u+v+w=1$, and that the tensor-product Bézier patch has the representation

$$\boldsymbol{X}_2(u,v) = \sum_{i=0}^{n}\sum_{j=0}^{n} \boldsymbol{a}_{ij} B_i^n(u) B_j^n(v). \tag{6.58b}$$

In order for the two patches to join continuously along the line $u = 0$, we need

$$\boldsymbol{X}_1(0,v,w) = \boldsymbol{X}_2(0,v),$$

which by (6.58) implies

$$\boldsymbol{b}_{0,k,n-k} = \boldsymbol{a}_{0k}, \qquad k = 0(1)n.$$

For C^1 continuity, we want the tangents along the parameter line to coincide, i.e.,

$$\frac{\partial \boldsymbol{X}_2}{\partial u}(0,v) = -\frac{D\boldsymbol{X}_1}{du}(0,v,w). \tag{6.59}$$

Here it is important to note that the two derivatives appearing here are, in general, of *different degrees*. For $u = 0$, the derivative of (6.58a) is of degree $n-1$ in v, while the derivative of (6.58b) is of degree n in v. In order to compare coefficients, we must first raise the degree of the derivative of (6.58a) from $n-1$ to n, using Lemma 6.3. By (6.53a) and (6.48), it follows that

$$\begin{aligned}\frac{D\boldsymbol{X}_1}{du}(0,v,w) &= n \sum_{j+k=n-1} (\boldsymbol{b}_{1jk} - \boldsymbol{b}_{0,j,k+1}) B_{0jk}^{n-1}(0,v,1-v)\\ &= n\sum_{j=0}^{n}\Bigg[\left(1-\frac{j}{n}\right)(\boldsymbol{b}_{1,j,n-1-j} - \boldsymbol{b}_{0,j,n-j})\\ &\qquad - \left(\frac{j}{n}\right)(\boldsymbol{b}_{0,j-1,n+1-j} - \boldsymbol{b}_{1,j-1,n-j})\Bigg] B_j^n(v).\end{aligned}$$

Comparing coefficients with those of the derivative of (6.58b) at $u = 0$ and using (6.59) and (6.11), we get

$$\boldsymbol{a}_{1j} = 2\left[\left(1-\frac{j}{2n}\right)\boldsymbol{b}_{0,j,n-j} + \left(\frac{j}{2n}\right)\boldsymbol{b}_{0,j-1,n+1-j}\right] - \boldsymbol{c}_j, \tag{6.60}$$

where

$$\boldsymbol{c}_j = \left(1-\frac{j}{n}\right)\boldsymbol{b}_{1,j,n-1-j} + \left(\frac{j}{n}\right)\boldsymbol{b}_{1,j-1,n-j},$$

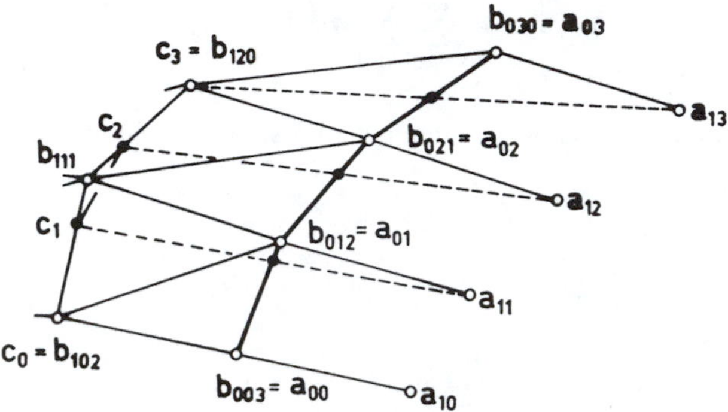

Fig. 6.31. Interpretation of GC^1 continuity for a tensor-product patch joining a triangular patch.

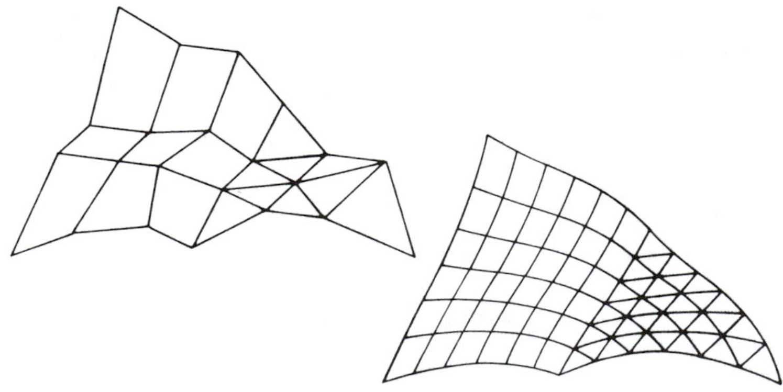

Fig. 6.32. GC^1 continuity between a tensor-product Bézier patch and a triangular Bézier patch; Bézier nets and parameter lines.

for $j = 0(1)n$. Fig. 6.31 provides a geometric interpretation of this condition, and Fig. 6.32 gives an example.

If we want to attach a *triangular Bézier patch* to a *tensor-product patch* with GC^1 continuity, then rearranging (6.60) leads to

$$\begin{aligned}\left(1-\frac{j}{n}\right)\boldsymbol{b}_{1,j,n-1-j}+\left(\frac{j}{n}\right)\boldsymbol{b}_{1,j-1,n-j}&\\ =-(\boldsymbol{a}_{1j}-\boldsymbol{a}_{0j})+\left(1-\frac{j}{n}\right)\boldsymbol{a}_{0j}&+\left(\frac{j}{n}\right)\boldsymbol{a}_{0,j-1}\end{aligned} \tag{6.61}$$

with $j = 0(1)n–1$. For $j = 0$,

$$\boldsymbol{b}_{102} = \boldsymbol{a}_{00} - (\boldsymbol{a}_{10} - \boldsymbol{a}_{00}).$$

The remaining $\boldsymbol{b}_{1jk}$ can be computed recursively directly from (6.61). Constructively, we may "reverse" the construction in Fig. 6.31. We now begin with the $\boldsymbol{a}_{1j}$, and use the corresponding relations to find the points $\boldsymbol{b}_{1,j,n-j-1}$. For a variety of approaches to this problem, see [Far 79].

The GC^1 continuity conditions between two tensor-product Bézier patches, between a tensor-product Bézier patch and a triangular Bézier patch, and between a triangular Bézier patch and a tensor-product Bézier patch can all be given a unified treatment which encompasses all of the conditions developed here as well as in Sect. 6.2.2.1, see [Far 82], [Kah 82], [Liu 89], and [Was 91].

6.3.4. Splines on Triangles

B-splines defined on triangles were first investigated by Sabin [Sab 76a] via convolutions. For this it is essential to have a *regular triangulation* in the parameter plane.

As observed in Chap. 4, B-splines can also be obtained as projections of simplices or by convolving with a unit impulse. In the same way, the domain in Fig. 6.34 can be considered to be the "shadows" of parallelepipeds (boxes), see Fig. 6.33, where various knot grids can be obtained by taking projections in various directions. These kinds of splines are called *box splines*. They were introduced by de Boor and Höllig [Boo 82], and later studied in great detail, see e.g., [Dah 84], [Pra 84, 85], [Sabl 85], [Dae 91]. If arbitrary polyhedra are projected, the "shadows" are *multivariate splines*, see [Pra 84], [Boo 88].

Given a (regular) triangulation of the (u, v) plane, suppose we have a corresponding set of B-spline basis functions N_j^n of degree n, and de Boor points $\boldsymbol{d}_j$. Then we can define a spline surface on the triangulation to any linear combination of the form

$$\boldsymbol{X}(u,v) = \sum_j \boldsymbol{d}_j N_j^n(u,v).$$

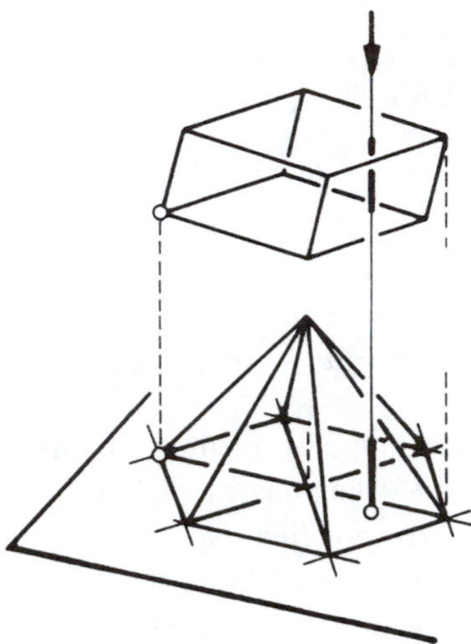

Fig. 6.33. Projection of a cube leads to a linear box spline.

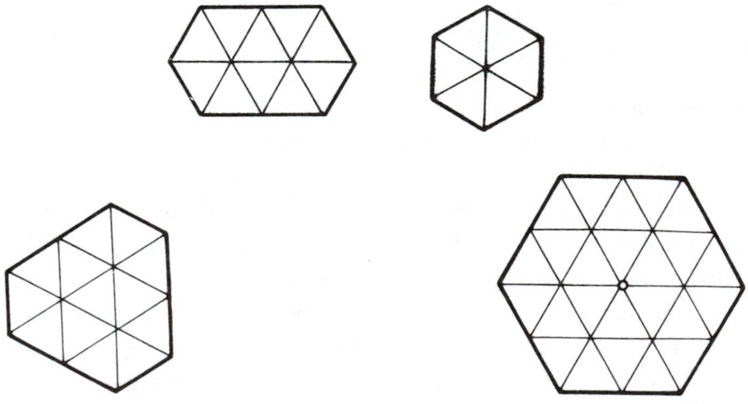

Fig. 6.34. Support of linear, quadratic, cubic and quartic B-splines.

Unfortunately, spline spaces defined on a triangular grid are not as easy to handle as tensor-product splines. The maximal possible order of continuity between patches depends on the geometry of the triangulation, and in general for a regular triangulation we can get C^r continuity only for $3r < 2n - 1$, see [Far 79]. In addition, B-spline functions on triangulations often do not form a basis. For example, cubic B-spline functions are linearly dependent, while quartic B-spline functions do not span the full space of C^2 continuous piecewise quartic functions. In fact, in some cases, the dimension of these kinds of spline spaces depends on the geometry of the triangulation, see e.g., [Schu 84, 88], [Alf 87].

A simple approach to splines defined on triangles is to use a direct generalization of the process of constructing B-spline basis functions via convolution with a unit impulse, see [Boeh 87b]. Suppose we have a regular triangular grid in the plane, see Fig. 6.35. Consider the base triangle T whose vertices are given in barycentric coordinates by

$$\boldsymbol{a} = (1,0,0), \qquad \boldsymbol{b} = (0,1,0), \qquad \boldsymbol{c} = (0,0,1).$$

Then

$$\boldsymbol{u} = \boldsymbol{b} - \boldsymbol{c}, \qquad \boldsymbol{v} = \boldsymbol{c} - \boldsymbol{a}, \qquad \boldsymbol{w} = \boldsymbol{a} - \boldsymbol{b}$$

are the three directions in the grid. Suppose G is the index set consisting of all $\boldsymbol{I} = (i,j,k)$ with $i+j+k=1$. For each s, we can define a set of *refined grid points* by $\boldsymbol{J}/s$ for $\boldsymbol{J} \in G$.

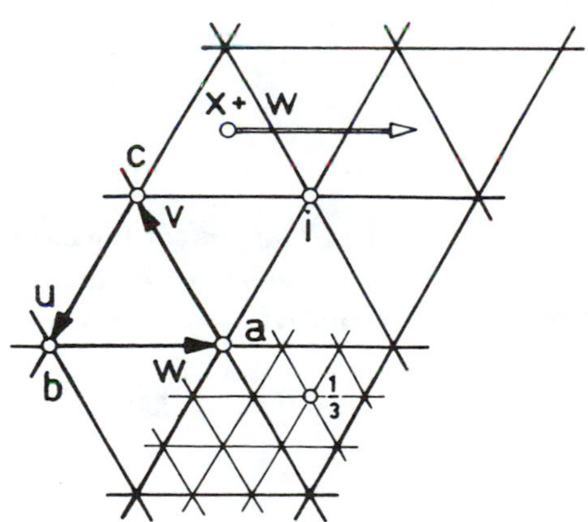

Fig. 6.35. Barycentric coordinates and refinement of a regular triangular grid.

The graph $N_{\boldsymbol{I}}$ of a piecewise linear box spline is a hexagonal pyramid with center point at $\boldsymbol{I}$ and height 1. For $\boldsymbol{p} = (p,q,r)$, let $N_{\boldsymbol{I}}^{\boldsymbol{p}}$ be the box spline constructed from $N_{\boldsymbol{I}}$ by $(p-1, q-1, r-1)$-fold convolution of the unit impulse in the directions $\boldsymbol{u}, \boldsymbol{v}, \boldsymbol{w}$. In this notation, the piecewise linear basis function

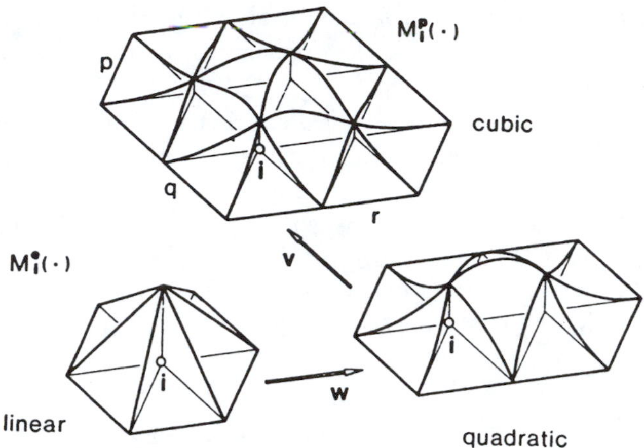

Fig. 6.36. Construction of quadratic and cubic box splines by convolution of a unit impulse with a piecewise linear box-spline.

N_I can be written as $N_I^{\boldsymbol{e}}$ with $\boldsymbol{e} = (1, 1, 1)$. Fig. 6.36 (provided by W. Boehm) shows two such box splines.

In general, we can write

$$N_I^{\boldsymbol{p}}(\cdot) = \int_{-1}^{0} N_I^{\boldsymbol{p}-\boldsymbol{c}}(\cdot + t\boldsymbol{w})dt, \tag{6.62}$$

with $\boldsymbol{p} - \boldsymbol{c} = (p, q, r - 1)$, where the $(\cdot)$ stands for the corresponding variables. Here $\boldsymbol{p}$ determines the lengths of the sides of the domain of definition. For each fixed $\boldsymbol{p}$, the B-splines $N_I^{\boldsymbol{p}}(\cdot)$ and $\boldsymbol{N}_I^{\boldsymbol{p}}(\cdot + \boldsymbol{J})$ are simply translates of each other, *i.e.*, we have

$$N_I^{\boldsymbol{p}}(\cdot) = N_{I+J}^{\boldsymbol{p}}(\cdot + \boldsymbol{J}).$$

Taking a linear combination

$$\boldsymbol{X}^{\boldsymbol{p}}(\cdot) = \sum_{I \in G} \boldsymbol{c}_I N_I^{\boldsymbol{p}}(\cdot) \tag{$*$}$$

of the B-spline functions $N_I(\cdot)$ with a fixed $\boldsymbol{p}$ leads to a B-spline surface with control points $\boldsymbol{c}_I$.

Since the direct computation of the value of the B-spline functions constructed above is very expensive, various authors have suggested transforming the control net $\boldsymbol{c}_i$ to Bézier points so that the computation can be done in terms of Bernstein polynomials. Sabin [Sab 76a] used Bernstein polynomials over triangles with $n := p + q + r - 2$ and constructed a Bézier formula

$$\boldsymbol{X}^{\boldsymbol{p}}(\cdot) = \sum_{J \in G} \boldsymbol{b}_J^n B_J^n(\cdot)$$

corresponding to the B-spline surface $(*)$. In this case, the piecewise linear function $B_i^1(\cdot)$ coincides with the pyramid $N_I(\cdot)$, and thus the Bézier points $\boldsymbol{b}_I^1$ are identical with the control points $\boldsymbol{c}_i$, *i.e.*, we have

$$\sum_{I \in G} \boldsymbol{c}_I B_I^1(\cdot) = \sum_I \boldsymbol{c}_I N_I^{\boldsymbol{e}}(\cdot) = \boldsymbol{X}(\cdot).$$

The formula (4.38) for integrating Bernstein polynomials leads to a simple way of computing the convolutions of $B_J^m(\cdot)$, $m = 1(1)n{-}1$, with the unit impulse in some net direction. *Filling and averaging algorithms* have been proposed for the recursive construction of the Bézier points $\boldsymbol{b}_J^m$ from $\boldsymbol{b}_I^{m-1}$, see [Sab 76a], [Pra 84], and [Boeh 87b]. If the Bézier points are known, then the de Casteljau algorithm can be used to compute points on the B-spline surface. Refinement algorithms for finding points on a B-spline surface using the control points $\boldsymbol{c}_I$ directly have also been developed. They are based on the fact that refining the Bézier nets corresponds to refining the projected boxes, and that the control points of the refinement converge to the desired surface, see e.g., [Boeh 83, 83a, 84a, 85], [Pra 85].

6.4. General Parameter Domains

Unfortunately, barycentric coordinates cannot be defined for general parameter domains such as five- or six-sided polygons in $\mathbb{R}^2$, since then each point can have more than one representation in terms of the vertices. For such parameter domains, we have to take a different approach. So far, there is no general mathematical method for handling arbitrary parameter domains, and the continuity conditions associated with polynomials of arbitrary degree on neighboring sets. We must therefore restrict ourselves to several special examples.

Sabin [Sab 83] has obtained blending patches for three-sided (*suitcase corners*) and five-sided domains connecting to Bézier patches and B-spline patches of degree 2, see Fig. 6.37. We consider first a three-sided patch. Since the neighboring patches are to be quadratic, we have to choose the form of the blending patch in such a way that it reduces to quadratic Bézier curves along the boundaries. These Bézier points, along with an *additional middle control point*, will generate the patch. If the points $\boldsymbol{A}, \boldsymbol{B}, \ldots, \boldsymbol{F}$ are the Bézier points of the boundary curves, see Fig. 6.37, and $\boldsymbol{G}$ is the additional control point, then Sabin chooses

$$\begin{aligned}\boldsymbol{X}(u,v,w) = u^2(1-2vw)\,\boldsymbol{A} + 2uv(1-w)\,\boldsymbol{B} \\ + v^2(1-2wu)\,\boldsymbol{C} + 2vw(1-u)\,\boldsymbol{D} + w^2(1-2uv)\,\boldsymbol{E} \\ + 2uw(1-v)\,\boldsymbol{F} + 4uvw\,\boldsymbol{G},\end{aligned} \tag{6.63}$$

subject to the *additional condition*

$$u + v + w - 2uvw = 1, \qquad 0 \le u, v, w \le 1.$$

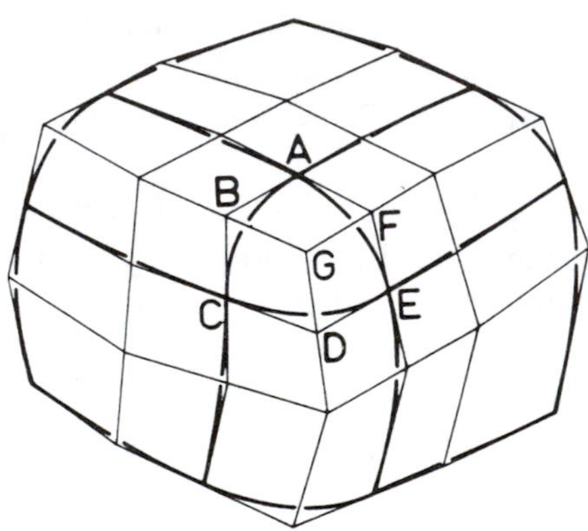

Fig. 6.37. Control points for a three-sided blending patch.

It is easy to see that (6.63) provides quadratic Bézier curves along the boundaries. If we write

$$\boldsymbol{X}(1,0,0) = \boldsymbol{A}, \qquad \boldsymbol{X}(0,1,0) = \boldsymbol{C}, \qquad \boldsymbol{X}(0,0,1) = \boldsymbol{E}$$

for the values at the vertices, then the boundary curves correspond to $u = 0$, $v = 0$ and $w = 0$. For example, for $w = 0$, (6.63) reduces to

$$\boldsymbol{X}(u,v,0) = u^2\,\boldsymbol{A} + 2uv\,\boldsymbol{B} + v^2\,\boldsymbol{C}, \qquad v = 1 - u,$$

which is a quadratic Bézier curve.

We now show that the blending patch (6.63) meets the neighboring Bézier patches with C^1 continuity. To see this, we again consider the boundary curve $w = 0$, and compute the tangent in the direction described by a curve

$u = u(t), v = v(t), w = w(t)$. Setting $w = 0$ and using the chain rule and the side condition in (6.63), we get

$$\frac{du}{dt} + \frac{dv}{dt} + \frac{dw}{dt} - 2uv\frac{dw}{dt} = 0.$$

Now if we choose t so that $\frac{dw}{dt} = 1$, it follows that

$$\frac{du}{dt} + \frac{dv}{dt} = 2uv - 1. \qquad (*)$$

Since $u + v = 1$, it follows that $(*)$ holds for

$$\frac{du}{dt} := -u^2, \qquad \frac{dv}{dt} := -v^2.$$

This gives the derivative

$$\frac{d\boldsymbol{X}}{dt}(u, v, 0) = 2u^2(\boldsymbol{F} - \boldsymbol{A}) + 4uv(\boldsymbol{G} - \boldsymbol{B}) + 2v^2(\boldsymbol{D} - \boldsymbol{C}),$$

and since $u = v - 1$, we have the boundary derivative of a biquadratic Bézier patch, cf. (6.11).

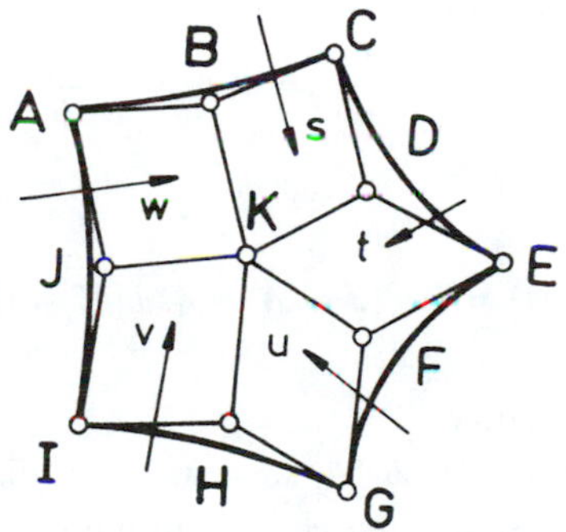

Fig. 6.38. A five-sided patch.

This method can be carried over to construct a *five-sided patch* (see Fig. 6.38) of the form

$$\begin{aligned}\boldsymbol{X}(s,t,u,v,w) = {} & u^2v^2w^2(1 - 2vstuvw)\,\boldsymbol{A} \\ & + 2tu^2v^2w(1 - 2stuvw)\,\boldsymbol{B} + t^2u^2v^2(1 - 2ustuvw)\,\boldsymbol{C} \\ & + 2st^2u^2v(1 - 2stuvw)\,\boldsymbol{D} + s^2t^2u^2(1 - 2tstuvw)\,\boldsymbol{E} \\ & + 2ws^2t^2u(1 - 2stuvw)\,\boldsymbol{F} + w^2s^2t^2(1 - 2sstuvw)\,\boldsymbol{G} \\ & + 2vw^2s^2t(1 - 2stuvw)\,\boldsymbol{H} + v^2w^2s^2(1 - 2wstuvw)\,\boldsymbol{I} \\ & + 2uv^2w^2s(1 - 2stuvw)\,\boldsymbol{J} + 4stuvw(1 - 2stuvw)\,\boldsymbol{K},\end{aligned} \qquad (6.64)$$

where

$$\begin{aligned} &0 \le s, t, u, v, w \le 1; \qquad s = 1 - uv, \qquad t = 1 - vw, \\ &u = 1 - ws, \qquad v = 1 - st, \qquad w = 1 - tu. \end{aligned} \tag{6.64a}$$

Along a boundary curve we have

$$s = 0 \qquad \text{with} \qquad t + w = 1, \quad u = 1, \quad v = 1,$$

which leads to the Bézier curve

$$\boldsymbol{X}(0, t, 1, 1, w) = w^2\boldsymbol{A} + 2tw\boldsymbol{B} + t^2\boldsymbol{C}.$$

For example, to compute the derivative with respect to a parameter r along the boundary curve $s = 0$, we note that by (6.64a), $t + w = 1 + stw$, and since $s = 0$,

$$\frac{dt}{dr} + \frac{dw}{dr} = wt\frac{ds}{dr}, \qquad \frac{du}{dr} = -w\frac{ds}{dr}, \qquad \frac{dv}{dr} = -t\frac{ds}{dr}.$$

Now taking

$$\frac{ds}{dr} := 1, \qquad \frac{dt}{dr} := t^2 w, \qquad \frac{dw}{dr} := tw^2,$$

gives the derivative

$$\frac{d\boldsymbol{X}}{dr}(0, t, 1, 1, w) = 2w^2(\boldsymbol{J} - \boldsymbol{A}) + 4tw(\boldsymbol{K} - \boldsymbol{B}) + 2t^2(\boldsymbol{D} - \boldsymbol{C})$$

which assures C^1 continuity.

Hosaka and Kimura [Hosa 84] have used similar methods to construct three- and five-sided blending patches of polynomial degree $n = 3$; but see [Sab 85, 86].

Sarraga [Sar 87] (see also [Sar 89, 90]) has taken a different approach to the problem of filling *multi-sided holes* based on ideas involving a cubic net and going back to Bézier [Bez 74], see also [Hosa 78], [Kah 82], and [Bee 86]. First the n-gon is divided into rectangular pieces by choosing one interior point C and connecting it to one point on each side. This reduces the problem to dealing with n boundary curves which meet at the point C, see Fig. 6.39. We want the patches to have a common tangent plane at the point C, so that for any two neighboring patches, $\boldsymbol{r}_1, \boldsymbol{r}_2$

$$\frac{\partial \boldsymbol{r}_1}{\partial u}(u, 0) + a(u)\frac{\partial \boldsymbol{r}_2}{\partial v_2}(u, 0) + b(u)\frac{\partial \boldsymbol{r}_2}{\partial u}(u, 0) = 0, \tag{6.65}$$

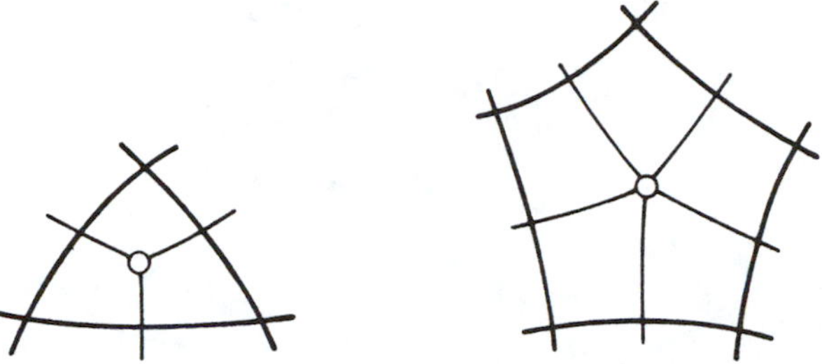

Fig. 6.39. Partition of an n-sided hole for $n = 3, 5$.

where u is the parameter of the boundary curve between $\boldsymbol{r}_1$ and $\boldsymbol{r}_2$. The coefficients of the polynomials a and b, as well as the twist vectors of the patches meeting at $\boldsymbol{C}$, are used to compute the boundary Bézier points of the corresponding patches. The degrees of the polynomials a, b depend on the geometry at the corner. The method can be used for both rational and integral surfaces, see Chap. 7.

Hahn, Gregory ([Hah 89], [Gre 86a, 87], and [Char 84]) have extended the approach of Sarraga to arbitrary n-gons. A way to fill n-sided holes using *functional splines* was suggested by [Li 90].

Some n-sided surface patches, called *S-patches of depth d*, associated with convex n-sided polygonal domains (n-gons) have been defined in [Loo 89] by means of an imbedding E of the n-gon in the domain of an $(n-1)$-dimensional Bézier simplex B of degree d via the mapping $S(\boldsymbol{u}^p) = B \circ E(\boldsymbol{u}^p)$. Here E is defined so that (see Fig. 6.40)

i) E maps edges K^p_{ij} of the n-gon to edges K^s_{ij} of the "Bézier-grid" of the $(n-1)$-dimensional simplex:

$$E : K^p_{ij} \to K^s_{ij} \qquad \text{for all } i, j \text{ with } i \neq j,$$

ii) E maps vertices E^p_i of the n-gon to vertices E^s_i of the "Bézier grid" of the $(n-1)$-dimensional simplex:

$$E : E^p_i \to E^s_i \qquad \text{for all } i,$$

iii) E maps points $\boldsymbol{u}^p$ in the interior of the n-gon to points $\boldsymbol{u}^s$ in the interior of the simplex:

$$E : \boldsymbol{u}^p \to \boldsymbol{u}^s.$$

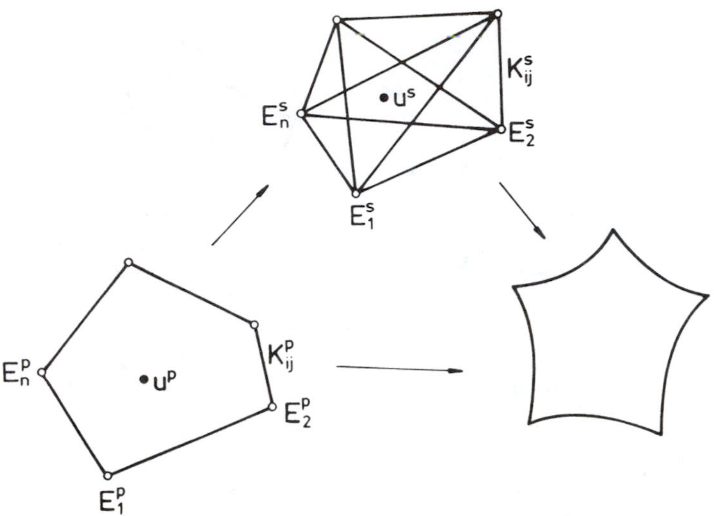

Fig. 6.40. Mapping a polyhedron to an S-patch.

An example of such a mapping E can be found in [Loo 89], based on results of [Char 84]. It involves a rational polynomial map of degree $n - 2$, which for $n = 3$ even reduces to an affine mapping.

In view of the definition of $S(\boldsymbol{u}^p)$, many properties of S-patches follow immediately from corresponding properties of Bézier simplices. For example, [Loo 89] gives a de Casteljau type construction algorithm, and discusses among other things, both C^k and GC^k continuity conditions (see Chap. 7) for neighboring triangular Bézier patches. The case of a GC^1 imbedding in a collection of rectangular surface patches (for the purpose of filling an n-sided hole), was considered in [Loo 90]. He also considered generalizations of biquadratic and bicubic B-spline surfaces based on S-patches which permit the modelling of general topologies such as general closed surfaces and surfaces with handles and branches.

6.5. Rational Tensor-product Surfaces

Suppose we choose the Bézier points (de Boor points) of a Bézier surface (B-spline surface) as points in E^4. Then if we interpret the components as homogeneous coordinates, we get *rational Bézier or B-spline surfaces* defined in E^3, see e.g., [Pie 86a, 87a, 87b], [Til 83], and [Vers 75].

We begin by discussing *rational Bézier surfaces*, with Bézier points

$$\boldsymbol{B}_{ij} = \begin{cases} (\beta_{ij}, \beta_{ij}\xi_{ij}, \beta_{ij}\eta_{ij}, \beta_{ij}\zeta_{ij})^T = (\beta_{ij}, \beta_{ij}\boldsymbol{b}_{ij})^T, & \beta_{ij} \neq 0, \quad \boldsymbol{b}_{ij} \in E^3 \\ (0, \xi_{ij}, \eta_{ij}, \zeta_{ij})^T = (0, \vec{\boldsymbol{b}}_{ij})^T, & \beta_{ij} = 0, \quad \vec{\boldsymbol{b}}_{ij} \in E^3. \end{cases} \tag{6.66}$$

Here the β_{ij} are (*homogenizing*) weight factors. If $\beta_{ij} = 0$, then $\boldsymbol{B}_{ij}$ is a point at infinity.

In terms of (6.66), we have the following parametric representation in E^3 of a *rational Bézier surface* of degree (n, m):

$$\boldsymbol{X}(u, v) = \frac{\sum_{i=0}^{n}\sum_{k=0}^{m}\beta_{ik}\boldsymbol{b}_{ik}B_i^n(u)B_k^m(v)}{\sum_{i=0}^{n}\sum_{k=0}^{m}\beta_{ik}B_i^n(u)B_k^m(v)}. \tag{6.67}$$

If the Bézier point $\boldsymbol{B}_{rs}$ is a *point at infinity*, then (6.67) becomes

$$\boldsymbol{X}(u, v) = \frac{\sum_{i=0}^{n}\sum_{k=0(i,k\neq r,s)}^{m}\beta_{ik}\boldsymbol{b}_{ik}B_i^n(u)B_k^m(v)}{N} + \frac{\vec{\boldsymbol{b}}_{rs}B_r^n(u)B_s^m(v)}{N}, \tag{6.67a}$$

where

$$N = \sum_{i=0}^{n}\sum_{k=0}^{m}{}_{(i,k\neq r,s)}\ \beta_{ik}B_i^n(u)B_k^m(v).$$

One of the weights (e.g., β_{00}) can be normalized to be 1. Moreover, on the boundary curves $u = 0$ and $v = 0$, the parametrization can be chosen so that e.g., $\beta_{n0} = 1$ and $\beta_{0m} = 1$. All other weights directly influence the shape of the surface.

It is common to refer to (6.67) as a *tensor-product surface*, although this is not entirely correct: a tensor-product surface has the form

$$\boldsymbol{X}(u, v) = \sum_{i=0}^{n}\sum_{k=0}^{m}\boldsymbol{C}_{ik}\ F_{ik}^{nm}(u, v),$$

where the basis functions $F_{ik}^{nm}(u, v)$ can be written as products

$$F_{ik}^{nm}(u, v) = G_i^n(u)H_k^m(v),$$

cf. Sect. 6.2. But the basis functions in (6.67) have the form

$$F_{ik}^{nm}(u, v) = \frac{\beta_{ik}B_i^n(u)B_k^m(v)}{\sum_{i=0}^{n}\sum_{k=0}^{m}\beta_{ik}B_i^n(u)B_k^m(v)},$$

and because of the presence of the denominator, cannot in general be written as the product of two factors. Thus, (6.67) is, strictly speaking, not really a tensor-product surface.

The use of the term "tensor-product surface" for (6.67) stems from the fact that it can be interpreted as the projection of a tensor-product surface in E^4 into E^3. This observation implies that (6.67) exhibits many of the properties of a "true" tensor-product surface, and that many tensor-product algorithms can be utilized, particularly when they can first be applied to the 4D representation. A similar situation persists for B-spline surfaces and for multivariate formulae.

The properties of rational Bézier curves presented in Sect. 4.1.4 all hold here. In particular, the de Casteljau algorithm applies provided that the Bézier points are multiplied by weights, and provided that the weights themselves are also processed by the algorithm. Moreover, we have the convex hull property (if all $\beta_{ij} > 0$), and the fact that the surface converges to the Bézier net as we increase the weights. In addition, by defining appropriate "continuity conditions", we can also construct rational spline surfaces.

Fig. 6.41 shows the effect of changing the weights. The first surface (Fig. 6.41a) has all weights equal to 1, *i.e.*, it is an ordinary biquadratic Bézier surface. Figs. 6.41b,c,d show surfaces corresponding to different weights.

In Fig. 6.41b the weights on the left and right boundaries are chosen to be 10.0 and 0.1, while in Figs. 6.41c,d the middle Bézier point has weights equal to -1.5 (Fig. 6.41c) and 10 (Fig. 6.41d), respectively. Clearly the choice of weights influences the shape of the parameter lines: they tend to bunch up in the neighborhood of higher weights. This effect can be reduced by introducing a piecewise linear rational reparametrization (see Chap. 4).

Fig. 6.42 illustrates the effect of using Bézier points at infinity. This surface is obtained from the one in Fig. 6.41 by replacing the middle Bézier point on the front boundary curve by a point at infinity.

Remark. In applying the de Casteljau algorithm to compute a point on the surface, the vanishing weights corresponding to points at infinity cannot simply be ignored, but must be carried along in the recurrence.

We now consider the special *rational biquadratic Bézier surfaces*. This class of surfaces contains the *quadrics*, but not all biquadratic rational Bézier surfaces are quadrics. Indeed, if the parameters (u, v) in (6.61) are eliminated, we get surfaces of degree eight in (x, y, z) in general, which are quadrics only if some additional conditions on the coefficients of (6.67) hold. The construction of Bézier curves, tensor-product patches, and triangular patches on quadrics was treated in [Hos 92, 92b].

Suppose now that $(m, n) = (2, 2)$ and the weight factors are chosen to be $\beta_{00} = \beta_{20} = \beta_{02} = \beta_{22} = 1$, (normalizing the boundary curves). Then if all remaining β_{ij} are chosen to be positive, the Bézier surface lies entirely inside the convex hull of the Bézier points. On the other hand, if the remaining β_{ij}

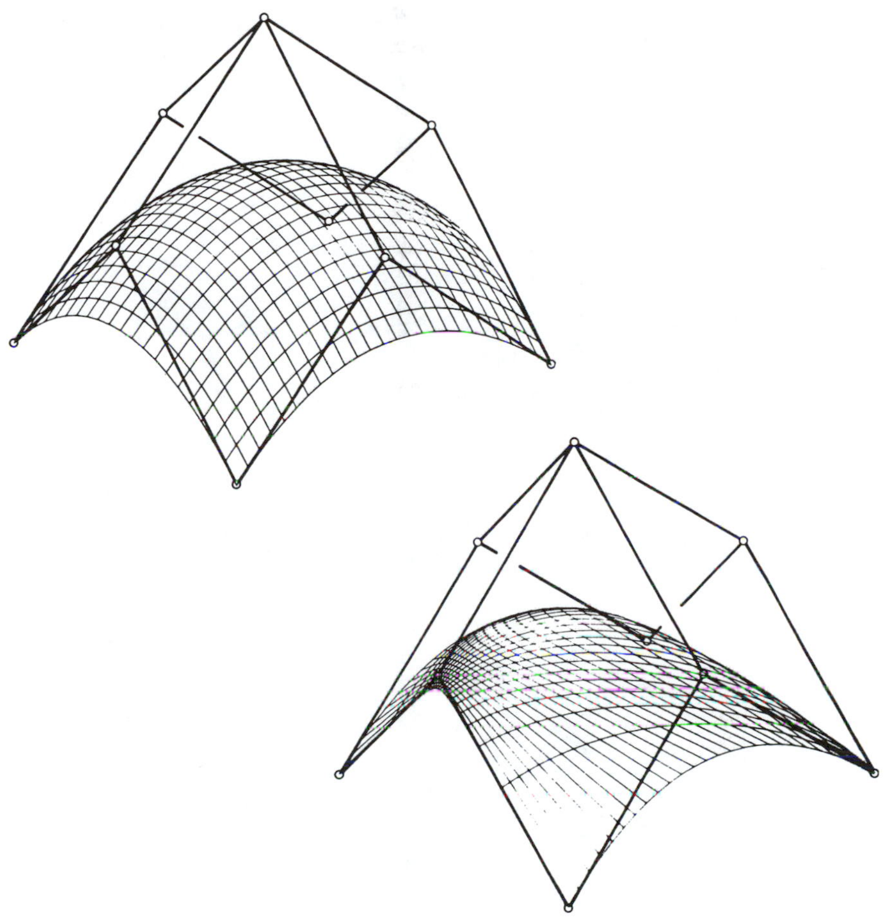

Fig. 6.41 a,b. Effect of the weights on rational Bézier surfaces.

are chosen to be negative, then the Bézier surface lies entirely outside of the convex hull of the Bézier points.

In the following we shall show how a *circular cylinder*, a *torus*, a *surface of revolution*, and a *sphere* can all be exactly represented using rational Bézier surfaces with appropriately chosen points at infinity, see [Pie 87, 87b]. In order to represent a (half) *cylinder* as a rational Bézier surface, we choose the polynomial degree (2,1) and in view of Fig. 6.43, take the Bézier points $\boldsymbol{B}_{10}$ and $\boldsymbol{B}_{11}$ to be points at infinity (*i.e.*, $\beta_{10} = \beta_{11} = 0$) with directions $\vec{\boldsymbol{b}}_{10}$ and $\vec{\boldsymbol{b}}_{11}$. Figs. 6.43 – 6.46 were kindly furnished by L. Piegl.

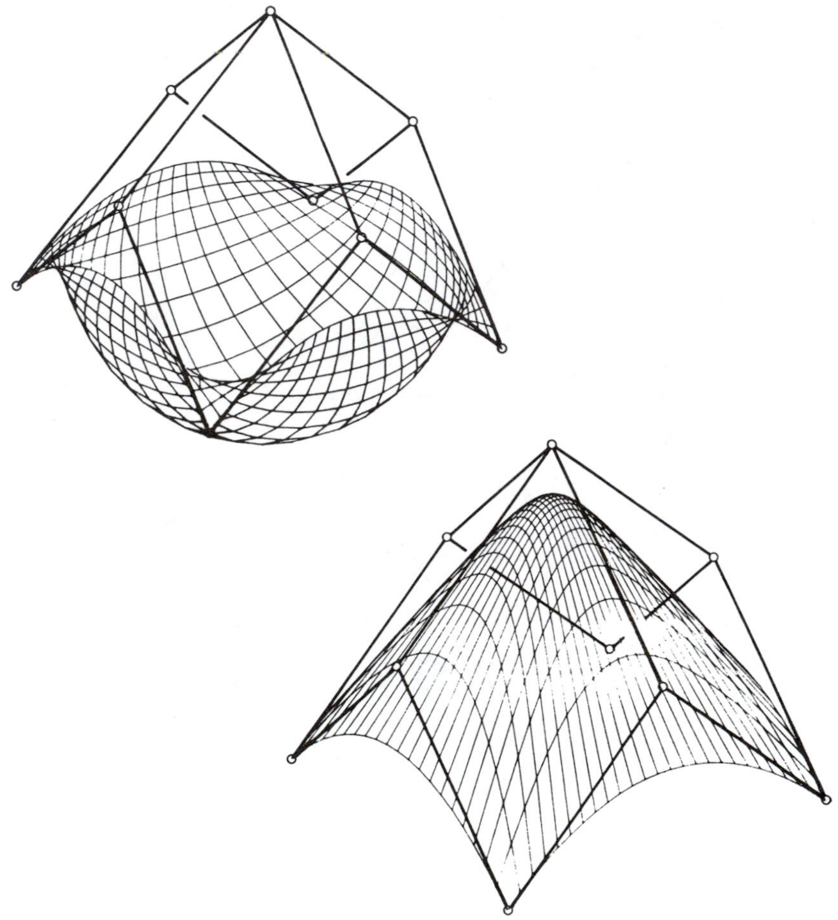

Fig. 6.41 c,d. Effect of the weights on rational Bézier surfaces.

The parametric lines $v =$ constant are straight lines, and because of the choice of points at infinity, the parametric lines $u =$ constant are ellipses, in general, see Chap. 4. Thus, (if all $\beta_{ij} = 1$) the (half) cylinder can be written as the rational Bézier surface

$$\begin{aligned} \boldsymbol{X}(u,v) = \tfrac{1}{N}\Big[&\boldsymbol{b}_{00}B_0^2(u)B_0^1(v) + \boldsymbol{b}_{20}B_2^2(u)B_0^1(v) + \boldsymbol{b}_{01}B_0^2(u)B_1^1(v) \\ &+ \boldsymbol{b}_{21}B_2^2(u)B_1^1(v) + \vec{\boldsymbol{b}}_{10}B_1^2(u)B_0^1(v) + \vec{\boldsymbol{b}}_{11}B_1^2(u)B_1^1(v)\Big], \end{aligned} \tag{6.68}$$

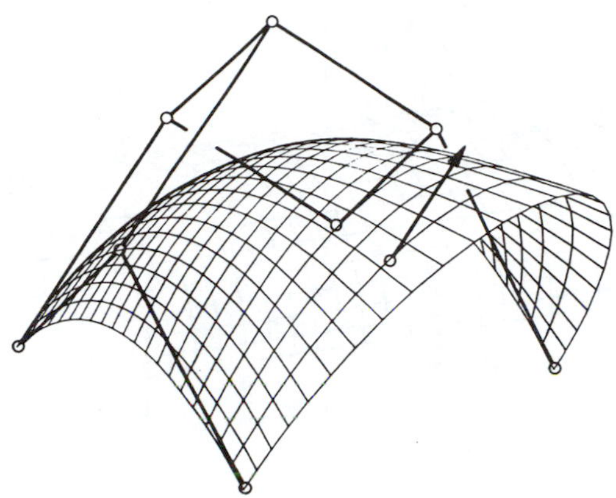

Fig. 6.42. The surface in Fig. 6.41 with one point moved to infinity.

with

$$N = B_0^2(u)B_0^1(v) + B_2^2(u)B_0^1(v) + B_0^2(u)B_1^1(v) + B_2^2(u)B_1^1(v).$$

In general, formula (6.68) represents an elliptical cylinder. We get a circular cylinder of radius r (and height h) by choosing the Bézier points to be (cf. Fig. 6.43):

$$\begin{aligned} \boldsymbol{b}_{00} &= (r,0,0)^T, & \boldsymbol{b}_{01} &= (r,0,h)^T, & \boldsymbol{b}_{20} &= (-r,0,0)^T, \\ \boldsymbol{b}_{21} &= (-r,0,h)^T, & \vec{\boldsymbol{b}}_{10} &= (0,r,0)^T, & \vec{\boldsymbol{b}}_{11} &= (0,r,0)^T. \end{aligned} \tag{6.68a}$$

Inserting these Bézier points in (6.68), it follows that the parametric formula for a half circular cylinder is given by

$$x = r\frac{(1-u)^2 - u^2}{(1+u)^2 + u^2}, \qquad y = r\frac{2u(1-u)}{(1+u)^2+u^2}, \qquad z = hv, \tag{6.68b}$$

for $u \in [0,1]$, $v \in [0,1]$.

If in (6.68b) we choose $\vec{\boldsymbol{b}}_{10} = \vec{\boldsymbol{b}}_{11} = (0,-r,0)^T$, then (6.68) describes the second half of the cylinder which, according to (6.68a), lies outside of the convex hull of the Bézier points.

In order to represent a *torus*, we choose $(m,n) = (2,2)$ and points at infinity with directions

$$\vec{\boldsymbol{b}}_{10}, \vec{\boldsymbol{b}}_{21}, \vec{\boldsymbol{b}}_{11}, \vec{\boldsymbol{b}}_{01}, \vec{\boldsymbol{b}}_{12},$$

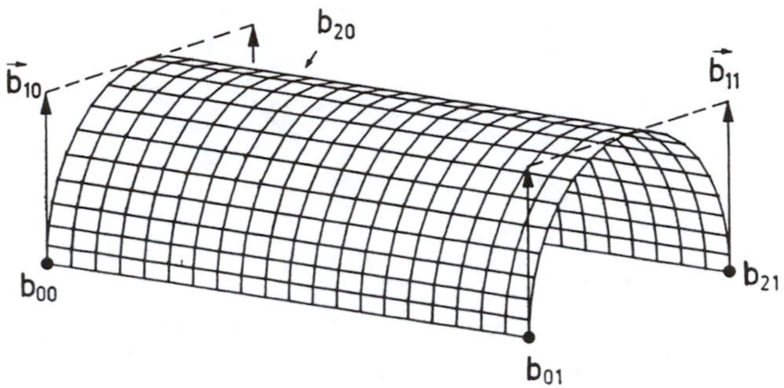

Fig. 6.43. Half a circular cylinder as a rational Bézier surface. The direction of the point at infinity is marked with an arrow.

i.e.,

$$\beta_{10} = \beta_{21} = \beta_{11} = \beta_{01} = \beta_{12} = 0.$$

We choose all other $\beta_{ij} = 1$, see Fig. 6.44a. This gives the representation

$$\begin{aligned} \boldsymbol{X}(u,v) = \tfrac{1}{N}\big[&\boldsymbol{b}_{00}B_0^2(u)B_0^2(v) + \boldsymbol{b}_{20}B_2^2(u)B_0^2(v) + \boldsymbol{b}_{02}B_0^2(u)B_2^2(v) \\ &+ \boldsymbol{b}_{22}B_2^2(u)B_2^2(v) + \vec{\boldsymbol{b}}_{10}B_1^2(u)B_0^2(v) + \vec{\boldsymbol{b}}_{21}B_2^2(u)B_1^2(v) \\ &+ \vec{\boldsymbol{b}}_{11}B_1^2(u)B_1^2(v) + \vec{\boldsymbol{b}}_{01}B_0^2(u)B_1^2(v) + \vec{\boldsymbol{b}}_{12}B_1^2(u)B_2^2(v)\big], \end{aligned} \tag{6.69}$$

with

$$N = B_0^2(u)B_0^2(v) + B_2^2(u)B_0^2(v) + B_0^2(u)B_2^2(v) + B_2^2(u)B_2^2(v)$$

for a torus-type surface patch. In order to get the quarter of a circular torus shown in Fig. 6.44b from (6.69), we choose the following Bézier points and points at infinity:

$$\begin{aligned} \boldsymbol{b}_{00} &= (0,-a,-r)^T, & \boldsymbol{b}_{20} &= (0,-a,r)^T, & \boldsymbol{b}_{02} &= (0,a,-r)^T, \\ \boldsymbol{b}_{22} &= (0,a,r)^T, & \vec{\boldsymbol{b}}_{10} &= (0,-r,0)^T, & \vec{\boldsymbol{b}}_{21} &= (a,0,0)^T, \\ \vec{\boldsymbol{b}}_{11} &= (r,0,0)^T, & \vec{\boldsymbol{b}}_{01} &= (a,0,0)^T, & \vec{\boldsymbol{b}}_{12} &= (0,r,0)^T. \end{aligned} \tag{6.69a}$$

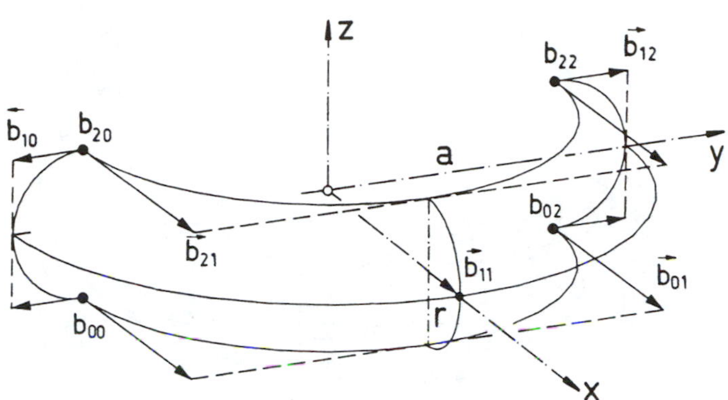

Fig. 6.44a. Bézier points and Bézier points at infinity (arrows) for the representation of a torus.

For $a > r$ this leads to a regular torus patch, for $a = 0$ to a *simple covered sphere*, see [Pie 87].

The choice (6.69a) of the Bézier points results in a representation of only the (exterior) quarter of the surface of the torus, see Fig. 6.44b.

If the direction vectors $\vec{b}_{10}, \vec{b}_{11}, \vec{b}_{12}$ in (6.69) are reversed, then we get the associated interior of the surface of the torus. On the other hand, if the direction vectors $\vec{b}_{21}$ and $\vec{b}_{01}$ are reversed, we get the corresponding back half of the torus. These four (C^2 continuous) Bézier surface patches can easily be combined into a closed formula using *B-spline basis functions*, see Sect. 4.2 and [Pie 87, 87b], [Til 83].

We now discuss the representation of surfaces of revolution by rational Bézier surfaces using Bézier points at infinity. We suppose that

- the z-axis is to be the axis of rotation,
- the meridian of the surface is prescribed as a (rational) Bézier curve of arbitrary degree lying in the (y, z) plane.

An example is given in Fig. 6.45, where the meridian is given by a quadratic rational Bézier curve.

In order to find the Bézier formula for a *surface of revolution*, we first select a point

$$\boldsymbol{P} = r\boldsymbol{e}_2 + z\boldsymbol{e}_3$$

in the meridian plane. We then rotate this point about the z-axis to obtain a

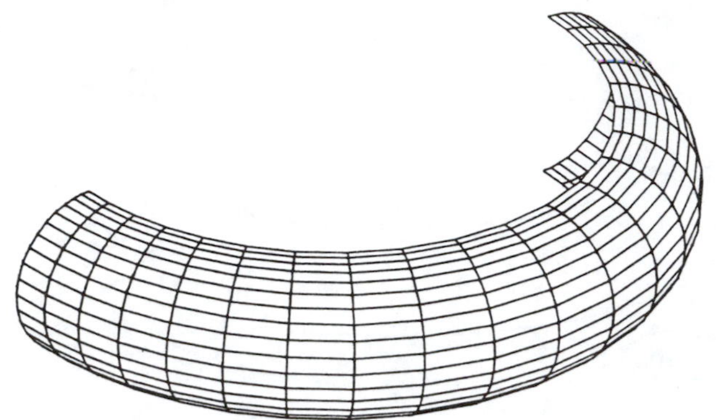

Fig. 6.44b. Piece of a torus generated by the Bézier points in Fig. 6.44a.

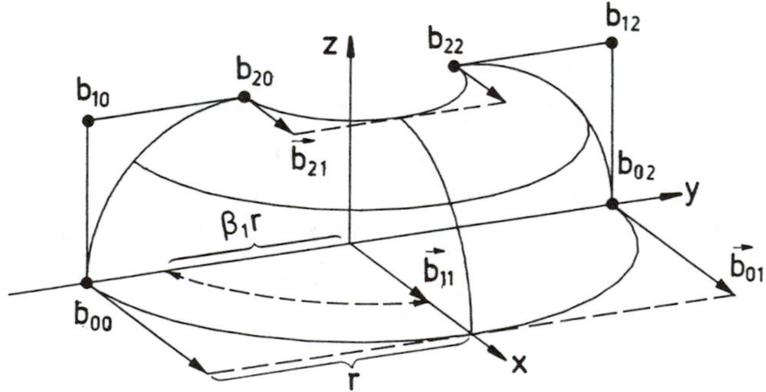

Fig. 6.45. Bézier points and Bézier points at infinity (arrows) for a surface of revolution with a quadratic rational meridian.

circle

$$\boldsymbol{K}(v) = \frac{\boldsymbol{P}_0 B_0^2(v) + \boldsymbol{P}_2 B_2^2(v)}{B_0^2(v) + B_2^2(v)} + \frac{\vec{\boldsymbol{P}}_1 B_1^2(v)}{B_0^2(v) + B_2^2(v)}, \qquad v \in [0,1], \tag{6.71}$$

where

$$\boldsymbol{P}_0 = \boldsymbol{P}, \qquad \boldsymbol{P}_2 = -r\boldsymbol{e}_2 + z\boldsymbol{e}_3, \quad \vec{\boldsymbol{P}}_1 = r\boldsymbol{e}_1.$$

If the meridian of the surface of revolution in the (y, z) plane is given by the

Bézier curve

$$X(u) = \sum_{i=0}^{n} \boldsymbol{b}_i B_i^n(u), \qquad \text{with } \boldsymbol{b}_i = r_i \boldsymbol{e}_2 + z_i \boldsymbol{e}_3,$$

then the Bézier points $\boldsymbol{b}_0$ and $\boldsymbol{b}_n$ lie on the meridian, and the corresponding parallels of latitude on the surface of revolution satisfy the equation (6.71). This means that the Bézier points $\boldsymbol{b}_{00}, \vec{\boldsymbol{b}}_{01}, \boldsymbol{b}_{02}, \boldsymbol{b}_{n0}, \vec{\boldsymbol{b}}_{n1}, \boldsymbol{b}_{n2}$ for the surface of revolution to be constructed can be found as in (6.71). But then the set of curve points corresponding to the parameter value $u = u_0$ must lie on the circle corresponding to (6.71). If we think of the point $u = u_0$ as being obtained using the de Casteljau algorithm, then rotation of points on the curve reduces to rotation of the Bézier points. It follows that the other (interior) Bézier points of the desired surface of revolution can be chosen to be

$$\boldsymbol{b}_{i0} = \boldsymbol{b}_i, \qquad \boldsymbol{b}_{i2} = -r_i \boldsymbol{e}_2 + z_i \boldsymbol{e}_3, \qquad \vec{\boldsymbol{b}}_{i1} = r_i \boldsymbol{e}_1.$$

To see that the set of points associated with the parameter value u form a half circle, we consider the Bézier formula for the surface:

$$\begin{aligned} X(u,v) &= \frac{\sum_{i=0}^{n} \sum_{k=0}^{2} \boldsymbol{b}_{ik} B_i^n(u) B_k^2(v)}{B_0^2(v) + B_2^2(v)} \\ &= \Big[\sum_{i=0}^{n} (r_i \boldsymbol{e}_2 + z_i \boldsymbol{e}_3) B_i^n(u) B_0^2(v) + \sum_{i=0}^{n} r_i \boldsymbol{e}_1 B_i^n(u) B_1^2(v) \\ &\qquad + \sum_{i=0}^{n} (-r_i \boldsymbol{e}_2 + z_i \boldsymbol{e}_3) B_i^n(u) B_2^2(v) \Big] \Big[B_0^2(v) + B_2^2(v) \Big]^{-1} \qquad (6.72a) \\ &= \frac{\boldsymbol{A}_0 B_0^2(v) + \boldsymbol{A}_2 B_2^2(v) + \vec{\boldsymbol{A}}_1 B_1^2(v)}{B_0^2(v) + B_2^2(v)}. \end{aligned}$$

Then as in (6.71), we get the Bézier points

$$\boldsymbol{A}_0 = \sum_{i=0}^{n} (r_i \boldsymbol{e}_2 + z_i \boldsymbol{e}_3) B_i^n(u), \qquad \boldsymbol{A}_2 = \sum_{i=0}^{n} (-r_i \boldsymbol{e}_2 + z_i \boldsymbol{e}_3) B_i^n(u),$$

$$\vec{\boldsymbol{A}}_1 = \sum_{i=0}^{n} r_i \boldsymbol{e}_1 B_i^n(u). \qquad (6.72b)$$

A comparison with (6.71) verifies our claim that the set of points associated with $u = u_0$ describe a circle.

If the meridian is prescribed as a rational curve, then we can rotate the point $\widetilde{\boldsymbol{P}} = (\beta, \beta \boldsymbol{P})$ to obtain, as in (6.71), the equation of a circle in homogeneous coordinates

$$K(v) = \frac{\boldsymbol{P}_0 B_0^2(v) + \boldsymbol{P}_2 B_2^2(v)}{B_0^2(v) + B_2^2(v)} + \frac{\lambda \vec{\boldsymbol{P}}_1 B_1^2(v)}{\beta (B_0^2(v) + B_2^2(v))}, \qquad (6.73a)$$

with Bézier points

$$\widetilde{\boldsymbol{P}}_0 = \widetilde{\boldsymbol{P}} = \begin{pmatrix} \beta \\ \beta(r\boldsymbol{e}_2 + z\boldsymbol{e}_3) \end{pmatrix}, \quad \widetilde{\boldsymbol{P}}_2 = \begin{pmatrix} \beta \\ \beta(-r\boldsymbol{e}_2 + z\boldsymbol{e}_3) \end{pmatrix}, \quad \widetilde{\boldsymbol{P}}_1 = \begin{pmatrix} 0 \\ \lambda r\boldsymbol{e}_1 \end{pmatrix}.$$

Comparing with (6.71), it follows that we must choose $\lambda = \beta$. Now if in addition to the Bézier points $\boldsymbol{b}_i = r_i\boldsymbol{e}_2 + z_i\boldsymbol{e}_3$, we are also given weights β_i, then it follows that the Bézier points of the surface of revolution must be

$$\boldsymbol{b}_{i0} = \boldsymbol{b}_i, \quad \beta_{i0} = \beta_i, \quad \boldsymbol{b}_{i2} = -r_i\boldsymbol{e}_2 + z_i\boldsymbol{e}_3, \quad \beta_{i2} = \beta_i, \quad \vec{\boldsymbol{b}}_{i1} = \beta_i r_i \boldsymbol{e}_1. \quad (6.73b)$$

If we apply these considerations to the example of a quadratic rational meridian in Fig. 6.45, we see the vector $\vec{\boldsymbol{b}}_{11}$ must have the length $\beta_1 r$.

If a meridian curve consists of several *Bézier spline segments*, then we must process each of the segments in an analogous way. We may treat a collection of Bézier segments as a B-spline curve, see e.g., [Pie 87b]. Fig. 6.46 shows a surface of revolution generated in this way, along with its associated meridian.

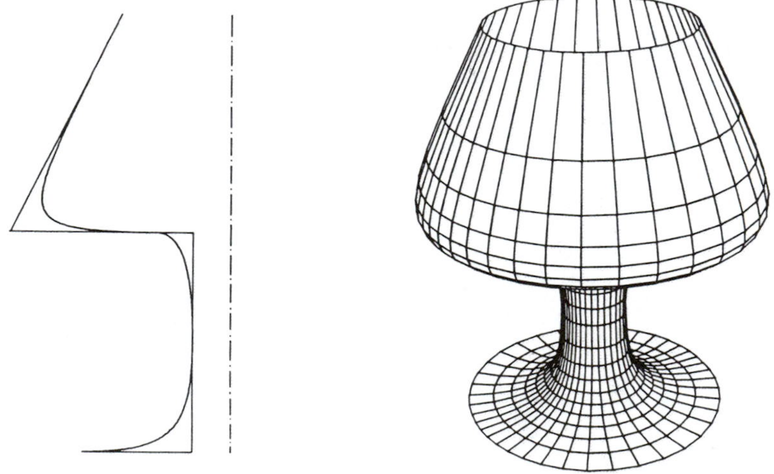

Fig. 6.46. Meridian and associated surface of revolution.

As already mentioned, we can of course also construct *rational B-spline surfaces*. The algorithms carry over directly, and the weights play an analogous role as for rational Bézier surfaces. For applications, the nonuniform tensor-product B-spline surfaces (NURBS) are of special importance, since they combine the advantages of the B-spline theory with the shape modelling abilities of rational surfaces. Methods for manipulating rational B-spline surfaces are described in detail in [Pie 89b]. Fig. 6.47 shows several examples

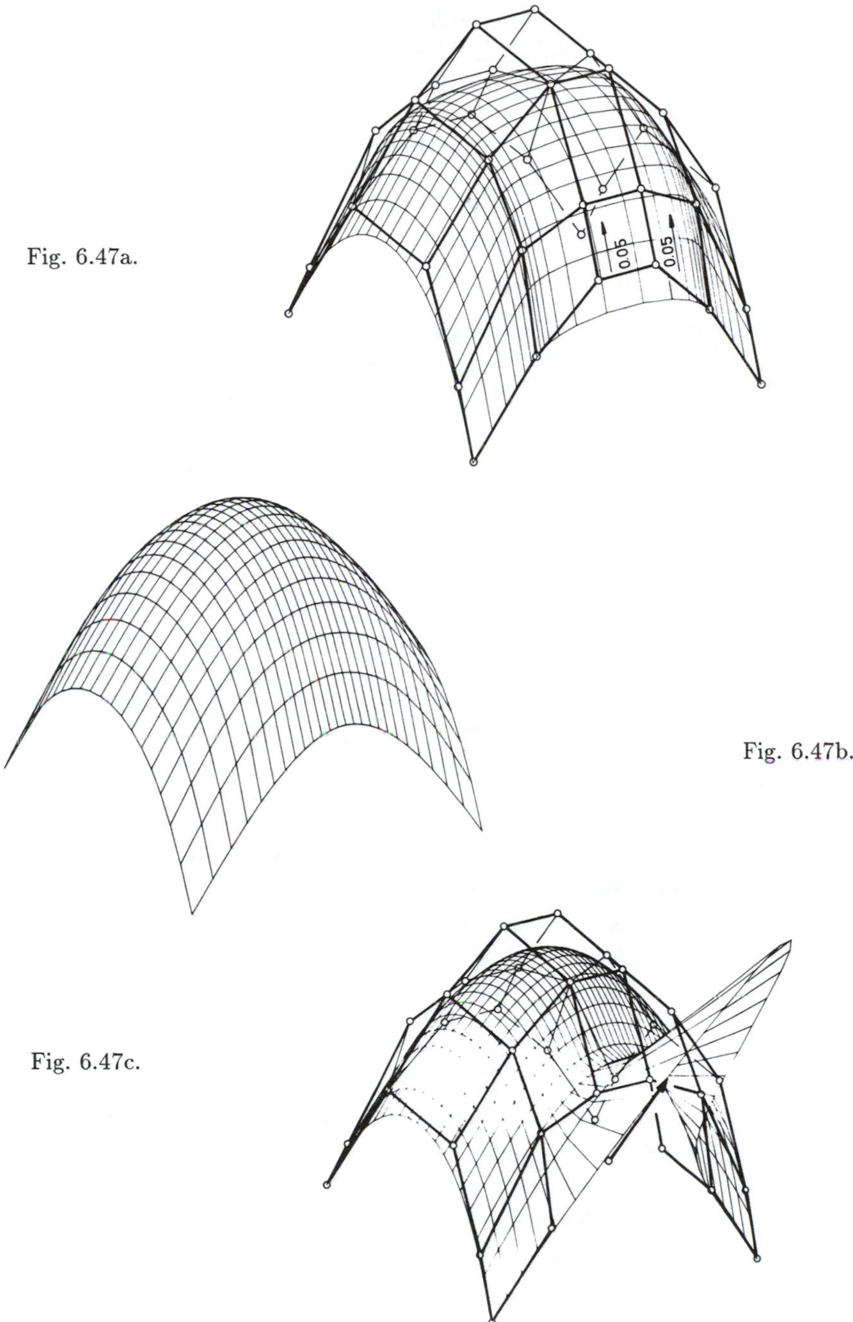

Fig. 6.47a.

Fig. 6.47b.

Fig. 6.47c.

Fig. 6.47. Rational B-spline surfaces.

of rational B-spline surfaces. Fig. 6.47a presents a tensor-product B-spline surface of order (4,4), where all weights are set equal to one. In Fig. 6.47b, the middle row of de Boor points (marked in the figure) are given weights 0.05, which clearly flattens out the surface. In Fig. 6.47c one of the two middle de Boor points on the front boundary curve is chosen to be a point at infinity, which leads to a deformation of the surface in the vicinity of the boundary.

6.5.1. Approximation with Rational B-spline Surfaces

The discussion in Sect. 4.4.3 of approximation by rational curves can be extended as in (4.74) to rational tensor-product B-spline surfaces, see [Schn 92]. The superiority of rational methods coupled with parametric correction is especially evident in the surface case. For example, approximating the surface in Fig. 6.20a using rational B-splines, we achieve the same error tolerance as in the third figure of Fig. 6.20a after only three iterations. For the surface shown in Fig. 6.48, the desired error bound cannot be attained using integral surfaces. Fig. 6.48a shows the approximation of a set of points using an integral quadratic B-spline surface: after 10 iterations we have a maximal error of 0.019. Fig. 6.48b shows the approximation of the same set of points using a quadratic rational B-spline surface: after 5 iterations the maximal error is 0.0008. The difference in the effectiveness of the two approximation methods is clearly visible if we compare Figs. 6.48a and 6.48b.

6.6. Rational Triangular Surfaces

The above discussion of rational tensor-product Bézier surfaces can be extended to *rational triangular Bézier surfaces*. If we index the weights in the same way as the Bézier points, then instead of (6.49) we have the following parametric formula for a *rational triangular Bézier surface*:

$$\boldsymbol{X}(u,v,w) = \frac{\sum_{i+j+k=n} \beta_{ijk} \boldsymbol{b}_{ijk} B^n_{ijk}(u,v,w)}{\sum_{i+j+k=n} \beta_{ijk} B^n_{ijk}(u,v,w)}. \tag{6.74}$$

The de Casteljau algorithm can also be generalized to cover these splines, provided the algorithm is applied to the weights as well as to the Bézier points multiplied by the weights. If the weights β_{ijk} are positive, then the properties of ordinary triangular Bézier surfaces (such as the convex-hull property) also extend to rational surfaces. The weights can also be chosen to be negative or zero, of course. In [Far 87a], rational triangular patches are used to represent octants of a sphere. A method for constructing triangular surfaces on arbitrary quadrics is developed in [Hos 92].

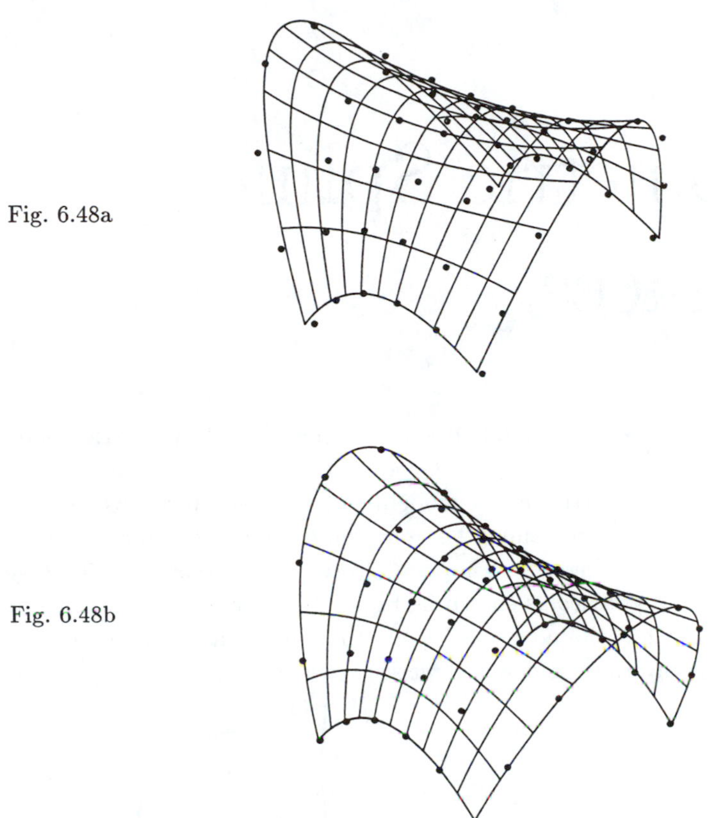

Fig. 6.48a,b. Approximation by a rational B-spline surface.

7 Geometric Spline Surfaces

In this chapter we extend to surfaces the idea of geometric continuity discussed in Chap. 5 for curves. Geometric spline surfaces are very useful in practice, and in particular for modelling various situations where ordinary C^r continuous surfaces cannot be constructed, e.g., for star-shaped patch configurations as in Fig. 7.1. They are also very useful for blending together various kinds of patches as shown in Fig. 7.2. More importantly, geometric continuity also has the advantage that, since it is based on the concept of contact, it is invariant under parametric transformations, in contrast to C^r continuity.

Fig. 7.1. Star-shaped patch configurations.

We begin by defining GC^r continuity in general, and later discuss Bézier representations for the cases of GC^1 and GC^2 in more detail. In Sect. 7.5 we consider some special problems relating to multi-patch surfaces. An overview and comparison of the various schemes can be found in Sect. 7.6, and in the last section, we discuss B-spline representations.

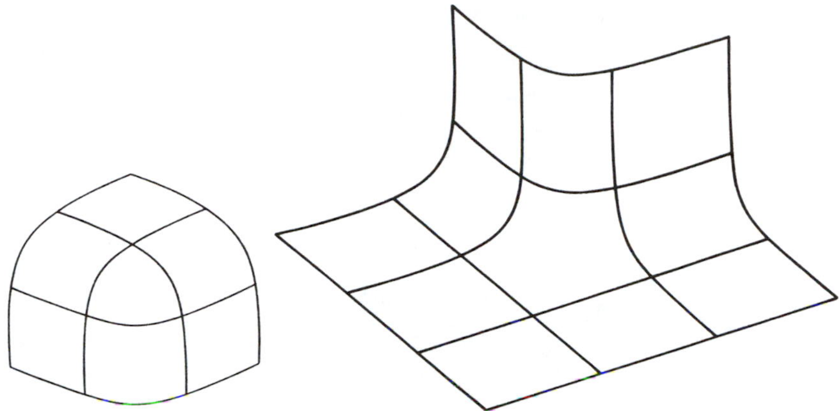

Fig.7.2a. Suitcase corner. **Fig. 7.2b.** House corner.

7.1. GC^r Continuous Surfaces

The concept of contact introduced in Chap. 5 for curves can be extended immediately to surfaces $\boldsymbol{X}(u,v)$ and $\boldsymbol{Y}(s,t)$. To derive GC^r continuity conditions, we assume that the two surfaces are to meet at a regular point $\boldsymbol{P} = \boldsymbol{X}(\bar{u},\bar{v}) = \boldsymbol{Y}(\bar{s},\bar{t})$ (i.e., at $\boldsymbol{P}$ we have $\boldsymbol{X}_u \times \boldsymbol{X}_v \neq 0, \boldsymbol{Y}_s \times \boldsymbol{Y}_t \neq 0$) with contact of order r, *i.e.*, $\boldsymbol{X}(u,v)$ and $\boldsymbol{Y}(s,t)$ should meet with C^r continuity at $\boldsymbol{P}$ after a reparametrization by an admissible, orientation-preserving transformation to common parameters, e.g., $u \mapsto u(s,t)$ with $a_{00} \equiv u(\bar{s},\bar{t})$, and $v \mapsto v(s,t)$ with $b_{00} \equiv v(\bar{s},\bar{t})$. Then as in Sect. 5.2, the GC^r continuity conditions follow by comparing the Taylor series expansions of the two surfaces. Here (cf. with Sect. 5.2) we need to apply the chain rule (in this case for derivatives of functions of several variables). For a mathematically rigorous derivation which takes account of the differential-geometric aspects and makes use of the theory of manifolds, see [Dero 85], [Gre 87, 89a], and [Hahn 89a].

Introducing the abbreviations,

$$a_{\mu\nu} = \frac{\partial^r}{\partial s^\mu \partial t^\nu} u(s,t)\Big|_{(s,t)=(\bar{s},\bar{t})} \qquad \text{with } r = \mu + \nu, \tag{7.1a}$$

$$b_{\mu\nu} = \frac{\partial^r}{\partial s^\mu \partial t^\nu} v(s,t)\Big|_{(s,t)=(\bar{s},\bar{t})} \qquad \text{with } r = \mu + \nu, \tag{7.1b}$$

then as a generalization of (5.6)ff, the GC^r *continuity condition* for $r = 1$ is

$$\begin{pmatrix} \boldsymbol{Y}_s \\ \boldsymbol{Y}_t \end{pmatrix} = \begin{pmatrix} a_{10} & b_{10} \\ a_{01} & b_{01} \end{pmatrix} \begin{pmatrix} \boldsymbol{X}_u \\ \boldsymbol{X}_v \end{pmatrix}. \tag{7.2}$$

For $r = 2$ we must also have

$$\begin{pmatrix} \boldsymbol{Y}_{ss} \\ \boldsymbol{Y}_{st} \\ \boldsymbol{Y}_{tt} \end{pmatrix} = \begin{pmatrix} a_{20} & b_{20} \\ a_{11} & b_{11} \\ a_{02} & b_{02} \end{pmatrix} \cdot \begin{pmatrix} \boldsymbol{X}_u \\ \boldsymbol{X}_v \end{pmatrix}$$

$$+ \begin{pmatrix} a_{10}^2 & 2a_{10}b_{10} & b_{10}^2 \\ a_{10}a_{01} & a_{10}b_{01} + a_{01}b_{10} & b_{10}b_{01} \\ a_{01}^2 & 2a_{01}b_{01} & b_{01}^2 \end{pmatrix} \cdot \begin{pmatrix} \boldsymbol{X}_{uu} \\ \boldsymbol{X}_{uv} \\ \boldsymbol{X}_{vv} \end{pmatrix}, \tag{7.3}$$

and for $r = 3$ additionally

$$\begin{pmatrix} \boldsymbol{Y}_{sss} \\ \boldsymbol{Y}_{sst} \\ \boldsymbol{Y}_{stt} \\ \boldsymbol{Y}_{ttt} \end{pmatrix} = \begin{pmatrix} a_{30} & b_{30} \\ a_{21} & b_{21} \\ a_{12} & b_{12} \\ a_{03} & b_{03} \end{pmatrix} \cdot \begin{pmatrix} \boldsymbol{X}_u \\ \boldsymbol{X}_v \end{pmatrix} \tag{7.4}$$

$$+ \begin{pmatrix} 3a_{10}a_{20} & 3(a_{20}b_{10} + a_{10}b_{20}) & 3b_{10}b_{20} \\ 2a_{10}a_{11} + a_{01}a_{20} & 2a_{10}b_{11} + 2a_{11}b_{10} + a_{20}b_{01} + a_{01}b_{20} & 2b_{10}b_{11} + b_{01}b_{20} \\ 2a_{01}a_{11} + a_{10}a_{02} & 2a_{01}b_{11} + 2a_{11}b_{01} + a_{02}b_{10} + a_{10}b_{02} & 2b_{01}b_{11} + b_{10}b_{02} \\ 3a_{01}a_{02} & 3(a_{02}b_{01} + a_{01}b_{02}) & 3b_{01}b_{02} \end{pmatrix} \cdot \begin{pmatrix} \boldsymbol{X}_{uu} \\ \boldsymbol{X}_{uv} \\ \boldsymbol{X}_{vv} \end{pmatrix}$$

$$+ \begin{pmatrix} a_{10}^3 & 3a_{10}^2b_{10} & 3a_{10}b_{10}^2 & b_{10}^3 \\ a_{10}^2a_{01} & a_{10}^2b_{01} + 2a_{10}a_{01}b_{10} & a_{01}b_{10}^2 + 2a_{10}b_{10}b_{01} & b_{10}^2b_{01} \\ a_{10}a_{01}^2 & a_{01}^2b_{10} + 2a_{10}a_{01}b_{01} & a_{10}b_{01}^2 + 2a_{01}b_{10}b_{01} & b_{10}b_{01}^2 \\ a_{01}^3 & 3a_{01}^2b_{01} & 3a_{01}b_{01}^2 & b_{01}^3 \end{pmatrix} \cdot \begin{pmatrix} \boldsymbol{X}_{uuu} \\ \boldsymbol{X}_{uuv} \\ \boldsymbol{X}_{uvv} \\ \boldsymbol{X}_{vvv} \end{pmatrix},$$

etc., [Cohe 82]. By the regularity assumption, $a_{10} > 0$, $b_{10} > 0$ and $a_{10}b_{01} - b_{10}a_{01} \neq 0$ must hold. A recurrence formula was given in [Was 91].

Conversely, if given two surfaces $\boldsymbol{X}(u, v)$ and $\boldsymbol{Y}(s, t)$ meeting at a point $\boldsymbol{P} = \boldsymbol{X}(\bar{u}, \bar{v}) = \boldsymbol{Y}(\bar{s}, \bar{t})$, there exist real-valued numbers $a_{\mu\nu}$ and $b_{\mu\nu}$ satisfying equations (7.2)–(7.4), then there exists a reparametrization to common parameters that relative to this parametrization, $\boldsymbol{X}(u, v)$ and $\boldsymbol{Y}(s, t)$ meet at $\boldsymbol{P}$ with C^r continuity, and thus $\boldsymbol{X}$ and $\boldsymbol{Y}$ meet with GC^r continuity. In this case, bivariate Taylor series expansions could be used to find the parameter transformation satisfying (7.1).

Let

$$\boldsymbol{X}^{(r)} = (\boldsymbol{X}^{(r,0)}, \boldsymbol{X}^{(r-1,1)}, \ldots, \boldsymbol{X}^{(1,r-1)}, \boldsymbol{X}^{(0,r)})^T$$

be the vector of partial derivatives of $\boldsymbol{X}(u,v)$, with a similar notation for $\boldsymbol{Y}$. Then, as in Sect. 5.2, we can write the above GC^r continuity conditions in the compact form

$$\begin{aligned} \boldsymbol{Y}' &= \Omega_{11}\boldsymbol{X}' \\ \boldsymbol{Y}'' &= \Omega_{22}\boldsymbol{X}'' + \Omega_{12}\boldsymbol{X}' \\ \boldsymbol{Y}''' &= \Omega_{33}\boldsymbol{X}''' + \Omega_{23}\boldsymbol{X}'' + \Omega_{13}\boldsymbol{X}' \end{aligned} \tag{7.5}$$

etc., where the matrices Ω_{ij} are those appearing in (7.2)–(7.4). As in the curve case (see (5.6) and following), the partial derivatives of $\boldsymbol{Y}$ of order r can be expressed as linear combinations of the partial derivatives of $\boldsymbol{X}$ up to order r.

The entire set of continuity conditions can now be written in matrix form

$$Y = AX,$$

where the connection matrix A is a block lower triangular matrix with blocks Ω_{ij}.

The GC^r continuity condition (7.5) can be given a geometric interpretation. For example, two surfaces meeting with contact of order one at a regular point $\boldsymbol{P}$ have the same tangent plane there (or equivalently a common normal vector). Surfaces with contact of order 2 at a regular point $\boldsymbol{P}$ have the same tangent plane, the same Dupin indicatrix, and the curvature has the same sign (or equivalently, the same principal curvature directions and radii, or the same normal curvatures, or the same second fundamental forms). GC^2 continuity implies the equality of the Gaussian and mean curvatures, see e.g., [Vero 76], [Kah 82], [Cohe 82], [Dero 85], [BEZ 86], and [Herr 87].

The geometric interpretation of GC^2 continuity is somewhat inconvenient for applications, since it requires equality of the normal curvatures at $\boldsymbol{P}$ for all possible tangent lines to the surface at $\boldsymbol{P}$. But the *3-Tangents Theorem* (see [Peg 92]) asserts that if two C^2 continuous surfaces meet at a point $\boldsymbol{P}$ with GC^1 continuity, then the equality of the normal curvatures corresponding to three tangent lines at $\boldsymbol{P}$, any two of which are always linearly independent, implies the equality for all tangents, and thus is equivalent to GC^2 continuity at this point. In this form, the criterion can be used as a tool for checking the smoothness of a surface.

The conditions for smooth contact of surfaces at a point can also be applied for constructing piecewise *GC^r continuous Hermite interpolants*. Such interpolants can, of course, be expressed in Bézier form, and are useful in connection with the conversion problem, see e.g., [Hos 88c, 89], [Was 91], and Chap. 10.

To construct a GC^r Hermite interpolant (in Bézier form, where for simplicity we assume the degree is n in both parameters) interpolating prescribed data on a rectangular grid, we may use the following two-step (or possibly four-step) method, (see Fig. 7.3):

1) construct a curve net such that the curve segments meet with GC^r continuity at the data points (see Fig. 7.3a),

2) determine the interior Bézier points near the patch corners of the Hermite interpolant in such a way that the patches meet at the data points with GC^r continuity (see Fig. 7.3b).

If the polynomial degree n of the patches is sufficiently large so that $n > 2r+1$, then we must also carry out the following two steps:

3) determine the interior Bézier points along the patch boundaries of the Hermite interpolant in such a way that adjoining patches meet with GC^r continuity along their common boundary curves (see Fig. 7.3c and step 4) of the algorithm in Sect. 7.5.1),

4) determine the remaining interior Bézier points of the interpolant by some optimization or design criterion (see Fig. 7.3d).

Step 3 involves GC^r continuity of two patches along a common boundary curve, and not just contact of order r at a single point. We encounter the same problem in the construction of spline surfaces, *i.e.*, we need to join a patch $\boldsymbol{Y}$ to a given patch $\boldsymbol{X}$ in such a way that the continuity conditions are satisfied along their common edges.

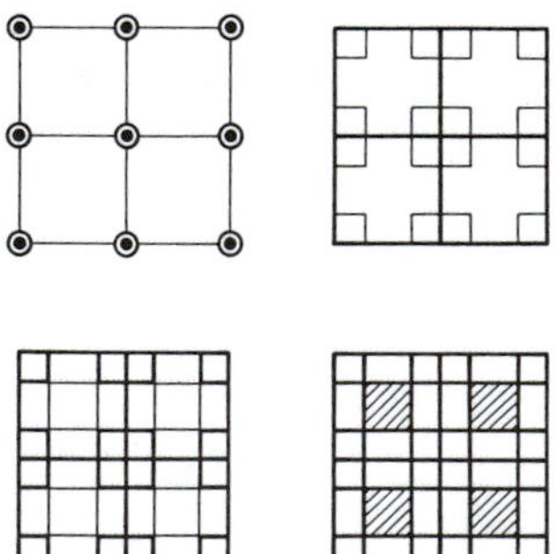

Fig. 7.3. Construction of a GC^r continuous Hermite interpolant.

We say that two surfaces $\boldsymbol{X}(u,v)$ and $\boldsymbol{Y}(s,t)$ join with GC^r continuity along a common *regular boundary curve* (also called a *linkage curve*), provided that they meet with GC^r continuity at every point on the common boundary curve. However, since the coefficients $a_{\mu\nu}$ and $b_{\mu\nu}$ can change from point to point, they must now be chosen to be functions of the parameter describing the common boundary curve. On the other hand, the conditions (7.5) are simplified, since the contact of two surfaces along a common boundary curve is in a certain sense more restrictive than contact at a single point. From now on, we assume that the common boundary curve has the same representation (e.g., in Bézier form) and the same parametrization (*i.e.*, $v = t$ for every point on the boundary curve) for both surfaces, see [Deg 90]. In this case, it follows immediately from the continuity condition

$$\boldsymbol{Y}(0,t) = \boldsymbol{X}(1,v), \qquad v = t,$$

that all partial derivatives in the direction of the common boundary curve are equal, *i.e.*,

$$\boldsymbol{Y}^{(0,\rho)}(0,t) = \boldsymbol{X}^{(0,\rho)}(1,v), \qquad v = t, \quad \rho = 1(1)r.$$

The condition (7.2) for *first order contact* (common tangent plane) at every point on the boundary curve thus simplifies to

$$\boldsymbol{Y}_s(0,t) = b(v)\boldsymbol{X}_u(1,v) + c(v)\boldsymbol{X}_v(1,v), \qquad v = t, \tag{7.6}$$

where $b(v) = a_{10}(v)$ with $b(v) > 0$ and $c(v) = b_{10}(v)$, see e.g., [Vero 76], [Sab 77], and [Hosa 78]. The coplanarity condition (7.6) is frequently also written in the form

$$\alpha(t)\boldsymbol{Y}_s(0,t) + \beta(v)\boldsymbol{X}_u(1,v) + \gamma(v)\boldsymbol{X}_v(1,v) = 0, \qquad v = t, \tag{7.7a}$$

with $\alpha(t) \neq 0$, $\beta(v) \neq 0$, and also in the determinant form

$$\det\left(\boldsymbol{Y}_s(0,t), \boldsymbol{X}_u(1,v), \boldsymbol{X}_v(1,v)\right) = 0, \qquad v = t. \tag{7.7b}$$

The condition (7.3) simplifies in the same way. Under the above assumptions, differentiating the GC^1 condition (7.6) with respect to t gives the GC^2 condition on $\boldsymbol{Y}_{st}$ automatically (see e.g., [Vero 76]). Thus, (7.3) reduces to the equation

$$\begin{aligned} \boldsymbol{Y}_{ss}(0,t) = {} & b(v)^2\boldsymbol{X}_{uu}(1,v) + 2b(v)c(v)\boldsymbol{X}_{uv}(1,v) + c(v)^2\boldsymbol{X}_{vv}(1,v) \\ & + e(v)\boldsymbol{X}_u(1,v) + f(v)\boldsymbol{X}_v(1,v), \qquad v = t, \end{aligned} \tag{7.8}$$

see [Kah 82, 83], [Höll 86], [Las 87]. Here $e(v) = a_{20}(v)$, and $f(v) = b_{20}(v)$. Equation (7.8) can also be written in a form analogous to (7.7a,b), [Kah 82,83]:

$$\delta(t)\boldsymbol{Z}(t,v) + \epsilon(v)\boldsymbol{X}_u(1,v) + \phi(v)\boldsymbol{X}_v(1,v) = 0, \qquad v = t, \tag{7.9a}$$

with $\delta(t) \neq 0$ and

$$Z(t,v) = Y_{ss}(0,t) - \left[b(v)^2 X_{uu}(1,v) + 2b(v)c(v)X_{uv}(1,v) + c(v)^2 X_{vv}(1,v)\right].$$

It can also be written in the determinant form

$$\det\left[Z(t,v), X_u(1,v), X_v(1,v)\right] = 0, \qquad v = t. \tag{7.9b}$$

A very practical criterion for two GC^1 continuous patches to join with GC^2 continuity is given in [Peg 92]. In particular, the *Linkage Curve Theorem* given there states that the equality of the normal curvatures with respect to just one surface tangent direction transversal to the linkage curve is equivalent to GC^2 continuity. Here the linkage curve need only be C^1 (and so can have polynomial degree one less than the patch), and the tangent direction need not be a continuous function of the curve parameter. The requirement of equality of all asymptote directions [Kah 82, 83] turns out to be redundant. On the other hand, the idea in the proof of the Linkage Curve Theorem is sufficiently general to cover the important case of complex asymptotic directions (*i.e.*, Gaussan curvatures positive), see [Peg 92], which was excluded in [Kah 82, 83].

For *contact of higher order* along a common boundary curve, we can also reduce the number of GC^r conditions to a single equation for each r, and indeed to an equation for the derivative $Y^{(r,0)}$. For example, for *third order contact* we need

$$\begin{aligned} Y_{sss}(0,t) &= b(v)^3 X_{uuu}(1,v) + 3b(v)^2 c(v) X_{uuv}(1,v) + 3b(v)c(v)^2 X_{uvv}(1,v) \\ &+ c(v)^3 X_{vvv}(1,v) + 3\Big[b(v)e(v)X_{uu}(1,v) + (c(v)e(v) + b(v)f(v))X_{uv}(1,v) \\ &+ c(v)f(v)X_{vv}(1,v)\Big] + h(v)X_u(1,v) + i(v)X_v(1,v), \qquad v = t, \end{aligned}$$

with

$$h(v) = a_{30}(v) \qquad \text{and} \qquad i(v) = b_{30}(v).$$

In the next section we discuss GC^1 and GC^2 continuity in more detail for triangular and rectangular patches given in Bézier form.

7.2. GC^1 Continuous Bézier Surfaces

In this section we discuss two applications. We begin with the problem of connecting a patch $Y(s,t)$ to a given patch $X(u,v)$ to build a GC^1 spline surface. Some early sources in the CAGD literature dealing with this problem include [BEZ 72], [Vero 76], [Sab 76a], [Hosa 78], and [FAU 81]. A geometric construction of the Bézier points for the two surface patches was given in [Far 82] and in [Kah 82].

We restrict ourselves first to *rectangular patches* $\boldsymbol{X}(u,v)$ and $\boldsymbol{Y}(s,t)$. Suppose two such patches share the common boundary curve $\boldsymbol{X}(1,v) = \boldsymbol{Y}(0,t)$, $v = t$, (see Fig. 7.4). Following [Far 82], in equation (7.7) we choose the $\alpha(v)$ and $\beta(v)$ to be constant and $\gamma(v)$ to be linear in v. This permits the use of patches of the same polynomial degree in v, and allows us to use comparison of coefficients to express the GC^1 continuity conditions between $\boldsymbol{X}(u,v)$ and $\boldsymbol{Y}(s,t)$ in terms of the Bézier points of the two surface patches. If $\boldsymbol{X}$ and $\boldsymbol{Y}$ do not have the same degree in v and t, respectively, then we apply degree raising to one of them, see e.g., [Liu 89] and [Du 90]. We suppose our two patches are written in the form

$$\boldsymbol{X}(u,v) = \sum_{i=0}^{m}\sum_{j=0}^{n} \boldsymbol{b}_{ij} B_i^m(u) B_j^n(v), \qquad u,v \in [0,1], \tag{7.10}$$

$$\boldsymbol{Y}(s,t) = \sum_{i=0}^{\bar{m}}\sum_{j=0}^{n} \bar{\boldsymbol{b}}_{ij} B_i^{\bar{m}}(s) B_j^n(t), \qquad s,t \in [0,1], \tag{7.11}$$

where without loss of generality, we may assume $\alpha(v) = -1$, $\beta(v) = \beta$ (with $\beta > 0$ by regularity), and $\gamma(v) = (1-v)\gamma_0 + v\gamma_1$.

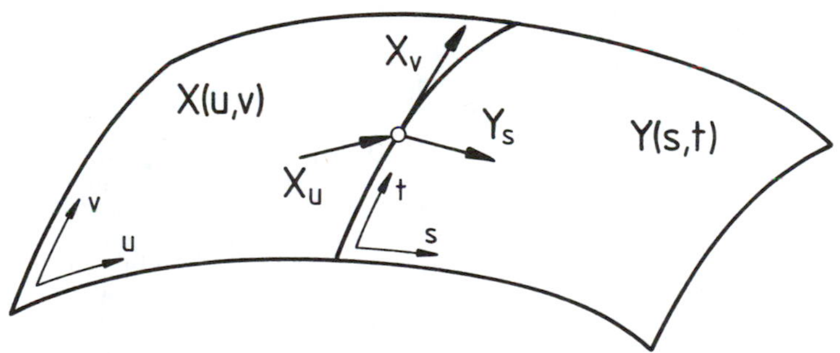

Fig. 7.4. Two rectangular patches meeting along a common boundary curve with GC^1 continuity.

Inserting the formulae (7.10) and (7.11) in (7.7a), using the identity

$$vB_j^{n-1}(v) = \frac{j+1}{n} B_{j+1}^n(v), \qquad (1-v)B_j^{n-1}(v) = \frac{n-j}{n} B_j^n(v),$$

and shifting the index j, we get

$$\bar{m}\sum_{j=0}^{n}\Delta^{10}\bar{\boldsymbol{b}}_{oj}B_j^n(v) = m\beta\sum_{j=0}^{n}\Delta^{10}\boldsymbol{b}_{m-1,j}B_j^n(v)$$

$$+\gamma_0\sum_{j=0}^{n}(n-j)\Delta^{01}\boldsymbol{b}_{mj}B_j^n(v)+\gamma_1\sum_{j=0}^{n}j\Delta^{01}\boldsymbol{b}_{m,j-1}B_j^n(v).$$

Comparing coefficients, it follows that

$$\bar{m}\Delta^{10}\,\bar{\boldsymbol{b}}_{0j} = m\beta\Delta^{10}\,\boldsymbol{b}_{m-1,j}+(n-j)\gamma_0\Delta^{01}\,\boldsymbol{b}_{mj}+j\gamma_1\Delta^{01}\,\boldsymbol{b}_{m,j-1}, \qquad (7.12a)$$

for $j = 0(1)n$. This can be written in the following equivalent form going back to Farin [Far 82] (see also [Kah 82, 83]) for the case $\bar{m} = m$:

$$\begin{aligned}\bar{\boldsymbol{b}}_{1j} &= (1-\frac{j}{n})\cdot\Big[(1+\frac{m}{\bar{m}}\beta-\frac{n}{\bar{m}}\gamma_0)\boldsymbol{b}_{mj}-\beta\frac{m}{\bar{m}}\boldsymbol{b}_{m-1,j}+\frac{n}{\bar{m}}\gamma_0\boldsymbol{b}_{m,j+1}\Big]\\ &+\frac{j}{n}\cdot\Big[(1+\frac{m}{\bar{m}}\beta+\frac{n}{\bar{m}}\gamma_1)\boldsymbol{b}_{mj}-\beta\frac{m}{\bar{m}}\boldsymbol{b}_{m-1,j}-\frac{n}{\bar{m}}\gamma_1\boldsymbol{b}_{m,j-1}\Big].\end{aligned} \qquad (7.12b)$$

Fig. 7.5a gives a geometric interpretation of equation (7.12), and Fig. 7.5b presents an alternative interpretation of this *sufficient condition* due to [Far 82].

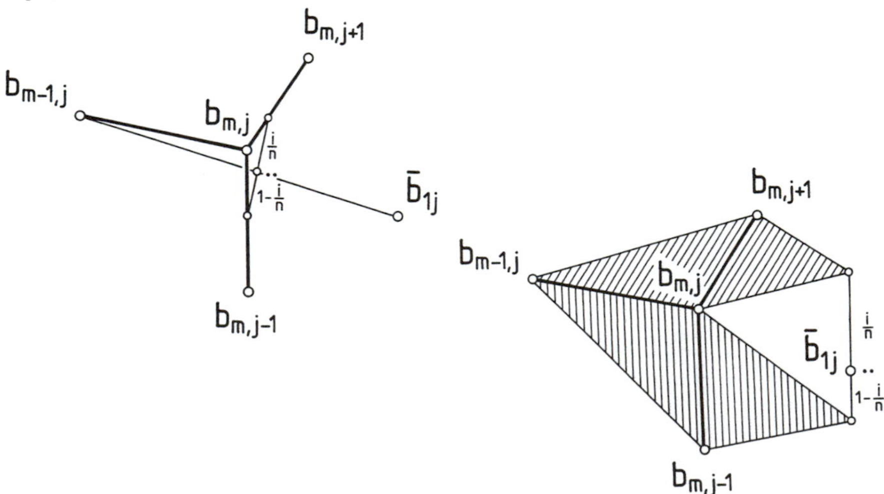

Fig. 7.5. Construction of the $\bar{\boldsymbol{b}}_{1j}$ for GC^1 continuity between two Bézier surfaces.

GC^1 continuity between two *triangular patches* or between a triangular and a rectangular patch can be handled in the same way. However, we must take into consideration that for triangular Bézier patches, the partial derivatives have to be replaced by directional derivatives, see Sect. 6.3.3, [Far 82], and [Kah 82, 83]. [Far 82] gives a unified treatment of some simple *sufficient conditions* for GC^1 continuity between three and four-sided patches.

In general, for GC^1 continuity, the parametric lines meeting along the linkage curve do not join together smoothly, but instead exhibit a bend.

The construction corresponding to (7.12) for geometric continuity between two rectangular (resp., two triangular) surface patches can also be obtained with the help of a parameter transformation corresponding to a de Casteljau extrapolation, cf. Sects. 4.1 and 6.3, and also see [Far 83a, 86], and [Las 87]. GC^r continuous extrapolation for curves and surfaces by means of reflection in the normal plane has been treated in [Shet 91].

It follows from (7.12), and also from the general equations (7.7), that for $v = t = 0$, *i.e.*, for $j = 0$, the Bézier points $\bar{\boldsymbol{b}}_{10}, \boldsymbol{b}_{m0}, \boldsymbol{b}_{m-1,0}$, and $\boldsymbol{b}_{m1}$ lie in the tangent plane of $\boldsymbol{X}(u,v)$ and $\boldsymbol{Y}(s,t)$ corresonding to $u = 1$, $v = 0$ and $s = t = 0$. In terms of barycentric coordinates, we can express this as

$$\bar{\boldsymbol{b}}_{10} = \alpha_1 \boldsymbol{b}_{m0} + \alpha_2 \boldsymbol{b}_{m-1,0} + \alpha \boldsymbol{b}_{m1}, \qquad \text{with } \alpha_1 + \alpha_2 + \alpha = 1. \tag{7.13a}$$

Similarly, we have

$$\bar{\boldsymbol{b}}_{1n} = \alpha_3 \boldsymbol{b}_{mn} + \alpha_4 \boldsymbol{b}_{m-1,n} + \alpha \boldsymbol{b}_{m,n-1}, \qquad \text{with} \qquad \alpha_3 + \alpha_4 + \alpha = 1. \tag{7.13b}$$

Equation (7.12) along with the equations given in Sect. 6.3.3 are sufficient for GC^1 continuity, but not necessary! Equation (7.7) allows very *general factors* $\alpha(v)$, $\beta(v)$ and $\gamma(v)$. For example, in [Hosa 78, 80], to get GC^1 continuity for bicubic and biquintic surfaces, the authors set $\alpha(v) = 1$ and choose $\gamma(v)$ to be cubic and $\beta(v)$ to be quadratic. [Liu 86] considers the case of two tensor-product Bézier surfaces of arbitrary degree, and to get GC^1 continuity takes the factors $\alpha(v)$, $\beta(v)$ and $\gamma(v)$ to be polynomials in Bézier form. [Liu 89] also gives some simple, special sufficient conditions along with *necessary and sufficient conditions* for GC^1 continuity between tensor-product and triangular Bézier patches. To get a solution by comparing coefficients, the degree of $\gamma(v)$ must be chosen to be at most one more than the larger of the degrees of $\alpha(v)$ and $\beta(v)$. Moreover, $\alpha(v)$, $\beta(v)$, and $\gamma(v)$ cannot have arbitrarily high degrees. The polynomial degrees of the two surfaces provide an upper bound on the degrees of these factors [Liu 86, 89, 90], [Du 90], and [Pet 91].

Since (7.7a) is vector-valued (and thus can be separated into coordinates), while in (7.7b) the three coordinates are multiplied together, (7.7a) is usually easier to use than (7.7b). However, if the right-hand side of (7.7b) is interpreted as a zero polynomial $\Phi(t) \equiv 0$, then a degree-wise separation of (7.7b) is possible, [Dero 90]. Indeed, suppose $\boldsymbol{S}_i$ denotes the Bézier points of the common boundary curve $\boldsymbol{S}(t)$, and $\boldsymbol{R}_j$ and $\boldsymbol{T}_k$ are the neighboring Bézier points to the $\boldsymbol{S}_i$ lying on the two patches, which we assume are either triangular or rectangular Bézier surfaces, see Fig. 7.8. Suppose the cross-boundary derivatives along the boundary are given by

$$\boldsymbol{X}'(t) = \sum_{i=0}^{R} \boldsymbol{R}'_i B_i^R(t), \qquad \boldsymbol{R}'_i = \boldsymbol{R}_i - \boldsymbol{S}_i,$$

$$\boldsymbol{Y}'(t) = \sum_{k=0}^{T} \boldsymbol{T}'_k B_k^T(t), \quad \boldsymbol{T}'_k = \boldsymbol{T}_k - \boldsymbol{S}_k,$$

and that the derivatives in the direction of the boundary curve are

$$\boldsymbol{S}'(t) = \sum_{j=0}^{S} \boldsymbol{S}'_j B_j^S(t), \quad \boldsymbol{S}'_j = \boldsymbol{S}_{j+1} - \boldsymbol{S}_j.$$

Inserting these derivatives in (7.7b) and using (4.2d), the multilinearity of the determinant leads to the following Bézier representation of the zero polynomial $\Phi(t) = \det(\boldsymbol{X}'(t), \boldsymbol{S}'(t), \boldsymbol{Y}'(t)) = 0$:

$$\Phi(t) = \sum_{\ell=0}^{D} \Phi_\ell B_\ell^D(t),$$

with

$$\Phi_\ell = \sum_{i+j+k=\ell} \frac{\binom{R}{i}\binom{S}{j}\binom{T}{k}}{\binom{D}{\ell}} \det\left(\boldsymbol{R}'_i, \boldsymbol{S}'_j, \boldsymbol{T}'_k\right),$$

where $0 \leq i \leq R$, $0 \leq j \leq S$, $0 \leq k \leq T$ and $R + S + T = D$. This means that the coefficients Φ_ℓ are the weighted averages of all terms of the form $\det(\boldsymbol{R}'_i, \boldsymbol{S}'_j, \boldsymbol{T}'_k)$ with $i + j + k = \ell$. From this it follows that (7.7b) is equivalent to the following set of $D + 1$ equations which have been separated according to degree:

$$\sum_{i+j+k=\ell} \det(\boldsymbol{R}^*_i, \boldsymbol{S}^*_j, \boldsymbol{T}^*_k) = 0, \qquad \ell = 0(1)D, \tag{7.14}$$

where

$$\boldsymbol{R}^*_i = \binom{R}{i}\boldsymbol{R}'_i, \quad \boldsymbol{S}^*_j = \binom{S}{j}\boldsymbol{S}'_j, \quad \boldsymbol{T}^*_k = \binom{T}{k}\boldsymbol{T}'_k.$$

For example, for two cubic triangular Bézier patches, (7.14) leads to seven equations (here the degrees of $\boldsymbol{X}', \boldsymbol{Y}'$, and $\boldsymbol{S}'$ are $R = S = T = 2$, which implies $D = 6$).

The conditions (7.14) are minimal in the sense that they are independent of each other [Dero 90]. This is an advantage over the above approach which may lead to a large set of design parameters, see e.g., [Liu 89, 90] [Was 91]. Their interdependence is hard to overlook, and has so far not been studied in depth (see e.g., 'An Example' in [Deg 90]). The above approach avoids this problem, since the conditions (7.14) are derived directly from the control points.

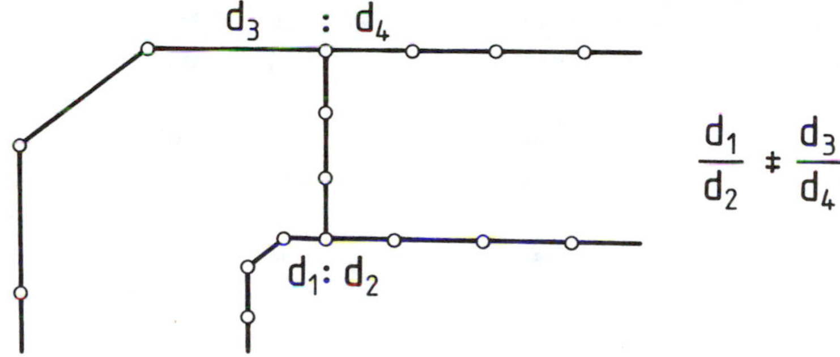

Fig. 7.6. Very large differences in the distances between Bézier points imply incompatible surfaces.

The special choice of $\alpha(v)$, $\beta(v)$ and $\gamma(v)$ used in the derivation of (7.12) leads to a simple direct way of joining two tensor-product Bézier surfaces of the same degree, but it can be too restrictive for a practical design process. In particular, problems can arise in the case of the *second application*. There we want a GC^1 join of two surfaces prescribed in terms of point, derivative, or tangent plane information at surface points, or along common boundary curves. Because of the special choice of the factors in (7.7), we may not have a sufficient number of free parameters in order to satisfy all of the boundary conditions. For example, if the two surfaces differ sharply in shape, or if along the common boundary they have very different "radii" (see Fig. 7.6), then the above choice of the parameters $\alpha(v)$, $\beta(v)$, and $\gamma(v)$ can be too restrictive, see [Sar 87] (but take note of [Sar 89]). If, however, $\beta(v)$ is chosen to be a cubic polynomial, then the above problem does not arise, and, moreover, the twist vectors of the surfaces at the end points of the common boundary curve decouple. This means that the solution of the system of equations for the twist vectors needed in (7.12) is localized with respect to the vertex of interest in the sense that for any pair of neighboring vertices, the corresponding GC^1

continuity conditions share no common variables, see Sect. 7.5.2. We note, however, that choosing $\beta(v)$ to be cubic increases the polynomial degree of $\boldsymbol{Y}$ as a function of v to $\bar{n}+3$, see [Sar 87, 89, 90].

This idea was implemented (see [Sar 90]) to construct *multi-patch surfaces* consisting of *tensor-product patches*, and allowing N-patch vertex configurations, see Sect. 7.5.2 and [Sar 90]. This requires however, the strong restriction that all even order vertices must involve $N = 4$ edges, and that such vertices must be of x-form in the sense that they are given by pairwise tangent continuous boundary curves [Sar 87, 90]. The polynomial degrees of the factors have to be chosen carefully to account for the particular vertex situation in order to produce the required continuity. Starting the construction with bicubic patches, the resulting patch complex thus includes not only bicubic patches, but also patches of degree (3,6), (6,3) and (6,6) [Sar 87, 89, 90]; see also Sect. 7.5.2.

We remark that increasing the polynomial degrees of the factors $\alpha(v)$, $\beta(v)$, and $\gamma(v)$ does not really increase the number of available degrees of freedom! It does increase the number of variables in the sense that the twist vectors at neighboring vertices to be determined from the GC^1 conditions can be decoupled, but at the same time, an increase in the polynomial degrees of $\alpha(v)$ and $\beta(v)$ by one implies a restriction on the choice of the Bézier points in the sense that (see [Du 90]):

- the number of equations corresponding to (7.12) increases by one, but they are interdependent,
- for every index j on the common boundary curve, three additional Bézier points are involved in the GC^1 continuity condition.

This means that if $\alpha(v)$ and $\beta(v)$ are chosen to be constants (see above), then (7.12) follows. This gives a total of $n+1$ equations, and for every index j (with the exception of the boundary values), in addition to $\boldsymbol{b}_{mj}$, four additional neighboring Bézier points will be involved in the GC^1 continuity conditions. In the same way, we see that for linear $\alpha(v)$ and $\beta(v)$, we have a total of $n+2$ conditions for GC^1 continuity, each of which involves an additional three Bézier points. Fig. 7.7 shows the number and kind of Bézier points which are influenced for each j (except for boundary values) by the GC^1 continuity conditions corresponding to constant, linear, and quadratic functions $\alpha(v)$ and $\beta(v)$.

The difficulties discussed above associated with special choices of the functions $\alpha(v)$, $\beta(v)$, and $\gamma(v)$ for given point and derivative/tangent plane information at the corners of the surfaces also arise in the case of *triangular patches*. Indeed, for the interpolation problem described above, two triangular Bézier patches of degree n with a common boundary curve of degree $n-1$ can

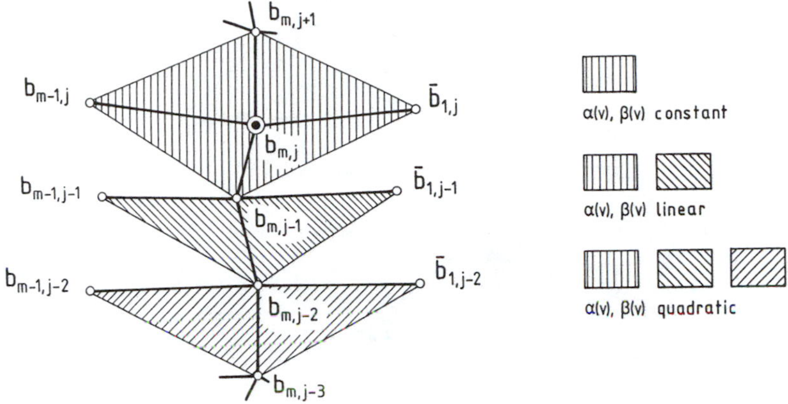

Fig. 7.7. The Bézier points affected by GC^1 conditions for a fixed j for the cases of constant, linear, and quadratic functions $\alpha(v)$ and $\beta(v)$.

only meet with GC^1 continuity provided that the prescribed $\boldsymbol{T}_0, \boldsymbol{R}_0, \boldsymbol{S}_0$ and $\boldsymbol{S}_1$ and the $\boldsymbol{T}_{n-1}, \boldsymbol{R}_{n-1}, \boldsymbol{S}_{n-1}$ and $\boldsymbol{S}_{n-2}$ satisfy equations of the form (7.12), where at $\boldsymbol{S}_0$ and $\boldsymbol{S}_n$ (for the notation, see Fig. 7.8), conditions of the form (7.13) must hold [Far 83a], see also [Shi 87] (but take note of [Shi 91]). In this sense, Farin interprets (7.13) as the following condition on the areas of the triangular patches:

$$\frac{\text{area}\{\boldsymbol{R}_0, \boldsymbol{S}_0, \boldsymbol{S}_1\}}{\text{area}\{\boldsymbol{T}_0, \boldsymbol{S}_0, \boldsymbol{S}_1\}} = \frac{\text{area}\{\boldsymbol{R}_{n-1}, \boldsymbol{S}_{n-1}, \boldsymbol{S}_{n-2}\}}{\text{area}\{\boldsymbol{T}_{n-1}, \boldsymbol{S}_{n-1}, \boldsymbol{S}_{n-2}\}}. \tag{7.15a}$$

The algebraic equivalent of this condition is that

$$\alpha_1 + \alpha_2 = \alpha_3 + \alpha_4, \tag{7.15b}$$

where α_i are as in (7.13), see [Shi 87].

[Far 83a] provides an algorithm for interpolating 3D data points using cubic Bézier curves, which are then degree-raised and spanned by quartic triangular Bézier patches. Each of these macro patches is subdivided into three pieces (micro patches). The GC^1 continuity between the macro patches is assured by (7.12), (7.15). The internal C^1 continuity of the micro patches is achieved by working from the outside to the inside, determining internal control points as the centroids of three surrounding points.

For the function-valued case, which is for example of great importance in scattered data fitting (see Chap. 9), (7.15) does not imply an additional

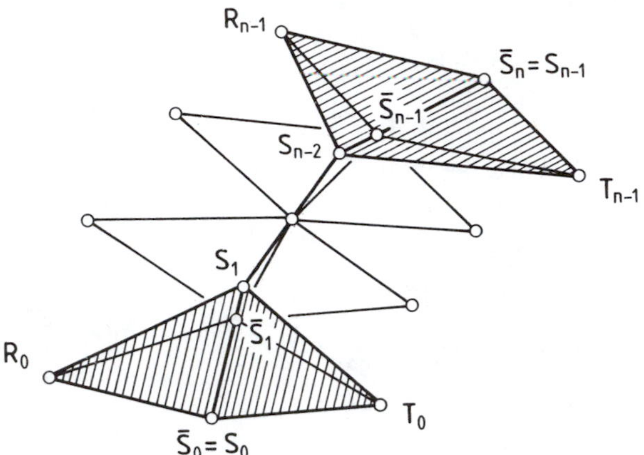

$\bar{s}_i$ degree n boundary Bézier points
s_i degree $n-1$ boundary Bézier points
R_i, T_i degree n surface Bézier points

Fig. 7.8. The Bézier points entering into Farin's GC^1 conditions for two triangular Bézier patches.

condition since it is already satisfied because of the planarity of the two quadrilaterals and the fact that the abscissae of the Bézier points in the two domain triangles form regular triangular grids.

[Pip 87] (see also [Far 86]) gives the following condition for GC^1 continuity between two triangular Bézier patches along a common boundary curve $w = 0$:

$$E(u,v)\boldsymbol{I}(u,v) + F(u,v)\boldsymbol{J}(u,v) + G(u,v)\boldsymbol{K}(u,v) + H(u,v)\boldsymbol{L}(u,v) = 0, \tag{7.16}$$

where

$$E(u,v) + F(u,v) + G(u,v) + H(u,v) = 0,$$

and where

$$\boldsymbol{I}(u,v) = \sum \boldsymbol{S}_{i+1} B_{ij0}^{n-1}(u,v,0), \qquad \boldsymbol{J}(u,v) = \sum \boldsymbol{S}_i B_{ij0}^{n-1}(u,v,0),$$
$$\boldsymbol{K}(u,v) = \sum \boldsymbol{T}_i B_{ij0}^{n-1}(u,v,0), \qquad \boldsymbol{L}(u,v) = \sum \boldsymbol{R}_i B_{ij0}^{n-1}(u,v,0).$$

The example [Pip 87]

$$\boldsymbol{S}_0 = (0,0,0), \quad \boldsymbol{S}_1 = (1,0,0), \quad \boldsymbol{S}_2 = (2,0,1), \quad \boldsymbol{S}_3 = (4,0,1),$$
$$\boldsymbol{T}_0 = (1,1,0), \quad \boldsymbol{T}_2 = (3,1,1), \quad \boldsymbol{R}_0 = (1,-1,0), \quad \boldsymbol{R}_2 = (3,-1,1),$$

(see Fig. 7.9) shows that in the case of two cubic triangular Bézier patches, (7.16) cannot always be satisfied in the above sense for prescribed $\boldsymbol{S}_0$, $\boldsymbol{S}_1$, $\boldsymbol{T}_0$, $\boldsymbol{R}_0$, $\boldsymbol{S}_{n-1}$, $\boldsymbol{S}_n$, $\boldsymbol{T}_{n-1}$, $\boldsymbol{R}_{n-1}$ if the factors are chosen to be too special. This happens, for example, if they are chosen to be constants or to be *degree-raising polynomials* (polynomials whose multiplication with the given patch lead to a degree raised surface in the sense of Sect. 6.3.2). This assertion again holds only in the parametric case; as discussed earlier, for two function-valued cubic triangular Bézier patches, the interpolation problem is always solvable.

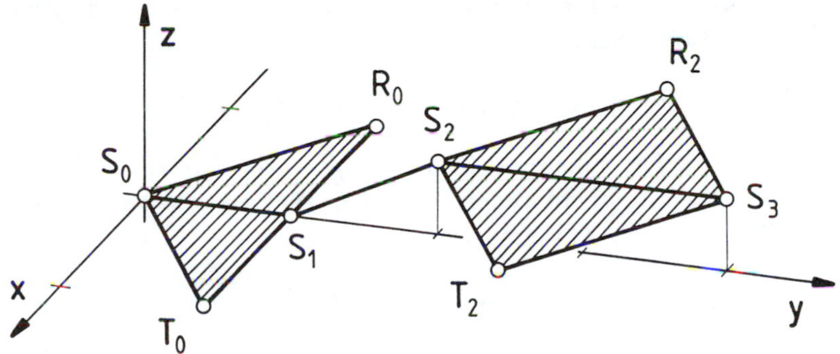

Fig. 7.9. An interpolation problem for which no solution using cubic triangular Bézier patches exists [Pip 87].

However, for quartic patches, (7.16) is always solvable, even for simple factors [Pip 87]. [Pip 87] gives an algorithm which first constructs a cubic triangular Bézier surface for each facet, and then subdivides each of them into three pieces about their centroids. After degree raising of the subpatches to degree 4, (7.16) guarantees GC^1 continuity along the boundaries of the macro patches. C^1 continuity is attained on the interior as in [Far 83a]. These surfaces tend to oscillate if all of the factors $E(u, v)$, $F(u, v)$, $G(u, v)$, $H(u, v)$ in (7.16) (which in [Pip 87] are at most linear) are nearly constant [Mann 92].

A solution using cubic patches is also possible, if we drop the usual restriction that the factors be constant or degree-raising polynomials. An example is presented in [Pet 90a, 91] where the factors are chosen to be linear functions, and none of them is a degree raising polynomial. We remark that in this case, the interior quadrilaterals formed by the Bézier net along the common boundary are no longer planar, in general.

Starting with the Bézier points of a given patch, the methods described above can be used to construct the Bézier points of a neighboring patch to join the given patch with GC^r continuity. We now discuss a method due to Chiyokura and Kimura [Chi 83] in which a *boundary curve* and a *tangent plane field* along the boundary curve is first constructed from the interpolation data. Then in a second stage, working "*to the right and to the left*", two adjoining patches are constructed. For Bézier surfaces, this means that we have to determine the unknown Bézier points for each of the two patches.

We can proceed as follows. Suppose the common boundary curve $\boldsymbol{S}(t)$, $t \in [0,1]$, is given or can be determined from vertex data using some appropriate interpolation method. Suppose $\boldsymbol{Q}(t)$ is a tangent-vector field transversal to $\boldsymbol{S}(t)$. Then together, $\boldsymbol{Q}(t)$ and $\boldsymbol{S}'(t)$ define a tangent-plane field along $\boldsymbol{S}(t)$. By the results of Sect. 7.1, a patch $\boldsymbol{Y}$ meets this tangent-vector field with GC^1 *continuity* precisely when

$$D_{\boldsymbol{q}}\boldsymbol{Y} = k(t)\boldsymbol{Q}(t) + h(t)\boldsymbol{S}'(t), \tag{7.17a}$$

where $\boldsymbol{q}$ is a direction transversal to $\boldsymbol{S}(t)$, $D_{\boldsymbol{q}}\boldsymbol{Y}$ is the directional derivative, and $k(t)$ and $h(t)$ are appropriately chosen so that we also have regularity. Similarly, for the patch $\boldsymbol{X}$, we must have

$$D_{\bar{\boldsymbol{q}}}\boldsymbol{X} = \bar{k}(t)\boldsymbol{Q}(t) + \bar{h}(t)\boldsymbol{S}'(t), \tag{7.17b}$$

with direction vector $\bar{\boldsymbol{q}}$ and functions $\bar{k}(t)$ and $\bar{h}(t)$, which can be different from $\boldsymbol{q}$, $k(t)$ and $h(t)$, in general, see [Deg 90]. Here we assume that $\boldsymbol{S}(t)$ and $\boldsymbol{Q}(t)$ are regular, i.e., $\boldsymbol{S}'(t) \neq 0$, $\boldsymbol{Q}'(t) \neq 0$. In addition, we require that $\boldsymbol{S}'(t)$ and $\boldsymbol{Q}(t)$ be linearly independent for all t: $\boldsymbol{S}'(t) \times \boldsymbol{Q}(t) \neq 0$, and $\boldsymbol{Q}(t), k(t), \bar{k}(t) \neq 0$, so that the tangent planes for $\boldsymbol{X}$ and $\boldsymbol{Y}$ are defined at every point on the common boundary curve.

Using this approach, where in contrast to the method first discussed in this section, the two patches are treated equally, and are constructed simultaneously, it is in principle possible to smoothly join all types of patches, even when they are of different patch types. This is done in [Chi 83], [Tak 91] for example for Gregory patches (see Sect. 7.5.3) and tensor-product Bézier patches. It has also been used for triangular Bézier patches (see e.g., [Shi 87], but take note of [Shi 91]), for triangular and rectangular Bézier patches (see e.g., [Jen 87]), and for Coons patches (see Sect. 8.1.2). If we work with the Bézier representation, then the cross derivatives (7.17) immediately lead to explicit values for the Bézier points of the surfaces $\boldsymbol{X}$ and $\boldsymbol{Y}$ which lie in the first rows parallel to the common boundary curve $\boldsymbol{S}(t)$.

The form of the patches, the choice of the factors, and also the required values or possible choices for the $\boldsymbol{Q}(t)$ and $\boldsymbol{S}(t)$, determine the possible degrees of freedom. We now discuss this in more detail for the important case of polynomial surfaces $\boldsymbol{X}$ and $\boldsymbol{Y}$.

If $\boldsymbol{X}$ and $\boldsymbol{Y}$ are polynomials, it follows that we can choose $\boldsymbol{S}(t)$, $\boldsymbol{Q}(t)$, $k(t)$, and $h(t)$ as polynomials, or as appropriate rational functions. Moreover, comparing coefficients in (7.17) leads to conditions on the polynomial degrees of $\boldsymbol{Q}(t)$, $\boldsymbol{S}(t)$, $k(t)$ and $h(t)$. In [Chi 83] (see also [Herr 85] and Sect. 8.1.2) $\boldsymbol{Q}(t)$ is defined by two vectors $\boldsymbol{Q}_0$ and $\boldsymbol{Q}_1$ transversal to $\boldsymbol{S}(t)$ at $\boldsymbol{S}(0)$ and $\boldsymbol{S}(1)$, whose signs are chosen according to whether the right or the left patch is to be constructed. In particular, he takes

$$\boldsymbol{Q}(t) = (1-t)\boldsymbol{Q}_0 + t\boldsymbol{Q}_1, \qquad t \in [0,1]. \tag{7.18}$$

Then $\boldsymbol{Q}_0$ and $\boldsymbol{Q}_1$ are chosen to be unit vectors which are orthogonal to the common boundary curve $\boldsymbol{S}(t)$ at $\boldsymbol{S}(0)$ and $\boldsymbol{S}(1)$. [Chi 86] also uses information on the surfaces in the choice of $\boldsymbol{Q}_0$ and $\boldsymbol{Q}_1$.

For $t = 0$ and $t = 1$, it follows from (7.17a) that

$$D_{\boldsymbol{q}}\boldsymbol{Y}\Big|_{t=0} = k_0\boldsymbol{Q}_0 + h_0\boldsymbol{S}_0', \tag{7.19a}$$

$$D_{\boldsymbol{q}}\boldsymbol{Y}\Big|_{t=1} = k_1\boldsymbol{Q}_1 + h_1\boldsymbol{S}_1', \tag{7.19b}$$

where $k_i = k(i)$, $h_i = h(i)$, and $\boldsymbol{S}_i' = \boldsymbol{S}'(i)$. This means that $k(t)$ and $h(t)$ must be at least linear in order to satisfy these boundary conditions. If they are linear, then by (7.19) they are already uniquely defined to be

$$k(t) = k_0(1-t) + k_1 t, \tag{7.20a}$$

$$h(t) = h_0(1-t) + h_1 t. \tag{7.20b}$$

Suppose now that $D_{\boldsymbol{q}}\boldsymbol{Y}$ is cubic, and that the boundary curve $\boldsymbol{S}(t)$ is also constructed to be a cubic. This happens, for example, for a quartic triangular Bézier patch and also for a tensor-product Bézier patch whose degree transversal to $\boldsymbol{S}(t)$ is four. Then with $h(t)$ as in (7.20b), $h(t)\boldsymbol{S}'(t)$ is also cubic. On the other hand, if $\boldsymbol{Q}(t)$ is constructed as in (7.18), then $k(t)$ cannot be selected as in (7.20a), but instead must be quadratic. This leads to a further scalar degree of freedom. An example of this can be found in the paper of [Jen 87]. However, if $k(t)$ is chosen as in (7.20a), then $\boldsymbol{Q}(t)$ cannot be constructed by (7.18); it has to be quadratic, leading to a vector-valued degree of freedom.

7.3. GC² Continuous Bézier Surfaces

Suppose $\boldsymbol{X}(u,v)$ and $\boldsymbol{Y}(s,t)$ are two rectangular patches which meet with GC^1 continuity as in (7.6). As in Sect. 7.2, we again assume that $\boldsymbol{X}(u,v)$ and

$\boldsymbol{Y}(s,t)$ are of the same polynomial degree in v and t, respectively, and are written in Bézier form as in (7.10)–(7.11).

If we choose $b(v)$ and $c(v)$ of the form $b(v) = b$, $c(v) = (1-v)c_0 + vc_1$, then (cf. Sect. 7.2) we can determine the Bézier points $\bar{\boldsymbol{b}}_{1j}$ by comparing coefficients. For GC^2 continuity between $\boldsymbol{X}(u,v)$ and $\boldsymbol{Y}(s,t)$, we must also require that equation (7.8) (or equivalently (7.9)) be satisfied. A *sufficient condition* for (7.8) is that $e(v)$ and $f(v)$ be constant and linear in v, respectively, *i.e.*, $e(v) = e$, $f(v) = (1-v)f_0 + vf_1$. Now we can compare coefficients as before. Inserting (7.10) and (7.11) in (7.8), it follows in the same way as in the GC^1 case that

$$\begin{aligned}\bar{m}(\bar{m}-1)\Delta^{20}\bar{\boldsymbol{b}}_{0j} &= me\Delta^{10}\boldsymbol{b}_{m-1,j} + (m-j)f_0\Delta^{01}\boldsymbol{b}_{mj} + jf_1\Delta^{01}\boldsymbol{b}_{m,j-1}\\ &\quad + m(m-1)b^2\Delta^{20}\boldsymbol{b}_{m-2,j} + 2mb[(n-j)c_0\Delta^{11}\boldsymbol{b}_{m-1,j}\\ &\quad + jc_1\Delta^{11}\boldsymbol{b}_{m-1,j-1}] + (n-j)(n-j-1)c_0^2\Delta^{02}\boldsymbol{b}_{mj}\\ &\quad + 2j(n-j)c_0c_1\Delta^{02}\boldsymbol{b}_{m,j-1} + j(j-1)c_1^2\Delta^{02}\boldsymbol{b}_{m,j-2}.\end{aligned} \tag{7.21}$$

The large number of design parameters and the complexity of the (sufficient) conditions implies that a practical implementation of these conditions requires a great deal of work. For $\bar{m} = m$, $e(v) = 0$, and $f(v) = 0$ (*i.e.*, $f_0 = f_1 = 0$), the sufficient conditions (7.21) reduce to a somewhat simpler form given in [Kah 82, 83]. But since even in this special case these GC^2 conditions are still very difficult to handle, [Kah 82] in fact only discusses some simple examples.

The case $d(v) = 0$, $e(v) = 0$ was also treated in [Vero 76] for surfaces expressed in monomial form, under the additional restriction that the $c(v)$ as well as the $b(v)$ are constant.

The cases of joining two triangular Bézier patches (or a triangular and a rectangular Bézier patch) with GC^2 continuity can be handled in the same way [Kah 82, 83] (see also [Was 91]), where, of course, for triangular Bézier patches, the partial derivatives must be replaced by directional derivatives.

The problem of finding *necessary and sufficient conditions* for GC^2 continuity between two Bézier patches analogous to those discussed in [Liu 89] for the GC^1 case has been treated in [Was 91], which also contains some special sufficient conditions. Our remarks in the previous section about the number of free parameters and their dependencies in the GC^1 case have analogs here, although the situation is considerably more complicated. In the polynomial Bézier case with GC^2 continuity, [Was 91] has carried out numerical calculations similar to those in [Dero 90], starting with the equivalent determinantal form (7.9b) of (7.8).

Given a prescribed boundary curve $\boldsymbol{S}(t)$ and a regular tangent-plane field along $\boldsymbol{S}(t)$, suppose $\boldsymbol{X}$ and $\boldsymbol{Y}$ are patches which join along $\boldsymbol{S}(t)$ with GC^1 continuity. Then the construction based on (7.17) can be generalized to the GC^2 case giving

$$D^2_{\boldsymbol{q}}\boldsymbol{Y} = k(t)^2\boldsymbol{P}(t) + h(t)^2\boldsymbol{S}''(t) + 2k(t)h(t)\boldsymbol{Q}'(t) + r(t)\boldsymbol{S}'(t) + p(t)\boldsymbol{Q}(t), \quad (7.22a)$$

$$D^2_{\bar{\boldsymbol{q}}}\boldsymbol{X} = \bar{k}(t)^2\boldsymbol{P}(t) + \bar{h}(t)^2\boldsymbol{S}''(t) + 2\bar{k}(t)\bar{h}(t)\boldsymbol{Q}'(t) + \bar{r}(t)\boldsymbol{S}'(t) + \bar{p}(t)\boldsymbol{Q}(t), \quad (7.22b)$$

where $\boldsymbol{q}$, $\bar{\boldsymbol{q}}$ are direction vectors, and $k(t)$, $\bar{k}(t)$, $h(t)$, $\bar{h}(t)$, $r(t)$, $\bar{r}(t)$, $p(t)$, and $\bar{p}(t)$ are appropriate functions [Deg 90]. Here $\boldsymbol{P}(t)$ is a vector field prescribing second derivatives transversal to $\boldsymbol{S}(t)$ and not lying in the tangent-plane.

The construction of GC^2 continuous patches using (7.17) and (7.22) is illustrated in [Deg 90] for two biquadratic patches. [Tak 91] treats C^2 continuous Gregory patches, and their GC^2 integration into a rectangular Bézier patch network so that (7.17) and (7.22) are satisfied.

The problem of interpolating transfinite and discrete curvature data given along the boundary of a triangular patch is treated in [Hag 86a, 89] using Nielson's side-vertex blending method (see [Nie 79], [Barn 77]). Starting with a prescribed regular curve network, [Web 90a] constructed GC^2 continuous Coons patches using biquintic blending functions. General curve networks are discussed in [Web 90], and blending methods (in the sense of Chap. 14) are also used for GC^r continuous functional splines in [Li 90], [Hos 91a].

7.4. Rational Geometric Spline Surfaces

If we use homogeneous coordinates, then most of the above results on GC^r continuity can be directly carried over to the rational case. This holds in particular for the results of Sect. 5.7.2 on GC^r continuous rational curves going back to [Deg 88]. Using the same argument as in Sect. 5.7.2, it follows that two surface patches $\boldsymbol{X}(u,v)$ and $\boldsymbol{Y}(s,t)$ join continuously at a *regular point* $\boldsymbol{P}$ corresponding to parameter values $(u,v) = (\bar{u},\bar{v})$ or $(s,t) = (\bar{s},\bar{t})$, provided that their $\mathbb{R}^4$ representations $\mathcal{X}(u,v)$ and $\mathcal{Y}(s,t)$ in terms of homogeneous coordinates satisfy

$$\mathcal{Y}(\bar{s},\bar{t}) = \alpha_{00}\mathcal{X}(\bar{u},\bar{v}), \quad (7.23)$$

where $\alpha_{00} = \alpha(\bar{s},\bar{t}) \neq 0$. To derive GC^r *continuity conditions*, we apply the chain rule to the following equation

$$\frac{\partial^\rho \mathcal{Y}}{\partial s^\mu \partial t^\nu}(s,t) = \frac{\partial^\rho}{\partial s^\mu \partial t^\nu}\Big(\alpha(s,t)\mathcal{X}(u(s,t),v(s,t))\Big), \qquad \rho = 0(1)r \quad (7.24)$$

at $(s,t) = (\bar{s},\bar{t})$ on the left, and $(u,v) = (\bar{u},\bar{v}) = (u(\bar{s},\bar{t}),v(\bar{s},\bar{t}))$ on the right. Here $\mathcal{X}(u,v)$ is reparametrized to the parameters s,t by $u \mapsto u(s,t)$,

$v \mapsto v(s,t)$ with $a_{00} = u(\bar{s},\bar{t}) = \bar{u}$ and $b_{00} = v(\bar{s},\bar{t}) = \bar{v}$. In addition, $\rho = \mu+\nu$, and for triangular Bézier surfaces, as usual, the partial derivatives must be replaced by the directional derivatives.

Introducing the abbreviations

$$\alpha_{\mu\nu} = \frac{\partial^\rho}{\partial s^\mu \partial t^\nu}\alpha(s,t)\Big|_{(s,t)=(\bar{s},\bar{t})} \qquad \rho = \mu + \nu, \tag{7.25}$$

and letting $a_{\mu\nu}$ and $b_{\mu\nu}$ be as in (7.1), we get the following GC^1 *continuity conditions* generalizing (7.2)ff (cf. (5.42)):

$$\begin{pmatrix} \mathcal{Y}_s \\ \mathcal{Y}_t \end{pmatrix} = \begin{pmatrix} \alpha_{10} \\ \alpha_{01} \end{pmatrix} \mathcal{X} + \alpha_{00} \begin{pmatrix} a_{10} & b_{10} \\ a_{01} & b_{01} \end{pmatrix} \begin{pmatrix} \mathcal{X}_u \\ \mathcal{X}_v \end{pmatrix}, \tag{7.26}$$

and the following GC^2 *continuity conditions*:

$$\begin{pmatrix} \mathcal{Y}_{ss} \\ \mathcal{Y}_{st} \\ \mathcal{Y}_{tt} \end{pmatrix} = \begin{pmatrix} \alpha_{20} \\ \alpha_{11} \\ \alpha_{02} \end{pmatrix} \mathcal{X}$$

$$+ \begin{pmatrix} \alpha_{00}a_{20} + 2\alpha_{10}a_{10} & a_{00}b_{20} + 2\alpha_{10}b_{10} \\ \alpha_{00}a_{11} + \alpha_{01}a_{10} + \alpha_{10}a_{01} & \alpha_{00}b_{11} + \alpha_{01}b_{10} + \alpha_{10}b_{01} \\ \alpha_{00}a_{02} + 2\alpha_{01}a_{01} & \alpha_{00}b_{02} + 2\alpha_{01}b_{01} \end{pmatrix} \begin{pmatrix} \mathcal{X}_u \\ \mathcal{X}_v \end{pmatrix}$$

$$+ \alpha_{00} \begin{pmatrix} a_{10}^2 & 2a_{10}b_{10} & b_{10}^2 \\ a_{10}a_{01} & a_{10}b_{01} + a_{01}b_{10} & b_{10}b_{01} \\ a_{01}^2 & 2a_{01}b_{01} & b_{01}^2 \end{pmatrix} \begin{pmatrix} \mathcal{X}_{uu} \\ \mathcal{X}_{uv} \\ \mathcal{X}_{vv} \end{pmatrix}, \tag{7.27}$$

etc. Regularity conditions follow as in Sect. 7.1.

For GC^r continuity along a common *regular boundary curve*, given by $s = 0$, $u = 1$, and by $v = t$ (see Sect. 7.1), the equations (7.23)ff can be simplified as follows. The *continuity condition* (7.23) becomes

$$\mathcal{Y}(0,t) = \alpha(0,t)\mathcal{X}(1,v), \qquad v = t, \quad \alpha(0,t) \neq 0. \tag{7.28}$$

Using (7.28), it follows immediately that the partial derivatives with respect to t must satisfy

$$\mathcal{Y}^{(0,\rho)}(0,t) = \frac{d^\rho}{dt^\rho}\Big(\alpha(0,t)\mathcal{X}(1,v)\Big), \qquad v = t, \quad \rho = 1(1)r, \tag{7.29}$$

which represents a structural simplification relative to $\mathcal{Y}_t$, $\mathcal{Y}_{tt}$, etc. in (7.26)ff. Condition (7.26) for *first order contact* reduces to the equation

$$\mathcal{Y}_s(0,t) = \alpha(0,t)\Big[b(v)\mathcal{X}_u(1,v) + c(v)\mathcal{X}_v(1,v)\Big] + \alpha_{10}(0,t)\mathcal{X}(1,v), \tag{7.30}$$

with $v = t$ and $b(v)$, $c(v)$ as in (7.6). Equation (7.30) can also be written in the equivalent form

$$A(t)\mathcal{Y}_s(0,t) + B(v)\mathcal{X}_u(1,v) + C(v)\mathcal{X}_v(1,v) + \Omega(t)\mathcal{X}(1,v) = 0, \qquad (7.31a)$$

with $v = t$ and functions $A(t) \neq 0$, $B(v) \neq 0$, $C(v)$ and $\Omega(t)$, resp., and also in the equivalent determinantal form

$$\det\left(\mathcal{Y}_s(0,t), \mathcal{X}_u(1,v), \mathcal{X}_v(1,v), \mathcal{X}(1,v)\right) = 0, \qquad v = t. \qquad (7.31b)$$

Condition (7.27) for *second order contact* can also be simplified in an analogous way (cf Sect. 7.1) to

$$\begin{aligned}\mathcal{Y}_{ss}(0,t) = {} & \alpha(0,t)\left[b(v)^2\mathcal{X}_{uu}(1,v) + 2b(v)c(v)\mathcal{X}_{uv}(1,v) + c(v)^2\mathcal{X}_{vv}(1,v)\right] \\ & + \left[\alpha(0,t)e(v) + 2\alpha_{10}(0,t)b(v)\right]\mathcal{X}_u(1,v) \\ & + \left[\alpha(0,t)f(v) + 2\alpha_{10}(0,t)c(v)\right]\mathcal{X}_v(1,v) \\ & + \alpha_{20}(0,t)\mathcal{X}(1,v), \qquad v = t,\end{aligned}$$

with $e(v)$, $f(v)$ as in (7.8). This can again be written in an equivalent form of type (7.31), cf. (7.9).

If two rational patches meet with GC^r continuity, then by (7.23)ff, the same need not necessarily be true for their homogeneous representations (see also page 249, Sect. 5.7.2). This was noticed first in [Vin 89] for $r = 1$.

Of course, equations (7.23)ff include the affine case (see the remark at the end of Sect. 5.7.2), and for $\alpha_{00} = 1$ ($\alpha(s,t) = 1$), (7.23)ff formally become (7.2)ff.

To give concrete versions of the above equations, we need to choose particular patch representations and insert them in (7.23)ff. There are a number of partial results for the Bézier case. [Sar 87] (but see also [Sar 89]!) works with affine coordinates, *i.e.*, in $\mathbb{R}^3$. Starting with (7.6) and using the quotient rule and polynomials $b(v)$ and $c(v)$, he gives *sufficient conditions* for GC^1 continuity between two rational tensor-product patches. Based on [Sar 87], [Shet 91] discusses GC^1 and GC^2 continuous extrapolations of rational curves and surfaces.

[Liu 90] has generalized his earlier work [Liu 89] on polynomial triangular and rectangular patches to give general *necessary and sufficient* conditions for GC^1 continuity for the rational case. He works with homogeneous coordinates, but carries over the GC^1 condition (7.7b) with the help of the quotient rule and properties of the determinant, leading him first to (7.31b). Then from the equivalent form (7.31a) of (7.31b), he derives the desired GC^r continuity conditions (by inserting the homogeneous Bézier representations). They contain a very large number of free parameters, leading [Liu 90] to give a more practical special *sufficient* condition.

Similarly, [Vin 89] works with an equation of the form (7.31a), where in order to simplify things, the factors are chosen to be constant and linear, respectively. This again leads to a special *sufficient* condition.

[Dero 90] showed that the *necessary and sufficient* condition arising from the determinant form of the GC^1 continuity condition (cf. (7.14) for the polynomial case), are in general independent and minimal. Instead of starting with (7.31b), [Dero 90] uses a corresponding equation which is defined for rational curves of the kind $\boldsymbol{I}(t)$, $\boldsymbol{J}(t)$, $\boldsymbol{K}(t)$, and $\boldsymbol{L}(t)$ in (7.16).

Working with affine coordinates, [Jie 90] discussed a special *sufficient* condition for a pair of rational triangular Bézier patches. To get GC^1 continuity conditions for triangular and rectangular patches, he extends the conversion equations between polynomial triangular and rectangular patches (see [Brü 80], [Gol 87]) to the rational case.

The construction of Chiyokura and Kimura [Chi 83] (see also [Deg 90]) is also applicable to rational representations. For example, the Gregory patches used in [Chi 83] can be considered as special rational patches of degree (7,7), (see Sect. 7.5.3). [Chi 91] constructs general rational Gregory patches (the so-called *rational boundary (RB) Gregory patches*), and applies the method described above in (7.6) and (7.17)ff to connect them in a GC^1 continuous way to a surrounding rectangular patch network. Curves in the mesh are represented by rational curves, thus including conic sections. In the general case, [Chi 91] chooses the $b(v)$ in (7.6) to be linear, and the $c(v)$ to be cubic.

7.5. Multi-patch Surfaces

Multi-patch surfaces (also called *sculptured surfaces*) are made up of collections of patches. They are often constructed in two steps: 1) first, find a (as smooth and regular as possible) (rectangular) net of curves, see e.g., [Ree 83], [Sar 87, 90], [Reu 89], and [Renn 91]. Then 2) find individual Coons, Bézier, or other patches interpolating the curve network.

On the other hand, in CAD/CAM volume modelling systems, the first step usually involves finding a Constructive Solid Geometry (CSG) representation of the object (see e.g., [MOR 85]), followed by a smoothing of the edges and corners, after which the resulting curve network is filled in with patches, see e.g., [Chi 86, 87]. For example, in [Bee 86] it is shown how to construct a GC^1 continuous patch network from a given polyhedral net using an appropriate smoothing operator, see also [Brun 91]. Repeated "corner cutting" can also be applied to a given polyhedral net, and under appropriate conditions, is known to converge to a smooth (*i.e.*, C^r continuous) surface. The result is, however, not a surface in parametric form, and deviations in the polyhedron from a regular structure can lead to reduced smoothness of the resulting surface, e.g., a drop from C^2 to C^0 continuity. Examples of these kinds of algorithms include the de Casteljau and degree raising algorithms. Further examples are given in [Doo 78], [Cat 78], [Brun 88]; see also [Mic 87], [Cav 89]. Corner cutting methods do not produce surfaces interpolating the original points. For methods which do, see [Dyn 87b, 90a, 92a], [Nas 87, 91, 91a].

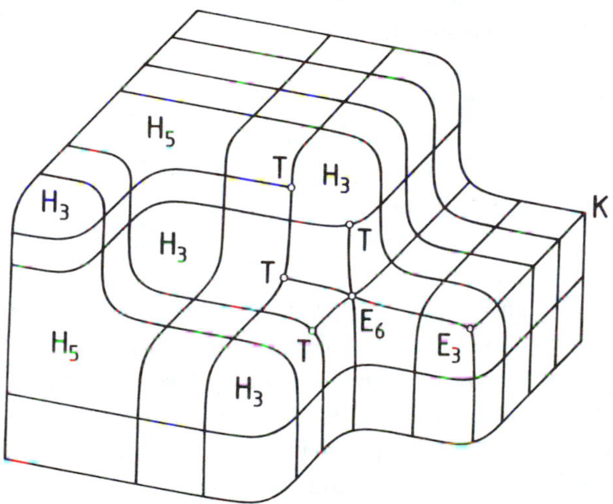

Fig. 7.10. Multi-patch surface.

For any of the approaches to building a polyhedral net, it can happen (see e.g., Fig. 7.10) that there are

- N-sided holes H_N in the net,
- N-sided vertex configurations E_N,
- T-node surface configurations T.

In addition, frequently the design requires sharp edges K in the interior of the patch network; they can be obtained for example if we require just C^0 continuity along a common boundary curve.

In the following we discuss the first two problems in detail. *T-nodes* T are more difficult to treat, and the approach depends essentially on what derivative information is given at T. Some methods are proposed in [Chi 86], [Renn 91], and [Herm 92].

7.5.1. N-sided Holes

A typical example of an N-sided hole in the interior of a rectangular curve network is provided by the *suitcase corner* configuration shown in Fig. 7.2a, where we have a three-sided "hole" in the rectangular net. Another example is provided by the *house corner* configuration shown in Fig. 7.2b, which involves a five-sided hole.

The problem of N-sided holes can be solved using an N-sided patch which meets the rectangular patches surrounding the hole with appropriate continuity. In Sects. 6.3 and 6.4 we have already discussed some methods for this problem, and in particular, in Sect. 6.3 triangular patches were treated (see also [Gre 86a], [Sto 89], and [Var 87] for surveys). However, many CAD/CAM systems do not make use of triangular patches, but rather work only with rectangular patches in tensor-product form.

We now present a method for filling an N-sided hole in a rectangular net using N rectangular patches due to Bézier [BEZ 72], [Bez 74] (see also [BEZ 86] and [Herm 89]). This requires a partitioning of the N patches which come into contact with the N sides of the hole; cf., Fig. 6.39 and Fig. 7.10. It is also possible to make the fill without this partitioning (see e.g., [Hah 89]), but then for each edge of the hole, there are two rectangular patches in the interior of the fill and only one rectangular patch in the exterior making contact there, *i.e.*, we get T-node patch configuration. Our aim is to construct a fill such that

- exterior boundary curves join together smoothly,
- every pair of patches in the fill join together smoothly along their common boundary,
- all N patches join together smoothly at the corner point $\boldsymbol{P}$,
- the patches join together smoothly along the boundary curves of the hole with the rectangular patches of the original surface.

While the first of the above requirements can be easily satisfied by construction, it is clear from the geometry that there can be no C^r continuous surface which satisfies both the second and the third requirements; to satisfy these, we have to work with geometric continuity.

For $N = 3$, a GC^1 continuous solution was given by [Hosa 78]; see also [Kah 82], and [Las 87] for the corresponding trivariate case. For arbitrary odd

N, the GC^2 continuous case was treated in [Jon 88]. Here we shall only sketch an algorithm for constructing a GC^r continuous fill for a triangular hole (see Fig. 7.11a), based on a general approach of [Hah 89] using rectangular patches for general N.

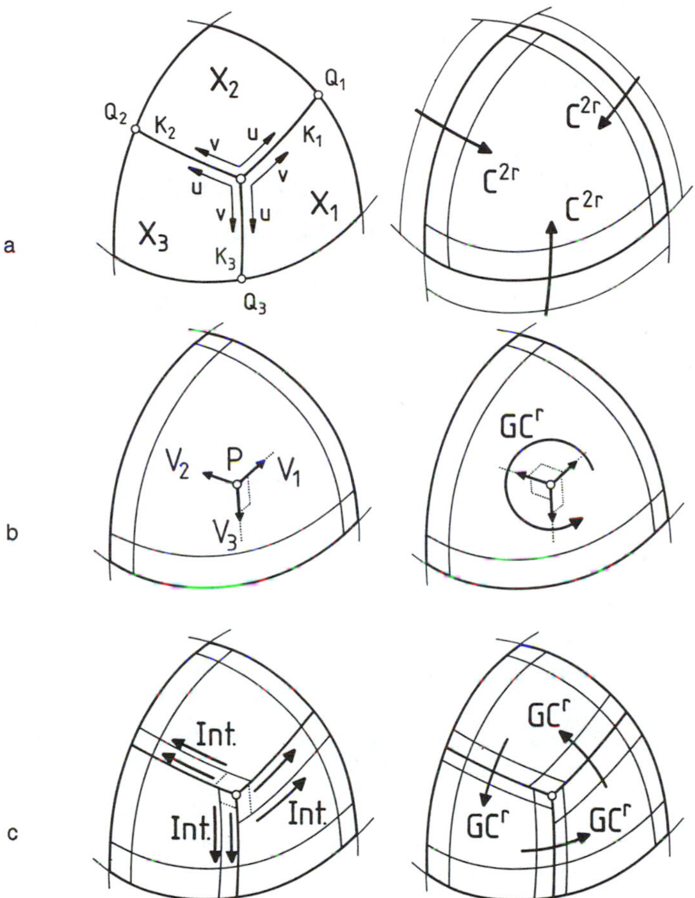

Fig. 7.11. GC^r continuous fill of a triangular hole by three rectangular patches.

Starting with the N rectangular patches surrounding the hole, the method proceeds stepwise as follows:

- for each of the surrounding N rectangular patches, compute its cross boundary derivatives up to order $2r$ along the edge, see Fig. 7.11b,
- choose a partition point $\boldsymbol{P}$ and a tangent plane $\boldsymbol{t}$ at $\boldsymbol{P}$. Also choose N vectors $\boldsymbol{V}_i$ which lie in $\boldsymbol{t}$, and which are to be the tangents at $\boldsymbol{P}$ to each

of the N curves meeting at $\boldsymbol{P}$. Finally, for one of the patches, choose all derivatives up to order $2r$ at $\boldsymbol{P}$, see Fig. 7.11c,

- compute derivatives for the other patches meeting at $\boldsymbol{P}$ in such a way that they all join with GC^r continuity at $\boldsymbol{P}$, see Fig. 7.11d,
- determine the boundary curves $\boldsymbol{K}_i = \boldsymbol{X}_i(0, v_i)$, and for each $j = 0(1)r$, find the cross boundary derivatives for the patches $\boldsymbol{X}_i(u_i, v_i)$ along the edges $\boldsymbol{K}_i$ as curves of degree $3r + 1 - j$ which interpolate the derivatives at $\boldsymbol{P}$ and $\boldsymbol{Q}_i$, see Fig. 7.11e,
- compute cross boundary derivatives of the patches $\boldsymbol{X}_i(u_i, v_i)$ along the boundary curves $\boldsymbol{X}_i(u_i, 0)$ to guarantee GC^r continuity, Fig. 7.11f,
- construct the patches $\boldsymbol{X}_i(u_i, v_i)$ from the derivative data along patch boundaries, e.g., as Coons patches (see Sect. 8.1), Bézier patches, or Gregory patches (see Sect. 7.5.3), where all undetermined interior Bézier points can now be considered to be free parameters to be used either for design or optimization purposes.

A generalization of the method discussed in Sect. 3.2 to the problem of constructing a GC^r continuous fill of an N-sided hole using functional splines was suggested by [Li 90], see Sect. 14.1.

7.5.2. N-vertex Problem

For the N-sided hole problem of the previous section, we must match given boundary curves and derivatives, but the partition point $\boldsymbol{P}$, the tangent plane $\boldsymbol{t}$ at $\boldsymbol{P}$, and the boundary curves $\boldsymbol{K}_i$ meeting at $\boldsymbol{P}$ are all free to choose. This assures the existence of a solution. For the N-vertex problem, however, $\boldsymbol{P}$ and the curves $\boldsymbol{K}_i$ meeting at $\boldsymbol{P}$ are all prescribed in advance. Consequently, this problem is much more difficult to handle.

We shall restrict our discussion to the use of N rectangular patches which are to join with GC^1 continuity. Even for this special case, it can happen that the "*vertex enclosure problem*" cannot be solved. According to [Pet 91], the existence of a vertex enclosure problem does not depend on the type of patches being used, but is in principle present for any combination of three, four or even N-sided patches of either polynomial or rational type, unless the term in the denominator vanishes at all of the data points (as for example happens for the Gregory patches, see Sect. 7.5.3). Partial solutions can be found in [BEZ 72, 86], [Hosa 78], [Sar 87] (note [Sar 89]!), [Sar 90], and [Bee 86] (see also [Höll 86]). As observed by [Wij 86] (see also [Watk 88a], [Du 88]), we have to distinguish between two cases: N even and N odd. The reason involves the twist vectors, which cannot always be defined to guarantee GC^1 continuity conditions at $\boldsymbol{P}$. Indeed, if all tangent vectors at $\boldsymbol{P}$ of the curves

$\boldsymbol{K}_i$ meeting at $\boldsymbol{P}$ happen to lie in the tangent plane at $\boldsymbol{P}$, then for odd N it is possible to choose appropriate twist vectors, but not always in the even case (*parity phenomenon*). We now discuss this situation in more detail, cf., [Watk 88a], [Sar 90]. Consider the N-vertex shown in Fig. 7.12, and suppose we choose the GC^1 continuity conditions between the patches $\boldsymbol{X}_i(u_i, v_i)$ and $\boldsymbol{X}_{i+1}(u_{i+1}, v_{i+1})$ to be

$$\frac{\partial \boldsymbol{X}_{i+1}}{\partial v_{i+1}}\bigg|_{v_{i+1}=0} = \beta_i(v_i)\frac{\partial \boldsymbol{X}_i}{\partial u_i} + \gamma_i(v_i)\frac{\partial \boldsymbol{X}_i}{\partial v_i}\bigg|_{u_i=0},$$

with $\beta_i(v_i) < 0$, by the regularity.

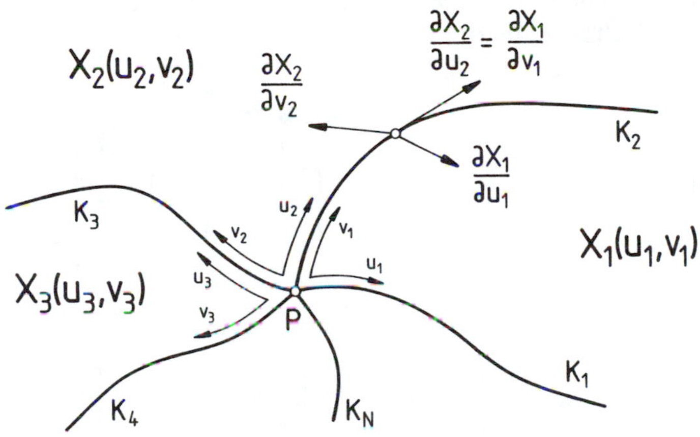

Fig. 7.12. On GC^1 continuity for the N-vertex problem.

Differentiating in the direction of the common boundary curve between $\boldsymbol{X}_i$ and $\boldsymbol{X}_{i+1}$, we get

$$\frac{\partial}{\partial u_{i+1}}\left(\frac{\partial \boldsymbol{X}_{i+1}}{\partial v_{i+1}}\right)\bigg|_{v_{i+1}=0} = \frac{\partial}{\partial v_i}\left(\beta_i(v_i)\frac{\partial \boldsymbol{X}_i}{\partial u_i} + \gamma_i(v_i)\frac{\partial \boldsymbol{X}_i}{\partial v_i}\right)\bigg|_{u_i=0},$$

and it follows that

$$\frac{\partial^2 \boldsymbol{X}_{i+1}}{\partial u_{i+1}\partial v_{i+1}} = \frac{\partial \beta_i(v_i)}{\partial v_i}\frac{\partial \boldsymbol{X}_i}{\partial u_i} + \beta_i(v_i)\frac{\partial^2 \boldsymbol{X}_i}{\partial u_i \partial v_i} + \frac{\partial \gamma_i(v_i)}{\partial v_i}\frac{\partial \boldsymbol{X}_i}{\partial v_i} + \gamma_i(v_i)\frac{\partial^2 \boldsymbol{X}_i}{\partial v_i^2}.$$

Here we have assumed that the $\boldsymbol{X}_i(u_i, v_i)$ are given as C^2 continuous patches, for example as Bézier patches. Now writing $\boldsymbol{T}_i$ for the twist vectors of

$\boldsymbol{X}_i(u_i, v_i)$ at $\boldsymbol{X}_i(0,0)$, and introducing the shorthand notation β_i and γ_i for the values of $\beta_i(v_i)$ and $\gamma_i(v_i)$ at $v_i = 0$, and β_i' and γ_i' for their derivatives at $v_i = 0$, the above equation becomes

$$-\beta_i \boldsymbol{T}_i + \boldsymbol{T}_{i+1} = \boldsymbol{R}_i, \tag{7.32a}$$

with

$$\boldsymbol{R}_i = \beta_i' \frac{\partial \boldsymbol{X}_i}{\partial u_i} + \gamma_i' \frac{\partial \boldsymbol{X}_i}{\partial v_i} + \gamma_i \frac{\partial^2 \boldsymbol{X}_i}{\partial v_i^2}\Big|_{(u_i,v_i)=(0,0)}.$$

This is a set of N vector-valued equations for the N unknown twist vectors $\boldsymbol{T}_i$. Writing (7.32a) in matrix form

$$\begin{pmatrix} -\beta_1 & 1 & & & \\ & -\beta_2 & 1 & & \\ & & \ddots & \ddots & \\ & & & \cdot & 1 \\ 1 & & & & -\beta_N \end{pmatrix} \cdot \begin{pmatrix} \boldsymbol{T}_1 \\ \boldsymbol{T}_2 \\ \cdot \\ \cdot \\ \cdot \\ \boldsymbol{T}_N \end{pmatrix} = \begin{pmatrix} \boldsymbol{R}_1 \\ \boldsymbol{R}_2 \\ \cdot \\ \cdot \\ \cdot \\ \boldsymbol{R}_N \end{pmatrix}, \tag{7.32b}$$

we see that

$$\det \mathbf{B} = (-1)^N [\prod_{i=1}^{N} \beta_i - 1],$$

where $\mathbf{B}$ is the coefficient matrix in (7.32b). Since $\beta_i < 0$, it follows that for N odd, $\det \mathbf{B} \neq 0$, and there is always a unique solution. On the other hand, for N even, the determinant may vanish (parity phenomenon). In this case there is a solution of the vertex enclosure problem only under additional conditions on the $\boldsymbol{R}_i$ (cf. (7.39)).

To make the situation more concrete, we now assume that the patches are given in Bézier form and are each of degree (m, m); the case $m = 3$ is also treated in [Lia 88], [Lee 91]. We rename the Bézier points which enter into the GC^1 continuity conditions as follows: $\boldsymbol{P}$ denotes the common vertex of the patches, $\boldsymbol{C}_i$ and $\boldsymbol{D}_i$ denote the Bézier points on the boundary curves $\boldsymbol{K}_i$ which determine the first and second derivatives along $\boldsymbol{K}_i$ at $\boldsymbol{P}$, and $\boldsymbol{T}_i$ denotes the Bézier point of $\boldsymbol{X}_i$ which determines its twist vector at $\boldsymbol{P}$, see Fig. 7.13.

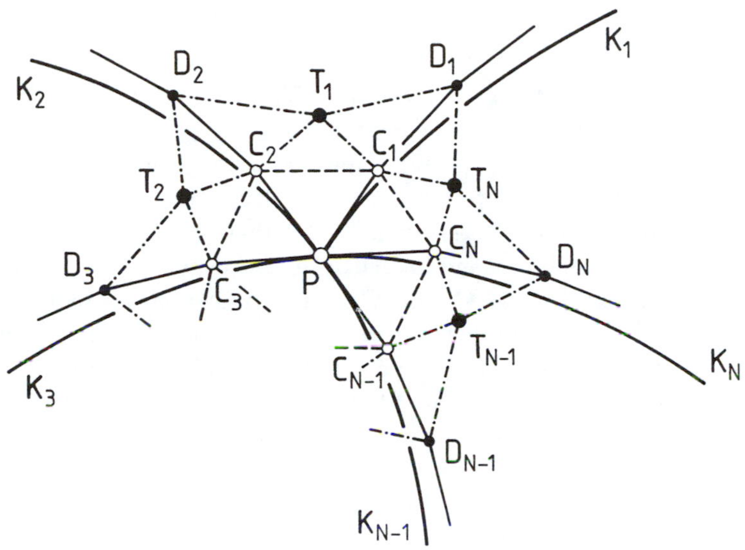

Fig. 7.13. N Bézier patches meeting at a point $\boldsymbol{P}$.

In view of the geometry of the problem, the first two equations (corresponding to $j = 0, 1$) of the GC^1 conditions (7.12) are coupled. Indeed, the Bézier points $\boldsymbol{C}_i$, $i = 1(1)N$, on $\boldsymbol{K}_i$ are involved (for $j = 0$) in the GC^1 conditions along $\boldsymbol{K}_{i-1}$, $\boldsymbol{K}_i$, and $\boldsymbol{K}_{i+1}$. Moreover, the Bézier points $\boldsymbol{T}_i$, $i = 1(1)N$, of $\boldsymbol{X}_i$ are involved (for $j = 1$) in the GC^1 conditions along $\boldsymbol{K}_i$ and $\boldsymbol{K}_{i+1}$, *i.e.*, using our new notation, for $j = 0$, (7.12) implies the system of equations [Du 88]

$$(\boldsymbol{C}_2 - \boldsymbol{P}) + \beta_1(\boldsymbol{C_N} - \boldsymbol{P}) - \gamma_1^0(\boldsymbol{C}_1 - \boldsymbol{P}) = 0, \tag{7.33.1}$$

$$(\boldsymbol{C}_{i+1} - \boldsymbol{P}) + \beta_i(\boldsymbol{C}_{i-1} - \boldsymbol{P}) - \gamma_i^0(\boldsymbol{C}_i - \boldsymbol{P}) = 0, \tag{7.33.i}$$

$$(\boldsymbol{C}_1 - \boldsymbol{P}) + \beta_N(\boldsymbol{C}_{N-1} - \boldsymbol{P}) - \gamma_N^0(\boldsymbol{C}_N - \boldsymbol{P}) = 0. \tag{7.33.N}$$

For $j = 1$, we get the system of equations

$$(\boldsymbol{T}_1 - \boldsymbol{C}_1) + \beta_1(\boldsymbol{T}_N - \boldsymbol{C}_1) - \gamma_1^0\frac{m-1}{m}(\boldsymbol{D}_1 - \boldsymbol{C}_1) + \gamma_1^1\frac{1}{m}(\boldsymbol{P} - \boldsymbol{C}_1) = 0, \tag{7.34.1}$$

$$(\boldsymbol{T}_i - \boldsymbol{C}_i) + \beta_i(\boldsymbol{T}_{i-1} - \boldsymbol{C}_i) - \gamma_i^0\frac{m-1}{m}(\boldsymbol{D}_i - \boldsymbol{C}_i) + \gamma_i^1\frac{1}{m}(\boldsymbol{P} - \boldsymbol{C}_i) = 0, \tag{7.34.i}$$

$$(\boldsymbol{T}_N - \boldsymbol{C}_N) + \beta_N(\boldsymbol{T}_{N-1} - \boldsymbol{C}_N) - \gamma_N^0 \frac{m-1}{m}(\boldsymbol{D}_N - \boldsymbol{C}_N) + \gamma_N^1 \frac{1}{m}(\boldsymbol{P} - \boldsymbol{C}_N) = 0. \quad (7.34.N)$$

This means that if the boundary curves $\boldsymbol{K}_i$ are given such that the Bézier points $\boldsymbol{P}, \boldsymbol{C}_i, \boldsymbol{D}_i$, and $\boldsymbol{T}_i$ are not compatible with (7.33) and (7.34), then no GC^1 continuous patch complex surrounding $\boldsymbol{P}$ can be constructed.

First, $\boldsymbol{P}$ is uniquely determined by the choice of the boundary curves $\boldsymbol{K}_i$. It then follows immediately from (7.33) that the N control points $\boldsymbol{C}_i$ and $\boldsymbol{P}$ must all be coplanar, and in fact lie in the tangent plane at the vertex $\boldsymbol{P}$. If two successive points $\boldsymbol{C}_i$ and $\boldsymbol{C}_{i+1}$ are fixed, then all remaining $\boldsymbol{C}_j$, $j \neq i, i+1$, can be determined successively from the system of equations (7.33). The design parameters β_i and γ_i^0 can certainly no longer be selected arbitrarily and independently. Indeed, if we go around the vertex $\boldsymbol{P}$, we have to get back to where we started. If the construction starts e.g., with $\boldsymbol{C}_1$ and $\boldsymbol{C}_2$, then crossing the boundary curves $\boldsymbol{K}_2$, $\boldsymbol{K}_3$, ..., $\boldsymbol{K}_{N-1}$, it follows from (7.33) that the vectors $\boldsymbol{C}_3 - \boldsymbol{P}$, $\boldsymbol{C}_4 - \boldsymbol{P}$, and so forth must all be linear combinations of the vectors $\boldsymbol{C}_1 - \boldsymbol{P}$ and $\boldsymbol{C}_2 - \boldsymbol{P}$. Now when we cross over the last two boundaries $\boldsymbol{K}_1$ and $\boldsymbol{K}_N$, the GC^1 constraints (7.33.N) and (7.33.1) define new values for the two vectors $\boldsymbol{C}_1 - \boldsymbol{P}$ and $\boldsymbol{C}_2 - \boldsymbol{P}$. We conclude that they must also be linear combinations of the original $\boldsymbol{C}_1 - \boldsymbol{P}$ and $\boldsymbol{C}_2 - \boldsymbol{P}$, leading to the following compatibility conditions [Du 88]:

$$(\boldsymbol{C}_1 - \boldsymbol{P}) = F_1(\beta_i, \gamma_i^0)(\boldsymbol{C}_1 - \boldsymbol{P}) + G_1(\beta_i, \gamma_i^0)(\boldsymbol{C}_2 - \boldsymbol{P}), \quad (7.35a)$$

and

$$(\boldsymbol{C}_2 - \boldsymbol{P}) = F_2(\beta_i, \gamma_i^0)(\boldsymbol{C}_1 - \boldsymbol{P}) + G_2(\beta_i, \gamma_i^0)(\boldsymbol{C}_2 - \boldsymbol{P}). \quad (7.35b)$$

In the general case where $\boldsymbol{P}$ and the $\boldsymbol{C}_1$ and $\boldsymbol{C}_2$ are not collinear, the factors F_j and G_j in (7.35) are linear polynomials in β_i and γ_i^0, and must satisfy the constraints

$$F_1(\beta_i, \gamma_i^0) = 1, \quad F_2(\beta_i, \gamma_i^0) = 0, \quad G_1(\beta_i, \gamma_i^0) = 0, \quad G_2(\beta_i, \gamma_i^0) = 1. \quad (7.36)$$

From (7.36) we get a relationship between the β_i which is of importance for the determination of the $\boldsymbol{T}_i$. If $\boldsymbol{N}$ denotes the normal vector to the tangent plane at $\boldsymbol{P}$, taking the vector product of (7.33.i) with $\boldsymbol{C}_i - \boldsymbol{P}$, then the scalar product with $\boldsymbol{N}$, and finally eliminating β_i leads to the equation

$$\beta_i = \frac{[(\boldsymbol{C}_{i+1} - \boldsymbol{P}) \times (\boldsymbol{C}_i - \boldsymbol{P})] \cdot \boldsymbol{N}}{[(\boldsymbol{C}_i - \boldsymbol{P}) \times (\boldsymbol{C}_{i-1} - \boldsymbol{P})] \cdot \boldsymbol{N}},$$

and it follows that the product of the β_i is given by

$$\prod_{i=1}^{N} \beta_i = \prod_{i=1}^{N} \frac{[(\boldsymbol{C}_{i+1} - \boldsymbol{P}) \times (\boldsymbol{C}_i - \boldsymbol{P})] \cdot \boldsymbol{N}}{[(\boldsymbol{C}_i - \boldsymbol{P}) \times (\boldsymbol{C}_{i-1} - \boldsymbol{P})] \cdot \boldsymbol{N}} = 1. \tag{7.37}$$

Assuming that the $\boldsymbol{C}_i$, β_i, and γ_i^0 have been chosen, we can now select all $\boldsymbol{D}_i$ and then the $\boldsymbol{T}_i$ in order to make sure conditions (7.34) are satisfied.

If we write (7.34) in the matrix form

$$\begin{pmatrix} 1 & 0 & & \cdots & 0 & \beta_1 \\ \beta_2 & 1 & & & 0 & 0 \\ 0 & \beta_3 & 1 & & 0 & 0 \\ \vdots & & & & & \vdots \\ 0 & 0 & & \beta_{N-1} & 1 & 0 \\ 0 & 0 & & \cdots & \beta_N & 1 \end{pmatrix} \begin{pmatrix} \boldsymbol{T}_1 \\ \boldsymbol{T}_2 \\ \boldsymbol{T}_3 \\ \vdots \\ \boldsymbol{T}_{N-1} \\ \boldsymbol{T}_N \end{pmatrix} = \begin{pmatrix} \boldsymbol{R}_1 \\ \boldsymbol{R}_2 \\ \boldsymbol{R}_3 \\ \vdots \\ \boldsymbol{R}_{N-1} \\ \boldsymbol{R}_N \end{pmatrix}, \tag{7.38}$$

with

$$\boldsymbol{R}_i = (1 + \beta_i)\boldsymbol{C}_i - \gamma_i^0 \frac{m-1}{m}(\boldsymbol{C}_i - \boldsymbol{D}_i) + \gamma_i^1 \frac{1}{m}(\boldsymbol{C}_i - \boldsymbol{P}),$$

then we see that for odd N, the system is always uniquely solvable since by (7.37), the determinant associated with the system (7.38) is equal to 2. For even N, however, this determinant vanishes in view of (7.37), and the system (7.38) is singular. In this case, (7.38) can only have a solution when the additional *twist compability conditions* [Du 88]

$$\sum_{i=1}^{N} (-1)^{i-1} \left[\prod_{k=1}^{i} \beta_k \right]^{-1} \boldsymbol{R}_i = 0 \tag{7.39}$$

are satisfied by the $\boldsymbol{R}_i$. This condition corresponds to the linear dependence of the rows of the coefficient matrix, see e.g., [Wij 86], [Lia 88], and [Sar 90]. In [Pet 91] this vector-valued condition is replaced by a scalar condition, and it is shown that a solution exists if all curves $\boldsymbol{K}_i$ meeting at $\boldsymbol{P}$ interpolate second derivative data, *i.e.*, if they are compatible with a prescribed second fundamental form at $\boldsymbol{P}$.

The general case where $\alpha_i \neq 1$ and the Bézier surfaces $\boldsymbol{X}_i(u_i, v_i)$ are of general degree (m_i, m_{i+1}) can be handled in a completely analogous way (see [Du 90]), and in fact, the considerations in [Pet 91] are even entirely independent of the particular Bézier form.

In order to be sure that the influence of the GC^1 conditions remains local, we should use tensor-product surfaces which are at least biquintic. For biquartic patches, the GC^1 conditions at $\boldsymbol{P}$ can be coupled with the GC^1

conditions at neighboring vertices. For bicubic patches, such coupling usually happens, making the problem very difficult to handle.

The only case where for bicubic patches compability conditions (7.39) can be satisfied, and the computation remains local, is the special case of a four-patch corner of a rectangular grid. In this case, all γ_i^0 and all γ_i^1 can be taken to be zero, cf. [Du 88, 90] and Sects. 6.2.2 and 6.2.3. However, if a four-patch corner (consisting of four curves meeting at $\boldsymbol{P}$ with pairwise collinear tangents) is imbedded in a general grid, and $\gamma_i^0 = 0$ at $\boldsymbol{P}$ for all i, then the factors $\beta(v)$ have to be chosen of higher polynomial degree in order to assure twist compability and locality. See [BEZ 72, 86], [Sar 87, 90], [Du 90], and also [Pet 91].

For N-vertices with $N \neq 4$, we cannot set all γ_i^0 to be zero since this can lead to undesirable behavior of patches associated with neighboring four-vertices, and in fact even to a singularity [Bee 86], [Du 90] where the Bézier net becomes degenerate (several Bézier points collapse to one). Then derivatives at the point $\boldsymbol{P}$ can become zero or collinear, causing difficulties with the definition of the tangent plane at this point. Thus, GC^1 continuity at such points can only be guaranteed under additional conditions, see [Du 90]. As a consequence, there is a coupling with nearby vertices, and the local nature of the scheme is lost at this point.

To insure the construction of tensor-product multi-patch surfaces using bicubic patches remains local, there are two possibilities:

1) subdivide certain surface patches,
2) construct surface patches of higher degree by degree raising.

For both strategies it is recommended to first consider subdividing or degree raising only the boundary curve at hand, and then later to extend the process to the rest of the surface if necessary, see [Sar 90], [Du 90]. Otherwise, method (1) always leads to a division into four subpatches, while method (2) always leads to a surface of degree (6,6), see Sect. 7.1.1. Thus with respect to the overall complexity, the second method appears to be superior. The first method only uses bicubic surfaces, which simplifies the data structure and results in surfaces which are easier to evaluate, but the subdivision process leads to more surface patches, and keeping track of the various cases which arise at each step of the subdivision is difficult and expensive, see [Du 90].

7.5.3. Gregory Patches

Gregory patches provide another way of solving both the localness and twist-vector problems for a multi-patch surface, and in fact in a very simple way.

The twist-vector problem came up already in connection with C^1 continuous Coons patches generated by cubic blending functions, see Sect. 8.1.2.

A solution was provided by Gregory [Gre 74] in the form of so-called *Gregory squares* which are a modification of the Coons patch. The four constant twist vectors at the corners of the patch are replaced by four variable twist vectors which are expressed as convex (*i.e.*, barycentric) combinations of the two mixed second derivatives, see Sect. 8.1.2. This allows the first order cross boundary derivatives to two neighboring boundaries to be prescribed independently.

The *Gregory patch* was introduced by Chiyokura and Kimura [Chi 83, 84, 86] as a corresponding modification of a bicubic Bézier patch. Thus, we get a Gregory patch from a bicubic Bézier patch by replacing the four interior constant control points by four variable control points, each of which is a convex combination of two given points. The patch is then defined by a total of 20 control points via (7.40) (cf. the equations for the commutative twist vectors for Gregory squares in Sect. 8.1.2 and see Fig. 7.14):

$$\boldsymbol{X}(u,v) = \sum_{i=0}^{3}\sum_{j=0}^{3} \boldsymbol{P}_{ij} B_i^3(u) B_j^3(v) = (1-u-uE)^3(1-v-vF)^3 \boldsymbol{P}_{00}, \quad (7.40)$$

$u, v \in [0,1]$, where the control points on the boundary are constant, while the four interior control points satisfy

$$\boldsymbol{P}_{11}(u,v) = \frac{u\boldsymbol{P}_{110} + v\boldsymbol{P}_{111}}{u+v}, \qquad \boldsymbol{P}_{12}(u,v) = \frac{u\boldsymbol{P}_{120} + (1-v)\boldsymbol{P}_{121}}{u+1-v},$$

$$\boldsymbol{P}_{21}(u,v) = \frac{(1-u)\boldsymbol{P}_{210} + v\boldsymbol{P}_{211}}{1-u+v}, \qquad \boldsymbol{P}_{22}(u,v) = \frac{(1-u)\boldsymbol{P}_{220} + (1-v)\boldsymbol{P}_{221}}{1-u+1-v},$$

where the *shift operators* E, F are defined by (cf. Sect.4.1)

$$E\boldsymbol{P}_{ij} = \boldsymbol{P}_{i+1,j}, \qquad F\boldsymbol{P}_{ij} = \boldsymbol{P}_{i,j+1}.$$

If the eight interior control points satisfy $\boldsymbol{P}_{ij0} = \boldsymbol{P}_{ij1}$, then (7.40) reduces to a bicubic Bézier patch. In general, a Gregory patch has a rational nature, which is a certain disadvantage. More precisely, a Gregory patch is a special kind of rational Bézier patch of degree (7,7), where the numerator polynomial is of degree (7,7), and the denominator is a biquartic polynomial [Taka 90]. The expression (7.40) avoids this complication very neatly: to compute a point $\boldsymbol{X}(u_0, v_0)$ on the surface, we first find the four interior control points, and the remaining computation reduces to the evaluation of a bicubic polynomial Bézier patch.

Triangular Gregory patches can be defined similarly, see e.g., [Lon 87], [Mann 92], and [Fol 92]. They can be interpreted as special rational triangular Bézier patches of degree seven with cubic polynomial denominator.

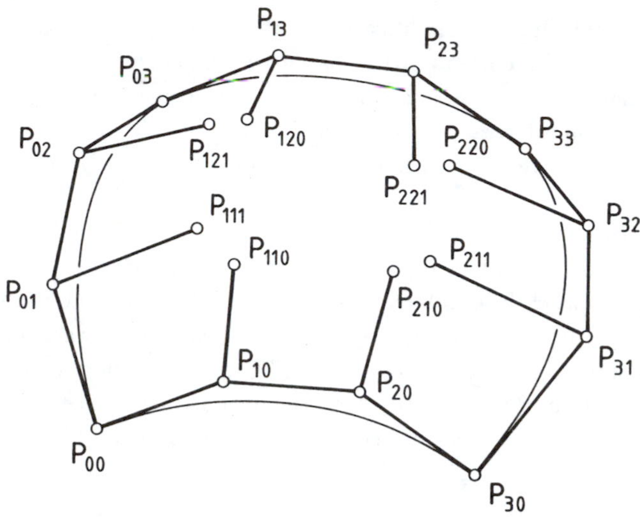

Fig. 7.14. Gregory patch.

In addition to the *convex hull property*, Gregory patches also have the property that the first derivatives transversal to the boundaries can be prescribed independently from one another. Thus, GC^1 continuous multi-patch surfaces can be easily constructed using linear factors $\beta(v)$ and $\gamma(v)$. The twist-vector problem does not come up, and neither does the problem of localness. Moreover, there is no difficulty in constructing patches which connect with GC^1 continuity to neighboring tensor-product Bézier patches, see Sect. 7.2.

The construction of a multi-patch surface using either bicubic Bézier patches or Gregory patches begins with a curve network. The algorithm proceeds as follows:

- fill all rectangular holes which are connected to neighboring patches at a 4-vertex using bicubic Bézier patches,
- divide N-sided holes by inserting a partition point and corresponding curves to get N-vertices,
- for all N-vertices either in the original net or constructed in the previous step, construct a GC^1 continuous surface using Gregory patches.

Gregory patches are used in solid modelling, in CAD/CAM systems, e.g., for blending (see Chap. 14) and [Chi 83, 84, 86, 87]. They are also used for free-form surface design, see [Tak 88] and [Renn 91]. Interpolation with Gregory patches generally gives good results regarding smoothness, form, and visual appearance, which explains their popularity. For example, an algorithm of [Shi 87] starts with cubic boundary curves, and raises them to degree four. Then a quartic triangular Gregory patch is used to fill three-sided holes. But, the Gregory patch does not really have to be constructed, because it is replaced by three quartic triangular Bézier patches. Internal GC^1 continuity of the micro patches is assured by (7.13), (7.15), and GC^1 continuity of the macro patches by (7.17) using linear factors. The modified algorithm of [Shi 91] constructs the Bézier patch so that it approximates the original Gregory patch as well as possible, and also maintains the smoothness properties. This requires that the normals coincide at four interior surface points. [Jen 87] proceeds in a similar way to [Shi 87], but gets C^1 continuity internally, where $k(t)$ in (7.17) is chosen to be quadratic.

[Shi 90] called the degrees of freedom involved in determining the derivatives transversal to the boundaries *tilt*, *bulge* and *shear*, and studied their suitability as form parameters for surface design.

A generalization of Gregory patches to the *rational case* was described in [Chi 91]. It is also possible (see [Tak 91]) to construct a C^2 continuous Gregory patch in a similar way to the modification of a biquintically blended C^2 continuous Coons patch used to get the C^2 *continuous Gregory square* [Barn 83] (see also Sect. 8.1.2). It is defined analogous to (7.40), starting with a biquintic Bézier patch. If for all 32 interior control points, $\boldsymbol{P}_{ij0} = \boldsymbol{P}_{ij1}$, then the C^2 continuous Gregory patch reduces to a biquintic Bézier patch; otherwise it is equivalent to a rational Bézier patch of degree 13 (in both parameters). The first and second derivatives transversal to the four boundaries can be chosen independently. This is useful for GC^2 continuous interpolation of a curve network [Tak 91], see also Sect. 7.3. This patch also has the convex hull property.

7.6. Multi-patch Schemes - Overview and Comparison

Available multi-patch schemes can be dividied into three groups:

- single patch schemes,
- blending patch (convex combination) schemes,
- splitting schemes.

Some of these schemes use rectangular patches, some use triangular patches, and some even use both. Only a few allow the integration of N-sided patches.

In this section we discuss and compare the methods treated in the previous sections (and the following). It is convenient to present our analysis in the form of three tables, see [Pet 90], [Mann 92]. Each of these tables is organized as follows:

- The *data* column describes the interpolation data using capital letters: $\boldsymbol{P}$ polyhedra, $\boldsymbol{N}$ normal, $\boldsymbol{T}$ tangent, $\boldsymbol{II}$ curvature, $\boldsymbol{M}$ curve net (mesh), $\boldsymbol{A}$ derivative, and $\boldsymbol{K}$ curvature transversal to the curves of the curve net. Small letters denote restrictions on the data: $\boldsymbol{k}$ compatible twist vectors required, $\boldsymbol{g}$ no closed surface possible, $\boldsymbol{e}$ for restrictions on the variables $\boldsymbol{N}$ or $\boldsymbol{II}$. The interpolation data satisfy the following chain of implications: $\boldsymbol{K} \Rightarrow \boldsymbol{A} \Rightarrow \boldsymbol{M} \Rightarrow \boldsymbol{T} \Rightarrow \boldsymbol{N}$.
- The *sides* column gives the number of sides a facet of the multi-patch surface may have. Here $\boldsymbol{N}$ means that patches with an *arbitrary* number of sides are allowed.
- The *degree* column contains the (maximally appearing) polynomial degree of the patches. For the rational case, the degrees of the numerator and denominator are separated with a "/ ". For blending methods, we give the degree of the data polynomials and the degree of the (possibly rational) partial interpolants separated by a "+".
- The *factor* column contains the polynomial degrees of the factor functions $\alpha(v)$, $\beta(v)$, and $\gamma(v)$ in (7.7a). For blending methods, we give the degree of the blending function in square brackets, and the degree of the rational parameters (in terms of the original coordinates) in round brackets.

7.6.1. Single Patch Methods

In these methods, exactly one patch is constructed for each grid facet. Table 7.1 (following [Pet 90]) summarizes some of the available methods of this type.

Data	Sides	Degree	Factors	Literature
Pe	4	(3,3)	0, 0, 1	[Bee 86]
M	4	(6,6)	0, 3, 2	[Sar 87, 89]
M	3, 4	4, (4,4)	1, 1, 3 resp. 0, 2, 1	[Pet 91]
IIe	4	(3,3)	$\boldsymbol{X}_v(\boldsymbol{N}_0 \times \boldsymbol{N}_1)$	[Sab 68]
Ne	3, 4	3, (3,3)	$(1-t)\boldsymbol{N}_0 + t\omega\boldsymbol{N}_1$	[Pet 90b]

Table 7.1. Single patch methods.

In [Sab 68] and [Pet 90b], equation (7.7a) is expressed by the condition $\boldsymbol{N} \cdot \boldsymbol{Y}_s = 0$, $\boldsymbol{N} \cdot \boldsymbol{X}_u = 0$, $\boldsymbol{N} \cdot \boldsymbol{X}_v = 0$ on the normal vectors $\boldsymbol{N} = \boldsymbol{N}(t)$ along the common boundary curve. In these papers, the definition of $\boldsymbol{N}(t)$ follows from the expressions given in the factor column of Table 7.1, where $\boldsymbol{N}_0$ and $\boldsymbol{N}_1$ are the prescribed normals at the ends of the boundary curve.

Other examples of single patch methods are given in [Hosa 84], [Lia 88], [Sto 88], [Loo 89, 90], [Li 90], [Hartm 90], and [Hos 91a].

7.6.2. Blending Patch Methods

These methods first construct N surface patches for each N-sided grid facet. Each of these patches interpolates a part of the boundary data, usually point and derivative data along an edge. Taking convex combinations leads to a single surface patch which interpolates all boundary data of an N-sided grid facet.

For C^{r-1} data, taking convex combinations leads to C^r discontinuities at the corners. This means that for prescribed boundary curves and tangent planes along the boundary curves, the mixed second derivatives at the corners need not coincide, *i.e.*, we do not have a twist problem. Table 7.2 (following [Pet 90]) compares several blending patch methods.

Data	Sides	Degree	Factors	Literature
Agk	4	(3,3)		[Coo 67]
Ag	3	2(3+3)	(1/1)	[Barn 73]
Ag	4	(4,4)/1		[Gre 74]
M	4	(4,4)/1	1, 1, 1	[Chi 83]
M	3	1+2	[4/2] 1, 1, 1	[Herr 85]
Ak	5	3+2	[6/6] (1/1)	[Char 84]
Kk	N	3+4/8	[3(n-2)/3(n-2)] (1/1)	[Gre 89a]
A	3	3+3	[2/2] (1/1)	[Nie 87c]
K	3	7+5	[4/4] (1/1)	[Hag 89]

Table 7.2. Blending patch methods.

Other examples of blending patch methods can be found in [Hag 86a], [Con 87], [Gre 86a, 87], [Lon 87], [Web 90, 90a], [Sait 90], [Chi 91], [Tak 91], and [Ham 91a].

7.6.3. Splitting Methods

These methods construct N non-overlapping triangular or rectangular surface patches for every N-sided grid facet. Thus, with each facet we have a macro patch, see Fig. 7.15. The subdivision allows the interpolation of boundary data along an edge of the macro patch by a subpatch which does not depend on the prescribed data along the other boundaries. The remaining degrees of freedom can be used to force internal C^1 or GC^1 continuity, *i.e.*, the construction of patches joining each other with C^1 or GC^1 continuity; compare the subdivision schemes for function-valued scattered data interpolation in Sect. 9.3.2 and in [Clo 65], [Pow 77]. The various algorithms are compared in Table 7.3 (following [Pet 90]).

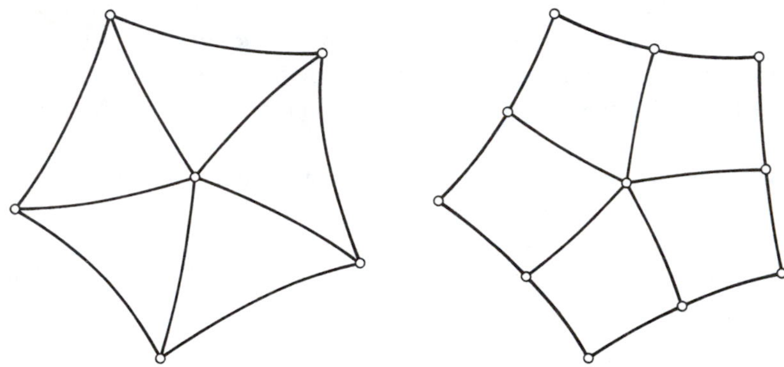

Fig. 7.15. Subdividing an N-sided hole (here $N = 5$) by triangular and rectangular surface patches.

Data	Sides	Degree	Factors	Literature
T	3	3	0, 0, 0	[Clo 65]
T	3	2	0, 0, 0	[Pow 77]
N	3	4	0, 0, 1	[Far 83a]
T	3	4	1, 1, 2	[Pip 87]
M	3, 4	4	1, 1, 1	[Shi 87, 91]
M	3	4		[Jen 87]
M	3, 4	3	1, 1, 2 and 1, 1, 1	[Pet 90a]
M	$N > 4$	4	2, 2, 3 and 1, 1, 1	[Pet 90a]
N	N	5	0, 0, 1	[Jon 88]

Table 7.3. Splitting methods.

Other examples of these kinds of methods can be found in [Bez 74], see also [BEZ 72, 86], [Hosa 78], [Kah 82]), [Hah 89], [Herm 89], and [Lee 91].

By their nature, splitting schemes generate more surface patches than the single patch schemes. But on the other hand, the degrees of the patches are usually lower, and fewer restrictions are required on the data. Blending patch methods lead to the same number of subpatches, but they are often of higher degree, and moreover, frequently rational. Generally, polynomial representations are preferable to rational ones, since point, derivative, and curvature calculations are significantly simpler for polynomial representations. The curvatures are discontinuous along the interior boundaries of macro patches, which doesn't happen for single patch and blending patch methods.

[Mann 92] presents a comparison of visual properties of splitting methods [Pip 87], [Jen 87], [Shi 87] and blending patch methods [Nie 87c], [Lon 87], [Herr 85]. In particular, he looks at images of the surfaces in terms of parametric lines, shading, and Gaussian curvature. All of the tested methods produced less than satisfactory results, and seemed to suffer from the same problems. (We should note that the surfaces constructed by the method of [Shi 91] exhibit significantly better visual properties than those in [Shi 87]). For example, most of these surfaces suffer from oscillation, *i.e.*, large variations in the Gaussian curvature, even in the interior of a single patch. Moreover, flat points often appear, and along the boundaries between patches there are even discontinuities in the sign of the Gaussian curvature.

Another problem is that curvatures tend to concentrate near the vertices and boundaries of the patches, while the interior remains relatively flat. This problem seems to originate in the form and parametrization of the boundary curves, since for all of the methods, a nonuniform distribution of curvature and the existence of flat points on the boundary curves carries over to the interior. An attempt to overcome this problem with a GC^1 method was presented in [Hans 91], where the boundary curves are chosen to be conic sections in rational Bézier form, and triangular patches are constructed using a Clough-Tocher partion and rational quartic triangular Bézier patches. This method has not yet been compared with the other methods. Methods for constructing curves which interpolate point, tangent, and curvature data with relatively uniform curvature (such as the one described by [Boo 87]) could lead to improved surface methods. The method in [Boo 87] is not so good for generating a curve network for a multi-patch surface, see [Mann 92]. Generally, a global curve optimization method (see Sect. 9.3.3, and Chap. 13) is more appropriate. In addition to the boundary curves, the free design parameters also have a major influence on the form of the surface. It is a very difficult problem to find a reasonable way to select these parameters.

7.7. B-Spline Representations

The (geometric) continuity conditions discussed above lead to a large number of restrictions which have to be imposed on the Bézier points of a surface, and which can be very difficult to keep track of. Geometric B-splines possess GC^r continuity as an intrinsic property, and so this property is automatically inherited by spline surfaces which are constructed as linear combinations of such B-splines.

For example, geometric spline surfaces in B-spline form can be expressed in terms of tensor-product geometric B-Splines $G_i^m(u)$ und $G_j^n(v)$ (see Chap. 5) defined over possibly different knot vectors and with different design parameters:

$$\boldsymbol{X}(u,v) = \sum_i \sum_j \boldsymbol{d}_{ij} G_i^m(u) G_j^n(v). \tag{7.41}$$

It follows from this definition that every parametric line corresponding to constant u inherits the properties of the $G_j^n(v)$, while every parametric line corresponding to constant v value inherits the properties of $G_i^m(u)$. These would include tangent, curvature, torsion, and GC^3 continuity. Moreover, it can be shown that for an appropriate choice of the basis functions, every planar slice of the surface also has these properties [Boeh 85a, 87], [Pot 88] (we recall that every planar slice is torsion-continuous since all planar curves are torsion free). It also follows that the design parameters are not associated with single control points, but rather with whole knot curves of the curve network, which in a certain sense is a disadvantage. Another serious disadvantage is the fact that (7.41) can only be formulated for rectangular curve networks. More general curve networks cannot be handled with surfaces defined as in (7.41).

In definition (7.41), $\boldsymbol{X}(u,v)$ can in principle be constructed using any of the curve representations discussed in Chap. 5. These would include β-spline surfaces [Bars 81], γ-spline surfaces [Boeh 85a] (see also [Boeh 87]), ν-spline surfaces [Nie 86], rational geometric spline surfaces [Boeh 87a], GC^3 spline surfaces [Pot 88], rational β-spline surfaces [Bars 88], geometric spline surfaces [Eck 89], $\beta\nu$-spline surfaces [Che 91], GC^1 und GC^2 continuous rational extrapolation surfaces using B-spline bases [Shet 91], and τ-spline surfaces [Neu 92].

The 'rectangular' ν-spline surfaces defined in [Nie 86] using Gordon-Coons methods (see Chap. 8) allow the specification of two tension parameters at each control point which affect only that control point.

8

Gordon-Coons Surfaces

Thus far, we have described surfaces in $\mathbb{R}^3$ using

- appropriately selected basis functions,
- tensor-product maps on rectangular parameter domains,
- special patches built on N-sided parameter domains.

In all of these methods, the vertices of the surface patches are given, while the boundary curves between patches are approximated.

In some applications, boundary curves between surface patches are prescribed, and it is useful to have methods which produce surfaces which interpolate these boundary curves. Throughout this section, we shall assume that the given boundary curves form nondegenerate triangles, quadrilaterals, or more generally N-gons. Sometimes we will assume we have additional information along the boundary curves, for example tangent planes (*i.e.*, the first derivatives along the boundary), or even higher derivatives.

The idea is to span the given curve complex using appropriate *blending functions*. This idea goes back to Coons [Coo 64, 67, 77], see also [Barn 76, 77, 82, 83]. Gordon [Gor 69, 71] provided a rigorous mathematical treatment using algebraic tools. Thus, we refer to these kinds of surfaces as *Gordon-Coons surfaces*. From the standpoint of classical numerical analysis, they can be considered as *Hermite interpolation surfaces*.

In the literature, Gordon-Coons surfaces are frequently referred to as *transfinite interpolants* since they interpolate a continuum of data, *i.e.*, all points along the prescribed boundary curves. In contrast, most other methods (and all of the methods discussed above) interpolate only to a set of *discrete data*.

The concept of a blending function also makes sense in $\mathbb{R}^2$. Suppose we are given function values $F(0)$ and $F(1)$. These values can be interpolated by the line

$$F(x) = (1 - x)F(0) + xF(1) \tag{8.1}$$

defined on $x \in [0,1]$. Here

$$f_0(x) := 1 - x, \qquad f_1(x) := x, \qquad x \in [0,1], \tag{8.2}$$

can be thought of as blending functions. In order to assure that the interpolant takes on the correct values at the ends of the interval $[0,1]$, these two blending functions must satisfy the conditions

$$f_i(k) = \delta_{ik} = \begin{cases} 1, & i = k \\ 0, & i \neq k. \end{cases} \tag{8.3}$$

In (8.1) we interpolated the prescribed function values using linear blending functions. We could instead have used cubic blending functions by replacing the polynomials in (8.2) by Hermite polynomials, see (3.12). As illustrated in Fig. 8.1, this leads, of course, to a different shape for the blended curve associated with the function values $F(0)$ and $F(1)$.

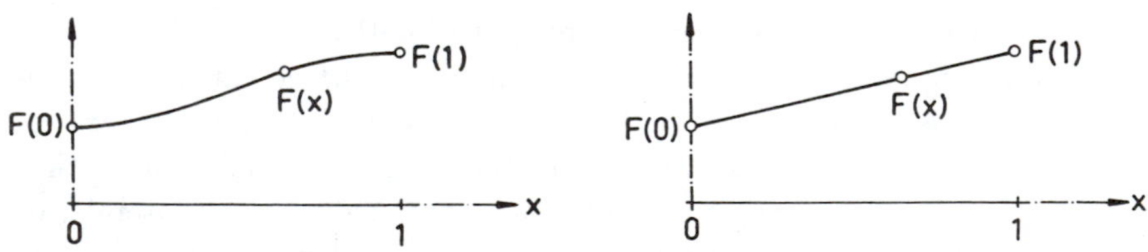

Fig. 8.1. Linear and cubic Hermite interpolation of $F(0)$, $F(1)$.

8.1. Gordon-Coons Surfaces on Rectangles

8.1.1. C^0 Continuous Patch Complexes

In this section we discuss how to construct a surface patch which interpolates given data along the boundary of a quadrilateral. Such patches can then be used to construct a C^0 complex of patches which interpolates data given on a rectangular network of space curves. To discuss the construction of such surface patches, we restrict our attention first to the relatively simple case where the boundary curves are straight lines defined by four corner values $F(i,k)$, $i,k = 0,1$, which we may think of as being associated with the four corners of the unit square. Thus, each of these lines lies in one of the planes $x = 0$, $x = 1$, $y = 0$, $y = 1$, see Fig. 8.2. As blending functions, we take the linear functions (8.2) defined on $[0,1] \times [0,1]$.

So far only the vertices of the interpolating surface patch are prescribed, and so we first construct the boundary curves by linear interpolation. Thus, for example, in the planes $y =$ const., we have (see Fig. 8.2)

$$P_1 = (1-x)F(0,0) + xF(1,0), \qquad \widetilde{P}_1 = (1-x)F(0,1) + xF(1,1). \quad (8.4)$$

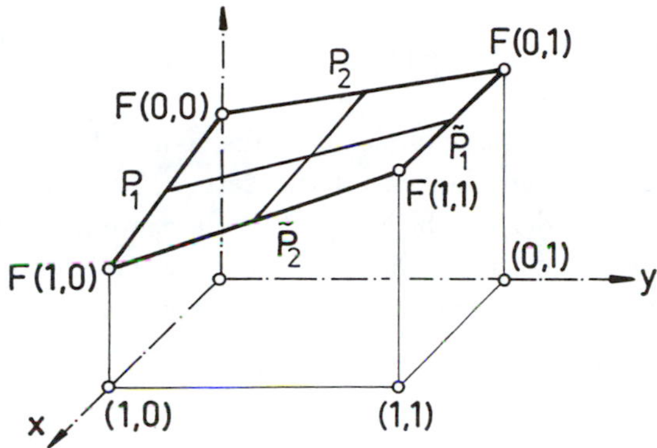

Fig. 8.2. Linear interpolation of a quadrilateral in space.

Next we linearly interpolate in the y direction. This leads to the surface

$$Q_1 = (1-y)\Big[(1-x)F(0,0)+xF(1,0)\Big] + y\Big[(1-x)F(0,1)+xF(1,1)\Big]. \quad (8.5)$$

We could, of course, also have interpolated first along the boundaries where x is constant, giving

$$P_2 = (1-y)F(0,0) + yF(0,1), \qquad \widetilde{P}_2 = (1-y)F(1,0) + yF(1,1), \quad (8.4a)$$

and then in the x direction, leading to

$$Q_2 = (1-x)\Big[(1-y)F(0,0)+yF(0,1)\Big] + x\Big[(1-y)F(1,0)+yF(1,1)\Big]. \quad (8.5a)$$

If this construction is to make sense, the final result should be independent of the order in which we interpolate, and indeed, it can be easily checked directly that (8.5) and (8.5a) agree, *i.e.*,

$$Q_1 = Q_2.$$

The problem becomes more difficult when the boundaries are no longer straight lines, but instead are curves $F(x,0)$, $F(x,1)$, $F(0,y)$, and $F(1,y)$, see

Fig. 8.3. If, analogous to (8.5), we linearly connect the boundaries $y = \text{const.}$, then we get the interpolation formula

$$P_1F(x,y) = (1-y)F(x,0) + yF(x,1), \tag{8.6a}$$

while linearly blending the boundaries corresponding to $x = \text{const.}$ leads to

$$P_2F(x,y) = (1-x)F(0,y) + xF(1,y). \tag{8.6b}$$

But now, in contrast to the simpler case discussed above, the two surfaces (8.6a) and (8.6b) are not in general the same. Since we want a method independent of the order in which we interpolate, we need to modify (8.6a) and (8.6b) in some way. To see how, we consider the error corresponding to (8.6a) on the boundaries $x = 0$ and $x = 1$, see Fig. 8.4.

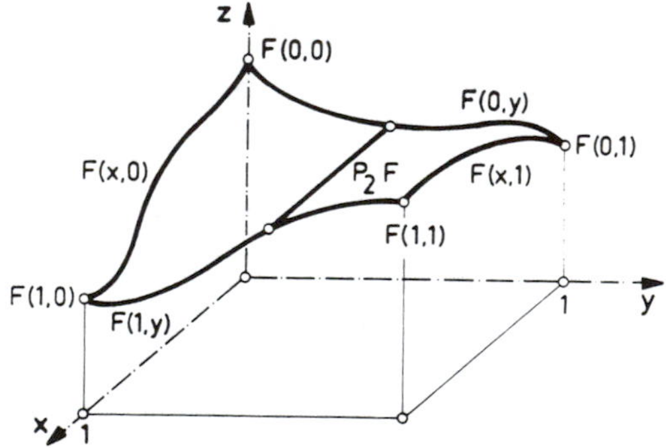

Fig. 8.3. Linear interpolation of a quadrilateral in space with curved boundaries.

In the plane $x = 0$, we have to correct P_1F by

$$F(0,y) - [(1-y)F(0,0) + yF(0,1)], \tag{8.7a}$$

(see Fig. 8.4), while in the plane $x = 1$, the needed correction is

$$F(1,y) - [(1-y)F(1,0) + yF(1,1)]. \tag{8.7b}$$

These correction terms can be written symbolically in the form

$$F - P_1F. \tag{8.8a}$$

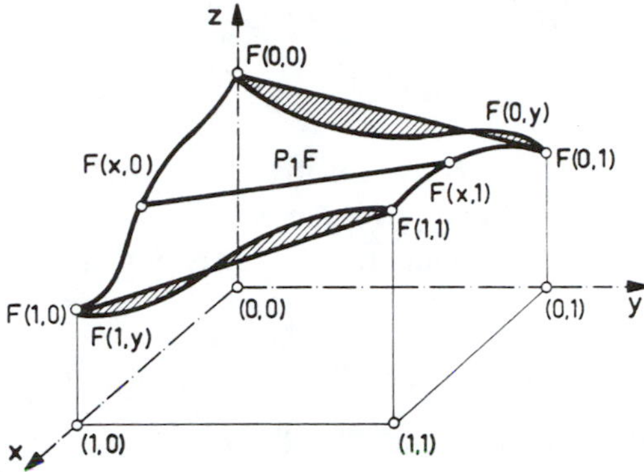

Fig. 8.4. Construction of correction terms for linear interpolation.

We want this correction to hold not only on the boundaries, but along all of the straight lines determined by P_2F. We can write this symbolically as

$$P_2(F - P_1F) = P_2F - P_2P_1F. \tag{8.8b}$$

Adding this correction, we find the following formula for the interpolating surface:

$$Q = P_1F + P_2F - P_2P_1F. \tag{8.9}$$

Again, we can work in the opposite direction. The correction in the plane $y = 0$ must be

$$F(x, 0) - [(1 - x)F(0, 0) + xF(1, 0)], \tag{8.10a}$$

and in the plane $y = 1$,

$$F(x, 1) - [(1 - x)F(0, 1) + xF(1, 1)]. \tag{8.10b}$$

Symbolically, this is

$$F - P_2F. \tag{8.11a}$$

If this correction is to hold along each of the lines determined by P_1, then we must have

$$P_1(F - P_2F) = P_1F - P_1P_2F, \tag{8.11b}$$

which leads to the following formula for the interpolating surface:

$$Q = P_2F + P_1F - P_1P_2F. \tag{8.12}$$

In order for (8.9) and (8.12) to describe the same surface, the operators must commute in the sense that

$$P_2P_1F = P_1P_2F.$$

The representations for the correction terms can be read off from (8.7) for $x = 0$ and $x = 1$, and from (8.10) for $y = 0$, $y = 1$. Combining these boundary values into one (commutative) correction term, and using (8.6a), (8.9), and (8.12), we get the following formula for the interpolating surface:

$$\begin{aligned} Q(x,y) &= (1-y)F(x,0) + yF(x,1) + (1-x)F(0,y) + xF(1,y) \\ &- \Big[(1-x)\left((1-y)F(0,0) + yF(0,1)\right) + x\left((1-y)F(1,0) + yF(1,1)\right)\Big]. \end{aligned} \tag{8.13}$$

The term in (8.13) in the square brackets is the desired correction term! Introducing the blending functions f_0, f_1 of (8.2), we can rewrite (8.13) in matrix form as

$$\begin{aligned} Q(x,y) &= (F(x,0), F(x,1))\begin{pmatrix} f_0(y) \\ f_1(y)\end{pmatrix} + (f_0(x), f_1(x))\begin{pmatrix} F(0,y) \\ F(1,y)\end{pmatrix} \\ &- (f_0(x), f_1(x))\begin{pmatrix} F(0,0) & F(0,1) \\ F(1,0) & F(1,1)\end{pmatrix}\begin{pmatrix} f_0(y) \\ f_1(y)\end{pmatrix}, \end{aligned} \tag{8.13a}$$

where the blending functions f_i can, of course, also be as in (3.12).

Here we note a certain disadvantage of the Coons formula: once we have chosen the blending functions, the formula (8.13a) does not leave any additional freedom for design purposes!

We can immediately carry (8.13a) over to parametrized boundary curves. We replace x and y by u and v, respectively, where $u, v \in [0,1]$, and assume that the boundary curves are given as parametric space curves. Then in place of (8.13a), we have the following parametric formula for the interpolating surface:

$$\begin{aligned} \boldsymbol{Q}(u,v) &= (\boldsymbol{P}(u,0), \boldsymbol{P}(u,1))\begin{pmatrix} f_0(v) \\ f_1(v)\end{pmatrix} + (f_0(u), f_1(u))\begin{pmatrix} \boldsymbol{P}(0,v) \\ \boldsymbol{P}(1,v)\end{pmatrix} \\ &- (f_0(u), f_1(u))\begin{pmatrix} \boldsymbol{P}(0,0) & \boldsymbol{P}(0,1) \\ \boldsymbol{P}(1,0) & \boldsymbol{P}(1,1)\end{pmatrix}\begin{pmatrix} f_0(v) \\ f_1(v)\end{pmatrix}, \end{aligned} \tag{8.13b}$$

where $\boldsymbol{P}(i,v)$ and $\boldsymbol{P}(u,i)$ are the parametric representations of the boundary curves for $i = 0, 1$. Given global coordinates (u,v) with the knots (u_k, v_j), $k = 0(1)n$ and $j = 0(1)m$, we can introduce local coordinates via

$$r = \frac{u - u_k}{\Delta u_k}, \qquad s = \frac{v - v_j}{\Delta v_j}, \qquad r, s \in [0,1], \tag{8.13c}$$

where $\Delta u_k = u_{k+1} - u_k$, and Δv_j is defined analogously. Now (8.13b) takes the form

$$Q(u,v) = (P(u,v_j), P(u,v_{j+1})) \begin{pmatrix} f_0(s) \\ f_1(s) \end{pmatrix} + (f_0(r), f_1(r)) \begin{pmatrix} P(u_k, v) \\ P(u_{k+1}, v) \end{pmatrix}$$

$$- (f_0(r), f_1(r)) \begin{pmatrix} P(u_k, v_j) & P(u_k, v_{j+1}) \\ P(u_{k+1}, v_j) & P(u_{k+1}, v_{j+1}) \end{pmatrix} \begin{pmatrix} f_0(s) \\ f_1(s) \end{pmatrix}. \tag{8.13d}$$

Since we have made no assumptions on the derivatives of the blending functions, if we build a piecewise surface using patches as in (8.13), we can only expect the overall surface to be continuous, in general. In particular, we cannot expect that two neighboring surface patches have the same tangent planes along a common boundary curve.

Remark. From the *algebraic* standpoint, (8.12) can also be interpreted as a Boolean sum [Gor 69] (see Chap. 9.4):

$$(P_1 \oplus P_2)F := P_1 F + P_2 F - P_1 P_2 F.$$

8.1.2. C^1 Continuous Patch Complexes

The Coons patches constructed in the previous section were completely determined by blending functions satisfying condition (8.3). Using them, we were able to construct C^0 continuous surfaces. If we want to get C^1 continuous surfaces, then the derivatives of the blending functions for points on the boundary of the patch must not have any influence on the derivatives of the patch, see [Gre 80]. In fact, the blending function must satisfy the additional condition

$$f_i'(k) = 0, \qquad i, k = 0, 1. \tag{8.14}$$

As we saw in (3.12), (3.13), the classical cubic Hermite polynomials

$$f_0(t) = 1 - 3t^2 + 2t^3, \qquad f_1(t) = 3t^2 - 2t^3$$

satisfy both conditions (8.3) and (8.14). Thus, if we construct our Coons patch (8.13b) using these blending functions, then by (8.14), derivatives of the patch along the boundary are given by

$$Q_u(i,v) = P_u(i,0) f_0(v) + P_u(i,1) f_1(v),$$

$$Q_v(u,i) = P_v(0,i) f_0(u) + P_v(1,i) f_1(u),$$

for $i = 0, 1$. This shows that the tangent vector to the curve $\boldsymbol{Q}(u_\ell, v)$ at the point $\boldsymbol{Q}(u_\ell, 0)$ (with u_ℓ constant) depends only on the tangent vectors $\boldsymbol{P}_v(0,0)$ and $\boldsymbol{P}_v(1,0)$ at the end points $\boldsymbol{P}(0,0)$, and $\boldsymbol{P}(1,0)$. A similar situation persists for the other boundary curves, see Fig. 8.5.

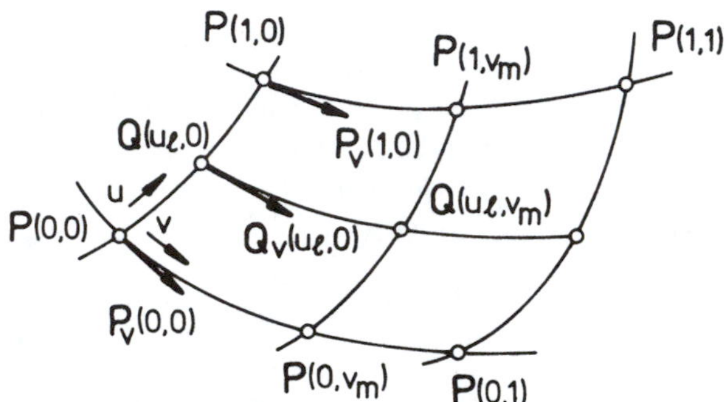

Fig. 8.5. Boundary derivatives perpendicular to a patch boundary.

By (8.15), we see that it is possible to join two Coons patches along a common parametric line in such a way that they share the same tangents along this line. Indeed, suppose we denote the two interpolating patches by I and II, and suppose that the two patches are to meet along the curves $\boldsymbol{P}_I(1, v)$ and $\boldsymbol{P}_{II}(0, v)$ with

$$\boldsymbol{P}_I(1, v) = \boldsymbol{P}_{II}(0, v).$$

It follows from (8.15) that the tangents along this common edge have the same directions provided that at the boundary points $v = 0$, $v = 1$, we have

$$\boldsymbol{P}_{Iu}(1, i) = c\boldsymbol{P}_{IIu}(0, i) \qquad \text{with } c > 0, \quad i = 0, 1, \tag{8.16}$$

i.e., the boundary curves $v = 0$ and $v = 1$ have the same tangent directions.

The above construction of two patches which join together with C^1 continuity was accomplished without making use of derivatives along the boundary curves. But, in fact we can even construct patches which interpolate given derivatives along the boundaries. Here we discuss the case where we are given the first derivatives along the boundary curves. In order to find a solution, we need to work with blending functions:

- which vanish along the boundary,
- whose derivatives are able to reproduce the prescribed derivatives along the boundary.

Again we can use the classical cubic Hermite polynomials

$$g_0(t) = t - 2t^2 + t^3, \qquad g_1(t) = -t^2 + t^3 \tag{8.17}$$

of (3.12), (3.13). Given boundary data as in Fig. 8.6, then analogous to (8.6a), (8.6b) we can construct the interpolants

$$\begin{aligned} \boldsymbol{P}_1\boldsymbol{F}(u,v) &= f_0(v)\boldsymbol{P}(u,0) + f_1(v)\boldsymbol{P}(u,1) + g_0(v)\boldsymbol{P}_v(u,0) + g_1(v)\boldsymbol{P}_v(u,1), \\ \boldsymbol{P}_2\boldsymbol{F}(u,v) &= f_0(u)\boldsymbol{P}(0,v) + f_1(u)\boldsymbol{P}(1,v) + g_0(u)\boldsymbol{P}_u(0,v) + g_1(u)\boldsymbol{P}_u(1,v). \end{aligned}$$

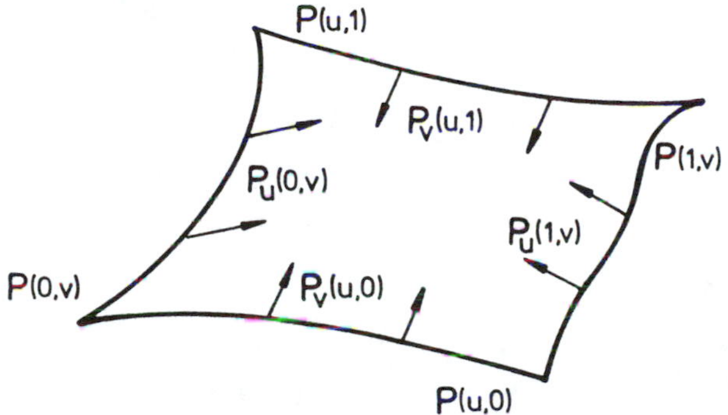

Fig. 8.6. Cross boundary data on the patch boundaries.

As in (8.12) above, we again need to introduce a correction term $\boldsymbol{P}_1\boldsymbol{P}_2\boldsymbol{F}$ or $\boldsymbol{P}_2\boldsymbol{P}_1\boldsymbol{F}$. As in (8.13a) we can write it in the matrix product form

$$\boldsymbol{P}_1\boldsymbol{P}_2\boldsymbol{F} := (f_0(u), f_1(u), g_0(u), g_1(u))\,\boldsymbol{B} \begin{pmatrix} f_0(v) \\ f_1(v) \\ g_0(v) \\ g_1(v) \end{pmatrix}. \tag{8.18}$$

Here $\boldsymbol{B}$ is a 4×4 matrix. The 2×2 submatrix in the upper left corner of $\boldsymbol{B}$ must be chosen to correspond to (8.13b), and provides the correction to values along the boundary. The 2×2 matrices in the upper right and lower left corners of $\boldsymbol{B}$ must correct the boundary derivatives with respect to u and v, respectively. Finally, as a direct computation of the first derivatives shows, the 2×2 submatrix in the lower right corner involves the mixed second derivatives at the boundary points. Thus, the matrix $\boldsymbol{B}$ should have the form

$$\boldsymbol{B} := \begin{pmatrix} \boldsymbol{P}(0,0) & \boldsymbol{P}(0,1) & \boldsymbol{P}_v(0,0) & \boldsymbol{P}_v(0,1) \\ \boldsymbol{P}(1,0) & \boldsymbol{P}(1,1) & \boldsymbol{P}_v(1,0) & \boldsymbol{P}_v(1,1) \\ \boldsymbol{P}_u(0,0) & \boldsymbol{P}_u(0,1) & \boldsymbol{P}_{uv}(0,0) & \boldsymbol{P}_{uv}(0,1) \\ \boldsymbol{P}_u(1,0) & \boldsymbol{P}_u(1,1) & \boldsymbol{P}_{uv}(1,0) & \boldsymbol{P}_{uv}(1,1) \end{pmatrix}, \tag{8.19}$$

and the corresponding interpolant takes the following form:

$$\boldsymbol{Q}(u,v) = (1, f_0(u), f_1(u), g_0(u), g_1(u))\, \bar{\boldsymbol{B}} \begin{pmatrix} 1 \\ f_0(v) \\ f_1(v) \\ g_0(v) \\ g_1(v) \end{pmatrix}, \tag{8.20}$$

where

$$\bar{\boldsymbol{B}} = \begin{pmatrix} 0 & \boldsymbol{P}(u,0) \quad \boldsymbol{P}(u,1) \quad \boldsymbol{P}_v(u,0) \quad \boldsymbol{P}_v(u,1) \\ \boldsymbol{P}(0,v) & \\ \boldsymbol{P}(1,v) & \\ \boldsymbol{P}_u(0,v) & -\boldsymbol{B} \\ \boldsymbol{P}_u(1,v) & \end{pmatrix}.$$

The correction defined by (8.18) is commutative if the order of the mixed derivatives in the matrix $\boldsymbol{B}$ can be reversed, *i.e.*, if

$$\boldsymbol{P}_{uv}(i,j) = \boldsymbol{P}_{vu}(i,j), \tag{$*$}$$

for $i, j = 0, 1$.

For C^2 surfaces, this requirement is theoretically trivially satisfied. But since in our case, the mixed derivatives at the corner points are not known, but instead must be estimated, the second mixed derivatives are, in general, *noncommutative*.

The elements in the lower right 2×2 submatrix of the matrix $\boldsymbol{B}$ in (8.19) are called *twist vectors*. In general, the components of the twist vectors must be estimated. Several suggestions for doing so can be found in the references [Barn 78, 85, 88], [Brun 85].

One reasonable approach is the following: first we set all components of the twist vectors equal to zero, and then modify them interactively to achieve the desired surface form.

It is, however, also possible to estimate the twist vectors from the given boundary data itself. For example, to get the *Adini twist vectors*, we construct a bilinear patch (8.13d) which interpolates the data at the four neighboring points corresponding to u_{k-1}, u_{k+1} and v_{j-1} and v_{j+1}. Since we are now working on a different domain, we must replace u_k and v_j in (8.13c,d) by u_{k-1} and v_{j-1}, respectively. Then the twist vector at (u_k, v_j) is taken to be the mixed partial derivative of this bilinear patch at (u_k, v_j). This gives the following formula (see [Barn 78, 88], [Far 92]) for the Adini twist:

$$\boldsymbol{Q}_{uv}(u_k, v_j) = \frac{-\boldsymbol{Q}_v(u_k, v_{j-1}) + \boldsymbol{Q}_v(u_k, v_{j+1})}{\Delta v_j}$$
$$+ \frac{-\boldsymbol{Q}_u(u_{k-1}, v_j) + \boldsymbol{Q}_u(u_{k+1}, v_j)}{\Delta u_k}$$
$$- \frac{\boldsymbol{Q}(u_{k-1}, v_{j-1}) + \boldsymbol{Q}(u_{k+1}, v_{j+1}) - \boldsymbol{Q}(u_{k-1}, v_{j+1}) - \boldsymbol{Q}(u_{k+1}, v_{j-1})}{\Delta u_k \Delta v_j},$$

where $\Delta u_k = u_{k+1} - u_{k-1}$ and $\Delta v_j = v_{j+1} - v_{j-1}$.

Alternatively, we can compute the *Bessel twist*. Here we construct bilinear Coons patches as in (8.13b) interpolating all combinations of four neighboring points (see Fig. 8.7), and compute their mixed derivatives at the point $\boldsymbol{P}(u_k, v_j)$. Thus, for example, for the bilinear patch in the upper right corner of Fig. 8.7, we have

$$\boldsymbol{Q}_{uv}(u_k, v_j) = \frac{-\boldsymbol{P}(u_k, v_j) + \boldsymbol{P}(u_k, v_{j+1}) + \boldsymbol{P}(u_{k+1}, v_j) - \boldsymbol{P}(u_{k+1}, v_{j+1})}{\Delta u_k \Delta v_j}.$$

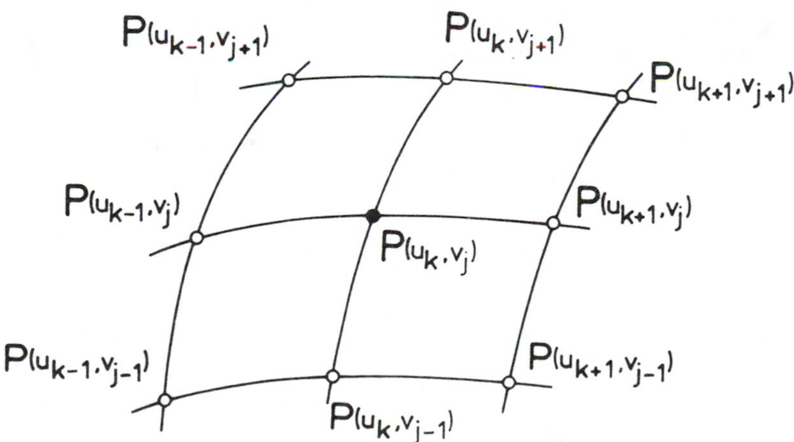

Fig. 8.7. Neighboring vertices used to estimate the twist vectors.

Similar formulae hold for the mixed partial derivatives of the other bilinear patches. We now take a linear combination of this collection of mixed derivatives to get the following formula for the Bessel twist:

$$\boldsymbol{X}_{uv}(u_k, v_j) = (1-\alpha_k, \alpha_k) \begin{pmatrix} \boldsymbol{Q}_{uv}(u_{k-1}, v_{j-1}) & \boldsymbol{Q}_{uv}(u_{k-1}, v_j) \\ \boldsymbol{Q}_{uv}(u_k, v_{j-1}) & \boldsymbol{Q}_{uv}(u_k, v_j) \end{pmatrix} \begin{pmatrix} 1-\beta_j \\ \beta_j \end{pmatrix},$$

where

$$\alpha_k = \frac{\Delta u_{k-1}}{u_{k+1} - u_{k-1}}, \qquad \beta_k = \frac{\Delta v_{k-1}}{v_{k+1} - v_{k-1}}.$$

It is also possible to get *commutative twist vectors* using various methods. For example, we can take

$$\begin{pmatrix} \dfrac{u\boldsymbol{P}_{vu}(0,0) + v\boldsymbol{P}_{uv}(0,0)}{u+v} & \dfrac{u\boldsymbol{P}_{vu}(0,1) + (1-v)\boldsymbol{P}_{uv}(0,1)}{u+1-v} \\ \dfrac{(1-u)\boldsymbol{P}_{vu}(1,0) + v\boldsymbol{P}_{uv}(1,0)}{1-u+v} & \dfrac{(1-u)\boldsymbol{P}_{vu}(1,1) + (1-v)\boldsymbol{P}_{uv}(1,1)}{1-u+1-v} \end{pmatrix}$$

as the matrix in the lower right-hand corner of (8.19). This matrix comes from a Gregory square, see [Barn 78], [Gre 74] and also (7.40). Other approaches to estimating the twists can be found e.g., in [CHI 88], [Gre 83], [Barn 88], and [Far 92].

Fig. 8.8a,b shows the effect of choosing different twist vectors, where we have taken $\boldsymbol{T} := \boldsymbol{P}_{uv}(0,0) = \boldsymbol{P}_{uv}(0,1) = \boldsymbol{P}_{uv}(1,0) = \boldsymbol{P}_{uv}(1,1)$. In Fig. 8.8a we chose $\boldsymbol{T} = (0,0,0)$. In Fig. 8.8b, we took $\boldsymbol{T} = (150,0,0)$ for the first two twist vectors, and $-\boldsymbol{T}$ for the third and fourth twist vectors. The boundary of the patch consists of straight lines and circular arcs connected together (with a total length 12). The bulging of the surface can be clearly seen.

Remark. It is also possible to construct C^2 continuous patches analogous to (8.20). This requires additional blending functions satisfying

$$h_i(k) = h_i'(k) = 0, \qquad h_i''(k) = \delta_{ik}, \qquad i,k = 0,1.$$

8.1.3. Bicubic Patches

A frequently applied special case of the interpolant (8.20) based on the cubic blending functions (3.12) and (8.17) arises when we are given the values and the tangents at the four corner points of the patch (discrete interpolant). We use these values along with the blending functions to construct approximate

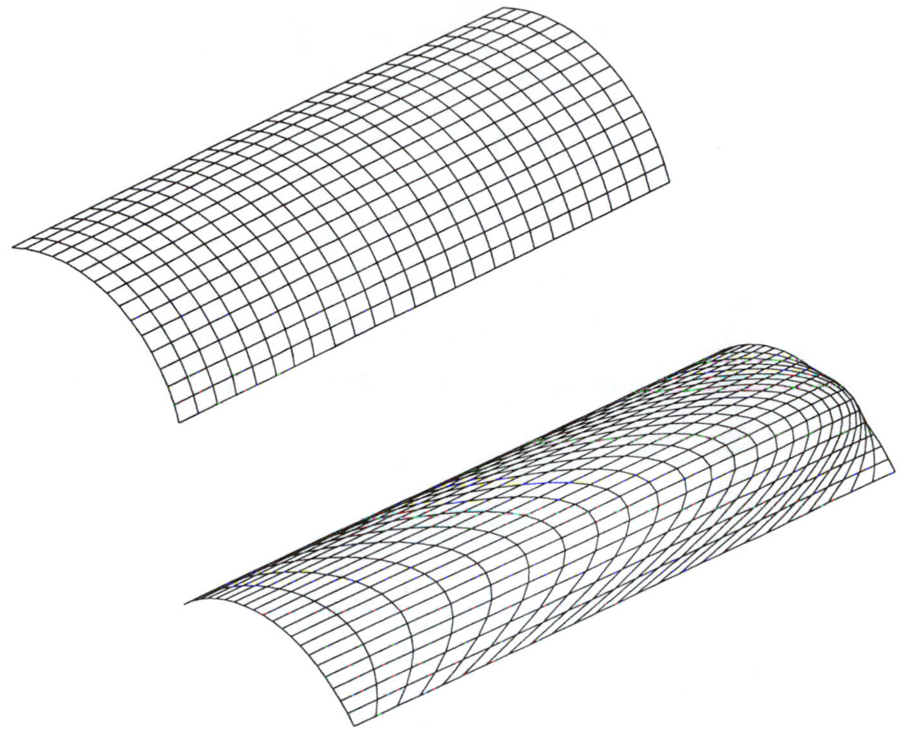

Fig. 8.8a,b. Effect of different twist vectors.

boundary curves, after which we can construct our discrete bicubic interpolating patch (8.20). In particular, we take the boundary curves to be

$$\boldsymbol{P}(u,i) := (f_0(u), f_1(u), g_0(u), g_1(u)) \begin{pmatrix} \boldsymbol{P}(0,i) \\ \boldsymbol{P}(1,i) \\ \boldsymbol{P}_u(0,i) \\ \boldsymbol{P}_u(1,i) \end{pmatrix}, \qquad i = 0, 1,$$

and

$$\boldsymbol{P}(i,v) := (\boldsymbol{P}(i,0), \boldsymbol{P}(i,1), \boldsymbol{P}_v(i,0), \boldsymbol{P}_v(i,1)) \begin{pmatrix} f_0(v) \\ f_1(v) \\ g_0(v) \\ g_1(v) \end{pmatrix}.$$

Using these, (8.20) simplifies to the following *bicubic patch*

$$\boldsymbol{Q}^*(u,v) = \boldsymbol{F}(u)^T \boldsymbol{P} \boldsymbol{F}(v), \tag{8.21}$$

with

$$\boldsymbol{F}(u) := (f_0(u), f_1(u), g_0(u), g_1(u))^T,$$
$$\boldsymbol{F}(v) := (f_0(v), f_1(v), g_0(v), g_1(v))^T,$$

and

$$\boldsymbol{P} := \begin{pmatrix} \boldsymbol{P}(0,0) & \boldsymbol{P}(0,1) & \boldsymbol{P}_v(0,0) & \boldsymbol{P}_v(0,1) \\ \boldsymbol{P}(1,0) & \boldsymbol{P}(1,1) & \boldsymbol{P}_v(1,0) & \boldsymbol{P}_v(1,1) \\ \boldsymbol{P}_u(0,0) & \boldsymbol{P}_u(0,1) & \boldsymbol{P}_{uv}(0,0) & \boldsymbol{P}_{uv}(0,1) \\ \boldsymbol{P}_u(1,0) & \boldsymbol{P}_u(1,1) & \boldsymbol{P}_{uv}(1,0) & \boldsymbol{P}_{uv}(1,1) \end{pmatrix}.$$

By (3.12) and (8.17), we can also express the blending functions in matrix form as

$$(f_0(u), f_1(u), g_0(u), g_1(u)) = (u^3, u^2, u, 1) \begin{pmatrix} 2 & 2 & 1 & 1 \\ -3 & 3 & -2 & -1 \\ 0 & 0 & 1 & 0 \\ 1 & 0 & 0 & 0 \end{pmatrix} =: \boldsymbol{u}^T \boldsymbol{K},$$

so that $\boldsymbol{Q}^*$ becomes

$$\boldsymbol{Q}^* = \boldsymbol{u}^T \boldsymbol{K} \boldsymbol{P} \boldsymbol{K}^T \boldsymbol{v}. \tag{8.22}$$

Remark. For bicubic patches, it is easy to see how to transform the Coons patch to a Bézier patch. Writing the patch $\boldsymbol{Q}^*$ as a bicubic tensor-product Bézier patch

$$\boldsymbol{Q}^* = \sum_{k=0}^{3} \sum_{i=0}^{3} \boldsymbol{b}_{ik} B_i^3(u) B_k^3(v) = \boldsymbol{u}^T \boldsymbol{L} \boldsymbol{B} \boldsymbol{L}^T \boldsymbol{v}, \tag{$*$}$$

where $\boldsymbol{B} = \{\boldsymbol{b}_{ik}\}$ is the matrix of Bézier points and

$$\boldsymbol{L} := \begin{pmatrix} -1 & 3 & -3 & 1 \\ 3 & -6 & 3 & 0 \\ -3 & 3 & 0 & 0 \\ 1 & 0 & 0 & 0 \end{pmatrix},$$

then comparing with (8.22) gives

$$\boldsymbol{P} = \boldsymbol{K}^{-1} \boldsymbol{L} \boldsymbol{B} \boldsymbol{L}^T (\boldsymbol{K}^T)^{-1}.$$

8.1.4. Gordon Surfaces

The use of the method of Coons to construct a system of patches leads to so-called *Gordon surfaces* [Gor 69, 71]. Suppose we are given a curve network consisting of the curves

$$\boldsymbol{f}(u_i, v), \quad i = 0(1)n \qquad \boldsymbol{f}(u, v_k), \quad k = 0(1)m.$$

Then we can construct a Gordon surface interpolating this curve network as the Boolean sum

$$(\boldsymbol{P}_1 \oplus \boldsymbol{P}_2)\boldsymbol{f} = \boldsymbol{P}_1\boldsymbol{f} + \boldsymbol{P}_2\boldsymbol{f} - \boldsymbol{P}_1\boldsymbol{P}_2\boldsymbol{f}, \tag{8.23}$$

where

$$\boldsymbol{P}_1\boldsymbol{f} = \sum_{i=0}^{n} \boldsymbol{f}(u_i, v) g_i(u), \qquad \boldsymbol{P}_2\boldsymbol{f} = \sum_{k=0}^{m} \boldsymbol{f}(u, v_k) h_k(v),$$

with blending functions $g_i(u)$ and $h_k(v)$. Here the product $\boldsymbol{P}_1\boldsymbol{P}_2\boldsymbol{f}$ is again just the tensor-product which interpolates the data at the corners of the curve network. For $n = m = 1$, the Gordon surfaces are just the Coons patches.

Nielson [Nie 86] has extended the concept of Gordon surfaces to ν-spline surfaces (rectangular ν-splines, see Chaps. 3, 7). Here the net lines are taken to be ν-spline curves with the corner points of the net chosen to be the knots.

8.2. Gordon-Coons Surfaces on Triangles

The Gordon-Coons method can also be used for *three-sided* and *five-sided parameter domains*, which means that Gordon-Coons surfaces can be used to solve the "vertex problem" discussed in Chaps. 6, 7, see also [Barn 73, 75, 81], [Gre 75].

We work with the standard triangle with vertices (1,0), (0,1), (0,0). First we interpolate parallel to the x-axis (see Fig. 8.9a):

$$\boldsymbol{P}_1\boldsymbol{F} = \frac{1 - x - y}{1 - y}\boldsymbol{F}(0, y) + \frac{x}{1 - y}\boldsymbol{F}(1 - y, y). \tag{8.24a}$$

On the boundaries we have

$$\boldsymbol{P}_1\boldsymbol{F} = \boldsymbol{F}(1 - y, y), \qquad \text{for } x + y = 1,$$

$$\boldsymbol{P}_1\boldsymbol{F} = \boldsymbol{F}(0, y), \qquad \text{for } x = 0,$$

$$\boldsymbol{P}_1\boldsymbol{F} = (1 - x)\boldsymbol{F}(0, 0) + x\boldsymbol{F}(1, 0), \qquad \text{for } y = 0.$$

Thus, the interpolant reduces to a straight line along the boundary $y = 0$.

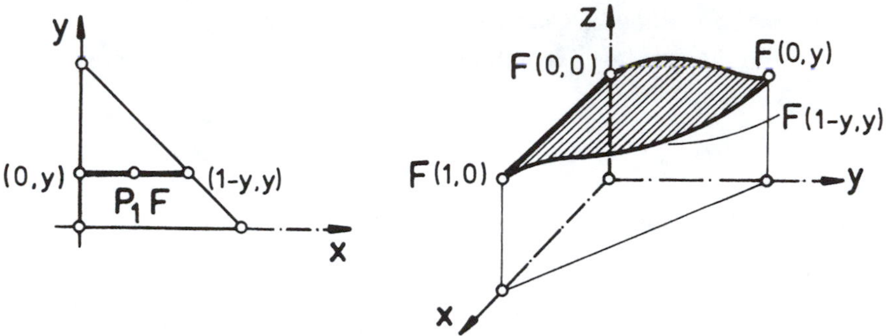

Fig. 8.9a. Interpolation parallel to the x-axis.

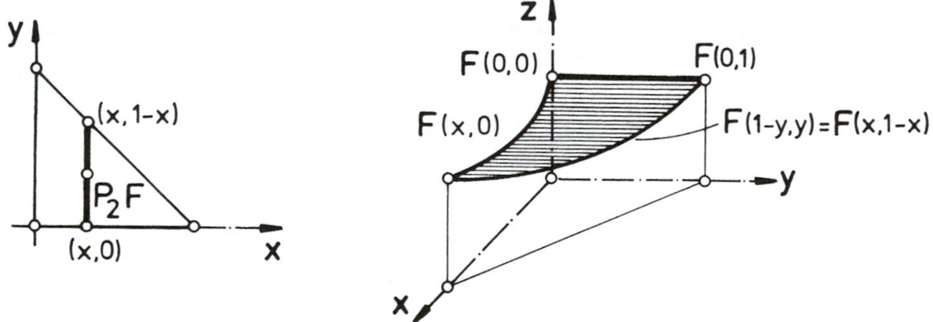

Fig. 8.9b. Interpolation parallel to the y-axis.

Similarly, we can construct the interpolant parallel to the y-axis (see Fig. 8.9b):

$$\boldsymbol{P}_2\boldsymbol{F} = \frac{1-x-y}{1-x}\boldsymbol{F}(x,0) + \frac{y}{1-x}\boldsymbol{F}(x,1-x), \tag{8.24b}$$

where now on the boundaries we have

$$\boldsymbol{P}_2\boldsymbol{F} = \boldsymbol{F}(x,0), \qquad \text{for } y=0,$$

$$\boldsymbol{P}_2\boldsymbol{F} = (1-y)\boldsymbol{F}(0,0) + y\boldsymbol{F}(0,1), \qquad \text{for } x=0.$$

We can also perform interpolation parallel to the line $y = 1 - x$:

$$\boldsymbol{P}_3\boldsymbol{F} = \frac{x}{x+y}\boldsymbol{F}(x+y,0) + \frac{y}{x+y}\boldsymbol{F}(0,x+y).$$

We now combine these three interpolants into

$$\boldsymbol{Q}(x,y) = \tfrac{1}{2}(\boldsymbol{P}_1\boldsymbol{F} + \boldsymbol{P}_2\boldsymbol{F} + \boldsymbol{P}_3\boldsymbol{F} - \boldsymbol{L}\boldsymbol{F}), \tag{8.25}$$

where $\boldsymbol{LF}$ has the following values on the boundaries:

$$\boldsymbol{LF}\big|_{x=0} = (1-y)\boldsymbol{F}(0,0) + y\boldsymbol{F}(0,1), \qquad \boldsymbol{LF}\big|_{y=0} = (1-x)\boldsymbol{F}(0,0) + x\boldsymbol{F}(1,0),$$

$$\boldsymbol{LF}\big|_{x=1-y} = x\boldsymbol{F}(1,0) + y\boldsymbol{F}(0,1),$$

so that

$$\boldsymbol{LF} = (1-x-y)\boldsymbol{F}(0,0) + x\boldsymbol{F}(1,0) + y\boldsymbol{F}(0,1).$$

In summary, we have

$$\begin{aligned}\boldsymbol{Q}(x,y) = \tfrac{1}{2}\Big[\left(\frac{1-x-y}{1-y}\right)\boldsymbol{F}(0,y) &+ \frac{x}{1-y}\boldsymbol{F}(1-y,y) + \frac{1-x-y}{1-x}\boldsymbol{F}(x,0)\\ &+ \frac{y}{1-x}\boldsymbol{F}(x,1-x) + \frac{x}{x+y}\boldsymbol{F}(x+y,0) + \frac{y}{x+y}\boldsymbol{F}(0,x+y)\\ &- [x\boldsymbol{F}(1,0) + y\boldsymbol{F}(0,1) + (1-x-y)\boldsymbol{F}(0,0)]\Big].\end{aligned}$$

Similar interpolants can be constructed via radial interpolation, see [Barn 77]. An interpolation method for arbitrary triangles is developed in [Barn 83], based on the barycentric coordinates (see Chap. 4) of a triangle. The method requires estimating the directional derivatives along the sides of the triangle, see [Lit 83].

Gregory [Gre 78] developed a different triangular interpolant using convex combinations; it can be generalized to *five-sided domains*, see [Gre 85].

The surface representation methods based on blending functions discussed here also have analogs for volume representations, where the boundary curves now come from cube-shaped domains, see Chap. 11 and [Gor 69].

9 Scattered Data Interpolation

In many applications, for example in geology, meteorology, cartography, and in the digitalization of model surfaces, we frequently have to deal with nonuniformly spaced data (so-called *scattered data*) which have to be interpolated or approximated. Here we restrict our attention to scalar-valued problems. For the corresponding 3D case, see Chap. 7.

The *interpolation problem* can be posed as follows. Suppose we are given N abscissae $\boldsymbol{x}_i = (x_i, y_i) \in \mathbb{R}^2$, and associated ordinates (e.g., measured values) z_i, $i = 1(1)N$. Then we seek a function $f(\boldsymbol{x}) = f(x, y)$ such that $z_i = f(x_i, y_i)$, $i = 1(1)N$. The *approximation problem* can be posed as a (*weighted* or *moving*) *least squares problem*: we seek to minimize the function $I(f) = \sum \omega_i(\boldsymbol{x})[f(x_i, y_i) - z_i]^2$. We may also consider the *smoothing problem* of mimimizing $I(f) = \sum \omega_i(\boldsymbol{x})[f(x_i, y_i) - z_i]^2 + \lambda J(f)$, where λ is a smoothing parameter, and $J(f)$ is a "*physical term*" such as the bending energy of a loaded elastic thin plate. In contrast to the previous chapters, here we do not make any special assumptions about the data (x_i, y_i, z_i). In this chapter we discuss only the interpolation problem. The approximation problem was discussed above in Sects. 2.3, 4.4, and 6.2.5. For more on approximation, see [Die 81], [Farw 86], [Fol 87b], [Fra 87], [Hay 74], [Hu 86], [Mcla 74, 76], [Mcm 87], [Lan 79], [LAN 86], [Schm 79, 83, 85], and [Schu 76], as well as the references listed in [Fra 87a]. For smoothing, see Chap. 13.

Two and higher dimensional scattered data interpolation problems are much more difficult to solve than the corresponding univariate problem. Indeed, by a theorem of Haar (see [DAV 75]), even the *existence* of a solution to the bivariate interpolation problem for given basis functions cannot be guaranteed in general. Moreover, even in those cases where the corresponding system of equations does have a solution, it is often very poorly conditioned. (A numerical method is called *well conditioned* provided that a relatively small perturbation in the input results in a relatively small change in the output). This problem can be avoided by working with explicit methods which do not require solving systems of equations, or blending of local approximations based on small data sets with solvable systems of equations. Many methods, however, actually use data dependent basis functions which lead to solvable systems, see [Gor 78].

There are many different applications of scattered data interpolation and approximation, and just as many different methods. There is no one method which works well in all cases. The choice of which method to use in a particular case depends entirely on the problem at hand, and also on various side conditions (such as the available hardware), since the different methods require quite different amounts of computational time and memory.

In a *global method*, the interpolant or approximant depends on all of the data points at once in the sense that adding, changing, or removing one data point implies that we have to solve the problem again. This usually means solving a linear system of $N+1$ equations, where in practice N can easily be as large as ten thousand or even a million. Often, this system is not sparse, and usually it is not very well-conditioned.

In a *local method*, adding, changing, or removing a data point affects the interpolant or approximant only locally, *i.e.*, in a certain subset of the domain. However, this advantage is often essentially lost in practice since for real scattered data, which have to be assumed to be "randomly distributed", we first have to compare all data points with each other in order to decide which points influence which parts of the surface. Moreover, global methods frequently produce more "pleasing" surfaces than local ones, in particular near the boundaries of the domain [Fra 82].

For very large data sets, often local and global methods can be combined. For example, for appropriate subsets of data, we may be able to first apply a global method, and then use the result to construct a locally defined global interpolant or approximant.

To evaluate and compare various methods, we should take account of difficulty of implementation, accuracy, appearance of the surface, and sensitivity of the surface to changes in the parameters [Fra 82]. For survey articles on scattered data interpolation, see [Schu 76], [Barn 77], and [Fra 82, 87, 91], and also [LAN 86], [Alf 89]. An extensive bibliography is contained in [Fra 87a].

In Sects. 9.1–9.4 we discuss the most important scattered data interpolation methods. Sect. 9.5 compares the performance of several of these methods on an example. Sect. 9.6 is devoted to the question of affine invariance, and Sect. 9.7 treats scattered data interpolation on surfaces in $\mathbb{R}^3$ (surfaces on surfaces).

9.1. Shepard's Method

Perhaps the best known approach to solving the scattered data interpolation problem is Shepard's method [Shep 68]. It was developed by meteorologists and geologists [Cre 59], [Cra 67], and is still undergoing development. Shepard defined his interpolating function $f(\boldsymbol{x})$ to be a weighted mean of the ordinates f_i:

$$f(\boldsymbol{x}) = \sum_{i=1}^{N} \omega_i(\boldsymbol{x}) f_i, \tag{9.1}$$

with weight functions (basis functions)

$$\omega_i(\boldsymbol{x}) = \frac{\sigma_i(\boldsymbol{x})}{\sum_{j=1}^{N} \sigma_j(\boldsymbol{x})}, \tag{9.2a}$$

where

$$\sigma_i(\boldsymbol{x}) = \frac{1}{d_i(\boldsymbol{x})^{\mu_i}} = \frac{1}{[(x-x_i)^2 + (y-y_i)^2]^{\frac{\mu_i}{2}}}, \qquad \mu_i \in \mathbb{R}^+, \tag{9.3}$$

is a power of the inverse Euclidean distance $d_i(\boldsymbol{x}) = |\boldsymbol{x} - \boldsymbol{x}_i|$. This is a kind of *inverse distance weighted method*, i.e., the larger the distance of $\boldsymbol{x}$ to $\boldsymbol{x}_i$, the less influence f_i has on the value of f at the point $\boldsymbol{x}$. The functions $\omega_i(\boldsymbol{x})$ have the following properties:

- $\omega_i(\boldsymbol{x}) \in C^0$, continuity
- $\omega_i(\boldsymbol{x}) \geq 0$, positivity
- $\omega_i(\boldsymbol{x}_j) = \delta_{ij}$, interpolation property ($f(x_i) = f_i$)
- $\sum \omega_i(\boldsymbol{x}) = 1$, normalization.

Differentiating (9.1), it can be shown [Gor 78] that the Shepard function has

- cusps at $\boldsymbol{x}_i = (x_i, y_i)$ if $0 < \mu_i < 1$,
- corners at $\boldsymbol{x}_i = (x_i, y_i)$ if $\mu_i = 1$,
- flat spots at $\boldsymbol{x}_i = (x_i, y_i)$ if $\mu_i > 1$, *i.e.*, the tangent planes at such points are parallel to the (x, y) plane.

In order to minimize computational time, it is recommended to choose $\mu_i = 2$, since then we do not need to compute the root in (9.3). However, this choice causes distant points to have a rather large influence, which for large values of

N, can lead to numerical instability. In view of this, in programming (9.2a), we should always use the equivalent expression

$$\omega_i(\boldsymbol{x}) = \frac{\prod_{j\neq 1} d_j(\boldsymbol{x})^{\mu_i}}{\sum_{k=1}^{N} \prod_{j\neq k} d_j(\boldsymbol{x})^{\mu_i}}, \tag{9.2b}$$

which is numerically more stable. For a careful implementation of the method, see [Ren 88a].

Besides the obvious weaknesses of Shepard's method [Shep 68], clearly all of the $\omega_i(\boldsymbol{x})$ have to be recomputed as soon as just one data point is changed, or a data point is added or removed. This implies that Shepard's method is a global method, see [Farw 86a].

There are several ways to improve Shepard's method, for example by removing the discontinuities in the derivatives and the flat spots at the interpolation points. Even the global character of the method can be made local:

1) by multiplying the weight functions $\omega_i(\boldsymbol{x})$ by a "damping function" (mollifying function) $\lambda_i(\boldsymbol{x})$, with $\lambda_i(\boldsymbol{x}_i) = 1$, $\lambda_i(\boldsymbol{x}) \geq 0$ inside a given neighborhood B_i of $\boldsymbol{x}_i$ (e.g., a circular disk with radius R_i), and with $\lambda_i(\boldsymbol{x}) = 0$ for points outside of B_i, i.e., for points $\boldsymbol{x}_i$ with distance $d_i > R_i$. As an example of such a mollifying function, we mention the *Franke-Little weights* [Barn 77, 84b], [Alf 89]:

$$\lambda_i(\boldsymbol{x}) = g(\boldsymbol{x})_+^{\mu} = \left(1 - \frac{d_i}{R_i}\right)_+^{\mu},$$

where

$$g(\boldsymbol{x})_+^{\mu} = \begin{cases} g(\boldsymbol{x})^{\mu}, & \text{if } g(\boldsymbol{x}) \geq 0, \text{ i.e., } d_i \leq R_i \\ 0, & \text{if } g(\boldsymbol{x}) < 0, \text{ i.e., } d_i > R_i. \end{cases}$$

These generalize weights given by [Fra 82] for the case $\mu = 2$. A similar method was also presented in [Schu 76], and exponential weights are investigated in [Mcla 74].

2) by using weight functions which are locally supported like B-splines,

3) by using a recurrence formula, so that adding new data points to the data set does not require the recalculation of all of the $\omega_i(\boldsymbol{x})$, but only involves inserting additional terms. Such a recursive construction is comparable to that of the classical Newton interpolation formula (which can be derived from the global Lagrange interpolation formula, see Chap. 2). If $f_{n-1}(\boldsymbol{x})$ denotes the Shepard interpolant to the data $(x_i, y_i, z_i = f_i)$, $i = 1(1)n\text{–}1$,

and we add one new data point, then the *recurrence formula* [Barn 83a] for $f_n(\boldsymbol{x})$ is

$$f_n(\boldsymbol{x}) = f_{n-1}(\boldsymbol{x}) + E_n(\boldsymbol{x})F_n,$$

where

$$E_1(\boldsymbol{x}) = 1,$$
$$E_k(\boldsymbol{x}) = \frac{\prod_{i=1}^{k-1} d_i(\boldsymbol{x})^\mu}{\sum_{j=1}^{k}\prod_{i=1,i\neq j}^{k} d_i(\boldsymbol{x})^\mu},$$

and

$$F_1 = f_1$$
$$F_2 = f_2 - E_1(\boldsymbol{x}_2)F_1$$
$$F_k = f_k - \sum_{j=1}^{k-1} E_j(\boldsymbol{x}_k)F_j = f_k - f_{k-1}(\boldsymbol{x}_k).$$

To eliminate the discontinuities in the derivatives and the flat spots which arise because of (9.3), we can

1) use other weight functions. For example, we can define $\omega_i(\boldsymbol{x})$ as above, but with

$$\sigma_i(\boldsymbol{x}) = \frac{1}{d_i(\boldsymbol{x})^{\mu_i} + c_i} = \frac{1}{[(x-x_i)^2 + (y-y_i)^2]^{\frac{\mu_i}{2}} + c_i}, \tag{9.4}$$

   where $c_i \neq 0$. Other possibilities include

$$\sigma_i(\boldsymbol{x}) = \frac{1}{\exp[c_i(x-x_i)^2 + c_i(y-y_i)^2]} \tag{9.5}$$

   and

$$\sigma_i(\boldsymbol{x}) = \frac{1}{\cosh(c_i(x-x_i)) + \sinh(c_i(y-y_i))}, \tag{9.6}$$

2) interpolate the first n terms of the Taylor series $T_i^n(\boldsymbol{x})$ of $f(\boldsymbol{x})$ at $\boldsymbol{x}_i$ instead of just the function values f_i. For example, the function

$$f(\boldsymbol{x}) = \sum_{i=0}^{N} \omega_i(\boldsymbol{x})\left[f(\boldsymbol{x}_i) + f_x(\boldsymbol{x}_i)(x-x_i) + f_y(\boldsymbol{x}_i)(y-y_i)\right] \tag{9.7}$$

   interpolates the function values $f(\boldsymbol{x}_i)$ along with the first partial derivatives at $\boldsymbol{x}_i$ which define the tangent planes at the data points. While this eliminates flat spots and improves the approximation order, it does require derivative values which are *a priori* not available, and thus must first be approximated in some way from the given data points. A program based on (9.7) can be found in [SPÄ 91].

Sometimes, however, it is desirable to "*reproduce*" flat spots. In this case, we can use larger values of μ_i, since for large values of μ_i, (9.1) mimics a piecewise constant function (cf. Fig. 9.1a where $\mu_4 = 15$). Such Shepard methods in both one and two variables are very useful in approximating *histograms*, see e.g., [Gor 78].

In the framework of the general interpolation problem, very good results (with relatively low computational effort) can be obtained with the *modified Shepard method*

$$f(\boldsymbol{x}) = \sum_{i=1}^{N} \omega_i(\boldsymbol{x}) L_i(\boldsymbol{x}), \tag{9.8}$$

where $L_i(\boldsymbol{x})$ are local approximations of $f(\boldsymbol{x})$ such that $L_i(\boldsymbol{x}_i) = f(\boldsymbol{x}_i)$ (for the *quadratic case*, see [Fra 80]). Using this method with the Franke-Little weights as mollifying functions gives very good results, does not require too much memory or computational time, and is simple to implement [Fra 82].

The Shepard methods discussed above can be generalized by replacing the Euclidean metric by an arbitrary metric on $\mathbb{R}^2$, [Gor 78]. In fact, we have already done this in (9.4)–(9.6). The properties of weight functions listed above continue to hold, along with the following important *positivity property*:

$$f_i \geq 0 \text{ for all } i \quad \Rightarrow \quad f(\boldsymbol{x}) \geq 0,$$

and the *maximum-minimum property*:

$$m \leq f(\boldsymbol{x}) \leq M, \qquad \text{where } m := \min_i\{f_i\}, \quad M := \max_i\{f_i\}.$$

(For the modified method based on interpolation of partial Taylor series given in (9.7), there is a different maximum-minimum property, see [Barn 84b]).

Fig. 9.1 shows the results of applying (9.3)–(9.5) to a planar curve. Fig. 9.1a corresponds to Shepard's method (9.3), (9.4) with $c_i = 0$ for different values of μ_i. Fig. 9.1b illustrates the generalized Shepard's method (9.4) with $\mu_i = 1$ and various c_i. Fig. 9.1c shows the result of using (9.5) with various c_i.

Fig. 9.12c is based on (9.3), Fig. 9.12d on (9.3) with $\mu_i > 1$, and Fig 9.12e on the modified quadratic method of [Fra 80]. Other examples can be found in [Gor 78], [Barn 83a], and [SPÄ 91].

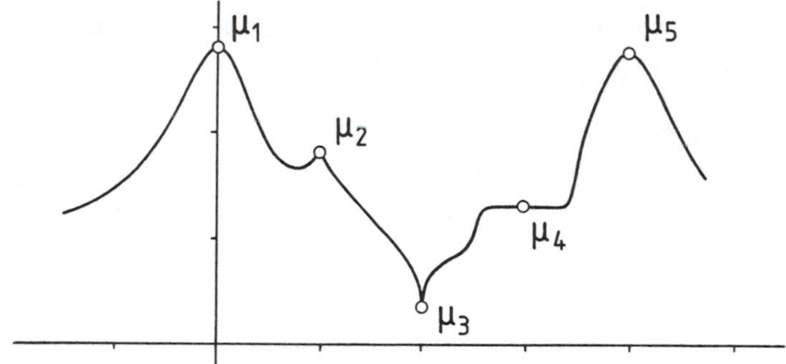

Fig. 9.1a. Shepard's method (9.3) with $\mu_1 = 2, \mu_2 = 1, \mu_3 = 0.5, \mu_4 = 15, \mu_5 = 2$.

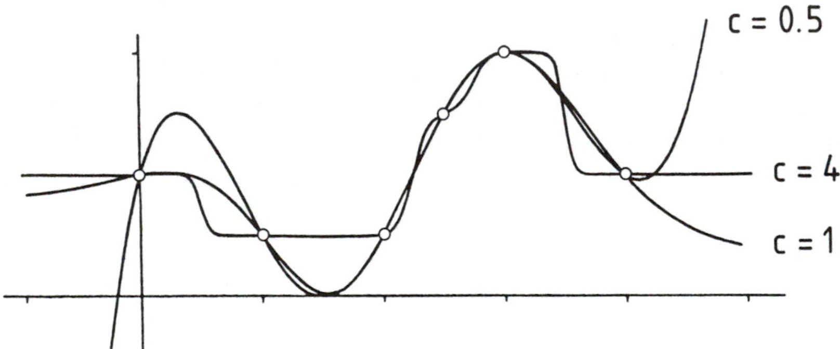

Fig. 9.1b. Shepard's method (9.4) with $\mu_i = 1$ and various c_i.

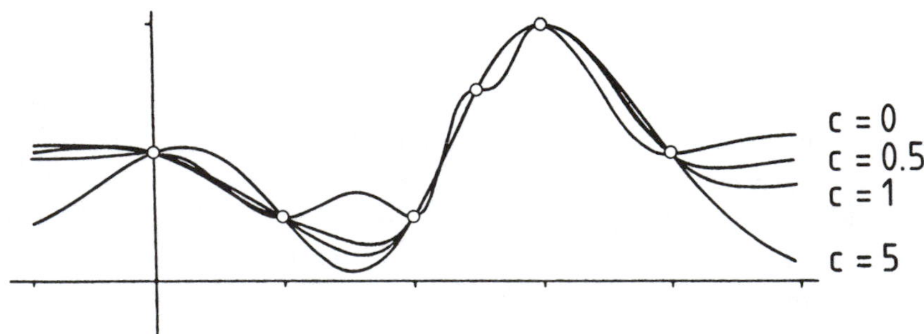

Fig. 9.1c. Shepard's method (9.5) for various c_i values.

9.2. Radial Basis Function Methods

Scattered data interpolation problems can also be solved using radial basis functions, in which case the interpolant has the form

$$f(\boldsymbol{x}) = \sum_{i=1}^{N} \alpha_i R(d_i(\boldsymbol{x})) + p_m(\boldsymbol{x}), \qquad p_m(\boldsymbol{x}) = \sum_{j=1}^{m} \beta_j p_j(\boldsymbol{x}). \tag{9.9}$$

Here the univariate basis functions $R(d_i(\boldsymbol{x}))$ are positive radial functions, *i.e.*, functions of the distance $d_i(\boldsymbol{x})$ of the point $\boldsymbol{x} = (x, y)$ to the interpolation point $\boldsymbol{x}_i = (x_i, y_i)$ in the parameter space, and $\{p_j(\boldsymbol{x})\}$ is a set of monomials of degree at most m. The unknown coefficients in (9.9) can be computed from the interpolation conditions $f(\boldsymbol{x}_i) = f_i$ together with the m side conditions

$$\sum_{j=1}^{N} \alpha_j p_i(\boldsymbol{x}_j) = 0, \qquad i = 1(1)m. \tag{9.10}$$

These conditions can be interpreted physically as equilibrium conditions (the sum of all forces and all moments is equal to zero), see [Fra 85]. It also can be thought of as guaranteeing that all polynomials in $\mathbb{P}_m$ are *reproduced* exactly (so the method has polynomial *precision*). Indeed, given $p(\boldsymbol{x}) \in \mathbb{P}_m$ and setting $f_i = p(\boldsymbol{x}_i)$, then for $\alpha_i = 0$, $i = 1(1)N$, (9.10) holds and we have $p_m(\boldsymbol{x}) = p(\boldsymbol{x})$, *i.e.*, the scheme reproduces elements in $\mathbb{P}_m$, see [Dyn 87a, 89], [Alf 89].

The coefficient matrix of the linear system of equations (9.9)–(9.10) is always full, and for more than about 200 interpolation points, quickly becomes very poorly conditioned [Fra 82], so that special ("*preconditioning*") methods are required [Dyn 86, 89]. Alternatively, we can localize the problem by dividing the data into overlapping subsets, then computing partial solutions for each data set, and finally, blending these solutions together to get a smooth locally-defined global interpolant, see e.g., [Fra 82, 82a], [Alf 89], and also Sect. 9.7.3.2. A theoretical justification for the method was recently found by [Mic 86], where the existence and uniqueness of a radial basis function solving the interpolation problem is established under appropriate conditions on the basis functions. The proof is based on the calculus of positive definite functions, see also [Dyn 87a, 89], [Fra 87].

In the following sections we describe three important examples of the class of radial basis functions. For other examples, see [Dyn 87a, 89].

9.2.1. Hardy's Multiquadrics

Multiquadrics (MQ) are among the most successful and most applied methods for interpolating scattered data. The MQ method is simple to implement, and for moderately sized data sets, leads to well-conditioned systems of equations. Surfaces constructed in this way usually are visually pleasing, and very smooth ($f(\boldsymbol{x})$ belongs to the class $C^\infty(\mathbb{R}^2)$). Hardy's MQ methods, see [Hardy 71], utilize rotation-symmetric basis functions of the general form

$$R(r_i) = \left(r_i^2 + R_i^2\right)^{\frac{\mu_i}{2}}, \qquad \mu_i \neq 0, \tag{9.11}$$

and are defined as in (9.9), but with $m = 0$, *i.e.*, these methods have no polynomial precision. The parameters r_i and R_i are free to be chosen by the user. The best choices seem to be $r_i = r(d_i) = d_i(\boldsymbol{x})$ and $R_i = R$, see [Fra 82], [Dyn 87a]. For an example, see Fig. 9.12f. These particular basis functions have both rotational invariance and translation invariance, and thus in this case the coefficient matrix of the system of equations is symmetric. Hardy himself chose $\mu_i = \mu = 1$, which corresponds to the upper part of a hyperboloid of rotation. The choice $\mu_i = \mu = -1$ leads to the so-called *reciprocal* (RMQ) method, which works just as well as the MQ method. The choice of R is critical, however, (especially in the case $\mu = -1$), since if R is chosen to be too small, cusps and dips appear at the interpolation points. This is due to the fact that for small R values, (9.11) becomes very similar to a cone, see [LAN 86], [Alf 89], [Fra 82]. Too large a choice of R quickly leads to a poorly conditioned system of equations whose solution can be difficult.

The Hadamard condition number

$$K = \frac{\det(\boldsymbol{A})}{\alpha_1 \cdots \alpha_n}, \qquad \alpha_i = \left(\sum_{k=1}^{n} a_{ik}^2\right)^{\frac{1}{2}},$$

provides a good measure of the conditioning of a linear system of equations with $n \times n$ matrix $\boldsymbol{A} = (a_{ik})$. The system is poorly conditioned if $K << 1$. Experience has shown that a problem is usually badly conditioned if $K < 0.01$, and that it is well conditioned if $K > 0.1$. For $0.01 < K < 0.1$, it is difficult to tell [ENG 85]. For other definitions of condition numbers, different values of K (e.g., $K > 10^6$) signal a badly conditioned problem, see [FORS 77].

The question of how to make an optimal choice for the constant R (or more generally for the R_i) is still an open problem. Some suggestions include the following:

- [Ste 84] recommends

$$R = \sqrt{\tfrac{1}{10} \max\{\max_{i,k} |x_i - x_k|, \max_{i,k} |y_i - y_k|\}}.$$

  This choice takes account of how dispersed the data is in both the x and y directions.

- [Hardy 71] used
$$R = 0.815d,$$
where d is the average distance of the data points to their nearest neighbors. The value of d can be estimated in various ways. For example, if all of the data points can be enclosed in a circle of diameter D, then $(\pi D^2)/(4N)$ represents the fraction $1/N$ of the area of the circle, which corresponds to a circle of diameter $D/\sqrt{N}$. We can use this value in defining d. For example, this leads to
$$R = 1.25\frac{D}{\sqrt{N}},$$
which was the value used in Franke's tests, see [Fra 82], [Carl 91]. This choice of R takes account of how dispersed the data is, and also of the fact that the condition number K depends on N.

Extensive testing by [Carl 91] using several different data sets and test functions with $\mu = \pm 1$ seems to verify the claim that K depends heavily on N and R. In particular, for $R^2 > \frac{1}{2}$, the condition of the linear system of equations seems to deteriorate rapidly.

RMS (root-mean-square) error calculations have shown that the optimal choice of the parameter R^2 depends almost exclusively on the ordinates (function values) $z_i = f_i$, and is essentially independent of both the distribution and number N of data points $\boldsymbol{x}_i = (x_i, y_i)$. For testing purposes, the RMS error can be found by comparing the test function $T(x, y)$ and its interpolant $f(x, y)$ at the points (x_k, y_k) of a dense $n \times m$ grid covering the domain of interest ([Carl 91] chooses $n = m = 33$). This gives

$$RMS = \sqrt{\frac{1}{nm}\sum_{i=1}^{nm}[T(x_i, y_i) - f(x_i, y_i)]^2}.$$

For relatively uniformly distributed data points whose ordinates do not vary too strongly, R^2 values between 0.1 and 1.0 produce good results. For large amounts of data, for data sets exhibiting a large degree of variation in the f_i, and for "track data" (such as the Monterey Coast/Big Sur data in [Fol 86, 87b]), smaller choices of R^2 (on the order of $10^{-6} \leq R^2 \leq 10^{-4}$) lead to better conditioned systems of equations, and seem to reduce undesired oscillations in the surfaces. The following algorithm from [Carl 91] often produces nearly optimal R^2 values:

1) scale the data points so that they lie in the unit cube by taking

$$\bar{x}_i = \frac{x_i - x_{\min}}{x_{\max} - x_{\min}}, \quad \bar{y}_i = \frac{y_i - y_{\min}}{y_{\max} - y_{\min}}, \quad \bar{z}_i = \frac{z_i - z_{\min}}{z_{\max} - z_{\min}},$$

2) compute a biquadratic least squares polynomial $Q(x, y)$ fitting the data points $(\bar{x}_i, \bar{y}_i, \bar{z}_i)$,

3) set $R^2 = (1+120V)^{-1}$, where $V = \sum_{i=1}^{N}[\bar{z}_i - Q(\bar{x}_i, \bar{y}_i)]^2/N$ is the variance,

4) compute the MQ or RMQ interpolant using $\mu = \pm 1$ and the above value of R^2 for the scaled data,

5) rescale the MQ or RMQ interpolant so that it is defined on the original domain.

For track data, e.g., for the Monterey Coast data, a "one-sided scaling" leading to a better distribution of data (as uniform as possible) can be very effective, see [Fra 91], [Eck 90].

[Hardy 90] contains an extensive overview of the history and applications of the MQ and RMQ methods. A wide variety of examples can be found in [LAN 86], [SPÄ 91].

9.2.2. Duchon's Thin Plate Splines

This method for interpolating scattered data goes back to the paper [Hard 72], and was later recognized to be the function interpolating the data points which minimizes the functional

$$I[f] = \int\int_{\mathbb{R}^2} \left[\left(\frac{\partial^2 f}{\partial x^2} \right)^2 + 2 \left(\frac{\partial^2 f}{\partial x \partial y} \right)^2 + \left(\frac{\partial^2 f}{\partial y^2} \right)^2 \right] dxdy \tag{9.12}$$

over an appropriate function space [Duc 77], [Mein 79, 79a] (see also [Schu 76], [Fra 82], [Dyn 87a, 89], and [Alf 89]).

According to (9.12), $I[f]$ measures the bending energy of a thin, infinite elastic plate which is constrained at the interpolation points, which explains the terminology *thin plate spline* (TPS). The solution of this variational problem involves the radial basis functions which are the solutions of the Laplace differential equation

$$\Delta^2 R = c\delta, \tag{9.13}$$

namely,

$$R(r_i) = r_i^2 \cdot \log r_i, \qquad r_i = d_i(\boldsymbol{x}). \tag{9.14}$$

Since for $m = 2$ and $R(d_i(\boldsymbol{x}))$ these kinds of interpolating functions are a generalization of the natural (cubic) spline curves, they are also referred to in the literature as *surface splines* [Hard 72].

Duchon's TPS [Duc 77] possess linear precision, *i.e.*, if the scattered data lie on a plane, then Duchon's TPS will give an exact fit. For an example of a TPS interpolant, see Fig. 9.12g.

Generalizations of the TPS involving the k-th iterated Laplace differential equation $\Delta^k R = c\delta$ with $k \in \mathbb{N}$ were discussed in [Dyn 87a, 89].

9.2.3. Franke's Thin Plate Splines in Tension

Thin plate splines in tension (TPST) are the two dimensional generalizations of splines in tension. They involve the radial basis functions

$$R(r_i) = \int_0^{\sqrt{r_i}} \rho \int_0^{\rho} \eta K_0(\alpha\eta)\, d\eta\, d\rho, \tag{9.15}$$

where K_0 is the Bessel function of second kind (the Macdonald function). These functions are the fundamental solutions of the differential equation

$$\Delta^2 R - \alpha^2 \Delta R = c\delta, \tag{9.16}$$

where α is a tension parameter, see [Fra 85] and also [Dyn 87a, 89]. They come from a variational problem involving the functional

$$I[f] = \int\int_{\mathbb{R}^2} \left[\left(\frac{\partial^2 f}{\partial x^2}\right)^2 + 2\left(\frac{\partial^2 f}{\partial x \partial y}\right)^2 + \left(\frac{\partial^2 f}{\partial y^2}\right)^2 + \alpha^2 \left(\frac{\partial f}{\partial x}\right)^2 + \alpha^2 \left(\frac{\partial f}{\partial y}\right)^2 \right] dx\, dy.$$

The solution of the scattered data interpolation problem can be written in the form (9.9), with $m = 1$, where $R(r_i)$ is given by (9.15). We note that as $\alpha \to 0$, (9.16) converges to (9.13), but not to a TPS, since a TPS must include the linear term in (9.9), while a TPST only involves a constant term.

9.3. FEM Methods

In this section we discuss a method for solving the scattered data problem which is completely different from the methods above. The method is based on constructing a triangulation of the convex hull of the set of data points $\boldsymbol{x}_i = (x_i, y_i) \in \mathbb{R}^2$, $i = 1(1)N$, where the vertices $\boldsymbol{P}_i$ of the triangulation coincide with the $\boldsymbol{x}_i$. Then for each triangle in the triangulation, we construct a surface patch which interpolates the given function values (and possibly also derivatives) at the vertices $\boldsymbol{P}_i$. In view of the form of the interpolant, this is a kind of *finite element method* (FEM). In the simplest case, we get a piecewise linear C^0 continuous surface which interpolates function values at the vertices. Often, however, we need to use higher degree polynomial pieces, and to enforce C^r continuity conditions with $r \geq 1$, between the patches.

This requires that either a certain number of derivatives are specified at the data points, or we will have to estimate them in a preprocessing step. There are a number of methods available for each of these steps, and they can be combined to produce a variety of methods. We discuss several of them in detail in this section.

9.3.1. Triangulation of Point Sets

9.3.1.1. Triangulation Methods

Let $\mathcal{P} = \{\boldsymbol{P}_i = (x_i, y_i)\}_{i=1}^N$ be a set of N points $\boldsymbol{P}_i$ in the x-y plane, and let Ω be the convex hull of $\mathcal{P}$.

Definition 9.1. *A set $\mathcal{T} = \{(\alpha_j, \beta_j, \gamma_j) : 1 \leq \alpha_j, \beta_j, \gamma_j \leq N\}$ consisting of M triples of integers $(\alpha_j, \beta_j, \gamma_j)$ defines a triangulation of $\mathcal{P}$ provided that*

- *i) for each $j = 1(1)M$, the points $\boldsymbol{P}_{\alpha_j}, \boldsymbol{P}_{\beta_j}, \boldsymbol{P}_{\gamma_j}$ are the vertices of a nondegenerate triangle T_j,*
- *ii) every triangle is defined by exactly 3 points in $\mathcal{P}$ which are the vertices of the triangle,*
- *iii) the intersection of the interior of two triangles T_j, T_k is empty whenever $j \neq k$,*
- *iv) the union of all of the triangles is the convex hull of $\mathcal{P}$.*

Since already for 4 points there is more than one triangulation, (see Fig. 9.2 and also the first Remark in Sect. 11.3.3.1), the question immediately arises of how to choose a "best" triangulation.

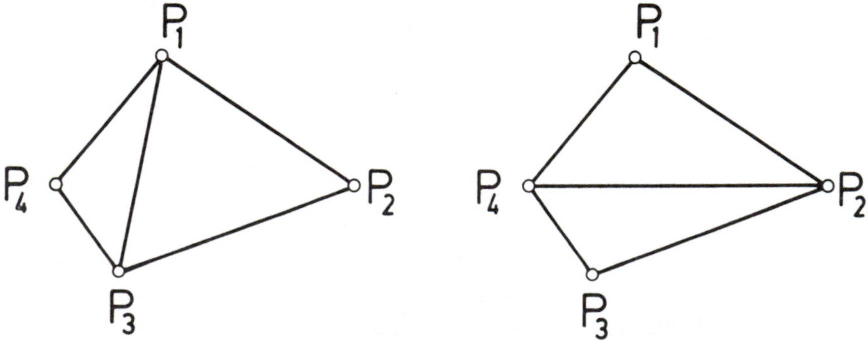

Fig. 9.2. Alternative triangulations of four points.

As an example of a simple criterion for comparing two triangulations $\mathcal{T}$ and $\widetilde{\mathcal{T}}$ of four points, we have (cf. [Mir 82], [Schu 87], [Wat 84], [Hua 89]):

Definition 9.2. (Shortest diagonal criterion). *The triangulation $\mathcal{T}$ of four points is better than the triangulation $\widetilde{\mathcal{T}}$ (w.r.t. the shortest diagonal criterion) provided that $d < \widetilde{d}$, where d is the length of the diagonal $\boldsymbol{P}_i\boldsymbol{P}_k$ of the triangulation $\mathcal{T}$, and $\widetilde{d}$ is the length of the diagonal $\boldsymbol{P}_j\boldsymbol{P}_\ell$ of $\widetilde{\mathcal{T}}$.*

While this criterion is simple to implement, it does not prevent selecting a triangulation with long thin triangles, which for interpolation should be avoided [Schu 87, 87a]. Indeed, such triangles are undesirable for the purposes of polynomial approximation since the smallest angle in the triangulation enters into the error estimate as a factor of the form $1/\sin^n \alpha$, see [Barn 85]. The following criterion of Lawson [Law 77], [Sib 78], [Lewi 78] is designed to avoid small angles:

Definition 9.3. (Max-min angle criterion). *The triangulation $\mathcal{T}$ is better than the triangulation $\widetilde{\mathcal{T}}$ (w.r.t. the max-min angle criterion) provided that $\alpha(\mathcal{T}) > \alpha(\widetilde{\mathcal{T}})$, where $\alpha(\mathcal{T}) = \min\{\alpha(T_j) : T_j \in \mathcal{T}\}$ and $\alpha(T_j)$ is the smallest angle in the triangle T_j, and $\alpha(\widetilde{\mathcal{T}})$ is defined similarly.*

Gregory [Gre 75] has shown that using the largest angle in the triangulation leads to a sharper error bound for polynomial approximation than does the smallest angle. This suggests that instead of trying to make the smallest angle as large as possible, it might be better to try to make the largest angle in the triangle as small as possible. This led Barnhill and Little [Barn 84a] to introduce the following criterion (see also [Nie 83]):

Definition 9.4. (Min-max angle criterion). *The triangulation $\mathcal{T}$ is a better triangulation than $\widetilde{\mathcal{T}}$ (w.r.t. the min-max angle criterion) provided that $\alpha(\mathcal{T}) < \alpha(\widetilde{\mathcal{T}})$, with $\alpha(\mathcal{T}) = \max\{\alpha(T_j) : T_j \in \mathcal{T}\}$, where $\alpha(T_j)$ is the largest angle in the triangle T_j, and $\alpha(\widetilde{\mathcal{T}})$ is defined similarly.*

Some other possible criteria for comparing triangulations include:

- *Max-min radius criterion*: We choose the triangulation for which the minimum of the radii of the set of circles inscribed in the triangles in the triangulation is maximized.
- *Min-max radius criterion*: We choose the triangulation for which the maximum of the radii of the set of circles inscribed in the triangles in the triangulation is minimized.
- *Max-min area criterion*: We choose the triangulation for which the minimum of the areas of the triangles in the triangulation is maximized.
- *Max-min height criterion*: We choose the triangulation for which the minimum of the heights of the triangles in the triangulation is maximized.

For still other criteria, see e.g., [Mcla 76], [Wat 84], and [PREP 85].

The criteria listed above are all different, but for any two of them, there are some choices of four points for which the two methods lead to the same triangulation.

9.3.1.2. Optimal Triangulations

The criteria introduced in the previous section for comparing the two alternative triangulations of a set of four points can also be used to construct *optimal* triangulations of larger point sets. One way to proceed is to start with four points, and then add one point at a time (e.g., the one which is closest to one of the sides of the current complex). This raises the question of whether the result of such an algorithm is an *optimal* triangulation, *i.e.*, whether using the criterion locally leads to a global optimum. This suggests (cf. [Schu 87]):

Definition 9.5. *A triangulation $\mathcal{T}$ is said to be locally optimal* w.r.t. *a criterion K provided that every quadrilateral defined by a pair of triangles in $\mathcal{T}$ which share a common edge is optimally triangulated* w.r.t. *K.*

For a given point set, there can, of course, be several locally optimal triangulations. For example, both of the triangulations in Fig. 9.3 are locally optimal w.r.t. the min-max angle test.

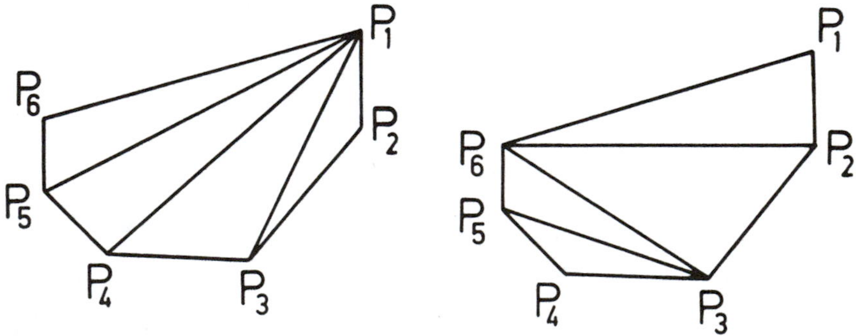

Fig. 9.3. Two triangulations both of which are locally optimal w.r.t. the min-max angle criterion (see [Schu 87]).

Starting with some local criterion for comparing triangulations, to get a global criterion, we first construct a vector with M components, where M is the number of triangles. In particular, if $\alpha(T_j)$ is a measure for the *numerical quality* of a triangle T_j (e.g. one of the quantities involved in the local criteria discussed above), then we define the vector $\boldsymbol{a}(\mathcal{T}) = (\alpha_1, \dots, \alpha_M)$ to be the result of sorting the quantities $\alpha(T_j)$ into increasing (or decreasing) order.

Then if $\mathcal{T}$ and $\widetilde{\mathcal{T}}$ are two triangulations, we can compare them using the *lexicographical order* for vectors: given two vectors $\boldsymbol{a}(\mathcal{T})$ and $\boldsymbol{a}(\widetilde{\mathcal{T}})$, we say $\boldsymbol{a}(\mathcal{T}) < \boldsymbol{a}(\widetilde{\mathcal{T}})$ provided that there exists some $k \in \mathbb{N}$ such that $\alpha_i = \tilde{\alpha}_i$ for $i = 1(1)k{-}1$ and $\alpha_k < \tilde{\alpha}_k$. Now following [Schu 87], we can introduce

Definition 9.6. *A triangulation $\mathcal{T}$ of a point set $\mathcal{P}$ is called globally optimal w.r.t. a criterion K provided that $\boldsymbol{a}(\mathcal{T}) \geq \boldsymbol{a}(\widetilde{\mathcal{T}})$ (resp. $\leq$) for every triangulation $\widetilde{\mathcal{T}}$ of $\mathcal{P}$.*

A globally optimal triangulation $\mathcal{T}$ is, of course, also locally optimal, and in fact is unique up to *neutral cases* where changing the triangulation does not change $\boldsymbol{a}(\mathcal{T})$. We should note, however, that the max-min angle criterion is the only known criterion for which a local optimum is also a global optimum in the sense of Definition 9.6. The following example is taken from the paper [Nie 87a]:

Example 9.1. Fig. 9.4 shows the 10 possible triangulations $\{\mathcal{T}^i\}_{i=1}^{10}$, associated with the six points

$$\boldsymbol{P}_1 = (5,9) \qquad \boldsymbol{P}_3 = (4,2) \qquad \boldsymbol{P}_5 = (8.5,4)$$
$$\boldsymbol{P}_2 = (2,5) \qquad \boldsymbol{P}_4 = (7,1.5) \qquad \boldsymbol{P}_6 = (5,8).$$

For the max-min and min-max criteria, the sorted vectors associated with these ten different triangulations are:

Max-min angle criterion	**Min-max angle criterion**
$\boldsymbol{a}(\mathcal{T}^1) = (0.04, 0.14, 0.35, 0.46, 0.62)$	$\boldsymbol{A}(\mathcal{T}^1) = (2.84, 2.36, 1.99, 1.77, 1.57)$
$\boldsymbol{a}(\mathcal{T}^2) = (0.02, 0.04, 0.35, 0.46, 0.50)$	$\boldsymbol{A}(\mathcal{T}^2) = (2.98, 2.84, 1.99, 1.91, 1.57)$
$\boldsymbol{a}(\mathcal{T}^3) = (0.02, 0.11, 0.42, 0.46, 0.50)$	$\boldsymbol{A}(\mathcal{T}^3) = (2.98, 2.42, 1.91, 1.88, 1.57)$
$\boldsymbol{a}(\mathcal{T}^4) = (0.04, 0.14, 0.35, 0.37, 0.66)$	$\boldsymbol{A}(\mathcal{T}^4) = (2.84, 2.36, 2.32, 1.99, 1.40)$
$\boldsymbol{a}(\mathcal{T}^5) = (0.11, 0.14, 0.42, 0.46, 0.62)$	$\boldsymbol{A}(\mathcal{T}^5) = (2.42, 2.36, 1.88, 1.77, 1.57)$
$\boldsymbol{a}(\mathcal{T}^6) = (0.02, 0.11, 0.50, 0.58, 0.88)$	$\boldsymbol{A}(\mathcal{T}^6) = (2.98, 2.42, 1.95, 1.91, 1.27)$
$\boldsymbol{a}(\mathcal{T}^7) = (0.11, 0.14, 0.37, 0.42, 0.66)$	$\boldsymbol{A}(\mathcal{T}^7) = (2.42, 2.36, 2.32, 1.88, 1.40)$
$\boldsymbol{a}(\mathcal{T}^8) = (0.11, 0.14, 0.37, 0.46, 0.70)$	$\boldsymbol{A}(\mathcal{T}^8) = (2.42, 2.36, 2.32, 1.50, 1.50)$
$\boldsymbol{a}(\mathcal{T}^9) = (0.11, 0.14, 0.57, 0.58, 0.70)$	$\boldsymbol{A}(\mathcal{T}^9) = (2.42, 2.36, 1.95, 1.74, 1.50)$
$\boldsymbol{a}(\mathcal{T}^{10}) = (0.11, 0.14, 0.58, 0.62, 0.88)$	$\boldsymbol{A}(\mathcal{T}^{10}) = (2.42, 2.36, 1.95, 1.77, 1.27)$

Comparing the vectors $\boldsymbol{a}(\mathcal{T}^i)$ corresponding to the max-min angle criterion lexicographically, we find that

$$\begin{aligned}\boldsymbol{a}(\mathcal{T}^2) < \boldsymbol{a}(\mathcal{T}^3) < \boldsymbol{a}(\mathcal{T}^6) < \boldsymbol{a}(\mathcal{T}^4) < \boldsymbol{a}(\mathcal{T}^1)\\ < \boldsymbol{a}(\mathcal{T}^7) < \boldsymbol{a}(\mathcal{T}^8) < \boldsymbol{a}(\mathcal{T}^5) < \boldsymbol{a}(\mathcal{T}^9) < \boldsymbol{a}(\mathcal{T}^{10}),\end{aligned}$$

and so the global optimum is given by $\mathcal{T}^{10}$. For this criterion, no matter where we start, if we always move to a better triangulation, we always end up at the global optimum, see Fig. 9.4 which was kindly provided by G. Nielson.

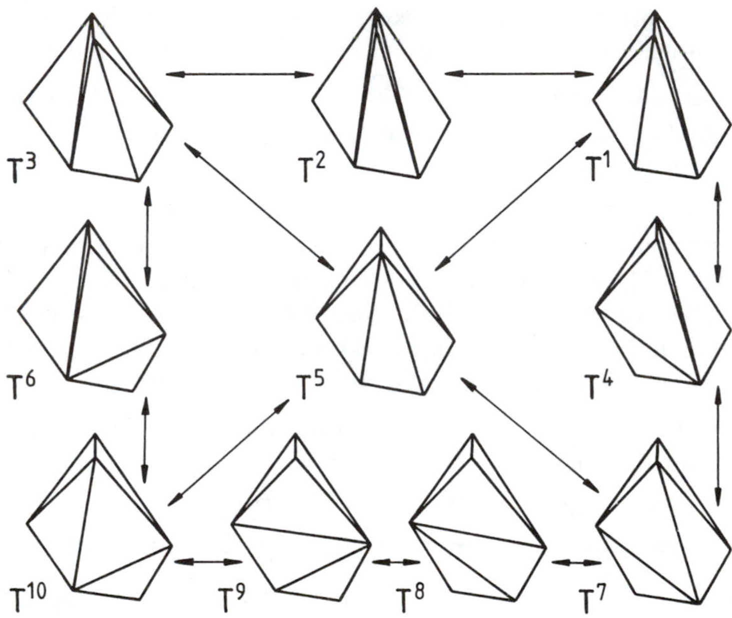

Fig. 9.4. Ten possible triangulations of 6 points.

For the min-max angle criterion, putting the vectors in lexicographical order gives

$$\begin{aligned}\boldsymbol{A}(\mathcal{T}^5) &< \boldsymbol{A}(\mathcal{T}^9) < \boldsymbol{A}(\mathcal{T}^{10}) < \boldsymbol{A}(\mathcal{T}^8) < \boldsymbol{A}(\mathcal{T}^7)\\ &< \boldsymbol{A}(\mathcal{T}^1) < \boldsymbol{A}(\mathcal{T}^4) < \boldsymbol{A}(\mathcal{T}^3) < \boldsymbol{A}(\mathcal{T}^6) < \boldsymbol{A}(\mathcal{T}^2).\end{aligned}$$

Here the global optimum is provided by $\mathcal{T}^5$. But in contrast to the case above, now it may happen that if we start at some triangulation and iteratively improve it by swapping edges, we may end up at the local optimum $\mathcal{T}^9$, see Fig. 9.4. Once we are there, it is impossible to reach the globally optimum triangulation $\mathcal{T}^5$ (w.r.t. the min-max angle criterion) since both $\mathcal{T}^{10}$ and $\mathcal{T}^8$ are worse than $\mathcal{T}^9$.

The situation in the above example can arise for all of the above criteria except for the max-min angle criterion. However, it has been observed that the max-min angle and min-max angle criteria often lead to the same triangulation. Moreover, comparison tests of [Nie 83] using randomly generated points showed that, in general, the triangulations associated with the two

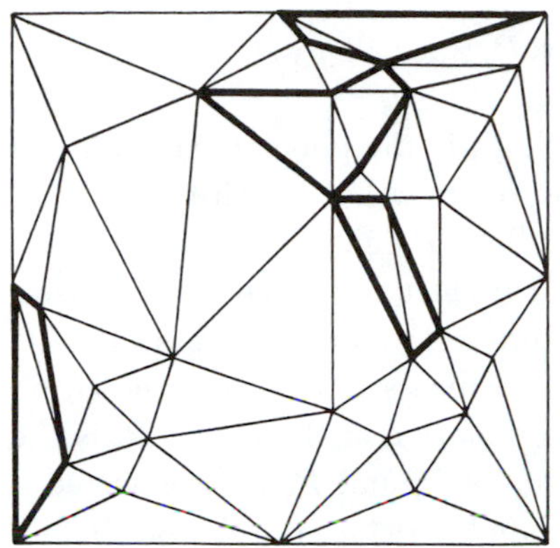

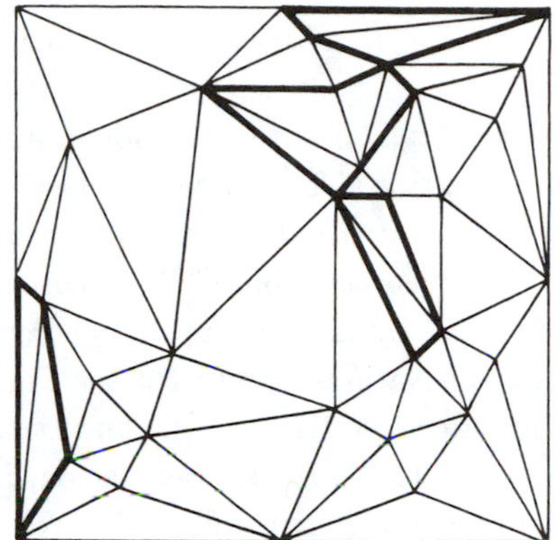

Fig. 9.5a. Max-min triangulation. **Fig. 9.5b.** Min-max triangulation.

criteria differ by at most a few triangles, on the order of 10%. Figs. 9.5a,b (taken from [Nie 83]) show a typical example.

Fig. 9.5c provides another comparison of the min-max angle and max-min angle criteria in terms of the loci of points $\boldsymbol{P}_\ell$ (the curves K_1 and K_2) which lead to the neutral cases for the associated quadrilaterals, see [Hans 90]. Clearly, neither of the sets enclosed by these curves contains the other, and so the two criteria can produce different triangulations, but only for points lying in the cross-hatched area, which is relatively small.

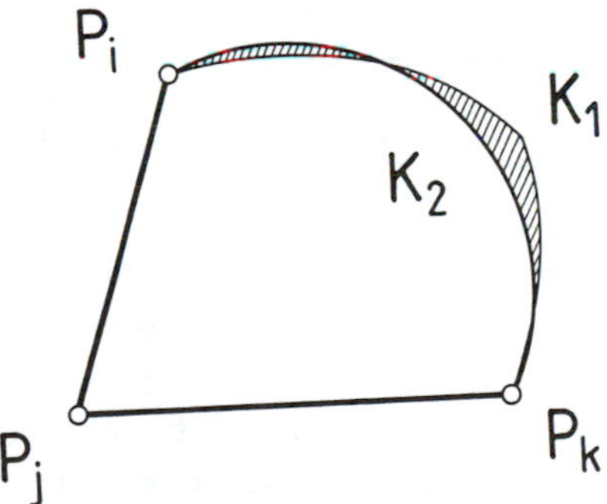

Fig. 9.5c. Loci of points leading to the neutral case for the min-max angle (thin line K_1) and the max-min angle criterion (thick line K_2).

It turns out that Lawson's local method actually delivers a global optimum, since the triangulation associated with the max-min angle criterion is a *Delaunay triangulation* of $\mathcal{P}$, see [Sib 78], [Law 72]. To explore this further, we recall that given a point set $\mathcal{P} = \{\boldsymbol{P}_i\}$, the corresponding *Dirichlet tessellation* (also called the *Thiessen* or *Voronoi tesselation*) is defined to be the partition of $\mathbb{R}^2$ into Dirichlet tiles

$$\boldsymbol{F}_i = \{\boldsymbol{x} \in \mathbb{R}^2 \ : d(\boldsymbol{x}, \boldsymbol{P}_i) \leq d(\boldsymbol{x}, \boldsymbol{P}_j) \ \text{ for all } j \neq i\}.$$

Here $d(\boldsymbol{x}, \boldsymbol{P}_k)$ is the Euclidean distance, *i.e.*, $\boldsymbol{F}_i$ is the polygon consisting of all points $\boldsymbol{x} \in \mathbb{R}^2$ which are closer to $\boldsymbol{P}_i$ than to any other $\boldsymbol{P}_j$ with $j \neq i$. The $\boldsymbol{F}_i$ are pairwise disjoint, and cover all of $\mathbb{R}^2$ [PREP 85].

Given the points $\boldsymbol{P}_i$, a corresponding Dirichlet tesselation can be constructed by finding the perpendicular bisectors to the line segments connecting the various points $\boldsymbol{P}_i$, see Fig. 9.6. (For a recursive algorithm based on this idea, see [Gree 78]). The Delaunay triangulation of the points $\boldsymbol{P}_i$ is the *dual* to the Dirichlet tesselation, *i.e.*, two points $\boldsymbol{P}_i$ and $\boldsymbol{P}_j$ are connected if and only if the tiles $\boldsymbol{F}_i$ and $\boldsymbol{F}_j$ of the associated Dirichlet tesselation share a (nontrivial) common edge, see [Law 77], [Gui 85], [PREP 85], and Fig. 9.6.

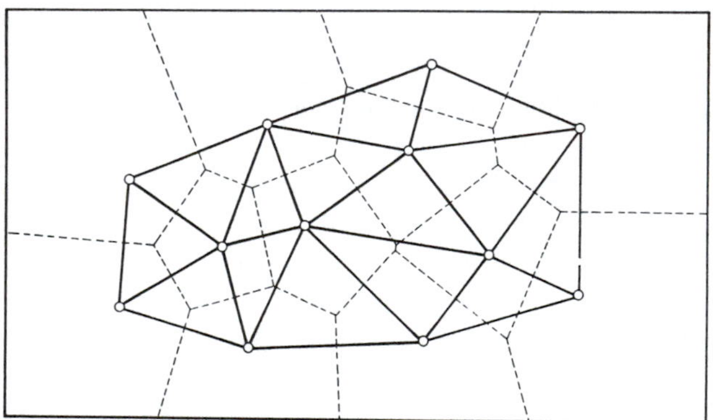

Fig. 9.6. Dirichlet tesselation and associated Delaunay triangulation.

Lawson [Law 77], [Sib 78] also showed that a Delaunay triangulation can be constructed using an appropriate "circle criterion". There are two versions of this criterion. We say that the *local circle criterion* is satisfied for a quadrilateral with vertices $\boldsymbol{P}_i$, $\boldsymbol{P}_j$, $\boldsymbol{P}_k$, $\boldsymbol{P}_\ell$ (see [Cli 84]) provided that the circumscribed circle associated with the triangle T_{ijk} with vertices $\boldsymbol{P}_i, \boldsymbol{P}_j, \boldsymbol{P}_k$ does not contain the vertex $\boldsymbol{P}_\ell$ of the triangle $T_{jk\ell}$ with vertices $\boldsymbol{P}_j, \boldsymbol{P}_k, \boldsymbol{P}_\ell$ which shares the edge $\boldsymbol{P}_j\boldsymbol{P}_k$, see Fig. 9.7.

If the local circle criterion is satisfied for every convex quadrilateral, then so is the "strong" global circle criterion which requires that for every triangle in the triangulation, the associated circumscribed circle contains no other data point [Law 77], [Sib 78]; see also [Law 86], [Alf 89].

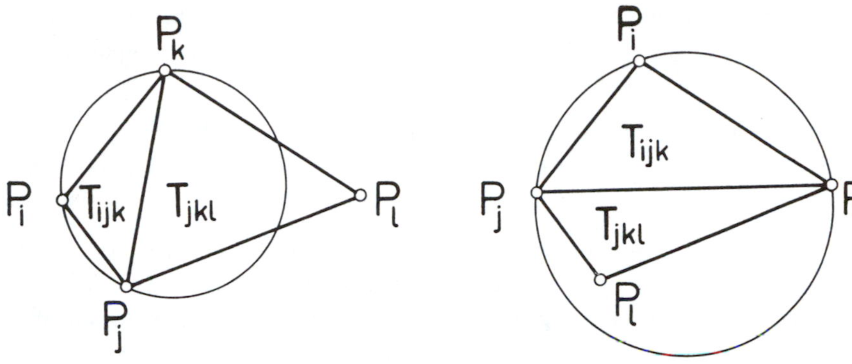

Fig. 9.7a. Local circle criterion is satisfied.

Fig. 9.7b. Local circle criterion is not satisfied.

Since a triangulation satisfying the global circle criterion is dual to the Dirichlet tesselation, and thus is equivalent to a Delaunay triangulation [Law 77], [Sib 78], it follows that the local circle criterion can be used to find a global optimum with respect to the max-min angle criterion via a series of local changes. This does not hold, however, for the min-max angle criterion, as shown in [Hans 90] by analyzing neutral cases.

Algorithms for constructing triangulations can be divided into three distinct classes, see [PREP 85], [Flo 87], and [Schu 87]:

- *post optimization algorithms*, where we first construct an initial triangulation, and then optimize using a local criterion, see e.g., [Law 72], [Lewi 78], and [Mir 82].
- *iterative (recursive) algorithms*, where we first construct an initial triangle, and then add one point at a time, always maintaining a locally optimal triangulation, see e.g., [Law 77], [Sib 78a], [Aki 78], [Gree 78], [Bow 81], [Boi 84], [Ren 84b], [Cli 84], and [Agi 91].
- *Divide-and-conquer algorithms*, where we first split the data set $\mathcal{P}$ into subsets, then construct locally optimal triangulations for each, and finally combine these triangulations, see e.g., [Lee 80], [Joe 86], [Gui 85], [Dwy 87], and [Hua 89].

For a comparison of the various triangulation algorithms [Law 77], [Gree 78], [Aki 78, 78a], [Ren 84b] with respect to memory and computation time, see [Cli 84]. The computational times of the three tested algorithms are essentially equal ([Law 77] is the fastest and [Aki 78, 78a] is the slowest). The memory requirements are quite different: the methods in [Ren 84b], [Law 77], and [Aki 78, 78a] require $7N$, $18N$, and $32N$ storage locations, respectively.

Summary. Since a good algorithm should always lead to a globally optimal triangulation, this suggests that if we want to construct triangulations using a local criterion, then we should use the max-min angle criterion (or equivalently the circle criterion) which leads to a Delaunay triangulation.

Remark. The triangulations corresponding to the criteria discussed above are not affinely invariant. The construction of (locally and globally optimal) affine invariant triangulations (cf. [Nie 89]) will be discussed in Sect. 9.6.

Remark. It is also possible to construct triangulations on curved surfaces using curved triangles (e.g., spherical triangles on the surface of a sphere). These can be useful for solving scattered data interpolation problems on such surfaces, see Sect. 9.7.3.

Remark. Algorithms for triangulating domains which are nonconvex, have holes in them, or are to include certain prescribed interior edges (so-called *barriers*), can also be constructed using the above criteria, see [Lewi 78], [Nie 83], [Barn 85], and also [Hua 89], [Cli 90].

Remark. Triangulations which depend not only on the location of the abscissae $\boldsymbol{x}_i = (x_i, y_i)$ of the data points, but also on the prescribed function values f_i at the $\boldsymbol{x}_i$ have been studied by [Cho 88] and [Dyn 90]. In [Dyn 90, 90b], [Rip 90, 90a] it was shown that for certain underlying functions with very specific behavior, taking account of the function values in constructing the triangulation often leads to substantially better piecewise linear approximations. In particular, this process allows long thin triangles, which can be very useful when there are strong variations in the size of the normal vector to the surface. Similar results can be found in [Qua 89, 90, 90a]; see also [Brow 91].

9.3.2. Triangular Interpolants

9.3.2.1. Nine Parameter Interpolant

Suppose we are given a function value and values for the first derivatives at each vertex of a triangulation. For each edge of the triangulation, this data determines the Bézier points of a cubic boundary curve. Thus, for each triangle, we have determined nine Bézier points along the edges which can be considered to be associated with a cubic patch defined on the triangle (cf. Sect. 4.1.4). This leaves only b_{111} to be determined.

If we choose

$$b_{111} = E = \tfrac{1}{3}(b_{300} + b_{030} + b_{003}), \tag{9.17}$$

then we get an interpolant which possesses linear precision, *i.e.*, if we take the nine pieces of data (three at each vertex) from a linear surface (plane), then the 9-parameter interpolant with b_{111} as in (9.17) reproduces the linear surface exactly. On the other hand, if we choose

$$b_{111} = \tfrac{3}{2}Q - \tfrac{1}{2}E, \tag{9.18}$$

where

$$Q = \tfrac{1}{6}(b_{201} + b_{102} + b_{021} + b_{012} + b_{210} + b_{120}),$$

then we get an interpolant which even has quadratic precision (meaning that quadratic surfaces are exactly reproduced).

One disadvantage of the 9-parameter interpolant is that it requires both function values and first derivatives, but produces a surface defined over the entire triangulation which is only C^0 continuous, see [STRA 73], [ZIE 77], and [Far 83a, 86]. This follows from the fact that the value of b_{111} depends on the data prescribed around the whole triangle rather than just data pertaining to only one edge. One can, of course, use the given function values at the vertices to estimate the needed derivatives as is done e.g., in the software package [Ren 84b], see also [SPÄ 91]. While the 9-parameter method can be used with any prescribed triangulation, it may be advantageous to adjust the triangulation to the data (cf. e.g. [Rip 90, 90a], [Dyn 90], [Qua 89, 90, 90a], [Brow 91]) rather than to simply use, for example, the Delaunay triangulation.

9.3.2.2. $\mathrm{C^r}$ Continuous Hermite Interpolants

Suppose we are given the values of a function and its first r derivatives at a set of points in the plane. In this section we are interested in interpolating this data using globally C^r piecewise polynomial functions defined on a triangulation of the data points. In order to avoid having to solve large systems of equations for the coefficients of an interpolant, we shall focus on methods which can be used to construct the interpolant on each triangle separately.

To analyze this problem in more detail, it is convenient to write each polynomial piece in Bézier form. Then it is clear (cf. Fig. 9.8 which illustrates the case $r = 3$) that for each pair of edges, there is a set of Bézier points (shown cross-hatched in the figure) which are affected by smoothness conditions involving neighboring patches, and this set involves Bézier points up to a distance of $2r$ from the common vertex. Thus, in order to keep the construction local, it is common practice to assign derivatives up to order $2r$ at each vertex. As the following theorem shows, this is only possible if $n \geq 4r + 1$.

Theorem 9.1. *Suppose we are given a triangulation. Then a necessary condition to assure the existence of a globally C^r piecewise polynomial of degree n which interpolates arbitrarily given function values and derivatives up to order $2r$ at each of its vertices $\boldsymbol{P}_i$ is that $n \geq 4r + 1$.*

Proof: Suppose we write the interpolant in Bézier form. Then the interpolation conditions immediately determine the Bézier points at the vertex and within a distance of $2r$ from the vertex. But if $n < 4r + 1$, these sets overlap at some Bézier point(s) along each edge (see Fig. 9.8), and hence the data at the vertices cannot be specified independently. ■

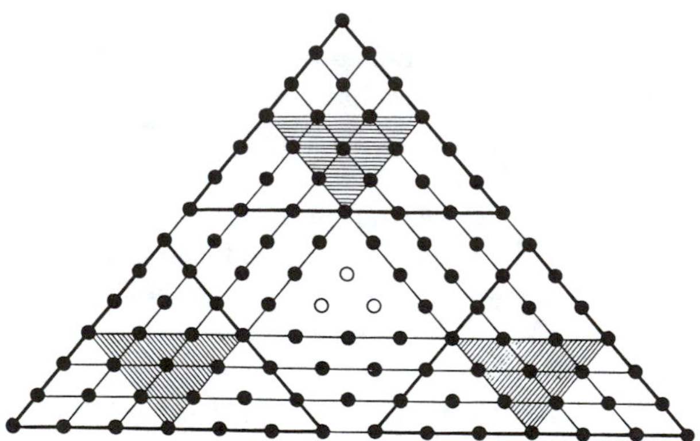

Fig. 9.8. An example for the proof of Theorem 9.1.

As already happens in the 9-parameter case, even if we prescribe all derivatives up to order r along the edges and up to order $2r$ at the vertices, the interpolant is not uniquely defined. In fact, there remain $r(r-1)/2$ Bézier ordinates in the interior of the Bézier net which are undefined by the data and which are not involved in any C^r continuity condition, cf. Fig. 9.8. These parameters can be used to make the polynomial precision of the interpolant as high as possible.

To handle those cases where we do not have enough cross-boundary derivative information to determine all Bézier coefficients within a distance of r of each edge, we can introduce a *condensation of the parameters*. To see how, we note that the p-th derivative of a Bézier patch of degree n transversal to an edge of the triangle can be written as a univariate Bézier polynomial of degree $n - p$. If we require that this derivative be a polynomial of degree $n - p - q$ instead of $n - p$, this introduces an additional q conditions for determining the Bézier ordinates in the rows of the Bézier net parallel to the edge.

A disadvantage of the C^r continuous Hermite interpolant is that to get C^r continuity, we need C^{2r} data, and moreover, have to use a rather high polynomial degree ($n \geq 4r+1$), see [Zen 70], [STRA 73], and [ZIE 77]. In particular,

- for C^1 continuity, we need to use quintic patches, see [Bel 69], [Aki 78, 78a] (for an example, see Fig. 9.12h), [Barn 81] (see also [Far 86]), and [Preu 84, 84a, 90, 90a],
- for C^2 continuity, we need to use nonic patches, see [Sabl 85a], [Whe 86], [Res 86], and [Preu 90, 90a].

We note that from the standpoint of visual appearance, there is very little difference between C^1 and C^2 continuous interpolants, see [Ren 84], [Preu 90a], but C^1 routines are about three times as fast as C^2 routines [Preu 90a].

One way of reducing the degrees of the patches, along with the number of derivatives needed from $2r$ to r, is to form so-called *macro elements* by subdividing each triangle in the triangulation into several subtriangles. Other possibilities include using rational representations (see e.g., [Bir 74], [Mans 74]), or weakening the C^r continutity to GC^r continuity, see Chap. 7. For example, to construct a GC^1 continuous Hermite interpolant, it suffices to use a polynomial of degree 4 instead of 5, cf. [Pip 87].

In the following two subsections, we discuss two of the most popular macro-element methods, the *Clough-Tocher* and the *Powell-Sabin* schemes. These methods construct polynomial interpolants, and thus are simpler to use than rational interpolation. Our analysis will be based on Bézier representations which will greatly simplify the construction.

9.3.2.3. Clough-Tocher Interpolants

To construct a C^1 continuous cubic Clough-Tocher interpolant [Clo 65], we divide each triangle into three micro-elements by connecting its vertices to a point lying in the interior, which to achieve symmetry is usually taken to be the center of gravity of the triangle. For each vertex of the triangulation, we prescribe C^1 data, *i.e.*, a function value and two first derivatives, along with some cross-boundary derivative at the center of each edge of the triangulation. The directions are usually chosen to be perpendicular to the edges, in which case we speak of cross-boundary normal derivatives. This gives a total of 12 data per macro-element, which we now use to construct a globally C^1 continuous, piecewise cubic interpolant, see e.g., [STRA 73], [Perc 76], [ZIE 77], and [LAN 86]. The Bézier ordinates of each micro-element are uniquely determined by the given data and the C^1 continuity conditions in a four step process, see Fig. 9.9, [Alf 84a], and [Far 86].

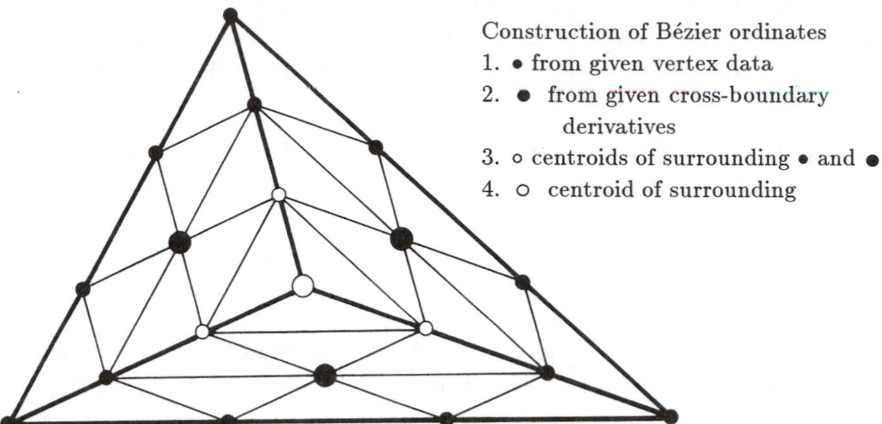

Construction of Bézier ordinates
1. • from given vertex data
2. ● from given cross-boundary derivatives
3. ∘ centroids of surrounding • and ●
4. ○ centroid of surrounding

Fig. 9.9. Clough-Tocher interpolant (top view).

It is of interest to note that the C^1 continuous cubic Clough-Tocher interpolant is actually C^2 continuous at the center point of each macro-element [Alf 84a], [Far 86]. For an algorithm, see [Law 77] (or also [Ren 84, 84b]). Fig. 9.12i shows an example of using the method. Other examples can be found in [SPÄ 91].

If we are not given the required cross-boundary derivatives at the centers of each edge of the triangulation, we can still define an interpolant by performing a condensation of the parameters. Alternatively, we can estimate the missing derivatives, for example by choosing them so that they are as close as possible to those of the 9-parameter interpolant [Far 83a], or by making the jumps in the second derivatives of the two adjoining patches as small as possible [Far 85a].

To construct C^2 continuous Clough-Tocher interpolants, we need C^3 data and polynomial degree 7 (C^2 data and polynomial degree 5 do not suffice) [Sabl 85a, 87], [Far 86]. Alternatively, one can form a different macro-element involving at least two interior lines connected to each vertex. [Alf 84] gave a construction based on C^2 data with $n = 5$ where three interior lines were used at each corner in forming the macro-elements.

9.3.2.4. Powell-Sabin Interpolants

Given prescribed C^1 data at the vertices of a triangulation, we now want to construct a globally C^1 continuous, piecewise quadratic interpolant. Clearly, in order to work with quadratic polynomials, we will have to use a larger number of micro-elements. If the largest interior angle of a given triangle is smaller than (the heuristically selected value of) 75 degrees, then we divide it into six micro-elements; otherwise we divide it into twelve. In the first case, we accomplish this by connecting the vertices and the midpoints of the sides with a point lying in the interior of the triangle ([Pow 77] chooses the center of the circumscribed circle). In the second case, we further subdivide the triangle by connecting the midpoints of the sides to each other. The reason for the two cases is that the center of the circumscribed circle of a given triangle can lie outside or very close to one of the sides of the triangle, which can lead to long thin triangles. In this regard, it is better to use the center of the inscribed circle, since this avoids the need to consider two cases.

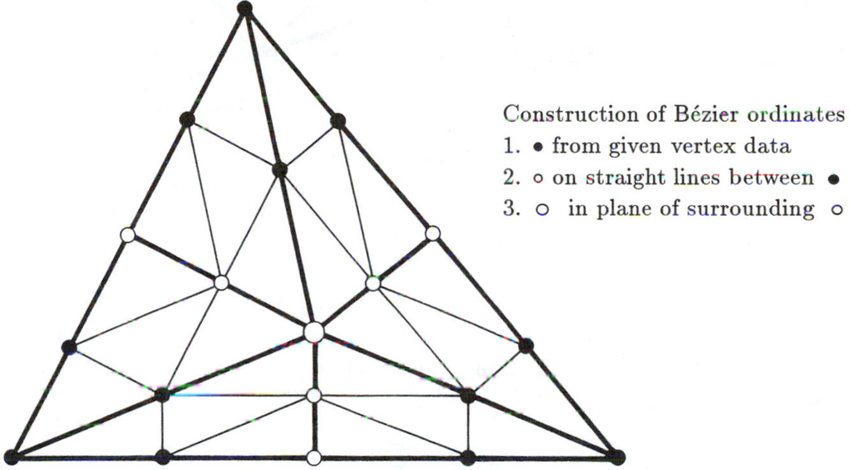

Fig. 9.10. Six-triangle Powell-Sabin interpolant (top view).

The Bézier ordinates of the quadratic Bézier polynomial associated with each micro-element can be determined by a three or four step process from the given data using the C^1 continuity conditions, see Figs. 9.10 and 9.11. The key to this is the fact that the normal cross-boundary derivatives along each edge of the triangles are linear instead of piecewise linear. This is automatically fulfilled in the first case in view of the way the partition is constructed.

An algorithm based on the six-micro-element partition of a triangle is described in [Cen 87]. Formulae for the twelve element case are given in [Chu 90]. Software and several examples can be found in [SPÄ 91]. C^k continuous Powell-Sabin interpolants with $k > 1$ were treated in [Sabl 85a] from the theoretical standpoint, and from a more constructive standpoint in the special case $k = 2$ in [Sabl 87].

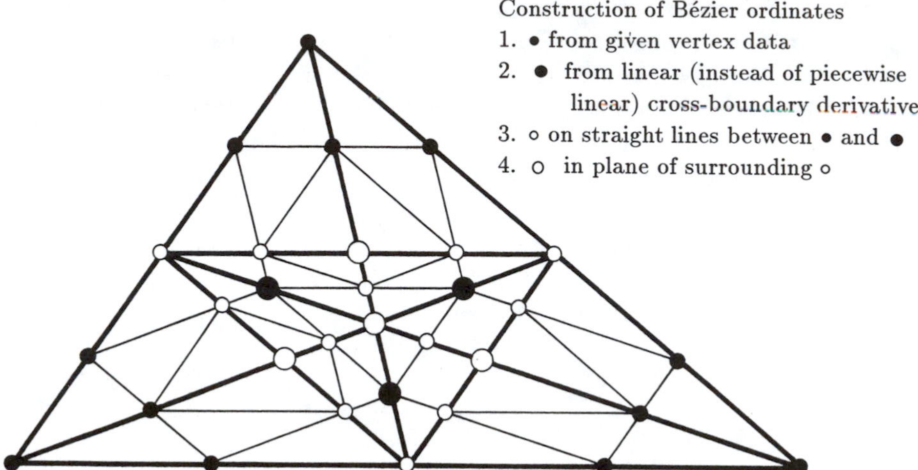

Fig. 9.11. Twelve-triangle Powell-Sabin interpolant (top view).

9.3.2.5. Rational Interpolants

The construction of rational interpolants is similar to that of the Shepard interpolants (9.1) and (9.2), but with the difference that now $f(\boldsymbol{x})$ is defined piecewise over the individual triangles T_{ijk} whose vertices are $\boldsymbol{x}_i$, $\boldsymbol{x}_j$ and $\boldsymbol{x}_k$. Suppose we define f on T_{ijk} by

$$f_{ijk}(\boldsymbol{x}) = \omega_i(\boldsymbol{x})Q_i(\boldsymbol{x}) + \omega_j(\boldsymbol{x})Q_j(\boldsymbol{x}) + \omega_k(\boldsymbol{x})Q_k(\boldsymbol{x}),$$

where in order to satisfy the interpolation conditions, we choose the weight functions ω_i so that $\omega_i(\boldsymbol{x}_k) = \delta_{ik}$, and the $Q_i(\boldsymbol{x})$ such that $Q_i(\boldsymbol{x}_i) = z_i$, see e.g., [Fra 82], [Lit 83], and [Agi 91]. Rational triangular interpolants can often be obtained by discretizing a transfinite triangular interpolant (see Sect. 8.2). This is the case, for example, for Nielson's minimum-norm network (MNN) interpolant, [Nie 83a, 84a], [Fra 80, 82] (see also Fig. 9.12j). For further examples, see [Barn 77, 84a], [Alf 84b, 89], and [Nie 87c].

9.3.2.6. Transfinite Interpolants

Transfinite interpolants (see Chap. 8) can, in principle, also be used to construct interpolating surfaces using the FEM method. Their main drawback is that they require data (such as function values and derivatives) at all of the points on the boundaries rather than just at the vertices.

9.3.3. Estimating Derivatives

Derivative data play an important role in most triangular interpolants. The quality of the corresponding interpolating surfaces (as measured in terms of visual appearance, smoothness, and accuracy, etc.) depends critically on the "*accuracy*" of this derivative data. Frequently, the derivatives have even more of an influence than the polynomial degree or continuity order [Ren 84], [Preu 90a]. Since we have already dealt with the problem in Sect. 8.1.2, here we may be brief. Programs for some of the methods can be found in [SPÄ 91].

9.3.3.1. Weighted Averages

The partial derivatives f_x and f_y at a data point $\boldsymbol{V}_i = (\boldsymbol{x}_i, z_i)$ can be estimated as a weighted average (convex combination) of the slopes $E_x(i,j,k)$ and $E_y(i,j,k)$ of the set of planes E_{ijk} which pass through the data point $\boldsymbol{V}_i$ and various pairs of neighboring points $\boldsymbol{V}_j$ and $\boldsymbol{V}_k$. This leads to the formulae

$$f_x(\boldsymbol{V}_i) = \sum \omega_{ijk} E_x(i,j,k), \qquad f_y(\boldsymbol{V}_i) = \sum \omega_{ijk} E_y(i,j,k), \tag{9.19}$$

with weights

$$\omega_{ijk} = \frac{\sigma_{ijk}}{\sum \sigma_{ijk}},$$

where the sums are to be taken over the set

$$N_i = \{(i,j,k) : j \neq k, \text{ where } \boldsymbol{V}_j, \boldsymbol{V}_k \text{ satisfy a "selection criterion"}\}.$$

The selection criterion might choose all of the points which are connected directly to $\boldsymbol{V}_i$, or it might involve the points which are within a prescribed distance of $\boldsymbol{V}_i$. There are a variety of possible choices for the weights σ_{ijk}. For example, we can take

$$\sigma_{ijk} = 1, \qquad \text{[Klu 78] (arithmetic mean)},$$

$$\sigma_{ijk} = d_i(\boldsymbol{x}_j)^{-2} d_i(\boldsymbol{x}_k)^{-2}, \qquad \text{[Lit 83]},$$

$$\sigma_{ijk} = \cos\gamma_{ijk} \cdot \text{area}\{\boldsymbol{V}_i, \boldsymbol{V}_j, \boldsymbol{V}_k\}, \qquad \text{[Aki 78, 84]},$$

where γ_{ijk} is the angle between the z-axis and the normal to the plane E_{ijk}, and where area$\{\boldsymbol{V}_i, \boldsymbol{V}_j, \boldsymbol{V}_k\}$ denotes the surface area of the triangle with vertices $\boldsymbol{V}_i$, $\boldsymbol{V}_j$, $\boldsymbol{V}_k$. Derivatives of order $n > 1$ can be estimated by repeated application of the above methods; once we have estimates for the $(n-1)$-st derivatives, we can use them to estimate the n-th derivatives.

9.3.3.2. Local Interpolation and Approximation

Another approach to estimating partial derivatives at the data point $\boldsymbol{V}_i$ is to replace them by the partial derivatives of some function which locally interpolates or approximates the data in a neighborhood of $\boldsymbol{V}_i$, see [Nie 83], [Alf 89]. In [Ste 84] several such methods are compared, including

- the method of Little described above,
- Shepard's interpolant (for an implementation, see e.g., [Ren 84]),
- Hardy's multiquadric interpolant,
- linear weighted least squares approximation,
- quadratic weighted least squares approximation (see e.g., [Law 77], [Nie 83], and [Ren 84]).

Each of these methods is localized by using only data points which lie in some appropriately defined neighborhood of the point of interest (cf. the discussion in the previous subsection). In Stead's tests [Ste 84], the Hardy MQ interpolant gave the best results.

Better estimates of derivatives can be obtained using global methods, in particular when they are based on some minimization process. We now discuss two such methods.

9.3.3.3. Nielson's Minimum Norm Network

Let s_{ij} denote the edge of the triangulation with endpoints $\boldsymbol{x}_i$ and $\boldsymbol{x}_j$, and suppose that $\alpha_{ij} > 0$ is an associated tension parameter. Then it is known [Nie 84a] (see also [Fra 87] and cf. [Mont 89]) that the solution of the problem of minimizing

$$\sum_{i,j} \int_{s_{ij}} \left[\left(\frac{\partial^2 f}{\partial s_{ij}^2} \right)^2 + \alpha_{ij}^2 \left(\frac{\partial f}{\partial s_{ij}} \right)^2 \right] ds_{ij}, \tag{9.20}$$

where the sum runs over all edges s_{ij}, is a C^1 function defined on the edge network whose restrictions to the edges are linear combinations of the basis functions 1, s, $\exp(\alpha_{ij}s)$, and $\exp(-\alpha_{ij}s)$, cf. Sect. 3.6.1. Here s denotes the Euclidean distance along the edge. For $\alpha_{ij} \to 0$, the solution of (9.20) becomes the piecewise cubic curve network of [Nie 80, 83a], while in the limiting case as $\alpha_{ij} \to \infty$, we get a piecewise linear curve network defined over the edges of the triangulation.

The problem of minimizing (9.20) leads to a sparse linear system of equations, which can be efficiently solved using an iterative method. The unique solution to (9.20) is a curve network (called the *minimum norm network*) defined on the edges of the triangulation. Then as discussed above, we can extend this data to the entire domain by means of a FEM interpolant [Nie 80, 83, 83a, 84a], [Fra 82, 87], [Ren 84], and [Mont 89]. The method works very well, as shown in the example in Fig.9.12j, which is based on the interpolant of [Nie 80, 83a].

A generalization to 3D data in the sense of Chap. 7 can be found in [Nie 88]. For an extension of the function-valued minimization problem (9.20) taking account of and determining derivatives transversal to the curves of the curve network, see [Pot 91].

9.3.3.4. Alfeld's Functional Minimization

Another approach to estimating derivatives is to minimize a functional of the form

$$F^k(Q) = \sum \int_{T_{ijk}} D^k Q,$$

where D^k denotes a derivative operator, and where Q is some global interpolation process. Here T_{ijk} denotes the triangle defined by the data points $\boldsymbol{V}_i$, $\boldsymbol{V}_j$, $\boldsymbol{V}_k$ which are being interpolated by Q. The summation is over all triangles in the triangulation, and the integration is over each individual triangle. We assume that positional data is given, and that the derivative values are parameters which can be varied so as to minimize the functional. If we use the Bézier representation, then the derivatives can be expressed in terms of Bézier points, which then become the variables of the minimization problem. However, depending on the application, other functionals and interpolants can be used. This method was suggested by [Alf 85], where Q was chosen to be a condensed quintic Hermite interpolant in Bézier form.

Now the desired partial derivatives can be computed by taking the corresponding derivatives of Q. It is possible to take Q itself to be an interpolant of the scattered data requiring only C^0 data, see [Alf 85, 89], and cf. [Ren 84].

9.3.3.5. Estimating Curvatures

Several triangular interpolants (see e.g., [Hag 86a]) require some "geometric data" such as curvatures. Curvatures can be computed from derivative data, provided they are available. Alternatively, we can take curvatures from some local interpolant or approximant, or estimate them directly by a method of [Hos 77], [Tod 86].

The method of Todd and McLeod is based on estimating the *Dupin indicatrix* at each data point $\boldsymbol{V}_i = (\boldsymbol{x}_i, z_i)$ by minimizing

$$\sum_{j=1}^{4} \left[A\bar{x}_j^2 + 2C\bar{x}_j\bar{y}_j + B\bar{y}_j^2 - \operatorname{sgn}(\kappa_{nj})\right]^2,$$

over A, B, and C. Here the sum is taken over four planes which are defined by $\boldsymbol{V}_i$ and eight neighbors, and $\bar{x}_j$ and $\bar{y}_j$ are the weighted components of the direction vectors $\boldsymbol{T}_j$ which span the planes at $\boldsymbol{V}_i$, and $\text{sgn}(\kappa_{nj})$ is the sign of the normal curvature at $\boldsymbol{V}_i$ in the direction $\boldsymbol{T}_j$. Then the principle curvatures and principle curvature directions at $\boldsymbol{V}_i$ are the eigenvalues and eigenvectors, respectively, of a 2×2 matrix defined by the A, B, and C.

9.4. Multi-stage Methods

Many scattered data methods require large amounts of memory, and are expensive to compute. In addition, often we need to calculate the values of a surface at a very large number of points. For example, rendering a surface requires many evaluations, often on a regular grid. For these purposes, tensor product methods are especially efficient. Methods for arbitrarily spaced data are often global, and they usually do not have a high degree of polynomial precision. Thus, it would be convenient to combine them somehow with local methods, and with methods which can be evaluated efficiently.

The idea of *multi-stage methods*, see [Schu 76, 79], [Barn 84b], [Fol 84, 86, 87b], and [Fra 82], is to combine the properties of several, possibly very different methods. In some ways, tensor-product methods are already simple examples of multi-stage methods since they can be considered to be the result of composing curve schemes.

One way to build two-stage methods is to use the Boolean sum

$$P \oplus Q = P + Q - PQ \tag{9.21}$$

formed from two operators P and Q, see Chap. 8. As shown by Barnhill and Gregory [Barn 75] (see also [Barn 84a]) the Boolean sum $P \oplus Q$ inherits the interpolation properties of P, and the precision of Q. Moreover, assuming P and Q produce surfaces of smoothness C^p and C^q, respectively, it turns out that $P \oplus Q$ produces a surface of smoothness $C^{\min\{p,q\}}$. Thus to get a method which both interpolates and has high precison, we may form the Boolean sum of some interpolation method P (which need not have any polynomial precision) with some method Q with high polynomial precision (which need not produce an interpolant).

The Boolean approach described above is restricted to the case where both methods P and Q are defined for the same type of data. Thus, a "*true*" scattered data method and a tensor-product method which requires gridded data cannot be combined as a Boolean sum. It would be nice, however, to have a way to combine the interpolation properties of scattered data methods with the efficiency of tensor-product methods. This can be done by forming the so-called *Delta sum* [Fol 80] (see also [Fol 84, 87b])

$$P\Delta Q = QP + P(I - QP), \tag{9.22}$$

where I is the identity. If P is chosen to be a scattered data intepolator, and Q is chosen to be a tensor-product interpolator, then $P\Delta Q$ interpolates scattered data, but has tensor-product form. This is a three-step method. In the first step we perform scattered data interpolation (with P), then form the tensor-product surface QP, and finally, since this no longer interpolates the original data, we compute and add the correction term $P(I - QP)$. To describe the properties of the delta sum, we note that since it is nothing more than the Boolean sum

$$P\Delta Q = P \oplus QP,$$

the Barnhill-Gregory Theorem [Barn 75] immediately implies that $P\Delta Q$ has the interpolation properties of P, the function precision of QP (it reproduces the intersection of the sets of functions reproduced by P and Q), and yields a surface in $C^{\min\{p,q\}}$, where p and q are the orders of smoothness associated with P and Q, respectively.

In principle, any number of methods can be combined. [Barn 84b] gives a table of the interpolation properties, smoothness properties, and polynomial precision for a variety of methods. Algorithms for multi-stage methods can be found in [Fol 84, 86, 87b].

9.5. An Example

To compare some of the scattered data interpolants discussed above, we consider the 33 scattered data points shown in Fig. 9.12a, along with the test function [Fra 80]

$$\begin{aligned} T(x,y) = {} & \tfrac{3}{4}e^{-\frac{1}{4}[(9x-2)^2+(9y-2)^2]} + \tfrac{3}{4}e^{-[\frac{1}{49}(9x+1)^2+\frac{1}{10}(9y+1)]} \\ & - \tfrac{1}{5}e^{-[(9x-4)^2+(9y-7)^2]} + \tfrac{1}{2}e^{-\frac{1}{4}[(9x-7)^2+(9y-3)^2]}, \end{aligned}$$

shown in Fig. 9.12b. Figures 9.12c to 9.12j, which were kindly provided by R. Franke, show the results of interpolating this scattered data using various methods.

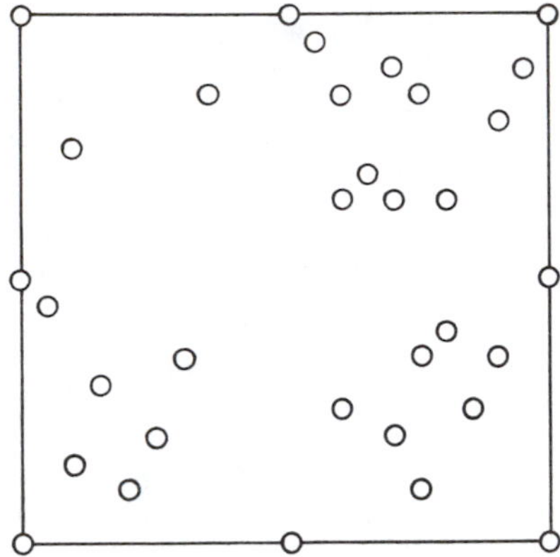

Fig. 9.12a. Data points.

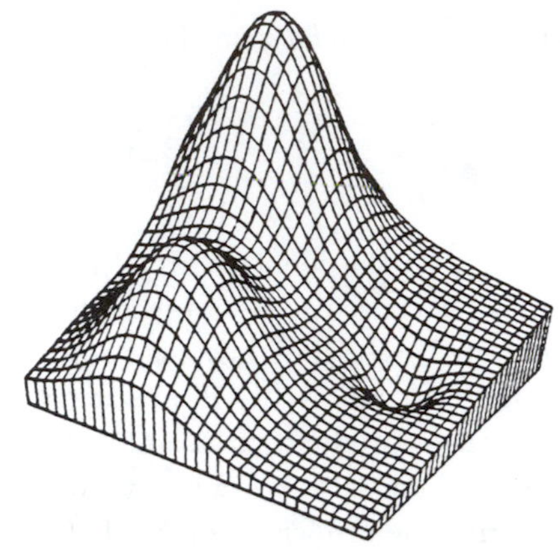

Fig. 9.12b. Test function.

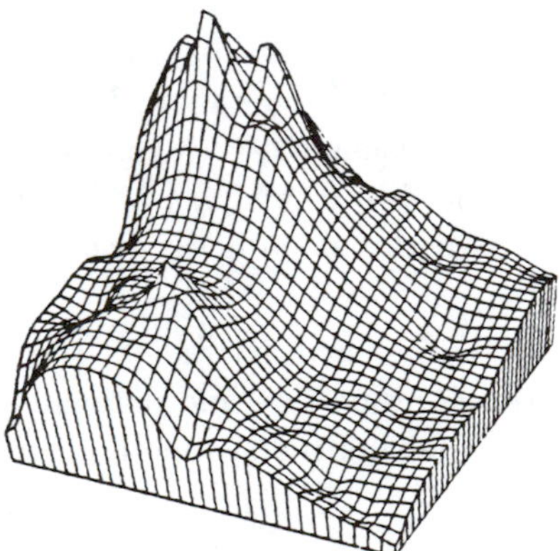

Fig. 9.12c. Shepard's method.

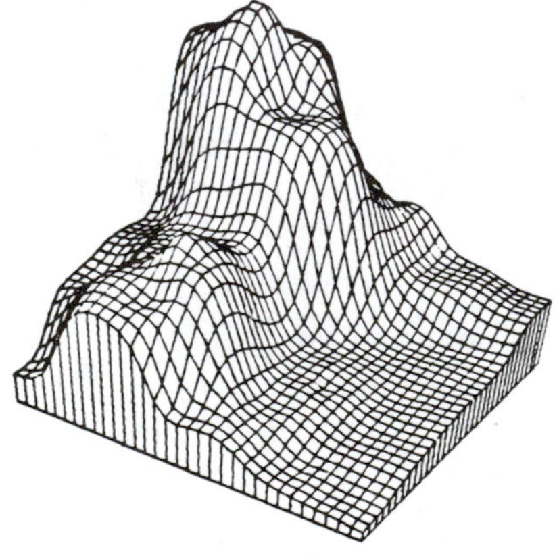

Fig. 9.12d. Modified Shepard's method of (9.7).

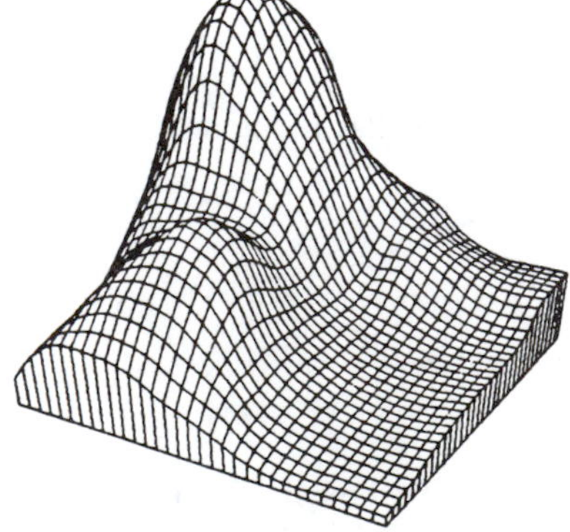

Fig. 9.12e. Modified quadratic Shepard's method [Fra 80].

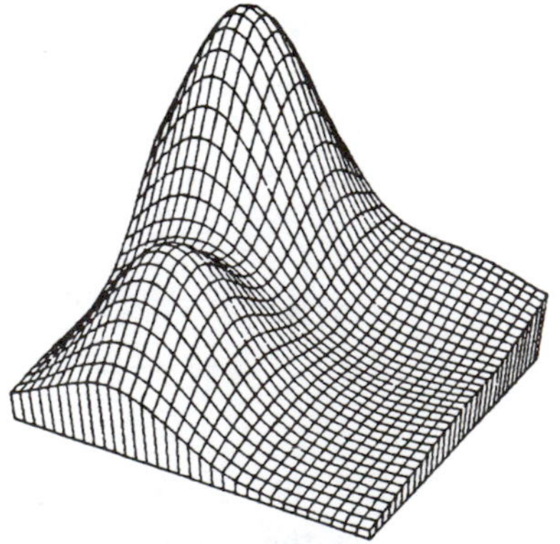

Fig. 9.12f. Hardy's MQ method.

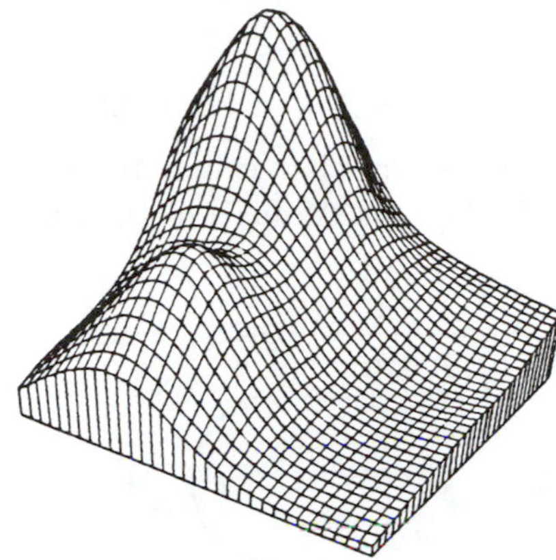

Fig. 9.12g. Duchon's TPS.

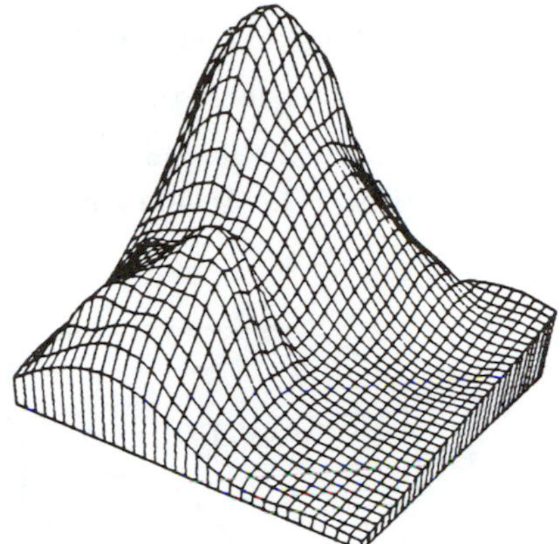

Fig. 9.12h. Akima's method.

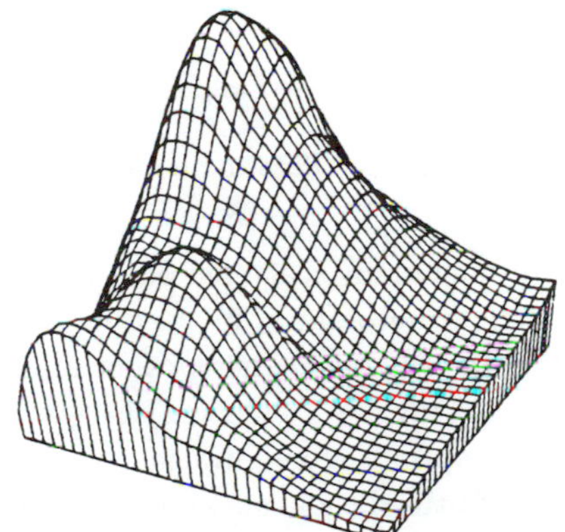

Fig. 9.12i. Lawson's method.

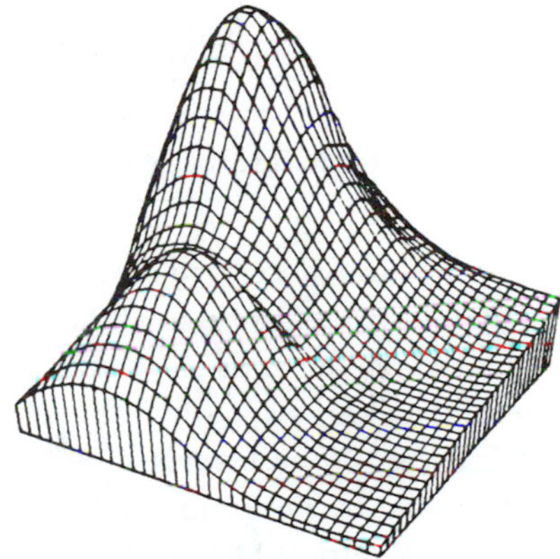

Fig. 9.12j. Nielson's MNN method.

9.6. Affine Invariance

In scattered data interpolation, usually the data are obtained by taking measurements in some units such as inches, feet, etc. Ideally, a good interpolation method should not depend on the choice of units and the placement of the origin of the coordinate system. But as shown by [Nie 87], most of the scattered data methods use some kind of basis functions which depend on the abscissae $\boldsymbol{x}_i = (x_i, y_i) \in \mathbb{R}^2$ in such a way that the space spanned by the basis functions is not closed with respect to affine transformations. Consequently, the associated interpolants are not affinely invariant.

Figure 9.13 describes the invariance of various scattered data interpolation methods in terms of four types of methods:

Type 1 : Hardy's multiquadric method [Hard 71],

Type 2 : Delta-sum method [Fol 80],

Type 3 : Franke's local thin plate spline [Fra 77, 82a],

Type 4 : Shepard's method [She 68], Duchon's thin plate spline [Hard 72], [Duc 77], McLain's method [Mcla 76], and Akima's method [Aki 78, 78a].

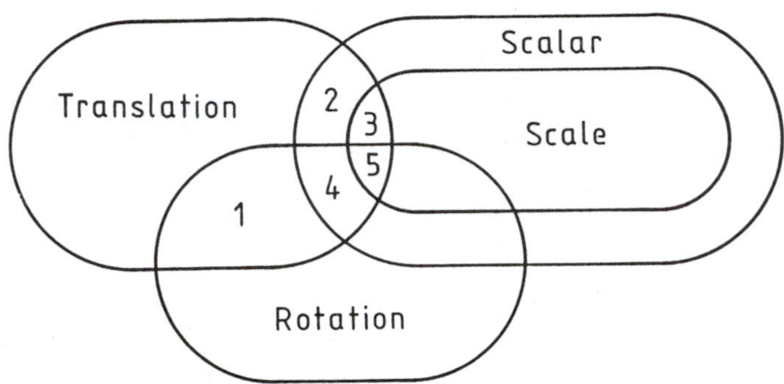

Fig. 9.13. Invariance properties of various scattered data methods.

[Nie 87, 89] described a method for modifying many scattered data methods in order to make them affinely invariant. His method can be used for any method which depends on the computation of the Euclidean distance, or more generally on the evaluation of some metric, and involves replacing the given metric by an *affine invariant metric*. The appropriate affine invariant metric can be found by performing a least squares fit to the data using an ellipse, see Sect. 4.4. Figure 9.14 (which along with Figs. 9.15-9.17 was kindly provided by G. Nielson) shows two affinely equivalent data sets, where the associated metrics are depicted by drawing lines with equal distances between them.

For data where the scaling of the x and y axes is very different, affine invariant interpolants can lead to significantly better error behavior than the original, non-modified interpolants; for ordinary data, the methods are comparable [Nie 87, 89].

FEM methods can also be made affinely invariant by using an *affine invariant triangulation* if the interpolant itself is already in an affine invariant form (e.g., in Bézier form). Since a (globally optimal) Delaunay triangulation is dual to the Dirichlet tesselation, which is defined in terms of the Euclidean metric, to get an affine invariant triangulation, we only need to construct an affine invariant Dirichlet tesselation. This can be done by using an affine invariant metric in place of the Euclidean metric. For example, we can modify Lawson's local circumscribed circle criterion (see Sect. 9.3.1.2 and

Fig. 9.7), which is equivalent to the max-min angle criterion, and which can be used to construct a globally optimal triangulation by repeatedly making local changes. Indeed, we simply replace the circle with an ellipse obtained from least squares fitting of the data. Figure 9.15 shows the Dirichlet tesselations of the two data sets of Fig. 9.14 which result from using the corresponding affine invariant metrics. Figure 9.16 shows the triangulations which arise as the duals to the tesselations in Fig. 9.15, while for comparison, Fig. 9.17 shows the triangulations obtained using the Euclidean norm.

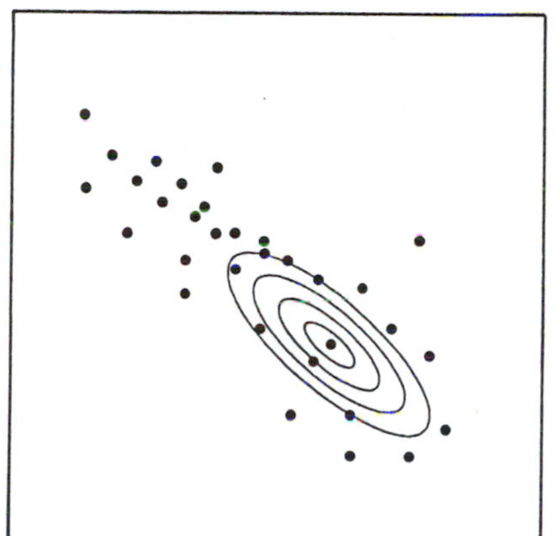

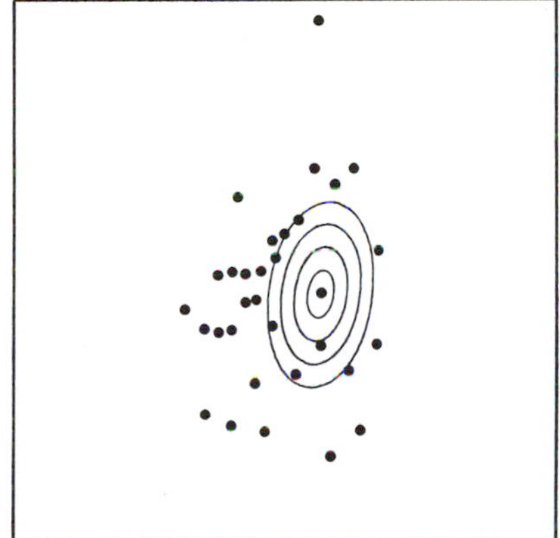

Fig. 9.14. Affinely equivalent data sets with associated norm.

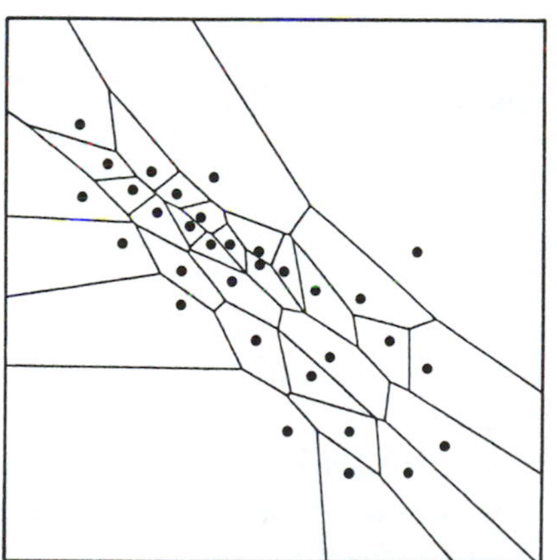

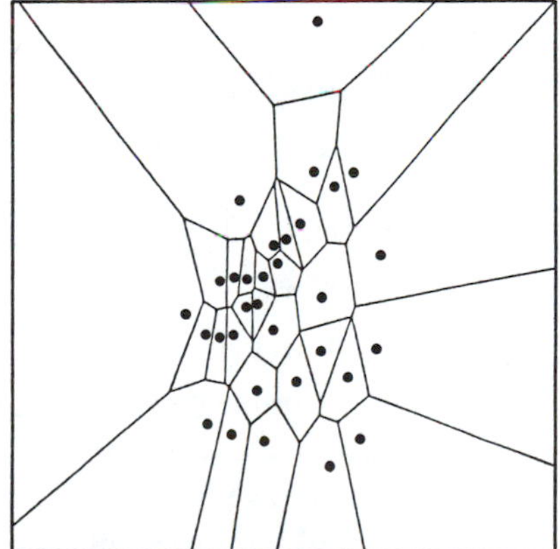

Fig. 9.15. The Dirichlet tesselation generated using the affine invariant norm for the data from Fig. 9.14.

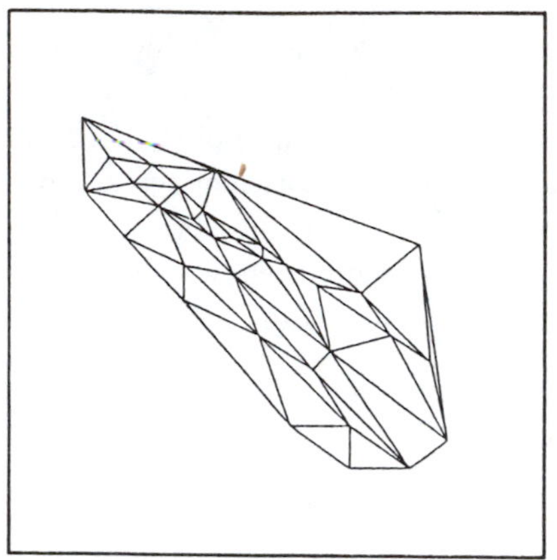
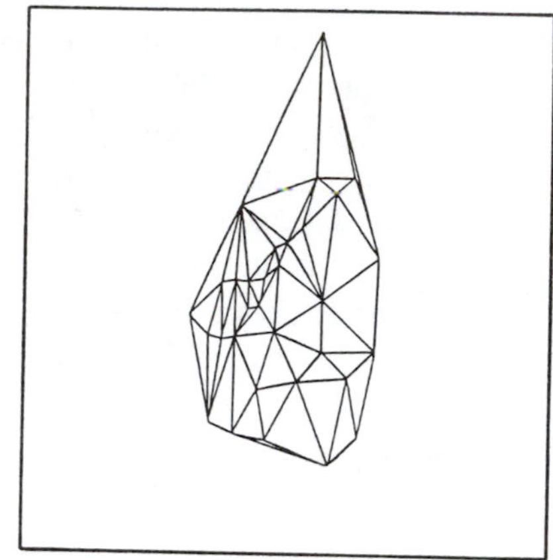

Fig. 9.16. The triangulation dual to the Dirichlet tesselation of Fig. 9.15.

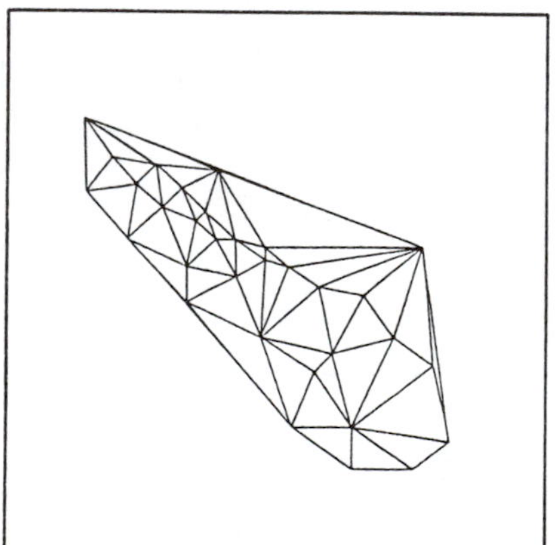
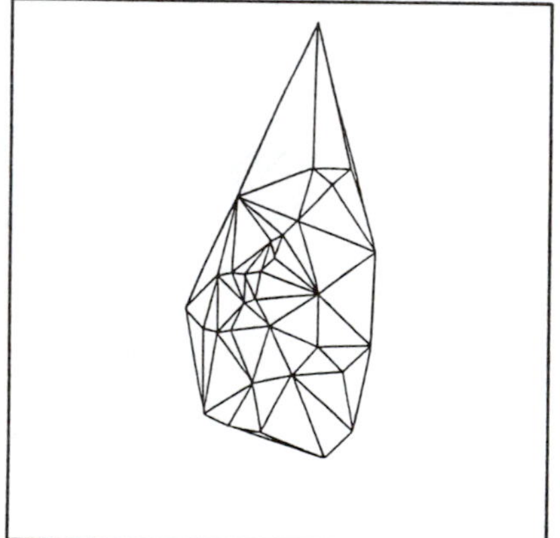

Fig. 9.17. Triangulation of the data of Fig. 9.14 using the Euclidean norm.

9.7. Scattered Data Methods in $\mathbb{R}^3$ – Surfaces on Surfaces

In many applications, we are given data measurements at points on a curved surface. For example, this data might come from the pressure distribution on the surface of an aircraft wing, or from measurements on precipitation, temperature, pressure, gravitational fields, or ozone levels at various points on the surface of the earth. In this case, we have the following scattered data *interpolation problem*: given N points $\boldsymbol{x}_i = (x_i, y_i, z_i) \in \mathbb{R}^3$, $i = 1(1)N$, on a surface $\boldsymbol{F}$ in $\mathbb{R}^3$ and associated real-valued measurements f_i, find a function $f(\boldsymbol{x})$ defined on $\boldsymbol{F}$ such that $f(\boldsymbol{x}_i) = f_i$. This problem can be treated either as a trivariate or bivariate problem.

9.7.1. The Trivariate Approach

The direct trivariate approach is to consider $f(\boldsymbol{x})$ to be the restriction to the surface $\boldsymbol{F}$ of some trivariate function $F(\boldsymbol{x})$ defined on $D \subset \mathbb{R}^3$, where $\boldsymbol{F} \subset D$. The idea is to construct $F(\boldsymbol{x})$ from the given data points $\boldsymbol{x}_i \in \boldsymbol{F}$, (taking account of the special geometry of the measurement surface $\boldsymbol{F}$) and the measured values f_i. The desired function is then given by the restriction $f(\boldsymbol{x}) = F(\boldsymbol{x})|_{\boldsymbol{x} \in \boldsymbol{F}}$. This approach leads to a real-valued function of three variables which cannot be directly displayed graphically. We discuss trivariate methods and ways of visualizing them in Chap. 11.

This direct trivariate approach goes back to [Barn 85]. It was further studied in [Pot 90a, 90b] for the case of multiquadric methods (and compared with the bivariate multiquadric approach). One conclusion of the study was that, for the problems considered there, the simple direct approach via trivariate functions was just as good as the much more complicated bivariate methods which we intend to discuss in more detail below.

To make the direct trivariate method work in practice, we have to impose *additional* conditions on the surface $\boldsymbol{F}$, and on the nature of the trivariate interpolant. The trivariate interpolation method must be able to deal with measurement points which lie on the special surface $\boldsymbol{F}$ in $\mathbb{R}^3$, *i.e.*, in very special subsets of their natural domain of definition. This essentially limits us to using the trivariate multiquadric (TMQ) method [Pot 90b] (see also [Carl 91] for remarks on the behavior of the MQ method on special data sets). [Pot 90a] also discusses a version of the TMQ method which makes use of additional points which do not lie on $\boldsymbol{F}$. The hybrid method of [Barn 87b] (see Sect. 9.7.2) also uses points $\boldsymbol{x}_i$ which in general do not lie on the surface $\boldsymbol{F}$.

In order to be able to work with the restriction $f(\boldsymbol{x})$ of $F(\boldsymbol{x})$ to $\boldsymbol{F}$, we must be sure that the surface $\boldsymbol{F}$ does not have any self-intersections. In addition, as happens for bivariate scattered data methods, if the data are subject to large variations in gradient, then the surface is likely to have large "*oscillations*". As a consequence, if we want to use the TMQ method, we should make sure that $d_r \geq K d_g$, with some reasonably large factor K (say $K > 0.5$), so that pairs of points $\boldsymbol{x}_i, \boldsymbol{x}_j$ on $\boldsymbol{F}$ which are far apart with respect to geodesic distance d_g (and thus could have very different associated function values) also are far apart with respect to Euclidean distance d_r, see [Pot 90b]. This holds, for example, for a circular cylinder and the sphere, but not for an aircraft wing! Consequently, we can get good results for the scattered data interpolation problem on a sphere or cylinder using the TMQ method [Pot 90a, 90b], [Eck 90]), but not such good results for an aircraft wing.

If $\boldsymbol{F}$ is such that a direct application of a trivariate method is not reasonable or possible, then an appropriate *domain mapping* may improve the situation [Pot 90b]. The idea is to map the surface $\boldsymbol{F}$ to a new surface $\bar{\boldsymbol{F}}$ using a mapping Φ in such a way that the geodesic distances on $\boldsymbol{F}$ are not too badly distorted (so that the intrinsic geometry is more or less preserved). Of course, we want $\bar{\boldsymbol{F}}$ to be such that we can apply a direct trivariate method. After applying the mapping Φ, the original data points $\boldsymbol{x} \in \boldsymbol{F}$ are mapped to points $\bar{\boldsymbol{x}}_i$ on $\bar{\boldsymbol{F}}$, which we associate with the original function values f_i. We now use the TMQ method to construct an interpolating function $\bar{F}(\bar{\boldsymbol{x}})$ on $\bar{\boldsymbol{F}}$ with $\bar{F}(\bar{\boldsymbol{x}}_i) = f_i$, $i = 1(1)N$. Then the desired function $f(\boldsymbol{x})$ on $\boldsymbol{F}$ is given by $f(\boldsymbol{x}) = \bar{F}(\Phi(\boldsymbol{x}))|_{\boldsymbol{x}\in\boldsymbol{F}}$. This idea of deforming the surface appeared already in [Fol 90b], although there he did not worry about maintaining the intrinsic geometry of the surface $\boldsymbol{F}$. Not doing so can cause a well-behaved (say uniformly distributed) set of data points on $\boldsymbol{F}$ to be mapped to a less well-behaved data set on $\bar{\boldsymbol{F}}$.

9.7.2. Hybrid Methods

In this section we discuss a hybrid method of [Barn 87b] (see also [Barn 91]) which involves a two-step process where first we construct a trivariate interpolant, and then use it to construct a surface interpolant.

In order to simplify the computation and *visualization* of the interpolant, the surface $\boldsymbol{F}$ is first approximated by a parametric C^1 continuous piecewise bicubic surface $\boldsymbol{w}(u,v)$. The final result is an approximation to $f(\boldsymbol{x})$ consisting of a parametric C^1 continuous piecewise bicubic surface $g(u,v)$, see Fig. 9.18. For visualization, [Barn 87b] evaluates both $\boldsymbol{w}(u,v)$ and $g(u,v)$ on a fine grid (u_p, v_q) for $p = 1(1)P$, $q = 1(1)Q$, in the parameter space. The function values $g(u_p, v_q)$ are then color coded by mapping the interval $[g_{min}, g_{max}]$ to a color code, and then the pixel corresponding to $\boldsymbol{w}(u_p, v_q)$ on the monitor is displayed with the associated color. Phong shading can be used for the pixels not associated with the grid, see Chap. 1 and [FOLE 82], [ROG 85], and [FEL 88].

To construct the piecewise bicubic *approximations* $\boldsymbol{w}(u,v)$ and $g(u,v)$, we need surface points $\boldsymbol{w}_{jk}$ and associated function values f_{jk} for points on a grid (u_j, v_k), with $j = 1(1)J << P$, $k = 1(1)K << Q$, see Fig. 9.18. We interpolate this data to construct the approximants $\boldsymbol{w}(u,v)$ and $g(u,v)$. [Barn 87b] assumes that the surface points $\boldsymbol{w}_{jk}$ are given. To get the f_{jk}, he uses a localized TMQ function $F_{jk}(\boldsymbol{x})$ which interpolates the data $(\boldsymbol{x}_i, f_i)$; i.e.,

$$F_{jk}(\boldsymbol{x}) = \sum_{i=1}^{M} a_{jk}^{i} \sqrt{d_i^2(\boldsymbol{x}) + R_{jk}^2}, \qquad \text{with } F_{jk}(\boldsymbol{x}_i) = f_i. \tag{9.23}$$

Then he takes $f_{jk} = F_{jk}(\boldsymbol{w}_{jk})$. Here $d_i(\boldsymbol{x}) = |\boldsymbol{x} - \boldsymbol{x}_i|$ is the three-dimensional Euclidean distance between $\boldsymbol{x}$ and $\boldsymbol{x}_i$, and the sum is taken over the M measurement points $\boldsymbol{x}_i$ on $\boldsymbol{F}$ which are closest to $\boldsymbol{w}_{jk}$, see Fig. 9.18. The coefficients a^i_{jk} are computed from the $M \times M$ linear system of equations $F_{jk}(\boldsymbol{x}_i) = f_i$, $i = 1(1)M$. Note that the original data $\boldsymbol{x}_i$ and f_i are no longer interpolated by the approximants $\boldsymbol{w}(u, v)$ and $g(u, v)$.

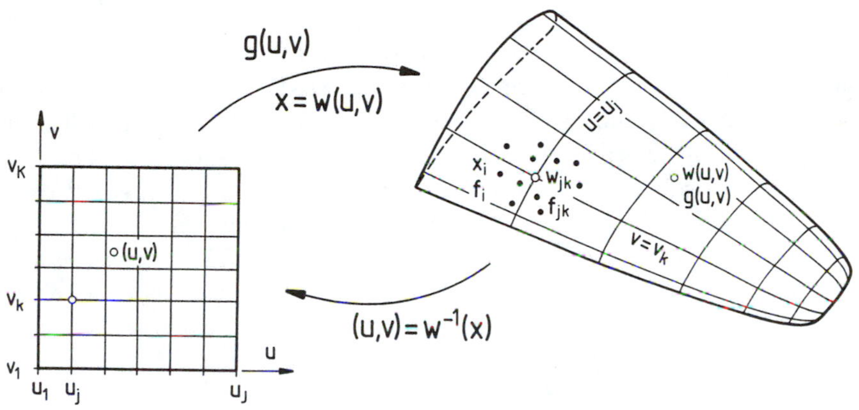

Fig. 9.18. Geometry and the approximations $\boldsymbol{w}(u, v)$ and $g(u, v)$ of the surface $\boldsymbol{F}$ and the function $f(\boldsymbol{x})$ defined on $\boldsymbol{F}$ for the scattered data problem.

For rendering purposes, the functions $\boldsymbol{w}(u, v)$ and $g(u, v)$ can be efficiently evaluated as described above, although the computation of $g(u, v)$ at a specified surface point $\boldsymbol{x} \in \boldsymbol{F}$ requires knowing the inverse $\boldsymbol{w}^{-1}(\boldsymbol{x})$, since $g(u, v) = g(\boldsymbol{w}^{-1}(\boldsymbol{x}))$.

9.7.3. The Bivariate Approach

The bivariate approach involves finding a bivariate function $f(\boldsymbol{x})$ defined on $\boldsymbol{F}$ which interpolates the data: $f(\boldsymbol{x}_i) = f_i$, $i = 1(1)N$. To construct f, we have to take account of the geometry and topology of $\boldsymbol{F}$. There are available methods for dealing with different types of surfaces $\boldsymbol{F}$, e.g., open, closed, convex, nonconvex, etc. In the following sections we discuss the Shepard, Hardy, and FEM methods.

9.7.3.1. Shepard's Method

[Barn 87b] (see also [Barn 90, 91]) used a modified version of Shepard's method for interpolating data on convex surfaces $\boldsymbol{F}$. The interpolant is defined by (9.8) with basis functions (weight functions) as in (9.2a), where now

$$\sigma_i(\boldsymbol{x}) = \frac{\lambda_i(\boldsymbol{x})}{g_i(\boldsymbol{x})^2}. \tag{9.24}$$

Here $\lambda_i(\boldsymbol{x})$ is a damping or mollifying function. The distance function $g_i(\boldsymbol{x})$ measures the geodesic distance from $\boldsymbol{x}$ to $\boldsymbol{x}_i$ along the surface $\boldsymbol{F}$.

To determine the $\lambda_i(\boldsymbol{x})$ and the local approximants (the so-called *nodal functions*) $L_i(\boldsymbol{x})$ of (9.8), we need to work with the points which are neighbors of $\boldsymbol{x}_i$. These can be determined from the *3D-Dirichlet tesselation* of the data points. There can be some difficulties when points are physically close together, but far apart on the surface $\boldsymbol{F}$ (which happens for example for two points on opposite sides of an aircraft wing). To avoid this problem, we erect a *barrier* inside of $\boldsymbol{F}$ which separates these kinds of points. The form of such a barrier depends on the shape of the surface, and e.g., could be a plane segment for an aircraft wing. Projecting the measurement points $\boldsymbol{x}_i$ onto the barrier leads to *barrier points*. Tesselation then produces two classes of neighboring points: barrier points and non-barrier points, see Fig. 9.19.

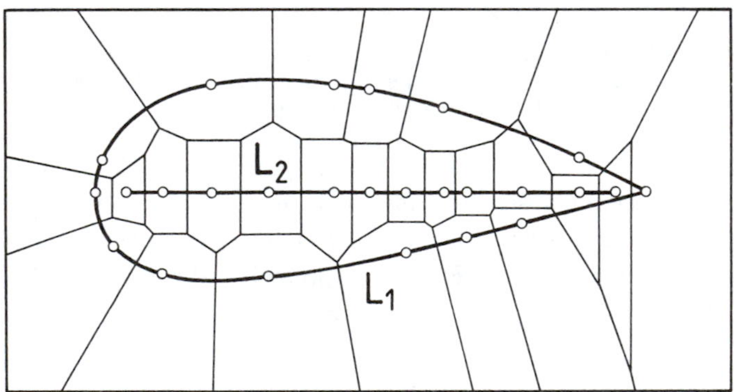

Fig. 9.19. Dirichlet tesselation of points on a profile (line $\boldsymbol{L}_1$) and points on a barrier (line $\boldsymbol{L}_2$).

To construct the $\lambda_i(\boldsymbol{x})$, we need points from each class, whereas the construction of the $L_i(\boldsymbol{x})$ only uses non-barrier points. We define the $\lambda_i(\boldsymbol{x})$ via cubic Hermite functions so that $\lambda_i(\boldsymbol{x}_i) = 1$, $\lambda_i(\boldsymbol{x}_k) = 0$, and the first derivative in the direction $\boldsymbol{v}_{ki} = \boldsymbol{x}_k - \boldsymbol{x}_i$ vanishes at both $\boldsymbol{x}_i$ and $\boldsymbol{x}_k$. They should be defined so that their support reflects the shape of the tile associated with $\boldsymbol{x}_i$, see Fig. 9.20. This construction assures that the support sets of two damping functions are disjoint whenever the two measurement points are separated by a barrier, see Fig. 9.20. We construct the $L_i(\boldsymbol{x})$ as a biquadratic least squares approximation to $f(\boldsymbol{x})$ in the neighborhood of $\boldsymbol{x}_i$, using (at least) five of the nearest neighbors of $\boldsymbol{x}_i$, and so that $L_i(\boldsymbol{x}_i) = f_i$. To this end, we approximate $\boldsymbol{F}$ in a neighborhood of $\boldsymbol{x}_i$ by a sphere which touches the surface $\boldsymbol{F}$ at the point $\boldsymbol{x}_i$. The function $g_i(\boldsymbol{x})$ measures the geodesic distance between $\boldsymbol{x}$ and $\boldsymbol{x}_i$. If $\boldsymbol{F}$ is not given explicitly, then $g_i(\boldsymbol{x})$ can be found approximately using the geodesics on the approximating sphere. This means that we can only expect to get good results for $L_i(\boldsymbol{x})$ and $g_i(\boldsymbol{x})$ when $\boldsymbol{F}$ is convex.

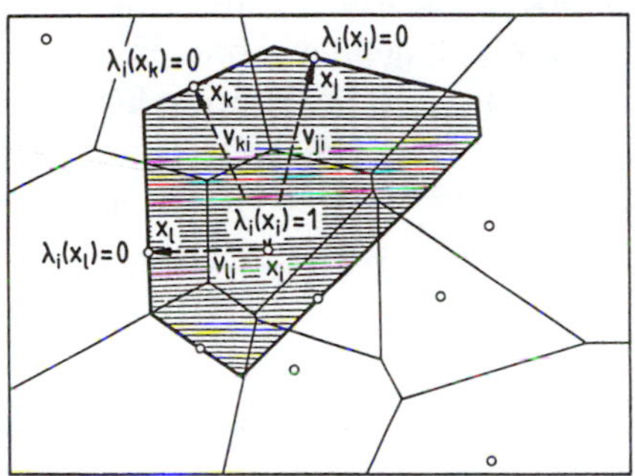

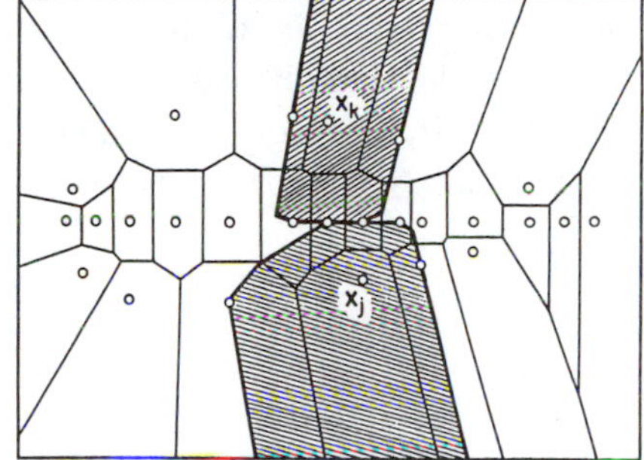

Fig. 9.20. Support (cross-hatched) of $\lambda_i(\boldsymbol{x})$ with no barrier points (left), and with several barrier points (right).

9.7.3.2. Multiquadric Methods

A very important special case of the interpolation problem on a surface is the case where $\boldsymbol{F}$ is a *sphere* $\boldsymbol{K}$ (of radius 1). In order to directly extend the planar MQ and RMQ methods

$$f(\boldsymbol{x}) = \sum_{i=1}^{N} \alpha_i R(d_i(\boldsymbol{x})), \qquad R(d_i(\boldsymbol{x})) = (d_i^2(\boldsymbol{x}) + R_i^2)^m, \quad m = \pm\tfrac{1}{2}, \tag{9.25}$$

to the sphere, the Euclidean distance $d_i(\boldsymbol{x})$ has to be replaced by the *geodesic distance* along the sphere

$$s_i(\boldsymbol{x}) = \phi_i(\boldsymbol{x}) = \arccos(\boldsymbol{x} \cdot \boldsymbol{x}_i). \tag{9.26}$$

We do not recommend this approach, however, since $s_i(\boldsymbol{x})$, which measures the great circle distance between $\boldsymbol{x}$ and $\boldsymbol{x}_i$, is not only singular at $\boldsymbol{x}_i$, but also at the antipodal point $\bar{\boldsymbol{x}}_i$. This means that $R(s_i(\boldsymbol{x}))$ has *discontinuous* derivatives at both $\boldsymbol{x}_i$ and $\bar{\boldsymbol{x}}_i$. Adding R_i^2 removes these singularities (discontinuities) at $\boldsymbol{x}_i$, but not at the antipodal point $\bar{\boldsymbol{x}}_i$. Moreover, when $R_i = 0$, there is a connection (*periodicity*) between the function values at $\boldsymbol{x}$ and at $\bar{\boldsymbol{x}}$, since $s_i(\bar{\boldsymbol{x}}) = \pi - s_i(\boldsymbol{x})$, and thus $f(\bar{\boldsymbol{x}}) = \sum \alpha_i s_i(\bar{\boldsymbol{x}}) = \sum \alpha_i(\pi - s_i(\boldsymbol{x})) = C - f(\boldsymbol{x})$. This relationship does not hold when $R_i \neq 0$, but the associated undesirable behavior of the interpolant is still present for small values of R_i.

We now discuss two different ways of solving this problem. [Fol 90c] described a C^2 *continuous modified RMQ method* which uses the basis function in (9.25), where $d_i(\boldsymbol{x})$ is replaced by the geodesic distance (9.26), and $m = -0.5$ when the geodesic distance satisfies $s_i(\boldsymbol{x}) \leq 3$, while if $3 < s_i(\boldsymbol{x}) < \pi$, the basis function is taken to be a fifth degree polynomial chosen in such a way that the overall basis function is C^2 continuous on the entire sphere, see Fig. 9.21. He chooses the parameter $R_i^2 = 28/N \approx 2.2\ \text{area}(\boldsymbol{K})/N$, which is approximately twice the surface area of the unit sphere divided by the number of points N.

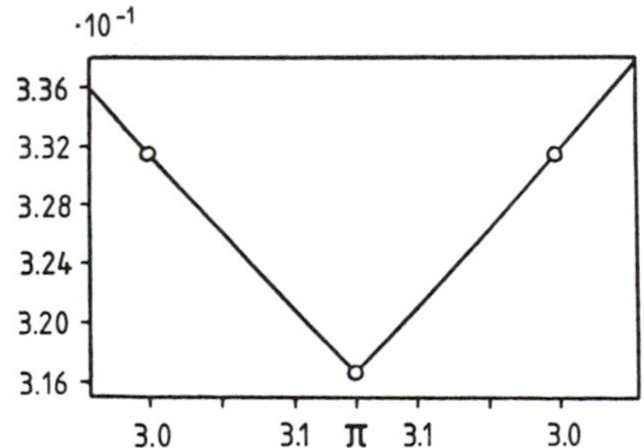

C^0 RMQ basis function (9.25), (9.26)

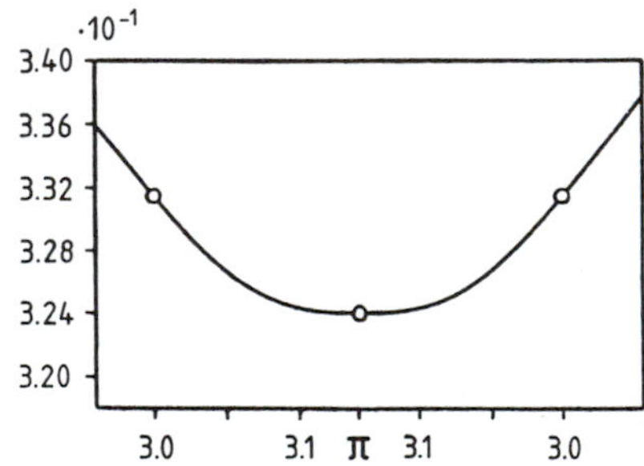

C^2 modified RMQ basis function

Fig. 9.21. RMQ basis functions ($R = 0.3$) near the antipodal point.

This interpolation method was applied in the domain mapping method in [Fol 90b], see also [Fol 90a], [Barn 91]. The basic idea is to find a one-to-one mapping Φ between the sphere $\boldsymbol{K}$ and a given surface $\boldsymbol{F}$ which is topologically equivalent to the sphere (but could be either convex or non-convex), thereby reducing the interpolation problem on $\boldsymbol{F}$ to one on the sphere. The interpolant is then constructed as in Sect. 9.7.1, but using a C^2 continuous modified RMQ method instead of a TMQ method, and then mapped back to $\boldsymbol{F}$ for a solution. This deformation method can, of course, also be used with other surfaces than a sphere, for example with a cone or a cylinder or even more general surfaces, see [Pot 90b].

We next discuss a method of [Hardy 75] (see also [Hardy 90]) called the *spherical RMQ method* which uses the basis functions

$$R_i(\boldsymbol{x}) = \left[1 + R^2 - 2R\cos(s_i(\boldsymbol{x}))\right]^m = \left[1 + R^2 - 2R\boldsymbol{x} \cdot \boldsymbol{x}_i\right]^m, \qquad (9.27)$$

with $0 < R < 1$ and $m = -0.5$. It can be seen from trigonometric identities that these functions measure the distance from the point $\boldsymbol{x} \in \boldsymbol{K}$ to a point $\boldsymbol{P}_i$ which lies at a distance of R from the center of the sphere on the ray passing through $\boldsymbol{x}_i$. The point $\boldsymbol{P}_i$ can be thought of as a source for a field, *i.e.*, as a mass [Hardy 75] (see also [Hardy 90]). [Pot 90a] studied the *spherical MQ method*, where m is replaced by $m = 0.5$ in (9.27), and found that it produced better results than the spherical RMQ method, and even worked well for values $R > 1$.

The spherical MQ method is easier to implement, and produces more accurate results than the C^2 continuous modified RMQ method.

In order to avoid poorly conditioned systems of equations, which often arise for *large data sets* ($N > 200$), [Pot 90a] suggested the following *localization scheme* extending the planar localization scheme of [Fra 82]. Using one of the above MQ methods, we construct M interpolants $f_i(\boldsymbol{x})$ defined on spherical caps $\boldsymbol{K}_i = \{\boldsymbol{x} \in \boldsymbol{K} : s_i(\boldsymbol{x}) \leq \rho_i\}$ with spherical midpoints at $\boldsymbol{M}_i$ and (geodesic) radii ρ_i. These M pieces are then combined by taking the convex combination

$$f(\boldsymbol{x}) = \sum_{i=0}^{M} \omega_i(\boldsymbol{x}) f_i(\boldsymbol{x}) \qquad \text{with} \qquad \omega_i(\boldsymbol{x}) = \frac{\lambda_i(\boldsymbol{x})}{\sum \lambda_i(\boldsymbol{x})},$$

using the Franke-Little weights (see page 391)

$$\lambda_i(\boldsymbol{x}) = \left(1 - \frac{s_i(\boldsymbol{x})}{\rho_i}\right)_+^{\mu}.$$

The choices $\mu = 2$ and $\mu = 3$ usually lead to sufficiently smooth surfaces. To construct $f_i(\boldsymbol{x})$, we use all measurement points $\boldsymbol{x}_i$ which lie in a somewhat larger spherical cap $\boldsymbol{K}_i'$ with spherical center $\boldsymbol{M}_i$ and radius $\rho_i' = \lambda\rho_i$, $\lambda \in [1.2, 1.4]$. The reason for using a larger $\boldsymbol{K}_i'$ for computing the $f_i(\boldsymbol{x})$ is to achieve a better behavior on the boundary of $\boldsymbol{K}_i$. Of course, we have to choose the spherical caps so that they completely cover the sphere. If we have very nonuniformly scattered data, we should choose $\boldsymbol{M}_i$ and ρ_i to depend on the data, for example in such a way that each of the caps has about the same number of points in it. If the measurement points are uniformly or almost uniformly distributed, then we can choose a regular or nearly regular covering of the sphere. In building such a covering, we can take the center points $\boldsymbol{M}_i$

to be the vertices of an inscribed regular polyhedron. This defines a spherical triangulation on the sphere, which can be refined if necessary (see Sect. 9.7.4). The selected radius ρ must then be larger than the supremum of the radii of the circles circumscribing the spherical triangles in this triangulation.

Other methods for localization are certainly possible. [Eck 90] described a *strip method* which can be applied when the measurement points $\boldsymbol{x}_i$ lie on (nearly) planar parallel closed convex tracks, which is for example the case for CAT-scan data. In this case the partial interpolants $f_i(\boldsymbol{x})$ are defined on parallel strips $\boldsymbol{S}_i = \{\boldsymbol{x} \in \boldsymbol{K} : z_i^b \leq z \leq z_i^t\}$ on the sphere, where without loss of generality, we can assume that the strips $\boldsymbol{S}_i$ are parallel to the x-y plane, see Fig. 9.22. Otherwise the method proceeds more or less as for the case of spherical caps.

The strip method also works very well on a *cylinder* $\boldsymbol{Z}$, see Fig. 9.22. [Eck 90] considered both the direct TMQ method, and a generalization of the spherical MQ method to a kind of *cylindrical MQ method*.

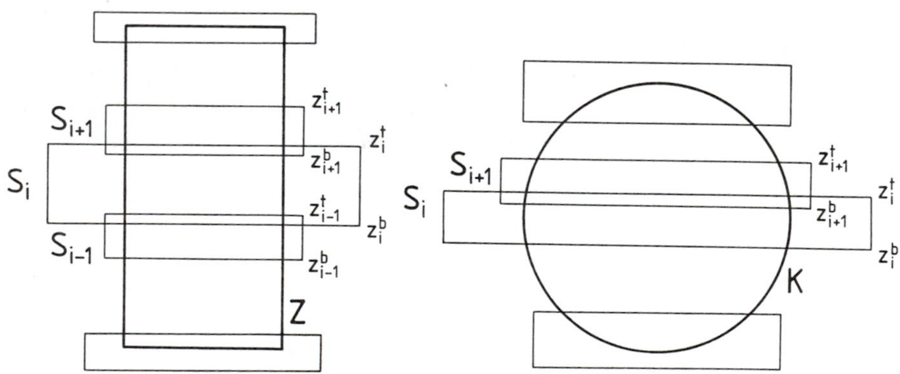

Fig. 9.22. Strip localization scheme on a sphere and cylinder.

The strip method can be applied to the problem of *reconstructing surfaces*, given a sequence of planar slices of the surface (usually described by contour lines). Reconstruction methods have numerous applications [Lin 89], for example in medical diagnostics, where we want to reconstruct the surface of an organ or a bone on the basis of CAT-scan data. [Eck 90] used the above methods to model the head and shaft of a femur, which have spherical and cylindrical shapes, respectively. His algorithm is a kind of *domain mapping method*, and involves fitting a least squares sphere (or cylinder) to the data to get surfaces $\boldsymbol{K}$ and $\boldsymbol{Z}$. The CAT-scan data points $\boldsymbol{p}_i$ are then radially projected onto measurement points $\boldsymbol{x}_i$ on these two surfaces, and the f_i are taken to be the radial Euclidean distance between the $\boldsymbol{x}_i$ and the $\boldsymbol{p}_i$. Then a

TMQ method is used to construct an interpolant to this data, see the remark in Sect. 9.7.1.

The most frequently used method for reconstructing surfaces from a set of planar slices involves representing the surface by piecewise linear triangular facets, see e.g., [Kep 75], [Chr 78], [Anj 87], [Zyd 87], [Boi 88], [Schu 90], and [Geig 91]. This can be accomplished by connecting the contours on neighboring slices by triangles, although this is not as simple as it sounds, since, in general, there are many different ways to connect any two given contours.

9.7.3.3. FEM Methods

The FEM method discussed in Sect. 9.3 for the bivariate case can be easily generalized to the case of curved surfaces in $\mathbb{R}^3$, *i.e.*, we again carry out the following three steps:

1) triangulate the set of measurement points $\boldsymbol{x}_i \in \boldsymbol{F}$,
2) estimate the required derivatives,
3) construct the triangular surfaces.

To describe how to construct a *triangulation*, we consider first the case where $\boldsymbol{F}$ is a *sphere* $\boldsymbol{K}$. Suppose that the $\boldsymbol{x}_i$ are not all on one side of a plane passing through the center of the sphere. The natural way to define a triangle on the sphere is to connect three vertices with (the shorter) segments of great circles, *i.e.*, with geodesics on the sphere. These *geodesic triangles* are the only ones which are intrinsic to the sphere. We shall require that our triangulation contain only *proper spherical triangles*, *i.e.*, geodesic triangles such that no pair of vertices is antipodal, and all three vertices do not lie in a plane through the center of the sphere, *i.e.*, on one great circle, see Fig. 9.23. Moreover, we require that the union of all triangles should provide a simple and complete covering of the sphere, and that every edge in the triangulation belongs to exactly two triangles. It can be shown that if we are given N data points $\boldsymbol{x}_i \in \boldsymbol{K}$, then there exist precisely $2(N-2)$ spherical triangles with $3(N-2)$ geodesic edges [Law 84], [Ren 84].

In principle, any one of the three types of algorithms listed on page 407 can be used to construct a triangulation of a given set of data points $\boldsymbol{x}_i$. However, the available software is all based on the idea of iteratively building the triangulation, since this is the most efficient. We are most interested in some kind of local criterion which can be used to iteratively make local changes in a given triangulation in such a way as to converge to a global optimum. Thus we are led to use the analogs of the usual max-min angle criterion and the circumscribed circle criterion.

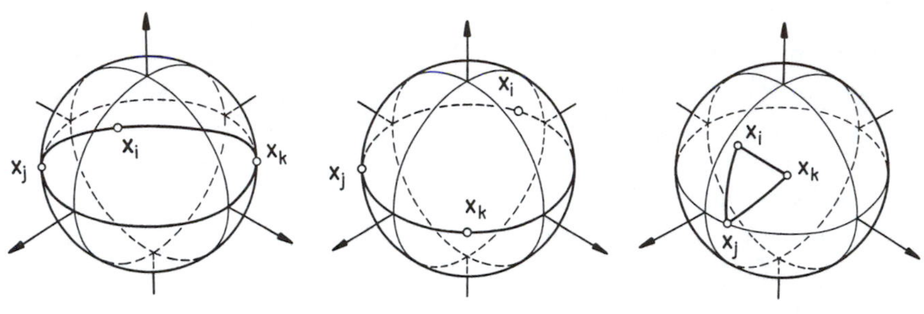

Non-admissible Non-admissible Admissible

Fig. 9.23. Admissible and non-admissible spherical triangles.

The *spherical circumscribed circle criterion* [Ren 84] checks whether a given point $\boldsymbol{x}_\ell$ lies inside or outside of the spherical circle $\boldsymbol{k}_{ijk}$ passing through the three points $\boldsymbol{x}_i, \boldsymbol{x}_j, \boldsymbol{x}_k$.

The *spherical max-min angle criterion* [Nie 87b] involves looking at the spherical quadrilateral $(\boldsymbol{x}_i, \boldsymbol{x}_j, \boldsymbol{x}_k, \boldsymbol{x}_\ell)$, and as in the planar case, compares the minimum angles Θ_{abc} of the two alternative triangulations, where the angle associated with a vertex of a triangle is defined to be the angle between the two planes passing through the center of the sphere and containing the edges $\boldsymbol{e}_{ab}$ and $\boldsymbol{e}_{bc}$, see Fig. 9.24. The criterion says

$$\begin{aligned}
&\textbf{if } \min\{\Theta_{kij}, \Theta_{ijk}, \Theta_{jki}, \Theta_{k\ell i}, \Theta_{\ell ik}, \Theta_{ik\ell}\} > \\
&\quad \min\{\Theta_{ij\ell}, \Theta_{j\ell i}, \Theta_{\ell ij}, \Theta_{jk\ell}, \Theta_{k\ell j}, \Theta_{\ell jk}\} \\
&\textbf{then } \text{choose edge } \boldsymbol{e}_{ik} \\
&\textbf{else } \text{choose edge } \boldsymbol{e}_{j\ell}.
\end{aligned}$$

The *3D polyhedron criterion* [Law 84] (cf. [Ren 84]) is different in the sense that it does not have a planar analog. If four points $\boldsymbol{x}_i, \boldsymbol{x}_j, \boldsymbol{x}_k, \boldsymbol{x}_\ell$ (ordered counter-clockwise) do not line in a plane, then the corresponding spherical quadrilateral, along with one of the two possible diagonals, forms a convex 3D polyhedron, while with the other it does not. Our criterion (which is equivalent to the spherical circumscribed circle criterion) for choosing one diagonal over the other now reads simply: choose the one which gives a convex 3D polyhedron:

$$\begin{aligned}
&\textbf{if } \det(\boldsymbol{x}_j - \boldsymbol{x}_i, \boldsymbol{x}_k - \boldsymbol{x}_i, \boldsymbol{x}_\ell - \boldsymbol{x}_i) > 0 \\
&\textbf{then } \text{choose edge } \boldsymbol{e}_{j\ell}, \\
&\textbf{else } \text{choose edge } \boldsymbol{e}_{ik}.
\end{aligned}$$

Tests [Nie 87b] have shown that all three criteria appear to lead to essentially the same spherical triangulations.

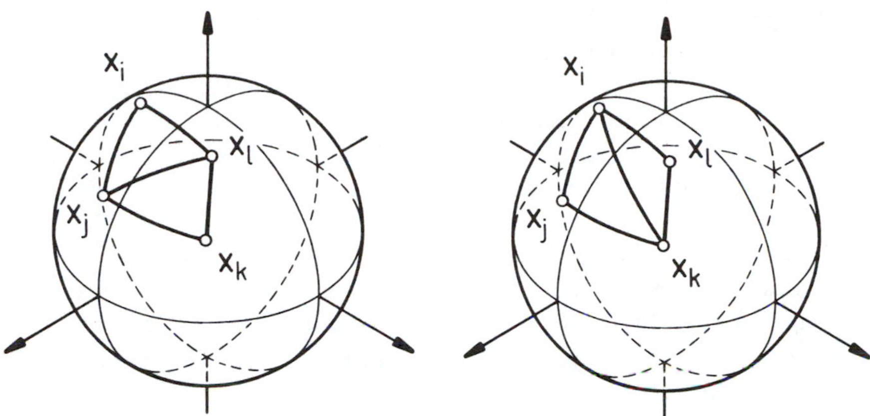

Fig. 9.24. Neighboring triangles with diagonals $\boldsymbol{e}_{j\ell}$ and $\boldsymbol{e}_{ik}$.

A triangulation method which can be applied to *general convex surfaces* was introduced in [Barn 90], see also [Barn 91]. The algorithm is based on the *one-sided property criterion*. A triangle $(\boldsymbol{x}_i, \boldsymbol{x}_j, \boldsymbol{x}_k)$ has the one-sided property provided that all other measurement points lie on the same side of a plane passing through the three points $\boldsymbol{x}_i, \boldsymbol{x}_j, \boldsymbol{x}_k$. A triangulation possesses the one-sided property provided that every triangle in the triangulation has this property. Now we can describe an iterative algorithm. We start with some initial triangle, then add one point after another to the triangulation, making sure that at each step, the one-sided property remains satisfied. The advantages of using this criterion are that 1) on the sphere, it leads to the same triangulations as the circumscribed circle criterion, 2) it avoids the swapping of diagonals required in the criteria above, and 3) it can be carried over directly from the sphere to a general convex surface.

The *derivative information* needed for constructing the triangular interpolant can be obtained with essentially the same methods as in the planar case, for example, by constructing a local quadratic least squares approximation [Law 84], [Ren 84], and [Barn 90, 91]. The local coordinate systems associated with each $\boldsymbol{x}_i \in \boldsymbol{F}$ can be defined using the tangent plane to the surface $\boldsymbol{F}$ at $\boldsymbol{x}_i$. If $\boldsymbol{F}$ is convex, then near $\boldsymbol{x}_i$, $\boldsymbol{F}$ can be approximated by an osculating sphere at $\boldsymbol{x}_i$, and the tangent plane can be replaced by the tangent plane to the osculating sphere, see e.g., [Barn 90, 91]. [Nie 87b] extended the idea of the MNN method (see Sect. 9.3.3.3) to the sphere, using the geodesic lines and directional derivatives. Using a piecewise linear triangulation of $\boldsymbol{F}$, [Pot 92] extended the method (9.20) to C^2 continuous surfaces which do not intersect themselves.

A simple extension of the *triangular surfaces* (discussed in Sect. 9.3.2) to surfaces on the sphere requires the existence of barycentric coordinates and the associated Bézier representation. So far, however, no one has succeeded in finding a sensible definition for triangular Bézier surfaces on a sphere or on a more general surface. This is due to the fact that a direct generalization of the barycentric coordinates from planar to spherical geodesic triangles does not seem to be possible [Brow 92]. However, barycentric coordinates in the plane can be projected onto the sphere. It would be reasonable to try to construct barycentric coordinates on the sphere using some kind of area preserving mapping (see e.g., [LAU 65]) between the plane and the sphere. But this leads to spherical triangles which are no longer geodesic, and which can exhibit very strong deformations as compared to geodesic spherical triangles when one or more of the vertices lie near one of the poles or near the *date line*. For this reason, so far only Coons-type triangular interpolants have been constructed. All of the papers [Law 84], [Ren 84], [Nie 87], [Barn 90, 91], and [Pot 92] prefer the *side-vertex method* of [Nie 79] over the BBG method because it involves constructing surface patches as convex combination of three partial interpolants, and consequently, no twist vector problem arises. The appropriate generalization involves carrying out the interpolation along great circles.

9.7.4. Visualization Methods

There are essentially three types of representation techniques.

Suppose $\boldsymbol{F}(\boldsymbol{x})$ is a surface in $\mathbb{R}^3$. In the *domain method* we display a function $f(\boldsymbol{x})$ defined over the domain $\boldsymbol{F}$ either by color coding of $\boldsymbol{F}$ according to the function values of $f(\boldsymbol{x})$ with Gouraud shading (see e.g., [FOLE 82], [ROG 85]), or by presenting a contour plot of $f(\boldsymbol{x})$ (i.e. lines of equal function values, see Sect. 12.4.1) [Barn 87b], [Nie 87b], and [Fol 90, 90a, 90b]. The best results are obtained using a combination of the two methods [Fol 90, 90a, 90b], [Pot 90b, 91a, 92], and [Nie 91a]. [Pot 90a, 91a] also employs isophotes. In implementing this method, often we need to approximate $\boldsymbol{F}(\boldsymbol{x})$ using planar triangles. The important special case of the sphere is discussed in [Ren 84], [Nie 87b], [Fol 90], and [Pot 90a].

In *bivariate methods*, we deal with the graph of $f(\boldsymbol{x})$ as a surface $\boldsymbol{f}(\boldsymbol{x})$ in $\mathbb{R}^3$ lying over $\boldsymbol{F}(\boldsymbol{x})$, and defined by $\boldsymbol{f}(\boldsymbol{x}) = \boldsymbol{F}(\boldsymbol{x}) + f(\boldsymbol{x})\boldsymbol{e}(\boldsymbol{x})$ [Pot 91a]. Here $\boldsymbol{e}(\boldsymbol{x})$ is usually chosen to be the unit normal vector $\boldsymbol{N}(\boldsymbol{x})$ to the surface $\boldsymbol{F}(\boldsymbol{x})$. However, for nonconvex surfaces, the use of the radial unit vector $\boldsymbol{R}(\boldsymbol{x})$ usually gives better results [Fol 90, 90a], [Nie 91a], and [Pot 91a]. The graph $\boldsymbol{f}(\boldsymbol{x})$ can be visualized using contour lines, isophotes, or color codings. Combinations of these methods and simultaneously displaying both $\boldsymbol{F}(\boldsymbol{x})$ and

$\boldsymbol{f}(\boldsymbol{x})$ (transparently) can often improve our understanding, see [Fol 90, 90a, 91a] and [Nie 91a].

Trivariate methods are based on trivariate respresentations. To display graphs lying in $\mathbb{R}^4$, we can use parallel or central projection to create graphs in $\mathbb{R}^3$, cf. [Pot 91a, 92], [Nie 91a], or we can construct contour plots, see Sect. 12.4.1. Other visualization techniques will be discussed in Sect. 11.7.2.

10 Basis Transformations for Curve and Surface Representations

There are a variety of modelling sytems in use for computer-aided design, based on different methods for mathematically describing free form curves and surfaces. These include, for example, monomials (ordinary polynomials) of degrees 3 to 19, Bernstein polynomials of various degrees, and B-spline basis functions of different orders [Boeh 84]. Frequently, Gordon-Coons surfaces and nonlinear basis functions are also used. This means that in order to work with data created by other modelling systems (for example, in dealing with a part created by another supplier), we need to perform a conversion or transformation of the provided curves and surfaces into ones we can work with on our own system. This can involve changing polynomial degrees, or even the form of the basis functions. Unfortunately, in general, such a transformation cannot be done exactly, and therefore we have to resort to approximation methods. This gives rise to the following related problem: given a prescribed error tolerance, replace a given set of spline surfaces by the smallest possible number of surface patches, either by merging patches or by splitting them.

Merging of surface patches can also be of independent interest in a modelling system. In developing a product, it often happens that in complicated regions, a large number of patches has been used. If so, it may be useful to try to reduce the number of patches by replacing clusters of smaller patches by a few larger ones (up to a given error tolerance), thus achieving a data compression.

10.1. Exact Basis Transformation

Conversion of curves and surfaces is simplest, of course, when it can be done exactly. Mathematically, the problem of transforming surface representations is simply a problem of transforming the associated bases.

The bases for (ordinary) polynomial representations are the monomials; for Bézier surfaces, the Bernstein polynomials; and for B-spline surfaces, the B-spline basis functions. The underlying structure in the case of Gordon-Coons surfaces consists of the description of the boundary and the form of the blending functions.

In general, exact basis transformation is possible for

- degree raising of monomials, Bézier polynomials, or B-splines, see Chap. 4,
- transformations involving monomials, Bernstein polynomials, or B-splines of the same polynomial degree.

A detailed treatment of various methods for exact basis transformation can be found in [Vri 91].

10.1.1. Exact Basis Transformation of Monomials and Bernstein Polynomials

Suppose we are given a Bézier curve

$$\boldsymbol{X}(t) = \sum_{\ell=0}^{n} \boldsymbol{b}_\ell B_\ell^n(t) \tag{10.1}$$

and a polynomial curve

$$\bar{\boldsymbol{X}}(t) = \sum_{i=0}^{n} \boldsymbol{a}_i t^i, \tag{10.2}$$

both defined on the same parameter interval $t \in [0, 1]$. It is known (see eg., [Cha 82]) that the Bernstein polynomials can be written as

$$B_\ell^n(t) = \sum_{k=0}^{n-\ell} (-1)^k \binom{n}{\ell} \binom{n-\ell}{k} t^{\ell+k}. \tag{10.3}$$

Substituting in (10.1), shifting the indices by $i := \ell + k$, and reordering the sum, (10.1) becomes

$$\boldsymbol{X}(t) = \sum_{i=0}^{n} \sum_{\ell=0}^{i} (-1)^{i-\ell} \binom{n}{i} \binom{i}{\ell} \boldsymbol{b}_\ell t^i.$$

Comparing coefficients with (10.2), it follows that

$$\boldsymbol{a}_i = \sum_{\ell=0}^{i} (-1)^{i-\ell} \binom{n}{i} \binom{i}{\ell} \boldsymbol{b}_\ell, \tag{10.4}$$

i.e., the associated transformation matrix $\boldsymbol{C}$ is a lower triangular matrix with entries

$$c_{i\ell} = \begin{cases} (-1)^{i-\ell}\binom{n}{i}\binom{i}{\ell}, & i \geq \ell \\ 0, & \text{otherwise.} \end{cases} \tag{10.5a}$$

To convert in the other direction, i.e., to compute the Bézier points from given polynomial coefficients, we may use the identity

$$\sum_{k=0}^{n} \boldsymbol{a}_k t^k = \sum_{k=0}^{n}\sum_{i=0}^{n-k} \binom{n-k}{i} t^{i+k}(1-t)^{n-(i+k)}\boldsymbol{a}_k.$$

Then expressing the right-hand side in terms of Bernstein polynomials leads to

$$\overline{c}_{\ell k} = \begin{cases} \binom{\ell}{k}/\binom{n}{k}, & \text{for } k \leq \ell \\ 0, & \text{otherwise,} \end{cases} \tag{10.5b}$$

cf. e.g., [Watk 88], [YAM 88], and [Vri 91].

To describe the transformation of tensor-product surfaces, it is convenient to use matrix notation. Suppose we are given a surface patch

$$\boldsymbol{X}(u,v) = \sum_{i=0}^{m}\sum_{j=0}^{n} \boldsymbol{a}_{ij} u^i v^j, \tag{10.6}$$

which we write in matrix form as

$$\boldsymbol{X} = \boldsymbol{u}^T \boldsymbol{A} \boldsymbol{v}, \tag{10.6a}$$

where $\boldsymbol{u} = (1, u, \ldots, u^m)^T$ and $\boldsymbol{v}$ is defined similarly, and where

$$\boldsymbol{A} = \begin{pmatrix} a_{00} & a_{01} & \ldots & a_{0n} \\ \vdots & & & \vdots \\ a_{m0} & a_{m1} & \ldots & a_{mn} \end{pmatrix}$$

is the corresponding coefficient matrix. In addition, suppose the same surface patch defined on the *same parameter set* is represented in terms of the Bernstein basis by

$$\boldsymbol{X}(u,v) = \sum_{i=0}^{m}\sum_{j=0}^{n} \boldsymbol{b}_{ij} B_i^m(u) B_j^n(v), \tag{10.7}$$

or in matrix form by

$$\boldsymbol{X}(u,v) = \boldsymbol{u}^T [BU] \boldsymbol{B} [BV]^T \boldsymbol{v}, \tag{10.7a}$$

where $\boldsymbol{B}$ is the coefficient matrix with entries $\boldsymbol{b}_{ij}$, while [BU] and [BV] are the coefficient matrices of the Bernstein polynomials in terms of monomials. By the assumption that both representations are defined over the same parameter set, a comparison of (10.6a) and (10.7a) implies

$$\boldsymbol{A} = [BU]\boldsymbol{B}[BV]^T. \tag{10.8}$$

As remarked in [Faro 87], [Dani 89], the numerical stability of this process depends heavily on the implementation.

10.1.2. Exact Basis Transformation of Monomials and B-spline Segments

Suppose we are given a B-spline curve

$$\boldsymbol{R}(t) = \sum_{i=0}^{n} \boldsymbol{d}_i N_{ik}(t) \tag{10.9}$$

with uniform knot vector $\boldsymbol{T}$, i.e., $t \in [t_0, t_{n+k}]$ with $t_\ell \in \mathbb{N}$. Our aim now is to convert this curve to monomial form

$$\bar{\boldsymbol{X}}_\ell(u) = \sum_{i=0}^{k-1} \boldsymbol{a}_{\ell,i} u^i, \tag{10.10}$$

interval by interval, using a local parameter $u \in [0, 1]$.

For each interval $[t_\ell, t_{\ell+1}]$, we can perform the transformation as follows:

1) compute all derivatives up to order $(k-1)$ of the B-spline curve $\boldsymbol{R}(t)$ at the point t_ℓ by repeated application of the formula (4.59) for differentiating a B-spline curve,
2) find the coefficients of the monomial expansion of $\bar{\boldsymbol{X}}_\ell(u)$ by computing the derivatives at $u = 0$:

$$\boldsymbol{a}_{\ell,i} = \frac{1}{i!} \frac{\partial^i \boldsymbol{R}(t)}{\partial t^i}\bigg|_{t=0}, \tag{10.11}$$

3) make a change of variables from the interval $[t_\ell, t_{\ell+1}]$ to the local interval $[0, 1]$.

Another approach is to work with the explicit matrix representation of B-spline coefficients. Suppose we are working with equally spaced knots. For t in the interval $t_r \le t \le t_{r+1}$, by Lemma 4.7, the B-spline curve (10.9) can be written as

$$\boldsymbol{X}(t) = \sum_{i=r-(k-1)}^{r} \boldsymbol{d}_i N_{ik}(t). \tag{$*$}$$

Now shifting indices so that $i := 0$, the upper limit on the sum becomes $r := k-1$. Then introducing $m := k+1$, we can express $(*)$ as

$$\boldsymbol{X}(\tau) = \boldsymbol{N}(\tau)^T \boldsymbol{D} = \boldsymbol{TCD}, \tag{10.12a}$$

where

$$\boldsymbol{D} = (\boldsymbol{d}_0, \boldsymbol{d}_1, \ldots, \boldsymbol{d}_m)^T, \qquad \boldsymbol{T} = (\tau^m, \tau^{m-1}, \ldots, 1)^T,$$
$$\boldsymbol{N}(\tau) = (N_{0,m+1}(\tau), N_{1,m+1}(\tau), \ldots, N_{m,m+1}(\tau))^T,$$

where $\tau \in [0,1]$ is a local parameter. The entries in the matrix $\boldsymbol{C}$ are given by (cf. e.g., [YAM 88])

$$c_{ij} = \frac{1}{n!}\binom{n}{i}\sum_{k=j}^{n}(n-k)^i(-1)^{k-j}\binom{n+1}{k-j}, \qquad i = 0(1)n, \quad j = 0(1)k\text{–}1. \tag{10.12b}$$

Matrix formulae in the case of nonuniform knots can be found, e.g., in [Gra 91].

10.1.3. Exact Basis Transformation of B-spline and Bézier Segments

One way to handle this case is to convert through monomials using a combination of the results in Sections 10.1.1 and 10.1.2. Here we discuss the *direct* transformation of a B-spline representation to the Bézier form. To simplify matters, we consider only uniformly spaced knots. The nonuniform case can be dealt with using appropriate matrices (see [Gra 91]), or by a generalization of the de Boor algorithm (where the knot vector itself is not transformed)

We begin by considering two examples where the monomial representation is used in an intermediate step. Let $k = 3$ so that we are working with quadratic basis functions. For a closed B-spline curve, (4.56) implies e.g., that the fourth segment can be written as

$$\boldsymbol{X}(w) = (\,\boldsymbol{d}_1,\ \ \boldsymbol{d}_2,\ \ \boldsymbol{d}_3\,)\begin{pmatrix} \frac{1}{2} & -1 & \frac{1}{2} \\ \frac{1}{2} & 1 & -1 \\ 0 & 0 & \frac{1}{2} \end{pmatrix}\begin{pmatrix} 1 \\ w \\ w^2 \end{pmatrix}, \qquad w \in [0,1]. \tag{10.13}$$

Similarly, using $n = 2$ in (4.2) and (4.4), we can write the Bézier curve as

$$\boldsymbol{X}(w) = (\,\boldsymbol{b}_0,\ \ \boldsymbol{b}_1,\ \ \boldsymbol{b}_2\,)\begin{pmatrix} 1 & -2 & 1 \\ 0 & 2 & -2 \\ 0 & 0 & 1 \end{pmatrix}\begin{pmatrix} 1 \\ w \\ w^2 \end{pmatrix}, \qquad w \in [0,1]. \tag{10.14}$$

The two formulae are supposed to be equal to each other for all parameter values $w \in [0,1]$, i.e.,

$$(\,\boldsymbol{d}_1,\quad \boldsymbol{d}_2,\quad \boldsymbol{d}_3\,)\begin{pmatrix} \frac{1}{2} & -1 & \frac{1}{2} \\ \frac{1}{2} & 1 & -1 \\ 0 & 0 & \frac{1}{2} \end{pmatrix} = (\,\boldsymbol{b}_0,\quad \boldsymbol{b}_1,\quad \boldsymbol{b}_2\,)\begin{pmatrix} 1 & -2 & 1 \\ 0 & 2 & -2 \\ 0 & 0 & 1 \end{pmatrix}. \qquad (10.15)$$

Multiplying by the inverse of the matrix on the right-hand side, leads to the following equation for the Bézier points in terms of the de Boor points:

$$\begin{aligned}(\,\boldsymbol{b}_0,\quad \boldsymbol{b}_1,\quad \boldsymbol{b}_2\,) &= (\,\boldsymbol{d}_1,\quad \boldsymbol{d}_2,\quad \boldsymbol{d}_3\,)\begin{pmatrix} \frac{1}{2} & -1 & \frac{1}{2} \\ \frac{1}{2} & 1 & -1 \\ 0 & 0 & \frac{1}{2} \end{pmatrix}\begin{pmatrix} 1 & 1 & 1 \\ 0 & \frac{1}{2} & 1 \\ 0 & 0 & 1 \end{pmatrix} \\ &= \left(\frac{\boldsymbol{d}_1+\boldsymbol{d}_2}{2}, \boldsymbol{d}_2, \frac{\boldsymbol{d}_2+\boldsymbol{d}_3}{2}\right). \end{aligned} \qquad (10.16)$$

This shows that the Bézier points $\boldsymbol{b}_0$ and $\boldsymbol{b}_2$ lie at the midpoints of the sides of the de Boor polygons $(\boldsymbol{d}_1, \boldsymbol{d}_2)$ and $(\boldsymbol{d}_2, \boldsymbol{d}_3)$, respectively, and that moreover, $\boldsymbol{b}_1 = \boldsymbol{d}_2$.

The case $k = 4$ is similar. By (4.50d), the parametric form of a curve segment of a closed B-spline curve can be written as

$$\boldsymbol{X}(w) = (\,\boldsymbol{d}_{l-3},\quad \boldsymbol{d}_{l-2},\quad \boldsymbol{d}_{l-1},\quad \boldsymbol{d}_l\,)\begin{pmatrix} \frac{1}{6} & -\frac{1}{2} & \frac{1}{2} & -\frac{1}{6} \\ \frac{2}{3} & 0 & -1 & \frac{1}{2} \\ \frac{1}{6} & \frac{1}{2} & \frac{1}{2} & -\frac{1}{2} \\ 0 & 0 & 0 & \frac{1}{6} \end{pmatrix}\begin{pmatrix} 1 \\ w \\ w^2 \\ w^3 \end{pmatrix}, \qquad (10.17)$$

for $w \in [0,1]$. By (4.4), a corresponding Bézier segment of degree 3 has the form

$$\boldsymbol{X}(w) = (\,\boldsymbol{b}_0,\quad \boldsymbol{b}_1,\quad \boldsymbol{b}_2,\quad \boldsymbol{b}_3\,)\begin{pmatrix} 1 & -3 & 3 & -1 \\ 0 & 3 & -6 & 3 \\ 0 & 0 & 3 & -3 \\ 0 & 0 & 0 & 1 \end{pmatrix}\begin{pmatrix} 1 \\ w \\ w^2 \\ w^3 \end{pmatrix}, \qquad w \in [0,1]. \qquad (10.18)$$

Setting these two expressions equal to each other, and multiplying by the inverse of the matrix in (10.18) leads to the following formula for the Bézier points $\boldsymbol{b}_i$ in terms of the de Boor points $\boldsymbol{d}_j$:

$$
\begin{aligned}
(\boldsymbol{b}_0,\ \boldsymbol{b}_1,\ \boldsymbol{b}_2,\ \boldsymbol{b}_3)
&= (\boldsymbol{d}_{l-3},\ \boldsymbol{d}_{l-2},\ \boldsymbol{d}_{l-1},\ \boldsymbol{d}_l)
\begin{pmatrix} \frac{1}{6} & -\frac{1}{2} & \frac{1}{2} & -\frac{1}{6} \\ \frac{2}{3} & 0 & -1 & \frac{1}{2} \\ \frac{1}{6} & \frac{1}{2} & \frac{1}{2} & -\frac{1}{2} \\ 0 & 0 & 0 & \frac{1}{6} \end{pmatrix}
\begin{pmatrix} 1 & 1 & 1 & 1 \\ 0 & \frac{1}{3} & \frac{2}{3} & 1 \\ 0 & 0 & \frac{1}{3} & 1 \\ 0 & 0 & 0 & 1 \end{pmatrix} \\
&= (\boldsymbol{d}_{l-3},\ \boldsymbol{d}_{l-2},\ \boldsymbol{d}_{l-1},\ \boldsymbol{d}_\ell)
\begin{pmatrix} \frac{1}{6} & 0 & 0 & 0 \\ \frac{2}{3} & \frac{2}{3} & \frac{1}{3} & \frac{1}{6} \\ \frac{1}{6} & \frac{1}{3} & \frac{2}{3} & \frac{2}{3} \\ 0 & 0 & 0 & \frac{1}{6} \end{pmatrix}.
\end{aligned}
\tag{10.19}
$$

Fig. 10.1a gives a geometric interpretation of formula (10.19) (the figure shows the nonuniform knot case, although (10.19) holds only for the uniform case). The analogous transformation formula for the nonuniform case can be read off directly from Fig. 10.1a. The geometric construction shown in Fig. 10.1a corresponds to the continuity construction of Fig. 4.12a. The auxiliary points introduced there correspond to the de Boor points! Fig. 10.1a also shows how to find the de Boor points, given the Bézier points. The associated transformation formulae correspond to equation (4.19).

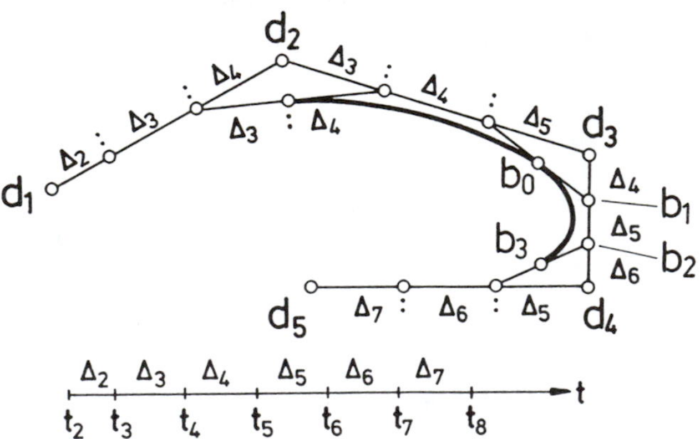

Fig. 10.1a. de Boor points $\boldsymbol{d}_i$ and Bézier points $\boldsymbol{b}_i$ of a cubic B-spline curve ($k = 4$).

The analogous construction for $k = 5$ is shown in Fig. 10.1b. The corresponding transformation formulae can be immediately read off from the geometry. A comparison of Fig. 10.1b with Fig. 4.12b shows how to handle

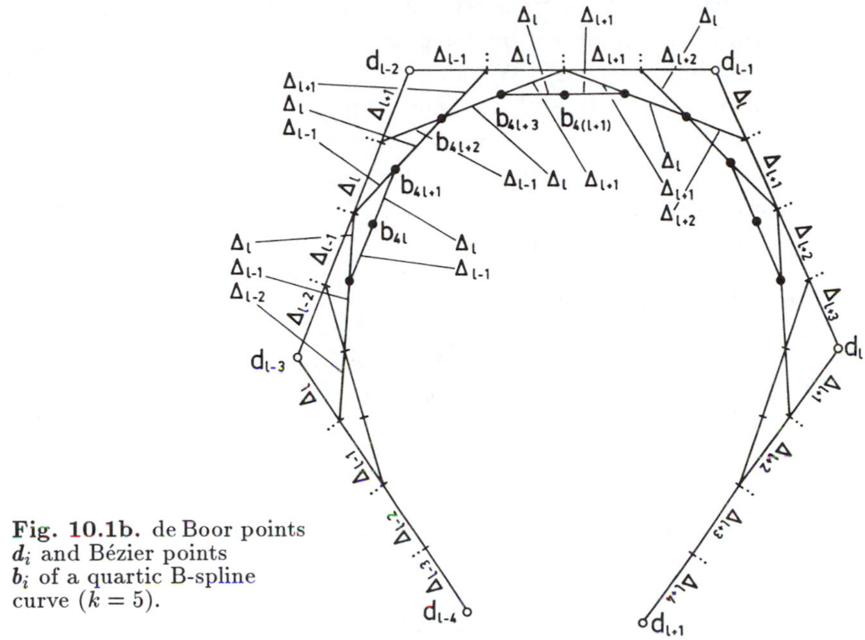

Fig. 10.1b. de Boor points d_i and Bézier points b_i of a quartic B-spline curve ($k = 5$).

Bézier curves. Fig. 10.2 illustrates the application of these transformation formulae. It shows a B-spline curve of order $k = 5$ (degree 4) along with both its Bézier and de Boor polygons.

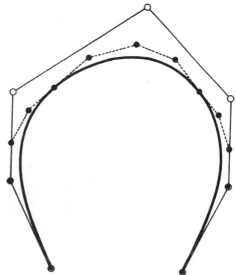

Fig. 10.2. Quartic B-spline curve ($k = 5$) and its Bézier and de Boor polygons.

Algorithmically, the transformation of a B-spline curve segment to Bézier form can be accomplished by using an extension of the de Boor algorithm due to Boehm [Boeh 81]. We assign a multiplicity of $k-1$ to each of the parameter values $t_\ell, t_{\ell+1}$ of the knot vector $\boldsymbol{T}$ associated with the B-spline curve. Then the set of B-splines of order k which do not vanish on the interval $[t_\ell, t_{\ell+1}]$ is precisely the set of Bernstein polynomials of degree $(k-1)$ associated with $[t_\ell, t_{\ell+1}]$, see Lemma 4.6. The de Boor algorithm then reduces to the de Casteljau algorithm.

Algorithm of Boehm:

Given:	de Boor points $\boldsymbol{d}_0, \ldots, \boldsymbol{d}_m$ of order $k-1$, knot vector $t_{k-1}, \ldots, t_{m+1}$.
For each if	$t_\ell < t_{\ell+1}$ from $t_{k-1}, \cdots, t_{m+1}$, $t_{\ell+1}$ has multiplicity $s < k-1$,
find where	$\boldsymbol{d}_\ell^p(t_{\ell+1})$ using the de Boor algorithm, $p = k-1-s$;
increase	m and for $j > l$ the index of $\boldsymbol{d}_j$ and t_j by p;
for	$i = 1, \ldots, p$
replace	$t_{\ell+i}$ by $t_{\ell+1}$, $\boldsymbol{d}_{\ell-p+i}$ by $\boldsymbol{d}_{\ell-p+i}^i$, $\boldsymbol{d}_{\ell+i}$ by $\boldsymbol{d}_\ell^{p-i}$.

To explain the idea behind this algorithm, we consider the transformation for order $k=4$. In this case the de Boor scheme at $t = t_\ell$ gives

$$\begin{array}{llll}
\boldsymbol{d}_{\ell-3} & & & \\
\boldsymbol{d}_{\ell-2} & \boldsymbol{d}_{\ell-2}^1 & & \\
\boldsymbol{d}_{\ell-1} & \boxed{\boldsymbol{d}_{\ell-1}^1 \quad \boldsymbol{d}_{\ell-1}^2} & & \\
\boldsymbol{d}_\ell & \boldsymbol{d}_{\ell-1} & \boldsymbol{d}_{\ell-1}^1 & \boldsymbol{d}_{\ell-1}^2 .
\end{array}$$

Comparing this with Fig. 10.1a and Fig. 4.12a shows that the points inside the box are Bézier points, cf. also Fig. 10.3.

The de Boor scheme corresponding to $t = t_{\ell+1}$ gives

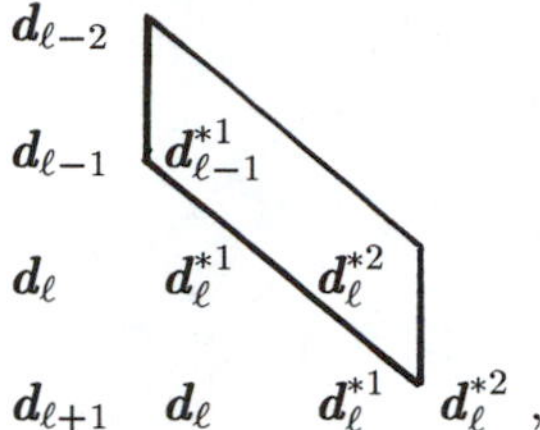

where again the points in the box are Bézier points.

These new (boxed in) Bézier points have the natural order

$$d_{\ell-2}^{1} \quad d_{\ell-1}^{2} \quad d_{\ell-1}^{1} \quad d_{\ell-1}^{*1} \quad d_{\ell}^{*2} \quad d_{\ell}^{*1}.$$

The points $d_{\ell-1}^{2}$ and d_{ℓ}^{*2} are curve points, and thus are boundary points of the corresponding Bézier polygon (compare Fig. 10.3 and Fig. 10.1a).

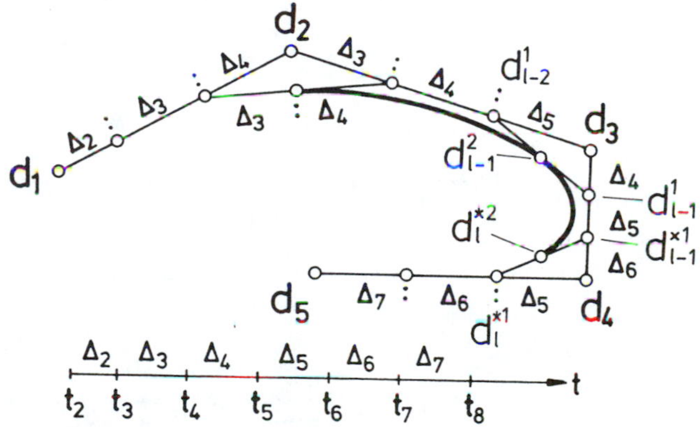

Fig. 10.3. Computation of the Bézier points of a B-spline curve using the de Boor algorithm.

Now consider $k = 5$. Then analogously, at $t = t_{\ell}$,

$$\begin{array}{lllll}
d_{\ell-4} & & & & \\
d_{\ell-3} & d_{\ell-3}^{1} & & & \\
d_{\ell-2} & d_{\ell-2}^{1} & d_{\ell-2}^{2} & & \\
d_{\ell-1} & d_{\ell-1}^{1} & \boxed{d_{\ell-1}^{2} \quad d_{\ell-1}^{3}} & & \\
d_{\ell} & d_{\ell-1} & d_{\ell-1}^{1} & d_{\ell-1}^{2} & d_{\ell-1}^{3}\ ,
\end{array}$$

while at $t = t_{\ell+1}$

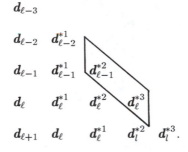

$$\begin{array}{lllll}
\boldsymbol{d}_{\ell-3} & & & & \\
\boldsymbol{d}_{\ell-2} & \boldsymbol{d}^{*1}_{\ell-2} & & & \\
\boldsymbol{d}_{\ell-1} & \boldsymbol{d}^{*1}_{\ell-1} & \boldsymbol{d}^{*2}_{\ell-1} & & \\
\boldsymbol{d}_{\ell} & \boldsymbol{d}^{*1}_{\ell} & \boldsymbol{d}^{*2}_{\ell} & \boldsymbol{d}^{*3}_{\ell} & \\
\boldsymbol{d}_{\ell+1} & \boldsymbol{d}_{\ell} & \boldsymbol{d}^{*1}_{\ell} & \boldsymbol{d}^{*2}_{l} & \boldsymbol{d}^{*3}_{l}.
\end{array}$$

The boxed points $\boldsymbol{d}^{3}_{\ell-1}, \boldsymbol{d}^{2}_{\ell-1}, \boldsymbol{d}^{*2}_{\ell-1}, \boldsymbol{d}^{*3}_{\ell}$ are Bézier points (cf. Fig. 10.1b and Fig. 10.4.). But there is *one point missing*. It lies between $\boldsymbol{d}^{2}_{\ell-1}$ and $\boldsymbol{d}^{*2}_{\ell-1}$, and by Fig. 10.4 can be computed as

$$\boldsymbol{b}_{4\ell+2} = \left(1 - \frac{\Delta_\ell + \Delta_{\ell+1}}{\Delta_{\ell-1} + \Delta_\ell + \Delta_{\ell+1}}\right) \boldsymbol{d}^{*1}_{\ell-1} + \frac{\Delta_\ell + \Delta_{\ell+1}}{\Delta_{\ell-1} + \Delta_\ell + \Delta_{\ell+1}} \boldsymbol{d}^{*1}_{\ell-2},$$

for $t = t_{\ell+1}$, or

$$\boldsymbol{b}_{4\ell+2} = \left(1 - \frac{\Delta_\ell + \Delta_{\ell-1}}{\Delta_{\ell-1} + \Delta_\ell + \Delta_{\ell+1}}\right) \boldsymbol{d}^{1}_{\ell-2} + \frac{\Delta_\ell + \Delta_{\ell-1}}{\Delta_{\ell-1} + \Delta_\ell + \Delta_{\ell+1}} \boldsymbol{d}^{1}_{\ell-1},$$

for $t = t_\ell$, respectively.

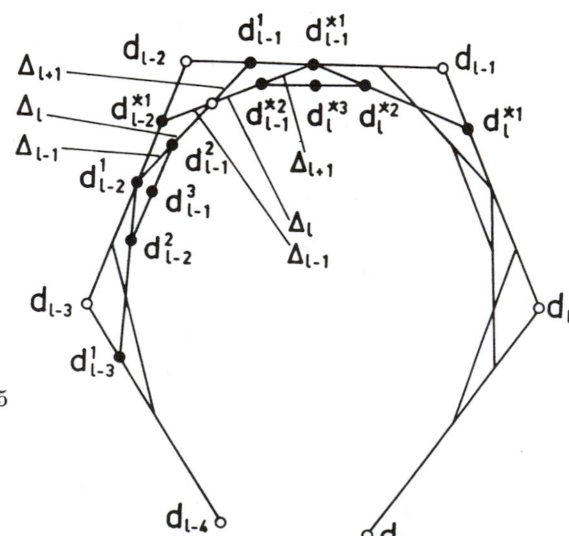

Fig. 10.4. Bézier and de Boor points for $k = 5$ (cf. Fig. 10.1b).

The Bézier points produced in this way correspond to those produced by the de Boor algorithm, see e.g., Fig. 4.28 and the diagram on page 179. We note that, obviously, repeated application of the de Boor algorithm to points in the de Boor scheme leads to the desired Bézier points. This repeated insertion step of the de Boor algorithm is done systematically in the algorithm of Boehm.

The basis transformation of B-spline segments to Bézier segments can also be accomplished using the so-called *Oslo Algorithm* of Cohen, Lyche, and Riesenfeld, see also [Coh 80], [BART 87].

These transformation algorithms can easily be extended to surfaces by applying them first in the u direction, then in the v direction. Fig. 10.5 shows the Bézier net of the B-spline surface of Fig. 6.13a. Note the increase in the number of control points.

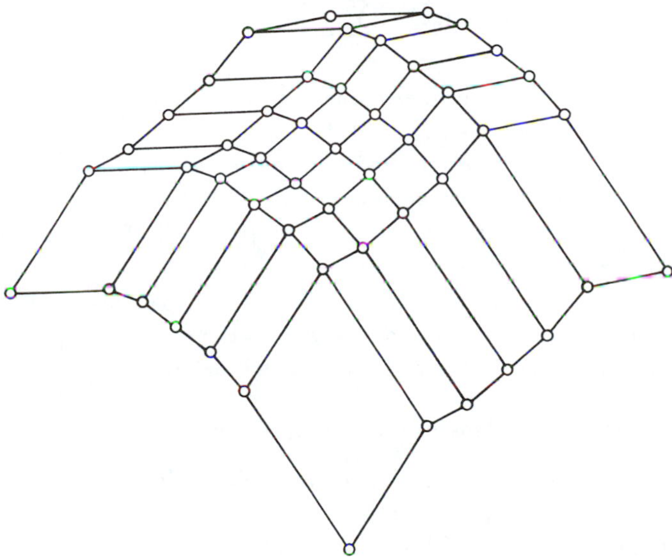

Fig. 10.5. Bézier net for the B-spline surface of Fig. 6.13a.

10.2. Approximate Basis Transformation

When exact basis transformation is not possible, for example in case of

- degree reduction,
- degree raising while simultaneously merging several patches,
- merging patches while maintaining the polynomial degree,

we have to use approximate methods.

As an example of this process, we briefly discuss the VDA conversion software of Dannenberg-Nowacki [Dan 85]. It is based on a method of Hölzle [Hölz 83] for locating new segment boundaries in such a way that the mean error of each segment is less than a prescribed tolerance. Then subject to prescribed Hermite boundary conditions, the following integral is minimized

$$Q = \int_0^1 \int_0^1 (\Delta \boldsymbol{p})^2 d\widetilde{u} d\widetilde{v}, \tag{10.20}$$

where $\boldsymbol{p}$ is the given surface, $\Delta\boldsymbol{p}(\widetilde{u},\widetilde{v}) = \boldsymbol{p}(u(\widetilde{u}), v(\widetilde{v})) - \boldsymbol{r}(\widetilde{u},\widetilde{v})$, and $\boldsymbol{r}$ is the desired surface. Here u, v are the local parameters of $\boldsymbol{p}$, while $\widetilde{u},\widetilde{v}$ are the local parameters of $\boldsymbol{r}$. The method assumes that the given surface is expressed in terms of monomials of degree (n, m):

$$\boldsymbol{p}(u,v) = \sum_{j=0}^{n}\sum_{k=0}^{m} a_{jk} u^j v^k.$$

Applying this method for degree reduction often leads to a surface with a large number of patches, see [Dan 85].

In [Lac 88] the surface conversion problem is solved using Chebycheff polynomials. Bézier curve conversion is treated by [Kal 87], [Watk 88] and B-spline curve transformation is discussed by [Patr 89]. The method of [Patr 89] was extended by [Bar 89] to B-spline surfaces.

In all these methods, the number of resulting patches of the approximating surface is always on the same order as the number of patches of the method in [Dan 85], see Table 10.1 on page 456 below.

10.2.1. Approximate Basis Transformation for Curves

Since the *conversion methods* discussed above always lead to a higher number of segments, in [Hos 87] a new method for curves was proposed which could also be extended to surfaces [Hos 89, 90, 92a].

The key idea behind these conversion methods is the use of reparametrization (see Sects. 4.4, 6.2.5) to reduce the number of curve or surface segments obtained. If one wants to use the parametrization as a tool for optimization, then we need to use parameter-invariant smoothness conditions, which leads us to consider GC^k or G^k continuity, see Chapters 5, 7.

Here we discuss reduction to cubic curves. Our approach is to use (5.6) and Bézier tools. We assume that the given curve is of degree n and can be written in Bézier form as

$$\boldsymbol{X} = \sum_{i=0}^{n} \boldsymbol{V}_i B_i^n(t), \qquad t \in [0,1], \tag{10.20a}$$

where $\boldsymbol{V}_i$ are prescribed Bézier points. We seek an approximating curve $\boldsymbol{Y}$ in parametric form

$$\boldsymbol{Y} = \sum_{i=0}^{3} \boldsymbol{W}_i B_i^3(t), \qquad t \in [0,1], \tag{10.20b}$$

where $\boldsymbol{W}_i$ are unknown Bézier points. Requiring that the endpoints of $\boldsymbol{X}$ and $\boldsymbol{Y}$ coincide, and that the two curves meet each other with first order smoothness at these points, it follows that coefficients in (10.20a, b) must satisfy the boundary conditions

$$\begin{aligned} \boldsymbol{W}_0 &= \boldsymbol{V}_0, & \boldsymbol{W}_1 &= \boldsymbol{V}_0 + \lambda_1(\boldsymbol{V}_1 - \boldsymbol{V}_0), \\ \boldsymbol{W}_m &= \boldsymbol{V}_n, & \boldsymbol{W}_{m-1} &= \boldsymbol{V}_n + \lambda_2(\boldsymbol{V}_{n-1} - \boldsymbol{V}_n). \end{aligned} \tag{10.21}$$

The new Bézier points are then determined by

- the parameters λ_1, λ_2,
- the parametrization of the Bézier segment $\boldsymbol{Y}$.

In order to find the best λ_1, λ_2, we choose at least $n+1$ (equally spaced) points $\boldsymbol{P}_i$ on the given curve $\boldsymbol{X}$. Assuming the corresponding parameter values are given by t_i, we can write

$$\boldsymbol{P}_i = \sum_{j=0}^{3} \boldsymbol{W}_j B_j^3(t_i) + \boldsymbol{\delta}_i, \tag{10.22}$$

where $\boldsymbol{\delta}_i$ is the error vector. Substituting (10.21) in (10.22), we can rewrite this equation as

$$\boldsymbol{D}_i = \lambda_1(\boldsymbol{V}_1 - \boldsymbol{V}_0)B_1^3(t_i) + \lambda_2(\boldsymbol{V}_n - \boldsymbol{V}_{n-1})B_2^3(t_i) + \boldsymbol{\delta}_i, \tag{10.23}$$

with given vector

$$\boldsymbol{D}_i := \boldsymbol{P}_i - \boldsymbol{V}_0 B_0^3(t_i) - \boldsymbol{V}_0 B_1^3(t_i) - \boldsymbol{V}_n B_2^3(t_i) - \boldsymbol{V}_n B_3^3(t_i).$$

It follows that the absolute value of the error vector in (10.23) is given by

$$\delta := \sum_{i=0}^{n} |\boldsymbol{\delta}_i|^2 = \sum_{i=0}^{n} \left[\boldsymbol{D}_i - \lambda_1(\boldsymbol{V}_1 - \boldsymbol{V}_0)B_1^3(t_i) - \lambda_2(\boldsymbol{V}_n - \boldsymbol{V}_{n-1})B_2^3(t_i)\right]^2.$$

The minimum of δ can be found by discrete least squares. The necessary conditions

$$\frac{\partial \delta}{\partial \lambda_1} = 0, \qquad \frac{\partial \delta}{\partial \lambda_2} = 0, \tag{10.24}$$

for a minimum lead to a linear system for λ_1 and λ_2. The result depends on the parametrization of the points $\boldsymbol{P}_i$, and hence after solving (10.24), we apply a parameter correction as discussed in Sect. 4.4 and then solve the system again. This process is iterated until nearly all error vectors are orthogonal. In general, this takes about three to four iterations. If the given error tolerance cannot be satisfied, the curve is split at the point where the error is maximal, and the algorithm is applied to each of the resulting two pieces.

This method can be extended to higher degrees of contact at the ends, and thus to higher degree polynomials as well as to rational Bézier curves, cf. e.g., [Hos 87, 88a, 88b]. It should be noted, however, that for higher degree polynomials, the minimization of the error leads to nonlinear systems of equations which have to be solved by appropriate optimization methods. Fig. 10.6 shows the reduction of a Bézier curve of degree 19 to a Bézier spline curve of degree 5 with 3 pieces and second order contact. The Bézier polygons of the original curve and the approximating curve are shown as a dashed line and a solid line, respectively. The corresponding curves cannot be distinguished visually.

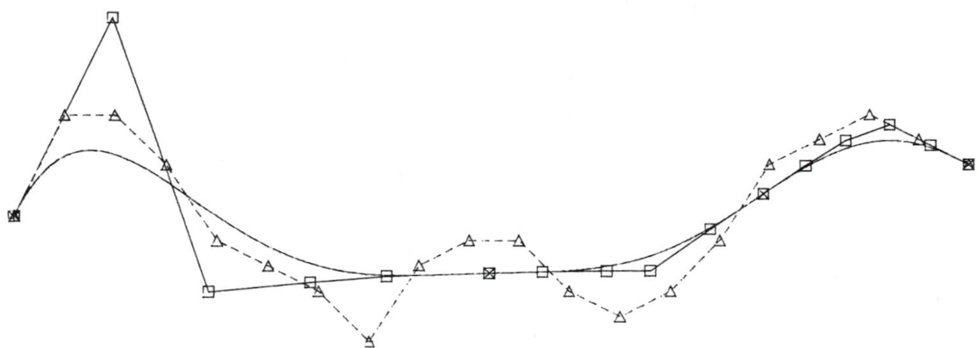

Fig. 10.6. Approximation of a Bézier curve of degree 19 by three quintic Bézier spline segments.

This method can also be used for *merging* several Bézier curves into a single Bézier curve, see [Hos 87].

10.2.2. Approximate Basis Transformation for Surfaces

Since basis transformations of surfaces also make essential use of *parameter correction*, as in the curve case we again describe the smoothness conditions in terms of order of contact, see Chap. 7.

Suppose the given surface is of degree (n, m) and has the parametric representation

$$\boldsymbol{X}(u,v) = \sum_{i=0}^{n}\sum_{k=0}^{m} \boldsymbol{V}_{ik} B_i^n(u) B_k^m(v), \qquad u,v \in [0,1], \tag{10.25a}$$

where the $\boldsymbol{V}_{ik}$ are the given Bézier points. Again we restrict our attention to the bicubic case (3,3), where the desired approximating Bézier patch has the form

$$\boldsymbol{Y}(u,v) = \sum_{i=0}^{3}\sum_{k=0}^{3} \boldsymbol{W}_{ik} B_i^3(u) B_k^3(v), \tag{10.25b}$$

where $\boldsymbol{W}_{ik}$ are the unknown Bézier points. Our approximation method consists of two steps:

1) approximate the boundary curves of the given surface patch $\boldsymbol{X}$ as in Sect. 10.2.1,
2) approximate the surface $\boldsymbol{X}$ on the interior.

We assume that the two surfaces $\boldsymbol{X}$ and $\boldsymbol{Y}$ have the same vertices, so that

$$\boldsymbol{X}(i,k) = \boldsymbol{Y}(i,k), \qquad i = 0,1 \quad k = 0,1. \tag{10.26}$$

This means that the vertices in the Bézier representations for the two surfaces also coincide. We shall also require that the two surfaces have common tangent planes at the corner points (GC^1 smoothness).

After constructing an approximation to the boundary curves, it follows that the Bézier points $\boldsymbol{W}_{0k}, \boldsymbol{W}_{i0}, \boldsymbol{W}_{3k}, \boldsymbol{W}_{i3}$ are determined for $i,k = 1,2$ by (10.21) in terms of the scalar parameters λ_i, see Fig. 10.7.

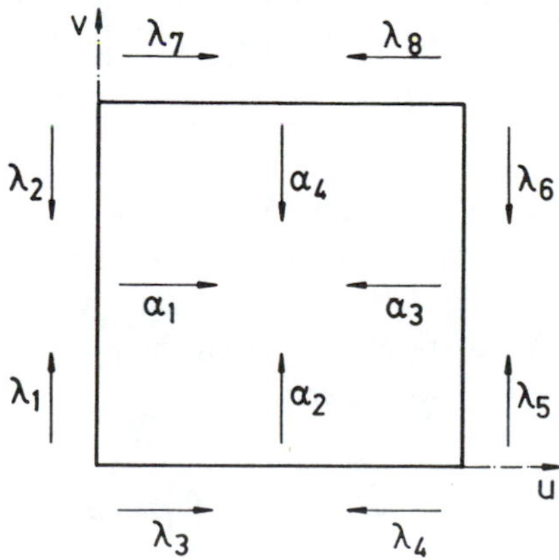

Fig. 10.7. Influence of the approximation parameters.

To find the unknown inner Bézier points $\boldsymbol{W}_{11}, \boldsymbol{W}_{12}, \boldsymbol{W}_{21}, \boldsymbol{W}_{22}$, we now assume that the following conditions for the cross derivatives on the boundary curves hold:

$$\begin{aligned}
&\text{boundary 1:} \qquad && \boldsymbol{Y}_u(0,v) = \alpha_1(v)\boldsymbol{X}_u(0,v),\\
&\text{boundary 2:} \qquad && \boldsymbol{Y}_v(u,0) = \alpha_2(u)\boldsymbol{X}_v(u,0),\\
&\text{boundary 3:} \qquad && \boldsymbol{Y}_u(1,v) = \alpha_3(v)\boldsymbol{X}_u(1,v),\\
&\text{boundary 4:} \qquad && \boldsymbol{Y}_v(u,1) = \alpha_4(u)\boldsymbol{X}_v(u,1),
\end{aligned}$$

where the unknown functions α_i are determined on the boundary curves by the parameters λ_k. Thus, we can take the α_i to be quadratic polynomials:

$$\begin{aligned}
\alpha_1(v) &= \frac{3}{n}\lambda_3 B_0^2(v) + \omega_1 B_1^2(v) + \frac{3}{n}\lambda_7 B_2^2(v),\\
\alpha_2(u) &= \frac{3}{m}\lambda_1 B_0^2(u) + \omega_2 B_1^2(u) + \frac{3}{m}\lambda_5 B_2^2(u),\\
\alpha_3(v) &= \frac{3}{n}\lambda_4 B_0^2(v) + \omega_3 B_1^2(v) + \frac{3}{n}\lambda_8 B_2^2(v),\\
\alpha_4(u) &= \frac{3}{m}\lambda_2 B_0^2(u) + \omega_4 B_1^2(u) + \frac{3}{m}\lambda_6 B_2^2(u).
\end{aligned} \tag{10.28}$$

This introduces new unknown parameters which are free and can be used to construct an optimal fit. Substituting (10.28) in (10.27), and taking account of the G^1 conditions at the corners, we obtain the following vector valued linear system of equations for the unknown Bézier points $\boldsymbol{W}_{11}, \boldsymbol{W}_{12}, \boldsymbol{W}_{21}, \boldsymbol{W}_{22}$:

$$\begin{pmatrix} B_1^3(v) & B_2^3(v) & 0 & 0\\ B_1^3(u) & 0 & B_2^3(u) & 0\\ 0 & 0 & B_1^3(v) & B_2^3(v)\\ 0 & B_1^3(u) & 0 & B_2^3(u) \end{pmatrix} \begin{pmatrix} \boldsymbol{W}_{11}\\ \boldsymbol{W}_{12}\\ \boldsymbol{W}_{21}\\ \boldsymbol{W}_{22} \end{pmatrix} = \begin{pmatrix} \frac{n}{3}\omega_1 \boldsymbol{N}_1(v) + \boldsymbol{Q}_1(v)\\ \frac{m}{3}\omega_2 \boldsymbol{N}_2(u) + \boldsymbol{Q}_2(u)\\ \frac{n}{3}\omega_3 \boldsymbol{N}_3(v) + \boldsymbol{Q}_3(v)\\ \frac{m}{3}\omega_4 \boldsymbol{N}_4(u) + \boldsymbol{Q}_4(u) \end{pmatrix}, \tag{10.29}$$

where $\boldsymbol{N}_i, \boldsymbol{Q}_i$ are abbreviations for known quantities. For each point (u_0, v_0), the matrix on the left has rank 3, so by eliminating variables, we can reduce the problem to finding four of them, say ω_1 and the three components of $\boldsymbol{W}_{11}$. Now we choose N points $\boldsymbol{P}_j(u_j, v_j)$ on the given surface, as uniformly distributed as possible. The overall error

$$d = \sum_{j=1}^{N} \left[\boldsymbol{P}_j - \sum_{i=0}^{3}\sum_{k=0}^{3} \boldsymbol{W}_{ik} B_i^3(u_j) B_k^3(v_j) \right]^2$$

is a function of these four variables. We can find a minimum by solving the linear system

$$\frac{\partial d}{\partial \boldsymbol{W}_{11}} = 0, \qquad \frac{\partial d}{\partial \omega_1} = 0, \tag{10.30}$$

for ω_1 and the three components of $\boldsymbol{W}_{11}$.

We now iterate between solving this system and correcting the parameters to find an optimal approximation surface of degree (3,3). This method can be extended to polynomials of higher degree, in which case we again need to solve a nonlinear system of equations (see [Hos 89], [Was 91]), or if higher mixed derivatives are used, we actually get a linear system, see [Was 91]. The method described here for bicubic surface patches produces a C^0 continuous surface, in general, although at the vertices the individual surface patches are joined with GC^1 smoothness. In the biquintic case, we have at least GC^1 continuity along the boundary curves. Fig. 10.8 shows the approximation of a Bézier surface of degree (15,15) by a Bézier surface of degree (3,3). It also shows the effect of parameter correction. The approximant constructed without parameter correction is shown in Fig. 10.8a, while the result after iterative parameter correction is shown in Fig. 10.8b. The dotted lines are the parameter lines of the approximating surface. The effect of the parameter correction is clear.

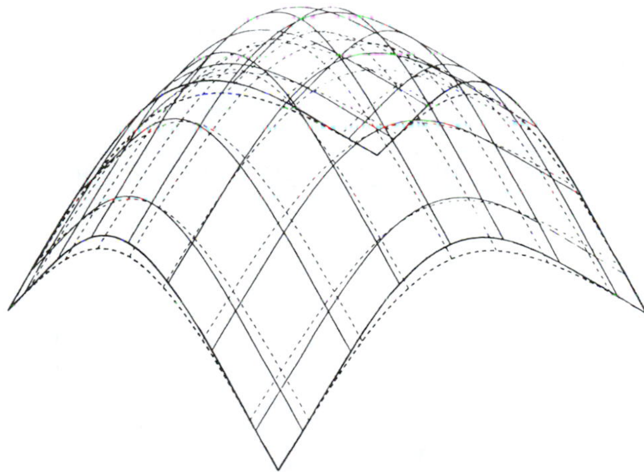

Fig. 10.8a. Approximation of a (15,15) Bézier surface by a bicubic Bézier surface without parameter correction.

A comparison between the methods of [Dan 85], [Bar 89], and the method [Hos 92a] presented here can be found in [Brod 90]. The test involved transforming an integral Bézier surface of degree (9,9) (see Fig. 10.9) into bicubic

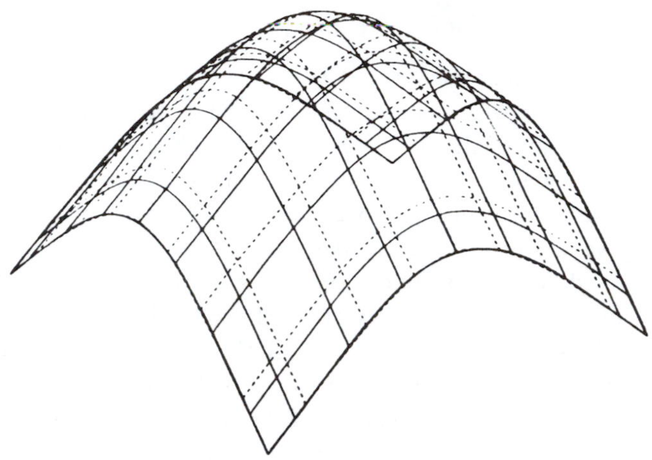

Fig. 10.8b. Approximation of the (15,15) Bézier surface in Fig. 10.8a by a bicubic Bézier surface with parameter correction.

patches. Table 10.1 clearly shows the superiority of the method using parameter correction (see Fig. 10.9) – it requires the smallest number of patches. Table 10.1 also includes information on the conversion of this test surface using the method developed by [Lac 88].

Method	Max. error	Number of patches	Smoothness
Bar 89	0.01	144	C^2
Dan 85	0.01	25	C^0
Hos 90	0.05	8	C^0
	0.01	12	C^0
Lac 88	0.10	24	C^0
	0.01	56	C^0
	0.10	40	C^1
	0.01	128	C^1

Table 10.1. Transformation of the (9,9) Bézier surface in Fig. 10.9 to a combination of bicubic Bézier patches.

Methods for basis transformation of Bezier and B-spline surfaces using *surface curves* (trimmed surfaces) were developed in [Hos 90, 92a]. There the surface curves are represented in B-spline form in the parameter plane.

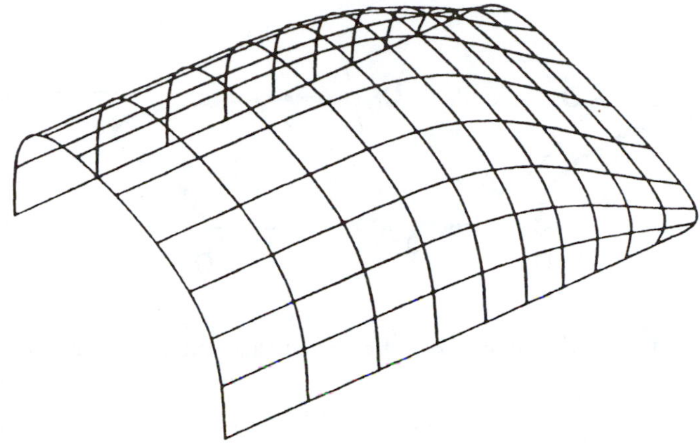

Fig. 10.9. A 9×9 Bézier surface.

10.3. Merging Bézier Surfaces

The methods developed in Sect. 10.2 can also be used to merge Bézier surfaces and thus, for data reduction. Suppose we are given a collection of Bézier patches whose boundaries have been constructed in the process of some design session. Our goal is to approximate this collection of Bézier patches to a prescribed accuracy, using a minimal number of different Bézier patches.

Since the original boundaries are artificial, we first look for some kind of natural boundaries. Our choice will be influenced by the fact that we intend to work with bicubic patches.

A *generic* (plane) cubic Bézier curve will generally not have more than one (interior) minimum of its curvature in the domain of interest because of the variation diminishing property. However, as shown in [SU 89], this kind of curve can have more than one point of minimum curvature, but such cases generally are of no interest for applications. Fig. 10.10 shows the distribution of the curvature minima of a generic cubic curve. For quintic curves, in general, there are no more than three minima of the curvature.

In view of the above observations, we cannot expect a cubic (or quintic) curve to be a good approximation of a curve with more than one minimum (or more than three minima) of curvature. Thus in order to get optimal results in approximating a given curve with more than one curvature minimum by a cubic curve, we generally will have to work with more than one segment.

In order to find these new boundary curves corresponding to a given system of spline segments, we first discretize the individual segments using

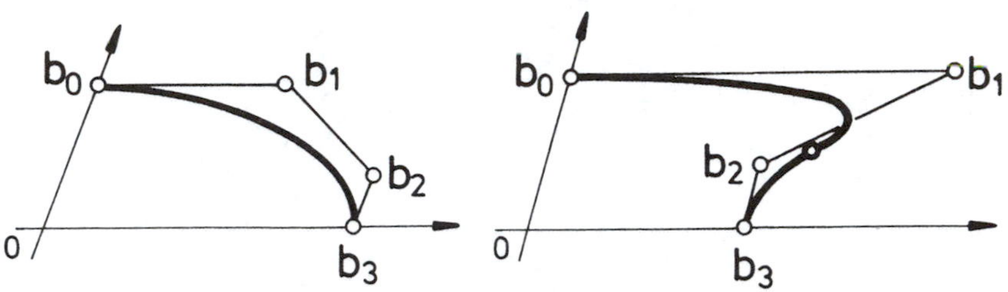

Fig. 10.10. Generic cubic Bézier curves and their curvature minima.

appropriately selected parameter nets (e.g., discretization with Δu). Afterwards, the parameter nets will be reparametrized with the parameter intervals Δt. We have a minimum of the curvature whenever

$$\kappa(t_i) - \kappa(t_{i-1}) < 0 \qquad \text{and} \qquad \kappa(t_{i+1}) - \kappa(t_i) \geq 0.$$

Thus, the goal of our subdivision strategy is

- separate the minimum points of curvatures of a parametric curve into distinct surface patches,
- using this criterion, reduce the number of surface patches to a minimum.

This segmentation process proceeds iteratively through several steps. We illustrate the method for bicubic patches and discretization in the u direction:

1) Starting with $u = 0$, find the first two minima of the curvature for each discretized parameter line $v =$ constant (see Fig. 10.11a) (if there exists only one minimum on a parameter line, the line can be cancelled);
2) Determine the local segmentation points by using the mean values of the parameter values found in step 1;
3) Find the smallest parameter value $u_{\kappa_{min}}$ for each curvature minimum located in the *second step*;
4) If $u_c \leq u_{\kappa_{min}}$, use the parametric value u_c corresponding to the average of all local segmentation points to find the first new boundary curve;
5) If $u_c > u_{\kappa_{min}}$, (see Fig. 10.11b), move the point u_c to u_c^* by successively removing the segmentation points with the largest parameter values, and computing the associated new average point. This step is repeated until the condition in Step 4 is satisfied (see Fig. 10.11c);
6) Once u_c^* is found, go back to Step 1 starting with u_c^*.

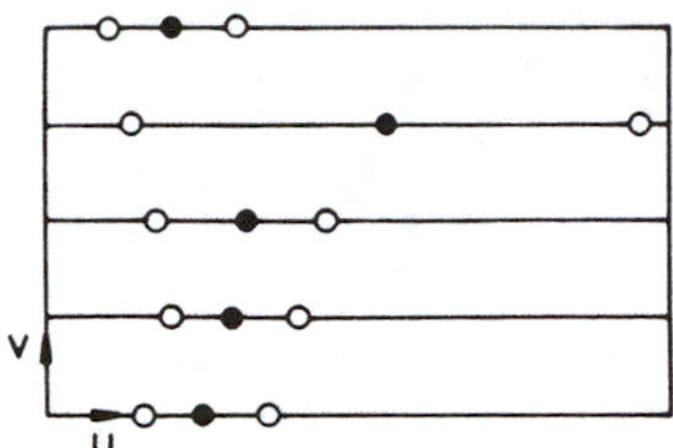

Fig. 10.11a. Local segmentation points in the parameter domain.

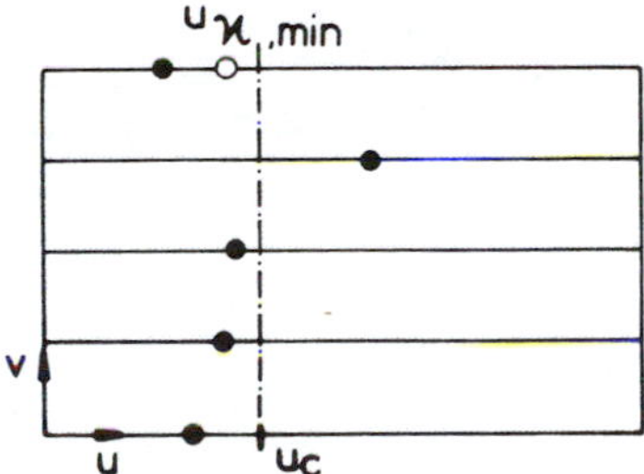

Fig. 10.11b. First segmentation line of a surface patch.

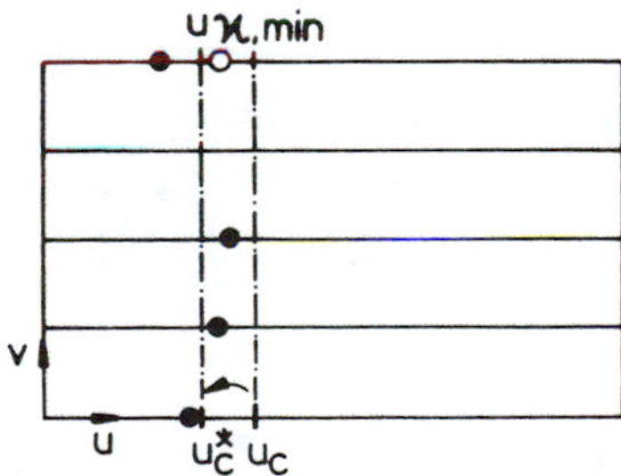

Fig. 10.11c. Removal of the largest parameter value of the local segmentation points.

Using this method leads to new surface patches which do not have more than one minimum of curvature on every discretized parameter line. Moreover, in the framework of the discretization used, we get a minimal number of such surface patches satisfying the above geometric conditions. Once we are done, we then analogously segment the surface in the v direction.

Once we have found the new patch boundaries, we fill in the surface patches as in Sect. 10.2. If the prescribed error tolerance cannot be satisfied, then we have to further subdivide the patches, e.g., by inserting new patch boundaries at those points with maximal error.

In order to illustrate the effectiveness of this method, we consider a collection of 220 bicubic patches, see Fig. 10.12 where the patch boundaries are shown as solid lines. This collection can be reduced to a set of 14 patches while maintaining a maximum error of at most 1 mm in the points, and at most 4 degrees in the angles between the corresponding normals. Here the underlying domain was 100 × 30 cm.

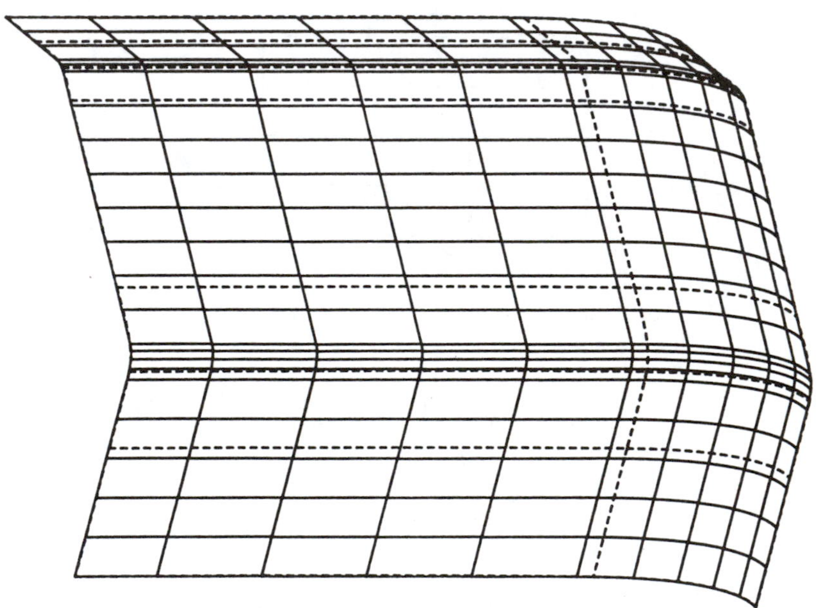

Fig. 10.12. Reduction of a collection of 220 bicubic Bézier patches (solid boundaries) to 14 patches (dotted boundaries).

Another approach to data reduction was proposed by [Lyc 88]. The idea is to remove unnecessary knots of a B-spline curve in such a way that the deviation of the simplified B-spline curve remains within a prescribed error bound. For applications of this method to data reduction, e.g., to characteristic curves, see [Wev 91] and Chap. 13.

10.4. Basis Transformation for Triangular Patches

The methods described so far for surface conversion deal only with the conversion of tensor-product surfaces. The approximation method of Sect. 10.3 can be extended to triangular Bézier patches, although in this case, certain additional problems arise: 1) closure problems when moving around a triangle, and 2) reduction in the number of unknowns due to the small number of Bézier points. This means that less geometric information can be carried over from the initial patches to the approximating patches. For details on the exact transformation of a monomial to a triangular Bézier patch, see [Wag 86].

Another type of basis transformation of interest for triangular patches is their conversion to four-sided patches. One way to (exactly) transform the Bézier points of a triangular patch into those of a four-sided patch is discussed in [Brü 80]. The (exact) decomposition of a four-sided patch into two triangular patches is treated by [Gol 87], where an (n, m)-tensor-product patch is converted to a triangular patch of degree $n + m$. An extension of the conversion equations between polynomial triangular and rectangular patches to the rational case can be found in [Jie 90].

It was noted above in Sect. 6.3.4 that for box splines, the control points can be converted to Bézier points, see [Boeh 83, 83a, 85], [Pra 85].

11 Multivariate Methods

While in the past, CAGD has been mostly concerned with curves and surfaces, more recently, there has been an increasing interest in higher dimensional multivariate objects such as volumes and hypersurfaces in $\mathbb{R}^n$, $n > 3$.

Some of the applications include

- the description of scalar or vector-valued physical fields, such as temperature, pressure, gravitation, velocity, electromagnetic fields, etc. as functions of several variables, e.g., the three positional coordinates and time, etc.,
- the description of spatial movement or deformation of a surface,
- the description of inhomogeneous materials,
- the construction and modification of homogeneous bodies which might arise in some design process involving geometric operations on a closed surface, as higher dimensional contour surfaces, or as parametric surfaces associated with hypersurfaces in $\mathbb{R}^n$.

For further details and examples, see [Alf 89], [Casa 85], [Faro 85a], [Sed 85b, 86a], and [Sanc 91].

Almost all of the surface methods described in Chapters 6 through 9 can be generalized to higher dimensional objects. Since visualizing objects in n variables with $n > 3$ is difficult, we shall frequently restrict our discussion in this chapter to the trivariate case.

In the bivariate case we used triangles and rectangles. In the trivariate case, the natural analog of these sets are tetrahedra, pentahedra, and hexahedra. They have been extensively used as domains for monomial, Lagrange, and Hermite basis functions, both in the FEM literature, see e.g., [ZIE 77], [Zen 73], [Grie 85], [GRIE 87], and [SABO 87], and in early papers in the CAGD literature, see [Fer 64], [Boo 62, 77], and for cubical solids, [MOR 85].

As in previous chapters, we again make heavy use of Bézier representations since they allow us to describe the geometrical connection between control points and the associated volumes.

11.1. Bézier Representations

11.1.1. Tensor-product Bézier Volumes

A *tensor-product Bézier volume* (TPB volume) of degree (ℓ, m, n) is defined to be

$$X(u) = \sum_{i=0}^{\ell} \sum_{j=0}^{m} \sum_{k=0}^{n} b_{ijk} B_i^{\ell}(u) B_j^m(v) B_k^n(w), \tag{11.1}$$

where $\boldsymbol{u} = (u, v, w)$, with $u, v, w \in I = [0, 1]$, and where we use the *ordinary Bernstein polynomials* of (4.2) as basis functions, see [Bez 78], [Casa 85], [Faro 85a], [Sed 86a], and [Las 85, 85a, 87].

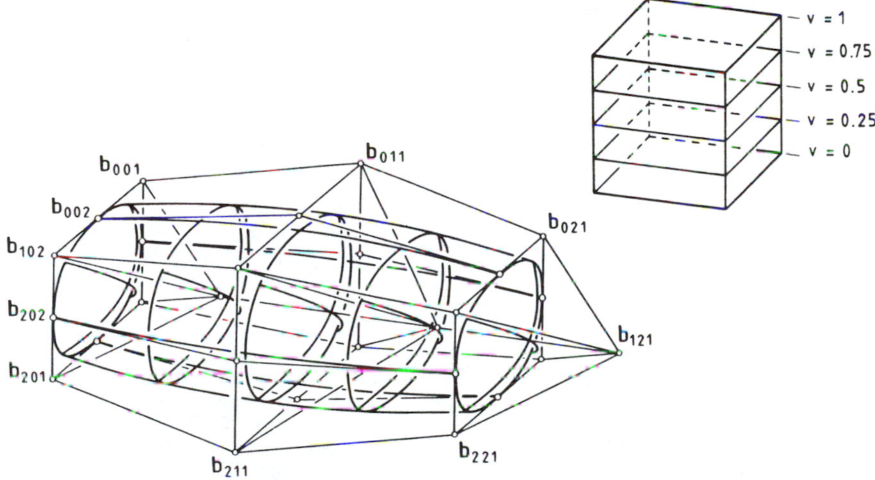

Fig. 11.1. Triquadratic TPB volume and its associated Bézier grid. The boundary surfaces of the TPB volume and the parametric surfaces corresponding to $v = 0, \frac{1}{4}, \frac{1}{2}, \frac{3}{4}, 1$ are shown. The parameter domain is shown in the upper right corner.

The vector-valued parametric formula (11.1) defines solid bodies in space. We call the $\boldsymbol{b}_{ijk} \in \mathbb{R}^3$ *Bézier points*, and if we connect them by lines according to the natural order of their indices, we get the associated *Bézier grid*, see Fig. 11.1. In the case where $b_{ijk} \in \mathbb{R}$, we have a function-valued nonparametric formula $X(\boldsymbol{u})$, where the b_{ijk} can be considered as ordinates associated with the abscissae $\boldsymbol{u}_{ijk} = (i/\ell, j/m, k/n) \in I^3$. In this case $(\boldsymbol{u}, X(\boldsymbol{u}))$ defines a *hypersurface* in $\mathbb{R}^4$ associated with the Bézier grid and the *Bézier points* $\boldsymbol{B}_{ijk} = (\boldsymbol{u}_{ijk}, b_{ijk}) \in \mathbb{R}^4$. Such a hypersurface could, for example, represent a spatial temperature distribution.

On the other hand, a function-valued formula with $\boldsymbol{b}_{ijk} \in \mathbb{R}^d$ can be used to describe vector-valued fields such as velocity fields. All of the following results for the parametric case carry over with appropriate minor modifications to the function-valued case.

The expression (11.1) can also be generalized to the case of *rational representations* in the same way as was done for Bézier curves and tensor-product Bézier surfaces [Las 91a]. Rational TPB volumes are useful for giving exact descriptions of 3D primitives such as the sphere, cylinder, torus, etc., with "*solid interiors*". Figs. 11.2 and 11.3 show two examples. The definition of *solids* using TPB volumes is also discussed in [Casa 85], [Faro 85a], and [Sai 87].

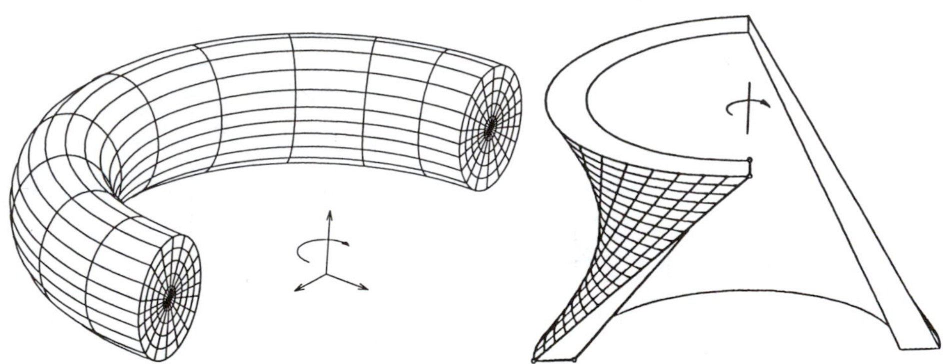

Fig. 11.2. Solid half-torus, described by a rational TPB volume of degree (1,2,2).

Fig. 11.3. Part of a solid hyperboloid of revolution described by a rational TPB volume of degree (1,1,2).

It follows immediately from the way in which (11.1) is defined that we can apply many of the Bézier curve operations discussed in Chap. 4 (such as degree raising, the de Casteljau algorithm, and various techniques derived from them such as subdivision) separately to the u, v, and w variables. This leads us to the following properties relating the behavior of the TPB volume and its Bézier grid (cf. Chaps. 4, 6, and [Las 87]):

- *Convex hull property*: All points defined by (11.1) lie inside the convex hull of the Bézier grid.
- *Parametric surfaces*: The parametric surfaces corresponding to constant u, v, or w are TPB surface patches of degree (m, n), (ℓ, n), and (ℓ, m), resp.
- *Parametric lines*: The parametric lines corresponding to constant u, v, or u, w, or v, w are Bézier curves of degree n, m, and ℓ, resp.
- *Boundary surfaces*: The boundary surfaces of a TPB volume are TPB surfaces. Their Bézier nets are the *boundary nets* of the Bézier grid.

- *Boundary curves*: The boundary curves of a TPB volume are Bézier curve segments. Their Bézier polygons are given by the *edge polygons* of the Bézier grid.
- *Vertices*: The vertices of a TPB volume coincide with the vertices of its Bézier grid.
- *Derivatives*: The partial derivatives of order (p, q, r) of a TPB volume of degree (ℓ, m, n) at the point $\boldsymbol{u}$ is given by (cf. (4.7b))

$$\frac{\partial^{p+q+r}}{\partial u^p \partial v^q \partial w^r} \boldsymbol{X}(\boldsymbol{u}) = \frac{l!}{(l-p)!} \frac{m!}{(m-q)!} \frac{n!}{(n-r)!} \cdot \\ \cdot \sum_{i=0}^{l-p} \sum_{j=0}^{m-q} \sum_{k=0}^{n-r} \Delta^{pqr} \boldsymbol{b}_{ijk} B_i^{l-p}(u) B_j^{m-q}(v) B_k^{n-r}(w), \tag{11.2}$$

where

$$\begin{aligned} \Delta^{000} \boldsymbol{b}_{ijk} &= \boldsymbol{b}_{ijk} \\ \Delta^{pqr} \boldsymbol{b}_{ijk} &= \Delta^{p00}[\Delta^{0q0}(\Delta^{00r} \boldsymbol{b}_{ijk})] \\ &= \Delta^{p-1,0,0}[\Delta^{0q0}(\Delta^{00r} \boldsymbol{b}_{i+1,j,k})] - \Delta^{p-1,0,0}[\Delta^{0q0}(\Delta^{00r} \boldsymbol{b}_{ijk})] \\ &= \Delta^{p00}[\Delta^{0,q-1,0}(\Delta^{00r} \boldsymbol{b}_{i,j+1,k}) - \Delta^{0,q-1,0}(\Delta^{00r} \boldsymbol{b}_{ijk})] \\ &= \Delta^{p00}[\Delta^{0q0}(\Delta^{0,0,r-1} \boldsymbol{b}_{i,j,k+1} - \Delta^{0,0,r-1} \boldsymbol{b}_{ijk})]. \end{aligned}$$

- *Degree raising*: If we write a TPB volume (11.1) of degree (ℓ, m, n) as one of degree $(\ell, m+\mu, n)$, then the new Bézier points $\bar{\boldsymbol{b}}_{iJk}^{\mu}$ are given by (cf. equation (4.14a))

$$\bar{\boldsymbol{b}}_{iJk}^{\mu} = \sum_{j=J}^{J-\mu} \boldsymbol{b}_{ijk} \frac{\binom{J}{j}\binom{m+\mu-J}{m-j}}{\binom{m+\mu}{m}}, \qquad \left\{ \begin{array}{l} J = 0(1)m{+}\mu \\ \text{all } i, k, \end{array} \right. \tag{11.3}$$

with corresponding formulae for degree raising in the u and w variables. If we take the limit as $\ell, m, n \to \infty$, then the points of the corresponding sequence of Bézier grids converge to the points of the TPB volume. The convergence rate is linear [Coh 85].

- *Degree reduction*: Suppose $\boldsymbol{X}(\boldsymbol{u})$ is a TPB volume of degree (ℓ, m, n) with $m = \bar{m} + \mu$. Then a necessary and sufficient condition for $\boldsymbol{X}(\boldsymbol{u})$ to be identical with a TPB volume of degree $(\ell, \bar{m}, n)$ is that (cf. equation (4.15))

$$\Delta^{0q0} \boldsymbol{b}_{ijk} = 0, \qquad \left\{ \begin{array}{ll} q = \bar{m}{+}1(1)m, & j = 0(1)m{-}q, \\ \text{all } i, k. \end{array} \right.$$

A similar assertion is valid for degree reduction in the u and w variables. If this condition holds, so that degree reduction of $\boldsymbol{X}(\boldsymbol{u})$ to $\bar{\boldsymbol{X}}(\boldsymbol{u})$ is possible, then we can think of $\boldsymbol{X}(\boldsymbol{u})$ as arising from degree raising of $\bar{\boldsymbol{X}}(\boldsymbol{u})$. Thus, if $\boldsymbol{X}(\boldsymbol{u})$ is a TPB volume of degree (ℓ, m, n) such that $\Delta^{0m0}\boldsymbol{b}_{i0k} = 0$ for all i, k, then we can reduce the polynomial degree with respect to the v variable to $m-1$, and the corresponding Bézier points of the representation $\bar{\boldsymbol{X}}(\boldsymbol{u})$ of degree $(\ell, m-1, n)$ are given by

$$\bar{\boldsymbol{b}}_{ijk} = \frac{1}{m-j}(m\boldsymbol{b}_{ijk} - j\bar{\boldsymbol{b}}_{i,j-1,k}), \qquad \left\{ \begin{array}{l} \text{all } i, k, \\ j = 0(1)m\text{–}1. \end{array} \right. \tag{11.4}$$

- *Point and derivative evaluation*: The point $\boldsymbol{X}(\boldsymbol{u}_0)$ can be found by repeated linear interpolation (the de Casteljau construction). Thus, for example, in the u direction we compute

$$\boldsymbol{b}_{ijk}^{i+1,j,k} = (1-u_0)\boldsymbol{b}_{ijk}^{ijk} + u_0\boldsymbol{b}_{i+1,j,k}^{i+1,j,k},$$

with similar recursions in v and w. Then starting with $\boldsymbol{b}_{ijk}^{ijk} = \boldsymbol{b}_{ijk}$, the desired point is given by $\boldsymbol{X}(\boldsymbol{u}_0) = \boldsymbol{b}_{000}^{\ell mn}$. This result stems from the fact that in (11.1), the de Casteljau algorithm can be applied separately to each of the parameter directions. The de Casteljau steps in different directions commute, and the result is independent of the order. Similarly, the derivative of order (p, q, r) of a TPB volume of degree (ℓ, m, n) at the point $\boldsymbol{u}$ can be found using the de Casteljau algorithm. Indeed, we have

$$\frac{\partial^{p+q+r}}{\partial u^p \partial v^q \partial w^r}\boldsymbol{X}(\boldsymbol{u}) = \frac{\ell!}{(\ell-p)!}\,\frac{m!}{(m-q)!}\,\frac{n!}{(n-r)!}\Delta^{pqr}\boldsymbol{b}_{000}^{\ell-p,m-q,n-r}, \tag{11.5}$$

where the forward differences $\Delta^{pqr}\boldsymbol{b}_{\alpha\gamma\epsilon}^{\beta\delta\zeta}$ are defined as in (11.2), but now operate on both the subscripts and superscripts.

- *Subdivision*: A TPB volume of degree (ℓ, m, n) can be subdivided into two TPB volumes of the same degree which join along the parametric surface corresponding to $u = u_0$ with ℓ continuous derivatives in the u direction. The Bézier points $\boldsymbol{b}_{0jk}^{ijk}$ and $\boldsymbol{b}_{ijk}^{\ell jk}$ of the two subsegments can be found by applying the de Casteljau algorithm for $u = u_0$ to all rows of the Bézier grid corresponding to $i = \text{const}$. Using the parameter transformations $\bar{u} = u/u_0$ for $u \in [0, u_0]$ and $\bar{u} = (u-u_0)/(1-u_0)$ for $u \in [u_0, 1]$, respectively, we can reparametrize the subsegments to again be defined on the unit cube I. A similar assertion holds for subdivision in the v and w directions. Again, the form of (11.1) allows us to apply the decomposition step to the individual variables separately. If subdivision is done properly, *i.e.*, if the set of partition points $\boldsymbol{u}_0 = (u_0, v_0, w_0)$ form a dense subset of the cube, then the Bézier points on the corresponding sequence of Bézier grids converge to points of the Bézier volume [Pra 84], see also [Lane 80] for the special case $u_0 = v_0 = w_0 = 0.5$. The convergence rate is quadratic, cf. [Coh 85], [Dah 86].

If we replace (11.1) by a d-fold tensor-product of univariate schemes, we get a *d-variate Bézier representation*, and it is clear that most of the above results carry over. The fact that everything essentially reduces to the univariate case leads to an enormous simplification. For example, taking advantage of this observation reduces the complexity of the interpolation problem from $\mathcal{O}((\prod_k n_k)^d)$ to $\mathcal{O}(\sum_k n_k^d)$, see [Boo 77], [BOO 78], [Alf 89], and compare with Sect. 6.2 and [Las 85, 85a, 87]. [Las 87] also discusses the approximation problem.

11.1.2. Tetrahedral Bézier Volumes

A *tetrahedral Bézier volume* (TB volume) of degree n is defined by

$$\boldsymbol{X}(\boldsymbol{u}) = \sum_{|\boldsymbol{i}|=n} \boldsymbol{b}_{\boldsymbol{i}} B_{\boldsymbol{i}}^n(\boldsymbol{u}), \tag{11.6}$$

where the sum is taken over all $\boldsymbol{i} = (i, j, k, \ell)$ which satisfy the conditions $|\boldsymbol{i}| = i + j + k + \ell = n$ and $i, j, k, \ell \geq 0$. As basis functions, we use the *generalized Bernstein polynomials* (cf. (6.44))

$$B_{\boldsymbol{i}}^n(\boldsymbol{u}) = \frac{n!}{i!j!k!\ell!} u^i v^j w^k t^\ell. \tag{11.7}$$

The four-tuples $\boldsymbol{u} = (u, v, w, t)$ with $|\boldsymbol{u}| = u + v + w + t = 1$ and $u, v, w, t \geq 0$ are the barycentric coordinates in $\mathbb{R}^3$, associated with a (nondegenerate) tetrahedron T in the parameter space [MÖB 67], [Far 86]. The coefficients $\boldsymbol{b}_{\boldsymbol{i}} \in \mathbb{R}^3$ are called *Bézier points*, and connecting them with straight lines according to their natural order leads to the corresponding *Bézier grid*, see Fig. 11.4.

If we take $b_{\boldsymbol{i}} \in \mathbb{R}$ in (11.6), then $X(\boldsymbol{u})$ is a function which defines a *hypersurface* $(\boldsymbol{u}, X(\boldsymbol{u}))$ in $\mathbb{R}^4$. In this case, we can consider the $b_{\boldsymbol{i}}$ to be *ordinates* associated with the (barycentric) *abscissae* $\boldsymbol{u}_{\boldsymbol{i}} = \boldsymbol{i}/n \in T \subset \mathbb{R}^3$, which can be combined to define the *Bézier points* $\boldsymbol{B}_{\boldsymbol{i}} = (\boldsymbol{u}_{\boldsymbol{i}}, b_{\boldsymbol{i}}) \in \mathbb{R}^4$. Connecting the Bézier points leads to a *Bézier grid* in $\mathbb{R}^4$.

All of the following results for the parametric case can be carried over to the function-valued case. Some special aspects of TB hypersurfaces have been investigated in [Gol 82, 83], [Alf 84a], [Sed 85b], [Dero 88a], see also [Far 86].

In the same way as for triangular Bézier patches, we can also define a version of (11.6) using *rational representations*, see [Las 91a].

As in the case of triangular Bézier surfaces (see Sect. 6.3 and [Boeh 84]), there is a close relationship between a TB volume and its corresponding Bézier grid, see e.g., [Far 86], [Boo 87a], [Las 87]. In particular, there are analogs of affine invariance, the convex hull property, parametric surfaces and lines, boundary surfaces and curves, properties of the vertices, etc. We list some of them here in more detail:

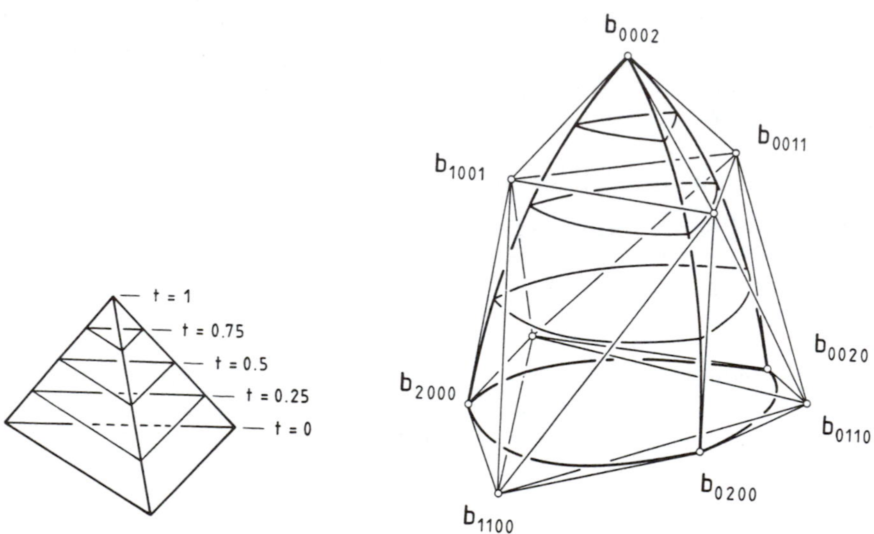

Fig. 11.4. TB volume of degree 2 and its associated Bézier grid. The figure shows the boundary surfaces and some parametric surfaces corresponding to $t = 0, \frac{1}{4}, \frac{1}{2}, \frac{3}{4}$. The parameter space domain is shown on the left.

- *Derivatives*: Because of the linear dependence of the barycentric coordinates, the partial derivatives of a TB volume do not have an obvious geometric interpretation; indeed, the partial derivatives with respect to u, v, w, and t do not coincide with the derivatives along parametric lines. Instead of working with partial derivatives, we must therefore use *directional derivatives*. Suppose σ denotes the parameter associated with a straight lines $\boldsymbol{u}(\sigma)$ in the three dimensional parameter space. Then $\dot{\boldsymbol{u}} = \frac{d}{d\sigma}\boldsymbol{u}(\sigma)$ describes a direction in parameter space, and the derivative of order α in the direction $\dot{\boldsymbol{u}}$ of a TB volume (cf. (6.53)) is given by

$$\begin{aligned} D^{\alpha}_{\dot{\boldsymbol{u}}}\boldsymbol{X}(\boldsymbol{u}) &= [\lambda D_{\dot{\boldsymbol{u}}_\lambda} + \mu D_{\dot{\boldsymbol{u}}_\mu} + \nu D_{\dot{\boldsymbol{u}}_\nu}]^{\alpha}\boldsymbol{X}(\boldsymbol{u}) \\ &= \frac{n!}{(n-\alpha)!}\sum_{|\boldsymbol{i}|=n-\alpha}[\lambda\Delta_{\dot{\boldsymbol{u}}_\lambda} + \mu\Delta_{\dot{\boldsymbol{u}}_\mu} + \nu\Delta_{\dot{\boldsymbol{u}}_\nu}]^{\alpha}\boldsymbol{b}_{\boldsymbol{i}}B^{n-\alpha}_{\boldsymbol{i}}(\boldsymbol{u}), \end{aligned} \tag{11.8}$$

where $\dot{\boldsymbol{u}} = \lambda\dot{\boldsymbol{u}}_\lambda + \mu\dot{\boldsymbol{u}}_\mu + \nu\dot{\boldsymbol{u}}_\nu$, and $\dot{\boldsymbol{u}}_\lambda$, $\dot{\boldsymbol{u}}_\mu$, $\dot{\boldsymbol{u}}_\nu$ are three linearly independent directions defined by edges of T. Here $\boldsymbol{e}_1 = (1,0,0,0)$, $\boldsymbol{e}_2 = (0,1,0,0)$, etc., and

$$\begin{aligned} \Delta^{0}_{\dot{\boldsymbol{u}}}\boldsymbol{b}_{\boldsymbol{i}} &= \boldsymbol{b}_{\boldsymbol{i}}, \\ \Delta^{\alpha}_{\dot{\boldsymbol{u}}}\boldsymbol{b}_{\boldsymbol{i}} &= \Delta^{\alpha-1}_{\dot{\boldsymbol{u}}}[\dot{u}\boldsymbol{b}_{\boldsymbol{i}+\boldsymbol{e}_1} + \dot{v}\boldsymbol{b}_{\boldsymbol{i}+\boldsymbol{e}_2} + \dot{w}\boldsymbol{b}_{\boldsymbol{i}+\boldsymbol{e}_3} + \dot{t}\boldsymbol{b}_{\boldsymbol{i}+\boldsymbol{e}_4}]. \end{aligned}$$

- *Degree raising*: The Bézier points $\bar{\boldsymbol{b}}^{\nu}_{\boldsymbol{I}}$ of the representation of degree $n+\nu$ of a TB volume of degree n are given by

$$\bar{\boldsymbol{b}}^{\nu}_{\boldsymbol{I}} = \sum_{|\boldsymbol{i}|=n}\boldsymbol{b}_{\boldsymbol{i}}\frac{\binom{I}{i}\binom{J}{j}\binom{K}{k}\binom{L}{l}}{\binom{n+\nu}{n}}, \qquad \text{where } \boldsymbol{I} = (I,J,K,L) \text{ with } |\boldsymbol{I}| = n+\nu. \tag{11.9}$$

  As n increases, the Bézier points on the corresponding sequence of Bézier grids converge to points on the associated TB volume.

- *Degree reduction*: Let $\boldsymbol{X}(\boldsymbol{u})$ be a TB volume of degree n with $n = \bar{n}+\nu$. Then a necessary and sufficient condition for $\boldsymbol{X}(\boldsymbol{u})$ to be identical with a TB volume of degree $\bar{n}$ is that

$$\Delta^{\beta}_{\dot{\boldsymbol{u}}_\lambda}\Delta^{\gamma}_{\dot{\boldsymbol{u}}_\mu}\Delta^{\delta}_{\dot{\boldsymbol{u}}_\nu}\boldsymbol{b}_{\boldsymbol{i}} = 0, \qquad \left\{ \begin{array}{ll} \text{all } |\boldsymbol{i}| = n-\alpha, & \beta+\gamma+\delta = \alpha, \\ \alpha = \bar{n}{+}1(1)n. & \end{array} \right.$$

  Here $\dot{\boldsymbol{u}}_\lambda, \dot{\boldsymbol{u}}_\mu, \dot{\boldsymbol{u}}_\nu$ are three linearly independent directions defined by edges of T. If this condition for degree reduction is satisfied, then we can think of it as being the converse of (11.9). Thus, if $\boldsymbol{X}(\boldsymbol{u})$ is a TB volume of degree n such that

$$\Delta^{\beta}_{\dot{\boldsymbol{u}}_\lambda}\Delta^{\gamma}_{\dot{\boldsymbol{u}}_\mu}\Delta^{\delta}_{\dot{\boldsymbol{u}}_\nu}\boldsymbol{b}_0 = 0 \quad \text{with} \quad \beta+\gamma+\delta = n,$$

  so that its polynomial degree can be reduced to $n-1$, then the Bézier points of the degree $n-1$ formula are given by

$$\bar{\boldsymbol{b}}_{\boldsymbol{i}-\boldsymbol{e}_1} = \frac{n}{i}\boldsymbol{b}_{\boldsymbol{i}} - \frac{j}{i}\bar{\boldsymbol{b}}_{\boldsymbol{i}-\boldsymbol{e}_2} - \frac{k}{i}\bar{\boldsymbol{b}}_{\boldsymbol{i}-\boldsymbol{e}_3} - \frac{l}{i}\bar{\boldsymbol{b}}_{\boldsymbol{i}-\boldsymbol{e}_4}, \qquad \text{all } |\boldsymbol{i}| = n, \ i \neq 0. \tag{11.10}$$

- *Point and derivative evaluation*: A volume point $\boldsymbol{X}(\boldsymbol{u}_o)$ can be found by repeated barycentric linear interpolation (the *de Casteljau algorithm*). This gives (cf. (6.52))

$$\boldsymbol{b}_{\boldsymbol{i}}^{\alpha+1} = u_0 \boldsymbol{b}_{\boldsymbol{i}+\boldsymbol{e}_1}^{\alpha} + v_0 \boldsymbol{b}_{\boldsymbol{i}+\boldsymbol{e}_2}^{\alpha} + w_0 \boldsymbol{b}_{\boldsymbol{i}+\boldsymbol{e}_3}^{\alpha} + t_0 \boldsymbol{b}_{\boldsymbol{i}+\boldsymbol{e}_4}^{\alpha}, \qquad \text{all } |\boldsymbol{i}| = n - \alpha - 1,$$

where $\boldsymbol{b}_{\boldsymbol{i}}^0 = \boldsymbol{b}_{\boldsymbol{i}}$, and $\boldsymbol{X}(\boldsymbol{u}_0) = \boldsymbol{b}_0^n$. The de Casteljau algorithm can also be used to compute derivatives. The derivative of order α in the direction $\dot{\boldsymbol{u}}$ of a TB volume is given by

$$D_{\dot{\boldsymbol{u}}}^{\alpha} \boldsymbol{X}(\boldsymbol{u}) = \frac{n!}{(n-\alpha)!} \left[\lambda \Delta_{\dot{\boldsymbol{u}}_\lambda} + \mu \Delta_{\dot{\boldsymbol{u}}_\mu} + \nu \Delta_{\dot{\boldsymbol{u}}_\nu}\right]^{\alpha} \boldsymbol{b}_0^n, \tag{11.11}$$

where $\dot{\boldsymbol{u}} = \lambda \dot{\boldsymbol{u}}_\lambda + \mu \dot{\boldsymbol{u}}_\mu + \nu \dot{\boldsymbol{u}}_\nu$, and $\dot{\boldsymbol{u}}_\lambda$, $\dot{\boldsymbol{u}}_\mu$, $\dot{\boldsymbol{u}}_\nu$ are three linearly independent directions defined by edges of T, and

$$\begin{aligned} \Delta_{\dot{\boldsymbol{u}}}^{0} \boldsymbol{b}_{\boldsymbol{i}}^{m} &= \boldsymbol{b}_{\boldsymbol{i}}^{m}, \\ \Delta_{\dot{\boldsymbol{u}}}^{\alpha} \boldsymbol{b}_{\boldsymbol{i}}^{m} &= \Delta_{\dot{\boldsymbol{u}}}^{\alpha-1} \left[\dot{u} \boldsymbol{b}_{\boldsymbol{i}+\boldsymbol{e}_1}^{m-1} + \dot{v} \boldsymbol{b}_{\boldsymbol{i}+\boldsymbol{e}_2}^{m-1} + \dot{w} \boldsymbol{b}_{\boldsymbol{i}+\boldsymbol{e}_3}^{m-1} + \dot{t} \boldsymbol{b}_{\boldsymbol{i}+\boldsymbol{e}_4}^{m-1}\right]. \end{aligned}$$

For some very efficient alternative algorithms based on a modified formula for TB volumes, see [Schu 86], and also [Schu 87a].

- *Subdivision*: A TB volume of degree n can be subdivided into an arbitrary number of TB volumes of degree n by repeatedly applying the de Casteljau algorithm in the same way as was done for triangular Bézier surfaces, see [Gol 83]. Generalizations to higher dimensional simplices can be found in [Pra 84]. In particular, if $\boldsymbol{u}_0, \ldots, \boldsymbol{u}_3$ are distinct points in the parameter space tetrahedron, and $\boldsymbol{a} = (a_0, a_1, a_2, a_3)$ with $\alpha = |\boldsymbol{a}| = a_0 + a_1 + a_2 + a_3$, then the de Casteljau recursion formula is

$$\boldsymbol{b}_{\boldsymbol{i}}^{\boldsymbol{a}+\boldsymbol{e}_j} = u_j \boldsymbol{b}_{\boldsymbol{i}+\boldsymbol{e}_1}^{\boldsymbol{a}} + v_j \boldsymbol{b}_{\boldsymbol{i}+\boldsymbol{e}_2}^{\boldsymbol{a}} + w_j \boldsymbol{b}_{\boldsymbol{i}+\boldsymbol{e}_3}^{\boldsymbol{a}} + t_j \boldsymbol{b}_{\boldsymbol{i}+\boldsymbol{e}_4}^{\boldsymbol{a}}, \qquad \left\{ \begin{array}{l} \text{all } |\boldsymbol{i}| = n - \alpha - 1 \\ j = 0(1)k \end{array} \right. \tag{11.12}$$

where $\boldsymbol{b}_{\boldsymbol{i}}^0 = \boldsymbol{b}_{\boldsymbol{i}}$. The recursions (11.12) for computing the Bézier points of the subsegments can be applied independently for the different points $\boldsymbol{u}_0, \ldots, \boldsymbol{u}_3$, i.e., the result does not depend on the ordering of the different de Casteljau steps. The subsegments can be redefined on T by appropriate linear transformations $\boldsymbol{u} \to \bar{\boldsymbol{u}}$.

Example 11.1. The subsegment $\bar{\boldsymbol{X}}(\bar{\boldsymbol{u}})$ of $\boldsymbol{X}(\boldsymbol{u})$ with $\boldsymbol{u}_0 = (1,0,0,0)$, $\boldsymbol{u}_1 = (0,1,0,0)$, $\boldsymbol{u}_2$ and $\boldsymbol{u}_3$ arbitrary, has the representation

$$\bar{\boldsymbol{X}}(\bar{\boldsymbol{u}}) = \sum_{\alpha=0}^{n} \sum_{\substack{|\boldsymbol{a}|=\alpha \\ i+j=n-\alpha}} \boldsymbol{b}_{ij00}^{\boldsymbol{a}} B_{ija_2a_3}^{n}(\bar{\boldsymbol{u}}),$$

i.e., the Bézier points of $\bar{\boldsymbol{X}}(\bar{\boldsymbol{u}})$ are $\boldsymbol{b}^{\boldsymbol{a}}_{ij00}$, $\boldsymbol{a} = (0,0,a_2,a_3)$, $i+j+a_2+a_3 = n$, and the parameter transformation is

$$\boldsymbol{u} = \begin{pmatrix} 1 & 0 & u_2 & u_3 \\ 0 & 1 & v_2 & v_3 \\ 0 & 0 & w_2 & w_3 \\ 0 & 0 & t_2 & t_3 \end{pmatrix} \bar{\boldsymbol{u}}.$$

Fig. 11.5a shows the subtetrahedron of T defined by $\boldsymbol{u}_0, \ldots, \boldsymbol{u}_3$. It is mapped via an affine mapping (equation (11.12)) to all upright subtetrahedra of the Bézier grid to form the next grid in the sequence. Fig. 11.5b shows $\bar{\boldsymbol{X}}(\bar{\boldsymbol{u}})$ as a subsegment of $\boldsymbol{X}(\boldsymbol{u})$. For $n = 2$, Fig. 11.5c shows the construction of the Bézier points $\boldsymbol{b}^{\boldsymbol{a}}_{ij00}$ of the subsegment $\bar{\boldsymbol{X}}(\bar{\boldsymbol{u}})$ of $\boldsymbol{X}(\boldsymbol{u})$ corresponding to $\boldsymbol{u}_1 = (0,1,0,0)$, $\boldsymbol{u}_0 = (1,0,0,0)$, and arbitrary $\boldsymbol{u}_3$, $\boldsymbol{u}_2$ as constructed by the above subdivision algorithm. The tetrahedra depict the sequence of Bézier grids arising in the algorithm. We mark only those Bézier points on the Bézier grids which define the subsegment $\bar{\boldsymbol{X}}(\bar{\boldsymbol{u}})$. These are the points on the Bézier grids which lie on the lower left edges of the tetrahedra (the edge of $\boldsymbol{X}(\boldsymbol{u})$ defined by $\boldsymbol{b}_{ij00}$ is where $\bar{\boldsymbol{X}}(\bar{\boldsymbol{u}})$ joins $\boldsymbol{X}(\boldsymbol{u})$). For other examples, see [Las 87].

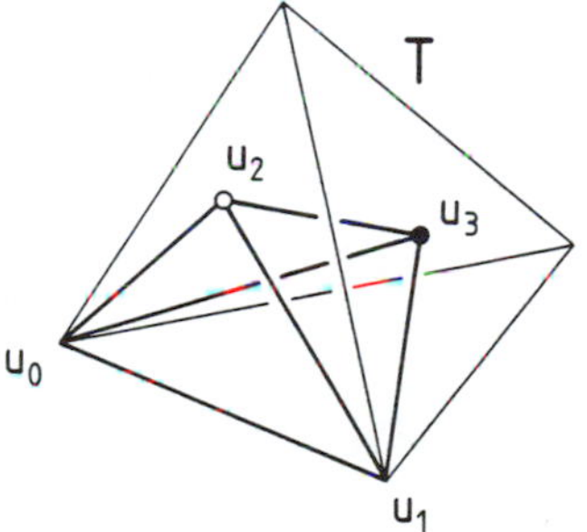

Fig. 11.5a. Parameter space.

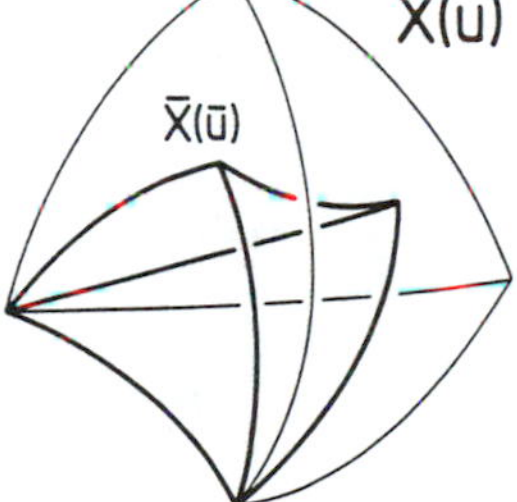

Fig. 11.5b. Coordinate space.

If repeated subdivision is done properly, the Bézier points converge to points of the TB volume.

Using d-variate Bernstein polynomials defined in terms of barycentric coordinates in $\mathbb{R}^d$, we can extend (11.6) to *d-variate representations*. Formally, we simply replace the index $\boldsymbol{i} = (i,j,k,\ell)$ in (11.6) by the multi-index $\boldsymbol{i} = (i_1, i_2, \ldots, i_{d+1})$ with $|\boldsymbol{i}| = \sum_k i_k = d+1$ and $i_k \geq 0$ (all k), and replace $\boldsymbol{u} = (u,v,w,t)$ by $\boldsymbol{u} = (u_1, u_2, \ldots, u_{d+1})$ with $|\boldsymbol{u}| = \sum_k u_k = 1$ and $u_k \geq 0$ (all k), see e.g. [Gol 83], [Far 86], and [Boo 87a].

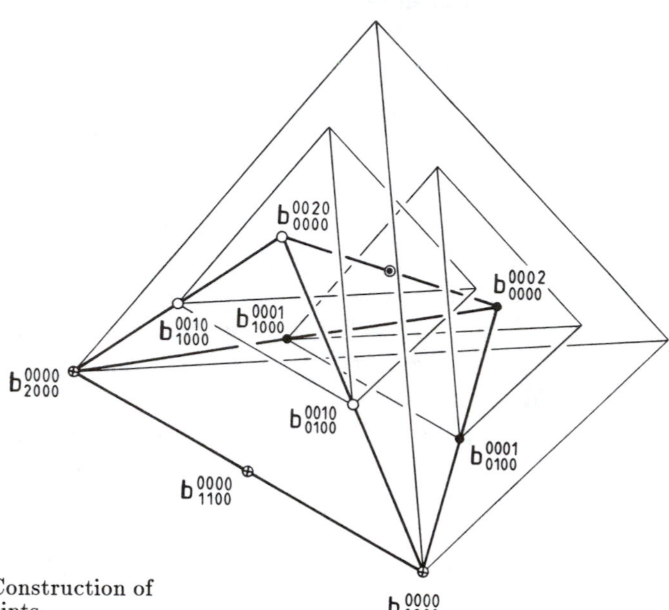

Fig 11.5c. Construction of the Bézier points.

11.1.3. Pentahedral Bézier Volumes

A *pentahedral Bézier volume* (PB volume) of degree $(m; n)$ is defined [Las 87] by

$$\boldsymbol{X}(\boldsymbol{u}; t) = \sum_{|\boldsymbol{i}|=m} \sum_{\ell=0}^{n} \boldsymbol{b}_{\boldsymbol{i};\ell} B_{\boldsymbol{i}}^{m}(\boldsymbol{u}) B_{\ell}^{n}(t), \tag{11.13}$$

where the first sum is taken over all $\boldsymbol{i} = (i, j, k)$ such that $|\boldsymbol{i}| = i + j + k = m$ with $i, j, k \geq 0$. Here $B_{\boldsymbol{i}}^{m}(\boldsymbol{u})$ denote the *generalized Bernstein polynomials* in (6.44), and the triples $\boldsymbol{u} = (u, v, w)$ with $|\boldsymbol{u}| = u+v+w = 1$ and $u, v, w \geq 0$, are the *barycentric coordinates* in $\mathbb{R}^2$. We use the *ordinary Bernstein polynomials* $B_{\ell}^{n}(t)$ of (4.2) as basis functions for the third parameter space direction of the (nondegenerate) pentahedron P in parameter space.

The coefficients can be either vector-valued, *i.e.*, $\boldsymbol{b}_{\boldsymbol{i};\ell} \in \mathbb{R}^3$, or real-valued, $b_{\boldsymbol{i};\ell} \in \mathbb{R}$, in which case they are to be considered as ordinates associated with the abscissae $(\boldsymbol{i}/m; \ell/n) \in P$. The *Bézier points* $\boldsymbol{b}_{\boldsymbol{i};\ell} \in \mathbb{R}^3$ (or $\boldsymbol{B}_{\boldsymbol{i};\ell} = (\boldsymbol{i}/m; \ell/n, b_{\boldsymbol{i};\ell}) \in \mathbb{R}^4$ in the case of real coefficients $b_{\boldsymbol{i};\ell} \in \mathbb{R}$, respectively) are the vertices of a *Bézier grid* lying in $\mathbb{R}^3$ ($\mathbb{R}^4$), see Fig. 11.6.

The expression (11.13) can be extended to *rational representations* [Las 91a]. Rational PB volumes are useful for constructing solid 3D primitives such as cones and tori. Figs. 11.7, 11.8 show two examples.

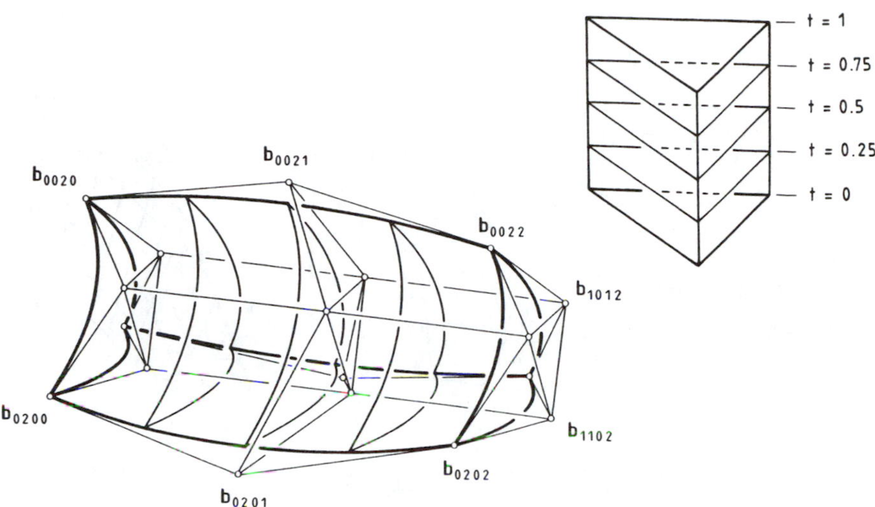

Fig. 11.6. PB volume of degree (2;2) and its associated Bézier grid. The boundary surfaces and parametric surfaces corresponding to $t = 0, \frac{1}{4}, \frac{1}{2}, \frac{3}{4}, 1$ are shown. The parameter space domain is also shown in the upper right corner.

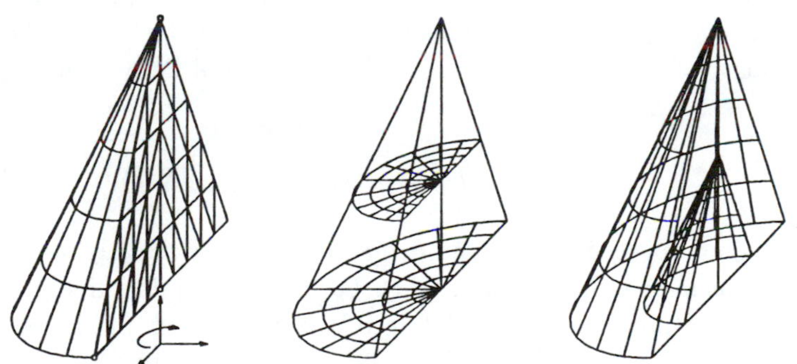

Fig. 11.7. Half of a solid cone defined as a rational PB volume of degree (1;2). The center and right figures show some boundary curves and parametric surfaces.

It follows immediately from (11.13) that the properties of triangular Bézier surfaces and Bézier curves carry over to PB volumes. In particular, there is a close relationship between the Bézier grid and the corresponding

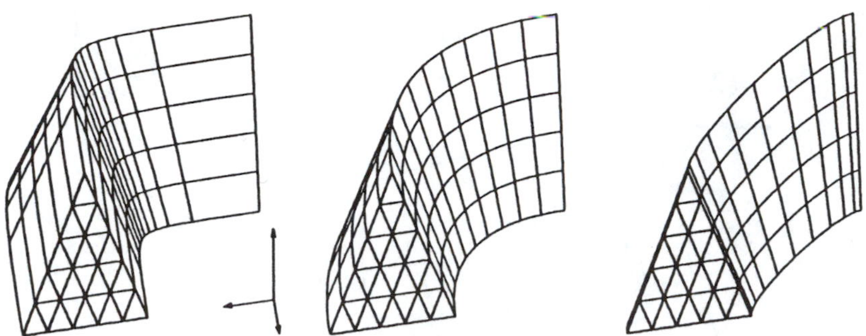

Fig. 11.8. Rational PB volumes of degree (1;2). The curved parametric lines are hyperbolic (left), parabolic (middle), and elliptic (right).

volume segment. We have affine invariance, the convex hull property, and various properties of parametric lines and surfaces, boundary curves and surfaces, and the vertices, cf. Sect. 11.1.1 and Sects. 4.1, 6.2.2, and 6.3.2. Also see [Las 87].

- *Derivatives*: Since in (11.13) we are using barycentric coordinates, we need to work with *directional derivatives* instead of partial derivatives. Let σ be the parameter for a line $\boldsymbol{g}(\sigma)$ in 3-dimensional parameter space, so that $\dot{\boldsymbol{g}} = \frac{d}{d\sigma}\boldsymbol{g}(\sigma)$ describes a direction in parameter space. Then the derivative of order α in the direction $\dot{\boldsymbol{g}} = (\dot{\boldsymbol{u}}; \dot{t})$ of a PB volume of degree $(m; n)$ is given by

$$D_{\dot{\boldsymbol{g}}}^{\alpha} \boldsymbol{X}(\boldsymbol{g}) = \left[\lambda D_{\dot{\boldsymbol{g}}_\lambda} + \mu D_{\dot{\boldsymbol{g}}_\mu} + \nu D_{\dot{\boldsymbol{g}}_\nu}\right]^{\alpha} \boldsymbol{X}(\boldsymbol{g}), \tag{11.14}$$

where $\dot{\boldsymbol{g}} = \lambda \dot{\boldsymbol{g}}_\lambda + \mu \dot{\boldsymbol{g}}_\mu + \nu \dot{\boldsymbol{g}}_\nu$, and $\dot{\boldsymbol{g}}_\lambda$, $\dot{\boldsymbol{g}}_\mu$, $\dot{\boldsymbol{g}}_\nu$ are three linearly independent directions defined by the edges of P. For the special direction $\dot{\boldsymbol{g}} = (\boldsymbol{0}; 1)$, which corresponds to the partial derivative w.r.t. t (cf. (4.7)), we have

$$D_{\dot{\boldsymbol{g}}}^{\alpha} \boldsymbol{X}(\boldsymbol{g}) = \frac{n!}{(n-\alpha)!} \sum_{|\boldsymbol{i}|=m} \sum_{\ell=0}^{n-\alpha} \Delta^{\alpha} \boldsymbol{b}_{\boldsymbol{i};\ell}\, B_{\boldsymbol{i}}^{m}(\boldsymbol{u}) B_{\ell}^{n-\alpha}(t).$$

For a direction corresponding to t =const. (cf. (6.53)) we get

$$D_{\dot{\boldsymbol{g}}}^{\alpha} \boldsymbol{X}(\boldsymbol{g}) = \frac{m!}{(m-\alpha)!} \sum_{|\boldsymbol{i}|=m-\alpha} \sum_{\ell=0}^{n} \Delta_{\dot{\boldsymbol{u}}}^{\alpha} \boldsymbol{b}_{\boldsymbol{i};\ell}\, B_{\boldsymbol{i}}^{m-\alpha}(\boldsymbol{u}) B_{\ell}^{n}(t).$$

Here Δ^α acts on ℓ and is defined as in (11.2) while $\Delta^\alpha_{\dot{\boldsymbol{u}}}$ acts on $\boldsymbol{i}$ and is defined as in (11.8).

- *Degree raising*: For degree raising in t, we have a formula analogous to (11.4), while for degree raising in $\boldsymbol{u}$, we have a formula analogous to (11.9) for finding the new Bézier points. As m and n go to infinity, the corresponding Bézier grid converges to the PB volume.

- *Degree reduction*: A necessary and sufficient condition for degree reduction from $n = \bar{n} + \nu$ to $\bar{n}$ in t is that (cf. Lemma 4.4 in Sect. 4.1.1)

$$\Delta^\alpha \boldsymbol{b}_{\mathbf{i};\ell} = 0, \qquad \begin{cases} \text{for all } |\boldsymbol{i}| = m \\ \alpha = \bar{n}{+}1(1)n, \quad \ell = 0(1)n{-}\alpha, \end{cases}$$

while for a reduction from $m = \bar{m} + \mu$ to $\bar{m}$ in $\boldsymbol{u}$ requires (cf. Sect. 11.1.2 and also [Far 79])

$$\Delta^\beta_{\dot{\boldsymbol{g}}_\mu} \Delta^\gamma_{\dot{\boldsymbol{g}}_\nu} \boldsymbol{b}_{\mathbf{i};\ell} = 0, \qquad \begin{cases} \text{all } |\boldsymbol{i}| = m - \alpha \text{ and } \beta + \gamma = \alpha \\ \alpha = \bar{m}{+}1(1)m, \end{cases}$$

where $\dot{\boldsymbol{g}}_\mu, \dot{\boldsymbol{g}}_\nu$ are two linearly independent directions defined by edges with $t =$ const. If degree reduction is possible, we can again carry it out by looking at the related degree raising problem, cf. (11.4), (11.10).

- *Point and derivative evaluation*: A volume point $\boldsymbol{X}(\boldsymbol{u}_0; t_0)$ can be found by repeated linear interpolation (de Casteljau algorithm) in both the $\boldsymbol{u}$ and t variables:

$$\begin{aligned} \boldsymbol{b}^{\gamma;\ell+1}_{\mathbf{i};\ell} &= (1 - t_0)\boldsymbol{b}^{\gamma;\ell}_{\mathbf{i};\ell} + t_0 \boldsymbol{b}^{\gamma;\ell+1}_{\mathbf{i};\ell+1} \\ \boldsymbol{b}^{\gamma+1;\ell}_{\mathbf{i};\ell} &= u_0 \boldsymbol{b}^{\gamma;\ell}_{\mathbf{i}+\mathbf{e}_1;\ell} + v_0 \boldsymbol{b}^{\gamma;\ell}_{\mathbf{i}+\mathbf{e}_2;\ell} + w_0 \boldsymbol{b}^{\gamma;\ell}_{\mathbf{i}+\mathbf{e}_3;\ell}, \end{aligned}$$

where $\boldsymbol{b}^{0;\ell}_{\mathbf{i};\ell} = \boldsymbol{b}_{\mathbf{i};\ell}$ and $\boldsymbol{X}(\boldsymbol{u}_0; t_0) = \boldsymbol{b}^{m;n}_{\mathbf{0};0}$. The derivative of order α in the direction $\dot{\boldsymbol{g}}$ of a PB volume of degree $(m; n)$ can also be computed using the de Casteljau algorithm. For the special direction $\dot{\boldsymbol{g}} = (\mathbf{0}; 1)$, we have

$$D^\alpha_{\dot{\boldsymbol{g}}} \boldsymbol{X}(\boldsymbol{u}; t) = \frac{n!}{(n-\alpha)!} \Delta^\alpha \boldsymbol{b}^{m;n-\alpha}_{\mathbf{0};0}(\boldsymbol{u}; t). \tag{11.15}$$

For a special direction with $t =$ constant, we get

$$D^\alpha_{\dot{\boldsymbol{g}}} \boldsymbol{X}(\boldsymbol{u}; t) = \frac{m!}{(m-\alpha)!} \Delta^\alpha_{\dot{\boldsymbol{u}}} \boldsymbol{b}^{m;n}_{\mathbf{0};0}(\boldsymbol{u}; t). \tag{11.16}$$

Here Δ^α and $\Delta^\alpha_{\dot{\boldsymbol{u}}}$ are defined in analogy with (11.5) and (11.11), respectively. In view of (11.14) – (11.16), we can use the de Casteljau algorithm to compute the directional derivative in an arbitrary direction.

- *Subdivision*: Subdivision w.r.t. the parameter t can be accomplished using the de Casteljau algorithm w.r.t. the parameter t, cf. Sect. 11.1.1. Similarly, subdivision w.r.t. $\boldsymbol{u}$ can be accomplished using the de Casteljau algorithm w.r.t. $\boldsymbol{u}$, cf. Sect. 11.1.2 and [Gol 83]. Because of the way in which (11.13) is defined, the two algorithms commute. If subdivision is appropriately repeated, then the sequence of Bézier grids converges to the associated volume.

11.1.4. Continuity Conditions

As for curves and surfaces, it is possible to join two volumes together. Conditions under which the two volumes join smoothly can be obtained as generalizations of the analogous results for curves and surfaces, see e.g., Sects. 6.2.2.1, 6.3.3, and also [Stä 76], [Far 79].

Suppose $\boldsymbol{X}$ and $\boldsymbol{Y}$ are two Bézier volumes which share a common boundary surface $\boldsymbol{Y}_0 = \boldsymbol{X}_0$. Then they connect with C^r *continuity* if and only if all of their directional derivatives $D^{\rho}_{\dot{\boldsymbol{x}}}$ of orders $\rho = 1(1)r$ in some direction $\dot{\boldsymbol{x}}$ across $\boldsymbol{Y}_0 = \boldsymbol{X}_0$ agree, *i.e.*, provided that

$$D^{\rho}_{\dot{\boldsymbol{x}}}\boldsymbol{Y}_0 = D^{\rho}_{\dot{\boldsymbol{x}}}\boldsymbol{X}_0, \qquad \rho = 1(1)r. \tag{11.17}$$

It is convenient to choose $\dot{\boldsymbol{x}}$ to be the normal direction [Alf 84a], since it allows a unified treatment of all volumes. Sometimes, this special choice of $\dot{\boldsymbol{x}}$ is actually necessary, for example in developing local schemes where two volumes are to be joined together, but we don't have any information on their actual geometric configurations. The vector $\dot{\boldsymbol{x}}$ can be written in terms of the *special directions* $\dot{\boldsymbol{x}}_\lambda$, $\dot{\bar{\boldsymbol{x}}}_\lambda$, $\dot{\boldsymbol{x}}_\mu$, etc., which correspond to the edges of the parameter space domains:

$$\dot{\boldsymbol{x}} = \lambda\dot{\boldsymbol{x}}_\lambda + \mu\dot{\boldsymbol{x}}_\mu + \nu\dot{\boldsymbol{x}}_\nu = \bar{\lambda}\dot{\bar{\boldsymbol{x}}}_\lambda + \bar{\mu}\dot{\bar{\boldsymbol{x}}}_\mu + \bar{\nu}\dot{\bar{\boldsymbol{x}}}_\nu.$$

Here the (possibly function-valued) factors $\lambda, \mu, \nu, \bar{\lambda}, \bar{\mu}$ and $\bar{\nu}$ are determined by the underlying parameter domains. We can also write $\dot{\boldsymbol{x}}$ in terms of local coordinates, see [Alf 84a]. By the linearity of the directional derivative operators, it follows that (11.17) can be written as

$$[\bar{\lambda}D_{\dot{\bar{\boldsymbol{x}}}_\lambda} + \bar{\mu}D_{\dot{\bar{\boldsymbol{x}}}_\mu} + \bar{\nu}D_{\dot{\bar{\boldsymbol{x}}}_\nu}]^{\rho}\boldsymbol{Y}_0 = [\lambda D_{\dot{\boldsymbol{x}}_\lambda} + \mu D_{\dot{\boldsymbol{x}}_\mu} + \nu D_{\dot{\boldsymbol{x}}_\nu}]^{\rho}\boldsymbol{X}_0.$$

Since $\boldsymbol{Y}_0 = \boldsymbol{X}_0$ (*i.e.*, the direction vectors $\dot{\bar{\boldsymbol{x}}}_\mu$ and $\dot{\boldsymbol{x}}_\mu$ as well as $\dot{\bar{\boldsymbol{x}}}_\nu$ and $\dot{\boldsymbol{x}}_\nu$ are identical), this is equivalent to

$$\bar{\lambda}^{\rho}D^{\rho}_{\dot{\bar{\boldsymbol{x}}}_\lambda}\boldsymbol{Y}_0 = [\lambda D_{\dot{\boldsymbol{x}}_\lambda} + \alpha D_{\dot{\boldsymbol{x}}_\mu} + \beta D_{\dot{\boldsymbol{x}}_\nu}]^{\rho}\boldsymbol{X}_0. \tag{11.18}$$

These C^r continuity conditions can often by simplified. [Las 87] gives a detailed treatment of continuity conditions for all possible segment configurations.

Example 11.2. Suppose $\boldsymbol{X}(\boldsymbol{x})$ is a TPB volume with $x \in [x_a, x_e]$, $y \in [y_a, y_0]$, $z \in [z_a, z_e]$, and $\boldsymbol{Y}(\boldsymbol{x})$ is another TPB volume with $x \in [x_a, x_e]$, $y \in [y_0, y_e]$, and $z \in [z_a, z_e]$ which are to join along the common face with continuous derivatives in the y-direction. Then condition (11.18) simplifies to

$$\frac{1}{(\Delta y_e)^\rho}\frac{\partial^\rho}{\partial \bar{v}^\rho}\boldsymbol{Y}(\bar{\boldsymbol{u}})\bigg|_{\bar{v}=0} = \frac{1}{(\Delta y_a)^\rho}\frac{\partial^\rho}{\partial v^\rho}\boldsymbol{X}(\boldsymbol{u})\bigg|_{v=1}, \quad \begin{cases} \text{all } \bar{u}=u, \ \ \bar{w}=w, \\ \rho = 1(1)r. \end{cases}$$

Here $\Delta y_a = y_0 - y_a$, $\Delta y_e = y_e - y_0$, $\boldsymbol{u} = (u,v,w)$ with $u,v,w \in [0,1]$, and $\bar{\boldsymbol{u}} = (\bar{u},\bar{v},\bar{w})$ where $\bar{u},\bar{v},\bar{w} \in [0,1]$ are local coordinates in the sense of (11.1).

Example 11.3. Suppose $\boldsymbol{X}$ and $\boldsymbol{Y}$ are two TB volumes defined over regular tetrahedra whose locally defined (barycentric) coordinates in (11.6) are chosen so that the common boundary corresponds to $u = \bar{u} = 0$. Then (11.18) simplifies to

$$D^\rho_{\dot{\bar{\boldsymbol{u}}}_\lambda}\boldsymbol{Y}\bigg|_{\bar{u}=0} = [D_{\dot{\boldsymbol{u}}_\lambda} + \tfrac{2}{3}D_{\dot{\boldsymbol{u}}_\mu} + \tfrac{2}{3}D_{\dot{\boldsymbol{u}}_\nu}]^\rho\boldsymbol{X}\bigg|_{u=0} \qquad \text{all } \boldsymbol{u} = \bar{\boldsymbol{u}}.$$

For applications of C^r continuous TPB volumes, see [Sed 86a] and [Coq 90], and for C^r continuous TB volumes, see e.g., [Alf 84a], [Dah 89], [Lod 90], and [Sed 90a]. For practical computations, it is convenient to express C^r continuity conditions in terms of Bézier points of the two volumes. For example, in the case where both TPB volumes are as in Example 11.2, inserting the Bézier representations in the C^r continuity conditions (cf. Sect. 6.2.2.1), the continuity conditions become:

- $\boldsymbol{X}$ and $\boldsymbol{Y}$ join with C^0 continuity if and only if $\bar{\boldsymbol{b}}_{i0k} = \boldsymbol{b}_{imk}$ for all i, k.
- $\boldsymbol{X}$ and $\boldsymbol{Y}$ join with C^1 continuity if in addition

$$\bar{\boldsymbol{b}}_{i1k} = \left(\frac{m}{\bar{m}}\frac{\Delta y_e}{\Delta y_a} + 1\right)\boldsymbol{b}_{imk} - \frac{m}{\bar{m}}\frac{\Delta y_e}{\Delta y_a}\boldsymbol{b}_{i,m-1,k}, \qquad \text{for all } i,k.$$

- $\boldsymbol{X}$ and $\boldsymbol{Y}$ join with C^2 continuity if in addition

$$\bar{\boldsymbol{b}}_{i2k} = \frac{\frac{m(m-1)}{\bar{m}(\bar{m}-1)}\left(\frac{\Delta y_e}{\Delta y_a}\right)^2 + 2\left(\frac{m}{\bar{m}}\frac{\Delta y_e}{\Delta y_a}\right) + 1}{\frac{m}{\bar{m}}\frac{\Delta y_e}{\Delta y_a} + 1}\bar{\boldsymbol{b}}_{i1k} - \frac{m}{\bar{m}}\frac{\Delta y_e}{\Delta y_a}\boldsymbol{d}_{ik}, \quad \text{all } i,k,$$

where

$$\boldsymbol{d}_{ik} = \frac{\frac{m(m-1)}{\bar{m}(\bar{m}-1)}\left(\frac{\Delta y_e}{\Delta y_a}\right)^2 + 2\left(\frac{m-1}{\bar{m}-1}\frac{\Delta y_e}{\Delta y_a}\right) + 1}{\frac{m}{\bar{m}}\frac{\Delta y_e}{\Delta y_a} + 1}\boldsymbol{b}_{i,m-1,k} - \frac{m-1}{\bar{m}-1}\frac{\Delta y_e}{\Delta y_a}\boldsymbol{b}_{i,m-2,k}.$$

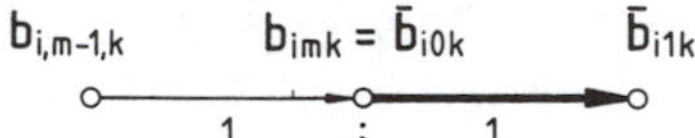

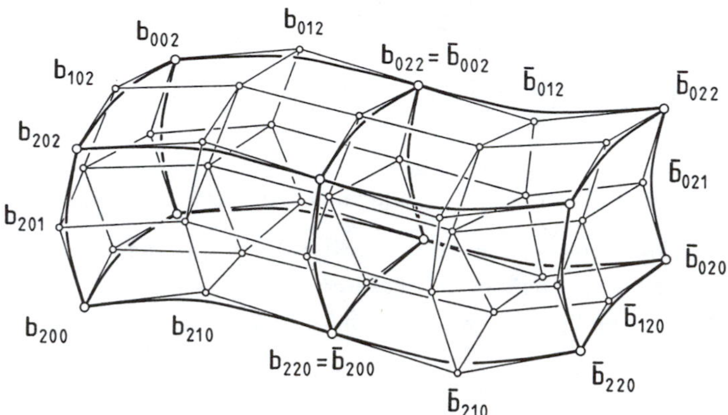

Fig. 11.9. Two C^1 continuous triquadratic TPB volumes parametrized over the unit cube.

For the case where the two TB volumes are as in Example 11.3, we have (cf. Sect. 6.3.3):

- $\boldsymbol{X}$ and $\boldsymbol{Y}$ join with C^0 continuity if and only if $\bar{\boldsymbol{b}}_{0jk\ell} = \boldsymbol{b}_{0jk\ell}$ for all $j + k + \ell = n$.
- $\boldsymbol{X}$ and $\boldsymbol{Y}$ join with C^1 continuity if we also have

$$\bar{\boldsymbol{b}}_{1jk\ell} = 2\boldsymbol{s}_{0jk\ell} - \boldsymbol{b}_{1jk\ell}, \qquad \text{for all } j + k + l = n - 1.$$

- $\boldsymbol{X}$ and $\boldsymbol{Y}$ join with C^2 continuity if in addition,

$$\bar{\boldsymbol{b}}_{2jk\ell} = 2\bar{\boldsymbol{s}}_{1jk\ell} - \boldsymbol{d}_{jk\ell}, \qquad \text{for all } j + k + l = n - 2,$$

with

$$\boldsymbol{d}_{jk\ell} = 2\boldsymbol{s}_{1jk\ell} - \boldsymbol{b}_{2jk\ell}.$$

Here the barycentrically weighted points $\boldsymbol{s}_{ijk\ell}$ are defined by

$$\boldsymbol{s}_{ijk\ell} = \frac{\boldsymbol{b}_{i,j+1,k\ell} + \boldsymbol{b}_{ij,k+1,\ell} + \boldsymbol{b}_{ijk,\ell+1}}{3}$$

with a corresponding definition for the $\bar{\boldsymbol{s}}_{ijk\ell}$.

If two volumes join with C^r continuity, it does not necessarily follow that their parametric lines and surfaces, or their boundary curves and surfaces, join with C^r continuity. Fig. 11.10 shows an example of what can happen. Moreover, for fixed parameter domains, forcing C^r continuity means that no free design parameters remain, since the continuity conditions uniquely determine all of the unknown Bézier points involved.

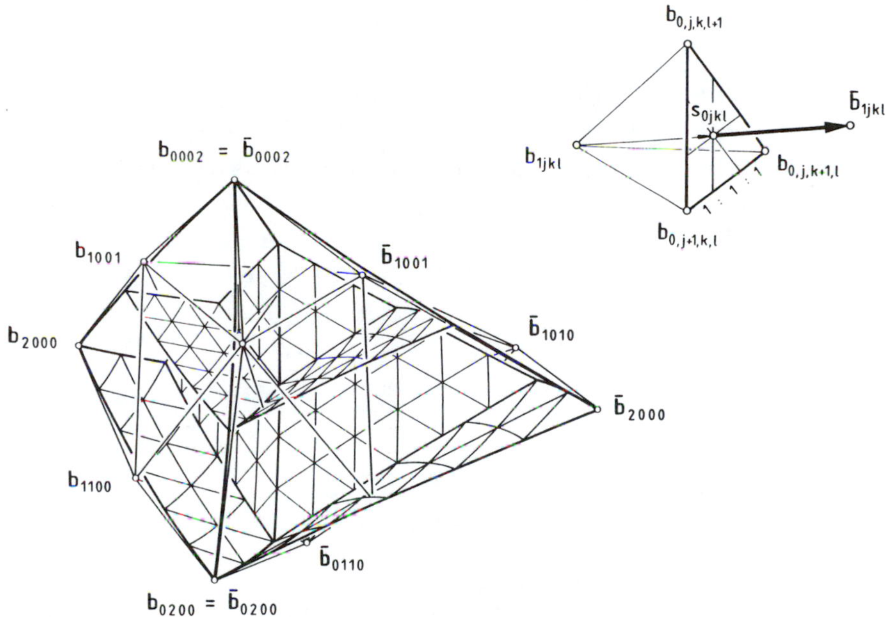

Fig. 11.10. C^1 continuity between two quadratic TB volumes defined over regular tetrahedra.

We can generalize (11.18) to geometric C^1 continuity (GC^1 continuity) by requiring the equality of the first terms in the Taylor series expansions of the two volumes along their common boundary surface, cf. Chap. 7 and [Gei 62]. This gives

$$\bar{\lambda} D_{\dot{\bar{\boldsymbol{x}}}_\lambda} \boldsymbol{Y}_0 = \left[\lambda D_{\dot{\boldsymbol{x}}_\lambda} + \alpha D_{\dot{\boldsymbol{x}}_\mu} + \beta D_{\dot{\boldsymbol{x}}_\nu}\right] \boldsymbol{X}_0. \tag{11.19}$$

Here the (possibly function-valued) factors $\bar{\lambda}$, λ, α, and β are no longer determined by the underlying parameter domain, but are now *free design parameters*. This eliminates both of the disadvantages of C^1 continuity mentioned above.

If we choose the factors in (11.19) to be polynomials, then because of the polynomial nature of the Bézier volumes, we can compare coefficients in (11.19) to get conditions in terms of Bézier points. This requires carrying out the differentiation, using identities of the form

$$(1-v)^\rho B_j^{m-\rho}(v) = \frac{\binom{m-\rho}{j}}{\binom{m}{j}} B_j^m(v), \qquad v^\rho B_j^{m-\rho}(v) = \frac{\binom{m-\rho}{j}}{\binom{m}{j+\rho}} B_{j+\rho}^m(v),$$

$$v^\rho B_{ijk}^{m-\rho}(\boldsymbol{u}) = \frac{\binom{m-\rho}{j}}{\binom{m}{j+\rho}} B_{i,j+\rho,k}^m(\boldsymbol{u}), \qquad v^\rho B_{ijk\ell}^{m-\rho}(\boldsymbol{u}) = \frac{\binom{m-\rho}{j}}{\binom{m}{j+\rho}} B_{i,j+\rho,k,\ell}^m(\boldsymbol{u})$$

(which follow from the definitions of binomial coefficients and the generalized Bernstein polynomials [Las 87]), shifting the indices, and finally comparing the coefficients.

In the case where both TB volumes are as in Example 11.3 (Fig. 11.10), we get

$$D_{\dot{\bar{\mathbf{u}}}_\lambda} \boldsymbol{Y}\Big|_{\bar{u}=0} = \left[\lambda D_{\dot{\mathbf{u}}_\lambda} + \alpha D_{\dot{\mathbf{u}}_\mu} + \beta D_{\dot{\mathbf{u}}_\nu}\right] \boldsymbol{X}\Big|_{u=0}$$

for all $u = \bar{u}$, where α, β, and $\lambda > 0$ are constant factors. This leads to the equation

$$\bar{\boldsymbol{b}}_{1jk\ell} = (1+\lambda)\boldsymbol{s}_{0jk\ell} - \lambda \boldsymbol{b}_{1jk\ell},$$

in terms of the barycentrically weighted points

$$\boldsymbol{s}_{0jk\ell} = \left(1 - \frac{\alpha}{1+\lambda} - \frac{\beta}{1+\lambda}\right)\boldsymbol{b}_{0,j+1,k\ell} + \frac{\alpha}{1+\lambda}\boldsymbol{b}_{0j,k+1,\ell} + \frac{\beta}{1+\lambda}\boldsymbol{b}_{0jk,\ell+1}.$$

With appropriate choices for the form parameters for GC^1 continuity, we can guarantee that the associated parametric lines and surfaces join with C^1 or GC^1 continuity. Fig. 11.11 provides an example.

Finally, we should remark that the results of Stärk on the use of the de Casteljau algorithm for smoothly joining Bézier curves and surfaces [Stä 76], [Far 83a] also hold here, *i.e.*, the Bézier points of the volume $\boldsymbol{Y}$ can be computed using de Casteljau extrapolation [Las 87].

11.2. B-spline Methods

A *tensor-product B-spline volume* (*B-spline volume*) is defined by

$$\boldsymbol{X}(\boldsymbol{u}) = \sum_{i=0}^{\ell}\sum_{j=0}^{m}\sum_{k=0}^{n} \boldsymbol{d}_{ijk} N_{ip}(u) N_{jq}(v) N_{kr}(w), \tag{11.20}$$

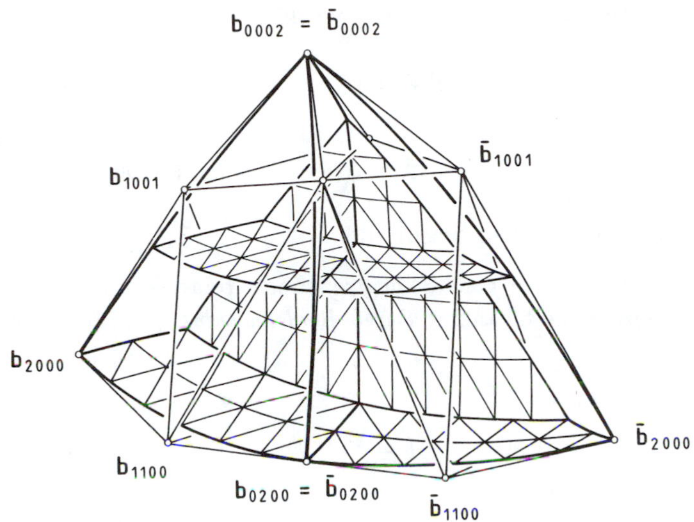

Fig. 11.11. GC^1 continuity with $\alpha = \beta = 0$, $\lambda = 1$.

where $\boldsymbol{u} = (u, v, w)$, $\boldsymbol{U} \times \boldsymbol{V} \times \boldsymbol{W}$ is a grid of knots with knot vectors $\boldsymbol{U}$, $\boldsymbol{V}$ and $\boldsymbol{W}$ as in Chap. 4.3, and where the N's are B-spline basis functions of orders p, q, and r as defined in Sect. 4.3.1. The $\boldsymbol{d}_{ijk} \in \mathbb{R}^3$ are called *de Boor points*, and form the *de Boor net*. We may also choose $d_{ijk} \in \mathbb{R}$, in which case we get a hypersurface lying in $\mathbb{R}^4$, which we call a *B-spline hypersurface*. We can, of course, also define rational B-spline volumes and hypersurfaces, including *trivariate NURBS*.

As a consequence of the tensor-product form of (11.20), all of the algorithms and properties of B-spline surfaces carry over to the trivariate case. We mention some useful references. [Sai 87] contains preliminary work on the integration of trivariate B-spline representations into a CSG-based solid modelling system. 3D primitives are generated using sweep, spin, and lofting operations. [Gries 89] describes the use of B-spline volumes in an FFD application, see Sect. 11.5 for surface modelling. [Patr 89a, 90] converted algebraic surfaces $F(x, y, z) = 0$ to B-spline form which can then be interpreted as contour surfaces $X(\boldsymbol{u}) = 0$ of trivariate function-valued B-spline hypersurfaces $(x, y, z, X(\boldsymbol{u}))$ with $(x, y, z) = (u, v, w)$, where $X(\boldsymbol{u})$ is given by (11.20) with $d_{ijk} \in \mathbb{R}$, see Sect. 11.6. They discuss the use of surfaces of the form $F(x, y, z) = 0$ for design and interpolation and approximation.

Taking d-fold tensor-products of univariate B-spline curve schemes leads to *d-variate B-spline representations*, see [Boo 77], [BOO 78] and the remark on d-variate Bézier representations on page 467.

11.3. Transfinite Methods

In analogy with the surface case, we can also use blending functions in the multivariate case. Here we work with hypercubes and simplices.

11.3.1. Transfinite Cube Segments

Following Sect. 8.1.1, we can define transfinite representations w.r.t. a d-dimensional hypercube by taking d-fold *Boolean sums*, [Coo 64]. For $d = 3$, we take

$$\begin{aligned} Q &= (P_1 \oplus P_2 \oplus P_3)F \\ &= P_1F + P_2F + P_3F - P_1P_2F - P_1P_3F - P_2P_3F + P_1P_2P_3F. \end{aligned} \tag{11.21}$$

Here

$$P_1F = (1-u)F(0,v,w) + uF(1,v,w)$$

interpolates the faces $F(0,v,w)$, $F(1,v,w)$. P_2F and P_3F have similar interpolation properties. The function

$$P_1P_2F = (1-u)[(1-v)F(0,0,w)+vF(0,1,w)]+u[(1-v)F(1,0,w)+vF(1,1,w)]$$

interpolates the edges $F(0,0,w)$, $F(0,1,w)$, etc., and P_1P_3F and P_2P_3F act similarly. Finally,

$$P_1P_2P_3F = (\,1-u, \quad u\,)\,G\begin{pmatrix} 1-v \\ v \end{pmatrix},$$

with

$$G = \begin{pmatrix} (1-w)F(0,0,0) + wF(0,0,1) & (1-w)F(0,1,0) + wF(0,1,1) \\ (1-w)F(1,0,0) + wF(1,0,1) & (1-w)F(1,1,0) + wF(1,1,1) \end{pmatrix}$$

interpolates the vertices $F(0,0,0)$, $F(0,0,1)$, ... of the cubical segment.

As in the bivariate case, there may be *twist incompatibilities*, *i.e.*, problems with the interchangeability of mixed higher derivatives. Every scheme which comes from combining several univariate schemes involving first derivatives in their definitions is plagued with this problem.

A corrected Coons' C^1 hyperpatch was constructed inductively in [Barn 84c]. For a C^2 hyperpatch, see [Wor 85].

11.3.2. Transfinite Tetrahedral Segments

By introducing barycentric coordinates, we can easily get transfinite simplicial representations which are generalizations of the methods discussed in Chap. 8. For example, [Barn 84a] constructed transfinite tetrahedral representations using Boolean sums of *BBG projectors* (which are cubic Hermite projectors along parallels to the sides). He also constructed transfinite d-dimensional simplicial representations using *radial projectors*, *i.e.*, cubic Hermite projectors along rays which join a vertex with the points on the opposite side. Since Boolean sums always lead to *twist incompatibilities*, the main problem is again to find appropriate corrections.

Since the problem of twist incompatibilities does not arise if we take *convex combinations*, Alfeld [Alf 84c] and Gregory [Gre 85] use this idea to construct simplicial interpolation schemes. [Alf 84c] defines

$$Q = \alpha_{12} P_{12} F + \alpha_{34} P_{34} F, \tag{11.22}$$

with

$$\alpha_{12} = \frac{u_3^2 u_4^2}{u_1^2 u_2^2 + u_3^2 u_4^2} \qquad \text{and} \qquad \alpha_{34} = 1 - \alpha_{12},$$

where P_{12} interpolates the function value and first derivatives along the boundary faces indexed by 1 and 2 of the tetrahedron, and P_{34} does the same for the boundary faces indexed by 3 and 4. Here u_k denote the barycentric coordinates.

[Gre 85] generalized his earlier symmetric methods for transfinite triangular interpolation [Gre 78] to $\mathbb{R}^d$ by taking convex combinations of Boolean sums of Taylor operators.

11.4. Scattered Data Methods

11.4.1. Shepard Methods

All of the Shepard interpolants discussed in Chap. 9 can easily be generalized to $\mathbb{R}^d$, see e.g., [Barn 84b, 87a], [Ren 88b], and [Alf 89]. These methods involve *distance-weighted interpolation*, *i.e.*, the interpolation data are weighted with functions $\omega_i(\boldsymbol{x})$ which depend on the (Euclidean) distance of the point of interest to each of the corresponding data points $\boldsymbol{x}_i$.

11.4.2. Radial Basis Function Methods

These *distance weighted interpolation* methods can also be easily generalized to $\mathbb{R}^d$, but so far most of the work has been on d-dimensional multiquadrics, see e.g., [Barn 84b, 87a], [Fol 87b] for $d = 3$, and also [Dyn 87a], [Alf 89].

11.4.3. FEM Methods

As in the bivariate case, multivariate scattered data interpolation by FEM methods requires a three step process [Barn 84a, 85, 87a]:

i) triangulate the scattered data,

ii) estimate the necessary derivatives,

iii) construct the FEM interpolant.

11.4.3.1. d-dimensional Triangulation

Given a set $\mathcal{P}$ of N points $\boldsymbol{x}_i \in \mathbb{R}^d$ with $d > 2$, we can define a d-dimensional triangulation in direct analogy with Definition 9.1. But there are some essential differences between triangulations in the plane and those in $\mathbb{R}^d$, for example [Alf 89]:

1) in the plane, $\mathcal{P}$ and its convex hull Ω uniquely determine the number of triangles M and the number of edges K of the triangulation. Indeed, it can be shown by induction that $M = 2N - E - 2$, $K = 3N - E - 3$, see [Law 77], [Chang 84], and also [Schu 87]. M and K also are bounded by $N - 2 \leq M \leq 2N - 5$ and $2N - 3 \leq K \leq 3N - 6$, cf. [Cli 84], [Ren 84], where E is the number of vertices of Ω. In higher dimensions we do not have such formulae, as can be seen (cf. [Barn 84a] and [Law 86]) already for a set of five data points in $\mathbb{R}^3$, see Fig. 11.12,

2) in $\mathbb{R}^d$ with $d > 3$ we cannot always distinguish triangulations on the basis of which data points are connected to each other [Law 86],

3) an iterative construction of the triangulation (cf. Sect. 9.3.1.2) is not always possible in the d-dimensional case with $d > 2$, see [Rud 58], [RUS 73], and [Alf 89].

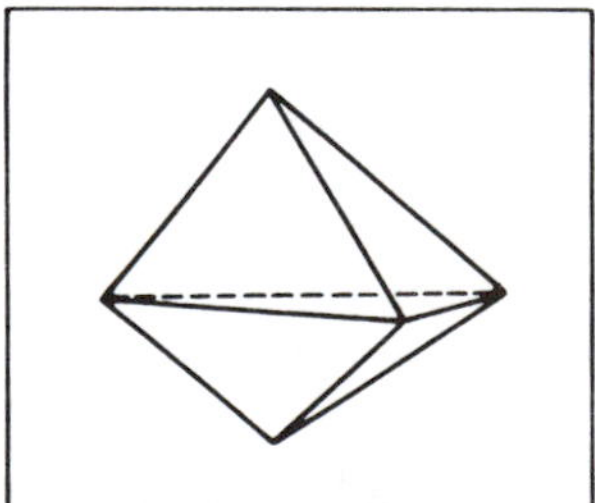
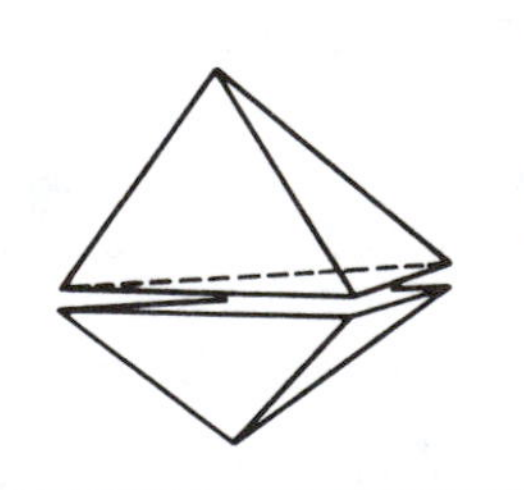
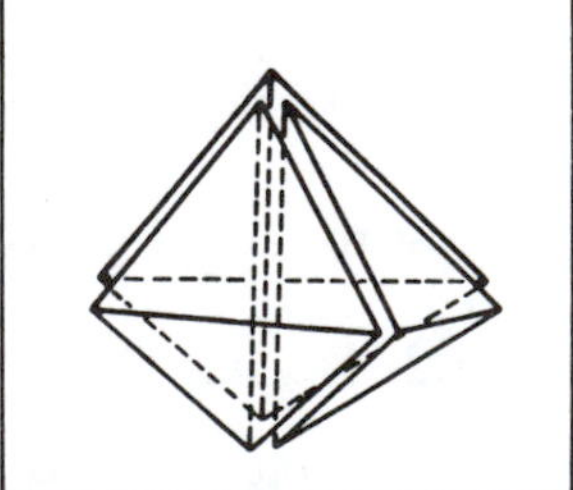

Fig. 11.12. Possible triangulations of five points in $\mathbb{R}^3$.

Some, but not all, of the criteria discussed in Sect. 9.3.1.1 for comparing plane triangulations can be generalized to the d-dimensional case. [Barn 84a] showed how to generalize the *max-min height criterion* as well as the *min-max radius quotient criterion*. This latter criterion involves computing the maximal quotient of the radii of the d-dimensional inscribed and circumscribed spheres over all triangles in the triangulation, and then minimizing over all possible triangulations.

The desire to construct a globally optimal triangulation suggests that we work with the d-dimensional version of the *Delaunay triangulation*. It arises as the dual to the *Dirichlet tessellation*, which is defined in the same way as in the planar case discussed in Sect. 9.3.1.2, see e.g., [Wat 81], [Avi 83] and also [Bruz 90], [Bow 81], [Barn 84a], and [Law 86].

While the max-min angle criterion does not generalize directly to $\mathbb{R}^d$, we can construct Delaunay triangulations by using a version of the *circumscribed circle criterion* involving (hyper)spheres [Law 86].

[Bruz 90] has given two data structures and certain Euler operators which can be used to construct a triangulation in $\mathbb{R}^3$, and has applied them to the algorithms of Watson [Wat 81] and Avis-Battacharya [Avi 83].

11.4.3.2. Interpolation

C^0 continuous cubic interpolants (generalized 9-parameter interpolants) require that we be given both function values and first derivatives at all the data points. Since cubic simplicial Bézier segments do not have any interior control points [Far 86], they are already defined by the 9-parameter interpolants corresponding to the boundary faces, provided they can be constructed.

To find C^r continuous *multivariate Hermite interpolants*, we need to use a very high polynomial degree n. For example, already for the trivariate case and $r = 1$ we need $n = 9$, while for $r = 2$ at least $n = 17$ is required [Zen 73]. The construction becomes more and more complicated, and higher and higher derivatives are needed. In the trivariate case, for example, for $r = 1$ we already need derivative data up to the fourth order at the vertices, along with function values and derivative data at the center of gravity and at several points on the boundary faces of the tetrahedron, for a total of 220 pieces of data, in order to determine the coefficients of the interpolant [Zen 73], [Res 87]. Thus, in the multivariate case, it is even more imperative than in the bivariate case (see Sect. 9.3.2) to split the domain into simplicial subdomains, and to construct the interpolant over these pieces. This allows us to use significantly lower polynomial degrees and derivatives of much lower order. We consider the Clough-Tocher and Powell-Sabin interpolants in more detail.

A C^1 *continuous trivariate Clough-Tocher interpolant* in Bézier form was described by [Alf 84a, 89]. Each macro-tetrahedron is split into four quintic micro-tetrahedra, see Fig. 11.13. In order to uniquely determine the Bézier points using the C^1 continuity conditions (see Sect. 11.1.4) from the prescribed function values and derivatives up to second order, we have to perform a condensation of parameters, cf. Sect. 9.3.2. Thus although we are using quintic pieces, the polynomial precision is only three.

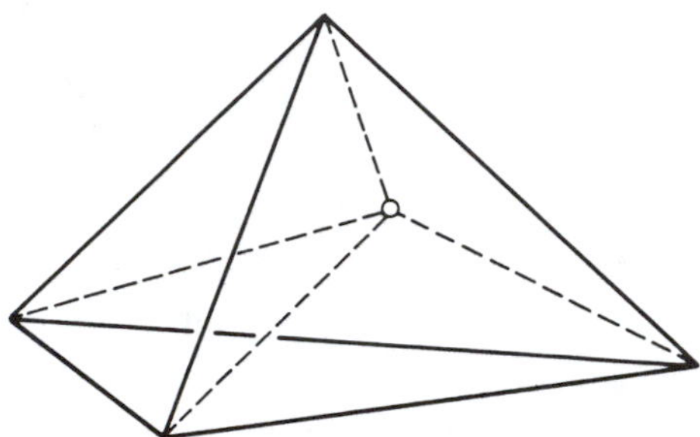

Fig. 11.13. Clough-Tocher split of a tetrahedron into four micro-tetrahedra.

A *d-dimensional* C^1 *continuous Clough-Tocher interpolant* with cubic accuracy was presented in [Far 86], [Wor 87]. In this case, every macro-simplex of the interpolant is broken into $(d+1)!/2$ micro-simplices. Since only cubic polynomials are used, the construction of the interpolant only requires function values and first derivatives, see also [Alf 89].

Powell-Sabin interpolants (cf. Sect. 9.3.2) are based on quadratic pieces, which are especially fast and simple to compute. As for the bivariate case, the use of such a small polynomial degree means that we have to split the domain into many pieces. For example, for $r = 1$, we need 24 micro-tetrahedra to construct one macro-tetrahedron, see Fig. 11.14. Besides this, the triangulation has to have some special properties before such a multivariate C^1 continuous Powell-Sabin interpolant can be constructed [Wor 88], see also [Alf 89].

Multivariate rational FEM interpolants can be obtained as generalizations of bivariate interpolants, mostly as discretized transfinite interpolants. For example, [Barn 84a] uses Boolean sums of BBG projectors (with an appropriate twist compatibility condition), while [Alf 84a, 89] uses convex combination of BBG projectors. [Alf 85a, 89] worked with *orthogonal projectors*, *i.e.*, projectors along perpendiculars to the "boundary facets" of the simplices. In principle, *transfinite interpolants* (see Sect. 11.3.2) can also be used as FEM interpolants.

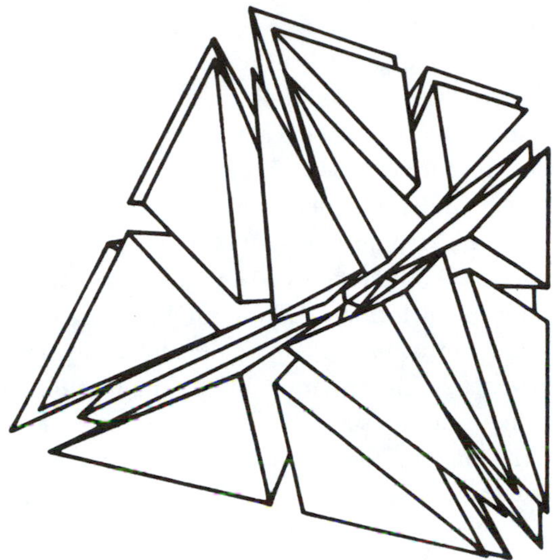

Fig. 11.14. Powell-Sabin split of a tetrahedron into 24 micro-tetrahedra ([Far 86], [Wor 88]).

11.4.3.3. Estimating Derivatives

Most of the methods discussed in Sect. 9.3.3, such as the method of weighted mean values, local interpolation and approximation methods, and Alfeld's functional minimization method, can be generalized to the d-variate case.

11.4.4. Multistage Methods

Multivariate multistage schemes can be built from the various multivariate methods discussed above in the same way as in the bivariate case. For example, we can use Boolean or Delta sums, see Sect. 9.4. For examples, see [Barn 84b] and [Fol 84, 87b].

11.5. Curve and Surface Modelling – FFDs

Free-form curves, surfaces, and volumes can be interpreted as deformations of the associated parameter domains under the mapping induced by their defining equations. In the 3D case, we are dealing, for example, with the deformation of a unit cube in the sense that we operate on the material forming the cube to alter its shape by pressing, drawing, or turning it, etc. Depending on the underlying *deformation prescription*, *i.e.*, as defined by a Bézier, B-spline, Coons, or other free-form formula, this deformation process can have either a global or a more local character. In general, the position of every point in the interior as well as those on the surface is altered, see e.g., Figs 11.1, 11.4, and 11.6 for the deformation of a cube, a tetrahedron, and a pentahedron by corresponding Bézier formulae. It is of particular interest for designers to be able to embed curves, surfaces, and volumes in the parameter domain of a free-form volume, since this allows concentrating on well-defined subsets of the parameter space. In simple cases, these subsets are in fact parametric lines and planes, which under the deformation are mapped to parametric curves and surfaces of the free-form volume, cf. Figs. 11.1, 11.4, and 11.6. The more interesting subsets, however, are non-isoparametric subsets, which might be given by an implicit formula. These might be 3D primitives such as a sphere, cone, cylinder, torus, etc. They might also be given by parametric formula, e.g., Bézier or B-spline forms, or by a combination or collection of the above. Finally, they might also be defined by solids such as those in the CSG or B-Rep-description. In the literature, in the context of these more general non-isoparametric subsets, the term FFDs (*Free form deformations*) is commonly used.

In CAGD, the idea of shaping a 3D object by first taking a "*detour*" to model a trivariate object goes back to Bézier [Bez 74, 78] (see also [BEZ 86]), where he used TPB volumes. TPB volumes were also used in [Sed 86a] and [Coq 90], where a number of illustrative examples are given. [Chad 89] used Bézier FFDs to deform muscles, and [Coq 91] used Bézier FFDs to perform animation. FFDs using B-spline bases were described by [Gries 89], while [Dero 88a] and [Las 92a] investigated the mathematical foundations of multivariate Bézier simplices and tensor-product Bézier representations.

In a *global FFD*, the entire object is embedded in the parameter domain of the volume, and subsequently deformed. On the other hand, in a *local FFD*, only those parts of an object are embedded in the parameter domain of a volume which are to be modelled.

In order to carry out an *FFD modelling process*, for every point of the object we must first find the associated parameter value in the parameter space of the volume. For this, we define a local coordinate system, the deformation domain, which includes all parts of the object to be deformed. For example, in the tensor-product case, this would be a box-shaped domain, which might be subdivided into sub-boxes corresponding to a segmented TPB volume or B-spline volume. By comparing coordinates, we can find the sub-box $\boldsymbol{Q}_0$, defined by $\boldsymbol{P}_0, \boldsymbol{U}, \boldsymbol{V}$, and $\boldsymbol{W}$ as in Fig. 11.15 which contains the given point $\boldsymbol{P}$ of the object. Then we can write $\boldsymbol{P}$ as

$$\boldsymbol{P} = \boldsymbol{P}_0 + u\boldsymbol{U} + v\boldsymbol{V} + w\boldsymbol{W},$$

where the local coordinates u, v, w are given by

$$u = \frac{\boldsymbol{V} \times \boldsymbol{W} \cdot (\boldsymbol{P} - \boldsymbol{P}_0)}{\boldsymbol{V} \times \boldsymbol{W} \cdot \boldsymbol{U}}, \quad v = \frac{\boldsymbol{U} \times \boldsymbol{W} \cdot (\boldsymbol{P} - \boldsymbol{P}_0)}{\boldsymbol{U} \times \boldsymbol{W} \cdot \boldsymbol{V}}, \quad w = \frac{\boldsymbol{U} \times \boldsymbol{V} \cdot (\boldsymbol{P} - \boldsymbol{P}_0)}{\boldsymbol{U} \times \boldsymbol{V} \cdot \boldsymbol{W}}.$$

Now we have to choose a deformation prescription, *i.e.*, a free-form volume formula. If we use a Bézier or a B-spline representation, this means we have to prescribe the polynomial degree and a corresponding control point grid which covers the deformation domain. Fig. 11.16 illustrates this process for a TPB formula as in (11.1). The control points of the grid are given by

$$\boldsymbol{b}_{ijk} = \boldsymbol{P}_0 + \frac{i}{\ell}\boldsymbol{U} + \frac{j}{m}\boldsymbol{V} + \frac{k}{n}\boldsymbol{W}.$$

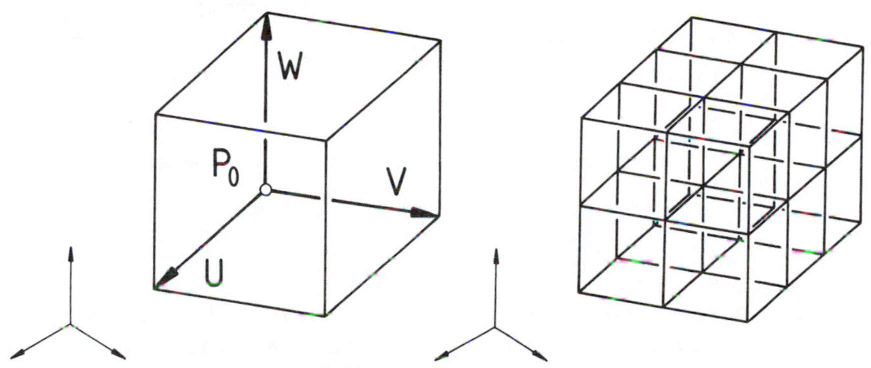

Fig. 11.15. Local coordinate system. **Fig. 11.16.** Control point grid.

The actual deformation of the object involves translation of the points $\boldsymbol{b}_{ijk} \to \bar{\boldsymbol{b}}_{ijk}$, and evaluation of the FFD defining equation with coefficients $\bar{\boldsymbol{b}}_{ijk}$ and the parameter values u, v, w of the object point $\boldsymbol{P}$ calculated above.

A *graphical image* of the deformed object can be obtained by mapping a piecewise linear approximation of some of the isoparametric lines. Present-day computers are very efficient at rendering surfaces consisting of piecewise linear triangulations (e.g., Phong shading), and so it is convenient to construct a triangulation of the object and then render it, see [Gries 89] and [Coq 90]. In any case, the deformed segments have to be approximated by linear or planar triangular segments for the purpose of rendering.

If the object is defined by control points, as it is for example in the case of Bézier curves, surfaces, and volumes, then we can restrict the mapping to the control points. The transformed Bézier points describe the FFD in compact form, although only approximately, and C^r continuity between segments is lost, thus requiring some kind of correction [Bez 78]. This approach produces a deformed object of the same degree as the original object. Fig. 11.17 illustrates this strategy in the 2D case for a global deformation of a planar Bézier curve using a TPB surface in $\mathbb{R}^2$.

From a mathematical point of view, the FFD construction involves composing two mappings [Bez 78], see Fig. 11.18. We discuss this point in more detail for the simple case of modelling a given curve in Bézier form using a tensor-product Bézier surface [Las 92a].

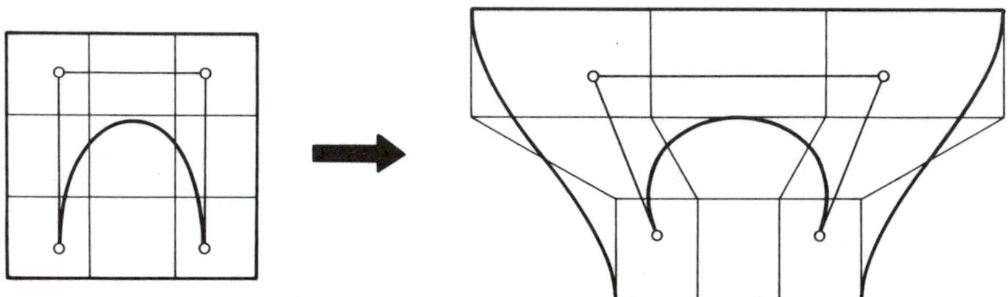

Fig. 11.17. The deformation of a Bézier polygon and a Bézier curve using an FFD mapping applied only to the Bézier points.

Theorem 11.1. *Let $\boldsymbol{K}(t) = (u(t), v(t)) : \mathbb{R} \to \mathbb{R}^2$ be a planar Bézier curve of degree N with Bézier points $\boldsymbol{k}_I = (u_I, v_I)$, and let $\boldsymbol{F}(u,v) : \mathbb{R}^2 \to \mathbb{R}^d$ with $d \geq 2$ be a Bézier surface of degree (ℓ, m) with Bézier points $\boldsymbol{b}_{ij}$. Then the surface curve $\boldsymbol{F}(t) = \boldsymbol{F}(\boldsymbol{K}(t)) = \boldsymbol{F}(u(t), v(t))$ can be represented as a Bézier curve of degree $(\ell + m)N$:*

$$\boldsymbol{F}(t) = \sum_{R=0}^{rN} \boldsymbol{B}_R(u^{\ell}_{\boldsymbol{I}^u}, v^{m}_{\boldsymbol{I}^v}) B^{rN}_R(t). \tag{11.23a}$$

Here $r = \ell + m$, and the Bézier points are

$$\boldsymbol{B}_R(u^{\ell}_{\boldsymbol{I}^u}, v^{m}_{\boldsymbol{I}^v}) = \sum_{|\boldsymbol{I}|=R} C^{\ell m}_R(N, \boldsymbol{I}) \boldsymbol{b}^{\ell m}_{00}(u^{\ell}_{\boldsymbol{I}^u}, v^{m}_{\boldsymbol{I}^v}), \tag{11.23b}$$

with constants

$$C^{\ell m}_R(N, \boldsymbol{I}) = \frac{\prod_{Q^u=1}^{\ell} \binom{N}{I^u_{Q^u}} \prod_{Q^v=1}^{m} \binom{N}{I^v_{Q^v}}}{\binom{rN}{R}}. \tag{11.23c}$$

The sum is taken over all $\boldsymbol{I} = (\boldsymbol{I}^u, \boldsymbol{I}^v)$, where $\boldsymbol{I}^u = (I_1^u, \ldots, I_\alpha^u)$ and $\boldsymbol{I}^v = (I_1^v, \ldots, I_\beta^v)$ are such that $0 \le I_1^u, \ldots, I_\alpha^u \le N$, $0 \le I_1^v, \ldots, I_\beta^v \le N$, and $|\boldsymbol{I}| = |\boldsymbol{I}^u| + |\boldsymbol{I}^v| = I_1^u + \cdots + I_\alpha^u + I_1^v + \ldots + I_\beta^v = R$.

For $\alpha = \ell$ and $\beta = m$, the $\boldsymbol{b}_{00}^{\ell m}(u_{\boldsymbol{I}^u}^{\ell}, v_{\boldsymbol{I}^v}^{m})$ can be computed recursively using the de Casteljau algorithm, *i.e.*, in the u-direction we use

$$\boldsymbol{b}_{ij}^{i+\alpha,j+\beta}(u_{\boldsymbol{I}^u}^{\alpha}, v_{\boldsymbol{I}^v}^{\beta}) = (1 - u_{I_\alpha^u})\boldsymbol{b}_{ij}^{i+\alpha-1,j+\beta}(u_{\boldsymbol{I}^u}^{\alpha-1}, v_{\boldsymbol{I}^v}^{\beta}) + u_{I_\alpha^u}\boldsymbol{b}_{i+1,j}^{i+\alpha,j+\beta}(u_{\boldsymbol{I}^u}^{\alpha-1}, v_{\boldsymbol{I}^v}^{\beta}),$$

and a corresponding formula for the v-direction.

The argument $(u_{\boldsymbol{I}^u}^{\alpha}, v_{\boldsymbol{I}^v}^{\beta})$ of $\boldsymbol{b}_{ij}^{i+\alpha,j+\beta}(u_{\boldsymbol{I}^u}^{\alpha}, v_{\boldsymbol{I}^v}^{\beta})$ indicates that this recurrence involves applying the de Casteljau algorithm α times in the u-direction (for the u parameter values $u_{I_1^u}, \ldots, u_{I_\alpha^u}$ corresponding to $\boldsymbol{I}^u = (I_1^u, \ldots, I_\alpha^u)$), and β times in the v-direction (for the v parameter values $v_{I_1^v}, \ldots, v_{I_\beta^v}$ corresponding to $\boldsymbol{I}^v = (I_1^v, \ldots, I_\beta^v)$). The construction points $\boldsymbol{b}_{ij}^{i+\alpha,j+\beta}(u_{\boldsymbol{I}^u}^{\alpha}, v_{\boldsymbol{I}^v}^{\beta})$ arise in the evaluation of the polar form of $\boldsymbol{F}(u,v)$. In view of the symmetry of the polar form w.r.t. permutations of the arguments and the tensor-product form of $\boldsymbol{F}(u,v)$, the order in which we carry out the steps has no influence on the result.

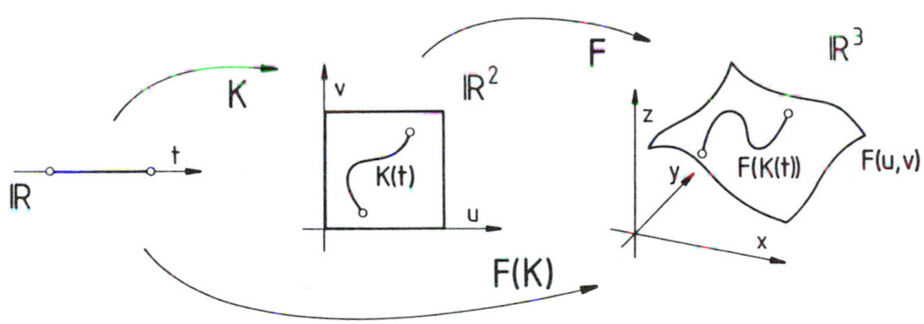

Fig. 11.18. FFD as a composition of two mappings: Deformation of a planar curve $\boldsymbol{K}(t)$ by a mapping $\boldsymbol{F} : \mathbb{R}^2 \to \mathbb{R}^3$.

The proof of Theorem 11.1 is based on induction on $r = \alpha + \beta$, where $\alpha \in \{0, \ldots, \ell\}$ and $\beta \in \{0, \ldots, m\}$, using the defining equation (4.1), the recurrence relation (4.2a), and the product formula (4.2d) for the Bernstein polynomials [Las 92a].

We now give a simple example to illustrate Theorem 11.1.

Example 11.4. The parametric lines of a Bézier surface $\boldsymbol{F}(u,v)$ of degree (ℓ, m) are Bézier curves of degree ℓ and m, respectively, see Sect. 6.2.2. For the special case $N = 1$, Theorem 11.1 immediately provides a generalization of this result to straight lines in general position in the domain, which are mapped to surface curves of degree $\ell + m$. For $\ell = m = 2$ (i.e., $N = 1$, $r = 4$), the Bézier points $\boldsymbol{B}_R$ of such a surface curve are given by

$$\begin{aligned}
\boldsymbol{B}_0 &= \boldsymbol{b}_{00}^{22}(u_0, u_0, v_0, v_0),\\
\boldsymbol{B}_1 &= \tfrac{1}{2}\left[\boldsymbol{b}_{00}^{22}(u_0, u_0, v_0, v_1) + \boldsymbol{b}_{00}^{22}(u_0, u_1, v_0, v_0)\right],\\
\boldsymbol{B}_2 &= \tfrac{1}{6}\left[\boldsymbol{b}_{00}^{22}(u_0, u_0, v_1, v_1) + 4\boldsymbol{b}_{00}^{22}(u_0, u_1, v_0, v_1) + \boldsymbol{b}_{00}^{22}(u_1, u_1, v_0, v_0)\right],\\
\boldsymbol{B}_3 &= \tfrac{1}{2}\left[\boldsymbol{b}_{00}^{22}(u_0, u_1, v_1, v_1 + \boldsymbol{b}_{00}^{22}(u_1, u_1, v_0, v_1)\right],\\
\boldsymbol{B}_4 &= \boldsymbol{b}_{00}^{22}(u_1, u_1, v_1, v_1).
\end{aligned}$$

Fig. 11.19a shows $\boldsymbol{K}(t) = (u(t), v(t))$ and its defining Bézier points $\boldsymbol{k}_I = (u_I, v_I)$ in the domain of $\boldsymbol{F}(u, v)$. Fig. 11.19b shows the Bézier net corresponding to the biquadratic TPB surface along with the affine images of the parametric domain on the individual quadrilaterals of the Bézier net, *i.e.*, the result of applying a de Casteljau step in the u and v-directions. The construction points $\boldsymbol{b}_{ij}^{i+1,j+1}(u_{I^u}^1, v_{I^v}^1)$ created in the process are marked. Fig. 11.19c shows the four bilinear Bézier nets corresponding to the $\boldsymbol{b}_{ij}^{i+1,j+1}(u_{I^u}^1, v_{I^v}^1)$, and the $\boldsymbol{b}_{00}^{22}(u_{I^u}^2, v_{I^v}^2)$ created by one more de Casteljau step in the u and v-direction from these bilinear Bézier nets. Fig. 11.19d shows the construction of the $\boldsymbol{B}_R$ by (11.23b) as convex combinations of the $\boldsymbol{b}_{00}^{22}(u_{I^u}^2, v_{I^v}^2)$.

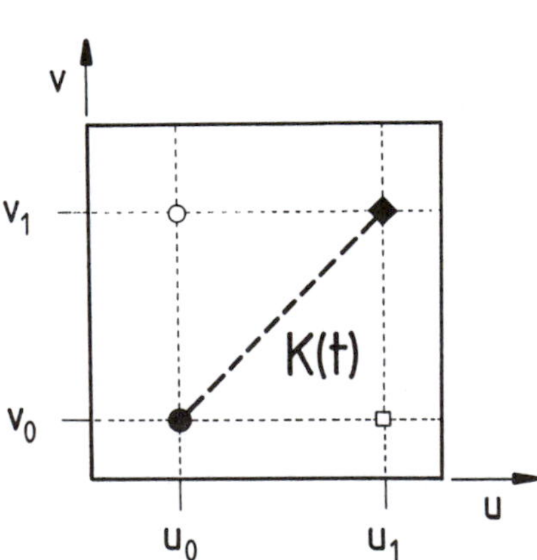

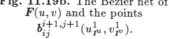

Fig.11.19a. The straight line as a curve in the parameter space of $\boldsymbol{F}(u, v)$.

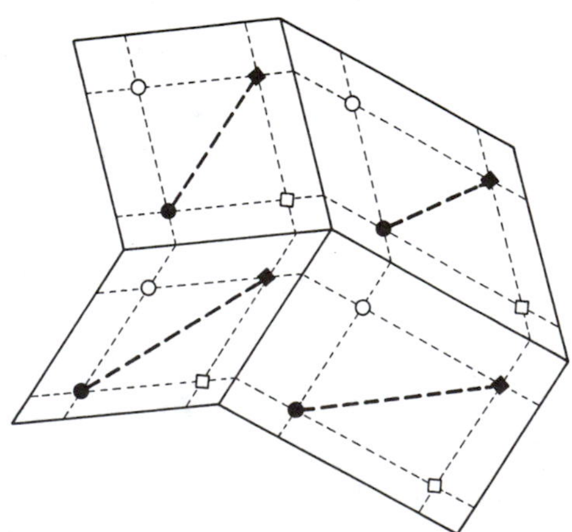

Fig. 11.19b. The Bézier net of $\boldsymbol{F}(u, v)$ and the points $\boldsymbol{b}_{ij}^{i+1,j+1}(u_{I^u}^1, v_{I^v}^1)$.

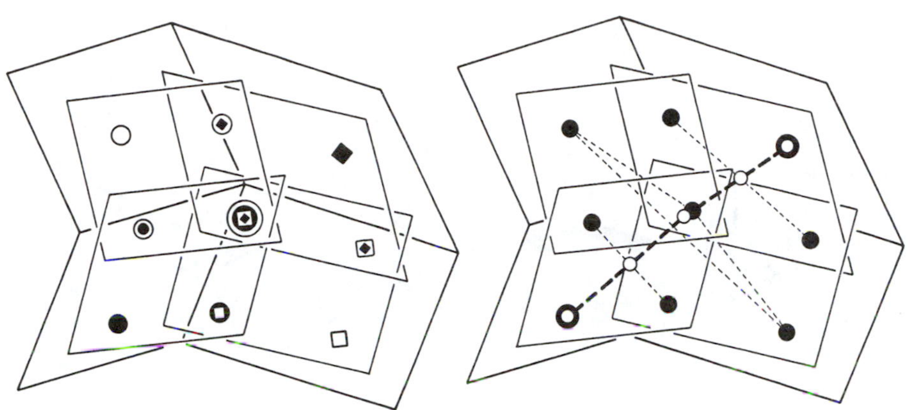

Fig. 11.19c. The Bézier net defined by $\boldsymbol{b}_{ij}^{i+1,j+1}(u_{I^u}^1, v_{I^v}^1)$ and the points $\boldsymbol{b}_{00}^{22}(u_{I^u}^2, v_{I^v}^2)$.

Fig. 11.19d. Construction of the Bézier points $\boldsymbol{B}_R$ as convex combinations of $\boldsymbol{b}_{00}^{22}(u_{I^u}^2, v_{I^v}^2)$.

Theorem 11.1 provides the mathematical foundation for the FFD approach to curve design via surface modelling. In addition, it provides the exact description of the non-isoparametric boundary curves of *trimmed surfaces*, which are surface patches whose boundaries are given by surface curves which are not in general isoparametric curves, see Fig. 11.20. Trimmed surfaces are of great importance in industrial applications, see [Mil 86], [Kob 86], [Cro 87], [Casa 87, 89, 92], [Shan 88], [Rock 89], [Hos 90, 91, 92a], and [Nis 90].

Theorem 11.1 can be generalized to the rational and multivariate case [Las 92a]. This allows an exact mathematical description of how to model a (rational) TPB surface by a (rational) TPB volume via the FFD idea. It also can be applied to find non-isoparametric boundary surfaces for *trimmed volumes*.

[Dero 88a] treated the simplicial case in a similar way, and gives several examples of applications.

It should be remarked, however, that because they involve composition of functions, the polynomial degree of FFDs can grow very quickly, especially in the important case of surfaces. Thus, the exact description of an FFD is not always the best approach. Fig. 11.21 compares an approximate description of an FFD with the exact description for an example where a quadratic Bézier curve is deformed using a quadratic triangular Bézier surface.

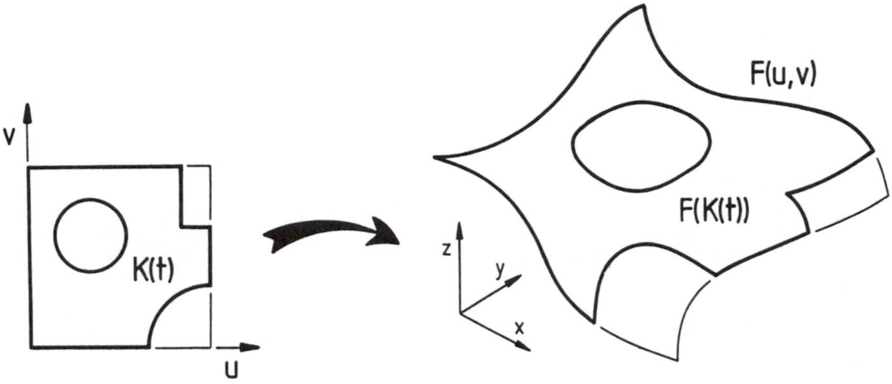

Fig. 11.20. Parametric and coordinate space representation of a trimmed surface.

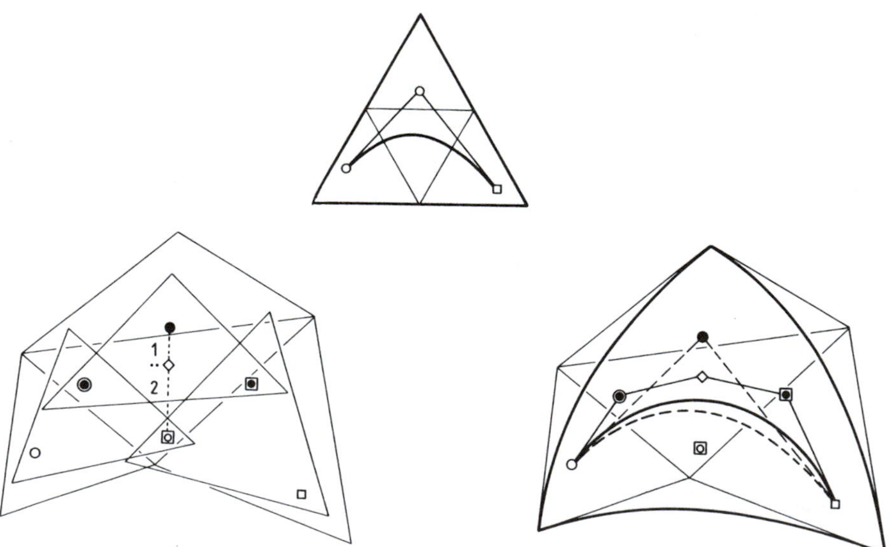

Fig. 11.21. A comparison of approximate and exact FFD methods. Top: an undeformed Bézier curve embedded in a triangle. Left: quadratic deformation of the triangular grid and the DeRose construction. Right: quadratic and quartic curves for the approximate and exact FFDs.

In contrast to Barr's *Jacobi matrix method* [Barr 84], which involves only operations such as twisting, bending, scaling, and tapering, the FFD method which is also described in [Far 90] allows much more general object deformations. However, as in [Barr 84], the method is purely mathematical in nature, with no physical basis. Alternatively, we can also employ *physically based methods*, which are playing an increasingly important role in *computer animation*. They involve physical variables and laws in the description of movement and deformation of an object. For some useful references to this complex area, see [BARR 87], [FOU 87], [Ter 88], [Wit 90], and [Wyv 86, 86a]. [Chad 89] combines the geometric aspects of FFDs with physical considerations.

11.6. Algebraic Curves and Surfaces

A *plane algebraic curve* is a curve which can be described by an implicit polynomial equation of the form $F(x, y) = 0$, and similarly, an *algebraic surface* is one which can be described by $F(x, y, z) = 0$. Polynomially parametrized curves and surfaces, e.g., in Bézier and B-spline form, constitute a subset of the sets of algebraic curves and surfaces, respectively; see Sect. 12.1.

Implicitly defined curves and surfaces have their specific *advantages* and *disadvantages*, just like parametric representations do, and indeed, the advantages of one are usually the disadvantages of the other, and vice versa.

Advantages of parametrized curve (segments) include: 1) they can be joined easily with C^r or GC^r continuity, 2) points on the curve can be efficiently computed (which is important for rendering purposes), 3) it is easy to describe subsegments in terms of the parameters, and 4) usually, as is the case for B-spline and Bézier representations, there is a close geometric relationship between the coefficients (control points) and the actual curve (e.g., the convex hull and variation diminishing properties hold).

On the other hand, some of the advantages of implicitly defined curves include: 1) it is easy to decide whether a point (x_0, y_0) lies on the curve where $F(x_0, y_0) = 0$, on one side where $F(x_0, y_0) < 0$, or the other where $F(x_0, y_0) > 0$; 2) the offset curve (surface), (see Chap. 15) is again an algebraic curve (surface); 3) the intersection (see Chap. 12) of two algebraic surfaces is an algebraic curve.

Since they are defined by equations of the form $F(x, y) = 0$, algebraic curves can be interpreted as contour lines of a function-valued, non-parametric surface $z = F(x, y)$ in $\mathbb{R}^3$, see Fig. 11.22. Similarly, an algebraic surface $F(x, y, z) = 0$ can be thought of as a contour surface of a function-valued, non-parametric hypersurface in $\mathbb{R}^4$. For more on calculating contour lines and surfaces, see Sect. 12.3.

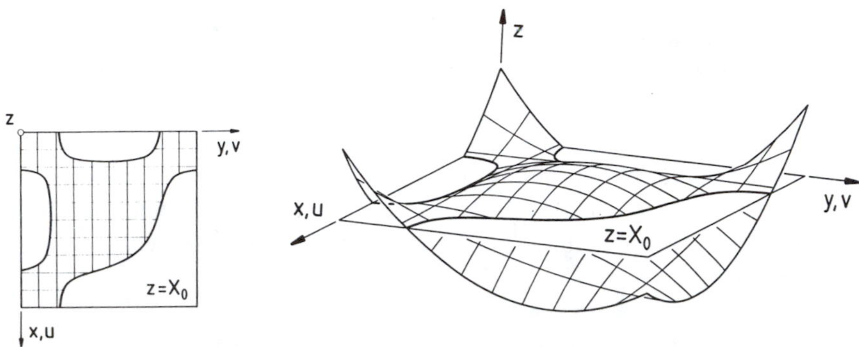

Fig. 11.22a. Plane algebraic curve defined by the implicit equation $F(x, y) = 0$.

Fig. 11.22b. Interpretation of a plane algebraic curve as the contour of a non-parametric function-valued surface $z = F(x, y)$.

Fig. 11.23 shows two examples of algebraic surfaces which are defined as contour surfaces $X(u, v, w) = 0$ of function-valued non-parametrized hypersurfaces in $\mathbb{R}^4$. They were computed using a modification of an algorithm in [Las 87, 90b].

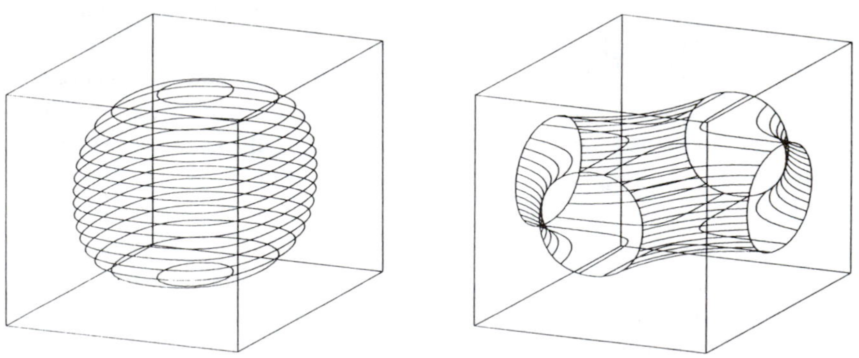

Fig. 11.23. Two algebraic surfaces which are contour surfaces of function-valued non-parametric hypersurfaces in $\mathbb{R}^4$.

As observed above, this method of representing an algebraic curve certainly requires the intersection of a surface $z = F(x, y)$ with the plane $z = 0$. But on the other hand, it follows that with the right kind of basis transformation, it is possible to give a geometric interpretation to the coefficients of the defining equation. Usually, the polynomial $F(\cdot)$ is written in terms of the monomial basis, so that an algebraic curve is described by

$$F(x, y) = \sum_{i=0}^{n} \sum_{k=0}^{n-i} a_{ik} x^i y^k. \tag{11.24}$$

The coefficients a_{ik} in (11.24) do not say anything about the geometric shape of the curve. But by means of a simple basis transformation (see Chap. 10), we can write F in terms of ordinary or generalized Bernstein polynomials or B-spline functions, where the (new) coefficients do have geometric significance. Indeed, conversion to ordinary Bernstein polynomials induces a description of $F(x, y) = 0$ in terms of a tensor-product Bézier surface, using generalized Bernstein polynomials leads to a triangular Bézier patch, and using B-spline functions leads to a tensor-product B-spline formula.

Example 11.5. $F(x, y) = x^2 + y^2 - 0.25 = 0$ describes a circle with center at the origin and radius 0.5. We can rewrite $F(x, y)$ in terms of generalized Bernstein polynomials as

$$\begin{aligned} -0.25\left[(1-x-y)^2\right] - 0.25\left[2x(1-x-y)\right] + 0.75\left[x^2\right] \\ - 0.25\left[2y(1-x-y)\right] - 0.25\left[2xy\right] + 0.75\left[y^2\right] = 0, \end{aligned} \tag{$*$}$$

where $u = x$, $v = y$, $w = 1 - x - y$ are the barycentric coordinates in $\mathbb{R}^2$, and where the expressions in the square brackets are the generalized Bernstein polynomials (6.44) for $n = 2$. It follows from $(*)$ that the Bézier ordinates of the function-valued non-parametric triangular Bézier patch $X(u, v, w)$, whose contour $X(u, v, w) = 0$ is described by the equation $x^2 + y^2 - 0.25 = 0$, are given by

$$b_{002} = b_{110} = b_{101} = b_{011} = -0.25, \qquad b_{200} = b_{020} = 0.75.$$

We associate these with the abscissae $(u, v, w) = (i/n, j/n, k/n)$ with $k = n - i - j$, $n = 2$, cf. Sect. 6.3.2. Since the triangular Bézier patch is defined only over the base triangle, the contour line $X(u, v, w) = 0$ corresponds to just the part of the algebraic curve (in this case a quarter circle) which lies in the triangle, see Fig. 11.24.

Other pieces of the algebraic curve can be constructed using other triangles and associated triangular Bézier patches. Fig. 11.25 (following [Sed 89a]) shows some examples. For the three examples on the left, we give the Bézier ordinates explicitly. The corresponding abscissae follow from the geometry of the regular subtriangulations of the individual triangles, cf. Sect. 6.3.2. In the

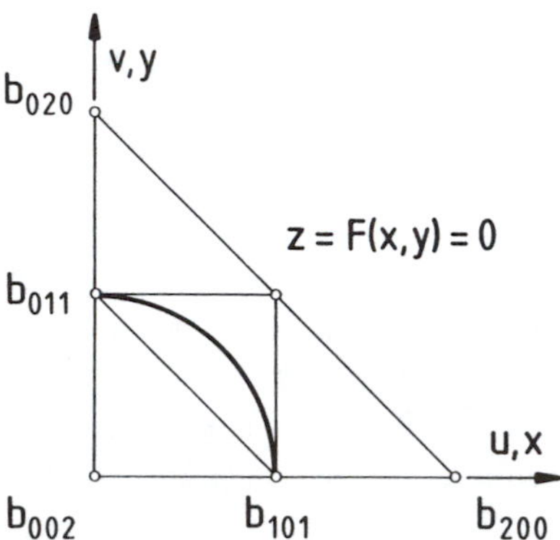

Fig. 11.24. A quarter circle defined as the contour line of a quadratic function-valued non-parametric triangular Bézier patch.

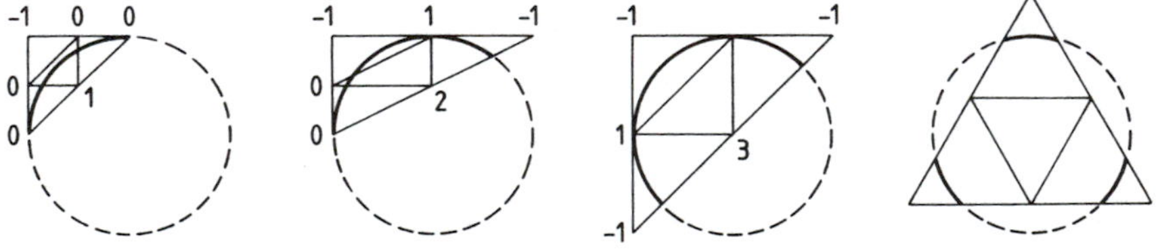

Fig. 11.25. Four circular segments defined by four different triangles.

literature, the Bézier ordinates in this context are frequently referred to as '*weights*', and the Bézier abscissae as '*control points*', see e.g., [Sed 87, 89a, 90a].

The indirect definition of an algebraic curve as the contour of a triangular Bézier patch gives us a way to use the geometric properties of the Bézier formula, see Sect. 6.3.2, to help in the design of algebraic curves. For example, if the algebraic curve is to pass through a vertex of a given triangle, say through $(u, v, w) = (0, 0, 1)$, then we must set the associated Bézier ordinate b_{00n} to zero. If the curve is to be tangent to one of the edges of the triangle at this point, say the edge $u = 0$, then we must also set $b_{1,0,n-1}$ to zero, and simultaneously $b_{0,1,n-1} \neq 0$ (otherwise we have a double-point of the algebraic curve at this vertex). Fig. 11.26 shows a cubic algebraic curve defined as the contour $X(u, v, w) = 0$ of a cubic triangular Bézier patch, and which interpolates at two vertices and has tangents matching the direction of the sides there. We can interpolate only at the vertices of the triangle because those are the points where the triangular Bézier patch interpolates.

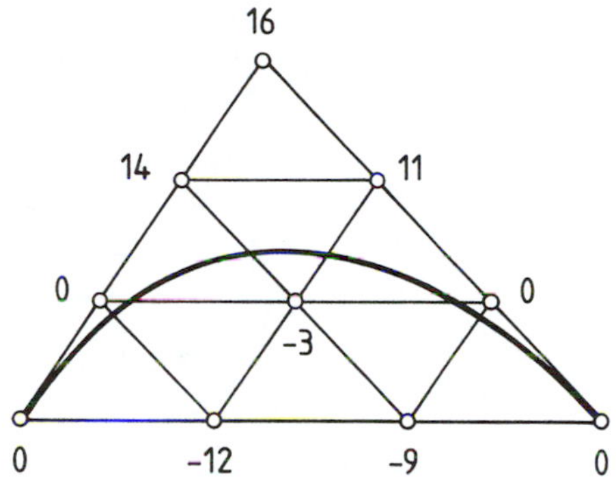

Fig. 11.26. Cubic algebraic curve interpolating prescribed boundary points and boundary tangents.

Similar properties hold for algebraic curves which come from tensor-product Bézier patches or B-spline surfaces, and also for algebraic surfaces which come from tensor-product Bézier, tetrahedral Bézier, and B-spline hypersurfaces.

[Sed 84a] studied algebraic curves corresponding to triangular Bézier patches, and found a monotonicity criterion which excludes loops and self-intersections, and also gave continuity conditions. In [Sed 90a], the same was done for algebraic surfaces associated with tetrahedral Bézier hypersurfaces. Using the ideas discussed here, [Sed 89a] also constructed approximate parametrizations of algebraic curves, *i.e.*, approximations to the algebraic curves by parametrized curves. Parametrization of algebraic surfaces is discussed in [Sed 87b, 90a], [Abh 87, 87a], see Sect. 12.1. Continuity conditions for algebraic curves and surfaces are treated in [Sed 84a, 85b, 87, 89a, 90a]. The papers [Dah 89], [Sed 90a], and [Lod 90] deal with *macro patches* (cf. Sects. 9.3.2.3, 9.3.2.4 and 11.4.3.2) to construct C^r continuous algebraic surfaces. In particular, [Lod 90] gives another *projective function-valued representation* for (quadratic) algebraic curves and surfaces, and compares it with the above methods, as well as the rational Bézier formulae in Sect. 4.1.4. [Patr 89a, 90] works with B-spline hypersurfaces, and presents a solution of the interpolation and approximation problems.

11.7. Visualization of Multivariate Objects

11.7.1. $\mathbb{R}^3$ Solids

Trivariate vector-valued formulae, *i.e.*, mappings from a cubical domain into $\mathbb{R}^3$, describe 3D objects in the usual sense. For simple patch configurations and/or small polynomial degrees, we can visualize such objects by displaying a set of *isoparametric surfaces* of the object. Thus, e.g., Fig. 11.27b shows several parametric surfaces corresponding to constant w values for the triquadratic TPB volume of Fig. 11.27a, see also the figures in Sect. 11.1.

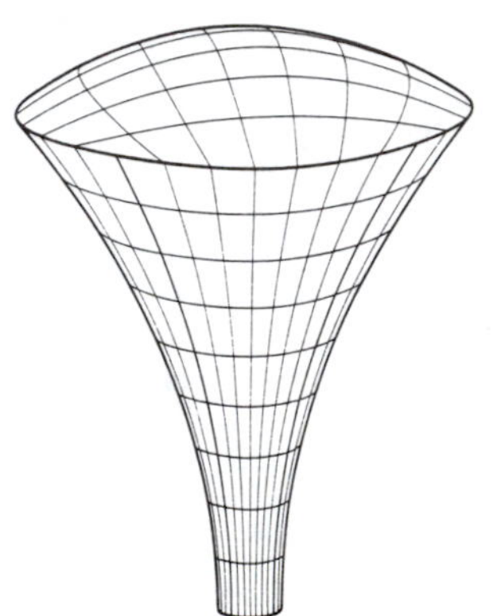

Fig. 11.27a. Triquadratic TPB-Volume.

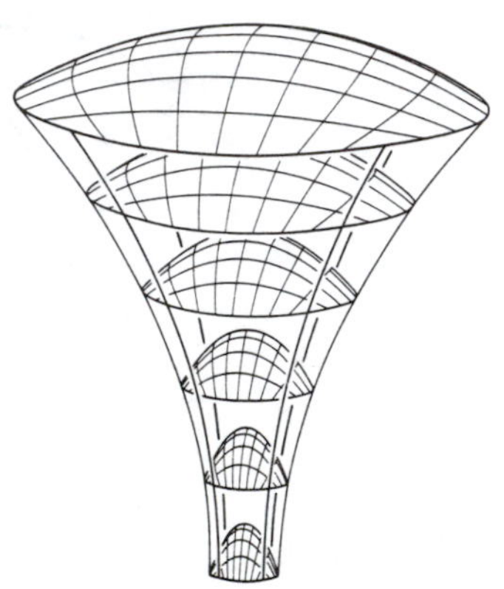

Fig. 11.27b. Representation of a family of parametric surfaces.

One way to get a good idea of the shape of an object is to display several different images. For example, for a trivariate tensor-product, we can look at three images displaying parametric surfaces corresponding to constant u, v, and w, respectively.

Such images can become complicated and confusing because of the limitations of two dimensional graphics, especially for objects consisting of many pieces or of pieces of high degree, for example spline volumes. In these cases we could proceed as follows: if we have a trivariate tensor-product formula, we display only the three parametric surfaces $\boldsymbol{X}(u_i, v, w)$, $\boldsymbol{X}(u, v_j, w)$, and $\boldsymbol{X}(u, v, w_k)$ which pass through the point $\boldsymbol{X}(u_i, v_j, w_k)$ in order to get a better impression of the "inside structure" of the object. This can be particularly useful if we let the point (and thus the corresponding three associated parametric surfaces) move around in the object.

Another way to render 3D objects is to display *slices* (e.g., cross or longitudinal sections) of the object. One way to do this is to choose a series of parallel slices, but as was done above with parametric surfaces, sometimes it can be useful to instead display three mutually orthogonal slices through a given point $\boldsymbol{X}(u_i, v_j, w_k)$ with $x = x(u_i, v_j, w_k)$, $y = y(u_i, v_j, w_k)$, and $z = z(u_i, v_j, w_k)$. More general slices (e.g., a *paddlewheel probe*) can also be useful, see e.g., [Spe 90]. Taking slices, along with finding parametric surfaces, can also be useful for other purposes, for example, when in addition to simply visualizing an object by perspective drawings, we also need an accurate scale reproduction of certain parts of the interior of the object. An algorithm for constructing such object slices can be found in [Las 87]. It is based on cutting the boundary surfaces and a given number of parametric surfaces using cutting planes selected by the user. Figs. 11.27c,d show a cross slice and a transversal slice of the object of Fig. 11.27a. The algorithm can also work with curved slices.

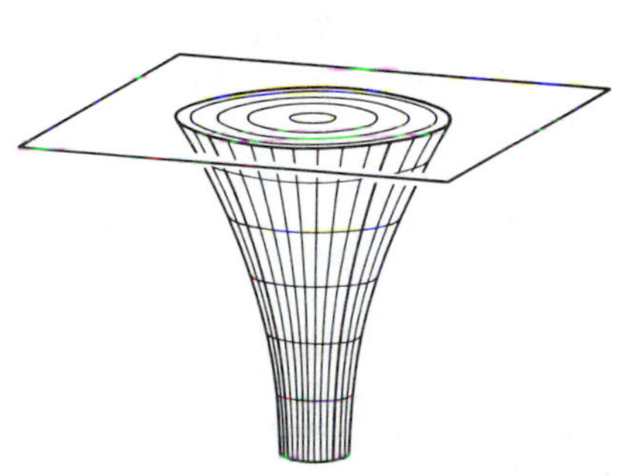

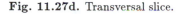

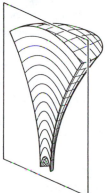

Fig. 11.27c. Cross section.

Fig. 11.27d. Transversal slice.

11.7.2. $\mathbb{R}^4$ Hypersurfaces

Trivariate function-valued representations describe hypersurfaces in $\mathbb{R}^4$. While we cannot render them directly, it is possible to visualize their behavior by displaying one or more projected surfaces in $\mathbb{R}^3$. For example, we can work with *isoparametric surfaces* which correspond to constant parameter values, and are the direct analogs of points and families of lines in the curve and surface cases, respectively. Alternatively, we can create *contour plots* of hypersurfaces (see Sect. 11.6) which correspond to constant function values. We are familiar with such plots from cartography and from temperature maps associated with weather reports, where they correspond to the surface of the earth.

Surfaces of constant parameter value are used in [Las 85a], [Fol 87b], and [Baj 90]. [Nie 90, 91a] suggest using a combination of three isoparametric surfaces $X(u_i, v, w)$, $X(u, v_j, w)$ and $X(u, v, w_k)$ associated with the parameter values corresponding to the function value $X(u_i, v_j, w_k)$. One implementation of this idea involves an axonometric view of the domain with the parameter planes $u = u_i$, $v = v_j$, and $w = w_k$ displayed, along with another copy of the domain with the graphs of the three parametric surfaces located on the faces of the cubical domain (decomposition technique). Another implementation, provides an axonometric view of the domain showing the three parameter planes and the contours of the associated isoparametric surfaces on them.

In addition to parallel projections, it is possible to work with central projections. [Pot 91a] used them to project hypersurfaces in $\mathbb{R}^4$ into $\mathbb{R}^3$ in order to display contour plots of trivariate scattered data interpolants. The visualization of trivariate scattered data interpolants by means of contour plots is also discussed in [Barn 84b, 85, 87a], [Nie 91a], and [Ham 91, 92]. The computation of contour plots for piecewise polynomials defined on tetrahedra is treated in [Pet 87] (generalizing [Pet 84]). The two examples in Sect. 11.6 come from TPB hypersurfaces. [Rock 90] also worked with TPB hypersurfaces, while [Dic 89] used B-spline hypersurfaces. The algorithm in [Sew 88] utilizes function values given on a uniform regular grid. Additional references can be found in Sect. 12.4.1, where the actual construction of contour lines and contour surfaces is discussed.

In principle, we could also use subsets of a hypersurface in $\mathbb{R}^4$ characterized in other ways for visualization purposes. For example, [Rat 88] considered *isophote surfaces* which are generalizations of isophote lines (lines with the same brightness on a surface in $\mathbb{R}^3$, see Sects. 1.6 and 13.4). For some examples, see [Pot 91a].

We also mention the papers [Nie 91a], [Pot 91a] which contain very good examples of visualizations.

Finally, we discuss several other methods for visualization. The *tiny cube method* of [Nie 90, 91a] (see also [Paj 90]) involves placing a number of small cubes (or boxes) in the 3D domain of the hypersurface. The images of the faces of these boxes are then displayed using a hidden surface algorithm, where the face is colored according to the function values of $X(u, v, w)$. If we use N_u, N_v, N_w boxes in the u, v, w directions, respectively, then there is a total of $N_u N_v N_w$ boxes, and thus $6N_u N_v N_w$ faces to be displayed. If we denote the side length in the u-direction by Δ_u, and assume that the separation between boxes in this direction is given by an integral multiple $\bar{M} = M\Delta_u$ of Δ_u, then (see Fig. 11.28)

$$\Delta_u = \frac{u_{max} - u_{min}}{N_u(M+1) - M},$$

and for each $i = 1(1)N_u$, the u coordinate of the left-hand lower vertex of the i-th box (see Fig. 11.28) is given by

$$u_i = u_0 + (i-1)\Delta_u(M+1).$$

Similar formulae hold in the v and w directions. For $N_u, N_v, N_w > 15$, the images produced by this method become difficult to interpret. It helps to draw in the lines connecting the centers of gravity of the boxes.

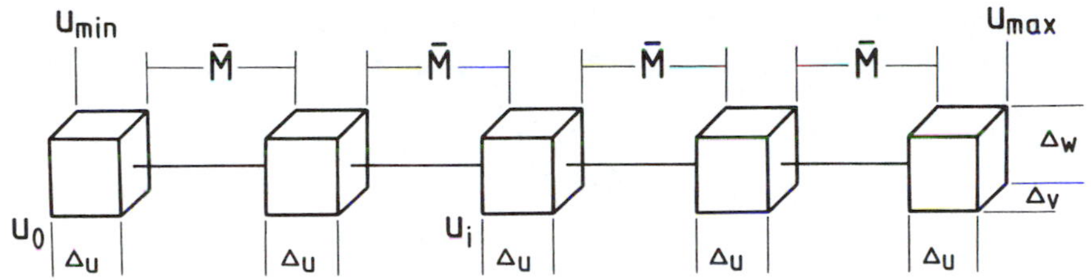

Fig. 11.28. Tiny cube method: computing u_i and Δ_u.

The *hexahedron method* [Paj 90] corresponds to the choice $M = 0$. Using this method, we use function values at the centers of gravity of some of the boxes, but not others (like turning light bulbs on or off). In this case, a box lying in the interior can be visible, provided those boxes between it and the viewer are "turned off". For this method, there is essentially no constraint on the size or number of boxes used.

In the *vanishing cube method* [Nie 90, 91a], we embed $N_u + 1$ equally spaced planes in the u direction, $N_v + 1$ in the v direction, and $N_w + 1$ in the w direction in the domain. This creates $3N_uN_vN_w$ rectangles, each of which is colored using bilinear interpolation of the values of $X(u, v, w)$ at its four corners. Using this method it is possible to view the interior if the exterior planes are rendered as partially transparent. In practice, the values of the N's cannot be too large.

Volume rendering [Levo 88, 90, 90a], [Dre 88], [Ups 88], and [Sabe 88] involves "shooting" rays through the object. We then "accumulate" information about the 3D field along the ray, and use it to define the color and intensity to be associated with the ray. To simplify the required summation/integration process, usually the object is divided into a large number of uniformly sized boxes called *voxels*, each of which is assigned a constant function value. We can improve on the method by using trilinear interpolation of the eight vertices within each voxel. The method is computationally expensive in any case.

[Krü 90] has suggested a method based on *linear transport theory* which is a generalization of the volume rendering method. Here the rays are replaced by *virtual particles* which can interact with the 3D field (which is represented by a fine point grid) in the sense of linear transport theory (e.g., absorption, scattering, sources, color shift, etc.) This method is also computationally very expensive.

Finally, we remark that interactively *rotating* the object (in order to see it from different viewpoints), *zooming* in on the object (in order to see the details better), and utilizing grey-scales or *color codes* (to differentiate and contrast between different regions) can greatly enhance our understanding of the image of an object. We would like to refer the reader to the proceedings [NIE 90, 91], [KAUF 90, 90a], [FRÜ 91], and the paper [Nie 91b].

12 Intersections of Curves and Surfaces

In many applications, we need to find the points (curves) where two curves (surfaces) intersect, for example

- in constructing a contour map to graphically represent a prescribed surface or hypersurface,
- in computing silhouettes to improve the graphical display of surfaces,
- in performing Boolean operations on solid bodies,
- in constructing smooth blending curves and surfaces for rounding off corners and edges between two curves or surfaces,
- in finding offset curves and surfaces, e.g., for NC-construction (theoretically defined offsets can have self-intersections), see Chap. 16.

A good (surface) intersection algorithm should have the following properties:

- *accuracy*, in the usual numerical sense,
- *robustness*, in the sense that all intersections are correctly identified, even when they decompose into pieces,
- *speed*, in particular in interactive settings,
- *self control*, in the sense that it does not require interactive help from a user.

Although presently there is no single algorithm which is optimal with respect to all four of these characteristics, there are many available algorithms. In this chapter we give an overview and, to the extent that tests on the different methods have been made, a comparison of several of the available intersection algorithms, see [Prat 86], [Sed 86], [Luk 89], [Dok 89], [Azi 90], [Barn 90a], and [Boe 91].

The difficulty of the intersection problem, and the type of algorithm which best suits it, depends on the general form of the curves or surfaces we are dealing with (implicit, explicit, parametrized), along with their particular properties (special or free-form). Thus, for example, the intersection of two planes is a straight line, while a plane and a quadric intersect in either a straight line or a conic section. In these simple cases, the calculation can be carried out directly. In contrast, the curve of intersection of two quadrics can already be an algebraic curve of fourth degree which is usually dealt with using numerical methods instead of Cardan's equation. Because of the special importance of quadrics for design purposes, the development of algorithms for determining intersections of quadrics has been heavily studied in the past by [Wei 66], [Woon 71], [Sab 76], [Lev 76, 79] [Sar 83], [Ock 84], and [Pfe 85], and more recently by [Mil 87] [Gei 88], [Pie 89, 92], [Faro 89a], [Gol 91], [She 91], and [Chi 91]. Most of these methods involve a clever algebraic reformulation of the problem to get a more convenient form for numerical computations. This is always possible, for example, if the surfaces are in special position relative to each other, e.g., when the axes of quadrics are coincident, parallel, orthogonal, or intersecting, i.e., when there is symmetry [Pfe 85], cf. the constructions in constructive geometry involving auxilliary planes and spheres; see e.g., [HOW 51], [MOR 52], [HAW 62], and [SCHÖ 77], and also [HARTM 88], [ADA 88]. In some cases, we can even find intersection curves directly in closed form, see Example 12.1.

In principle, the intersection problem always has a simple solution when both curves or surfaces are given as functions. For example, given two curves $y = f_1(x)$ and $y = f_2(x)$, then the set of intersection points is precisely the set of zeros of the equation

$$f_1(x) - f_2(x) = 0.$$

These zeros can be computed either directly, or by using an appropriate numerical method.

For two surfaces $z = f_1(x, y)$ and $z = f_2(x, y)$, the projection of the intersection curve in the (x, y) plane is given by

$$f_1(x, y) - f_2(x, y) = 0.$$

The projection of the intersection curve onto some other coordinate plane can be found by eliminating the corresponding coordinate, and using these results, the intersection curve itself can be explicitly expressed as a space curve in parametric form.

Example 12.1. Suppose we are given a circular cylinder and a sphere

$$(x + \tfrac{r}{2})^2 + y^2 = b^2, \qquad \text{and} \qquad x^2 + y^2 + z^2 = r^2, \qquad b > r.$$

The projection of the intersection curve into the (x, z) plane can be found by eliminating y from this pair of equations. This leads to the equation

$$z^2 - rx = \tfrac{5}{4}r^2 - b^2$$

for a parabola. The projection of the intersection curve into the (y, z) plane is given by

$$r^2(b^2 - y^2) = \left(\tfrac{3}{4}r^2 - b^2 - z^2\right)^2 .$$

The intersection curve can be written in parametric form as (see Fig. 12.1)

$$x = r\cos u \cos v, \quad y = r\sin u \cos v, \quad z = r\sin v,$$

with

$$\cos v = \frac{-\cos u}{2} + \sqrt{\frac{\cos^2 u}{4} + \left(\frac{b^2}{r^2} - \frac{1}{4}\right)}.$$

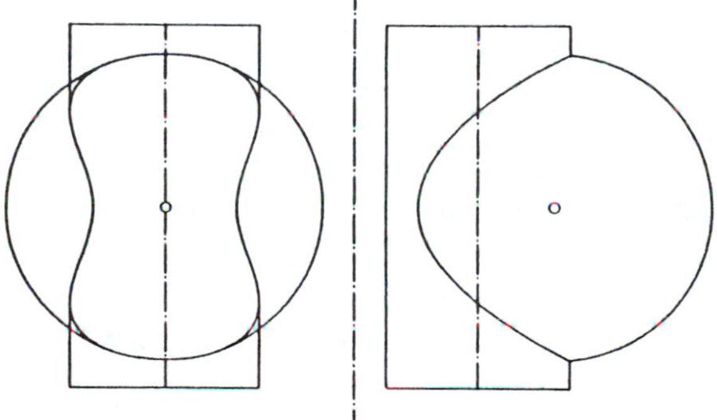

Fig. 12.1. Two projections of the intersection of a cylinder and sphere.

If both curves (surfaces) are given as non-decomposable implicit equations $f_1(x, y) = 0$ and $f_2(x, y) = 0$ ($f_1(x, y, z) = 0$ and $f_2(x, y, z) = 0$), then the intersection points (curves) can be found as solution of the corresponding nonlinear system of equations $f_1(x, y) = 0$, $f_2(x, y) = 0$ ($f_1(x, y, z) = 0$, $f_2(x, y, z) = 0$). To solve this system, we can use numerical methods such as the Newton-Raphson method (cf. e.g., [FAU 81]), differential geometric methods [Ast 88], or a combination of geometric and analytic methods (cf. e.g., [Owe 87]).

The intersection problem is also easy to solve when we are given one plane curve (or a surface) in implicit form $f(x, y) = 0$ ($f(x, y, z) = 0$), and the second in parametric form. In this case we can insert the parametric formula into the implicit form to get a (nonlinear) equation $f(x(t), y(t)) = 0$ $(f(x(u,v), y(u,v), z(u,v)) = 0)$ for the intersection, which, e.g., can be solved using Newton's method. Thus, in this case we have the methods:

a) curve $k_1 \cap$ curve k_2:

$$k_1 : f(x,y) = 0, \qquad k_2 : x = x(t), y = y(t)$$

leads to $f(x(t), y(t)) = 0$, whose zeros $t = t_i$ give us the intersection points $\boldsymbol{P}_i = (x(t_i), y(t_i))$.

b) surface $\Phi_1 \cap$ surface Φ_2:

$$\Phi_1 : f(x,y,z) = 0, \qquad \Phi : \boldsymbol{X} = (x(u,v), y(u,v), z(u,v))$$

leads to $f(x(u,v), y(u,v), z(u,v)) = 0$. Now e.g., we set $u = u_0$ and find the zeros $v = v_i$. We repeat this for a series of equally spaced u values, resulting in a set of intersection points $\boldsymbol{X}(u_k, v_i)$. If we sort these points and appropriately interpolate or approximate, we get approximations to the intersection curves, see Sect. 12.4.2.

A large collection of programs for finding intersections of simple surfaces and solid objects can be found in [ADA 88], [HARTM 88], and [END 89]. The relatively simple situations dealt with there do not often occur in practice, however. Usually both curves or surfaces are given in parametric form or as free-form curves or surfaces. Thus, in the following, we occupy ourselves primarily with intersection algorithms for this more general case. Our aim is to present some of the available methods and to compare them, see also [Prat 86], [Luk 89].

12.1. Algebraic Methods

As we already have seen above, the mathematical nature of the intersection problem is "very simple" when one plane curve or surface is given implicitly while the other is given in parametric form. In view of this, if both curves or surfaces are given in parametric form, it is natural to try to convert one of them to implicit form. In the literature [Sed 83, 84, 85a, 87], [Gol 85, 87a], this idea, which comes from algebraic geometry (cf. e.g., [SAL 85], [WALK 50]) is called *implicitization*. For a given plane curve, this involves determining the implicit polynomial formula $f(x, y) = 0$ by eliminating the parameter t from the given rational polynomial parametric representation

$$x = \frac{x(t)}{w(t)}, \qquad y = \frac{y(t)}{w(t)},$$

of the curve. For a surface, we find $f(x, y, z) = 0$ by eliminating both of the parameters s and t from the parametric representation

$$x = \frac{x(s,t)}{w(s,t)}, \quad y = \frac{y(s,t)}{w(s,t)}, \quad z = \frac{z(s,t)}{w(s,t)}$$

of the surface. Implicitization is always possible for rational polynomial representations, see e.g., [Sed 84], [Chio 92]. The converse of this assertion does not hold. An implicitly given curve or surface cannot generally be written exactly in rational polynomial parametric form. For example, exact conversion is only possible for: linear and quadratic curves and surfaces; for higher polynomial degrees only in very special cases; for example, for cubic curves of genus zero, where the genus of an algebraic curve depends on its degree n and the number of singular points, counting multiplicities, see [Abh 88], [Kat 88]. The problems of finding either exact or approximate parametrizations of algebraic curves have been studied by [Abh 87, 87a, 88], [Sed 89a], and [Wag 89].

Given a rational curve $\boldsymbol{X}(t) = (x(t), y(t))$, where

$$x(t) = \frac{\sum_{i=0}^{n} a_i t^i}{\sum_{i=0}^{n} c_i t^i}, \qquad y(t) = \frac{\sum_{i=0}^{n} b_i t^i}{\sum_{i=0}^{n} c_i t^i},$$

then it can be written in implicit form as (see e.g., [Sed 87])

$$f(x,y) = \begin{vmatrix} L_{n-1,n-1}(x,y) & \cdots & L_{0,n-1}(x,y) \\ \vdots & & \vdots \\ L_{n-1,0}(x,y) & \cdots & L_{0,0}(x,y) \end{vmatrix} = 0, \tag{12.1}$$

with

$$\begin{aligned} L_{ij}(x,y) &= \alpha_{i,j} x + \beta_{i,j} y + \gamma_{i,j} \\ &= \sum_{\substack{l \le \min(i,j) \\ l+m=i+j+1}} (b_m c_l - c_m b_l)x + (a_l c_m - a_m c_l)y + (a_m b_l - a_l b_m). \end{aligned} \tag{12.2}$$

The determinant $f(x, y)$ in (12.1)-(12.2) is called the (Bezout) resultant, cf. Chap. 13. From (12.2) we deduce that the implicit formula for a curve has the same polynomial degree as the original parametric formula.

Example 12.2. For the parabola $x = t^2 - 1$, $y = t^2 - 2t + 2$, we get the implicit representation

$$f(x,y) = \begin{vmatrix} -2 & x-y+3 \\ x-y+3 & -2x-2 \end{vmatrix} = -x^2 + 2xy - y^2 - 2x + 6y - 5.$$

Example 12.3. For the rational cubic parabola,

$$x = \frac{2t^3 - 18t^2 + 18t + 4}{-3t^2 + 3t + 1}, \qquad y = \frac{39t^3 - 69t^2 + 33t + 1}{-3t^2 + 3t + 1},$$

we get the implicit representation

$$f(x,y) = \begin{vmatrix} -117x + 69y + 564 & 117x - 6y - 636 & 39x - 2y - 154 \\ 117x - 6y - 636 & -69x - 2y + 494 & -66x + 6y + 258 \\ 39x - 2y - 154 & -66x - 6y + 258 & 30x - 6y - 114 \end{vmatrix}$$

$$= -156195x^3 + 60426x^2y - 7056xy^2 + 224y^3 + 2188998x^2 - 562500xy + 33168y^2 - 10175796x + 1322088y + 15631624.$$

In certain cases (see [Chio 92]), the implicitization of surfaces given in parametric form can be accomplished in the same way [Sed 84]. A triangular surface of degree n has an implicit formula of degree n^2, and a tensor-product surface of degree (m, n) has an implicit formula of degree $2mn$. A bicubic surface patch thus results in an equation of degree 18 with 1330 terms! Because of this algebraic complexity, implicitization for surfaces via parametric elimination can usually be done only with the help of formula manipulation software such as Mathematica or Reduce.

Usually, we need to know not only the coordinates of points of intersections of two curves or surfaces, but also the values of the corresponding parameters. Thus, it is frequently necessary to perform an *inversion*, i.e., given the coordinates of an intersection point on a curve (surface), we have to calculate the corresponding parametric value(s). For a $\boldsymbol{P}_0$ with coordinates (x_0, y_0) lying on a planar curve $\boldsymbol{X}(t)$, to find the parameter value $t = t_0$ with $\boldsymbol{X}(t_0) = \boldsymbol{P}_0$, we must solve the linear system of equations (see e.g., [Sed 87])

$$f(x,y) = \begin{pmatrix} L_{0,0}(x,y) & \cdots & L_{0,n-1}(x,y) \\ \vdots & & \vdots \\ L_{n-1,0}(x,y) & \cdots & L_{n-1,n-1}(x,y) \end{pmatrix} \begin{pmatrix} t^{n-1} \\ t^{n-2} \\ \vdots \\ t \\ 1 \end{pmatrix} = 0. \qquad (12.3)$$

Example 12.4. For the parabola in Example 12.2, we solve

$$\begin{pmatrix} -2 & x - y + 3 \\ x - y + 3 & -2x - 2 \end{pmatrix} \begin{pmatrix} t \\ 1 \end{pmatrix} = 0,$$

to get

$$t = \frac{x - y + 3}{2} \qquad \text{and} \qquad t = \frac{2x + 2}{x - y + 3}.$$

The two equations lead to the same parameter values for points lying on the curve.

Example 12.5. For the cubic rational parabola in Example 12.3, we have

$$t = \frac{\begin{vmatrix} 117x - 6y - 636 & 39x - 2y - 154 \\ -69x - 2y + 494 & -66x + 6y + 258 \end{vmatrix}}{\begin{vmatrix} -117x + 69y + 564 & 39x - 2y - 154 \\ 117x - 6y - 636 & -66x + 6y + 258 \end{vmatrix}}.$$

Under certain restrictions (see [Chio 92]), the inversion problem for surfaces can be solved in a similar way, see e.g., [Sed 83, 84].

An *algorithm* for computing the intersection points of two planar rational polynomial curves of degrees n_1 and n_2 given in parametric form by

$$\boldsymbol{X}_1(t) = \boldsymbol{X}_1(x(t), y(t)) : \qquad x(t) = \frac{x_1(t)}{w_1(t)}, \qquad y(t) = \frac{y_1(t)}{w_1(t)}, \qquad 0 \le t \le 1,$$

and

$$\boldsymbol{X}_2(\tau) = \boldsymbol{X}_2(\xi(\tau), \eta(\tau)) : \qquad \xi(\tau) = \frac{\xi_2(\tau)}{\omega_2(\tau)}, \qquad \eta(\tau) = \frac{\eta_2(\tau)}{\omega_2(\tau)}, \qquad 0 \le \tau \le 1$$

can be accomplished as follows [Sed 85a, 87]:

- compute the implicit representation of $\boldsymbol{X}_1(t)$ of the form $f_1(x, y, w) = 0$, where f_1 is a homogeneous polynomial and $w(t)$ is a homogeneous coordinate,
- insert the parametric formula for $\boldsymbol{X}_2(\tau)$ into the implicit formula for $\boldsymbol{X}_1(t)$. This gives $f_1(\xi(\tau), \eta(\tau), \omega(\tau)) = 0$, where the polynomial degree of f_1 is $n_1 n_2$,
- compute the real solutions of the polynomial $f_1(\xi(\tau), \eta(\tau), \omega(\tau))$ in the domain $0 \le \tau \le 1$,
- use the parametric formula $\boldsymbol{X}_2(\tau)$ and the parameter values τ_i computed above which correspond to the intersection points $\boldsymbol{S}_i = \boldsymbol{X}_2(\tau_i)$ to find the (x, y) coordinates of the intersection points,
- perform inversion on $\boldsymbol{X}_1(t)$ to get the parameter values t_i such that $\boldsymbol{X}_1(t_i) = \boldsymbol{S}_i$. Only those points with $0 \le t \le 1$ belong to the solution set.

We can perform preprocessing to greatly simplify the determinantal expansion in (12.1) and the inversion calculation (12.3). This involves introducing zeros in the first column and row of the determinant (12.1) (with the exception of L_{00}, L_{01} and L_{10}), see [Sed 86]. Nevertheless, the algebraic complexity of the method grows rapidly. In addition, for high polynomial degrees, the algebraic method leads to significant numerical problems. The expansion of the determinant (12.1) for a curve of degree n leads to an equation of degree n^2, whose zeros are to be found. Thus, the available software for finding zeros of polynomials has an essential influence on the *robustness*, *accuracy*, and *efficiency* of the algorithm. While the algorithm is still extremely effective for $n = 2, 3$, already for $n = 4$ we need to take advantage of some special tricks, even when the computation is all done in double precision. In particular, we expand the L_{ij} in terms of Bernstein polynomials because of their numerical stability, see [Sed 86]. Because of the associated numerical problems, results using the algorithm for $n \geq 5$ have not yet been obtained [Sed 90], but see also [Hob 91].

The above algorithm can also be used to compute the intersection points for curves in $\mathbb{R}^3$, see [Gol 85]. First we preprocess both curves by projecting them into the x-y plane. We then find the intersection of the resulting projected curves using the above algorithm. Since the projected curves can have more intersection points than the original space curves, we then need a postprocess (e.g., comparing the z-coordinates) to eliminate the pseudo intersection points, see [Gol 87a]. Since this algorithm almost always generates undesired pseudo intersection points, [Gol 87a] has suggested an alternative approach using *resolvents* in which pseudo intersection points can be recognized early.

By *Bezout's Theorem*, two planar algebraic curves of degrees n_1 and n_2 (which could be the planar projections of two space curves) have at most $n_1 n_2$ intersection points. This does not hold, however, for curves in $\mathbb{R}^n$ with $n \geq 3$. As shown in [Gol 85], [Chan 87a], for example, two cubic space curves can have at most 5 (and not 9) intersection points. Upper bounds on the maximum number of intersection points of two curves in $\mathbb{R}^n$ can be found in [Abh 89, 90a].

As remarked above, these algebraic methods can be generalized to surfaces, see e.g., [Sed 84]. For example, using a resultant going back to Sylvester, [Wei 66] computes the intersections of two quadrics, while [Woon 71] concentrates on finding their planar intersection curves. The computation of nonplanar intersection curves of quadrics was treated in [Mil 87], and [Ock 84] used algebraic decompositions to find intersections. [Faro 89a] considered degenerate (i.e., decomposable) quadric intersections, using a new factorization method. [Chio 91] also computed the intersection set of quadrics, using a resultant defined by Macaulay.

The intersection curves for general surfaces given in parametric form can be found using an algorithm of [Hos 87a] involving double parameter elimination. Suppose we are given two parametrized surfaces Φ_1, Φ_2, where Φ_1 is expressed in parametric form $\boldsymbol{X}_1(u,v) = \boldsymbol{X}_1(x(u,v), y(u,v), z(u,v))$, while Φ_2 is a tensor-product surface of degree (m,n) expressed in terms of a *monomial basis*, i.e., the parametric formula $\boldsymbol{X}_2(\mu,\nu) = \boldsymbol{X}_2(\xi(\mu,\nu), \eta(\mu,\nu), \zeta(\mu,\nu))$ for Φ_2 is given by

$$\boldsymbol{X}_2(\mu,\nu) = \sum_{i=0}^{m}\sum_{k=0}^{n} \boldsymbol{A}_{ik}\mu^i\nu^k, \qquad \mu,\nu \in [0,1].$$

Suppose the components of the (vector-valued) coefficients $\boldsymbol{A}_{ik}$ are denoted by a_{ik}, b_{ik}, c_{ik}, and suppose

$$\alpha_{1k}(\mu) = \sum_{i=0}^{m} a_{ik}\mu^i, \quad \alpha_{2k}(\mu) = \sum_{i=0}^{m} b_{ik}\mu^i, \quad \alpha_{3k}(\mu) = \sum_{i=0}^{m} c_{ik}\mu^i.$$

Then the intersection curve $\Phi_1 \cap \Phi_2$ is described by the following conditions:

$$x(u,v) - \sum_{k=0}^{n} \alpha_{1k}\nu^k = 0, \quad y(u,v) - \sum_{k=0}^{n} \alpha_{2k}\nu^k = 0, \quad z(u,v) - \sum_{k=0}^{n} \alpha_{3k}\nu^k = 0. \tag{12.4}$$

Now introducing the vectors

$$\boldsymbol{B}_0 = \begin{pmatrix} \alpha_{10}(\mu) - x(u,v) \\ \alpha_{20}(\mu) - y(u,v) \end{pmatrix}, \quad \boldsymbol{B}_k = \begin{pmatrix} \alpha_{1k}(\mu) \\ \alpha_{2k}(\mu) \end{pmatrix},$$

$$\boldsymbol{C}_0 = \begin{pmatrix} \alpha_{20}(\mu) - y(u,v) \\ \alpha_{30}(\mu) - z(u,v) \end{pmatrix}, \quad \boldsymbol{C}_k = \begin{pmatrix} \alpha_{2k}(\mu) \\ \alpha_{3k}(\mu) \end{pmatrix},$$

for $k = 1(1)n$, the intersection condition (12.4) becomes

$$\boldsymbol{P} = \boldsymbol{B}_0 - \sum_{k=1}^{n} \boldsymbol{B}_k\nu^k = 0, \quad \boldsymbol{Q} = \boldsymbol{C}_0 - \sum_{k=1}^{n} \boldsymbol{C}_k\nu^k = 0. \tag{12.4a}$$

The points on the intersection curve $\Phi_1 \cap \Phi_2$ are the common zeros of the equations in (12.4a). Using Bezout elimination, we can now eliminate the parameter ν from both equations in (12.4a). The associated determinant leads to two (real-valued) polynomials $\widetilde{P}$, $\widetilde{Q}$ of maximal degree $(2mn)$ in μ. The common zeros of the polynomials $\widetilde{P}$, $\widetilde{Q}$ can be found by another parameter elimination, or using a numerically less complicated method of [Hos 87a].

The intersection problem for curves and surfaces was treated in [Chan 87] with similar methods. [Faro 87a] used a combination of algebraic and analytic methods to compute the intersection curves for an implicit and a parametric surface. Intersections of algebraic and rational surfaces are discussed in [Kri 90] using a hybrid algorithm which combines subdivision (see Sect. 12.2), tracing (see Sect. 12.5), and numerical methods (Newton). [Garr 89] expresses the intersection curve for two algebraic surfaces by means of a plane projection. The problem is that one has to choose the plane and the direction of projection in such a way that the mapping is (almost everywhere) one-to-one.

The problems discussed above for curves, especially with inversion and implicitization, are even more severe for surfaces. This led [Kri 90] to state: "the high degree of the resulting polynomials and the characteristics of the elimination process, involving large scale computation, contribute to substantial loss of accuracy, making the method unattractive for practical application".

A quite different method of [Nef 89], [Bil 89] is based on the theory of *Gröbner bases*, see e.g., [Buch 85], [HOFF 89] for a very readable introduction. This approach allows us to not only solve the intersection, implicitization, and inversion problems, but, e.g., also to compute the singular points of a (planar, implicitly described) intersection curve of two surfaces [Buch 89]. However, the computational time, memory requirements, and especially the interaction rate of the algorithm are still very large, and so a solution of the intersection problem which possesses all four characteristics listed at the beginning of this section still remains elusive.

12.2. Subdivision Methods

A subdivision algorithm divides the objects to be intersected into (many) pieces, after which we look for intersections of the pieces. Often, piecewise linear approximations are used to find approximate intersection points or curves. This linearizes the problem.

Earlier subdivision algorithms divided both of the objects uniformly and completely. For curves this leads to a uniform binary tree, and for surfaces it leads to a uniform quadtree data structure, see Figs. 12.2 and 12.5. These *uniform subdivision algorithms* are quite costly, and require a great deal of memory [Grif 75], and thus are somewhat inefficient.

A significant reduction in computational time and memory requirements is possible if we modify our algorithms to include some kind of criterion to estimate where possible intersections can occur, *i.e.*, for each curve or surface we construct a *bounding box* which completely encloses the curve or surface. Then we can restrict our search for intersection points to these boxes using a *separability test*, see Fig. 12.3. When the boxes do not overlap, there is no

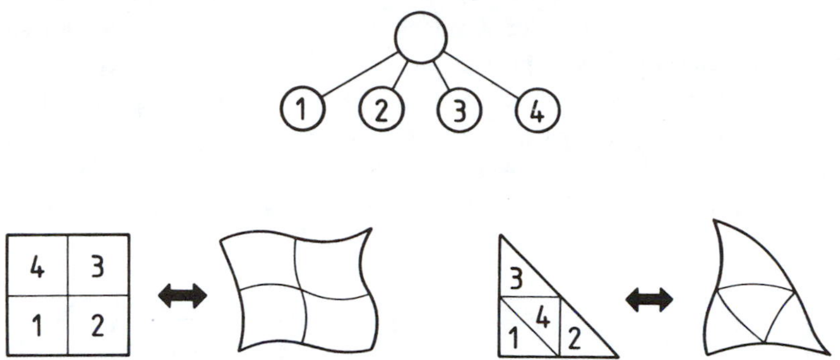

Fig. 12.2. Quadtree data structure for quadrilateral and triangular surfaces.

intersection, and the program stops. When the boxes do overlap, we subdivide both curves or surfaces, and repeat the process.

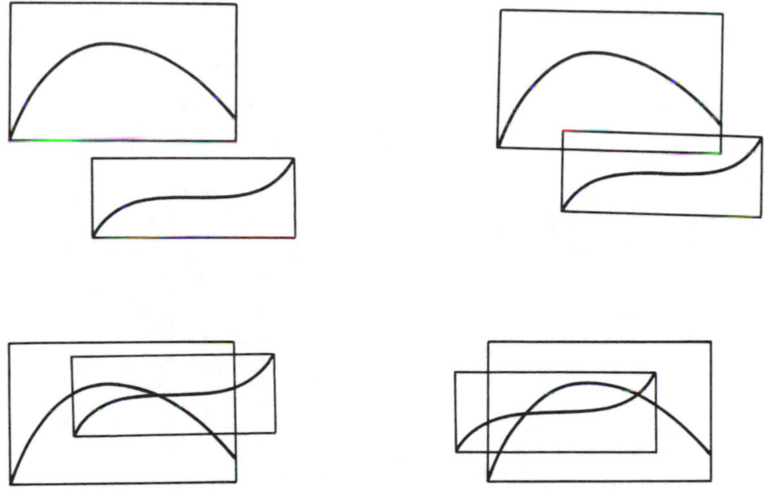

Fig. 12.3. Separability test for four curve pairs with associated bounding boxes.

This process is a non-uniform, adaptive subdivision algorithm. By successively refining the structure and removing the separable boxes (divide-and-conquer principle) we will eventually end up with a set of boxes which surround the intersection, and whose sizes can be made smaller than any prescribed tolerance. These boxes can be used to define a piecewise linear approximation of the curves or surfaces, respectively. Then we can easily construct a piecewise linear approximation to the desired intersection by intersecting the lines (resp. planes), see [Lane 80], [Coh 80]. The overall adaptive algorithm based on this *divide-and-conquer* principle is diagrammed in Fig. 12.4. For curves it generates a non-uniform binary tree, and for surfaces we get a non-uniform quadtree, see e.g., [Car 82] (also [Barn 90a], [Deh 91]) and Fig. 12.5.

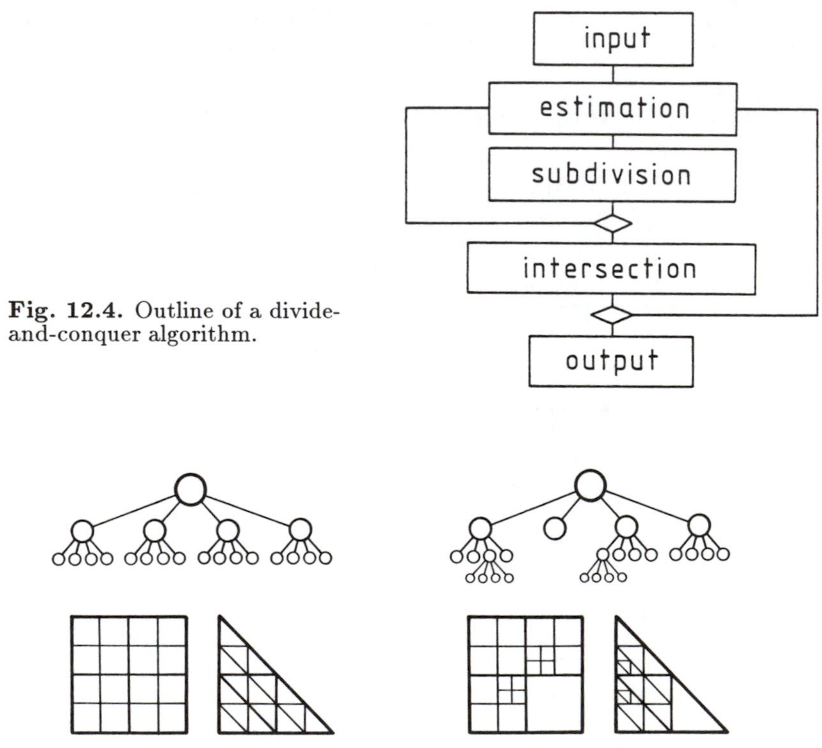

Fig. 12.4. Outline of a divide-and-conquer algorithm.

Fig. 12.5. Uniform and non-uniform quadtrees.

Divide-and-conquer algorithms are quite easy to program, since for example, they do not involve any starting point problem, such as arises for the *marching algorithms* discussed below. The divide-and-conquer strategy always produces all of the pieces of the intersection (up to the prescribed accuracy), without any interactive help from the user, see Fig. 12.6. Decomposable intersection curves are no problem for the subdivision algorithm, see Fig. 12.21. However, to *increase the accuracy*, we have to perform additional subdivision. Since in the case of two surfaces, for example, the intersecting curve segments are found in a (relatively) random order, *sorting* them so that we can construct a continuous curve is always required, although the quadtree structure can be a great help, see [Barn 90a]. This sorting should always be done in the parameter space of the surfaces, since sorting in coordinate space is more prone to errors.

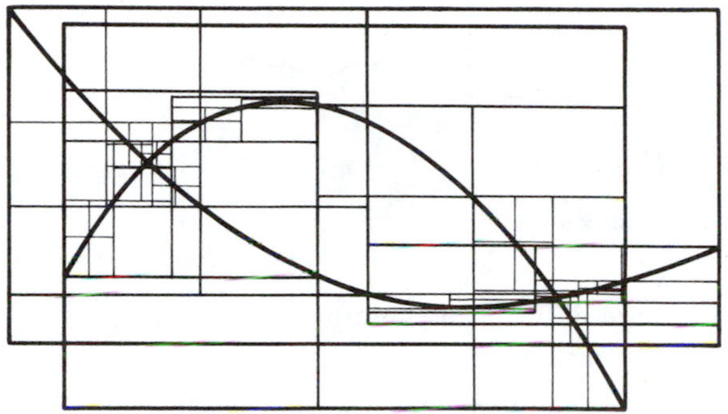

Fig. 12.6. Intersection of two curves using a divide-and-conquer algorithm utilizing min-max boxes.

The difficulty involved in programming a subdivision algorithm, and how fast it runs, depends entirely on the form of the curves or surfaces being intersected. Various representations allow different subdivision strategies and bounding box definitions, which of course lead to entirely different intersection criteria and convergence properties. For example, for representations which do not possess some kind of analog of the *convex hull property* of Bézier representations, it may be a problem to choose bounding boxes which are guaranteed to contain the intersecting curve(s) or surface(s). To solve this problem, some bounding box definitions require a great deal of *preprocessing*.

Almost all methods require an initial computation of several curve or surface points to be used in constructing the bounding boxes. The most commonly used bounding boxes are the *min-max boxes*. Constructing such a box for curves usually requires two end points, while for surfaces we need four points (vertices) describing a quadrilateral in space. A *min-max box* is then defined as the smallest box (with edges parallel to the axes) which contains the points, see Fig. 12.7.

It is also possible to use *oriented bounding boxes* which can be thought of as min-max boxes which are *rotated into position*, i.e., the edges of the boxes no longer have to be parallel to the axes, see [Bal 81], [Hou 85], [Barn 87, 90a], and Fig. 12.8. Using oriented boxes usually gives better results, but they are,

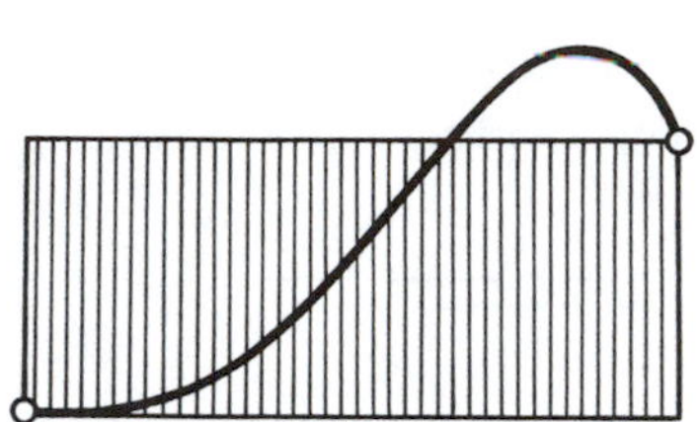

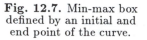

Fig. 12.7. Min-max box defined by an initial and end point of the curve.

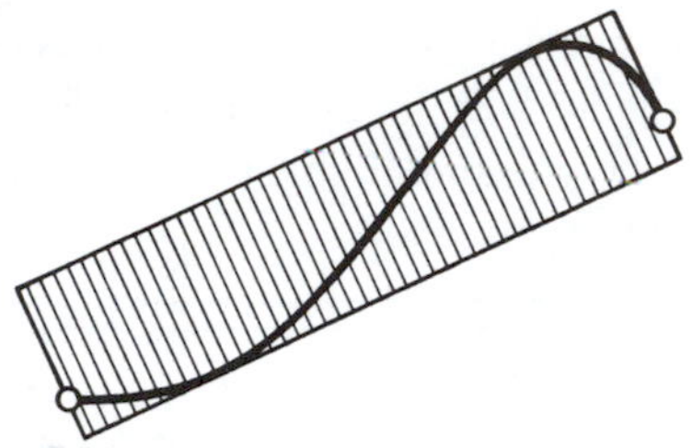

Fig. 12.8. Oriented bounding box.

of course, usually more difficult to construct and more effort is required for performing the separability tests.

It is also possible to use open neighborhoods, defined by up to 6 boundary conditions for objects in space, and up to four boundary conditions for objects in the plane. For example, for plane curves, we can use a pair of parallel lines g_1 and g_2 to define a *strip* or *fat line* of thickness δ, where δ is the distance between the two lines. Similarly, for surfaces, we can use two parallel planes to describe a *fat plane or slab* of thickness δ, see [Car 82], [Kay 86], and [Yen 91]. Slabs are uniquely defined by a normal vector and a thickness δ. By taking the intersection of several oriented slabs, we can very efficiently generate neighborhoods and carry out the separability test [Kay 86], [Yen 91]. Thus, for example, using three (possibly mutually orthogonal) slabs provides a very simple and fast way to construct oriented min-max boxes. The speed-up factor of using *slab boxing* as compared with the traditional way ([Hou 85], [Barn 90a]) of constructing oriented min-max boxes is around two [Yen 91].

Some of the construction methods discussed above do not automatically guarantee that the curve or surface lies entirely inside the associated neighborhood, see [Barn 90a] and also Fig. 12.7, where the min-max box is constructed using two end points of the curve. This means that to get a functioning algorithm, we still have to make certain modifications.

This problem is attacked heuristically in [Hann 83], [Hou 85], and [Barn 87] by introducing a user-selected *magnification factor* which is used to magnify the size of all of the (oriented) min-max boxes. [Lane 79] gave an analytical approach to estimating magnification factors using first derivatives, while both [Wan 84] and [Fil 86a], used second derivatives. First derivatives (the *Jacobi matrix*) were also used by [Herz 90] (and in the case of B-spline surfaces second derivatives by [Nat 90]) in the case of bounding boxes with edges parallel to the axes. [Barn 90a] finds "*almost safe*" magnification factors by using the angle between the normals (or the tangents) at the vertices of the surface segments.

By making a clever choice of curve or surface subdivision points, we can avoid this problem altogether. For example, [Kop 83] uses a *preprocessing step* to select curve points $\boldsymbol{X}(t_i)$ with either horizontal or vertical tangents (when they exist). The resulting pieces then have possible horizontal or vertical tangents only at their end points. We can use the x and y coordinates of the end points $\boldsymbol{X}(t_i)$ to construct *min-max boxes*, see Fig. 12.9.

[Faro 87a, 89] also proposed a *preprocessing step* which selects special curve points, *i.e.*, points where the derivative and differential-geometric characteristics such as curvatures and their derivatives are identically zero. These are then used to construct bounding *boxes* and *triangles* which completely enclose the (monotone) curve segments, see Fig. 12.10.

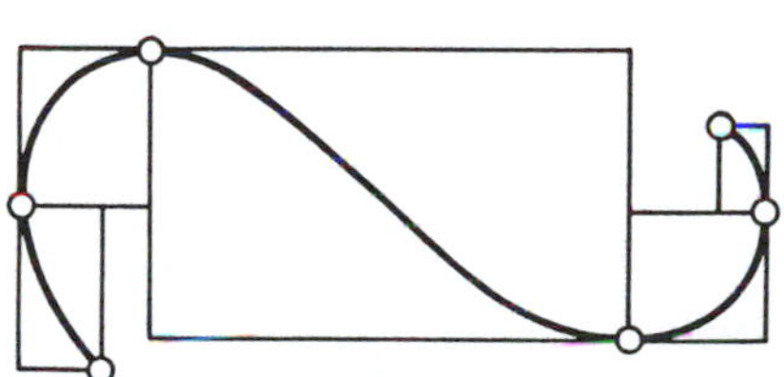

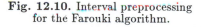

Fig. 12.9. Interval preprocessing for the Koparkar-Mudur algorithm.

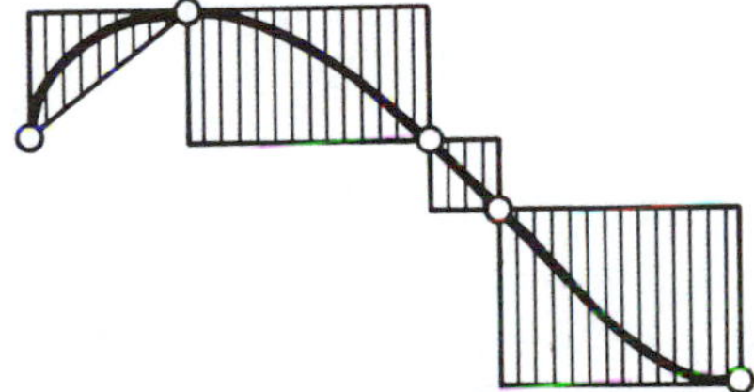

Fig. 12.10. Interval preprocessing for the Farouki algorithm.

Both methods require a relatively expensive preprocessing step to find the zeros of the differential-geometric quantities in [Faro 89], and to solve the nonlinear systems of equations arising in [Kop 83]. The same holds for the method of [Kri 90], where special *significant points* (to be used in the subdivision process) must be found for the parameter space representation $F(u, v) = 0$ describing the intersection curve of an algebraic and a rational polynomial surface. As in [Faro 87a, 89], the monotone pieces of the intersection curve are then computed using a tracing algorithm. These special curve points $F(u_i, v_j) = 0$ of [Kri 90] are defined as *border points*, *i.e.*, points where, e.g., $u_i = 0$ or $v_j = 0$, as *turning points*, *i.e.*, points where $F_u(u_i, v_j) = 0$ (horizontal tangent) or $F_v(u_i, v_j) = 0$ (vertical tangent), and as *singular points*, *i.e.*, points where $F_u(u_i, v_j) = F_v(u_i, v_j) = 0$ (note [Mon 91]).

We now discuss several other methods for constructing neighborhoods. For methods involving a control polygon and whose associated basis functions add up to one for every parameter value, [Gol 86] computes a *magnification factor* which guarantees that the curve or surface lies entirely inside the *magnified convex hull* by estimating the deviation of the control points from a straight line or plane.

[Herz 87] uses a Lipschitz condition to construct *ellipsoidal neighborhoods* which are guaranteed to contain the curve or surface. These are generalizations of the *bounding spheres* which are heavily used in computer graphics, see e.g., [Herz 90]. But since the intersection of two ellipsoids is relatively expensive, it can be advantageous to embed the ellipsoids in (oriented) min-max boxes, and to use these instead of the ellipsoids, even though we do end up with somewhat weaker estimates for the intersection, see also [Bou 85].

[Sab 76a], [Marc 84], and [Ross 84] (see also Sect. 4.1.4) apply circular arcs for the approximation of curves or the intersection curves of two surfaces. If we assign a given thickness δ to each circular arc, we end up with a set of circular *fat arcs* of thickness δ which enclose the intersection curve, see Fig. 12.16. In comparison with min-max boxes, circular fat arcs are very expensive to construct, and relatively expensive to compare with each other. However, they do have a very high convergence rate, see below. Fat arcs are defined by a center point $\boldsymbol{M}_m$, a radius ρ_m, a thickness $\delta = \rho_{max} - \rho_{min}$, and an angle ϕ.

Frequently, in CAGD it is convenient to work with parametrized curves and surfaces which are defined by control polygons with properties similar to those of the Bézier polygons of the *Bézier representation*. For these, we can construct neighborhoods of the type discussed above as follows. First, because of the nature of the Bézier representation, a natural choice for a neighborhood is the *convex hull* of the Bézier points, see Figs. 4.6 and 12.11. Algorithms for finding the convex hull of a point set can be found, e.g., in [Edd 77], [Prep 77, 79], [PREP 85], [Kao 90]. But in view of the complexity of using convex hulls (especially of the separability test), convex hulls are not particularly useful for defining neighborhoods in a subdivision algorithm.

The same holds for the so-called *small convex hulls*. These were defined in [Herr 89] by using special quadratic and cubic basis functions for curves which result in convex hulls containing the entire curve, but which have areas which are only about 77% (for $n = 2$) and 42% (for $n = 3$) of those for the convex hulls of the usual Bézier formulae, see Fig. 12.11.

For objects defined in Bézier form, a *min-max box* can be constructed as the smallest rectangular parallelepiped which contains the convex hull, see Fig. 12.12. Although this gives somewhat weaker estimates than the convex hull, the min-max boxes are extremely simple to implement, and in particular, their construction and the testing for intersections are much faster.

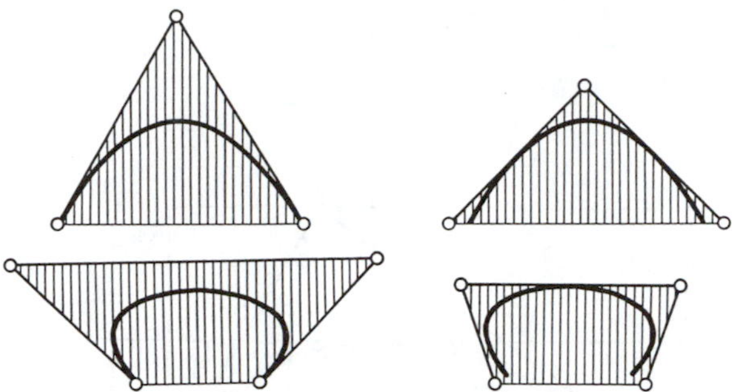

Fig. 12.11. Convex hulls (left) and small convex hulls (right) for quadratic and cubic curves.

It is still faster to construct min-max boxes defined directly from the corner Bézier points. However, to assure that this box entirely encloses the object, we have to enlarge it appropriately. [Fil 86a] discusses how to find a magnification factor using second derivatives. [Gol 86] gives estimates using the maximal deviation of the curve from the convex hull of its control points.

A rectangular *strip* can be defined as the smallest rectangle containing the convex hull of the Bézier points with one edge parallel to the line $\boldsymbol{b}_0\boldsymbol{b}_n$, see [Bal 81] and Fig. 12.13. In the quadratic and cubic cases, it is possible to find still smaller rectangular strips with a small amount of additional effort, see [Sed 89, 90], [Wan 91a], without losing the property that it is an enclosing neighborhood, see Fig. 12.14.

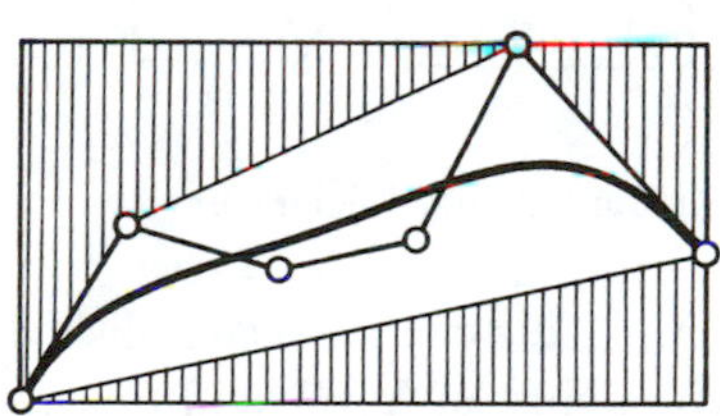

Fig. 12.12. Min-max box for a Bézier curve.

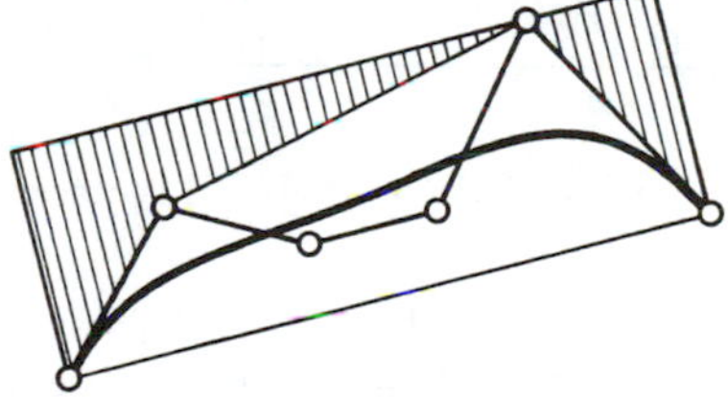

Fig. 12.13. Rectangular strip for a Bézier curve.

Strips of thickness δ (*fat lines*) were investigated for Bézier curves in [Sed 90]. This leads to a very efficient algorithm for finding intersections of curves, where first the Bézier curve $\boldsymbol{X}_1(t)$ is split into pieces using the strip around the second Bézier curve $\boldsymbol{X}_2(\tau)$. Then a new strip is constructed for the middle subsegment of $\boldsymbol{X}_1(t)$, and the clipping process is then applied to the second curve $\boldsymbol{X}_2$, etc. In *Bézier clipping*, subsets of the parameter domain are constructed where it is guaranteed that $\boldsymbol{X}_1$ and $\boldsymbol{X}_2$ do not intersect, e.g.,

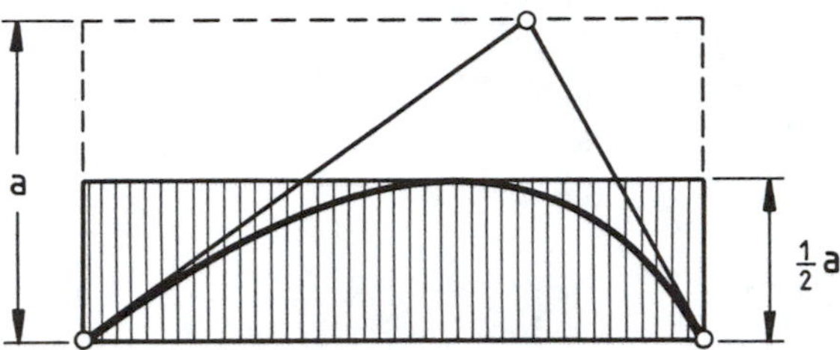

Fig. 12.14a. Reduced rectangular strips for a quadratic Bézier curve.

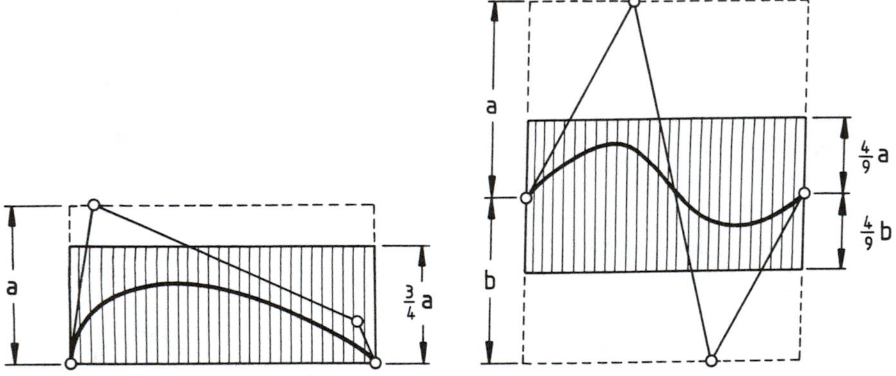

Fig. 12.14b. Reduced rectangular strip for a cubic Bézier curve.

because $\boldsymbol{X}_1$ lies outside of the strip around $\boldsymbol{X}_2$. This can be accomplished as follows: Suppose $\boldsymbol{b}_i = (x_i, y_i)$ denote the Bézier points of $\boldsymbol{X}_1(t)$, and that $\boldsymbol{B}_i = (X_i, Y_i)$ are those of $\boldsymbol{X}_2(\tau)$. Let L be the line joining $\boldsymbol{B}_0\boldsymbol{B}_n$, see Fig. 12.15a:

$$ax + by + c = 0, \qquad \text{with} \qquad a^2 + b^2 = 1.$$

Then the distance $d(t)$ of a point on the curve $\boldsymbol{X}_1(t)$ from L is given by

$$d(t) = \sum_{i=0}^{n} d_i B_i^n(t), \qquad \text{with} \qquad d_i = ax_i + by_i + c.$$

The d_i are the distances of the $\boldsymbol{b}_i$ from L. The function $d(t)$ can be interpreted as a function-valued Bézier curve. To display it graphically, we associate the

Bézier ordinates d_i with the abscissae i/n, see Sect. 4.1 and Fig. 12.15b. The zeros of $d(t) = 0$ correspond to the points where $\boldsymbol{X}_1(t)$ intersects the straight line L. At t's where $d(t) > d_{max}$ or $d(t) < d_{min}$, $\boldsymbol{X}_1(t)$ lies outside of the strip around $\boldsymbol{X}_2(\tau)$, and therefore cannot be a point of intersection with the curve $\boldsymbol{X}_2(\tau)$, see Fig. 12.15a. The points t_{min} and t_{max} where the convex hull of $d(t)$ intersects the strip $d_{min} \leq d \leq d_{max}$ produces a set in the parameter space which is guaranteed not to contain any intersection points of $\boldsymbol{X}_1(t)$ and $\boldsymbol{X}_2(\tau)$. The Bézier clipping process then uses the de Casteljau algorithm to split $\boldsymbol{X}_1(t)$ into three pieces, using $t = t_{min}$ and $t = t_{max}$. Only the middle subsegment of $\boldsymbol{X}_1(t)$ is retained, and its strip is used to split $\boldsymbol{X}_2(\tau)$.

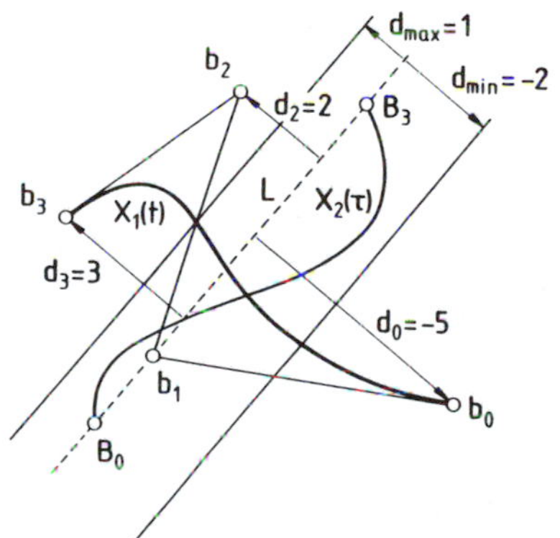

Fig. 12.15a. Bézier curves $\boldsymbol{X}_1, \boldsymbol{X}_2$ and strip for $\boldsymbol{X}_2$.

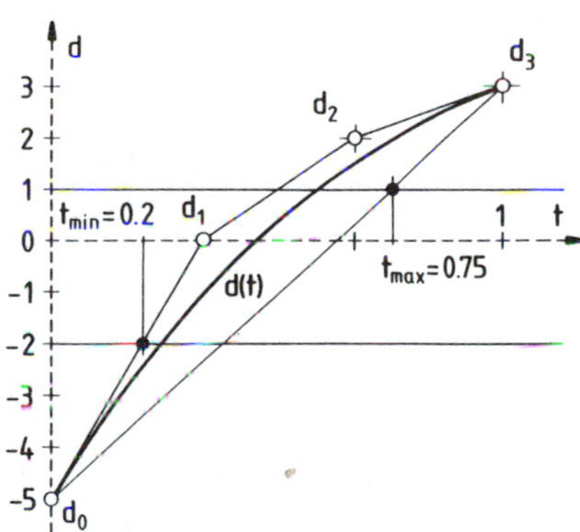

Fig. 12.15b. Function-valued Bézier curve $d(t)$.

Fig. 12.15. Sederberg-Nishita Bézier clipping algorithm.

When two curves intersect at more than one point, then we must subdivide the curves so that any pair of subsegments has at most one point of intersection. Then we can apply the above algorithm to find each intersection point separately.

A divide-and-conquer algorithm based on *fat planes* of thickness δ was presented in [Car 82] for Bézier surfaces.

[Sed 89] constructed *fat arcs* of thickness δ for curves written in Bézier form by taking the center $\boldsymbol{M}_m$ and radius ρ_m from the circle which interpolates the points $\boldsymbol{X}(0)$, $\boldsymbol{X}(\frac{1}{2})$, and $\boldsymbol{X}(1)$. If two of these points coincide, then we must first subdivide the Bézier curve. The exterior and interior radii are equal to the minimum and maximum of

$$d(t) = \|\boldsymbol{X}(t) - \boldsymbol{M}_m\|, \qquad t \in [0,1].$$

If $\boldsymbol{X}(t)$ is a Bézier curve of degree n, then

$$d(t)^2 = \sum_{k=0}^{2n} c_k \binom{2n}{k} (1-t)^{2n-k} t^k.$$

Since

$$\max\{d(t)\} = \sqrt{\max\{d(t)^2\}}, \qquad \min\{d(t)\} = \sqrt{\min\{\max\{d(t)^2, 0\}},$$

for $t \in [0,1]$, it follows from $0 \leq B_k^{2n}(t) \leq 1$ that

$$\rho_{max}^2 = \max\{d(t)^2\} \leq \max_{0 \leq k \leq 2n} \{c_k\}, \qquad \rho_{min}^2 = \min\{d(t)^2\} \geq \min_{0 \leq k \leq 2n} \{c_k\}.$$

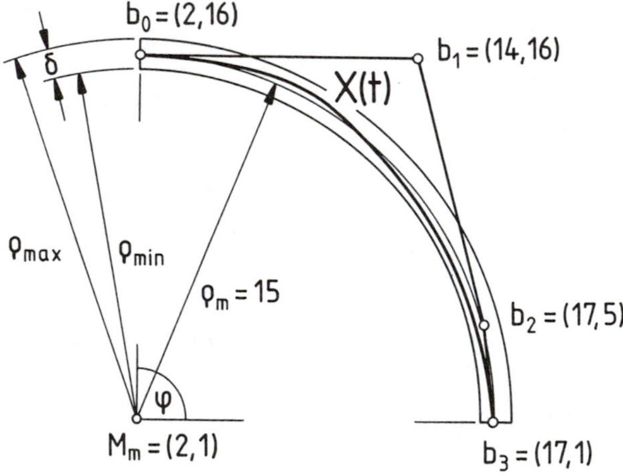

Fig. 12.16. Definition of a fat arc for a Bézier curve.

To compute ρ_{max} and ρ_{min}, we suggest translating the curve so that $\boldsymbol{M}_m$ coincides with the origin of the coordinate system. The angle ϕ (see Fig. 12.16) is defined by the lines $\boldsymbol{M}_m\boldsymbol{b}_0$ and $\boldsymbol{M}_m\boldsymbol{b}_n$, provided that the tangent to $\boldsymbol{X}(t)$ varies by at most 90° as t runs over $[0,1]$. This can be checked by examining the derivative curve $\boldsymbol{X}'(t)$ (the *hodograph* of $\boldsymbol{X}(t)$), see e.g., [For 72], [BEZ 86], and [Sed 88]. If this is not the case, then we must subdivide $\boldsymbol{X}(t)$.

With appropriate modifications, all of the approaches to defining neighborhoods described above can be generalized to space curves and surfaces in

$\mathbb{R}^3$. This is very simple for min-max boxes; we need only add the z-coordinate to get min-max boxes (rectangular parallelepipeds).

In *refining* a collection of bounding boxes for a curve or a rectangular surface, for simplicity, we usually bisect the parameter intervals. Thus, for example, for a curve we would choose the point with parameter value $t_H = (t_i + t_{i+1})/2$, see the left-hand side of Fig. 12.17, and then define the two new refined neighborhoods using the new point $\boldsymbol{X}(t_H)$. For rectangular surfaces, we find the surface curves corresponding to the parameter values u_H and v_H, and use them to construct four new refined rectangular segments, see the right-hand figure in Fig. 12.17 and e.g., [Las 86, 88], [Nat 90]. It is also possible to subdivide a rectangle into two subrectangles, see the center figure in Fig. 12.17, and [Deh 91].

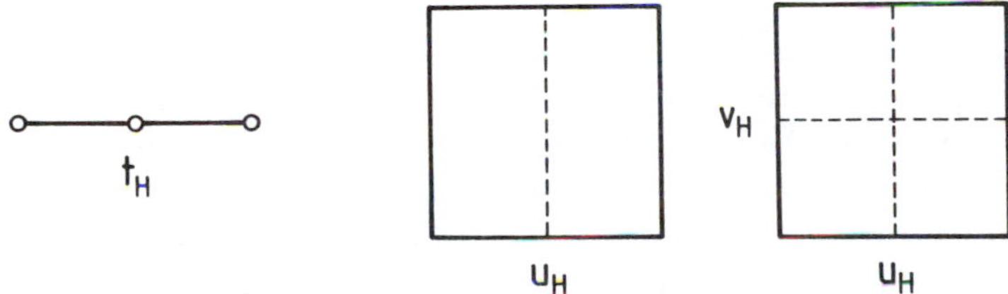

Fig. 12.17. Subdivision of curves and rectangular surfaces.

There are several different subdivision strategies for triangular surfaces. Fig. 12.18 shows subdivision using midpoints of one, two, or three sides of the triangle, or its center of gravity [Gol 83]. The most common methods are to bisect just the longest side of the triangle, to use the center of gravity, or to subdivide into four subtriangles using the midpoints of all three sides. As discussed in [Fil 86], it is better to use an adaptive strategy rather than to use one fixed subdivision method everywhere.

Recent results of [Sed 90], [Nis 90] show that a similar situation persists for curves; it is better to use an *intelligent adaptive subdivision* rather than simply bisecting every time at the midpoint of the interval. Nonuniform subdivision is also used in the *root algorithm* of [Rock 89] for finding the zeros of a given function-valued Bézier curve. The method proceeds sequentially as follows. If all b_i are positive or negative, then it follows from the convex hull property that the curve cannot have any zero. If this is not the case, then the first piece of the Bézier polygon which crosses the t axis is determined by the first pair of Bézier ordinates b_i and b_{i+1} with $b_i b_{i+1} < 0$. Thus, as a first

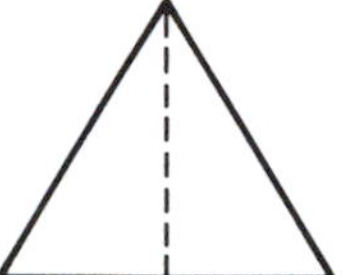
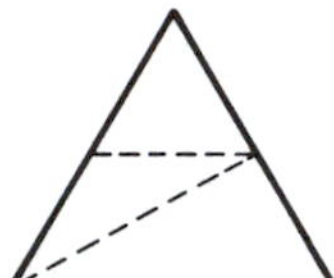
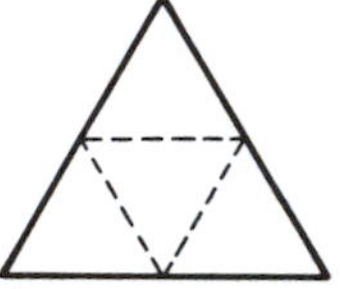
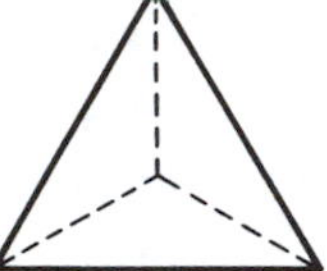

Fig. 12.18. Four possible subdivisions of a triangular surface.

approximation to the smallest zero of the curve, we can take

$$t = \frac{b_i}{n(b_i - b_{i+1})} + \frac{i}{n}.$$

We then subdivide the curve at this parameter value, and repeat the process by looking at the piece defined to the left of this point, and if there is no zero, then at the one on the right. Repeating this process, the algorithm produces a sequence of numbers which converges to the smallest zero. The next largest zero can be found by examining the remaining pieces on the right, etc. The efficiency of this algorithm can be increased by using a *deflation formula* due to R. Chang, see [Rock 90].

Lemma 12.1. (Deflation formula). *Suppose*

$$P(t) = \sum_{i=0}^{n} p_i B_i^n(t) \qquad \text{and} \qquad Q(t) = \sum_{j=0}^{n-1} q_j B_j^{n-1}(t)$$

are Bézier polynomials of degree n and $n-1$. In addition, suppose that $P(0) = 0$ and $P(t) = tQ(t)$. Then

$$q_j = \frac{n}{j+1} p_{j+1}, \qquad j = 0(1)n\text{–}1. \tag{$*$}$$

Proof: $P(0) = 0$ implies $p_0 = 0$. The summation index of $P(t)$ can thus be changed to $1 \le i \le n$, and after shifting the index by $j \to i = j+1$ for $Q(t)$, we get

$$P(t) = t\sum_{i=1}^{n} p_i \binom{n}{i} (1-t)^{n-i} t^{i-1} = t\sum_{i=1}^{n} q_{i-1} \binom{n-1}{i-1} (1-t)^{n-i} t^{i-1} = tQ(t).$$

Comparing coefficients and shifting the indices by $i \to j = i-1$ gives $(*)$. ■

We can now describe the very efficient *all roots algorithm* of [Rock 90]:

- find the smallest zero t_1 of $X(t)$,
- subdivide $X(t)$ at $t = t_1$,
- compute the new Bézier points of the right subsegment via

$$b_j^{new} = \frac{n}{j+1} b_{j+1}^{old}, \qquad j = 0(1)n\text{–}1,$$

- set $n^{new} = n^{old} - 1$,
- repeat until all n zeros are found.

As concerns refinement, we note that splitting an interval for a B-spline representation corresponds to a knot insertion [Coh 80], [Boeh 80, 81], and for a Bézier representation to a subdivision using the de Casteljau algorithm [Lane 80]. Recursive subdivision algorithms for polynomial representations are given in [Gol 86] (note also [Gol 84]).

If some parts of a surface are subdivided more than other parts, it can happen that the resulting linear approximation to the surface (*i.e.*, the Bézier net in the case of a Bézier formula) will exhibit discontinuities (*cracks* or *subdivision gaps*). For ways to eliminate such discontinuities, see e.g., [Hou 85], [Bars 87], and [Deh 91].

Refinement algorithms need some kind of *termination criterion.* [Cat 74] subdivides until the pieces are the size of pixels, but this is often too much. In [Nyd 72], the difference between the normals $\boldsymbol{N}_i$ at the vertices of the subsegments are used to define a termination criterion. This criterion is highly local (only the vertices are involved), and can therefore lead to bad results. [Herz 87] suggested also using the differences in the tangent vectors $\boldsymbol{T}_i$ at the vertices. Normals distributed uniformly over the surface (*flatness test*) and tangents at the vertices (*edge linearity test*) are used in [Hou 85] and also in [Barn 90a], see Fig. 12.19. The absolute value of the second partial derivatives is used by [Cla 79] to decide whether further subdivision is needed or not. Second derivatives are also used in [Wan 84], [Fil 86a], and [Nat 90]. Frequently, (see e.g., [Lane 79, 80, 80a], [Bars 87]) the flatness test is carried out by checking to see if the deviation from linear or planar form is within a prescribed tolerance. This can be done by computing the thickness δ of the neighborhood being used, *i.e.*, for Bézier representations the thickness of the convex hull of the Bézier points (*convex hull flatness test*), see Fig. 12.20 and [Car 82], [Las 86], and [Nas 87]. With a little additional effort, the termination criterion in Fig. 12.20 can be somewhat improved for rational quadratic and cubic Bézier curves, see [Sed 89, 90], [Wan 91a], and Fig. 12.14.

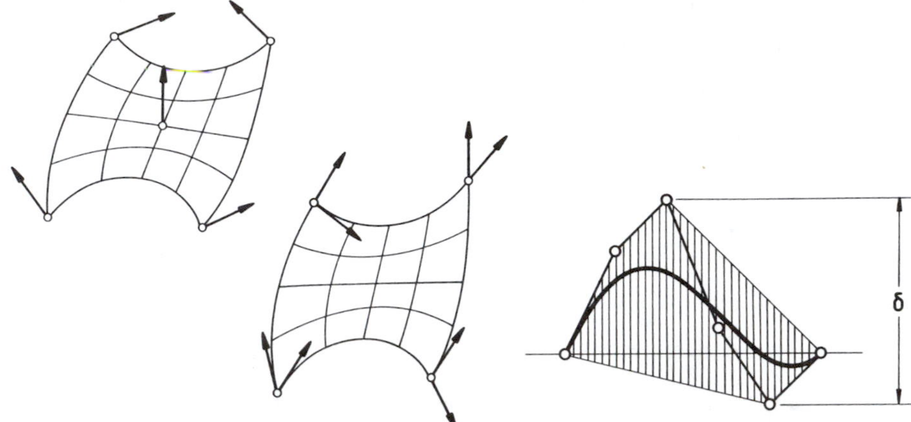

Fig. 12.19. Flatness and edge linearity test.

Fig. 12.20. Convex hull flatness test.

Both the flatness test and the subdivision steps can require a considerable amount of computation. For polynomial representations, it may be desirable to perform a *degree reduction of the segments*, *i.e.*, to approximate them by segments with smaller polynomial degrees. This reduces the overall computation significantly, and in particular that needed for the flatness test and the subdivision. Degree reduction is used, e.g., in [Pete 84, 87]. It should be mentioned though, that the test to decide whether degree reduction is possible can itself be expensive. Thus, degree reduction cannot be applied for all representations or with all subdivision methods [Fil 86]. For example, for Bézier representations, the convex hull of the degree-reduced formula is in general larger than the original one. Hence, degree reduction can even result in a *decrease in overall performance* [Sed 86], and so cannot be recommended in general.

It seems to be best not to perform a flatness test at first, but instead to find a preliminary estimate of how many subdivisions will be required to satisfy the flatness test, see [Wan 84], [Fil 86a], [Sed 86], and [Nat 90].

A *comparison* of several neighborhood definitions can be found in [Sed 89] for cubic Bézier curves. It turns out that circular fat arcs are clearly superior in terms of "*convergence rate*". In the limit, the area of the circular arc surface is reduced by about a factor of eight for each subdivision step. The factor for convex hulls or rectangular strips is about four, while for min-max boxes, it is only two. However, circular fat arcs require more work to construct and to intersect. On the other hand, while min-max boxes have the worst convergence rate, they are extremely simple to construct and compare with each other as shown in Table 12.1, which comes from [Sed 89]. For curves of degree $n > 3$, the amount of computation for min-max boxes and rectangular strips grows linearly, while for circular fat arcs and convex hulls, it grows faster. The numerical entries in the table give the number of $(*/, +-, <>)$ operations needed to compute (create) and intersect (separability test) the various neighborhoods for a planar cubic Bézier curve. The last row contains a comparison of the convergence rates.

	Fat arcs	Min-max box	Convex hulls	Strips
Computation	(54,42,10)	(0,0,12)	(12,30,3)	(15,10,4)
Intersection	(8,18,8)	(0,0,4)	expensive	(16,24,16)
Convergence	$\mathcal{O}(n^3)$	$\mathcal{O}(n)$	$\mathcal{O}(n^2)$	$\mathcal{O}(n^2)$

Table 12.1. Comparison of four neighborhood definitions.

A *comparison* [Rock 90] of various divide-and-conquer algorithms for curves shows that Rockwood's *all root algorithm* and Sederberg-Nishita's *Bézier clipping algorithm* are the best performers. For $n \leq 5$ they are between 2 and 10 times faster than Koparkar-Mudur's *interval bisection algorithm* [Sed 90], which in turn is somewhat faster than Lane-Riesenfeld's *min-max box algorithm* [Sed 86, 90]. These comparison times depend, of course, on the degree of the curves and also on the particular example being tested. For example, for $n \leq 5$, the interval bisection algorithm is faster than the min-max box algorithm, but because it requires extensive preprocessing, for larger values of n it becomes comparatively less efficient, [Sed 86]. In spite of the fact that the min-max box algorithm does not compare so favorably with other methods, it is nevertheless heavily used. This is due in part to the fact that it is independent of the polynomial degree n, and is easy to implement for curves as well as for surfaces (for both rational and non-rational representations).

A comparison of algebraic methods with subdivision methods [Sed 86, 90], [Rock 90] shows that the method based on resultants is indeed faster for $n \leq 3$, but for $n = 4$ is already slower than the all root algorithm (but is still 1 to 3 times faster than the Bézier clipping algorithm). For $n = 5$, even the Bézier clipping algorithm is faster than the algebraic method which is clearly very inefficient for $n > 5$, if it can be used at all, see [Gol 87a].

We should remark, that in general, the running time of an algorithm depends very strongly on the implementation. It can vary widely for different computers and compilers [Sed 86, 90]. We should also note the possibility of parallel processing (see e.g., [Kauf 89]), which can lead to significant increases in speed. For example, the de Casteljau algorithm, which normally exhibits $\mathcal{O}(n^2)$ behaviour, can be implemented in parallel as an $\mathcal{O}(n)$ algorithm [Dero

87, 89]. The degree of parallelizability is by nature higher for subdivision algorithms than for algebraic methods.

Figs. 12.21 and 12.22 show examples of curve intersections.

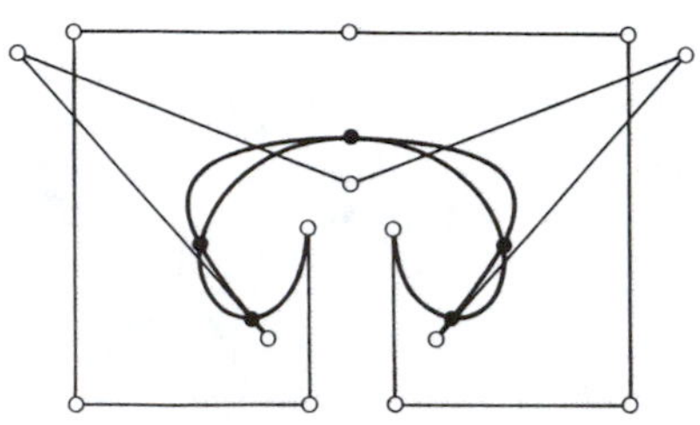

Fig. 12.21. Intersection of two Bézier curves using a divide-and-conquer algorithm.

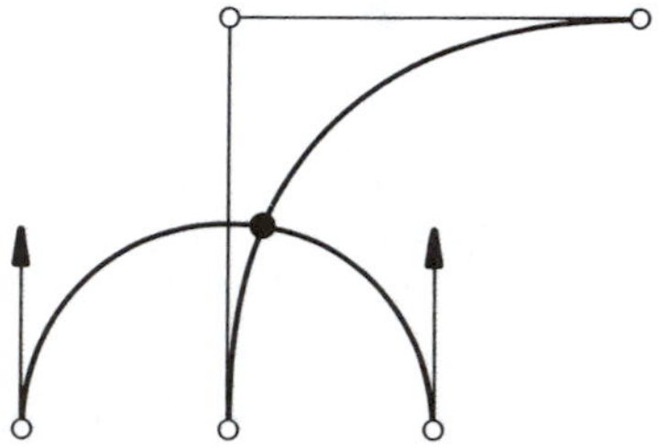

Fig. 12.22. Intersection of two rational Bézier curves using a divide-and-conquer algorithm.

The Bézier divide-and-conquer algorithm can also be used to find *self-intersection points* (see Fig. 12.24) of a Bézier curve [Las 89]. This can be done by introducing an *angle criterion* (see Fig. 12.23) which measures the changes in the tangent $\boldsymbol{X}'(t)$ as the parameter value t varies. Self-intersection can occur for offset curves and surfaces, see Chap. 15 and [Aom 90].

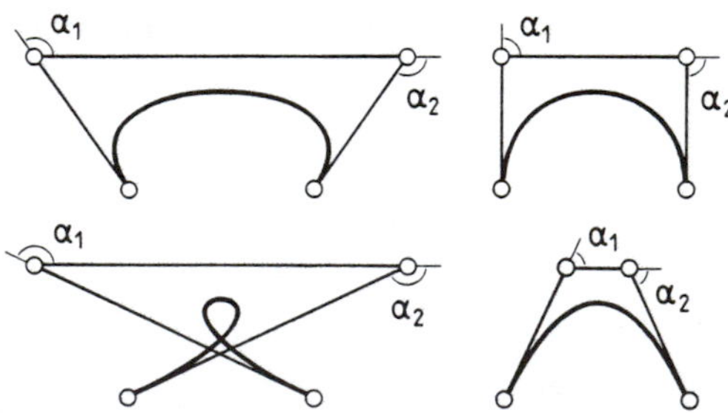

Fig. 12.23. Angle criterion for Bézier curves. A Bézier curve has no self-intersections, provided $\sum |\alpha_k| \leq \pi$, where the α_k are the exterior angles of the Bézier polygon.

Given two curves, we say that they have *collinear normals* at the common point $\boldsymbol{P} = \boldsymbol{X}_1(t) = \boldsymbol{X}_2(\tau)$, provided that the normal vectors $\boldsymbol{N}_1(t)$ and $\boldsymbol{N}_2(\tau)$

to $\boldsymbol{X}_1(t)$ and $\boldsymbol{X}_2(\tau)$ at $\boldsymbol{P}$ are parallel, [Sed 90] and also [Sed 89b]. In this case we say that the two curves *osculate* at $\boldsymbol{P}$. Similar definitions apply to surfaces. Such osculatory points cause considerable difficulty for intersection algorithms. In [Mark 89] they are called magic points, while in differential geometry (see e.g., [SPI 79]), they are referred to as *critical points*, see also [Kri 92].

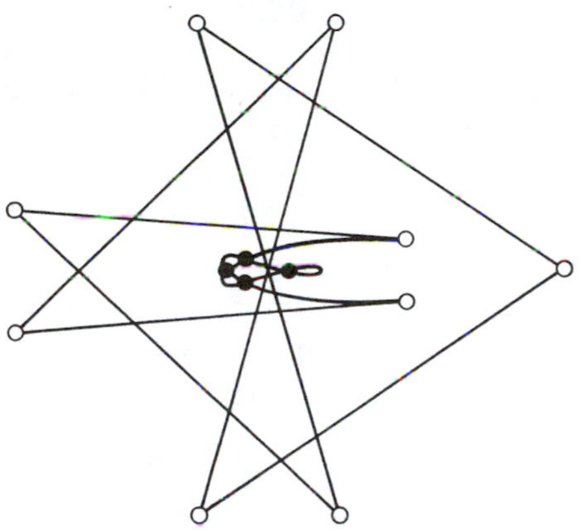

Fig. 12.24. Self-intersections of a Bézier curve.

This problem can be dealt with using a *vector field* vf in the parameter space D_1 of $\boldsymbol{X}_1$, whose zeros correspond to the points of tangential osculation, see [Mark 89], [Chen 89]. A critical part of the algorithm is the construction of a function $h : D_1 \rightarrow D_2$, *i.e.*, from one parameter space to the other, such that the associated points in coordinate space have minimal distance. Newton's method can be used to construct h, and to calculate the zeros of the vector field vf which is defined by h. The idea is developed further in [Kri 92]. For a discussion of the problem of finding the *minimal distance* between two points, curves or surfaces, see [MOR 85]. For curves, this problem is also treated in [Fri 92], [Deg 92], and [Eis 92].

[Sed 90] finds osculatory points for a pair of Bézier curves using the divide-and-conquer method. As in the Bézier clipping algorithm discussed above, this involves alternately computing *Bézier focus curves* of the form $\boldsymbol{F}(t) = \boldsymbol{X}(t) + c(t)\boldsymbol{N}(t)$ for the two Bézier curves, where $c(t) \in \mathbb{R}$ and $\boldsymbol{N}(t)$ is the normal vector (*i.e.*, $\boldsymbol{N}(t) \cdot \boldsymbol{X}(t) = 0$). The $c(t)$ are defined in such a way that by repeated subdivision of both curves, the focus curves corresponding to the subsegments converge to the centers of curvature of the subsegments. This divide-and-conquer strategy can be used to find curve points with first order osculation.

Subdivision algorithms are used for a wide variety of tasks: to compute intersections, to compute self-intersections of curves and surfaces [Las 88, 89], to determine silhouettes of a surface [Kop 86], [Elb 90], to construct contour maps [Pete 84, 87], in NC milling using parallel curves [Suh 90], in building hidden-line [Grif 75], [Li 88], ray-tracing [Kay 86], solid-modeling algorithms [Car 82], [Mil 86], [Casa 87], [Nas 87], and [Tur 88], and for visualization algorithms [Whi 78], [Lane 80a], and [Herz 87].

Divide-and-conquer algorithms can be used with any surface representation. For example, [Hann 83] gives an algorithm utilizing spline bases for surfaces of rotation, for translation surfaces, and for blending surfaces. [Owe 87] presents an algorithm for implicitly defined surfaces. In [Pete 84, 87] there is an algorithm for triangular Bézier surfaces and tetrahedral Bézier volumes, and [Car 82], [Las 86, 87, 90b] describe an algorithm for tensor-product Bézier surfaces and volumes. Algorithms for tensor-product B-spline surfaces can be found in [Pen 84], [Dok 85], and [Nat 90], for NURBS-surfaces in [Elb 90], and for parametric surfaces with no special form in [Hou 85].

12.3. Embedding Methods

Embedding methods are methods for determining all solutions (including multiple solutions) of a system

$$F(\boldsymbol{X}) = 0 \tag{12.5}$$

of N polynomial equations in N variables. The idea of the method is to solve (12.5) by starting with the known or easily found solutions $\boldsymbol{X}_i$ of a simpler system

$$G(\boldsymbol{X}) = 0. \tag{12.6}$$

Here $F(\boldsymbol{X})$ and $G(\boldsymbol{X})$ are related by

$$H(\boldsymbol{Z}) = tF(\boldsymbol{X}) + (1-t)G(\boldsymbol{X}) \qquad \text{with} \qquad \boldsymbol{Z} \equiv (\boldsymbol{X}, t), \quad t \in [0,1], \tag{12.7}$$

i.e., the systems of equations (12.5) and (12.6) are both embedded in a family of systems (12.7). Let $\boldsymbol{DH}(\boldsymbol{Z})$ be the matrix of partial derivatives of $H(\boldsymbol{Z})$, and suppose that $\{\boldsymbol{X}_i\}$ are the solutions of the simpler system (12.6). We then generate a homotopy path (or "flow") from the solutions $\boldsymbol{X}_i$ of (12.6) to the desired solutions $\bar{\boldsymbol{X}}_i$ of (12.5) given by $\boldsymbol{Z}(1) = (\bar{\boldsymbol{X}}_i, 1)$. Numerically, we do this by solving, for each solution $\boldsymbol{X}_i$ of (12.6), the initial value problem

$$\boldsymbol{DH}(Z)\frac{dZ}{dt} = 0 \qquad \text{with} \qquad \boldsymbol{Z}(t) \equiv (\boldsymbol{X}, t), \qquad \boldsymbol{Z}_i(0) = (\boldsymbol{X}_i, 0),$$

see [Drex 77], [Gar 79], and [Morg 83].

Embedding methods are also known as *homotopy, continuation* or *incremental loading methods* [Gar 79], [Morg 83], and [Wri 85]. They seem well suited for computing intersections of implicitly defined surfaces, but are still under development, and so far have been used in CAGD only for quadrics, see [Prat 86].

12.4. Discretization Methods

In general, the problem of finding intersections for surfaces leads to an underdetermined nonlinear system of equations, independent of the nature of the surface representations being used. Indeed, to compute the four parameters of an intersection point, we have only three coordinate equations at our disposal. Discretization methods solve this problem by reducing the number of degrees of freedom to three by discretizing the surface representation. This can be done in various ways.

12.4.1. The Grid Method – Contouring

The first step in the *grid method* is to discretize both surfaces. Thus, for each surface we get a "*matrix*" of points $\boldsymbol{P}_{ij}$ (usually surface points of the form $\boldsymbol{X}(u_i, v_j)$). We then use these in sets of three or four to define "*cells*" on which we construct either linear or bilinear approximations to the surfaces, see e.g., [Schu 76], [Hartw 83]. The quality of the approximation (and hence also the qualilty of resulting intersection curves) depends on the grid size and the nature of the points.

While grid methods are frequently used in pre- or post-processing stages of multi-step surface intersection algorithms, their main application is to the computation of *contours* (also called *potential lines, isolines*, or *level curves*) for given function-valued surfaces, see e.g., [Preu 84, 86], [Suf 84], and Fig. 12.25. Many of the applications involve *terrain modelling*, see e.g., [Mir 82], [Eva 87], [Petr 87], and [Hua 89], or *scattered data fitting*, see e.g., [Mcla 74], [Patt 78], [Sab 80], and [Schag 82], where the data are unevenly distributed, and therefore must first be triangulated, see Fig. 12.26.

There are similar algorithms for data describing a given function-valued surface defined over a regular rectangular grid, or over either regular or non-regular (scattered data) triangular grids. The basic idea is as follows (see e.g., [Schu 76], [Patt 78], [Sut 80], and [Petr 87]): Suppose we want to draw the level curve $f(x, y) = z = z_0$, and that z_0 is bounded from above and below by values of the surface at the endpoints of some edge. Then the contour line $z = z_0$ intersects this edge at some point which can be determined using *inverse interpolation*, see e.g., [Schu 76], [Patt 78].

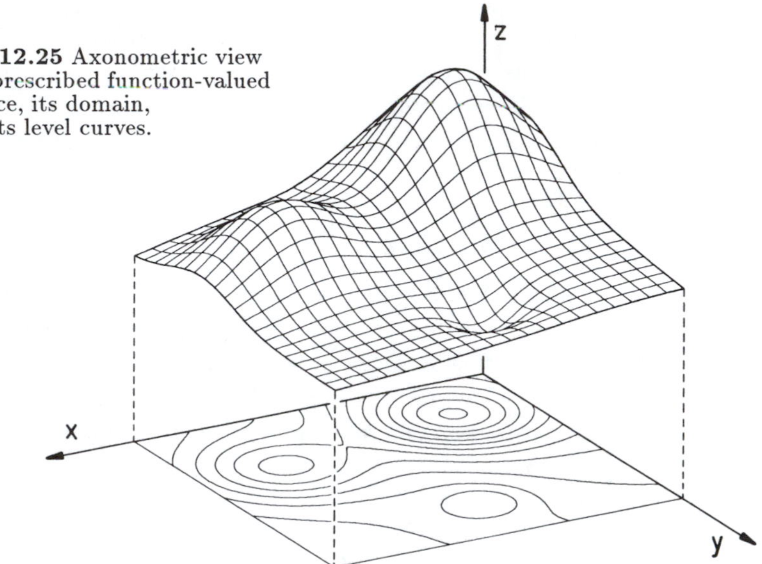

Fig. 12.25 Axonometric view of a prescribed function-valued surface, its domain, and its level curves.

An alternative approach is to interpolate the grid points with a network of curves lying on the surface of the form $\boldsymbol{X}(u_i, v)$ and $\boldsymbol{X}(u, v_j)$. Then we can find the points of intersection of the contour line $z = z_0$ with one of the curves $\boldsymbol{X}(u_i, v)$ by solving the equation $\boldsymbol{X}(u_i, v) = z_0$, see [Mcla 74].

A contour line which "enters" a cell by crossing one edge, must again "exit" the cell by crossing some (possibly another) edge. This exit point for the contour line can be determined by considering the other edges defining the cell. This exit point then serves as an entry point for the adjoining cell. We can then get an approximate contour line by simply connecting the entry and exit points, see Fig. 12.27 and [Patt 78], [Prat 86], [Petr 87], and [Hua 89]. Alternatively, we can use some kind of marching technique such as the very simple one described in [Mcla 74] (but see [Sut 76]!). Better results can be obtained by an appropriate *forward march* through a cell, starting at the entry point and going to the exit point, see e.g., [Preu 84, 86] and also Sect. 12.5.

To construct *smoother* level curves, we can refine the grid [Sut 76a], [Gold 77]. Another possibility is to interpolate or approximate the points found above using a spline. This can, however, lead to overlapping or intersection of different contour lines [Wrig 79], [Sut 80]. One way to avoid this is to use splines in tension with a sufficiently large tension parameter, see [Wrig 79].

The cases where the contour line passes through a vertex or lies on an edge of a cell (or when the whole cell is at the contour level) require special

Fig. 12.26. Level curves of a scattered data surface.

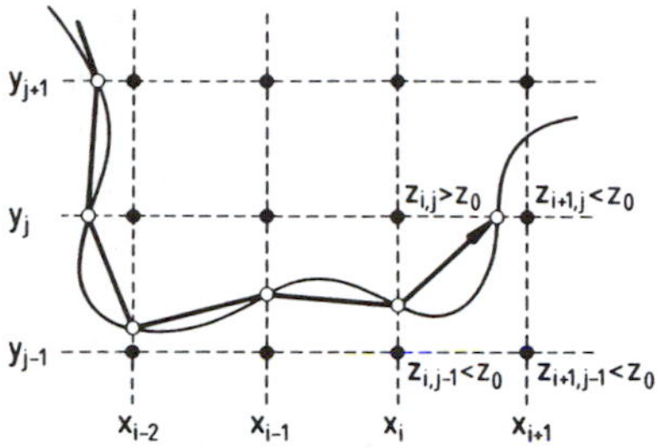

Fig. 12.27. Calculating contour lines via the grid method.

attention. Problems can arise if too coarse a grid is given or selected. For example, a closed loop can lie entirely inside of the cell, and thus be completely "*overlooked*". We also have problems if the entry and exit point of a contour line both lie on the same edge of a cell, or if a contour crosses all four edges of a four-sided cell. The problem which arises when a contour crosses all four edges of a four-sided cell is called the *saddle-point problem* or *four-point problem*, see Fig.12.28 and [Schu 76], [Sut 80], [Prat 86], and [Petr 87]. If we are dealing with a surface given by discrete data and whose local geometry is unknown, then it is useful to impose a consistency condition to guarantee that the algorithm gives the same result independent of the starting point and the order in which the cells are examined.

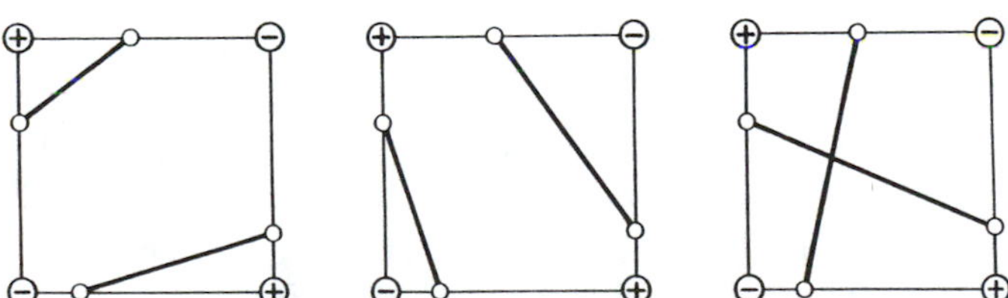

Fig. 12.28. Four-point problem: possible contour lines.

Such a *uniqueness condition* could be: given one point on each of the four edges, we always connect pairs of points which lie on edges sharing a common vertex where the function value $f(x, y)$ is larger than the level value

z_0. Fig. 12.29a shows an example where using this consistency condition, the pairs $\boldsymbol{P}_1$, $\boldsymbol{P}_4$ and $\boldsymbol{P}_2$, $\boldsymbol{P}_3$ should be connected with contour lines. An alternative approach [Day 63] is to *subdivide* the rectangle into four triangles by drawing in the diagonals. We associate the average of the four corner values with the new vertex. The problem is now reduced to the triangle case, where it is easy to construct a consistent solution, although triangular elements then have to be incorporated into the data structure, see Fig. 12.29b.

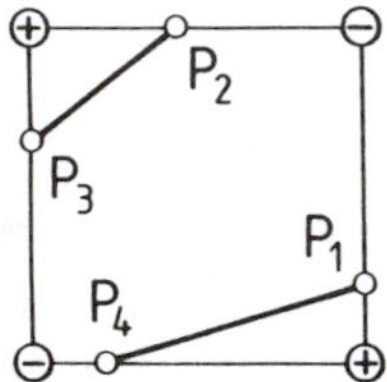

Fig. 12.29a. Consistency condition involving the common vertex between two points.

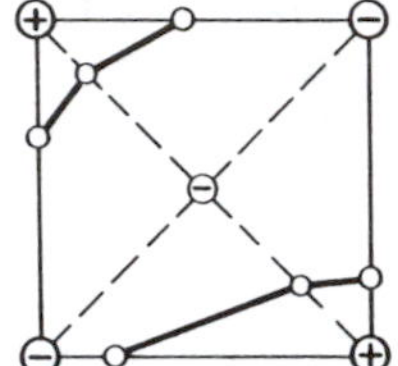

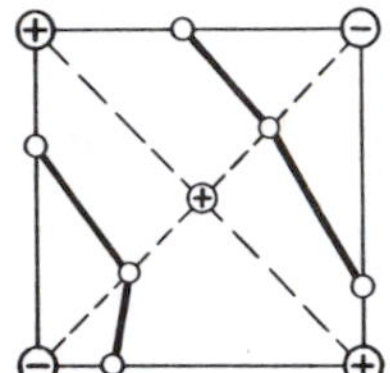

Fig. 12.29b. Consistency condition based on subdividing into four triangles.

Fig. 12.29. Solution of the four-point problem for rectangles.

These ideas can be generalized for use in finding *contours of hypersurfaces* based on cube or tetrahedral shaped cells [Wrig 79], [Gal 89], [Dob 90], [Ham 91, 92]. As before, there is a problem with deciding how to connect contour points. Simply comparing the level value with the function values at the eight vertices of a cell (*marching cube method* [Lor 87], [Gal 89], see Fig. 12.30) not only can lead to incorrect contours, but can even create holes in the surfaces [Dür 88], [Wil 90], and [Bri 91]. The problem can be overcome with an appropriate *consistency condition* such as a common vertex [Ham 91, 92] (cf. Fig. 12.29a), facial average [Wyv 86] (cf. Fig. 12.29b), use of additional intersection patterns [Bri 91], or the use of derivative data [Wil 90].

To find contour lines (surfaces) for *arbitrary curved* (*hyper*) *surfaces*, we have to make use of sophisticated subdivision, tracing, or algebraic algorithms which produce approximate level lines (surfaces). For example, [Fu 90] considered bicubic surfaces, [Pete 84] triangular Bézier surfaces, [Pete 87], [Dah 89] TB-hypersurfaces, [Sat 85], [Dic 89] B-spline surfaces, [Rock 90] TPB-hypersurfaces, and [Patr 89a, 90] B-spline hypersurfaces.

In addition to using piecewise linear surfaces with the grid method, we can also use scalar-valued, piecewise *quadratic triangular Bézier surfaces*. These are of special interest since their intersection with a contour plane can be

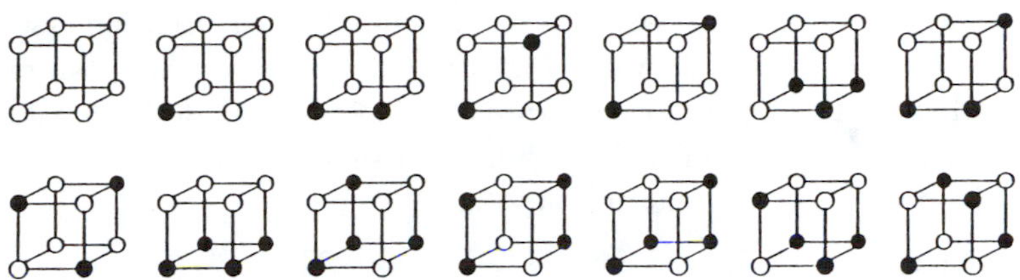

Fig. 12.30. The 14 principle cases of the marching cube method.

be computed exactly as a conic section. A scalar-valued quadratic triangular Bézier surface $X(u, v, w)$ can be written as a quadratic form

$$X(u,v,w) = (\,u, v, w\,) \begin{pmatrix} b_{200} & b_{110} & b_{101} \\ b_{110} & b_{020} & b_{011} \\ b_{101} & b_{011} & b_{002} \end{pmatrix} \begin{pmatrix} u \\ v \\ w \end{pmatrix}. \qquad (*)$$

Since the barycentric coordinates u, v, w are related to the Cartesian coordinates x, y by a linear transformation, $(*)$ represents a quadric surface, in particular, a paraboloid [Far 86]. A planar section of a paraboloid is a conic section. It can be expressed as a (piecewise) quadratic rational Bézier curve with pieces given by (12.8) (cf. (4.31), and see [Pie 86], [FAR 90]):

$$\boldsymbol{X}(t) = \frac{\boldsymbol{b}_0 B_0^2(t) + \omega_1 \boldsymbol{b}_1 B_1^2(t) + \boldsymbol{b}_2 B_2^2(t)}{B_0^2(t) + \omega_1 B_1^2(t) + B_2^2(t)}. \qquad (12.8)$$

To determine the Bézier points $\boldsymbol{b}_0$, $\boldsymbol{b}_1$, $\boldsymbol{b}_2$ and the weight ω_1 of a segment of the contour curve, [Far 86] suggested the following algorithm (see Fig. 12.31):

1) examine only those surface patches where the contour plane intersects the convex hull of the Bézier points,
2) find the intersection point of the three boundary curves with the contour plane. If there are not exactly two intersection points on different edges, go to 6,
3) let the intersection points be $\boldsymbol{b}_0$ and $\boldsymbol{b}_2$, and let $\boldsymbol{b}_1$ be the intersection of the tangent planes at $\boldsymbol{b}_0$ and $\boldsymbol{b}_2$ with the contour plane. If $\boldsymbol{b}_1$ lies at infinity or far outside of the triangle, go to 6,
4) find the shoulder point $\boldsymbol{S}$ as the intersection of the Bézier surface with the straight line passing through $\boldsymbol{b}_1$ and the midpoint $\boldsymbol{M}$ of the line segment $\boldsymbol{b}_0\boldsymbol{b}_2$. If $\boldsymbol{S}$ does not lie between $\boldsymbol{b}_1$ and $\boldsymbol{M}$, go to 6,

5) use inverse linear interpolation to find the ratio s between the lengths of $\boldsymbol{b}_1\boldsymbol{S}$ and $\boldsymbol{b}_1\boldsymbol{M}$, and set $\omega_1 = s/(1-s)$,

6) subdivide the triangle, and apply the algorithm to each subtriangle.

The subdivision of step 6) serves to reduce complicated situations (see Fig. 12.32) to the simple one shown in Fig. 12.31 (top view), where the process described above can be used to determine the weight and Bézier points. Unfortunately, the way in which this subdivision should be done (cf. Fig.12.18) cannot be prescribed in advance [Wor 90].

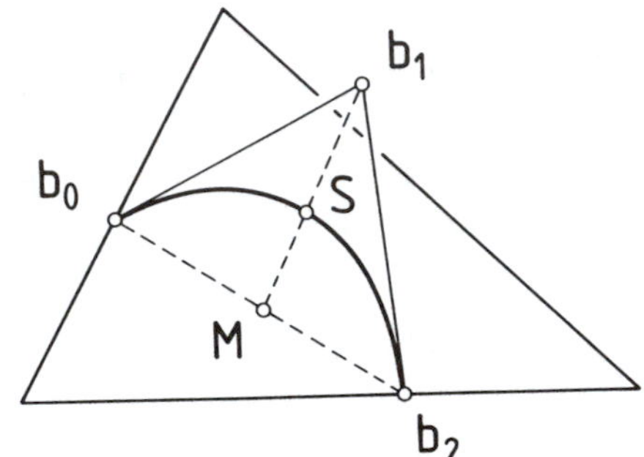

Fig 12.31. Contour calculation for a quadratic triangular Bézier patch.

An efficient, robust and completely deterministic algorithm for finding contours even in the cases of degenerate or decomposable conic sections is described in [Wor 90]. To determine the Bézier points and weights, it uses the above strategy in some cases, but does not make use of subdivision of the surface. The essential idea is to compute tangents to the contour line which are parallel to the sides of the triangles (see Fig. 12.32 where they are shown as dashed lines). The contour is then subdivided at these points of tangency, and the tangents themselves are used in the construction of the Bézier polygon. This approach solves the saddle point problem, the problem of contours which lie entirely inside a triangle or touch an edge, the problem of small closed loops, and even the problem when a contour enters and exits a triangle on the same edge.

A corresponding algorithm for contouring quadratic TB-hypersurfaces in $\mathbb{R}^4$ was given by [Barn 92a].

There are a number of ways to *graphically display* contour line images. The usual approach is to present elevations (as is done for maps), see Fig. 12.26. Axonometric images, which include contour lines or a combination of the two methods (see Fig. 12.25), can increase our understanding of the surface. Frequently, the surfaces between level curves are shaded or colored, see Fig. 12.33

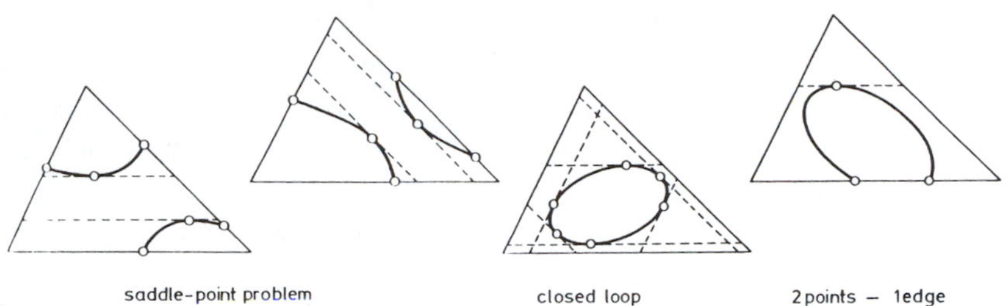

Fig. 12.32. Problems in finding contours for quadratic triangular Bézier surfaces.

and [Preu 86,89], [LUT 89]. Again using axonometric views improves understanding. Using a uniform shading in each region suggests that the surface has a stepwise shape. This led [Fol 90, 90a] to suggest varying the color intensity between two level curves linearly from a minimal intensity I_{min} to a maximal intensity I_{max}. [Fol 90, 90a] suggests choosing $I_{min} = 0$ corresponding to black. Another common method involves "*smearing*" the sharp boundaries between colors using half-tone methods, see e.g., [ENC 88].

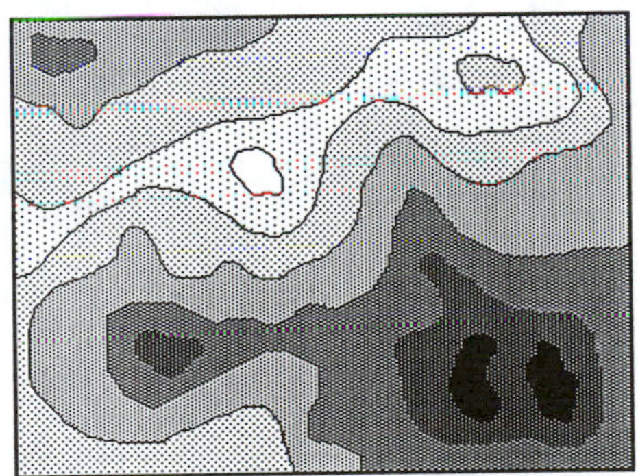

Fig. 12.33. Coded projection of a surface. The regions between the contour lines are coded with gray tones or colors from a palette.

Good overviews of various methods for constructing contour line images can be found in [Sut 80], [Sab 80, 85a], see also [Schu 76]. A number of contouring software packages are presented in [Petr 87], and some excellent graphical images can be found in [Fol 90, 90a], [Nie 91a], and [Pot 91a].

12.4.2. Parameter Discretization

Given two surfaces, parameter discretization involves selecting a prescribed number of parameter values for one of the four surface parameters to construct corresponding parametric lines on one surface. These lines are then intersected with the second surface, see Fig. 12.34. The problem is thus reduced to a problem involving three equations in three unknowns. Before we can find a parametric formula for the intersection curve $\boldsymbol{S}$ via interpolation or approximation, we must first sort the intersection points $\boldsymbol{S}_i$ arising from the discretization.

Suppose $\nu = \nu_i$ define a sufficiently dense set of parametric lines on the surface $\boldsymbol{X}_2(\mu, \nu)$. Then, for example, the intersection points of these space curves with the surface $\boldsymbol{X}_1$ can be found by an *iterative projection method*. Starting with some $\boldsymbol{P}_{i0}$ corresponding to $\mu = \mu_{i0}$, we construct a line passing through it and perpendicular to the surface $\boldsymbol{X}_1(u, v)$. We denote the foot of this perpendicular by $\boldsymbol{P}_{i1}$. Then the point $\boldsymbol{P}_{i1}$ is projected perpendicularly onto a point $\boldsymbol{P}_{i2}$ on the surface $\boldsymbol{X}_2(\mu, \nu)$, etc., see Fig. 12.35. The computation of the perpendicular projection of the point $\boldsymbol{P}_{i0}$ onto $\boldsymbol{X}_1(u, v)$ can be accomplished, for example, by minimizing $|\boldsymbol{P}_{i0} - \boldsymbol{X}_1(u, v)|$ using available optimization methods, see e.g., [HORS 79].

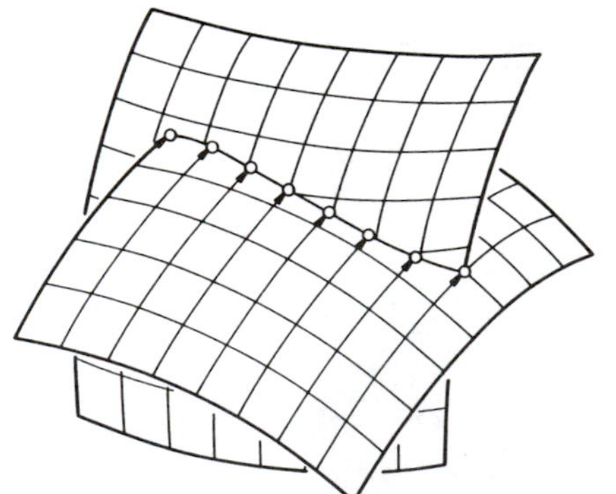

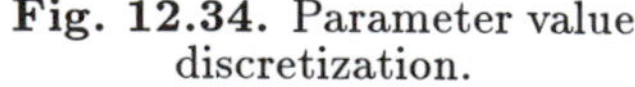

Fig. 12.34. Parameter value discretization.

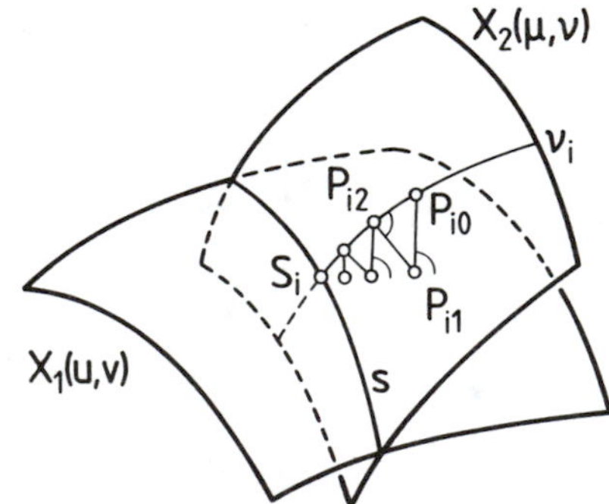

Fig. 12.35. Projection method.

[Barn 87] suggests replacing the parameter lines by *piecewise linear approximants*, which are then intersected with the piecewise linear approximation of the second surface. The points found in this way are then used as starting values for a numerical method (e.g., Newton's method) to quickly find the exact location of the intersection points. For more on the numerical treatment of the problem of finding points where a curve pierces a surface, see e.g., [FAU 81], [MOR 85], and [Che 88].

More on methods for iteratively finding intersection points for a line (curve) with a surface, or for iteratively improving the accuracy of estimates of such intersection points, can be found in [Hos 85], [Hou 85], [Casa 87], and [Barn 87, 90a]. The method of parameter value discretization is also implemented in the algorithms in [Hos 87a], see also Sect. 12.1 and [Catl 87].

The problem of finding the point where a straight line pierces an arbitrary curved surface is treated in great detail in the ray tracing literature, see [GLA 89], [MÜLL 88]. For a discussion of subdivision strategies, see e.g., [Whi 80], [ROG 85], [Kay 86], [Woo 89], and [Nis 90]; for numerically based methods, see [Tot 85], [Joy 86], and [Gig 89]; for algebraic methods, see [Kaj 82]; and for hybrid algorithms which combine subdivision and numerical methods, see [Swe 86], [Yan 87].

12.5. Tracing Methods

Discretization algorithms are often part of preprocessing for tracing algorithms. Such algorithms usually consist of a *search phase* (also called a *hunting phase*), a *tracing* or *marching phase*, and a *sorting phase*. The search phase serves to provide starting values for the tracing phase, and usually involves a discretization algorithm. The tracing phase computes pieces of the intersection curve in the form of point sequences. The sorting phase orders the pieces, *i.e.*, the point sequences, and divides them into disjoint pieces and curve loops.

The *discretization*, which is needed in the first step of the algorithm may involve finding a prescribed number of values for both parameters of one of the two surfaces, say $\boldsymbol{X}_1(u, v)$. This produces a finite number of parametric curves on this surface which form a curve network, see e.g., [Tim 77], [MOR 85], and [Che 88]. The mesh size should be as large as possible, but must be chosen small enough so that no parts of the contour can get lost. The surface patches generated by the curve network are then numbered, see the example in Fig. 12.36. Next, the parametric lines on $\boldsymbol{X}_1(u, v)$ are intersected with the second surface $\boldsymbol{X}_2(\mu, \nu)$, working our way in order through the lines corresponding to $u =$ const., and then those corresponding to $v =$ const. In this way contour points are found sequentially, see Fig. 12.36. The actual computation of these contour points can be done with any of the methods discussed above in connection with parameter value discretization. On the other hand, [Faro 87a] and [Kri 90] find points on the intersection curve which divide it into monotone pieces, and use these as the starting values for the following tracing phase.

In the *tracing phase*, we construct the contour curves piecewise, using the intersection points found in the search phase as starting points. This is done by looking at the surface patches one after another. We order the intersection points falling on the boundary of a patch according to their indices. Thus, for example, for patch number 3 in Fig. 12.36, we list the points in the order 5, 9, 13, and 15. Then for the given patch, we start with the intersection point with the smallest index, and trace the intersection curve of the two surfaces until it leaves the patch. For the example in Fig. 12.36, this corresponds to constructing the piece of the curve denoted by c which starts at the point 5 and goes to the point 13. Then the remaining points in the point list associated with this patch are treated. To insure that no part of the intersection curve is constructed twice, we drop the points we have already used. Thus in the example, points 5 and 13 are dropped from the list, leaving 9 and 15 to consider. Starting at point 9, we construct the curve segment d which exits at point 15, which completes the work on patch 3, and we move on to the fourth patch.

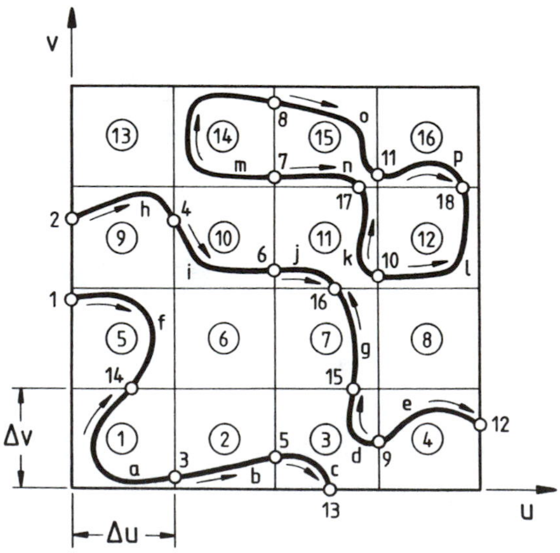

Fig. 12.36. A parametric curve network on a surface $\boldsymbol{X}_1(u, v)$ along with the intersection points $1, \ldots, 18$ of the isoparametric lines of $\boldsymbol{X}_1(u, v)$ with $\boldsymbol{X}_2(\mu, \nu)$ resulting from the search phase, and the composite intersection curve of $\boldsymbol{X}_1$ and $\boldsymbol{X}_2$ consisting of the pieces $a, \ldots, p$.

A tracing algorithm should construct a sequence of points which are on (or near) the intersection curve, starting at a given initial point, and ending at a given end point. The idea is to move from point to point by a prescribed amount in the direction of the intersection curve. The size of the steps and the direction to move can be computed from the local differential geometry of the two surfaces. For example, if we are currently at the point $\boldsymbol{S}_0$ with $\boldsymbol{X}_1(u_0, v_0) = \boldsymbol{X}_2(\mu_0, \nu_0)$, then we can compute the direction in which to move to get a new starting point by finding the tangent T_0 to the desired intersection curve $\boldsymbol{s}$. This tangent must be perpendicular to the surface normals $\boldsymbol{N}_1$, $\boldsymbol{N}_2$ to the surfaces $\boldsymbol{X}_1$, $\boldsymbol{X}_2$ at the point $\boldsymbol{S}_0$. Thus (see Fig. 12.37), it has the direction

$$\boldsymbol{T}_0 = \boldsymbol{N}_1(u_0, v_0) \times \boldsymbol{N}_2(\mu_0, \nu_0).$$

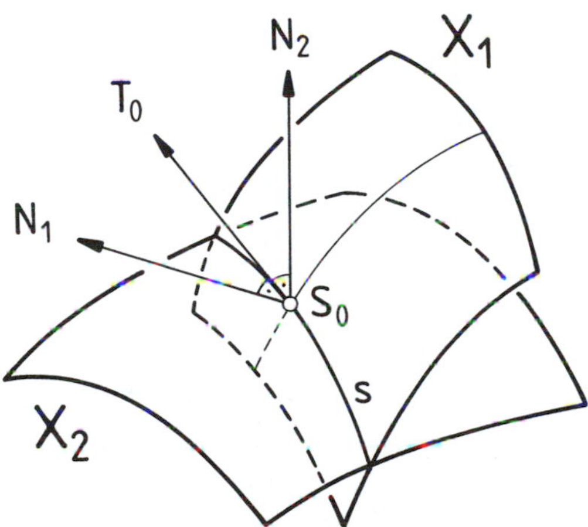

Fig. 12.37. Construction of a tangent to the intersection curve.

As remarked above, the intersection problem is in general under-determined, *i.e.*, it has too many degrees of freedom. This problem can be somewhat alleviated by imposing additional *arbitrary* conditions. For example, they can help determine the step size. Such conditions can be interpreted as a supplementary curve or third surface $\boldsymbol{X}_3$, which intersects the intersection curve at a point which is a prescribed distance d from the last found intersection point. A good choice for $\boldsymbol{X}_3$ is, e.g., a circle or sphere k_0 with center at the last found intersection point $\boldsymbol{S}_0$, and with radius $r_0 = r(\kappa_0)$, where κ_0 is the curvature of the intersection curve, see [FAU 81], [Prat 86] and Fig. 12.38.

Often, because of the resulting computational savings, a straight line (plane) g_0 is constructed perpendicular to the tangent T_0 at a distance $d_0 =$

$d(\kappa_0)$ from $\boldsymbol{S}_0$, see [FAU 81], [Schag 82], [MOR 85], [Barn 87], [Luk 89], and Fig. 12.38.

[Preu 86] constructs a *parabola* which passes through the last three constructed points $\boldsymbol{S}_{-2}$, $\boldsymbol{S}_{-1}$, and $\boldsymbol{S}_0$, see Fig. 12.38.

In [Suf 84], [Che 88], and [Ast 88], the next starting point $\boldsymbol{S}_{S1}$ is computed using a *circle of curvature* K_0 (instead of the tangent T_0) passing through the last found point. However, this requires second partial derivatives. If the angle of turning is not too large (say $\delta_0 \leq 5°$), then this approach has the advantage that the next starting point $\boldsymbol{S}_{S1}$ usually lies close to the true intersection curve, so that we do not need a Newton iteration step to improve it, see Fig. 12.38. Indeed, [Suf 84] found that iterative improvement is needed in only about 2% of the steps (and then only one iteration), while marching in the direction of the tangent requires the use of Newton iteration about 95% of the time.

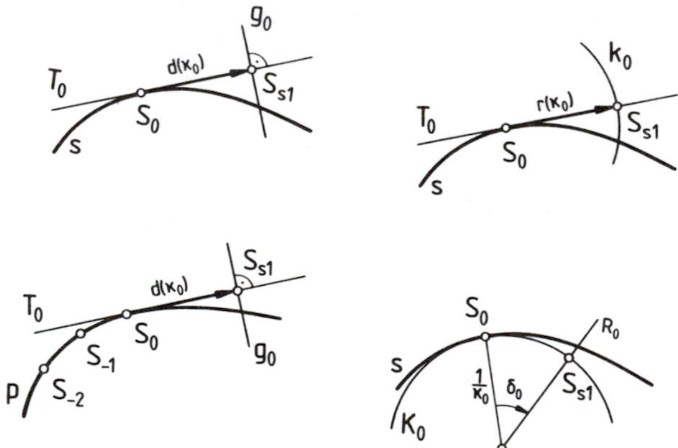

Fig. 12.38. Marching methods.

In [Barn 90a], the curvature circle K_0 is computed directly from $\boldsymbol{S}_0$ and two neighboring points to the right and left of $\boldsymbol{S}_0$, without using second partial derivatives. [Gei 84] also presents a curvature dependent step size selection algorithm using only the first derivatives and a polygonal approximation to a circle.

To assure that a marching algorithm works well in the presence of strong changes in the curvature of the intersection curve generally requires the computation of third and higher partial derivatives. The application of general *quadratic forms* Q_0, defined at the last found point $\boldsymbol{S}_0$, can also be used in this case to provide a relatively safe way to track the intersection curve without making use of third or higher derivatives [Dic 89]. An additional improvement can be made by adaptively choosing the "turning angle" δ_0, [Dic 89].

Farouki's algorithm [Faro 86, 87a] subdivides the intersection curve into pieces with monotonic behavior, whose endpoints are special curve points (see above), and then marches along a section of the curve by constructing the *Taylor series* expansion at the last found point, see also [Mon 86] and [Baj 88]. The *implicit function theorem* provides the mathematical foundation for this method, which provides a relatively safe way to deal with singularities.

A singularity is signaled by the degeneracy of the *Jacobi matrix* $\boldsymbol{J}$. A simple singularity can be identified using the *Hessian matrix* $\boldsymbol{H}$. If $\boldsymbol{H}$ has two negative (positive) eigenvalues, then we have a relative maximum (minimum), and if the two eigenvalues have different signs, then we have a saddle point. An estimate of the distance to a simple singularity is given in [Dic 89]. The *radius of convergence* of the Taylor series expansion can also be used to make assertions in the presence of singularities of higher order. In particular, the radius of convergence can be used to control the step size. Situations where parts of the intersection curve come very close together without touching, or where they cross (*branch points*), can also be dealt with using this method [Mon 86].

The *branch point problem* can also be treated as follows [Abh 83, 88], [Baj 88], see also [WALK 50]. First, the singularity of order n is translated into the origin (it may also be appropriate to rotate the coordinates), and then a series of n quadratic transformations is used to decompose the singularity into a number of simple points for which the direction in which to move can be uniquely determined. This is then retransformed to the original situation with the help of two points (one on the right and one on the left of the singularity).

The *sorting phase* orders the points or individual pieces of the intersection, and divides them into disjoint curve segments. In the simplest case this is done by comparing the inital and end points of the pieces, see e.g., [MOR 85] and the remark on sorting on page 517. However, more care is required for singularities, such as branch points and points where the intersection curve touches a boundary, in order to insure that the points are correctly connected. For recent work, see [Barn 90a], [Wag 89], and [Joh 90].

Another class of tracing algorithms are based on reformulating the problem of finding intersection curves as a *minimization problem*. Suppose the two surfaces Φ_1 and Φ_2 and the supplementary third surface are given by

$$f(x, y, z) = 0, \qquad g(x, y, z) = 0, \qquad h(x, y, z) = 0.$$

Then the intersection curve of Φ_1 and Φ_2 can be found by minimizing the function

$$F = f^2 + g^2 + h^2,$$

which is identically zero along the intersection curve, and is otherwise positive [Pow 72], [FAU 81], see also [Prat 86].

The intersection curve problem for two implicit surfaces has also been reformulated as a problem of finding the solution curves corresponding to a *system of ordinary differential equations* by [Phi 84] and [Pfe 85].

All tracing algorithms share the common problems of finding a *good starting point*, of locating all branches of the intersection curve (small interior closed loops should not be overlooked), and of preventing multiple copies of the same piece of the intersection curve. An additional problem is the *correct* choice of the direction in which to move, and also of the step size, so that, for example, when pieces of the intersection curve are close together, we don't jump from one curve to the other. We also have to find a way to automatically stop when we get near the starting point in the case of closed loops, and when approaching the boundary of the surface. Moreover, we want to assure the *correct* behavior when the intersection curve has self-intersections and cusps. Most marching algorithms display a very high (undesirable) sensitivity to local singularities.

This *list of problems* vastly complicates the *fine tuning* of a tracing algorithm. So far, not all problems have satisfactory solutions, and many are still the subject of current research. For example, an interesting approach to the starting point question using hodographs has been proposed by [Sed 88], while [Chan 87] has suggested using elimination methods (cf. Sect. 12.1) for determining starting values. A method for testing for interior closed loops by means of an angle criterion is described by [Sed 89b]. For further ideas, see [Hoh 91], [Malo 91], and [Kri 92].

Since the embedding methods described in Sect. 12.3 are still in the early stages of development, and probably also will not turn out to work too well, while the algebraic methods described in Sect. 12.1 are inefficient for surfaces of higher degree, tracing methods (in combination with discretization methods) are of great practical importance. In particular, by combining them with subdivision algorithms, as is done, e.g., in [Pen 84], [Dic 89], [Barn 87, 90a], [Azi 90], [Kri 90], and [Kop 86, 91a], they lead to very powerful algorithms which can even be applied to find self-intersections of a surface [Barn 87, 90a]. A further advantage of tracing algorithms is their general applicability to arbitrary surface types.

Advantages of the tracing algorithms as compared to the subdivision algorithms include the following (see also [Azi 90]): points on (pieces of) the intersection curve can be found sequentially, so that the amount of work needed for sorting is much less than for subdivision algorithms. The *accuracy* can be increased simply by taking a smaller step size, and/or by increasing the number of iterations used to compute the intersection point. Since Newton's method has a quadratic rate of convergence, for the same error tolerance

ϵ, tracing algorithms are in general much faster. In addition, for the same ϵ, they require fewer points to represent the intersection curve. This is because in using a subdivision algorithm to produce an approximation to the intersection curve within a prescribed accuracy, if there is some critical point on the curve, it will produce an *unnecessarily* large number of tiny segments even in areas where the curve behaves well. With an appropriate step size selection process, a tracing algorithm will only use small steps in areas where it is necessary to meet the tolerance. For further comparisons of the methods, see [Dok 89] and [Barn 90a].

12.6. Concluding Remarks

We saw in Sect. 12.1 that every tensor-product surface of degree (m, n) has an implicit representation of degree $2mn$, while in general, the intersection of any two algebraic surfaces of degrees N_1 and N_2 is an algebraic curve of degree $N_1 N_2$. This implies that the intersection of two tensor-product surfaces of degree (m_1, n_1) and (m_2, n_2) is an algebraic curve of degree $4m_1 m_2 n_1 n_2$. For two bicubic surfaces, we thus get an algebraic curve of degree 324! A curve with this high a degree can decompose into many pieces as illustrated in Fig. 12.39, where it is shown that a planar section of a bicubic surface can consist of eight pieces, see also [Fu 90]. In addition to this *global complexity*, the problem can also suffer from *local complexity* in the sense that it can have self-intersections (nodes), and other types of singularities appearing in the intersection curve, e.g., cusps or tangential touching (tacnodes), see Fig. 12.40, and also [Faro 86], [Prat 86], and [Baj 88].

So far, there still is no algorithm which meets all of the four criteria listed at the beginning of this chapter, and thus the problem of developing such an intersection algorithm remains one of the most difficult problems in CAGD. This is due in part to the fact that the four criteria are not completely compatible with each other. For example, accuracy and robustness must be traded off against speed. For these reasons, each algorithm has to be individually tuned by selecting certain *tolerances*. They also serve to improve the user-friendliness of the algorithms, and to reduce the amount of interactive control required. Such tolerances include SPT (*same point tolerance*), SRT (*search refinement tolerance*), CRT (*curve refinement tolerance*), and other OT's (*optimization tolerances*), and should be expressed in terms of angle sizes in order to avoid dealing with units of length such as inches, etc., [Barn 90a].

The intersection problem can be approached with a variety of different techniques, including numerical, analytic, geometric, and algebraic methods, and using various methods such as subdivision, tracing, and discretization

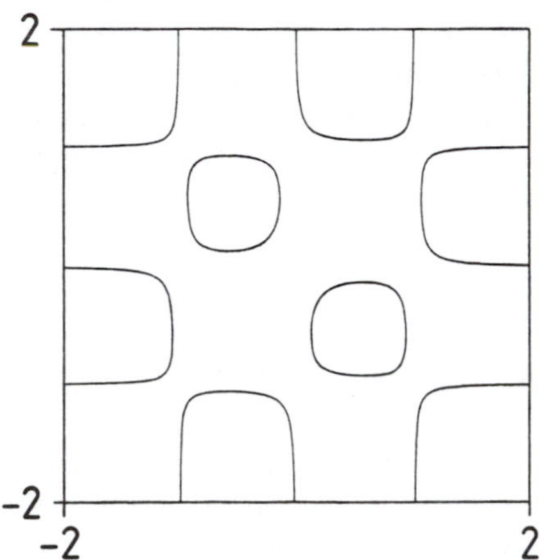

Fig. 12.39. Level curves $z = 0$ of a bicubic surface given by the implicit formula $z = (x-1)x(x+1)(y-1)y(y+1) + 0.05$ ([Prat 86]).

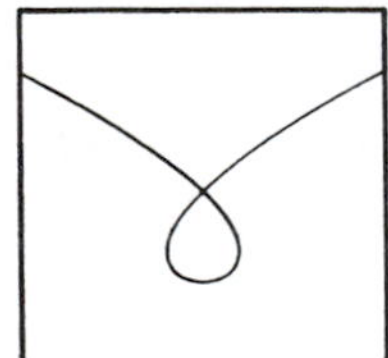
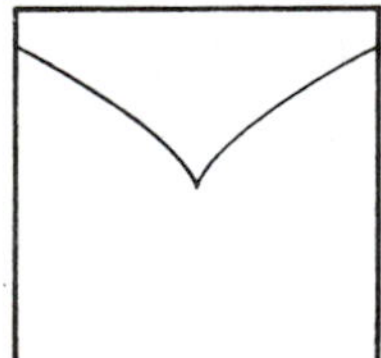

Fig. 12.40. Examples of local complexities of the intersection curve of two surfaces.

methods, etc. While the classic algorithms of the first and following generations are mostly based on just one method (with all its advantages and disadvantages), more recently the tendency is to combine two or more methods into a multi-step method or so-called *hybrid-algorithm*. The idea is to preserve the advantages of each of the ingredients, but to eliminate their disadvantages if possible. For some effective algorithms of this type, see [Barn 90a], [Dic 89], and [Kri 90].

Current research is focused on starting points, running time and convergence, singular points, tangential touching points and branch points, touching boundary curves, partial overlapping, near intersections, small loops, and self-intersections. These problems are discussed in [Nat 90], [Mülle 90, 91], [Hoh 91], [Mano 91], [Mark 91], [Malo 91], and [Kri 92].

13 Smoothing of Curves and Surfaces

Frequently, we are given measured data, and need to construct approximations to it using free-form curves and surfaces. Usually the data are subject to measurement errors. In designing curves and surfaces, we not only want a good approximation of the data, but we want the curve or surface to be "visually pleasing", in some functional or esthetic way. For example, the roof of a car should not have undesirable bumps or wiggles but should be convex, and the tail of an aircraft should not have any oscillations in the surface which could affect its aerodynamic properties. These kinds of undesirable features of curves and surfaces follow from corresponding properties of their curvatures. Removing them is referred to as *smoothing*, and is often performed at the end of a design session. Fig. 13.1 describes the overall procedure for generating free-form curves and surfaces, starting with the data and ending with the final product.

The process often begins by pre-smoothing the given data, after which we interpolate or approximate. This step may involve minimizing some form of bending energy (cf. Sects. 3.6 and 9.2), in which case we get some kind of special energy-minimizing spline. Other methods based on Bézier and B-spline representations were discussed in Sects. 4.4 and 6.3. Here we are again interested in minimizing certain expressions, but while staying with Bézier and B-spline forms. The next step is to test the resulting interpolant or approximant to decide whether it is acceptable or not. If it is not sufficiently "pleasing", we may then alter the parameters in the representation or in the optimality criterion, or we may apply an additional algorithm to either locally or globally smooth parts of the curve or surface.

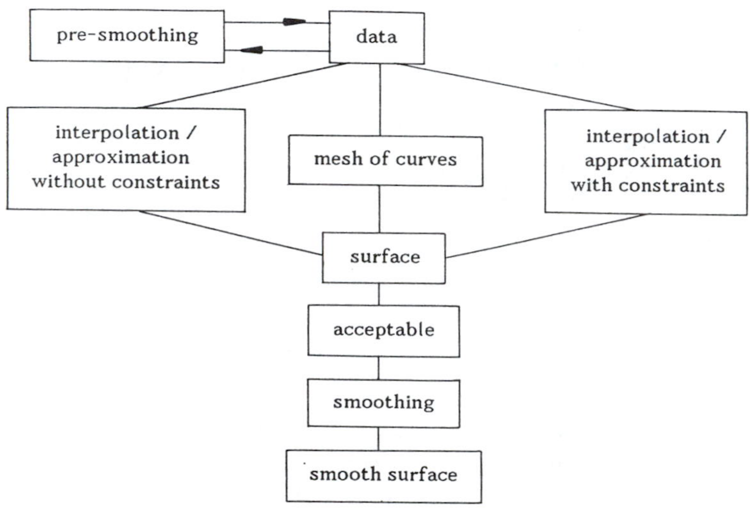

Fig. 13.1. Curve/surface modelling and smoothing process.

13.1. Pre-smoothing of Data

Methods for pre-smoothing data are based on using difference quotients to discretize various geometric invariants (curvatures) of curves or surfaces. For example, for functions, the curvature is proportional to the second derivative. Hence, oscillations in a planar function-valued curve can be detected by examining first and second differences of the data (x_i, t_i), see [Renz 82]. This involves forming the divided differences

$$\Delta_{i,i+1} = \frac{x_i - x_{i+1}}{t_i - t_{i+1}}, \qquad dt_i = \frac{t_i + t_{i+1}}{2}, \tag{13.1a}$$

$$\Delta_{i,i+1,i+2} = \frac{\Delta_{i,i+1} - \Delta_{i+1,i+2}}{t_i - t_{i+2}}, \qquad ddt_i = \frac{dt_i + dt_{i+1}}{2}, \tag{13.1b}$$

and looking at their associated graphs. Regions where there is a strongly oscillating second difference can be smoothed out by performing a formal integration of the corresponding difference quotients. In order to make sure that the resulting function matches up at the end points, we can correct it with a linear weight function.

[Sap 90] has suggested the following smoothing criterion for twice continuously differentiable curves. Given a curve $\boldsymbol{X}(s)$, we first calculate the jumps

$$z_i = |\dot{\kappa}(s_i^-) - \dot{\kappa}(s_i^+)| \tag{13.2a}$$

in the derivative of the curvature with respect to the arc length at the points $\boldsymbol{P}_i$ corresponding to the parameter values s_i, $i = 0(1)N$. The z_i characterize the local discontinuities in $\dot{\kappa}$, and summing them leads us to the global quantity

$$\xi = \sum_{i=1}^{N-1} z_i \tag{13.2b}$$

which we wish to minimize. The algorithm can be summarized as follows:

a) compute ξ^0 for the initial curve,

b) find the largest z_j, and smooth the curve in the region influenced by $\boldsymbol{P}_j$ by changing it to $\boldsymbol{P}_j^*$,

c) if the resulting ξ is smaller than it was, go to step b); otherwise, reverse the last step and stop.

This method was extended by [Eck 90] to consider sets of points, using a method of [SU 89] to approximate the curvature of the curve by the reciprocal of the radius of the circle which passes through the points $(\boldsymbol{P}_{i-1}, \boldsymbol{P}_i, \boldsymbol{P}_{i+1})$:

$$K_i = \frac{4\Delta_i}{|\boldsymbol{L}_i|\,|\boldsymbol{L}_{i+1}|\,|\boldsymbol{Q}_i|}, \tag{13.3a}$$

where

$$\boldsymbol{L}_i = \boldsymbol{P}_i - \boldsymbol{P}_{i-1}, \qquad \boldsymbol{Q}_i = \boldsymbol{P}_{i+1} - \boldsymbol{P}_{i-1}, \qquad \Delta_i = \tfrac{1}{2}\det(\boldsymbol{L}_i, \boldsymbol{L}_{i+1}).$$

The quantity K_i has a negative sign when $(\boldsymbol{P}_{i-1}, \boldsymbol{P}_i, \boldsymbol{P}_{i+1})$ are in clockwise order, and a positive sign if they are in counter-clockwise order. Then as in (13.2a) and (13.2b), we introduce the local measure

$$D_i = \frac{(K_{i+1} - K_i)}{|\boldsymbol{L}_{i+1}|} - \frac{(K_i - K_{i-1})}{|\boldsymbol{L}_i|}$$

of discrete curvature, and the corresponding global measure

$$\xi = \sum_{i=1}^{N-1} D_i. \tag{13.3b}$$

Then the algorithm described above is applied to reduce ξ.

This method can be extended to space curves by working with suitable differences for torsion values. An algorithm for smoothing surface data using the bi-Laplace equation was developed by [Langr 84].

In addition to data smoothing, it may also be useful to perform some kind of data reduction. The idea is to first smooth the data, and then to either thin it out or fit it with some interpolation or approximation method so that a curve can be constructed to within a prescribed error tolerance, see, e.g., [Wev 88, 91], [Eck 90]. Fig. 13.2 shows how the method works for some CAT-scan data involving a slice of a person's femur. Fig. 13.2a shows the original data and the associated curvature plot based on (13.3a). Fig. 13.2b shows the smoothed data obtained by using the method described above, and also shows the associated curvature plot. Fig. 13.2c shows an approximation to the 217 CAT-scan data points of Figs. 13.2a,b using a geometric B-spline curve with 32 Bézier segments.

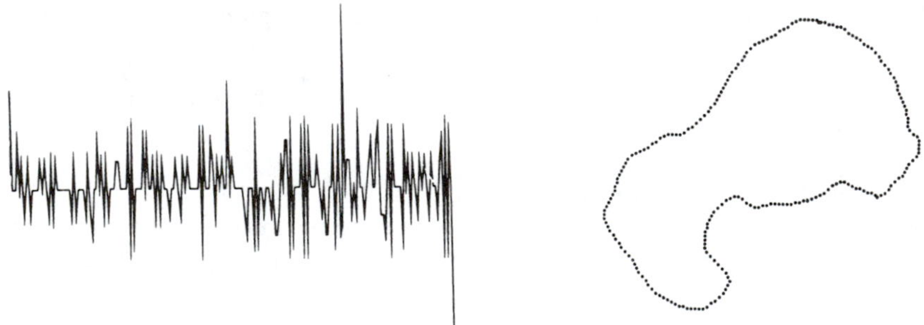

Fig. 13.2a. Original data and curvature plot.

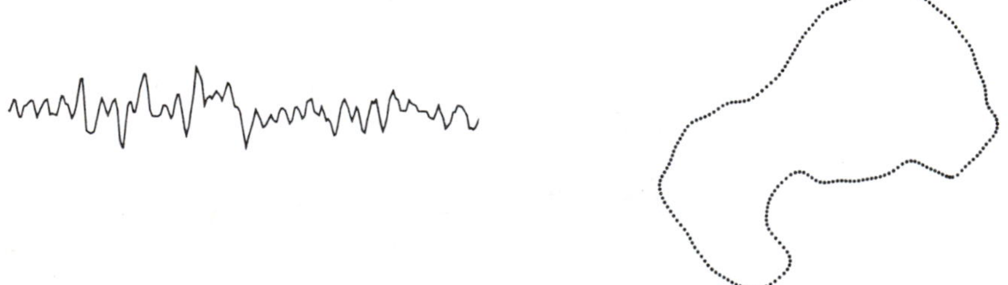

Fig. 13.2b. Smoothed data from Fig. 13.2a and curvature plot.

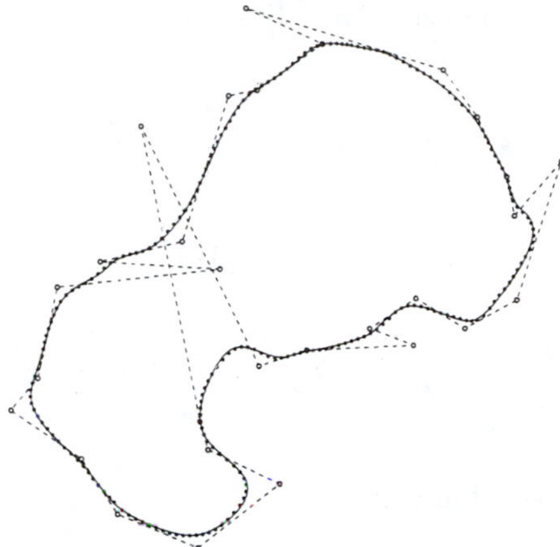

Fig. 13.2c. Geometric B-spline curve describing the CAT-scan data.

13.2. Constructing Smooth Curves and Surfaces by Optimization

In this approach to fitting data, the idea is to choose the curve or surface which is smoothest in the sense that it has the least bending energy. Since this often leads to nonlinear problems, we have to be careful about how we choose a functional to measure energy. We have already encountered several such criteria for curves. Here we consider

- minimal bending [Rei 67], see (3.2):

$$\int_a^b |f''(t)|dt,$$

- minimal bending energy [Meh 74], see (3.53):

$$E = \int_a^b \kappa^2 ds,$$

- minimal jerk [Mei 87], [Hag 90]:

$$R = \int_a^b |\dddot{X}(t)|dt,$$

- minimal tension [Schw 66], see (3.37):

$$\sigma = \int_a^b |f''(t)|^2 dt + \alpha^2 \int_a^b |f'(t)|^2 dt,$$

- discretized minimal tension [Nie 74], see (3.46):

$$\sigma = \int_a^b |f''(t)|^2 dt + \sum_{i=1}^{n-1} \nu_i |f'(t_i)|^2,$$

- the expression [Hag 85], [Pot 90], see (3.50):

$$\sigma = \int_a^b |f'''(t)|^2 dt + \sum_{i=1}^{n-1} \nu_i |f'(t_i)|^2 + \sum_{i=1}^{n-1} \eta_i^2 |f''(t_i)|^2,$$

- the expression, see [Ran 91]:

$$\sigma = \int_a^b \left(\frac{\tau^2}{\kappa^2} + \frac{\kappa'^2}{\kappa^4} \right)^{1/2} ds,$$

  where κ is the curvature and τ is the torsion.

For surfaces, the following measures have been proposed:

- minimal bending energy of a plate:

$$U = \int_a^b \int_c^d \left[(f_{xx} + f_{yy})^2 - 2(1-\nu)(f_{xx} f_{yy} - f_{xy}^2) \right] dx dy,$$

  see [Wal 71], [Rei 71], [Lot 88], and [Kal 90],
- approximate bending energy of a plate:

$$U = \int_a^b \int_c^d (\kappa_1^2 + \kappa_2^2) du dv, \tag{13.4}$$

  where κ_1 and κ_2 are principal curvatures, see [Wal 71], [Ree 83], and [Hag 87, 90],
- an extension of the minimal jerk to surfaces:

$$U = \alpha \int_a^b \int_c^d \left| \frac{\partial^3 \boldsymbol{X}(u,v)}{\partial u^3} \right|^2 du dv + \beta \int_a^b \int_c^d \left| \frac{\partial^3 \boldsymbol{X}(u,v)}{\partial v^3} \right|^2 du dv, \tag{13.5}$$

  see [Wör 91], [Hag 92],
- the smoothness measure

$$U = \int_a^b \int_c^d \left| \frac{\partial \boldsymbol{c}}{\partial u} \times \frac{\partial \boldsymbol{c}}{\partial v} \right| du dv, \tag{13.6}$$

where for the indicatrix $\boldsymbol{c}(u,v)$, [Ran 91] has suggested one of the following

$$\boldsymbol{c}(u,v) := \begin{cases} K(u,v)\boldsymbol{N}(u,v), & \text{or} \\ \boldsymbol{X}(u,v) + \frac{H(u,v)}{K(u,v)}\boldsymbol{N}(u,v), & \text{or} \\ [K(u,v) + H(u,v)^2]\boldsymbol{N}(u,v), & \end{cases}$$

with $\boldsymbol{N}(u,v)$ the unit normal vector to the surface $\boldsymbol{X}(u,v)$, K the Gaussian curvature, and H the mean curvature.

If we represent our curves or surfaces in either Bézier or B-spline form, then minimizing the measure of smoothness (13.4) for TP-B-spline surfaces can be accomplished using standard nonlinear optimization methods, see e.g., [Lot 88]. A number of authors have suggested combining the overall error with convex combinations of the energy expressions, see, e.g., [Now 89, 90, 92]. For example, for curves, we can minimize

$$(1-\alpha)\sum_{i=1}^{n}(\boldsymbol{X}(t_i) - \boldsymbol{P}_i)^2 + \alpha\big[(1-\beta)\int_a^b |\boldsymbol{X}''|^2 ds + \beta\int_a^b |\boldsymbol{X}'''|^2 ds\big]. \qquad (13.7)$$

For analogous measures for surfaces, see [Wör 91], [Hag 92]. The integrals appearing here must be calculated using some numerical quadrature rule.

Smoothing of data in connection with interpolation has proved to be very effective. [Hosa 69] combined the energy integrals with a measure of closeness of fit, leading to the problem of minimizing the following functional:

$$U = \tfrac{1}{2}EI\int_a^b \kappa^2(s)ds + \tfrac{1}{2}\sum_{i=0}^{n} c_i(\boldsymbol{P}_i - \boldsymbol{P}_i^*)^2, \qquad (13.8)$$

where E is the modulus of elasticity, I is the moment of inertia, c_i is the coefficient of stiffness, $\boldsymbol{P}_i$ are the given points, and $\boldsymbol{P}_i^*$ are the corrected points.

The expression (13.8) can be extended to surfaces by working with curve networks. Suppose we are given an ordered set of measurement points $\boldsymbol{P}_{ij}$ which are attached to movable grid points $\boldsymbol{Q}_{ij}$ by springs with stiffness α_{ij}. We define a curve network over the grid consisting of families of curves $\boldsymbol{C}_j$ and $\boldsymbol{D}_i$ which are themselves elastic splines with constant stiffness EI. The equilibrium position of this elastic network can be found by minimizing the energy functional

$$U = \tfrac{1}{2}\sum_{i=1}^{m}\sum_{j=1}^{n}\alpha_{ij}|\boldsymbol{P}_{ij} - \boldsymbol{Q}_{ij}|^2 + \tfrac{1}{2}EI\Big[\sum_{i=1}^{m}\int_{\boldsymbol{C}_i}\kappa_u^2 ds_u + \sum_{j=1}^{n}\int_{\boldsymbol{D}_j}\kappa_v^2 ds_v\Big],$$

([Hosa 69], [Now 89, 90, 92]), where κ_u, κ_v are the curvatures in the u and v directions, and ds_u and ds_v are the arc-length differentials in the u and v

directions, respectively. The behavior of the curves at the boundaries of the domain can be controlled by end conditions. The ratios α_{ij}/EI determine the relative influence of the various data points.

Another approach to smoothing curve networks with fixed grid points Q_{ij} was suggested by [Now 92]. Given a net

$$h = \{(x, y_i, h_{x_i}(x)), \quad i = 1(1)N, \qquad (x_j, y, h_{y_j}(y)), \quad j = 1(1)M\},$$

with $h_{x_i}, h_{y_j} \in C^2$ and $h_{x_i}(x_j) = h_{y_j}(y_i)$, the smoothness is measured by the functional

$$L_2(h) = \sum_{i=1}^{N} \int_{x_1}^{x_M} (h''_{x_i}(x))^2 dx + \sum_{j=1}^{M} \int_{y_1}^{y_N} (h''_{y_j}(y))^2 dy.$$

Combining this with the sum of squares of the errors, and applying methods of variational calculus, gives us the desired smooth curve network. To get a surface, we can then interpolate this net using G^1 or G^2 continuous patches, see [Now 89]. Similar methods were discussed by [And 88].

13.3. Detecting Undesirable Curve and Surface Behavior

We begin by describing when a curve or surface has some undesirable behavior which requires smoothing. We formulate our criteria in terms of expected curvature. A curve contains an *undesirable segment* provided

- the segment should be convex, but instead has points of inflection, see Fig. 13.3a,
- the segment should have just one point of inflection, but in fact has more, see Fig. 13.3b.

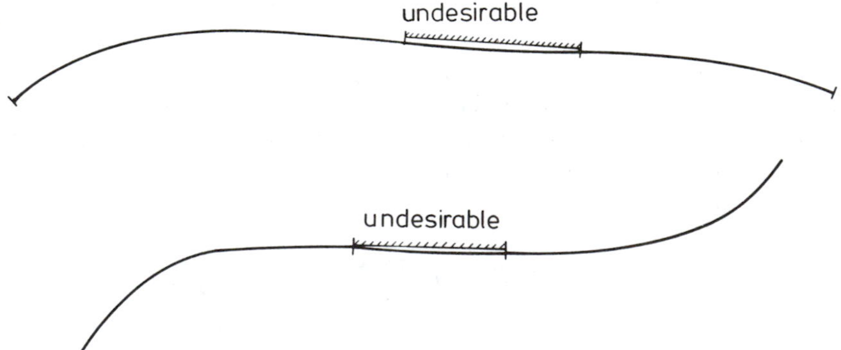

Fig. 13.3a,b. Undesirable segments of a curve.

A surface patch has *undesirable segments* provided

- the segment should be convex (positive Gaussian curvature), but the curvature changes sign,
- the segment should have negative Gaussian curvature, but the curvature changes sign,
- the Gaussian curvature of the segment should change sign at most once, but in fact has several sign changes.

We have already presented a method for detecting undesirable segments of curves in (13.2a) and (13.3a). Other methods which can be used for curves and surfaces include

- the *isoline method*, where we examine the surface on a network of curves corresponding to some geometric property such as constant height, constant angle of slope, constant Gaussian curvature, constant mean curvature, etc.,
- the *reflection line method*, which looks for irregularities in the reflected image corresponding to parallel linear light sources,
- the *mapping method*, where singularities in a certain image of the curve or surface correspond to undesirable segments.

The *isoline method* is computationally very expensive, since it almost always leads to a transcendental system of equations. Contour lines, lines of constant Gaussian curvature, etc., can always be described as the zero set of some appropriate function of the surface parameters (u, v). The problem then is to find zeros of these functions. One way to do this is to keep one variable constant, and look for zeros with respect to the other variable (parameter discretization, see Sect. 12.4.2). The determination of isolines can be simplified by dividing the surface into a number of smaller patches (cells) which can be examined individually, see [Hartw 83], [Ree 83], and Sect. 12.4.

To describe the *reflection line method* (see [Kla 80]), suppose we are given a continuously differentiable surface $\boldsymbol{X} = \boldsymbol{X}(u, v)$ in parametric form with normal vector $\boldsymbol{N} = \boldsymbol{N}(u, v)$ for $u, v \in [0, 1]$. Given a straight line $\boldsymbol{L}$ and a fixed observation point $\boldsymbol{A}$, we define the corresponding *reflection line* to be the image of the line $\boldsymbol{L}$ as reflected in the surface $\boldsymbol{X}$, see Fig. 13.4.

To examine the behavior of a surface patch, we look at the reflection lines corresponding to a family of lines. Irregularities in the reflection lines point to undesirable behavior. Fig. 13.5 shows a typical reflection line image which exhibits irregularities in the center of the image.

The reflection line method is a mathematical model of a method which is actually employed in the automobile industry for examining surfaces of auto

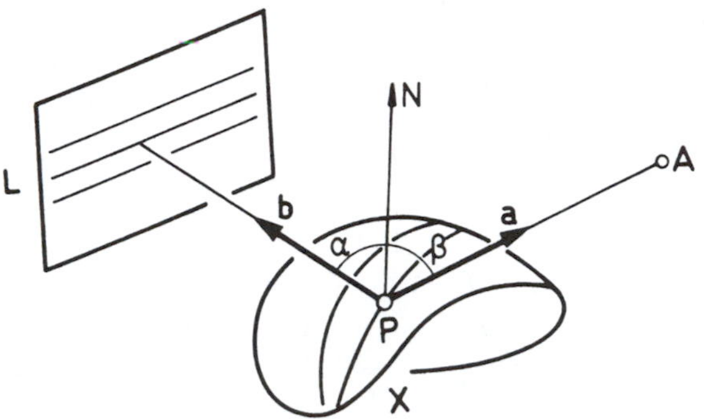

Fig. 13.4. Reflection of light lines.

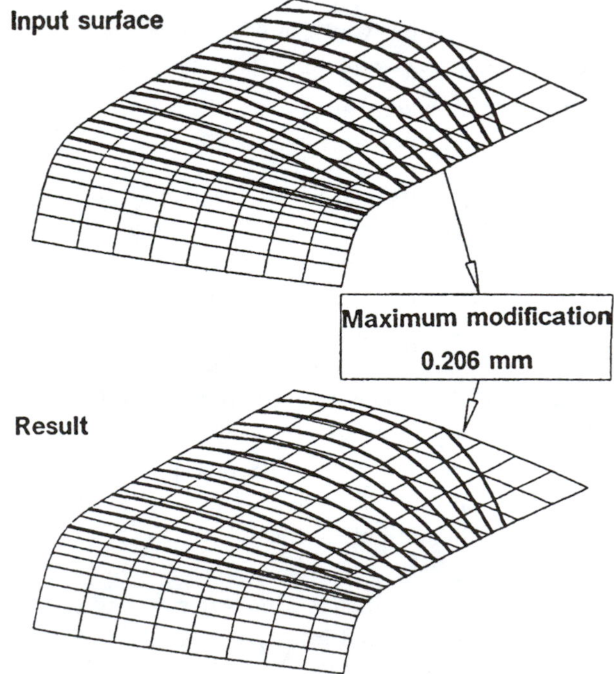

Fig. 13.5. Reflection line image with irregularities.

bodies using a cage of parallel light sources. A designer makes use of this tool to find irregularities in the surface and correct them, see Fig. 13.5 (which was kindly provided by Mercedes-Benz). Automobile customers are used to judging bodies of cars by reflection lines.

Recently, [Kau 88] has proposed an easy to use *modification of the reflection line method* which is very sensitive to irregularities in surfaces. First we construct a family of curves on the surface, for example by taking plane parametric lines if they exist, or by finding the intersection curves of the surface with a family of planes. Next, we select a light direction $\boldsymbol{a}$ and a set of reflection angles α_i, $i \in I$. Finally, we find the points $\boldsymbol{P}_j$ on the family of curves whose tangents form an angle α_ℓ, $\ell \in I$, with the light direction $\boldsymbol{a}$.

Another approach to detecting regions where the curvature exhibits some undesirable behavior is the method of *k-orthotomics*, which comes from optics and catastrophy theory, see [Bru 84] and [Hos 85a]. A method based on certain polarities was suggested in [Hos 84].

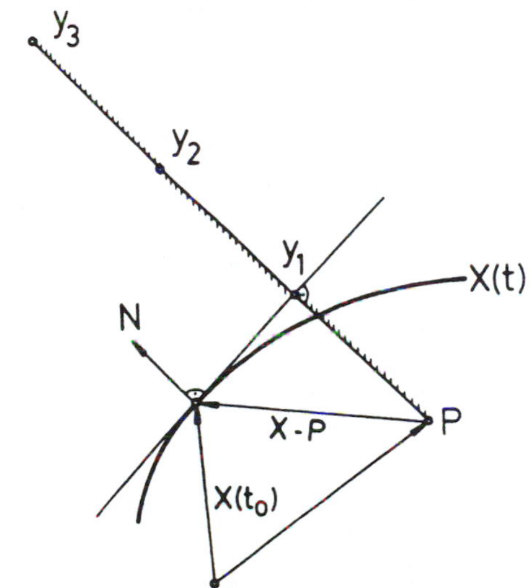

Fig. 13.6. Reflection of a point $\boldsymbol{P}$ in the tangent to a curve $\boldsymbol{X}(t)$ leads to $\boldsymbol{Y}_2$. Doubling the distance from the line gives $\boldsymbol{Y}_3$.

We explain the construction of k-orthotomics first for planar curves. Suppose $\boldsymbol{X}(t)$ is such a curve, and that $\boldsymbol{P}$ is a given point which does not lie on $\boldsymbol{X}(t)$ or on any tangent to $\boldsymbol{X}(t)$. For each point $\boldsymbol{X}(t_0)$ on the curve, let $\boldsymbol{Y}_2$ be the reflection of the point $\boldsymbol{P}$ in the line which is tangent to the curve at the point $\boldsymbol{X}(t_0)$, see Fig. 13.6. As $\boldsymbol{X}(t)$ runs over the curve, this process generates a curve $\boldsymbol{Y}_2(t)$, called the *2-orthotomic of $\boldsymbol{X}(t)$ with respect to $\boldsymbol{P}$*. The 2-orthotomic can also be interpreted as the *reflected wave front* corresponding to a point light source at $\boldsymbol{P}$, or as the *virtual image* of $\boldsymbol{P}$ in a mirror $\boldsymbol{X}(t)$. From Fig. 13.6, it follows that the parametric representation of $\boldsymbol{Y}_2(t)$ is given by

$$Y_2(t) = P + 2[(X(t) - P) \cdot N(t)]\, N(t), \tag{13.9a}$$

where $N(t)$ is the unit normal vector to the curve $X(t)$. If we replace the factor 2 by k, (13.9a) becomes the equation of the *k-orthotomic*. The next theorem is due to [Hos 85a].

Theorem 13.1. *Let $X(t)$ be a C^3 regular parametrized planar curve, and let P be a point which does not lie on the curve or on any tangent to the curve. Then the k-orthotomic with respect to the point P has a singularity (a cusp) at $t = t_0$ if and only if $X(t)$ has a point of inflection at $t = t_0$.*

Remark. If P lies on a tangent to $X(t)$, then (13.9a) implies that the k-orthotomic passes through P, and has a singularity there.

We illustrate this mapping method with an example. Consider a family of Bézier curves of degree 4 with Bézier points

$$b_0 = (-12, 0), \quad b_1 = (-6, 6), \quad b_2 = (0, p), \quad b_3 = (6, 3), \quad b_4 = (12, 0),$$

where p is a parameter running over $[-1.5, 2.5]$. It is clear from Fig. 13.7a that the top curve in the family is convex, while the bottom one is not. The associated 10-orthotomics are shown in Fig. 13.7b. They clearly indicate that the bottom two curves in Fig. 13.7a are not convex.

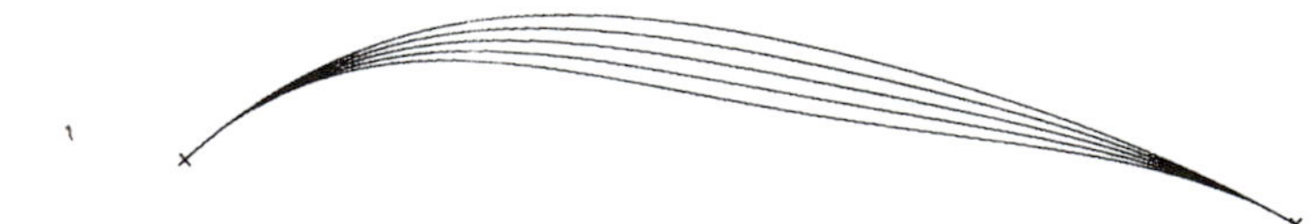

Fig. 13.7a. Family of Bézier curves.

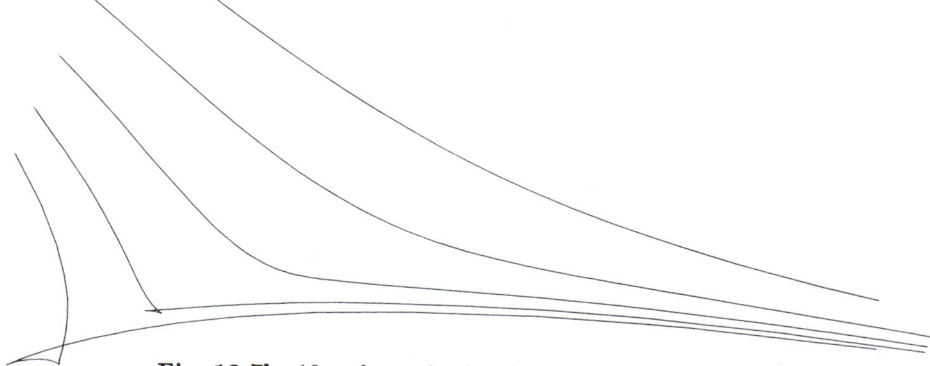

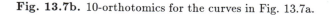

Fig. 13.7b. 10-orthotomics for the curves in Fig. 13.7a.

The principle of k-orthotomics can be extended to surfaces. Given a surface $\boldsymbol{X}(u,v)$ and a point $\boldsymbol{P}$, we reflect $\boldsymbol{P}$ in a tangent plane to the surface. As before, we can extend the reflected line segment by a factor of k. This leads to the following equation for the k-orthotomic:

$$\boldsymbol{Y}(u,v) = \boldsymbol{P} + k[(\boldsymbol{X}(u,v) - \boldsymbol{P}) \cdot \boldsymbol{N}(u,v)]\,\boldsymbol{N}(u,v). \tag{13.9b}$$

The following result is due to [Hos 85a].

Theorem 13.2. *Let $\boldsymbol{X}(u,v)$ be a regular parametrized surface of class C^3, and let $\boldsymbol{P}$ be a point which does not lie on the surface $\boldsymbol{X}$ or on any tangent plane to it. Then the k-orthotomic $\boldsymbol{Y}$ of $\boldsymbol{X}$ with respect to $\boldsymbol{P}$ has a singularity at (u_0, v_0) if and only if the Gaussian curvature of $\boldsymbol{X}$ is zero or changes sign at this point.*

To illustrate this result, consider the Bézier surface of degree $(3,3)$ in Fig. 6.3 with Bézier points

$$\begin{array}{llll}
\boldsymbol{b}_{00} = (0,0,0), & \boldsymbol{b}_{01} = (0,1,2), & \boldsymbol{b}_{02} = (0,2,1), & \boldsymbol{b}_{03} = (0,3,0), \\
\boldsymbol{b}_{10} = (1,0,1), & \boldsymbol{b}_{11} = (1,1,2), & \boldsymbol{b}_{12} = (1,2,2), & \boldsymbol{b}_{13} = (1,3,2), \\
\boldsymbol{b}_{20} = (2,0,2), & \boldsymbol{b}_{21} = (2,1,2), & \boldsymbol{b}_{22} = (2,2,2), & \boldsymbol{b}_{23} = (2,3,1), \\
\boldsymbol{b}_{30} = (3,0,0), & \boldsymbol{b}_{31} = (3,1,1), & \boldsymbol{b}_{32} = (3,2,2), & \boldsymbol{b}_{33} = (3,3,0).
\end{array}$$

All parametric lines on this surface are convex, but the surface itself is not, since as can be seen in Figure 13.8, the Gaussian curvature changes sign near the four corners.

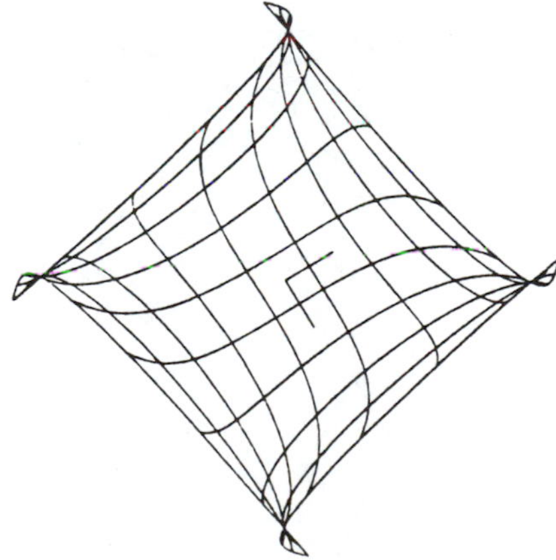

Fig. 13.8. Orthotomic image of the surface in Fig. 6.3.

The locations of the bad parts of the surface in Fig. 13.9 are easy to identify. In general, however, we can't draw any conclusions as to which parts are to be smoothed. In order to overcome this problem, we can map the singularities of the orthotomic surface onto the original surface. At singularities of the orthotomic surface $\boldsymbol{Y}$, the normal vector to $\boldsymbol{Y}$ vanishes, i.e.,

$$\boldsymbol{Y}_u \times \boldsymbol{Y}_v = 0. \tag{13.10}$$

To avoid computing the derivatives explicitly, we can discretize the parameter domain, e.g., using equally spaced intervals Δ, and replacing the derivatives in (13.10) by difference vectors. This leads to the approximations

$$\begin{aligned} \boldsymbol{N}_u(u_i, v_k) &\approx \boldsymbol{N}(u_i, v_k) - \boldsymbol{N}(u_i + \Delta, v_k), \\ \boldsymbol{N}_v(u_i, v_k) &\approx \boldsymbol{N}(u_i, v_k) - \boldsymbol{N}(u_i, v_k + \Delta). \end{aligned}$$

Now we mark those points $\boldsymbol{P}_{ij}$ on the original surface $\boldsymbol{X}$ where

$$|\boldsymbol{N}_u \times \boldsymbol{N}_v| < \epsilon. \tag{13.11}$$

This discretization process can, of course, falsely identify regions (where the Gaussian curvature is near zero) as being bad. In Fig. 13.9 we have marked the points on the surface corresponding to singular points of the orthotomic (which theoretically correspond to points where the Gaussian curvature changes sign).

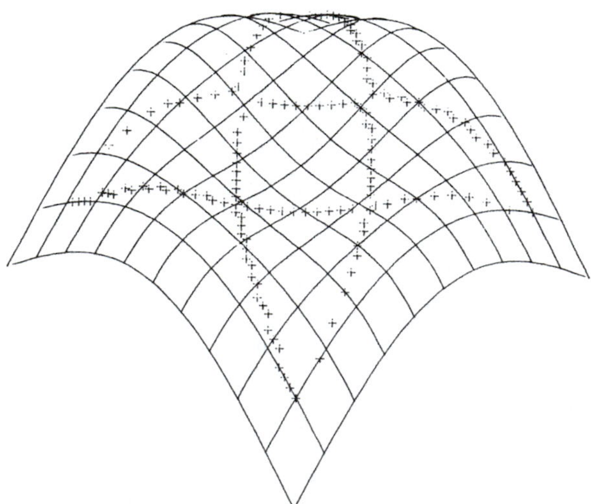

Fig. 13.9. Surface patch with points marked where the Gaussian curvature changes sign.

13.4. Fairing Curve and Surface Segments

It is more difficult to improve bad segments of curves and surfaces than it is to detect them. We would like to have methods which are local in the sense that any modifications we make will effect only the segment of interest, leaving the rest of the curve or surface unchanged. In some cases, this may require some additional subdivision using the algorithm of de Casteljau or de Boor. This leads to new smaller curve segments or patches which then have to be smoothed, taking account of continuity conditions.

13.4.1. Fairing Curve Segments

In simple cases, we can fair a curve *interactively* by modifying some of its control points. For example, if the bad part of the curve is in the middle, then we recommend splitting off the ends using the de Casteljau algorithm, and then altering a Bézier point in the middle to remove the offending point of inflection, see [Hos 84, 85a].

The above interactive approach is generally applicable only in very special cases, and so we now want to develop *algorithmic methods* for fairing.

We again consider first the case of *fairing curves*. [Kje 83] assumes that the bending energy of a cubic spline curve is reduced if the jumps in the third derivatives vanish or are made as small as possible. Thus, to fair a curve in some neighborhood of a given knot, we try to move it in one direction or the other to make the jump in the third derivative equal to zero. This condition on the third derivative is then added to the system of equations (3.18), where now the knot becomes one of the unknowns.

A similar approach was suggested by [Far 87], [Sap 90], based on minimizing the expression in (13.2b). By modifying the algorithm described there, we can define an automatic fairing method which is as close as possible to being convexity preserving. It proceeds as follows:

- Knot removal: remove the knot t_j corresponding to the largest variation in κ', and use appropriate symmetry properties to find the new control points. The new spline curve $\bar{\boldsymbol{X}}$ is defined over the knot vector $\bar{\boldsymbol{T}} = \boldsymbol{T}\backslash\{t_j\}$.
- Knot insertion: insert the knot t_j into the knot vector $\boldsymbol{T}$, and find the new control points so that the curve $\bar{\boldsymbol{X}}$ is defined over the new knot vector $\bar{\boldsymbol{T}}$. These new control points of $\bar{\boldsymbol{X}}$ are not the same as the original control points for the curve.

Bad segments of curves can also be faired by *eliminating* the parameter t using algebraic methods. This idea is based on the observation that for Bézier curves (or surfaces), the shape of the curve (or surface) depends exclusively on the Bézier points, while the parameter only serves as a means for finding individual points on the curve (or surface). For parameter elimination, we use the method of *Bézout elimination* (see Chap. 12), which is discussed in many books on algebra, e.g., [SAL 85], or [Gol 85], [Sed 84].

In order to describe fairing algebraically, we first need a *sufficient condition* for a *planar* curve to be fair. We want to alter the behavior of the curve in a neighborhood of undesired points of inflection in such a way that a pair of points of inflection is changed into a single point with stationary curvature.

Suppose we want the curvature κ to vanish at a point on the curve which is simultaneously an extreme value of κ. We call such a point a *stationary flat point*. At such a point we must have $\kappa = 0$ and $\kappa' = 0$, see Fig. 13.10. Thus, differentiating the formula for the curvature κ of the curve (see Chap. 2), we see that sufficient conditions for a point to be a stationary flat point are (see [Hos 84]) that

$$\boldsymbol{X}' \times \boldsymbol{X}'' = 0, \qquad \boldsymbol{X}' \times \boldsymbol{X}''' = 0. \tag{13.12}$$

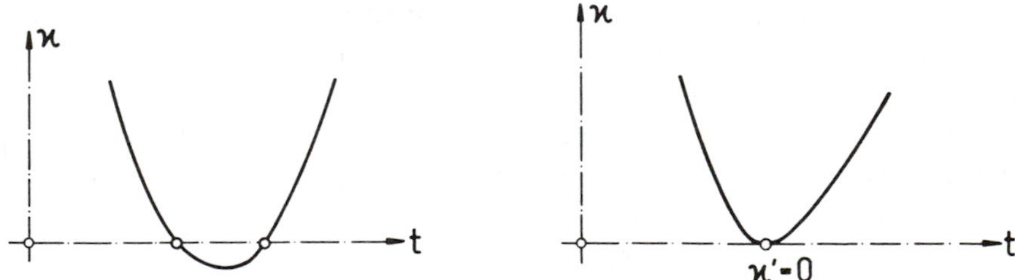

Fig. 13.10. Curvature diagram and stationary point with $\kappa' = 0$.

As shown in Chap. 4, the derivatives of a Bézier curve are given by

$$\begin{aligned} \boldsymbol{X}' &= \frac{n!}{(n-1)!}\sum_{i=0}^{n-1} B_i^{n-1}(t)\Delta^1 \boldsymbol{b}_i, \\ \boldsymbol{X}'' &= \frac{n!}{(n-2)!}\sum_{i=0}^{n-2} B_i^{n-2}(t)\Delta^2 \boldsymbol{b}_i, \\ \boldsymbol{X}''' &= \frac{n!}{(n-3)!}\sum_{i=0}^{n-3} B_i^{n-3}(t)\Delta^3 \boldsymbol{b}_i, \end{aligned} \tag{$*$}$$

where Δ^1, Δ^2, and Δ^3 denote the first, second, and third differences. Now transforming the Bernstein basis in t to a *standard basis* in w by dividing by

$(1-t)^n$ and substituting $w = t/(1-t)$, we get

$$\frac{B_k^n(t)}{(1-t)^n} = \binom{n}{k} w^k, \qquad w \in \mathbb{R}^+ .$$

But then $(*)$ is transformed into

$$\begin{aligned}
\boldsymbol{X}' &\hat{=} \textstyle\sum_{k=0}^{n-1} \boldsymbol{C}_k w^k, && \text{with} && \boldsymbol{C}_k := \binom{n-1}{k}\Delta^1 \boldsymbol{b}_k,\\
\boldsymbol{X}'' &\hat{=} \textstyle\sum_{k=0}^{n-2} \boldsymbol{D}_k w^k, && \text{with} && \boldsymbol{D}_k := \binom{n-2}{k}\Delta^2 \boldsymbol{b}_k,\\
\boldsymbol{X}''' &\hat{=} \textstyle\sum_{k=0}^{n-3} \boldsymbol{E}_k w^k, && \text{with} && \boldsymbol{E}_k := \binom{n-3}{k}\Delta^3 \boldsymbol{b}_k,
\end{aligned}$$

where we have dropped non-essential factors. If we substitute $(*)$ in (13.12), we get the following polynomials (expressed in terms of monomials):

$$\boldsymbol{X}' \times \boldsymbol{X}'' := \sum_{k=0}^{2n-3} \boldsymbol{A}_k w^k, \qquad \boldsymbol{X}' \times \boldsymbol{X}''' := \sum_{k=0}^{2n-4} \boldsymbol{B}_k w^k, \tag{13.13}$$

where the coefficients are given by

$$\boldsymbol{A}_k = \sum_m (\boldsymbol{C}_m \times \boldsymbol{D}_l) \quad \text{with} \quad \begin{cases} m + l = k, & m, l > 0, \\ m < n-1, & l < n-2, \end{cases}$$

and

$$\boldsymbol{B}_k = \sum_m (\boldsymbol{C}_m \times \boldsymbol{E}_l) \quad \text{with} \quad \begin{cases} m + l = k, & m, l > 0, \\ m < n-1, & l < n-3. \end{cases}$$

In order to be able to use vector algebra, we introduce

$$\boldsymbol{M}_k := (\boldsymbol{A}_k, \boldsymbol{B}_k).$$

Now both of the equations in (13.13) can be expressed in vector-valued polynomial form as

$$\boldsymbol{P}(w) := \sum_{k=0}^{N} \boldsymbol{M}_k w^k = 0, \quad \text{with} \quad N := 2n-3. \tag{13.14}$$

In order to eliminate the parameter w, we now proceed as follows. We split $\boldsymbol{P}(w)$ into two terms

$$\begin{aligned}
\boldsymbol{P}(t) = {} & w^{N-k}(\boldsymbol{M}_N w^k + \boldsymbol{M}_{N-1} w^{k-1} + \cdots + \boldsymbol{M}_{N-k}) \\
& + (\boldsymbol{M}_{N-k-1} w^{N-k-1} + \cdots + \boldsymbol{M}_0).
\end{aligned}$$

Setting

$$\boldsymbol{Q}_k(w) := \boldsymbol{M}_N w^k + \boldsymbol{M}_{N-1} w^{k-1} + \cdots + \boldsymbol{M}_{N-k},$$

it follows that the N vector products $\boldsymbol{Q}_0(w) \times \boldsymbol{P}(w), \dots, \boldsymbol{Q}_{N-1}(w) \times \boldsymbol{P}(w)$ are all scalar polynomials of degree $N-1$ in w ($\boldsymbol{P}$ and the $\boldsymbol{Q}_k$ are vectors of length 2):

$$\begin{pmatrix} \boldsymbol{Q}_0 \times \boldsymbol{P} \\ \boldsymbol{Q}_1 \times \boldsymbol{P} \\ \vdots \\ \boldsymbol{Q}_{N-1} \times \boldsymbol{P} \end{pmatrix} = \begin{pmatrix} R_{00} & \cdots & R_{0,N-2} & R_{0,N-1} \\ R_{10} & \cdots & R_{1,N-2} & R_{1,N-1} \\ \vdots & & & \vdots \\ R_{N-1,0} & \cdots & R_{N-1,N-2} & R_{N-1,N-1} \end{pmatrix} \begin{pmatrix} w^{N-1} \\ w^{N-2} \\ \vdots \\ w \\ 1 \end{pmatrix}, \tag{13.15}$$

where

$$R_{ik} = \sum_p (\boldsymbol{M}_p \times \boldsymbol{M}_q)$$

and

$$p + q = 2N - i - k - 1, \quad p \geq \max(N-i, N-k), \quad p \leq N.$$

The scalar coefficients in (13.15) depend only on the Bézier points. These polynomials vanish only if the vector polynomial $\boldsymbol{P}(w)$ vanishes. The vanishing of $\boldsymbol{P}(w)$ leads to the conditions (13.12) for points with stationary curvature. Thus, we have a criterion for the fairing of Bézier curves which depends only on the Bézier points:

$$\det |R_{ik}| = 0. \tag{13.16}$$

This determinant of order N is called the *Bézout determinant*.

In order to apply this Bézout determinant to the problem of fairing Bézier curves, we select a Bézier point $\boldsymbol{b}_k$, and choose the parameter p so that $\boldsymbol{b}_k = (x_k, p)$, where x_k is the given abscissa of $\boldsymbol{b}_k$. If we insert the Bézier point $\boldsymbol{b}_k$ into the Bézout determinant (13.16), we get a function depending on the parameter p. We now vary this parameter p in order to force $\det |R_{ik}(p)|$ to vanish. A value p where this happens defines a flat point of the associated Bézier curve.

This algebraic method can clearly be carried out with the help of *formula manipulation software* such as Reduce, Macsyma, Mathematica, etc.

In Chap. 4 we have seen that the *length of the parameter interval* has a strong influence on the shape of a curve. [Schel 84] has developed a fairing method for curves and surfaces based on a reparametrization. The mathematical basis for the method is provided by the following two theorems. The first deals with the problem of maintaining the boundary conditions on the derivatives while reparametrizing.

Theorem 13.3. *Let $\boldsymbol{X}(t)$ be a Bézier curve of degree n defined on $[a,b]$, where $n > 2j+1$. Suppose we scale the parameter interval $[a,b]$ by the factor $x := (\tilde{b}-a)/(b-a)$, and change the first $j+1$ and last $j+1$ Bézier points of $\boldsymbol{X}(t)$ to*

$$\widetilde{\boldsymbol{b}}_k(x) = \sum_{\ell=0}^{k} \binom{k}{\ell} \boldsymbol{b}_\ell (1-x)^{k-\ell} x^\ell, \qquad \widetilde{\boldsymbol{b}}_{n-k}(x) = \sum_{\ell=0}^{k} \binom{k}{\ell} \boldsymbol{b}_{n-\ell} (1-x)^{k-\ell} x^\ell,$$

for $0 \le k \le j$. Then the first j derivatives of $\boldsymbol{X}(t)$ and $\tilde{\boldsymbol{X}}(\tilde{t})$ agree on the boundary of the parameter interval.

A function $f = f(x_1, x_2, \ldots, x_n)$ is called *variation diminishing* provided that the number of sign changes $V(f)$ of the function f satisfies

$$V(f) \le V(x_1, x_2, \ldots, x_n),$$

where $V(x_1, x_2, \ldots, x_n)$ denotes the number of sign changes in the sequence $x_1, x_2, \ldots, x_n$.

Theorem 13.4. *Let $\boldsymbol{X}(t)$ be a Bézier curve of degree $n = 2q+1$, and suppose $\boldsymbol{B}$ is the vector of Bézier points $\boldsymbol{b}_0, \ldots, \boldsymbol{b}_n$ of $\boldsymbol{X}$. Let*

$$\tilde{\boldsymbol{B}} = \boldsymbol{X}_q \boldsymbol{B} \tag{13.17}$$

be the vector of transformed Bézier points $\widetilde{\boldsymbol{b}}_0, \ldots, \widetilde{\boldsymbol{b}}_n$ obtained by multiplying by the $(n+1) \times (n+1)$ transformation matrix

$$\boldsymbol{X}_q = \begin{pmatrix} 1 & 0 & \cdots & 0 & 0 & \cdots & & 0 \\ B_0^1(x) & B_1^1(x) & & 0 & 0 & & & \\ \vdots & & & \vdots & \vdots & & & \vdots \\ B_0^q(x) & & \cdots & B_q^q(x) & 0 & \cdots & & 0 \\ 0 & & & 0 & B_q^q(x) & \cdots & & B_0^q(x) \\ \vdots & & & 0 & 0 & & & \vdots \\ & & & & & & B_1^1(x) & B_0^1(x) \\ 0 & & \cdots & 0 & 0 & \cdots & 0 & 1 \end{pmatrix}.$$

Then this transformation is variation-diminishing for all $x \in [0,1]$.

Remark. The proof of Theorem 13.4 can be accomplished using a result of [Lane 77] which gives conditions for a mapping of the form (13.17) to be variation diminishing. If we choose $x > 1$, then the mapping (13.17) is variation increasing. Fig. 13.11 shows the effectiveness of the method of [Schel 84], where to enhance understanding, we have also included x values larger than 1.

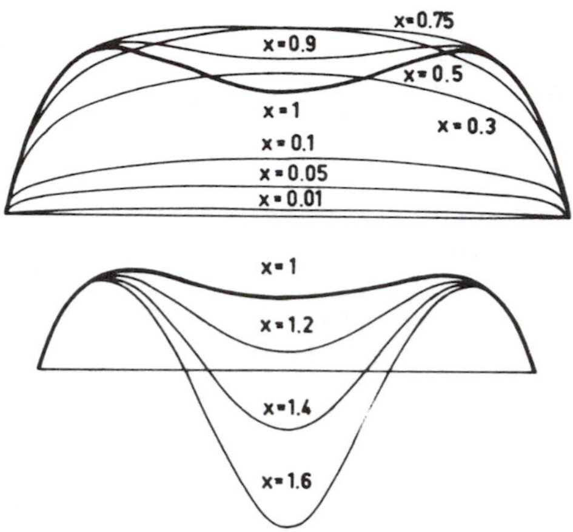

Fig. 13.11. Fairing effect of the parametric transformation of [Schel 84] (the original curve corresponds to $x = 1$).

13.4.2. Fairing Surfaces

[Kje 83] has suggested a method for fairing surfaces based on the assumption that the bending energy of a cubic spline surface is reduced if jumps in the third derivatives are reduced. In this section we discuss some other formula-based and interactive methods.

We consider first an interactive method which provides visual control using the mapping method. The first step is to mark the region on the surface which is to be faired (for example, by mapping singularities of the k-orthotomic onto the surface). Next, we select a point $\boldsymbol{P}$ in the middle of this area, and move this point in an appropriate direction. The control points of the new surface (e.g., Bézier points) are computed, and the orthotomic image is used to control whether the undesired behavior in this region of the surface has been removed. With appropriate software (see, e.g., [Hauc 88]), this process can be carried out directly on the computer screen. The parameter value (u, v) corresponding to the selected surface point $\boldsymbol{P}$ is computed by iteration, and then the true position in $\mathbb{R}^3$ of the translated point $\bar{\boldsymbol{P}} = \boldsymbol{P} + \boldsymbol{d}$ is found, where $\boldsymbol{d}$ is the translation vector. Finally, certain pre-selected interior control points of the surface (e.g., on the Bézier net) are moved in the direction $\boldsymbol{d}$ to force $\bar{\boldsymbol{P}}$ to lie on the new surface. The computation of the new Bézier points can be easily accomplished using formula manipulation software.

If reflection lines are used to identify surface irregularities, then these reflection lines themselves can be used to correct the surface. In particular, by introducing parameters d_i, we change the surface $\boldsymbol{X}(u, v)$ into a new surface $\bar{\boldsymbol{X}}(u, v)$ as follows:

$$\bar{\boldsymbol{X}}(u, v) = \boldsymbol{X}(u + d_1, v + d_2) + n(u + d_1, v + d_2)\boldsymbol{N}(u + d_1, v + d_2). \quad (13.18)$$

The transformation $\boldsymbol{X}(u, v) \to \bar{\boldsymbol{X}}(u, v)$ maps all points $\boldsymbol{P}$ on a reflection line $\boldsymbol{r}$ into points $\bar{\boldsymbol{P}}$ on an (infinitely close) reflection line $\bar{\boldsymbol{r}}$. A given point $\boldsymbol{P} \in \boldsymbol{r}$ can be mapped in more than one way to $\bar{\boldsymbol{P}} \in \bar{\boldsymbol{r}}$, i.e., the d_i are not unique. [Kla 80] suggested that the change in each point $\boldsymbol{P} \in \boldsymbol{r}$ should be perpendicular to $\boldsymbol{r}$. In addition, we have to be sure that $\boldsymbol{P}$ lies on the surface $\bar{\boldsymbol{X}}$ according to (13.18). To accomplish this, we use series expansions to transform (13.18) into a system of partial differential equations for the d_i. This system is then solved numerically.

For *reflection lines with common slopes*, the fairing process can be greatly simplified, see [Kau 88].

For a useful approach to fairing surfaces involving optimally altering the twist vector while minimizing bending energy, see [Ree 83] and [Hag 86a].

The *method of reparametrization* of [Schel 84] can also be extended to surfaces. In this case we can change the parameters in both the u and v directions, but we have to simultaneously change the parameter intervals of the entire network of splines, which means that the method is of a global nature.

In order to apply algebraic methods based on Bézout elimination to fairing surfaces, we suppose that the part of the surface where the undesired sign change in the Gaussian curvature appears is transformed so that it includes a *stationary point* where the Gaussian curvature satisfies $K = 0$. Thus, we get the following conditions for the Gaussian curvature,

$$K = 0, \qquad \frac{\partial K}{\partial u} = 0, \qquad \frac{\partial K}{\partial v} = 0, \quad (13.19)$$

and using (2.16a) we get

$$\begin{aligned} &(\boldsymbol{N}, \boldsymbol{N}_u, \boldsymbol{N}_v) = 0, \\ &(\boldsymbol{N}, \boldsymbol{N}_{uu}, \boldsymbol{N}_v) + (\boldsymbol{N}, \boldsymbol{N}_u, \boldsymbol{N}_{uv}) = 0, \\ &(\boldsymbol{N}, \boldsymbol{N}_{uv}, \boldsymbol{N}_v) + (\boldsymbol{N}, \boldsymbol{N}_u, \boldsymbol{N}_{vv}) = 0. \end{aligned} \quad (13.20)$$

Now we can use Bézout elimination as in (13.12)-(13.16) to remove the parameter u from the first two and last two equations. Then the parameter v

can be eliminated from the resulting polynomials. This leads to an equation which depends only on the Bézier points $\boldsymbol{b}_{ik}$ of the Bézier surface. Finally, we find a zero of this equation by making appropriate changes in individual Bézier points. This elimination process can be done "by hand" only in the simplest cases, and in general, again requires the use of formula manipulation software. Because of memory restrictions, the elimination process generally fails for nonsymmetric surfaces, and for polynomial degrees larger than three.

13.5. Detecting Incorrect Joins in Spline Surfaces

In the construction of C^k continuous spline surfaces, it can sometimes happen that an error can occur which causes a loss of C^k continuity between two patches. We need to have a way of detecting these kinds of irregularities in the spline surface, since they can cause problems later in using the surface in applications, e.g., in milling of surfaces. One approach to detecting lack of smoothness at patch boundaries is to use isophotes, which are lines of equal brightness, see [Rös 37] and Sect. 1.6. If $\boldsymbol{L}$ is the direction of the incident light, and $\boldsymbol{N}$ is the normal vector to the given surface, then along an isophote

$$\boldsymbol{N} \cdot \boldsymbol{L} = c, \tag{13.21}$$

where c is a real constant. The choice $c = 0$ corresponds to the silhouettes, see Chap. 1. The following result is due to [Pös 84].

Theorem 13.5. *If two surface patches join with C^k continuity along a curve $\boldsymbol{X}$, then the isophotes are only C^{k-1} continuous at points on $\boldsymbol{X}$.*

This property is well-known in constructive geometry. Fig. 13.12 illustrates such shadows for two model objects. The model on the left consists of a cone on top of a cylinder, joined with C^0 continuity. This means that its self-shadow boundaries are C^{-1}. The model on the right consists of a half sphere on top of a cylinder, joined with C^1 continuity, so that in this case, its self-shadow boundaries are C^0.

Figs. 13.13 a,b shows two spline surface patches which do not have any apparent irregularities. But since the isophotes in Fig. 13.13a have corners, we know that the first patch is only C^1 continuous. Similarly, the jumps in the isophotes in Fig. 13.13b indicate that the patch shown there is only C^0 continuous. The isophotes were constructed by choosing several different values of c in (13.21), and then solving the resulting transcendental equations numerically. From some view-points, an observer may not be able to detect any defects in the isophotes, and so a complete examination of the surface should include rotating it to look at it from different angles. An improvement of the isophote method can be found in [Pot 88a].

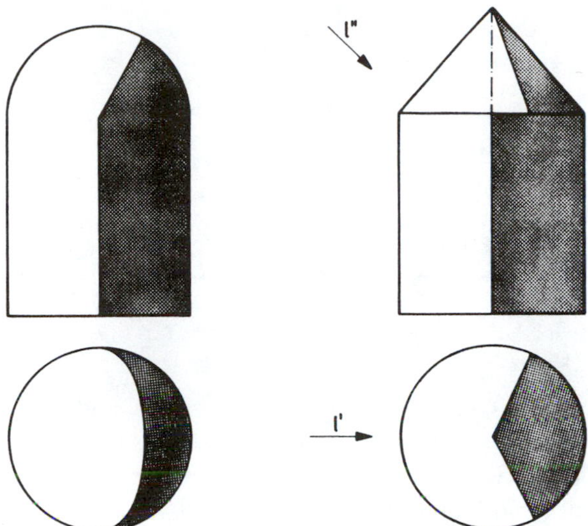

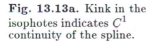

Fig. 13.12. Parts of a body shaded by itself under illumination by parallel light rays in the direction ℓ.

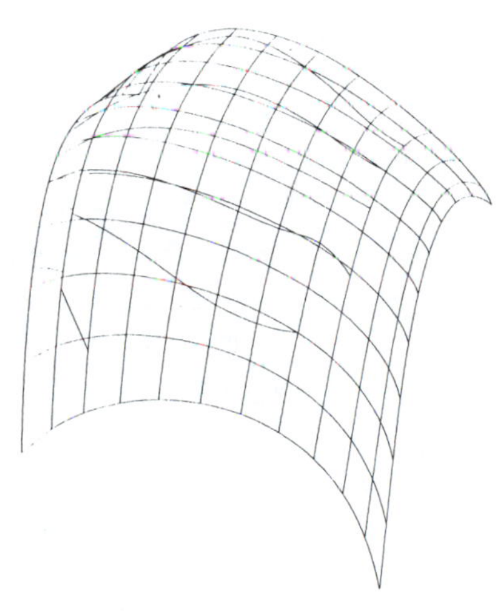

Fig. 13.13a. Kink in the isophotes indicates C^1 continuity of the spline.

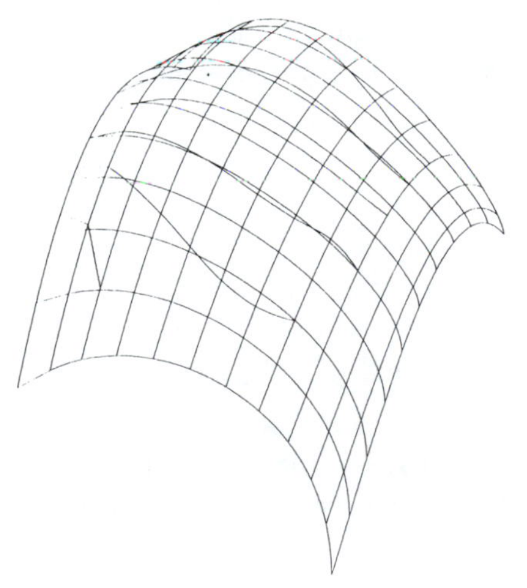

Fig. 13.13b. Jump in the isophotes indicates C^0 continuity of the spline.

14 Blending Methods

The construction of connecting curves and surfaces, and the rounding off of sharp corners or edges, commonly called *blending*, is of central importance in CAGD. We differentiate between exterior and interior blends and connections between (possibly non-intersecting) non-connected curves or surfaces. There are a variety of ways to define blends mathematically: in parametric form, in implicit form, by recursive processes, or by means of numerical methods. It is convenient [Var 89] to distinguish

- *superficial blending*, where there is no explicit mathematical formula for the blend. The blending corresponds to some procedure to be applied to a corner or edge such as "round off with radius R". The problems associated with this approach, and in particular the fact that the blending can only take place in the actual production process, are discussed in [Wel 84], see also [Wood 87], [Var 89],
- *surface blending*, where the blend is described by an additional surface which connects smoothly with the given surfaces. The surfaces to be blended may or may not intersect, and can be given in either implicit or parametric form,
- *polyhedral blending*, where the objects to be blended are defined by polyhedra. This involves taking account of the topology. Here the blending is accomplished by constructing a smooth connecting surface patch (i.e., surface blending [Chi 83, 87, 91], [Kim 84], [Fja 86], [Sait 90] or by defining a recursive subdivision process as is done in the corner cutting method, see e.g., [Cat 78], [Doo 78, 78a], and [Nas 87, 91],
- *volumetric blending*, which involves using a solid modelling system such as a CSG or a boundary representation system. Here the blending is just one of the system operations which may involve e.g., surface or polyhedral blending, and which can be combined with other operations. The surfaces can be given in either implicit or parametric form. The system automatically takes care of the problem of topology.

Blending curves and surfaces given in implicit and parametric form are treated in Sects. 14.1 and 14.2. In Sects. 14.3 and 14.4 we describe ways to construct blending curves and surfaces recursively, or by numerically solving a partial differential equation.

Because of space limitations, many important aspects of blending will be discussed only superficially or not at all. These include the problems associated with multiple blending, the blending of non-convex corners, topological considerations, and the question of automation and implementation in CAGD systems. Some of these problems are discussed in the survey paper of [Wood 87], and in the papers [Var 89], [Har 90].

14.1. Blends in Implicit Form

Planar curves and surfaces in implicit form are described by an equation $F(x, y) = 0$ or $F(x, y, z) = 0$, see Sect. 11.6. In Sect. 3.2 we have already seen how for $n = 2$, a planar implicit curve of degree n which joins given curves with GC^1 continuity can be constructed by enforcing derivative conditions on its $\binom{n+2}{2}$ coefficients. This permits the construction of blends using conic splines (i.e., quadratic blending curves). This approach can also be extended to curves of higher degree and to surfaces. This was done by [Baj 89, 92], who recently considered C^1 data interpolation using implicit surfaces. Our discussion of various other blending methods for implicitly defined curves and surfaces follows the survey article of [Wood 87]. In particular, we consider

- *unbounded blends*, where the blends globally influence the shape of the objects, see Fig. 14.1a,
- *volume-bounded blends*, where the spread of the blend is controlled by a prescribed volume, see Fig. 14.1b,
- *range-bounded blends*, where the spread of the blend is controlled by a given distance measure, see Fig. 14.1c,
- *curvature-bounded blends*, where the radius of curvature of the blend is fixed, see Fig. 14.1d.

14.1.1. Global Blends

Unbounded blending is suitable for the *design of molecules* [Bli 82], and for free-form modelling [Ric 73]. The basic idea is to replace the mathematical operations of union and intersection of sets (half spaces bounded by curves, surfaces, etc.) by algebraically defined approximants. The resulting *smoothed* object thus will always lie slightly inside or outside of the (exact) union or intersection of the individual objects.

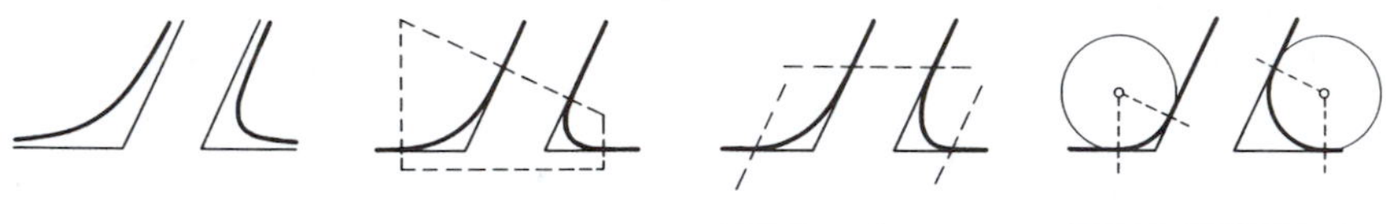

Fig. 14.1. Four possible ways of controlling the region in which a blend of two intersecting cylinders lies.

[Ric 73] assumes that the objects to be blended are described by non-negative functions $G_i = G_i(\boldsymbol{x})$ with $0 \leq G_i \leq 1$. The approximation of the intersection operator is given by

$$F = (G_1^{\mu} + \cdots + G_n^{\mu})^{1/\mu}, \tag{14.1}$$

while the approximation of the union operator is given by

$$F = (G_1^{-\mu} + \cdots + G_n^{-\mu})^{-1/\mu}. \tag{14.2}$$

Here μ is a smoothing parameter which can be used to control the degree of the blending. For small values of μ, the objects G_i are strongly blended (even if they were originally disjoint), while for larger values of μ, (14.1) and (14.2) are closer to the actual set theoretical operations.

In Blinn's *blobby (man) model* [Bli 82], the objects to be blended are described by half spaces defined by (field) functions $G_i = G_i(\boldsymbol{x})$ with $G_i \leq 0$. The approximation of the intersection operator is given by

$$F = \sum_{i=1}^{n} c_i e^{-a_i G_i}. \tag{14.3}$$

The parameters a_i describe the rate of decay of the fields G_i, while the parameters c_i weight the contribution of the G_i to the overall expression $F(\boldsymbol{x})$. Disjoint objects are again allowed; in [Bli 82] these are atoms. Some applications and a generalization to non-spherical objects G_i can be found in [Kalr 89], [Mur 91], and [Bid 92]. The integration of this method into a set theoretical based solid modeller is described in [Quar 86]. These global blending methods have proven to be relatively robust [Quar 86], and are quite efficient [Bli 82], see also [Wood 87].

Field functions are also employed in the *soft object method* [Wyv 86, 86a, 89], [Bloo 90], which is based on the same idea. We also mention the *convolution method* of [Bloo 91], which is another generalization of [Bli 82].

14.1.2. Volume-bounded Blends

Here the region in which the blend is to lie is prescribed by a volume which is usually described in terms of its bounding surfaces, and is used directly in defining the blend.

The first blending surface of this kind was suggested by Sabin, see [Wood 87]. The blend of the surfaces $G_1(\boldsymbol{x}) = 0$ and $G_2(\boldsymbol{x}) = 0$ has the form

$$F = (1-\lambda^2)G_1 + \lambda^2 G_2 + (1-\lambda^2)\lambda^2 M = 0, \qquad \text{with} \qquad \lambda = \frac{H_1}{H_1 + H_2}. \quad (14.4)$$

Here H_1 and H_2 are *supplementary surfaces* which are used to control the spread of the blend, and in particular to fix the boundary curves T_1 and T_2 of the blend lying on the surfaces G_1 and G_2. These are called *contact curves*, *link curves*, or *trim curves*. Without the *correction term* $(1-\lambda^2)\lambda^2 M$, with constant M, the blending surface would pass through the intersection curve $G_1 \cap G_2$, which would lead to a bulge or bump.

Many volume-bounded (and range-bounded) blending methods are closely related to the *conic section pencil methods* [LIM 44] discussed in Sect. 3.2. For example, if a GC^1 continuous blending curve between two curves K_1 and K_2 is to be constructed (see Fig. 14.2), then the tangents at the *trim points* $\boldsymbol{T}_1$ and $\boldsymbol{T}_2$ on the curves K_1 and K_2 can be interpreted as support lines ℓ_1 and ℓ_2 of the pencil of conic sections. Here $\boldsymbol{T}_1 \cup \boldsymbol{T}_2$ is the third support line $\ell_3 = \ell_4$. These three support lines determine all possible conic section blends

$$K = (1-\lambda)\ell_1\ell_2 + \lambda\ell_3^2 = 0, \qquad \lambda \in [0,1]. \quad (14.5)$$

They all lie in a region bounded by the three support lines. This type of blend is illustrated in Fig. 14.2 as a dashed line.

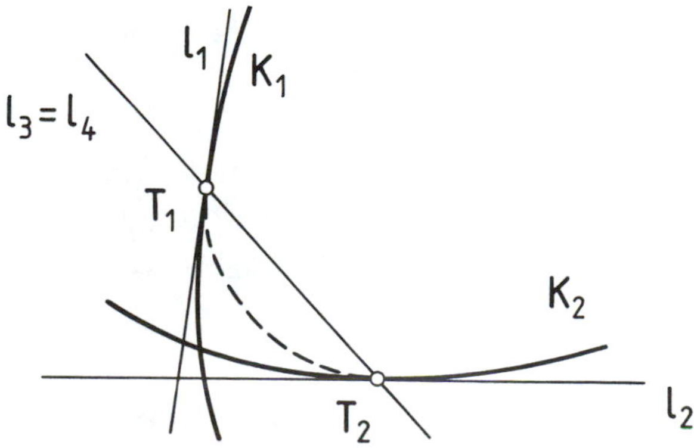

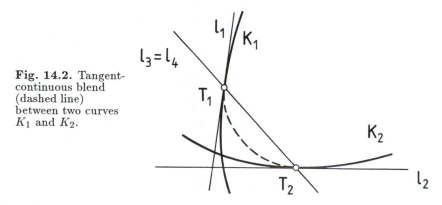

Fig. 14.2. Tangent-continuous blend (dashed line) between two curves K_1 and K_2.

If the straight lines ℓ_i in (14.5) are replaced by curves K_i, then this method can be used to construct more general tangent-continuous blends. The formula (14.5) can also be easily modified to produce surface blends. Using planes E_i then leads to cylindrical or conic blending surfaces. The trim lines are described by straight lines. In the case of more general surfaces, the problem is to construct the bounding surface H in such a way that it intersects both G_1 and G_2 along the desired trim lines, see e.g., [Pratt 87]. A simple way to achieve this is to take H to be the product of two surfaces H_1 and H_2, where H_1 defines the trim line T_1 on G_1, and H_2 defines the trim line T_2 on G_2, respectively. The equation of the blending surface is then given by ([Wood 87]):

$$F = (1-\lambda)G_1G_2 + \lambda H_1^2H_2^2 = 0, \qquad \lambda \in [0,1]. \tag{14.6}$$

More generally,

$$F = (1-\lambda)\prod_{i=1}^{m} G_i + \lambda \prod_{i=1}^{n} H_i^2 = 0, \qquad \lambda \in [0,1], \tag{14.7}$$

describes a surface blending together the m surfaces G_i, and which lies in the region defined by the n bounding surfaces H_i, see [Zha 86].

So far we have dealt with blending curves and surfaces which join with GC^1 continuity. To construct a blend with higher order of contact, we can use the following formula (see [Li 90], [Hos 91a] and cf. [Pratt 87]):

$$F = (1-\lambda)G - \lambda H^{\mu} = 0, \qquad \lambda \in [0,1], \quad \mu \geq 2. \tag{14.8}$$

Here G is called the *basis curve (surface)*, H the *transversal curve (surface)*, and F the *blending curve (surface)* or functional spline. The following result was established in [Li 90], [Hos 91a].

Theorem 14.1. *Let $\boldsymbol{P} = G \cap H$, (or $\boldsymbol{P} \in \Gamma = G \cap H$) with $G, H \in C^{\mu}(\boldsymbol{P})$, and suppose F is as in (14.8). Then at the point $\boldsymbol{P}$, the functional spline F meets the basis curve (surface) G with an order of contact of (at least) $\mu - 1$.*

For $\mu \geq 3$, this assures that F and G have the same curvature (Dupin indicatrix) at the point $\boldsymbol{P}$. In the case of planar curves, the first $\mu - 3$ derivatives of the curvature are actually equal when $\mu > 3$.

Example 14.1. Let $G = \ell_1\ell_2$ and $H = \ell_3$ where ℓ_1, ℓ_2, ℓ_3 are the straight lines defined by $y - 2(x+1) = 0$, $y + 2(x-1) = 0$, and $y = 0$, respectively. Then the functional spline

$$F = (1-\lambda)\ell_1\ell_2 - \lambda\ell_3^{\mu} = 0, \qquad \lambda \in [0,1], \quad \mu \geq 3,$$

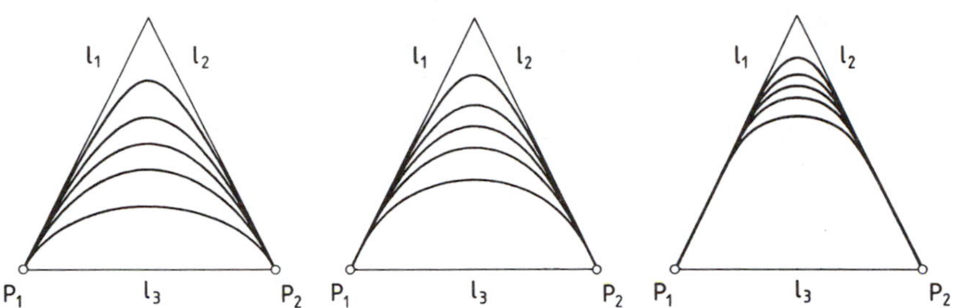

Fig. 14.3. Functional splines for $\mu = 2$ (left), $\mu = 3$ (middle) and $\mu = 8$ (right).

has zero curvature at the points $\boldsymbol{P}_1 = \ell_1 \cap \ell_3$ and $\boldsymbol{P}_2 = \ell_2 \cap \ell_3$. For increasing values of μ and/or decreasing values of λ, F approaches the intersection of the lines ℓ_1 and ℓ_2, see Fig. 14.3.

A blending curve or surface defined as a functional spline lies entirely inside the region bounded by G and H, and moreover, $F \cap G = F \cap H = G \cap H$, see [Li 90] and Fig. 14.3.

Convex blending curves and surfaces can be constructed if the function G in (14.8) is a product of lines (or planes) $G_i(\boldsymbol{x}) = \boldsymbol{N}_i \cdot \boldsymbol{x} + c_i$, $|\boldsymbol{N}_i| = 1$, $c_i \in \mathbb{R}$, $i = 0(1)m$, where the $G_i(\boldsymbol{x})$ are oriented so that the normals $\boldsymbol{N}_i$ point towards F. In this case we have (see [Hartm 92]):

Theorem 14.2. *The functional spline curve (surface)*

$$F = (1-\lambda) \prod_{i=1}^{m} G_i^{\mu_i} - \lambda G_0^{\mu_0} \tag{14.9}$$

is convex provided that span$(\boldsymbol{N}_1, \ldots, \boldsymbol{N}_m)$ *lies in* $\mathbb{R}^2$ ($\mathbb{R}^3$) *and* $\mu_0 > \sum_{i=1}^{m} \mu_i$, $\mu_i \in \mathbb{R}$.

Additional convexity criteria for curves and surfaces can be found in [Li 90], [Hos 91a], and [Hartm 92].

Functional splines can be used to construct very simple blending surfaces when the boundary curves of the blends and the surfaces are given implicitly. For example, to find the blending surface between two intersecting cylinders G_1, G_2 with the same radii and perpendicular axes, and which is bounded by circles on the cylinders, we choose the basis surface to be the product of the two cylinders, and the transversal surface to be the product of the four planes through the circles on the cylinders. If the axes of the two cylinders intersect at the origin, and if the four planes are all at the same distance from the origin, then the blending surface is described by

$$F = (x^2 + z^2 - 3)(y^2 + z^2 - 3) - \lambda\big((x-4)(x+4)(y-4)(y+4)\big)^{\mu}.$$

Fig. 14.4 shows the blending surface constructed in this way.

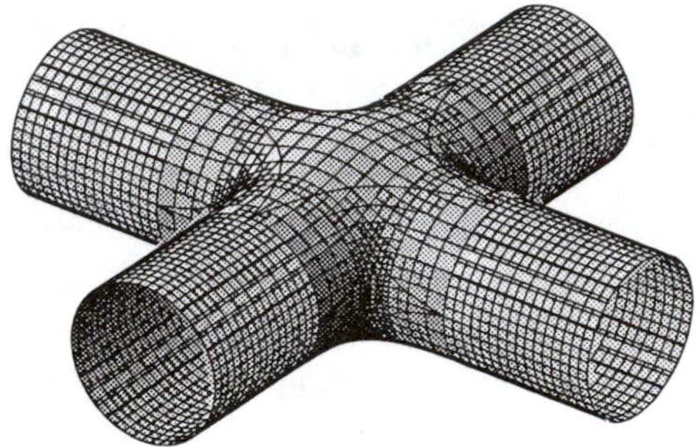

Fig. 14.4. Blending surface between two cylinders.

If the boundary curves of the blending surface cannot be given implicitly, we can use a construction (cross-section method) described by [Hartm 90]. Let G_1 and G_2 be two intersecting surfaces with intersection curve Γ. The first step is to construct two curves T_1 and T_2 on G_1 and G_2, respectively, which lie in a neighborhood of Γ, see Fig. 14.5 (left). The local distance between T_1 and Γ, and T_2 and Γ, respectively, should depend on the angle between the two surfaces G_1 and G_2 along the intersection curve. The contact points of a rolling sphere give these trim curves. Now we connect corresponding points on T_1 and T_2 with a ruled surface ψ whose generating lines form the basis curves for the desired blending curves, see Fig. 14.5 (middle). Now through each generating line of the ruled surface ψ, we construct a plane which cuts the intersection curve of the two given surfaces G_1 and G_2 orthogonally. If the ruled surface ψ at these intersections is described by the equation $H(\boldsymbol{x}) = 0$, and if the equations of the G_i have the form $G_i(\boldsymbol{x}) = 0$, then a G^2 continuous blending surface (with appropriate orientation) is given by the equation

$$F = (1-\lambda)G_1G_2 - \lambda H^3 = 0,$$

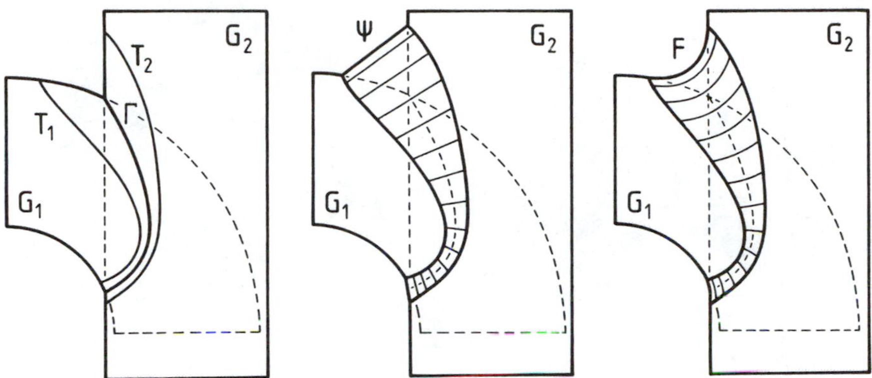

Fig. 14.5. Construction of a blending surface.

see Fig. 14.5 (right).

A construction of blending surfaces without prescribed boundary curves is described in [Schm 92]. He selects $\mu = 0$ in equation (14.8), leading to the following equation for the blending surface:

$$F = (1 - \lambda)G - \lambda = 0,$$

and so F can be interpreted as a contour surface of G. Such contour surfaces can be used for both blending and interpolation.

Example 14.2. The equation

$$F = (1 - \lambda)(x^2 + z^2 - 4)(y^2 + z^2 - 4) - \lambda$$

describes the blending surface between the two cylinders in Fig. 14.6. The equation

$$F = (1 - \lambda)(x^2 + z^2 - 4)(y^2 + z^2 - 4)(x^2 + y^2 - 4) - \lambda$$

gives the blending surface for the three cylinders in Fig. 14.7.

Blending surfaces between curves can be constructed as follows. In the (x, y)-plane we choose a curve $f(x, y) = 0$, and in the (y, z)-plane we choose a curve $b(y, z) = 0$. The equation of the blending surface is then

$$F(x, y, z) = f(x, b(y, z)) = 0.$$

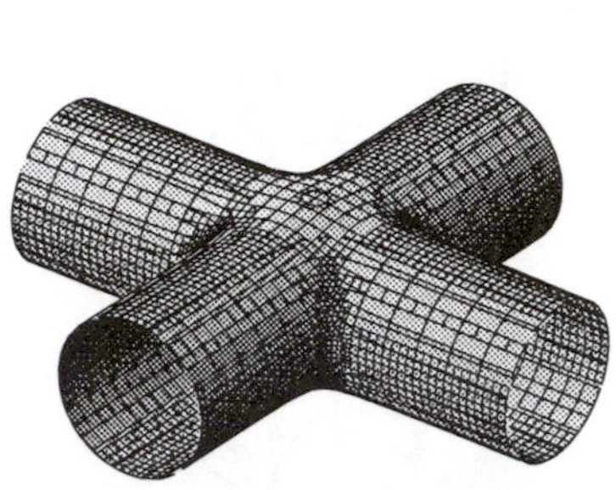

Fig. 14.6. Unbounded blending surface between two cylinders.

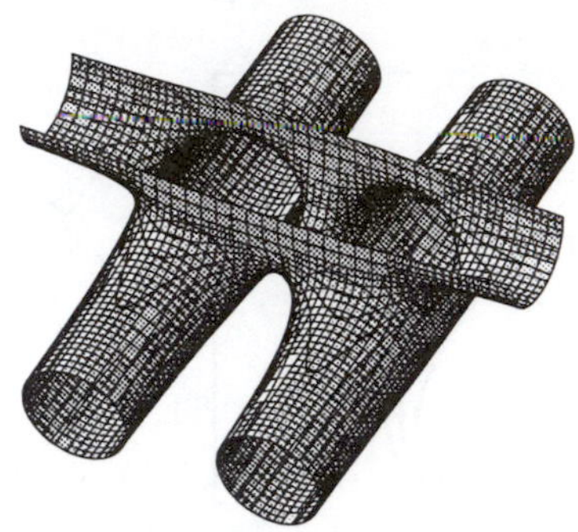

Fig. 14.7. Unbounded blending surface between three cylinders.

Example 14.3. Fig. 14.8 (see also the front cover) shows a blending surface constructed in this way. We take

$$f(x, y) = G - \lambda H = 0,$$

with basis and transversal curves G and H given by

$$G = (y - 2.5)(\sqrt{3}x - y - 0.5)(-\sqrt{3}x - y - 0.5),$$
$$H = (1 - y)(y + \sqrt{3}x - 2.5)(y - \sqrt{3}x - 2.5).$$

As a torus generating spin curve (directrix), we take

$$b(y, z) = (y^n + z^n)^{1/n}.$$

Fig 14.8. Blending surface between two curves with $n = 4$, $\lambda = -18.9525$.

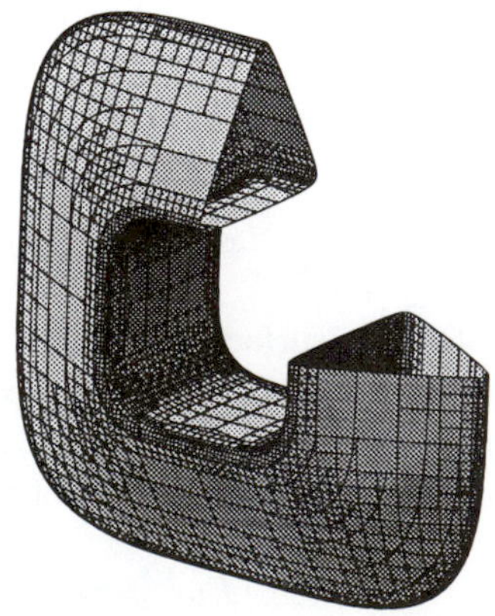

14.1.3. Range-controlled Blends

In this approach, the shape of the blend is controlled by distance parameters which are inserted directly into the defining equations. In this connection, we mention [Rock 84], [Mid 85], and [Hoff 85], the last of which has led to a number of follow-up papers.

The *superelliptic blend* of Rockwood and Owen [Rock 84, 87] is defined by

$$F = 1 - \left(1 - \frac{G_1}{r_1}\right)^{\mu} - \left(1 - \frac{G_2}{r_2}\right)^{\mu} = 0. \tag{14.10}$$

The parameters r_1 and r_2 determine the spread of the blend. In a certain sense, they can be compared to the (half length of the) major and minor axes of an ellipse: for $r_1 = G_1$, F reduces to the surface G_2, while for $r_2 = G_2$ it reduces to G_1. The shape parameter μ is referred to in [Rock 84, 87] as the *thumbweight*. It controls the tightness of the blend: for $\mu = 2$, the cross-section of the blend (14.10) is approximately a circle, while for larger values of μ, it is superelliptic, i.e., has a more pronounced tightness, see Fig. 14.9.

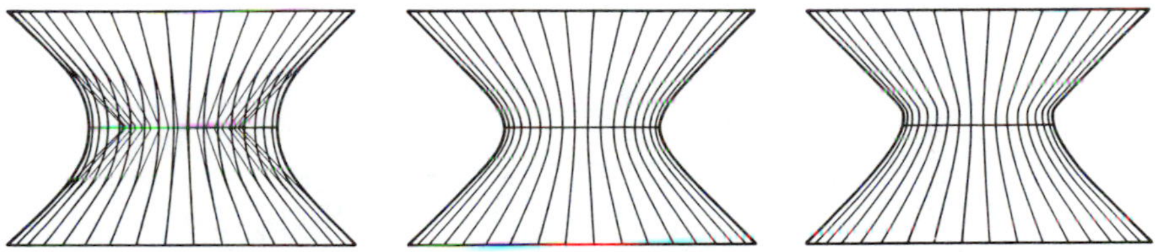

Fig. 14.9. Various blending surfaces between two cones [Rock 84]. Left: $\mu = 2$, middle: $\mu > 2$, right: $r_1 \neq r_2$.

In analogy to (14.7) with $m = 2$, $n = 1$, and $H_1 = G_1 + G_2 - r$, [Mid 85] has introduced the blending surface

$$(1 - \lambda)\hat{G}_1\hat{G}_2 + \lambda(\hat{G}_1 + \hat{G}_2 - r)^2 = 0, \qquad \lambda \in [0, 1]. \tag{14.11}$$

Here $\hat{G}_i$ is obtained from G_i by multiplication with a factor to compensate for the differences in the distances of G_i from the potential surface $G_i = r$ and the prescribed Euclidean distance r from the trim curve T_1 (T_2) to the surface G_2 (G_1) which determines the spread of the blend. These *normalization factors* have to be separately defined for each surface. For quadrics and for the torus, this is relatively straightforward, and for planes, cylinders, and the sphere, there is a direct connection.

A blending surface between two (intersecting) surfaces can be constructed as follows. First we choose trim curves on each of the surfaces (near the intersection curve). These are then connected with a blending function similar to those used for Coons surface interpolation to connect two opposite boundary edges of a patch. The *potential method* of Hoffmann and Hopcroft [Hoff 85, 86] proceeds in this way, see also [Hoff 87, 88]. This approach is simple to carry out, can be easily automated, and is generally applicable, since it works for arbitrary algebraic surfaces. Moreover, it is efficient in the sense that it produces a blending surface of smallest possible polynomial degree.

Suppose we are given algebraic surfaces G_1 and G_2 in implicit form $G_1(\boldsymbol{x}) = 0$ and $G_2(\boldsymbol{x}) = 0$. The potential surfaces $G_1 = s$, $(H_1 = G_1 - s = 0)$, with $s \in \mathbb{R}$, and $G_2 = t$, $(H_2 = G_2 - t = 0)$, with $t \in \mathbb{R}$, have shapes similar to the surfaces G_1 and G_2. However, depending on the sign of the constants s and t, they lie either entirely inside or entirely outside of G_1 and G_2, respectively. Suppose a is an appropriately selected maximal s value for which $H_1 \cap G_2$ defines a non-degenerate space curve lying on G_2 which is to be a trim curve T_2 for the blend. Similarly, suppose the trim curve T_1 on G_1 is the intersection of the potential surface H_2 with G_1 corresponding to an appropriately chosen maximal t value b, see Fig. 14.10.

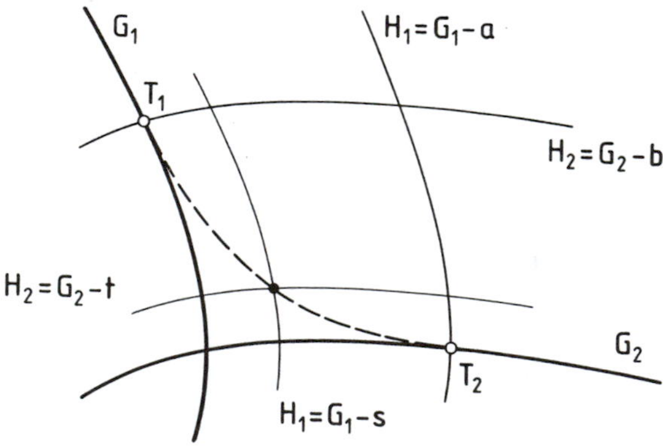

Fig. 14.10. Potential method for blending algebraic surfaces G_1 and G_2.

The connecting surface is now defined by blending these two trim curves using a blending function $f(s,t)$. The point $(s,t) = (a,0)$ in the (s,t)-parameter space corresponds to the trim curve T_2, while the point $(s,t) = (0,b)$ in the (s,t)-parameter space corresponds to the trim curve T_1. Moreover, the parameter values (s,t) with $0 < s < a$, $0 < t < b$ generate space curves which are the intersections $H_1 \cap H_2$ of the potential surfaces $H_1 = G_1 - s = 0$

and $H_2 = G_2 - t = 0$. The intersection curves form the connecting surface $F(\boldsymbol{x}) = f(G_1, G_2) = 0$. [Hoff 85] has established the following important

Theorem 14.3. *Suppose the blending function $f(s,t)$ approaches the s-axis tangentially at $(a,0)$, and similarly approaches the t-axis tangentially at $(0,b)$. Then F joins G_2 and G_1 with GC^1 continuity along the trim curves T_2 and T_1, respectively.*

If the blending function $f(s,t)$ meets the axes with higher order of contact, then the blending surface F joins G_1 and G_2 with a correspondingly higher degree of contact. The construction of such functions was carried out by [Kos 91], although it is quite complicated in comparison with the functional splines in [Li 90], [Hartm 90], and [Hos 91a].

By Theorem 14.3, the problem of blending two curved algebraic surfaces is reduced to the much simpler problem of blending the axes (planes) in the parameter space! For a GC^1 join, we have to satisy four boundary conditions, which means that $f(s,t)$ must be chosen to be at least quadratic. This means that if we want to keep the polynomial degree of the blending surface as small as possible, we should choose $f(s,t)$ to be quadratic, i.e., a conic section. Taking account of the boundary conditions (cf. with Sect. 3.2), it follows that we must choose

$$f(s,t) = b^2s^2 + 2\epsilon st + a^2t^2 - 2ab^2s - 2a^2bt + a^2b^2 = 0, \tag{14.12}$$

with $\epsilon \in \mathbb{R}$, and hence the blending surface $F(\boldsymbol{x}) = f(G_1, G_2)$ is given by

$$F = b^2G_1^2 + 2\epsilon G_1G_2 + a^2G_2^2 - 2ab^2G_1 - 2a^2bG_2 + a^2b^2 = 0. \tag{14.13}$$

Thus, the blending surface F has polynomial degree $\max\{2m, 2n\}$ when G_1 and G_2 are of degree m and n, respectively. When these are quadrics, we get a blending surface of degree 4.

The parameters a and b determine the spread of the blend. We should note that depending on the sign of a and b, we get entirely different blending surfaces, see [Hoff 85, 88]. The choice of the free parameter ϵ determines the form of the conic section (14.12), and thus influences the curvature distribution of the blending surface F. For $a, b > 0$, we have e.g.,

$\epsilon = -\infty$	$f(s,t)$ is a pair of lines $s = 0$, $t = 0$,
$-\infty < \epsilon < -ab$	$f(s,t)$ is a hyperbola,
$\epsilon = -ab$	$f(s,t)$ is a parabola,
$-ab < \epsilon < ab$	$f(s,t)$ is an ellipse (a circle for $a = b$ and $\epsilon = 0$),
$\epsilon = ab$	$f(s,t)$ is the line $bs + at - ab = 0$, counted twice.

Fig. 14.11 shows examples of various blending functions for the case $a = 7, b = 3$.

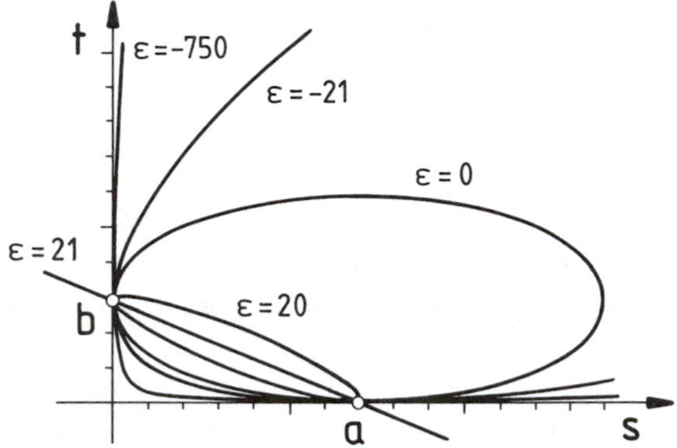

Fig. 14.11. Quadratic blending functions $f(s,t)$ for $a = 7, b = 3$.

Example 14.4. Let G_1 and G_2 be cylinders with radii 8 and 4, respectively, whose axes intersect at a right angle:

$$G_1(\boldsymbol{x}) = x^2 + y^2 - 8^2 = 0, \qquad G_2(\boldsymbol{x}) = y^2 + z^2 - 4^2 = 0.$$

The potential surfaces H_1 and H_2 are given by

$$H_1(\boldsymbol{x}) = x^2 + y^2 - 8^2 - s = 0, \qquad H_2(\boldsymbol{x}) = y^2 + z^2 - 4^2 - t = 0.$$

Suppose the two trim curves T_1 and T_2 correspond to the values $a = 36$ and $b = 20$, i.e., they are the intersection of the cylinder $H_2 = G_2 - 20 = 0$ (of radius 6) with G_1, and the intersection of the cylinder $H_1 = G_1 - 36 = 0$ (of radius 10) with G_2, respectively. Suppose the blending function $f(s,t)$ is taken to be an ellipse with $\epsilon = 0$:

$$f(s,t) = \frac{(s-36)^2}{36^2} + \frac{(t-20)^2}{20^2} - 1 = 0.$$

Then the quartic blending surface which joins G_1 and G_2 with GC^1 continuity along the trim curves T_1 and T_2 is given by

$$\begin{aligned} F(\boldsymbol{x}) = f(G_1, G_2) &= \frac{(G_1-36)^2}{36^2} + \frac{(G_2-20)^2}{20^2} - 1 \\ &= 81z^4 + 162y^2z^2 - 5832z^2 + 106y^4 + 50x^2y^2 \\ &\quad - 10832y^2 + 25x^4 - 5000x^2 + 322576 = 0. \end{aligned}$$

Of course, the blend is only that part of the surface $F(\boldsymbol{x}) = f(G_1, G_2)$ which corresponds to the arc $f(s,t) = 0$ where $bs + at - ab < 0$ holds, see Fig. 14.11.

The *affine potential method* described above is not completely general, since there are some quartic blending surfaces between two quadrics which cannot be derived by the affine method, see e.g., [Hoff 87]. On the other hand, the *projective potential method* employs the potential surfaces given by $H_1 = G_1 - sW = 0$ and $H_2 = G_2 - tW = 0$, where W is an arbitrary polynomial which is not identically 1. In this case, the blending surface can be obtained by the substitution of $s = G_1/W$ and $t = G_2/W$ in $f(s,t) = 0$.

In the case where $f(s,t)$ is chosen to be quadratic and G_1 and G_2 are quadrics, then under the following *hypotheses*

- the intersection curves $G_1 \cap G_2$, $T_1 = H_2 \cap G_1$, and $T_2 = H_1 \cap G_2$ are distinct and none of them is the union of algebraic curves of lower degree,
- F is not a union of algebraic surfaces of lower degree,
- the quadratic terms in G_1 and G_2 do not have a common factor,

we have

Theorem 14.4. (Uniqueness Theorem). *The projective potential method produces all possible blending surfaces of degree 4 which join two quadrics G_1 and G_2 along quartic trim curves T_1 and T_2 with GC^1 continuity.*

For a proof, see [Hoff 86]. The proof also implies the following

Corollary 14.1. *Given two quadrics G_1 and G_2, there is a GC^1 continuous blending surface of degree 4 if and only if the trim curves T_1 and T_2 lie on a common quadric Q.*

Applying the potential method, we get ([Hoff 86, 87, 88])

$$Q = bG_1 + aG_2 - ab, \tag{14.14}$$

which leads to the one parameter family of blending surfaces

$$F = \psi G_1 G_2 + Q^2, \qquad \text{with} \qquad \psi = 2\epsilon - 2, \tag{14.15}$$

where ϵ is as above. Equation (14.13) is clearly easier to manipulate and automate than (14.15).

The blends corresponding to (14.10) for $\mu = 2$ are a subset of those defined by (14.13) with $\epsilon = 0$. However, in general, (14.10) leads to blends of higher degree than those obtained with the potential method. For quadrics, the method of [Mid 85] also produces blending surfaces of degree four. By

Theorem 14.4, his method must be an equivalent formulation of the affine potential method. Indeed, for $H_1 = G_1/a$, $H_2 = G_2/b$, and $r = 1$, equation (14.11) can be transformed into the equation (14.13) by substituting $\epsilon = ab(1+\lambda)/2\lambda$. However, in the projective formulation, the potential method is more general. Moreover, it is much simpler to deal with, since the spread of the blend is controlled by the potential values a and b. In most cases there is no simple connection with the Euclidean distance.

Implementations of the three range-controlled blending methods discussed above are considered in [Rock 84, 87], [Mid 85], and [Hoff 85]. For a treatment of multiple blending and blending of non-convex corners, see [Rock 87], [Hoff 87, 88], [Hol 87], and [Kos 89, 91].

14.1.4. Rolling-ball Blends

Most blending methods do not allow the user to select the shape of the cross-section of the blending surface. However, to a certain extent, this is possible for rolling-ball blends, which have circular cross-sections, thus allowing e.g., the calculation of stress concentration factors. On the other hand, restricting the form leads to a certain loss in design freedom.

A rolling-ball blend can be realized by rolling a sphere along a (possibly fictitious) intersection curve of two surfaces in such a way that it touches both surfaces, see e.g., [Ross 84], [Peg 87, 90]. Mathematically, such a blend is a piece of a tubular surface which joins the two surfaces G_1 and G_2 along the trim lines T_1 and T_2 in a GC^1 continuous fashion. A *canal surface* is defined to be the envelope E of a sphere with variable radius r, whose midpoint moves along a (space) curve S which is sometimes referred to as the *spin curve* or *directrix* of the canal surface. In the case of a constant r value (which leads to *tubular surfaces* and *constant radius (rolling ball) blends*), the directrix is the intersection curve of the two offset surfaces O_1 and O_2 which run parallel to G_1 and G_2 at a distance r, see Chap. 15. The *directrix* can be written in the form $S(\boldsymbol{x}) = 0$ for all $\boldsymbol{x} = \boldsymbol{x}_S \in \mathbb{R}^3$ for which

$$O_1(\boldsymbol{x}_S) = 0, \tag{14.16a}$$

$$O_2(\boldsymbol{x}_S) = 0, \tag{14.16b}$$

are simultaneously satisfied. Thus, the *envelope* $E(\boldsymbol{x}) = 0$ of the moving spheres, which defines the tubular surface, is given by the system of equations consisting of (14.16a) – (14.16b), and the two additional equations

$$(x - x_S)^2 + (y - y_S)^2 + (z - z_S)^2 - r^2 = 0, \tag{14.16c}$$

$$(\boldsymbol{x} - \boldsymbol{x}_S) \cdot \boldsymbol{T}(\boldsymbol{x}_S) = 0, \tag{14.16d}$$

see [Hoff 90]. Here the direction vector $\boldsymbol{T}$ of the tangent T is defined as the cross product of the normals $\boldsymbol{N}_1$ and $\boldsymbol{N}_2$ to G_1 and G_2, respectively. These equations assert that the envelope points $\boldsymbol{x}$ correspond to those points of the sphere of radius r which lie in a plane perpendicular to the tangent T to the directrix S at the points $\boldsymbol{x}_S$. Thus, we have a *circular blending*, which also follows from the geometric interpretation of the *spherical blending* defined above. From the standpoint of production, however, there is a difference between using a 5-axis milling machine with a sliding cutter, and a 3-axis milling machine with a spherical cutter, see Chap. 16 and [Peg 87, 90].

In some simple cases (e.g., for a plane, cylinder, sphere, or cyclide), it is possible to give an explicit equation for the offset surface, but in general the *offset surface* O_F corresponding to an implicit surface F is described by the four equations

$$(x - x_F)^2 + (y - y_F)^2 + (z - z_F)^2 - r^2 = 0, \tag{14.17a}$$

$$F(\boldsymbol{x}_F) = 0, \tag{14.17b}$$

$$(x - x_F)F_y(\boldsymbol{x}_F) - (y - y_F)F_x(\boldsymbol{x}_F) = 0, \tag{14.17c}$$

$$(y - y_F)F_z(\boldsymbol{x}_F) - (z - z_F)F_y(\boldsymbol{x}_F) = 0, \tag{14.17d}$$

see e.g., [Hoff 90]. Equations (14.17c) – (14.17d) express the orthogonality of the direction vector $\boldsymbol{x} - \boldsymbol{x}_F$ to the two surface tangent vectors $(F_y, -F_x, 0)$ and $(0, F_z, -F_y)$, where $\boldsymbol{x}_F \in F$ is the foot of the perpendicular to the offset surface $O_F(\boldsymbol{x}) = 0$ passing through the offset point $\boldsymbol{x}$.

The systems of equations (14.16) and (14.17) are in general quite difficult to solve. A closed form solution can sometimes be found by eliminating variables, e.g., by means of resultants or Gröbner bases, although the computation time using available programs can be very large, see e.g., [Hoff 90].

One source of the difficulty in giving an exact description of a constant radius (CR) blend is the very high polynomial degree of a tubular surface. For example, in [Rock 87] the CR-blend of two cylinders is described by an equation with 969 terms. In light of this, Rossignac and Requicha [Ross 84, 86] construct an approximation by first defining several cross-sections of the blend (i.e., circular arcs) which are then connected with cylindrical, toroidal, and spherical pieces. The directrix is thus approximated by circular arcs and straight line segments. There are, however, several problems with the implementation of this idea, and indeed, it is, in general, not possible to enforce GC^1 continuity between the surface patches and the original surfaces. An algorithm based on the same ideas for blending the *natural quadrics* (plane, cone, sphere, and cylinder) and the torus is given in [Hol 88].

The approximation method of Rockwood and Owen [Rock 87] is based on the local approximation of the angle θ (and thus the angle $\psi = \pi - \theta$

of intersection between the two surfaces) which is obtained by taking the scalar product of the normal vectors $\boldsymbol{N}_1$ and $\boldsymbol{N}_2$ to the surfaces G_1 and G_2, respectively. From Fig. 14.12 we see that

$$G_1 = r - (r - F)\cos(\frac{\theta}{2} - \alpha), \qquad G_2 = r - (r - F)\cos(\frac{\theta}{2} + \alpha),$$

and thus

$$F = r - \frac{1}{\sin\theta}\left[(r-G_1)^2 + (r-G_2)^2 - 2(r-G_1)(r-G_2)\cos\theta\right]^{1/2} = 0. \quad (14.18)$$

The blending surface is

$$G_1^2 + G_2^2 - 2\cos\theta G_1 G_2 + 2r(\cos\theta - 1)(G_1 + G_2) + r^2(2 - 2\cos\theta - \sin^2\theta) = 0.$$

For $\psi = \theta = 90°$ this reduces to the superelliptic blend of equation (14.10) with $r_1 = r_2 = r$.

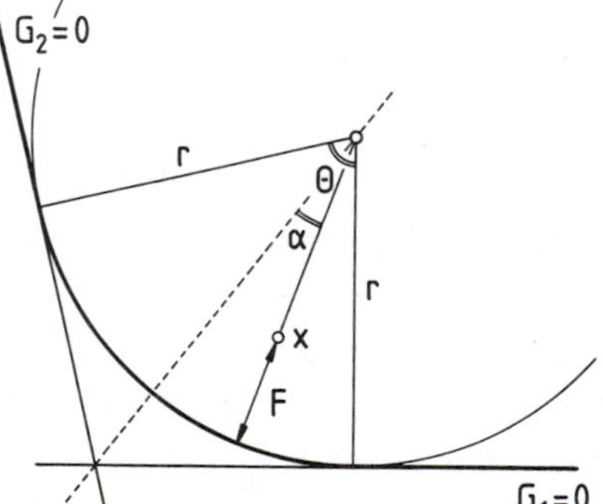

Fig. 14.12. Derivation of the Rockwood-Owen approximation of a constant radius rolling-ball blend.

If the radius of the sphere which generates a rolling-ball blend is allowed to vary, we get a so-called *variable radius (rolling-ball) blend*. Even for simple cases, the mathematical description of such blending surfaces can be very complicated, see [Chan 90]. For variable radius (VR) blends, we have to deal with parts of a canal surface. The problem is to determine the directrix, and to specify an appropriate radius function. While the directrix of a CR-blend is just the intersection of two offset surfaces, this is no longer the case for VR-blends.

In several cases of special importance to CSG modeling, VR-blends can in fact be constructed precisely as pieces of a cyclide. As shown in Forsyth [FORS 12] (see also [Prat 88, 90], [Zhou 91]), a *(Dupin) cyclide* (in standard position and orientation) can be written in either of the two following equivalent implicit forms:

$$(x^2 + y^2 + z^2 - \mu^2 + b^2)^2 = 4(ax - e\mu)^2 + 4b^2y^2, \qquad (14.19a)$$
$$(x^2 + y^2 + z^2 - \mu^2 - b^2)^2 = 4(ex - a\mu)^2 + 4b^2z^2, \qquad (14.19b)$$

and thus is a surface of fourth degree. In the case of the ring cyclide, the surface looks like a ring of nonuniform thickness. The parameters a, b, and e are given as the distances of the vertices and of the foci from the origin of an ellipse and a hyperbola which lie in mutually orthogonal planes. The ellipse and hyperbola are oriented such that the vertices of one are the foci of the other (so-called anti-conics, also referred to as focal conics). They are described by the equations

$$\frac{x^2}{a^2} + \frac{y^2}{b^2} = 1,\ z = 0, \qquad \frac{x^2}{e^2} - \frac{z^2}{b^2} = 1,\ y = 0, \qquad e^2 = a^2 - b^2,$$

see Figs. 14.13 and 14.14. The parameter μ describes the relationship between the radius and the location of the center of a sphere with variable radius which moves along the ellipse. Indeed, a cyclide can be constructed as the hull of a sphere whose center moves along the ellipse, and whose radius varies between $R_{\min}$ and $R_{\max}$, and thus is a special kind of canal surface! The radius function can be described in more detail using the thread construction of Maxwell [Max 68], see also [Boeh 89, 90a]. A thread of length $L = a + \mu$ is attached to one of the two foci of the ellipse, while the stretched out thread is moved along the ellipse. The radius of the sphere is then defined by the length of the part of the string which hangs freely. It follows immediately that for the two vertices A and B of the ellipse defined by the ends of the major and minor axes $R(A) = R_{\min} = L - (a + e) = \mu - e$ and $R(B) = R_{\max} = L - (a - e) = \mu + e$, see Figs. 14.13 and 14.14.

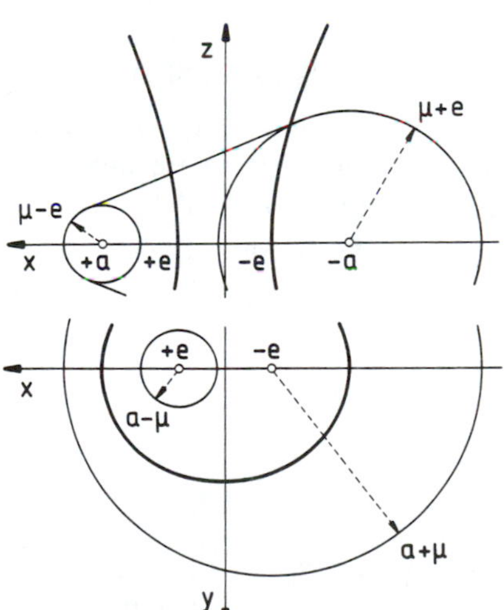

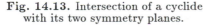

Fig. 14.13. Intersection of a cyclide with its two symmetry planes.

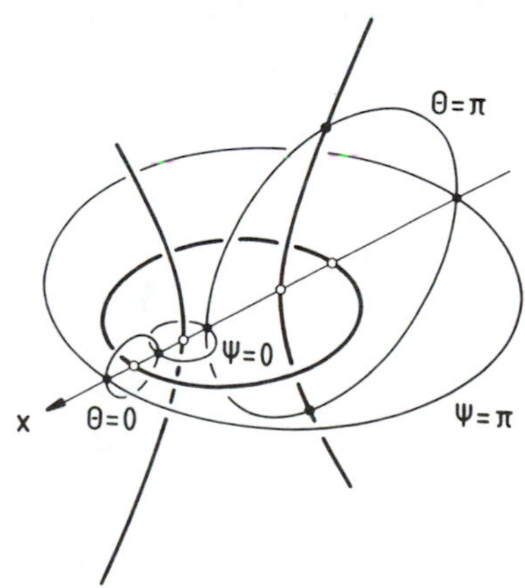

Fig. 14.14. A pair of anticonics defining the cyclide.

Further definitions and alternative constructions are given in [Chan 89], where properties of cyclides are also discussed, see also [Boeh 89, 90a], [Prat 90]. We note that cyclides are (the only) surfaces of fourth degree whose curvature lines are circles.

For integration into CAD-systems, it is important to note that in addition to the ring, horn, and spindel cyclides, the set of surfaces defined by (14.19) also includes natural quadrics and the torus, which are important basis surfaces for solid modelling. For a morphology of cyclide forms, see [Chan 89], [FLA 75].

The following general result is due to [NUT 88].

Theorem 14.5. *The angle Φ between the surface normal and the curve normal of a curvature line of a surface is constant along planar curvature lines.*

Since all curvature lines of a cyclide are circular, and thus planar, this theorem applies to cyclides. It leads to a simple explicit construction of a GC^1 continuous variable radius blend between a cyclide G_1 and, for example, a plane or a sphere G_2 [Prat 88, 90]. First we choose a trim line T_1 on G_1 to be a curvature line, i.e., a circle. We then compute the angle Φ_1 for this circle. Then for any cyclide blending surface which joins G_1 with GC^1 continuity along T_1, we must have $\Phi = \pi - \Phi_1$ (the curve normals along T_1 are equal, but the surface normals have opposite signs). The variable radius blend is then defined to be that part of the cyclide which joins G_2 along a circular trim line T_2 on the surface G_2 with GC^1 continuity. If there is some symmetry present, then the determination of this cyclide is relatively simple.

We now give a simple example.

Example 14.5. Suppose we want to construct a variable radius blend between a cylinder G_1 with radius R_1 and a plane G_2 which is inclined at an angle α to the axis of rotation of the cylinder. Suppose the trim line T_1 is to be a circle lying in the horizontal plane at a distance h from the point of intersection of the axis of the cylinder with the plane G_2, and that the trim line T_2 is to be a circle lying in the plane G_2 with radius R_2. Then as can be seen in Fig. 14.15,

$$\begin{aligned} R_{\min} &= (h - R_1 \cot\alpha)\tan\frac{\alpha}{2}, \\ R_{\max} &= (h + R_1 \cot\alpha)\cot\frac{\alpha}{2}, \\ R_2 &= h + R_1 \operatorname{cosec}\alpha. \end{aligned}$$

If the coordinate system is chosen as in Figs. 14.13 and 14.14, then the (x, y)-plane is a horizontal symmetry plane of the cyclide, and it follows (see Fig. 14.13) that

$$a - \mu = R_1, \qquad \mu + e = R_{\max}, \qquad \mu - e = R_{\min},$$

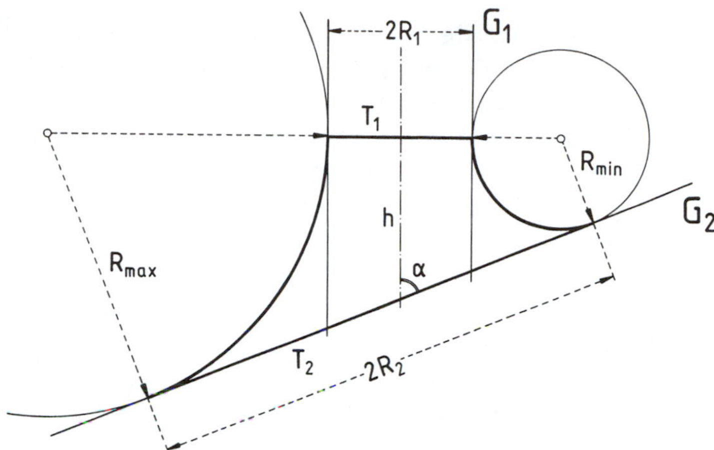

Fig. 14.15. Cross-section of a cyclide-defined variable radius blend of a cylinder and an inclined plane.

from which we can determine the cyclide parameters a, e, and μ, see [Prat 88, 90].

Other examples can be found in [Pr- 89, 90], [Boeh 89, 90a], and [Chan 90]. For more general cases, we need an appropriate generalization of the definition of the directrix of a constant radius blend. In this case, points on the directrix still are at equal distance from the two surfaces, but because of the radius function, S is no longer the intersection of two parallel surfaces at a constant offset distance r. Some suggestions on how to construct the directrix in this case can be found in [Peg 87, 90] and in [Chan 90], [Hoff 90a].

Pegna's approach [Peg 87, 90] is based on an interactively prescribed initial directrix S^0. Then an iteration is carried out to produce a sequence of directrices S^i which have as many points as possible satisfying the equidistance property. Here one of the given surfaces serves as a reference surface (but note [Chan 90]!). Orthogonal projection of S onto G_1 and onto G_2 leads to trim curves T_1 and T_2. [Ross 86] carries out the projection by minimizing $\|S - G_i\|$ using a Newton-Raphson algorithm; see also [Che 87], [Peg 87, 90, 90a], and take note of the discussion of distance versus orthogonal projection in [Peg 90a], and the algorithms presented there for implicit and for parametric representations.

[Chan 90] gives an explicit construction for the directrix of a variable radius blend with the help of *Voronoi* surfaces, also called *skeletal* or *medial axis* surfaces, see [Nac 82]. A Voronoi surface V_{12} associated with two given surfaces G_1 and G_2 is defined to be the set

$$V_{12} = \{\boldsymbol{x} \in \mathbb{R}^3 \; : d^{\perp}(\boldsymbol{x}, G_1) = d^{\perp}(\boldsymbol{x}, G_2)\}$$

of all points which are equidistant from both G_1 and G_2, where $d^{\perp}(\boldsymbol{x}, G_i)$ is the (shortest, i.e., perpendicular) Euclidean distance of $\boldsymbol{x}$ to the surface G_i. This concept is a natural generalization of the Voronoi diagrams discussed in Sect. 9.3.1. The Voronoi surface V_{12} associated with the two surfaces G_1 and G_2 is defined by the following eight equations in the ten unknowns $\boldsymbol{x}$, $\boldsymbol{x}_1$, $\boldsymbol{x}_2$, and r, where $\boldsymbol{x}_i$ is the point on the surface G_i corresponding to $\boldsymbol{x} \in V_{12}$:

$$G_1(\boldsymbol{x}_1) = 0, \tag{14.20a}$$

$$G_2(\boldsymbol{x}_2) = 0, \tag{14.20b}$$

$$(x - x_1)^2 + (y - y_1)^2 + (z - z_1)^2 - r^2 = 0, \tag{14.20c}$$

$$(x - x_2)^2 + (y - y_2)^2 + (z - z_2)^2 - r^2 = 0, \tag{14.20d}$$

$$\nabla K_1 \cdot \boldsymbol{t}_1 = 0, \tag{14.20e}$$

$$\nabla K_1 \cdot \boldsymbol{T}_1 = 0, \tag{14.20f}$$

$$\nabla K_2 \cdot \boldsymbol{t}_2 = 0, \tag{14.20g}$$

$$\nabla K_2 \cdot \boldsymbol{T}_2 = 0. \tag{14.20h}$$

Here ∇K_i denotes the gradient of the sphere K_i defined in (14.20c) and (14.20d), and $\boldsymbol{t}_i$ and $\boldsymbol{T}_i$ are two linearly independent tangent vectors to the surface G_i, $i = 1, 2$. Eliminating $\boldsymbol{x}_1$, $\boldsymbol{x}_2$, and r from (14.20), we get the equation of the Voronoi surface $V_{12}(\boldsymbol{x}) = 0$ in implicit form.

Example 14.6. The Voronoi surface corresponding to the two cylinders

$$G_1(\boldsymbol{x}) = x^2 + (y - 2)^2 - 1 = 0, \qquad G_2(\boldsymbol{x}) = z^2 + (y - 2)^2 - 1 = 0,$$

is the hyperbolic paraboloid

$$V_{12}(\boldsymbol{x}) = x^2 - 8y - z^2 = 0.$$

It follows immediately from the definition of the Voronoi surface that every curve lying on V_{12} can serve as the directrix for a VR-blend of the surfaces G_1 and G_2. The directrix can be determined as the intersection curve $S := V_{12} \cap R$ with the help of an appropriately chosen reference surface R, see [Chan 90]. Then as above, T_1 and T_2 can be determined by orthogonal projection onto G_1 and G_2.

We remark that the explicit determination of the equations for the Voronoi and canal surfaces, and for the directrices and trim curves, can only be carried out in some particularly simple (but important) cases where we have considerable geometric insight, e.g., when G_1 and G_2 are planes, cylinders, spheres, or cones. An approximate construction is thus required. In view of the above remarks on cyclides, it seems reasonable to try to construct the approximation from pieces of cyclides. This has been done in [Chan 90] based on the method of [Ross 84, 86] for finding an approximate CR-blend. We proceed in three steps:

- use the above method to compute the directrix of the variable radius blend,
- approximate the directrix by a GC^1 continuous curve consisting of elliptical arcs using the method of pencils of conic sections,
- for each elliptic arc, construct a corresponding cyclide segment whose directrix is the elliptic arc in such a way that these cyclide segments join smoothly, see [Prat 89, 90], [Chan 90].

14.2. Blends in Parametric Form

Blending objects (in parametric form) requires choosing trim lines (see Sect. 14.2.1), the computation of derivatives along and transversal to these curves, and the construction of blending surfaces which join smoothly along the trim lines, i.e., with C^r or GC^r continuity, see Sect. 14.2.2. Since C^r or GC^r joins have already been discussed in detail in Chapters 6, 7, and 8, we do not consider them any further here. The important special case of the rolling-ball blend is treated in Sect. 14.2.3.

14.2.1. Determination of Trim Lines

In the simplest cases, *trim lines* can be given in terms of parametric lines, in which case we get a constant-range blend in parameter space (but in general, only there). But almost always, trim lines have to be chosen as more general surface curves. This can be done either in the parameter or the coordinate space, and by some interactive or automatic process.

Interactive methods are employed in the algorithms in [Hansm 87, 88], [Bar 89a], and [Peg 87, 90]. Their advantages include high flexibility, which arises because of the interactive construction of the trim lines, and the fact that the existence and calculation of intersection curves for the two surfaces to be blended is not required.

In the interactive method of Hansmann [Hansm 87], the trim lines are defined interactively by the user by marking a sequence of points on the prescribed surfaces [Hansm 88]. These points are inverted onto the parameter planes, and approximated there using integral B-spline curves of fourth order. The resulting approximations are then mapped back onto the given surfaces.

[Bar 89a] uses rational B-spline curves to describe trim lines in parameter space.

The interactive method of Pegna and Wilde [Peg 87, 90, 90a], [Kop 91] allows the directrix to be freely prescribed by the user. This space curve is then projected orthogonally onto the two surfaces to produce the trim lines.

Automatic methods have been suggested by [Bar 89a] and [Peg 87, 90]. They are based mostly on the calculation of intersections of offset surfaces to the two given surfaces. The automatic method of Pegna and Wilde [Peg 87, 90, 90a], [Kop 91] proceeds in this way. Then the trim curves are taken to be the orthogonal projections of this curve onto the surfaces $\boldsymbol{G}_i(u_i, v_i)$. In the automatic method of Bardis and Patrikalakis [Bar 89a], it is assumed that the two surfaces intersect. The trim lines are then found by offsetting the intersection curve in the directions of the geodesics of the two surfaces. By cutting the surfaces with offset surfaces, the trim curves can similarly be determined automatically.

Trim lines can be described either in parameter or coordinate space. In *coordinate space* we get an explicit representation of the trim curves using (spline) interpolation of surface points, see e.g., [Har 90], [Kla 92]. These can be given by any of the three methods described above. Of course, the interpolation curves no longer lie exactly on the surfaces. This can lead to gaps between the prescribed surfaces and the blend whose boundary curves are the trim lines. Such gaps can be reduced by increasing the number of points being interpolated, i.e., we can make the gap smaller than any prescribed tolerance.

Describing trim lines in *parameter space* corresponding to surfaces $\boldsymbol{G}_i(u_i, v_i)$, e.g., interactively or with the help of an intersection algorithm, has the advantage that the trim curves can be given as true surface curves via functional composition. For Bézier representations, this can be done using the theorem in Sect. 11.5. The polynomial degree of such an explicit description can, of course, be high, cf. Theorem 11.1. Hence, in [Fil 89], instead of using an explicit coordinate space representation $\boldsymbol{T}_i(\tau) = \boldsymbol{G}_i(\boldsymbol{t}_i(\tau))$ for the trim curves given in the parameter space of $\boldsymbol{G}_i$ by $\boldsymbol{t}_i(\tau) = (u_i(\tau), v_i(\tau))$, $\tau \in [\tau_0, \tau_1]$, the author evaluates $\boldsymbol{T}_i(\tau) = \boldsymbol{G}_i(\boldsymbol{t}_i(\tau))$ only pointwise at individual τ values. Often, the surface points obtained in this way can be interpolated with lower order polynomial spline curves. As before, these approximations to the trim curves do not usually lie on the surfaces.

For a prescribed curve $\boldsymbol{T}_i(\tau) = \boldsymbol{G}_i(u_i(\tau), v_i(\tau))$, Koparkar [Kop 91, 91b] defines a *normal surface* $\boldsymbol{G}_i^{\perp}(\tau, \delta)$ parametrized over τ and δ by

$$\boldsymbol{G}_i^{\perp}(\tau, d) = \boldsymbol{G}_i(u_i(\tau), v_i(\tau)) + \delta \boldsymbol{N}_i(u_i(\tau), v_i(\tau)) = \boldsymbol{T}_i(\tau) + \delta \boldsymbol{N}_i(\tau). \quad (14.21)$$

The isoparametric lines $\delta =$ constant of $\boldsymbol{G}_i^{\perp}$ are the intersection curves of the normal surface with the δ offset surface of $\boldsymbol{G}_i$, whereas the lines defined via the surface normal vector $\boldsymbol{N}_i$ at the point $\boldsymbol{G}_i(u_i(\tau), v_i(\tau))$ correspond to the isoparametric lines $\tau = $ constant. It follows that $\boldsymbol{G}_i^{\perp}$ is a special ruled surface, called a *fanout surface* in [Kop 91, 91b]. Taking the intersection $\boldsymbol{G}_1^{\perp} \cap \boldsymbol{G}_2^{\perp}$ leads to the directrix $\boldsymbol{S}$ which can then be used with the cross-section method described below to construct the blending surface. In view of the definition of $\boldsymbol{G}_i^{\perp}$, $\boldsymbol{S}$ has the useful property that the orthogonal projection of $\boldsymbol{S}$ onto $\boldsymbol{G}_i$ is again $\boldsymbol{T}_i(\tau)$.

14.2.2. Definition of the Blending Surface

If we are given an *explicit formula* for the trim curves in coordinate space, then it is possibile to give the blending surface in closed form. For example, if the trim curves are (rational) B-spline or Bézier (spline) curves, then the blend can be constructed as a (rational) tensor-product B-spline or Bézier (spline) surface, see e.g., [Bar 89a], [Cho 89], [Kla 92], and [Chi 87, 91]. In [Har 90], (RB) Gregory patches are used, while [Szi 91] uses bicubic monomial forms (Ferguson patch); cf. the example in Fig. 6.11.

In the *cross-section method*, the trim curves need not be given in closed form, but can instead be prescribed in terms of sets of points. However, the points on the two trim curves $\boldsymbol{T}_1$ and $\boldsymbol{T}_2$ have to be somehow uniquely identified with each other, possibly by a joint parameter τ. The blend is then described as a family of curves: for each pair of corresponding trim curve points $\boldsymbol{T}_1(\tau)$ and $\boldsymbol{T}_2(\tau)$, $\tau \in [\tau_0, \tau_1]$, we define a (planar) (cross-section) curve $\boldsymbol{Q}_\tau(\sigma)$, $\sigma \in [\sigma_0, \sigma_1]$ of the blend. The blend can also be described as a sweep of the (varying) curve $\boldsymbol{Q}_\tau(\sigma)$, the generatrix, along the boundary curves $\boldsymbol{T}_1(\tau)$ and $\boldsymbol{T}_2(\tau)$, see e.g., [Fil 89]. The construction of rolling-ball blends (cf. Sect. 14.2.3) is almost always done in this way.

In both cases, *derivatives* along the trim curves are required for the construction. If the blending boundary curves are given in closed form, then the derivatives in the direction of the trim lines can be computed from the representation of these curves. Otherwise, we have to go back to the representation of the surfaces. Cross derivatives always have to be obtained from the formulae for the surfaces. There are several different strategies for choosing a direction $\boldsymbol{R}_i$ across a given trim line $\boldsymbol{T}_i$. Assuming that $\boldsymbol{N}_i$ denotes the normal to the surface at the point $\boldsymbol{T}_i(\tau) \in \boldsymbol{G}_i$, we can choose $\boldsymbol{R}_i(\tau)$ as:

- the tangent to the surface $\boldsymbol{G}_i$ perpendicular to the tangent $\boldsymbol{T}_i'(\tau)$ to the trim curve $\boldsymbol{T}_i(\tau)$ (see e.g., [Hansm 87, 88], [Fil 89], [Bar 89a], and [Kop 91]), i.e., $\boldsymbol{R}_i = \pm\, \boldsymbol{N}_i \times \boldsymbol{T}_i'$,
- the orthogonal projection (onto the tangent plane of the given surface $\boldsymbol{G}_i$ at the point $\boldsymbol{T}_i(\tau) \in \boldsymbol{G}_i$) of the segment connecting the point $\boldsymbol{S} = \boldsymbol{S}(\tau)$

on the directrix, and its orthogonal projection $\boldsymbol{S}_k^{\perp}(\tau) = \boldsymbol{T}_k(\tau)$ onto the other surface $\boldsymbol{G}_k$, see [Peg 90a], [Raj 91],

- the direction of the intersection line of the tangent plane to the surface at the point $\boldsymbol{T}_i(\tau) \in \boldsymbol{G}_i$ and the plane $\boldsymbol{E}(\tau)$ with normal vector $\boldsymbol{N}_E = (\boldsymbol{S} - \boldsymbol{T}_1) \times (\boldsymbol{S} - \boldsymbol{T}_2)$ determined by the points $\boldsymbol{T}_1(\tau)$, $\boldsymbol{T}_2(\tau)$ and the corresponding point $\boldsymbol{S} = \boldsymbol{S}(\tau)$ on the directrix, see [Kop 91], i.e., $\boldsymbol{R}_i = \pm \boldsymbol{N}_i \times \boldsymbol{N}_E$,
- the projection of the difference vector $\boldsymbol{V}(\tau) = \boldsymbol{T}_2(\tau) - \boldsymbol{T}_1(\tau)$ onto the tangent plane of the surface $\boldsymbol{G}_i$ at the point $\boldsymbol{T}_i(\tau) \in \boldsymbol{G}_i$, see [Fil 89], i.e.,

$$\boldsymbol{R}_i = \boldsymbol{V} - \frac{\boldsymbol{V} \cdot \boldsymbol{N}_i}{\boldsymbol{N}_i \cdot \boldsymbol{N}_i} \boldsymbol{N}_i.$$

Then, for example, we get a GC^1 continuous cubic Hermite-blend (by the cross-section method) by choosing

$$\boldsymbol{F}(\sigma, \tau) = \boldsymbol{T}_1(\tau) H_{00}(\sigma) + \boldsymbol{T}_2(\tau) H_{01}(\sigma) + \boldsymbol{Q}_1(\tau) H_{10}(\sigma) + \boldsymbol{Q}_2(\tau) H_{11}(\sigma),$$

with the cubic Hermite polynomials as in Sect. 3.3 and the cross derivatives

$$\boldsymbol{Q}_i(\tau) = \delta_i \ell_i(\tau) \frac{\boldsymbol{R}_i(\tau)}{|\boldsymbol{R}_i(\tau)|},$$

see also [Fil 89]. The factors $\ell_i(\tau)$ and the design parameters δ_i allow the construction of various blend cross-sections.

If we want a C^2 or GC^2 blending surface, we can get the necessary second derivatives from the surface normal curvature in the direction of the selected $\boldsymbol{R}_i$, see [Hansm 87], [Peg 90a], and [Raj 91].

14.2.3. Rolling-ball Blends

The explicit computation of a *constant radius* (*rolling-ball*) *blend* of two surfaces given in parametric form $\boldsymbol{G}_1(u_1, v_1)$ and $\boldsymbol{G}_2(u_2, v_2)$ can be carried out using equations analogous to (14.16) – (14.17), see e.g., [Hoff 90, 90a]. For example, for a directrix $\boldsymbol{S}$ and trim curves $\boldsymbol{T}_1$ and $\boldsymbol{T}_2$ we can proceed as follows.

If the offset surfaces $\boldsymbol{O}_1(u_1, v_1)$ and $\boldsymbol{O}_2(u_2, v_2)$ to the surfaces $\boldsymbol{G}_1(u_1, v_1)$ and $\boldsymbol{G}_2(u_2, v_2)$ are given by the equations

$$\boldsymbol{O}_1(u_1, v_1) = \boldsymbol{G}_1(u_1, v_1) + r\boldsymbol{N}_1(u_1, v_1), \tag{14.22a}$$

$$\boldsymbol{O}_2(u_2, v_2) = \boldsymbol{G}_2(u_2, v_2) + r\boldsymbol{N}_2(u_2, v_2), \tag{14.22b}$$

(see Chap. 15), then the directrix $\boldsymbol{S}$ can be found immediately from the intersection condition

$$\boldsymbol{O}_1(u_1, v_1) = \boldsymbol{O}_2(u_2, v_2).$$

If $\boldsymbol{S}$ is parametrized by $\tau \in [\tau_0, \tau_1]$, then $u_i = u_i(\tau)$, $v_i = v_i(\tau)$, and we get

$$\boldsymbol{S}(\tau) = \boldsymbol{O}_1(u_1(\tau), v_1(\tau)) = \boldsymbol{O}_2(u_2(\tau), v_2(\tau)),$$

and (14.22) implies that the trim curves $\boldsymbol{T}_1$ and $\boldsymbol{T}_2$ are given by

$$\boldsymbol{T}_1(\tau) = \boldsymbol{G}_1(u_1(\tau), v_1(\tau)), \qquad \boldsymbol{T}_2(\tau) = \boldsymbol{G}_2(u_2(\tau), v_2(\tau)).$$

In practice, the determination of explicit formulae for the directrix, the trim curves, and the rolling-ball blend can be very expensive, as it was for implicitly defined surfaces, provided it can be done at all. [San 90] shows that a solution is possible for the case of natural quadrics and a torus. However, general quadrics lead to equations or systems of equations which have no analytic solution. This situation also occurs, of course, for free-form surfaces, see [Hoff 90, 90a]. Thus, numerous *approximation methods* have been suggested. Usually this involves blends based on sweep surfaces which include design parameters which can be chosen to give (approximately) circular arcs as cross-sections. Thus, for example, in [Fil 89] the cross-sections are constructed by cubic Hermite interpolants, while [Cho 89] uses quadratic rational Bézier curves. [Har 90] uses cubic rational Bézier curves, and [Kla 92] uses cubic polynomial Bézier curves. [Peg 87, 90] applies a non-polynomial circular formula depending on the directrix and its Frenet frame.

In certain important cases arising in solid modelling with parametric boundary representations, *variable radius (rolling-ball) blends* can be given explicitly in terms of pieces of cyclides. By [FORS 12] (see also [Prat 88, 90], [Zhou 91]), a *(Dupin) cyclide* (in standard position and orientation) can be parametrized as

$$\begin{aligned} x(\theta,\psi) &= \frac{1}{\omega}(\mu(e - a\cos\theta\cos\psi) + b^2\cos\theta), \\ y(\theta,\psi) &= \frac{b}{\omega}\sin\theta(a - \mu\cos\psi), \\ z(\theta,\psi) &= \frac{b}{\omega}\sin\psi(e\cos\theta - \mu), \end{aligned} \tag{14.23}$$

with

$$\omega(\theta,\psi) = a - e\cos\theta\cos\psi,$$

and $\theta, \psi \in [0, 2\pi]$. Both the isolines $\theta =$ const. and the isolines $\psi =$ const. of this parametrization are curvature lines, and thus are circular, see Fig. 14.14.

Because of this, it is easy to connect them with a natural quadric or with a torus along a circle. After the reparametrization $\theta, \psi \to u = u(\theta, \psi)$, $v = v(\theta, \psi)$, (14.23) becomes a rational biquadratic, see [Prat 88, 90], [Boeh 90a], and [Zhou 91]. Bézier representations of this type are given in [Prat 88, 90] and [Boeh 90a], although those in the former appear not to be practical. [Zhou 91] uses the blossoming principle to get a NURBS formula associated with the knot vector $\boldsymbol{T} = (-1, -1, -1, 0, 0, 1, 1, 1)$ in both the u and v directions, which consists of four Bézier segments, where the one corresponding to $[-1, 0] \times [0, 1]$ is the one in [Boeh 90a].

As in Sect. 14.1.4, the directrix of the blending surface is defined as the intersection of an appropriately chosen reference surface with the Voronoi surface $\boldsymbol{V}_{12}$ corresponding to the surfaces $\boldsymbol{G}_1$ and $\boldsymbol{G}_2$. The system of equations analogous to (14.20) consists of six equations in eight unknowns, and is given in [Hoff 90a], along with the complete system for describing the VR-blend. For non-trivial surfaces, we again have to work with approximations to the blending surface. This can be accomplished with an algorithm based on parametric representations similar to the one given at the end of Sect. 14.1.4 for implicit representations. In general, algorithms based on the cross-section method can similarly be extended to the case of VR-blends, see e.g., [Peg 87, 90], [Har 90].

14.3. Recursively Defined Blends

Subdivision algorithms were originally developed to provide fast ways to evaluate curves and surfaces, but they also can be applied to blending polyhedrally defined objects. The idea is to smooth out edges and corners by a stepwise refinement of the polyhedral net structure. This can be done with repeated subdivision of the polyhedral facets, and is equivalent to cutting off the edges and corners at each step (*corner cutting*), see the second paragraph of Section 7.5.

In each step of Chaikin's algorithm [Chai 74], the polygon edges are divided in the ratio 1 : 2 : 1, producing two new points per edge:

$$\boldsymbol{P}_{2i}^{k+1} = \tfrac{3}{4}\boldsymbol{P}_i^k + \tfrac{1}{4}\boldsymbol{P}_{i+1}^k, \qquad \boldsymbol{P}_{2i+1}^{k+1} = \tfrac{1}{4}\boldsymbol{P}_i^k + \tfrac{3}{4}\boldsymbol{P}_{i+1}^k. \tag{14.24}$$

The process starts with the initialization $\boldsymbol{P}_i^0 = \boldsymbol{P}_i$, $i = 0(1)N$, see Fig. 14.16. Repeatedly applying the rule (14.24) produces a sequence of polygons which converges to a C^1 continuous curve which is equivalent to a quadratic B-spline curve with $\boldsymbol{d}_i = \boldsymbol{P}_i$ and a uniform knot vector, see [Rie 75]. If the recurrence

$$\boldsymbol{P}_{2i}^{k+1} = \tfrac{1}{2}\boldsymbol{P}_i^k + \tfrac{1}{2}\boldsymbol{P}_{i+1}^k, \qquad \boldsymbol{P}_{2i+1}^{k+1} = \tfrac{1}{8}\boldsymbol{P}_i^k + \tfrac{3}{4}\boldsymbol{P}_{i+1}^k + \tfrac{1}{8}\boldsymbol{P}_{i+2}^k, \tag{14.25}$$

is used, then we get a sequence of polygons (see Fig. 14.17) which converges to a C^2 continuous cubic B-spline curve with uniform knot vector, see [Dyn 89a, 90a]. In general, all B-spline curves can be generated by recursive subdivision, and similarly, B-spline surfaces can be obtained from regular rectangular grids.

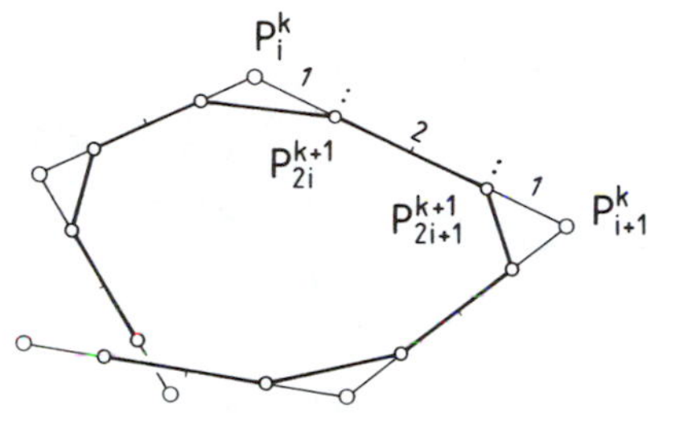

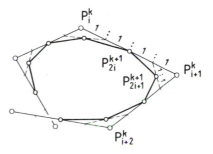

Fig. 14.16. A subdivision step of the algorithm of [Chai 74].

Fig. 14.17. A subdivision step of the cubic algorithm of [Dyn 89a].

Catmull and Clark [Cat 78] have extended the above idea to arbitrary polyhedra for the recursive construction of biquadratic and bicubic B-spline surfaces. The Doo-Sabin algorithm [Doo 78], see also [Brun 88, 91], [Nas 91], leads to a biquadratic B-spline surface, and is a generalization of the Chaikin algorithm. For a given N-sided facet, it constructs N new vertices as the midpoints of the lines connecting the original vertices of the facet to the point which is the average of these vertices. There are three different types of new facets which reproduce in each step. If the original polyhedron had N-sided corners, these yield N-sided facets in the first refinement, but not in any later refinements, see Fig. 14.18. [Rock 84], [Brun 88] give several examples, and [Vee 82] describes an implementation.

The resulting surface can possess *singular points* (also called *extraordinary points*), where only GC^1 continuity holds. These points correspond to the N-sided vertices with $N \neq 4$, or to N-vertex facets with $N \neq 4$ of the original polyhedron. A similar behavior is exhibited by the algorithm of [Cat 78], which leads to surfaces which do not behave as well in the neighborhood of singular points. A study of the smoothness properties and of C^1 continuous modifications of the algorithm of [Cat 78] can be found in [Doo78], [Ball 86, 88]. In general, it is impossible to achieve C^2 or GC^2 behavior at the singular points, see [Ball 90]. Additional results on convergence and smoothness behavior of subdivision algorithms can be found in [Pra 87], [Mic 87, 89], and [Dyn 91]. [Brun 88] generalized the biquadratic algorithm of [Doo 78]

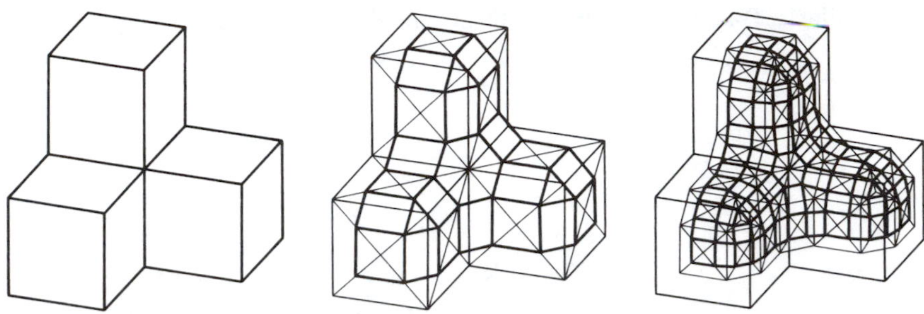

Fig. 14.18. Defining polyhedron (left) and two refinement steps of the quadratic Doo-Sabin algorithm.

by assigning a tension parameter to each vertex of the original polyhedron. Recursively defined curves and surfaces which interpolate the vertices have been treated in [Nas 87, 91, 91a], [Dyn 87b, 90a]. Those in [Dyn 87b, 90a] allow the use of tension parameters for design purposes. The papers [Sab 86a], [Dyn 92] provide a good overview of recursive subdivision and corner cutting, see also [ABR 91].

The polyhedral representation of objects described above have the advantage of being much faster for computing a graphic representation, i.e., for rendering purposes, but the lack of a parametrization or a closed formula can be a disadvantage in some applications.

14.4. Numerically Defined Blends

The blending problem can be interpreted as a boundary value problem. Bloor and Wilson [Blo 89] therefore construct blends as surfaces which take on the desired values on the boundaries (i.e., match up with the trim lines), and which solve the system of (Laplace) partial differential equations

$$\Big(\frac{\partial^2}{\partial u^2} + \frac{\partial^2}{\partial v^2}\Big)\boldsymbol{X}(u,v) = 0. \tag{14.26}$$

However, solutions of (14.26) are not really well suited for the purposes of CAGD, since by the nature of the boundary conditions, or because of an inappropriately selected parametrization, they can have a constriction. Moreover, blends constructed in this way are only C^0 continuous.

The *constriction effect* can be eliminated with the help of a parameter a which gives different weights to u and v, i.e., instead of (14.26), we solve the modified system

$$\left(\frac{\partial^2}{\partial u^2} + a^2 \frac{\partial^2}{\partial v^2}\right) \boldsymbol{X}(u, v) = 0, \tag{14.27}$$

see Fig. 14.19.

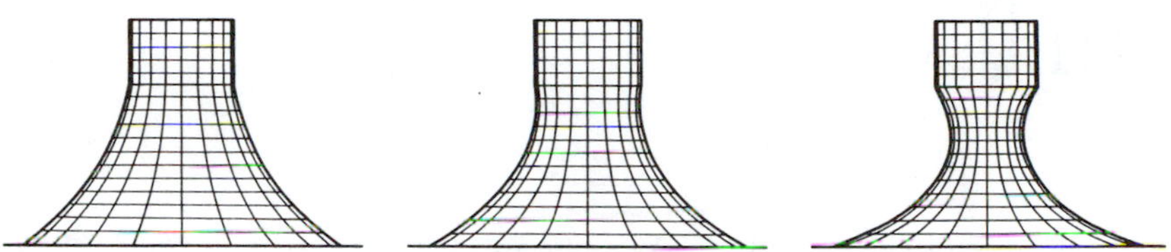

Fig. 14.19. Influence of the design parameter a (increasing from left to right) on the form of the solution of the boundary value problem (14.27) for a blend between a cylinder standing perpendicular to a horizontal plane.

To get a blend which joins the given surfaces with a higher degree of smoothness, a solution of a differential equation of higher order must be computed so that we can also match the derivatives along the trim lines. For C^1 continuity, for example, we have to solve the system of partial differential equations of fourth order

$$\left(\frac{\partial^2}{\partial u^2} + a^2 \frac{\partial^2}{\partial v^2}\right)\left(\frac{\partial^2}{\partial u^2} + a^2 \frac{\partial^2}{\partial v^2}\right) \boldsymbol{X}(u, v) = 0, \tag{14.28}$$

while for C^2 continuity, we get a corresponding system of equations of sixth order. We note however, that it is not always necessary to solve a differential equation of fourth order for every component of $\boldsymbol{X}(u, v)$ to get GC^1 continuity of the blending surface, see [Bloo 89, 89a].

In general, it is not possible to find solutions of the above systems in closed or analytic form, and thus numerical methods have to be applied, see e.g., [SMI 78]. Since many such methods produce a solution only on a grid, in such cases the blending surface has to be constructed by interpolation. In order to get surfaces which are compatible with CAD systems, [Blo 90a] uses B-spline surfaces for the interpolation.

The method has been applied in [Blo 89b, 90], [Low 90] for filling N-sided holes, and for the construction of functional free-form surfaces.

15 Offset Curves and Surfaces

Offset curves and surfaces are curves and surfaces which are a constant distance d from a given initial curve or surface. Thus the offset of a parametric curve is given by

$$\boldsymbol{X}_d(t) = \boldsymbol{X}(t) + \boldsymbol{N}(t)d, \tag{15.1a}$$

while the offset of a parametric surface is given by

$$\boldsymbol{X}_d(u, v) = \boldsymbol{X}(u, v) + \boldsymbol{N}(u, v)d, \tag{15.1b}$$

where $\boldsymbol{N}$ is the unit normal vector with $|\boldsymbol{N}| = 1$. For a given orientation of $\boldsymbol{N}$, d can be either positive or negative, so that the offset can lie on either side of the initial curve or surface, see Fig. 15.1. For plane curves and for surfaces, the normal vector $\boldsymbol{N}$ is uniquely defined. For space curves, all vectors in the plane defined by $\dot{\boldsymbol{X}} \cdot \boldsymbol{N} = 0$ are normal vectors.

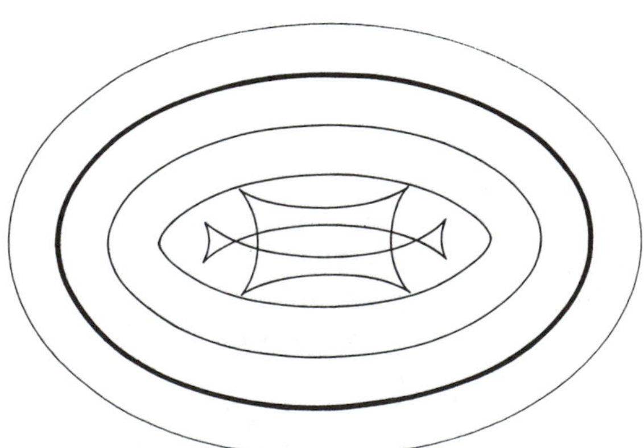

Fig. 15.1. Offsets of a plane curve.

Offset curves and surfaces are of great importance in applications [Til 84], [Ross 86]. For example, the interior skin of a ship or automobile can be described by an offset surface corresponding to a distance equal to the thickness of the material. Pieces of the surface can be cut, milled or polished using a laser-controlled device to follow the offset. Applications of offset curves in road construction are discussed in [Mee 90].

15.1. Analytic Properties of Plane Offset Curves

By (15.1), the parametric formula for an offset curve at distance d from a given plane curve is

$$\begin{aligned} x_d(t) &= x(t) \pm d\frac{\dot{y}(t)}{\sqrt{\dot{x}^2(t)+\dot{y}^2(t)}}, \\ y_d(t) &= y(t) \mp d\frac{\dot{x}(t)}{\sqrt{\dot{x}^2(t)+\dot{y}^2(t)}}, \end{aligned} \tag{15.2}$$

assuming that the initial curve $\boldsymbol{X}(t)$ is *differentiable.*

For each parameter value t, the initial curve and its offset have parallel tangents, since differentiating (15.2) gives

$$\dot{\boldsymbol{X}}_d(t) = \dot{\boldsymbol{X}}(t)(1+\kappa(t)d). \tag{15.3}$$

It follows immediately from (15.3) that even if the initial curve is regular, the offset curve can have cusps at points where

$$1+\kappa d = 0. \tag{15.4}$$

Equation (15.4) depends on the orientation of the curve, and thus is not too suitable for finding cusps of the offset; [Hos 85] has given a stable method.

From (15.3) we see that the curvature of the offset curve is given by

$$\kappa_d = \frac{\kappa}{1+\kappa d},$$

and its radius of curvature by

$$\rho_d = \rho + d, \tag{15.5}$$

where κ and ρ are the quantities associated with the initial curve. A detailed discussion of offset curves from the view point of differential geometry can be found in [Faro 90a].

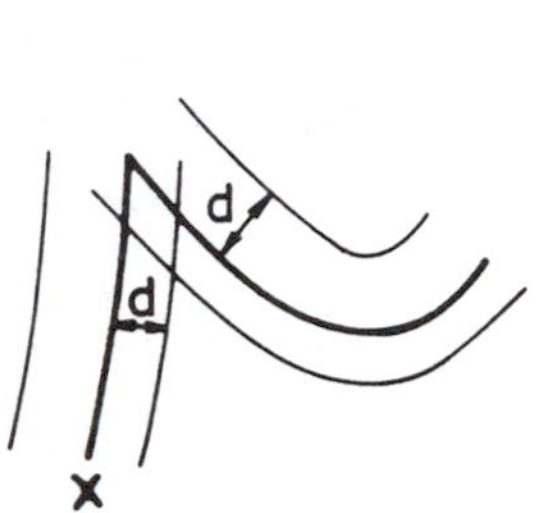

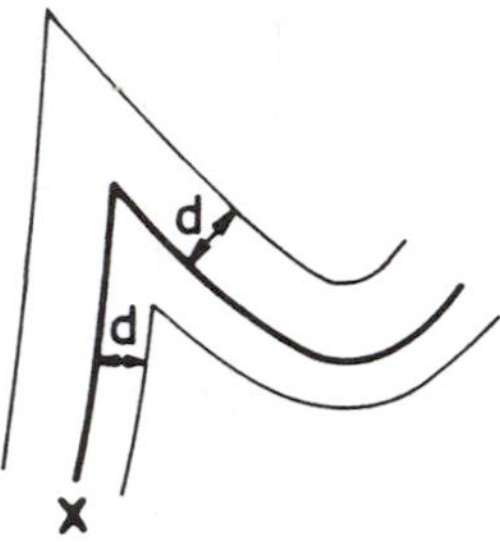

Fig. 15.2a. Possible offset curves near a cusp.

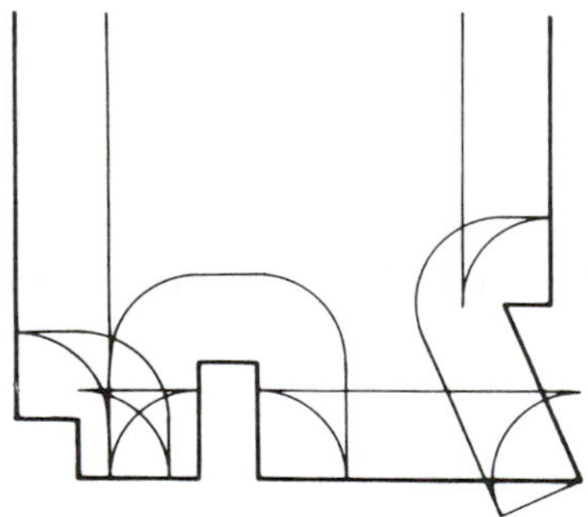

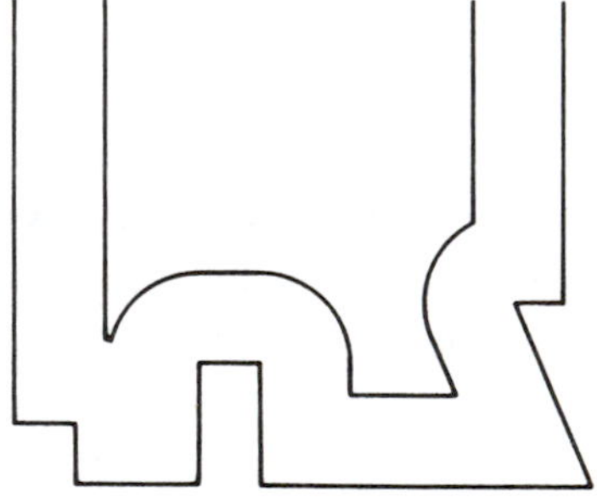

Fig. 15.2b. Possible offset curves at corners.

If the initial curve is not differentiable at some points, then the corresponding offset curve has to be specially defined at such points. This can be done, for example, using lines or circular arcs, see Fig. 15.2a,b.

It is clear from the figures that, depending on the shape of the initial curve, its offset curve can come closer than d to the initial curve, thus causing problems with *collisions* when steering a tool. To avoid this, we need to remove certain segments of the curve which start and end at self-intersections, see Fig. 15.3 and [Sae 88], [Stei 90]. These can be found by first estimating their location, and then using an iterative Newton like method. Given two curves $\boldsymbol{X}_1(t)$ and $\boldsymbol{X}_2(\tau)$ in parametric form, the process proceeds as follows:

1) choose an appropriate $t = t_0$, and find the equation of the tangent T_1 at the point $\boldsymbol{P}_0 = \boldsymbol{X}_1(t_0)$:

$$\frac{x - x(t_0)}{y - y(t_0)} = \frac{\dot{x}(t_0)}{\dot{y}(t_0)},$$

or equivalently

$$x\dot{y}(t_0) - y\dot{x}(t_0) - [x(t_0)\dot{y}(t_0) - y(t_0)\dot{x}(t_0)] = 0, \qquad (*)$$

2) find the point $\boldsymbol{P}_1 = \boldsymbol{X}_2(\tau_1)$ where T_1 intersects $\boldsymbol{X}_2$ by solving the equation

$$\xi(\tau)\dot{y}(t_0) - \eta(\tau)\dot{x}(t_0) - [x(t_0)\dot{y}(t_0) - y(t_0)\dot{x}(t_0)] = 0,$$

for $\tau = \tau_1$, see Fig. 15.4,

3) find the tangent T_2 to $\boldsymbol{X}_2$ at $\boldsymbol{P}_1 = \boldsymbol{X}_2(\tau_1)$, and then the point $\boldsymbol{P}_2$ where T_2 intersects $\boldsymbol{X}_1$. Repeat this process until $|\boldsymbol{P}_i - \boldsymbol{P}_{i+1}| < \epsilon$, which gives $\boldsymbol{P}_i = \boldsymbol{X}_1(t_i) \approx \boldsymbol{S} \approx \boldsymbol{X}_2(\tau_i) = \boldsymbol{P}_{i+1}$, see Fig. 15.4.

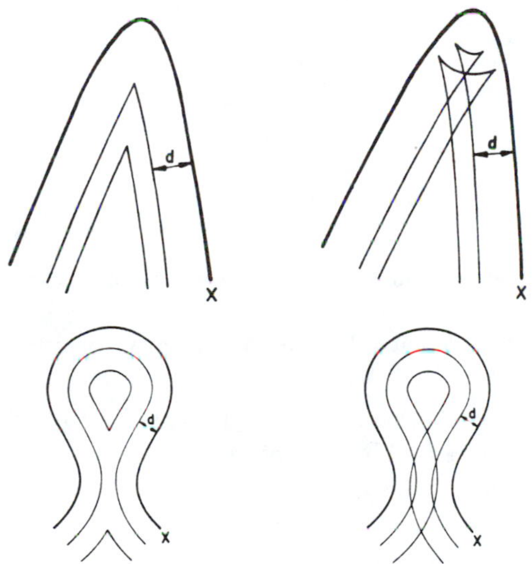

Fig. 15.3. Collision of offset curves.

This method converges whenever the tangents intersect the curves at each step, i.e., provided that the starting point $\boldsymbol{P}_0 \in \boldsymbol{X}_1(t)$ lies in the cross-hatched area in Fig. 15.5.

If the initial curve itself has self-intersections, then additional problems arise. A global criterion for identifying self-intersections of Bézier curves is given in [Las 89], see also Fig. 12.23. We should note, however, that offset curves of Bézier curves are not in general Bézier curves. The criterion can still

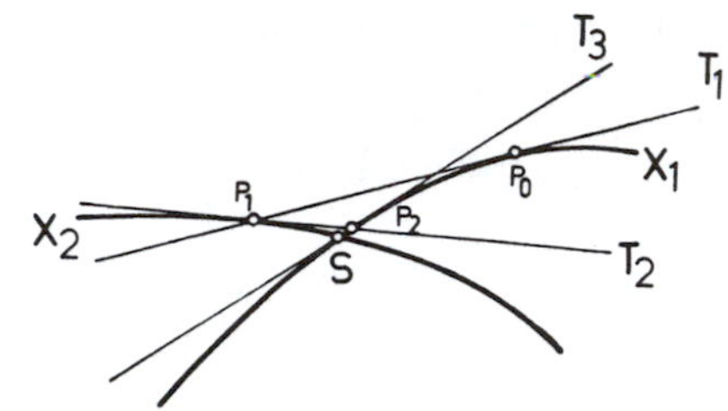

Fig. 15.4. Intersection algorithm for plane curves.

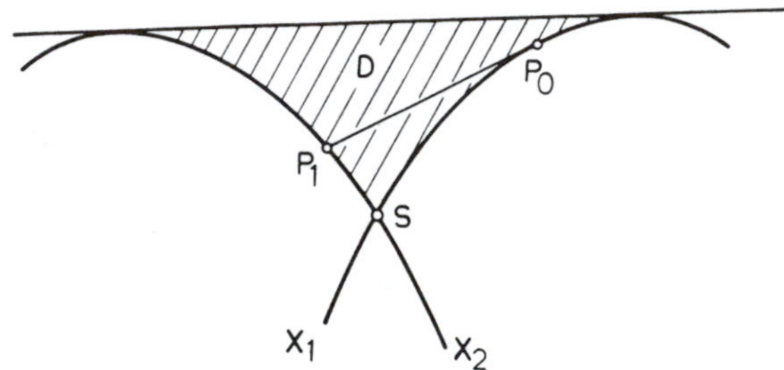

Fig. 15.5. Admissible starting points for the intersection algorithm.

be used though, if we replace the exact offset curve by a Bézier approximation to it, see Sect. 15.4.

Collision problems arise in many applications. Special methods for dealing with them in the case of interior offsets (as are used for example in milling holes or pockets in an object, cf. Fig. 15.6), can be found in [Per 78] for the case of simply connected regions, and in [HEL 91] for the case of multiply connected regions. An application of offset curves for the design of a pattern is shown in Fig. 15.7, see also [Hos 85].

Another approach to offset curves is to define them to be the envelope corresponding to moving the center of a circle of radius d along the initial curve. This simultaneously defines both an "inside" and "outside" offset curve, and has immediate applications in milling. For an initial curve $\boldsymbol{X}(t)$ given in parametric form, the circle of radius d and center at $\boldsymbol{X}(t)$ is given by

$$\boldsymbol{X}_d(t,\tau) = \begin{pmatrix} x(t) \\ y(t) \end{pmatrix} + \begin{pmatrix} \cos\tau \\ \sin\tau \end{pmatrix} d, \tag{15.6}$$

where $\tau \in [0, 2\pi]$. The desired offset curves are defined by the envelopes of this

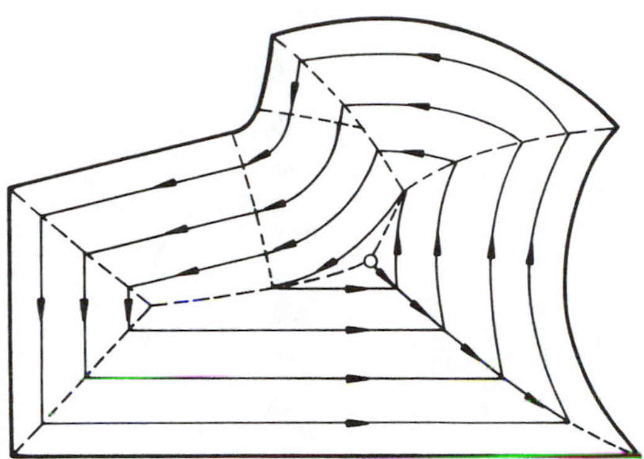

Fig. 15.6. Cutting paths for milling a hole.

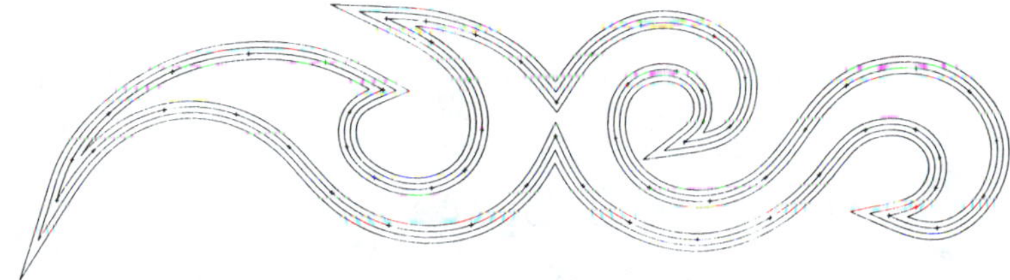

Fig. 15.7. Decorative pattern for a boot.

circle. At points on the envelope, the partial derivatives of $\boldsymbol{X}_d$ with respect to t and τ must be parallel, *i.e.*,

$$\frac{\partial \boldsymbol{X}_d}{\partial \tau} = \lambda \frac{\partial \boldsymbol{X}_d}{\partial t}, \qquad \lambda \in \mathbb{R}, \tag{15.7a}$$

which leads to the envelope condition

$$\tan \tau = \frac{-\dot{x}}{\dot{y}}. \tag{15.7b}$$

If the initial curve has a cusp at a point $\boldsymbol{S}$, then we have $\dot{x} = \dot{y} = 0$ there, and so τ is undetermined. In this case we have to insert a piece of a circle of radius d into the offset curve, see Fig. 15.8 and [HOFF 89].

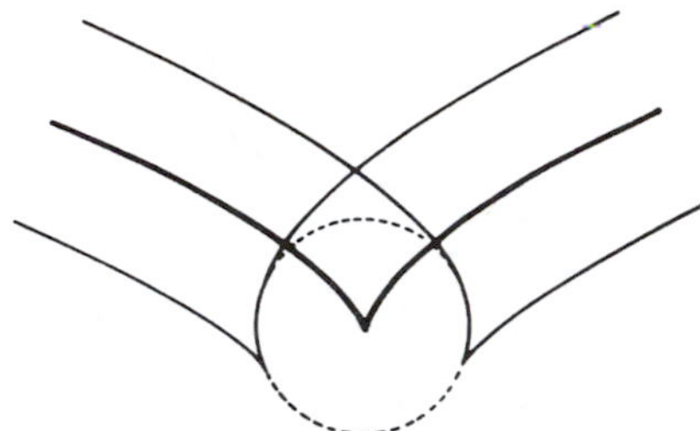

Fig. 15.8. Complete offset curve for an initial curve with a cusp.

The offset curve problem can be formulated in algebraic terms by eliminating τ from (15.6). Then in an (ξ, η) system, we have

$$(\xi - x(t))^2 + (\eta - y(t))^2 - d^2 = 0,$$

along with the envelope condition

$$\dot{x}(t)(\xi - x(t)) + \dot{y}(t)(\eta - y(t)) = 0.$$

Eliminating the parameter t leads to the algebraic formula for the (complete) offset curve.

15.2. Offset Curves in Space

Given a curve in space, we cannot define a single unique associated offset curve. However, we can still generate offsets by moving a circle with radius d along the curve, keeping the osculating plane which contains the circle orthogonal to the given curve.

Let s be the arc length parameter for the initial curve, and suppose $\boldsymbol{v}_1 = \boldsymbol{X}'/|\boldsymbol{X}'|$ is the unit tangent to the curve. Then both the principal normal vector $\boldsymbol{v}_2 = \boldsymbol{X}''/|\boldsymbol{X}''|$ and the binormal vector $\boldsymbol{v}_3 = \boldsymbol{v}_1 \times \boldsymbol{v}_2$ are perpendicular to $\boldsymbol{v}_1$. Thus, any curve of the form

$$\boldsymbol{X}_d(s, \tau) = \boldsymbol{X}(s) + d(\boldsymbol{v}_2 \cos\tau + \boldsymbol{v}_3 \sin\tau) \tag{15.8}$$

is a distance of d from the initial curve $\boldsymbol{X}$ at every value of s. The parameter τ can either be fixed, or can be a function of s.

Offset curves in space have applications to the control of ball end cutting tools used to mill free form surfaces, see Fig. 15.9. In this case, we have to select the parameter τ so that $\boldsymbol{X}_d(s)$ remains on the offset surface at a distance d from the given free form surface.

In general, it is not easy to find the arc length parametrization of a curve on a free form surface. Thus, we suggest another approach. If $\boldsymbol{X}(t)$ is a curve on the patch $\boldsymbol{X}(u,v)$, then the plane ϵ defined by the equation

$$(\boldsymbol{Y} - \boldsymbol{X}(u(t_0), v(t_0))) \cdot \dot{\boldsymbol{X}}(t_0) = 0 \qquad (*)$$

is orthogonal to $\boldsymbol{X}(t)$ at the point $t = t_0$ (here $\boldsymbol{Y}$ is a point running over the plane ϵ). We are looking for the point on the intersection curve of the plane ϵ with the given patch $\boldsymbol{X}(u,v)$ which lies at a distance d from the point $\boldsymbol{X}(t)$, *i.e.*, in addition to $(*)$ we want

$$|\boldsymbol{X}(u,v) - \boldsymbol{X}(t_0)| - d = 0. \qquad (**)$$

The equations $(*)$ and $(**)$ constitute a nonlinear system of equations which can be solved using the Newton-Raphson method, although frequently there are problems with convergence. The numerical computation required can be significantly reduced if we do not work with the exact normal intersection curve, *i.e.*, if we give up $(*)$ and require only $(**)$. As a possible trajectory orthogonal to the initial curve $\boldsymbol{X}(t)$, we can take the curve $\boldsymbol{X}_q$ lying on the patch which is orthogonal to $\boldsymbol{X}(t)$ at the point $\boldsymbol{X}(t_0)$; this curve does not necessarily lie on the plane $(*)$. The tangent to $\boldsymbol{X}_q$ at the point $\boldsymbol{X}(t_0)$ does lie in the tangent plane to the patch, and thus

$$\dot{\boldsymbol{X}}_q = \frac{\dot{\boldsymbol{X}} \times \boldsymbol{N}}{|\dot{\boldsymbol{X}} \times \boldsymbol{N}|}.$$

Now with the decomposition of $\dot{\boldsymbol{X}}_q$ into components of the partial derivatives, $\dot{\boldsymbol{X}}_q = \Delta u \boldsymbol{X}_u + \Delta v \boldsymbol{X}_v$, we find that the image of the orthogonal direction vector in the parameter plane is given by $\boldsymbol{q} = (\Delta u, \Delta v)^T$. If we assume that the orthogonal curve in the parameter plane is the image of a straight line with the direction $\boldsymbol{q}$, then the equation for the direction of the line is

$$\boldsymbol{u}(\lambda) = \begin{pmatrix} u(t_0) \\ v(t_0) \end{pmatrix} + \lambda \begin{pmatrix} \Delta u \\ \Delta v \end{pmatrix}, \qquad \lambda > 0,$$

(with an appropriate orientation of $\boldsymbol{q}$). The condition $(**)$ now leads to the equation

$$|\boldsymbol{X}(u(t_0) + \lambda \Delta u, v(t_0) + \lambda \Delta v) - \boldsymbol{X}(t_0)| - d = 0.$$

This is a one parameter equation in λ, whose zeros can easily be found using regula falsi or Newton's method.

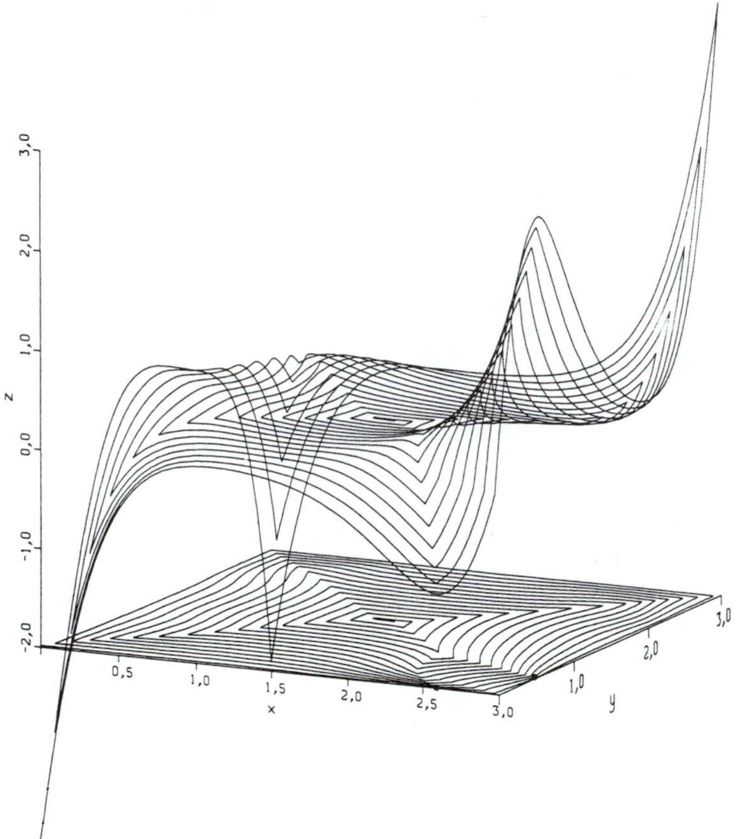

Fig. 15.9. Offset curves as milling paths on a patch.

Fig. 15.9 shows offset curves lying on a parametric patch as computed by this method. We start with the patch boundary as an initial curve, then use its offset as an initial curve, etc. The method has problems at points where the initial curve is not differentiable, see [Fels 91].

Offset curves on a patch can also be constructed as geodesic offset curves by finding a trajectory orthogonal to $\boldsymbol{X}(t)$ which is simultaneously a geodesic at each point $\boldsymbol{X}(t_0)$ on the initial curve. This requires solving the differential equation for the geodesic using the orthogonal direction as an initial value. This geodesic line is followed until we are at a distance d from the initial point, measured along the geodesic. An algorithm for constructing geodesic offset curves along with examples of applications can be found in [Patr 89b],

[Stei 90]. If a trim curve is used to cut a hole in a patch, then by uniformly expanding the hole, we get an offset curve at a distance d from the boundary of the hole. The distance d can be either spatial distance or geodesic distance along the patch.

15.3. Offset Surfaces

For each point on an offset surface, the tangent plane there is parallel to the tangent plane at the corresponding point on the initial surface. If we differentiate (15.1b), we get

$$\frac{\partial \boldsymbol{X}_d}{\partial u} = \boldsymbol{X}_u + \boldsymbol{N}_u d, \qquad \frac{\partial \boldsymbol{X}_d}{\partial v} = \boldsymbol{X}_v + \boldsymbol{N}_v d,$$

and thus the normal to the offset surface is given by

$$\boldsymbol{N}_d = \frac{\partial \boldsymbol{X}_d}{\partial u} \times \frac{\partial \boldsymbol{X}_d}{\partial v} = \boldsymbol{X}_u \times \boldsymbol{X}_v = \boldsymbol{N}\, |\boldsymbol{X}_u \times \boldsymbol{X}_v|, \tag{15.9}$$

since $\boldsymbol{N} \cdot \boldsymbol{N}_u = 0$, $\boldsymbol{N} \cdot \boldsymbol{N}_v = 0$, and $|N| = 1$, *i.e.*, $\boldsymbol{X}$ and $\boldsymbol{X}_d$ have a common normal vector at corresponding points.

In analogy to (15.5), the principal radii of curvature $\rho_{i,d}$ of an offset surface are given by

$$\rho_{1,d} = \rho_1 + d, \qquad \rho_{2,d} = \rho_2 + d, \tag{15.10}$$

where ρ_i are the principal radii of curvature of the given patch, see [STR 50], [FAU 81], and [Faro 85, 86a, 90a, 90b]. Since $\kappa = \rho^{-1}$, this implies that the Gaussian and mean curvatures of the offset surface are given by

$$K_d = \frac{K}{1 + 2Hd + Kd^2}, \qquad H_d = \frac{H + Kd}{|Kd^2 + 2Hd + 1|}, \tag{15.11}$$

respectively, see (2.15 a,b). Singularities of offset surfaces can be located by finding the zeros of the outer product of the partial derivatives w.r.t. u and v. After some rearrangement (see [Faro 86a]), (15.9) implies

$$\left| \frac{\partial \boldsymbol{X}_d}{\partial u} \times \frac{\partial \boldsymbol{X}_d}{\partial v} \right| = |\boldsymbol{X}_u \times \boldsymbol{X}_v|\, |(1 + 2Hd + Kd^2)|. \tag{15.12}$$

If the parametrization is regular, then singularities appear whenever the second factor in (15.12) vanishes. By (15.10), the second factor in (15.12) vanishes for $-\rho_1^{-1} = -\kappa_1 = d$ and/or $-\rho_2^{-1} = -\kappa_2 = d$.

Singularities at a point arise when the distance d of the smallest absolute value of the principal curvature is attained at $\boldsymbol{P}$. Singularities of offset surfaces

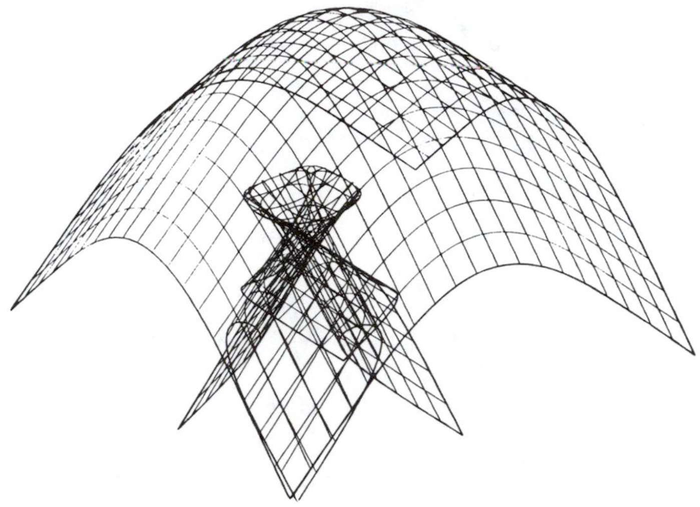

Fig. 15.10. Singularities of the offset surface of a convex patch.

can be of many different types. For example, they can involve cusps, sharp edges, or self-intersections, see [Faro 86a] and Fig. 15.10. If only one of the principal curvatures of an offset surface vanishes, then the surface has a sharp edge. If both principal curvatures vanish simultaneously at some point, then the offset surface has a cusp. Self-intersections of offset surfaces are also a problem, see [Aom 90].

If the initial surface has sharp edges, cusps, or other singularities, then the offset surface must be specially constructed near such points. Spheres, cylinders, and planar patches can be used to connect pieces of the offset surface together, see e.g., [Faro 85].

If the offset surface is constructed as the envelope of a moving sphere, then we get a uniquely defined offset surface at vertices, along edges, etc. Suppose the initial surface is $\boldsymbol{X}(u, v)$. Then the equation of a sphere of radius d whose center lies on the initial surface is given by

$$F = (\xi - x(u,v))^2 + (\eta - y(u,v))^2 + (\zeta - z(u,v))^2 - d^2 = 0, \qquad (15.13a)$$

in terms of a (ξ, η, ζ) system. As (u, v) runs over the parameter domain, (15.13a) describes the envelope of the family of spheres.

The points of the envelope can be determined as follows. The tangent planes to F must satisfy the equation

$$\frac{\partial F}{d\xi}\Delta\xi + \frac{\partial F}{d\eta}\Delta\eta + \frac{\partial F}{d\zeta}\Delta\zeta = 0.$$

On the other hand, the complete differential of (15.13a) is

$$\frac{\partial F}{d\xi}\Delta\xi + \frac{\partial F}{d\eta}\Delta\eta + \frac{\partial F}{d\zeta}\Delta\zeta + \frac{\partial F}{du}\Delta u + \frac{\partial F}{dv}\Delta v = 0.$$

At points on the envelope, the complete differential reduces to the tangent plane, *i.e.*, the two expressions must coincide, resulting in the *envelope condition* (see [HOFF 89])

$$\frac{\partial F}{\partial u} = 0, \qquad \frac{\partial F}{\partial v} = 0. \tag{15.13b}$$

Eliminating the parameter u and v from (15.13a) – (15.13b), we get an algebraic formula for the offset surface.

15.4. Approximation of Offset Curves and Surfaces

If we are given initial curves or surfaces in terms of some basis (e.g., the Bernstein basis or B-spline basis), then because of the square root which appears in the normalization of the normal vectors, an offset curve or surface cannot be represented in terms of the same basis system. In order to use the same bases in a CAD system for offset curves or surfaces, we have to replace the exact offsets with approximations to them.

Based on geometric properties of the offset, [Kla 83] has discussed approximations to plane offset curves using cubic spline curves. [Arn 86] uses cubic Bézier splines and appropriate measures of the approximation error. [Coq 87] and [Pha 88] discuss approximating offset curves in terms of B-spline curves. Approximations of offset surfaces by bicubic spline surfaces are developed in [Faro 86a]. [Wol 92] presents an effective algorithm for approximating geodesic offset curves.

As a supplement to the methods for approximate basis transformation of spline curves and surfaces presented in Chap. 10, we now discuss the use of geometric G^1 Bézier splines for the approximation of offset curves and surfaces. We first consider the approximation of a prescribed offset curve $\boldsymbol{X}_d$ by a cubic Bézier curve $\boldsymbol{Y}$. We assume that the given curve $\boldsymbol{X}$ is a Bézier curve of degree n, given in parametric form as

$$\boldsymbol{X}(t) = \sum_{i=0}^{n} \boldsymbol{V}_i B_i^n(t), \qquad t \in [0,1],$$

where $\boldsymbol{V}_i$ are the (given) Bézier points. The associated offset curve $\boldsymbol{X}_d$ is given in (15.2).

We are seeking to approximate the offset curve $\boldsymbol{X}_d$ by a a cubic integral Bézier curve of the form

$$\boldsymbol{Y}(t) = \sum_{i=0}^{3} \boldsymbol{W}_i B_i^3(t),$$

with unknown Bézier points $\boldsymbol{W}_i$.

By the construction of the offset curve, it follows immediately that the endpoints of the approximate Bézier curve $\boldsymbol{Y}$ are given by

$$\boldsymbol{W}_0 = \boldsymbol{V}_0 + d\,\boldsymbol{n}(0), \qquad \boldsymbol{W}_3 = \boldsymbol{V}_n + d\,\boldsymbol{n}(1), \tag{15.14}$$

where $\boldsymbol{n}$ is the unit normal to the given Bézier curve $\boldsymbol{X}(t)$, see Fig. 15.11.

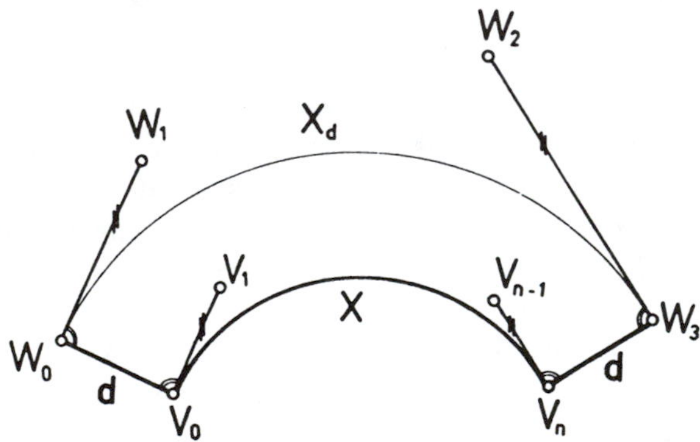

Fig. 15.11. End points of the approximating curve to an offset curve $\boldsymbol{X}_d$ corresponding to a given Bézier curve $\boldsymbol{X}$.

For a geometric C^1 match of the exact offset curve $\boldsymbol{X}_d$ and the approximating curve $\boldsymbol{Y}$, we must have (cf. (5.1))

$$\boldsymbol{W}_1 = \boldsymbol{W}_0 + \lambda_1(\boldsymbol{V}_1 - \boldsymbol{V}_0), \qquad \boldsymbol{W}_2 = \boldsymbol{W}_3 + \lambda_2(\boldsymbol{V}_n - \boldsymbol{V}_{n-1}), \tag{15.15a}$$

for some arbitrary parameters λ_i.

In order to approximate the offset curve, we choose $k+1$ equally-spaced points $\boldsymbol{P}_i$ on the exact offset curve $\boldsymbol{X}_d$, corresponding to the parameter values t_i, $i = 0(1)k$, with $k > n$. Evaluating the approximating curve at these points, we get

$$\boldsymbol{P}_i = \sum_{j=0}^{3} \boldsymbol{W}_j B_j^3(t_i) + \boldsymbol{\delta}_i, \tag{15.15b}$$

with errors $\boldsymbol{\delta}_i$. As in Chap. 10, we now minimize the error by applying a least-squares method and parameter correction.

Fig. 15.12 shows an example of a geometric Bézier spline curve with four cubic segments, and two offset curves, one on each side of the initial curve. The Bézier points defining the initial curve are marked with dots, while those of the approximating curves are marked with squares. The approximating offset curves involve 7 segments, and the maximal error is given by $\epsilon_0 = 0.01$ cm. The cusps were computed using the algorithm discussed on page 604, see also [Hos 85]. The exact offset curves and their approximations are drawn on top of each other.

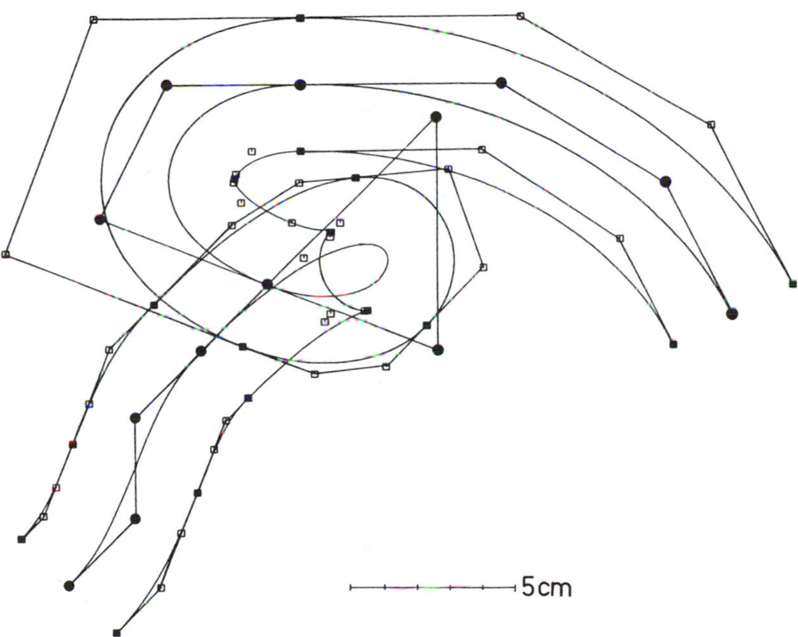

Fig. 15.12. A cubic Bézier spline curve with 7 segments (with a loop) and its approximate offsets defined by cubic G^1 Bézier spline curves.

Offset curves can also be approximated using Bézier curves of higher degree, see the discussion in Chap. 10 and [Hos 88a, 90]. If we are approximating offset curves by rational Bézier curves, then the additional degrees of freedom provided by the weights result in curves with fewer pieces than those obtained using integral Bézier curves, see [Hos 88a].

To approximate offset surfaces using Bézier patches, we assume that the initial surface is given in the form

$$\boldsymbol{X}(u,v)=\sum_{i=0}^{n}\sum_{k=0}^{m}\boldsymbol{V}_{ik}B_i^n(u)B_k^m(v),$$

and that the approximate offset surface is given by

$$\boldsymbol{Y}(u,v)=\sum_{i=0}^{p}\sum_{k=0}^{q}\boldsymbol{W}_{ik}B_i^p(u)B_k^q(v).$$

The corners of the offset surface are given by

$$\begin{aligned}\boldsymbol{W}_{00}&=\boldsymbol{V}_{00}+\boldsymbol{N}(0,0)\,d, & \boldsymbol{W}_{p0}&=\boldsymbol{V}_{n0}+\boldsymbol{N}(1,0)\,d,\\ \boldsymbol{W}_{0q}&=\boldsymbol{V}_{0m}+\boldsymbol{N}(0,1)\,d, & \boldsymbol{W}_{qp}&=\boldsymbol{V}_{nm}+\boldsymbol{N}(1,1)\,d.\end{aligned} \tag{15.16}$$

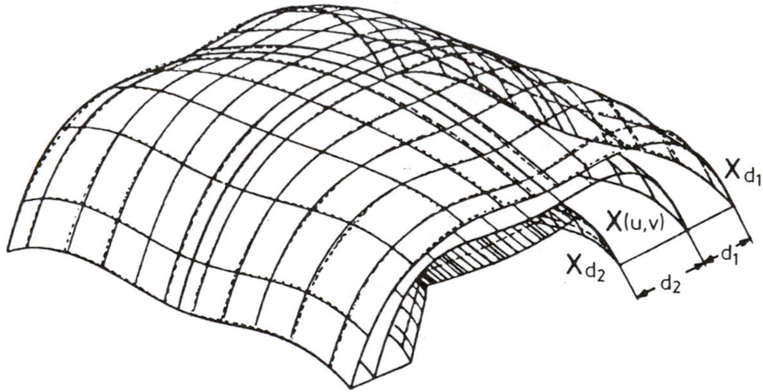

Fig. 15.13. Bicubic patch approximation of the offset surfaces of a given patch $\boldsymbol{X}(u,v)$ corresponding to distances d_1, d_2.

Now as in Chap. 10, we can require that the offset surface $\boldsymbol{X}_d$ and the approximating surface $\boldsymbol{Y}$ meet with first order continuity at these corner points in the bicubic case, and with second order continuity in the biquintic case. This leads to the same problem as in Chap. 10, and we can apply the approximation methods discussed there. Fig. 15.13 shows an example of a given Bézier patch of degree (4,4), and the bicubic approximations of its offset surfaces at oriented distances d_1 and d_2, see also [Hos 89]. In order to reduce the approximation error, both offset surfaces are subdivided. The surface $\boldsymbol{X}_{d_2}$ consists of 8 patches, while $\boldsymbol{X}_{d_1}$ involves only four patches. As can be seen from the boundary curves, there is a considerable approximation error between $\boldsymbol{X}_{d_1}$ and $\boldsymbol{Y}$. If $\boldsymbol{X}_{d_1}$ is also subdivided into eight segments, the error can be greatly reduced.

16 Mathematical Modelling of Cutting Paths

Numerically controlled milling machines (*NC-machines*) are an essential tool for manufacturing free-form surfaces. For example, the dies and injection molds used to make automobile parts are manufactured using milling machines. The task of milling free-form surfaces has given rise to a number of new mathematical and technical problems, see e.g., [Zir 89], [Kla 91]. In this section we give a short overview of the mathematical modelling of the milling process.

There are a number of different milling machines available:

- *two-axis milling machines*, which can move in both the x and y directions, see [HEL 91], [Per 78], and Fig. 16.1a,
- *two-and-one-half-axis milling machines*, which can also move by a constant amount in the z direction,
- *three-axis milling machines*, which can move in all three of the x, y, and z directions, see [Bob 85], [Lone 87], [Cho 88a], [Oli 90], [Lee 91a], [Cui 91], and Fig. 16.1b,
- *four-axis milling machines*, which in addition to moving in all three spatial directions, can swing in a plane,
- *five-axis milling machines*, which in addition to moving in all three spatial directions, require two angles to describe the motion, see [Dam 76], [Hen 76, 76a], [Pri 86], [Vie 89], [Zir 89], and Fig. 16.1c.

Two-axis milling machines are used to cut holes (pockets) in plane profiles, see [Suh 90], [HEL 91]. Three-, four-, and five-axis milling machines are used to mill free-form surfaces, see [Zir 89].

The differences can be seen clearly in Fig. 16.1. For a two-axis milling machine, the tool has a fixed position (e.g., parallel to the z-axis), and can be moved in only two coordinate directions. In three-axis milling, the tool has a fixed position (e.g., parallel to the z-axis of the machine), and can be moved in any of the three coordinate directions. For a five-axis milling machine, the tool is positioned according to the normal vector to the surface to be milled. It is clear that the five-axis machine can do a better job of milling a surface.

Fig. 16.1 depicts so-called *face milling*. Fig. 16.2 gives an example of *side milling*, where the shaft of the cutting tool does the cutting. This type of milling is especially useful for ruled surfaces [Sie 81].

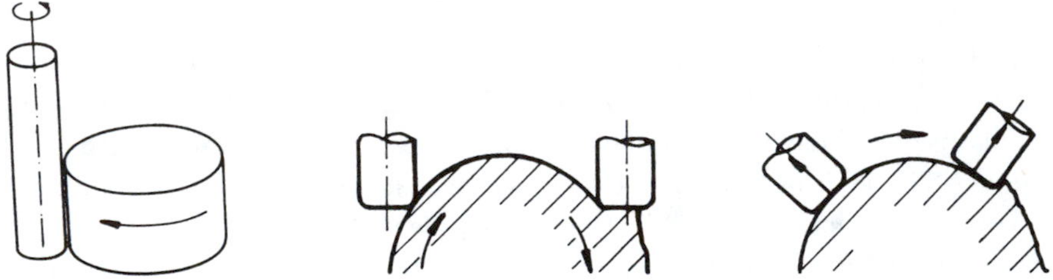

Fig. 16.1. Tool position for two, three, and five-axis milling.

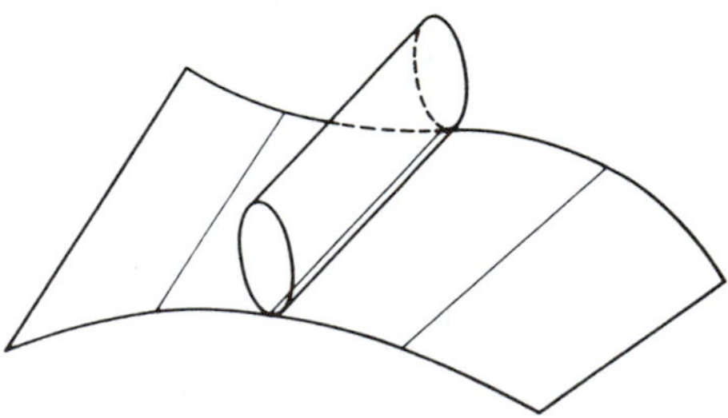

Fig. 16.2. Side milling.

The numerical control of milling machines leads to the following mathematical problems:

a) determination of the milling coordinates and the milling axis relative to the desired surface depending on the type of milling,

b) moving the tool along special surface curves in order to control the shape, size, and location of scallops, see Fig. 16.3,

c) transforming the desired control curves to machine coordinates,

d) computation of the scallop profile,

e) collision checking, *i.e.*, checking to make sure that the "back side" of the milling tool does not disturb the previously milled surface or its boundary, see [Cho 89a], [Hen 76], [Wal 82], [Zir 88], and Fig. 16.4.

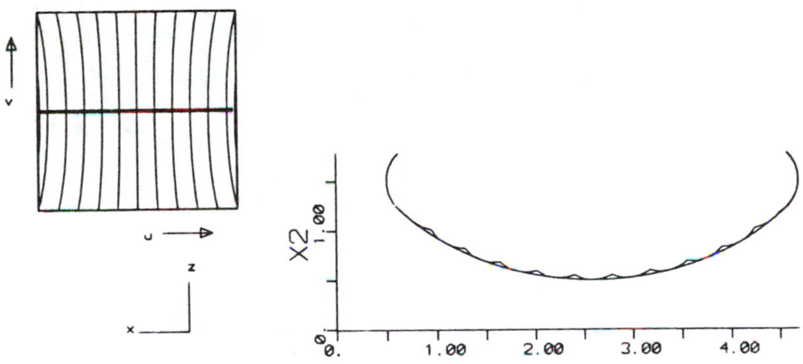

Fig. 16.3. Top and side views of a ridge profile.

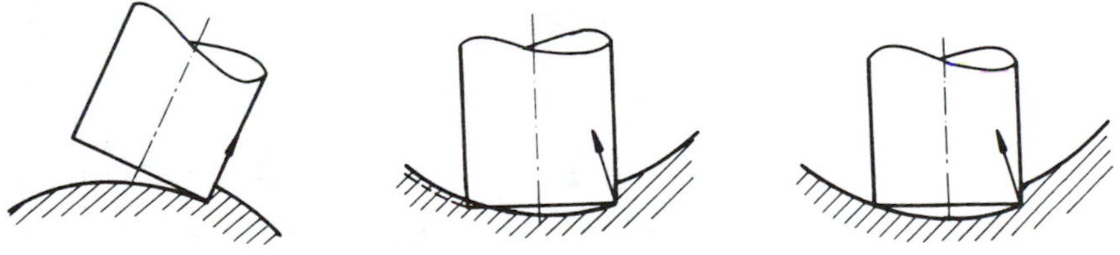

Fig. 16.4. Collision check for five-axis milling.

The basic idea for milling strategies for three-axis or five-axis milling machines is to determine which parts of a surface will be effected as we move the milling tool. This requires studying the coverings of spherical normal images of the surface, see [Tse 91], [Che 92]. Fig. 16.5 shows three critical situations for the three-axis cutting tool. The surface in Fig. 16.5a can be milled with a three-axis machine provided it is properly positioned. On the other hand, the surface in Fig. 16.5b cannot be milled with such a machine. Both surfaces can be milled with a five-axis machine with an appropriate cutting tip. Fig. 16.5c shows an example where part of the surface cannot be milled because of the size of the tool.

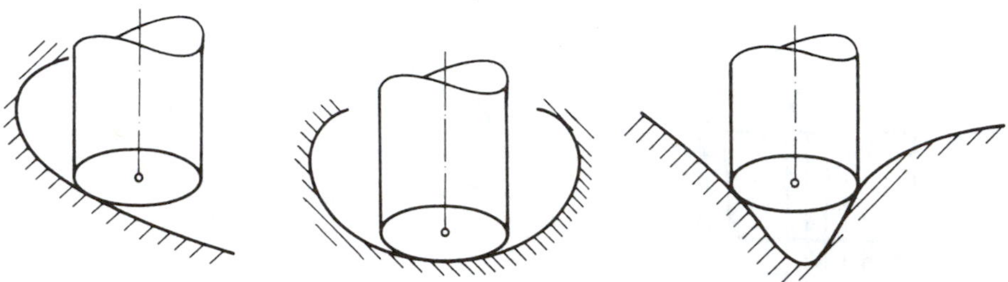

Fig. 16.5. Problem zones for three-axis milling.

Remark. Frequently, milling paths are controlled by micro-processors where the basic incremental movement is a straight line or circular arc. This means that we need to discuss discretization of cutting paths, see [Gas 90], [Hos 92c].

16.1. Mathematical Description of Milling Tools

The mathematical basis for controlling milling is a description of the outer surface of the cutting tool in an orthonormal basis. This basis is then moved along the surface to be milled. Thus, we begin by analytically describing the various tool surfaces in terms of a moving basis $(\boldsymbol{e}_1, \boldsymbol{e}_2, \boldsymbol{e}_3)$ with $|\boldsymbol{e}_i| = 1$. First we consider a *cylindrical end cutter*, see Fig. 16.6, whose surface is described by

$$\boldsymbol{Wz}(r,s) = (qR_W \cos r, qR_W \sin r, sS)^T. \tag{16.1}$$

Here $q \in [0,1]$, $s = 0$ corresponds to the tip of the tool, while $q = 1$, $s \in [0,1]$ corresponds to its shaft. Cylindrical tools are used primarily in two-axis milling and for side milling.

We turn now to ball-end cutters, which are used primarily in three-axis milling. Such a tool can be described as in Fig. 16.7. As can be seen from the figure, the boundary of a ball-end cutter consists of two essential shapes: a quarter circle, and a straight line. The quarter circle can be described by the following formula:

$$\boldsymbol{k}_1(v) = (R_W \sin t, 0, -R_W \cos t)^T, \qquad t \in [0, \pi/2], \tag{16.2a}$$

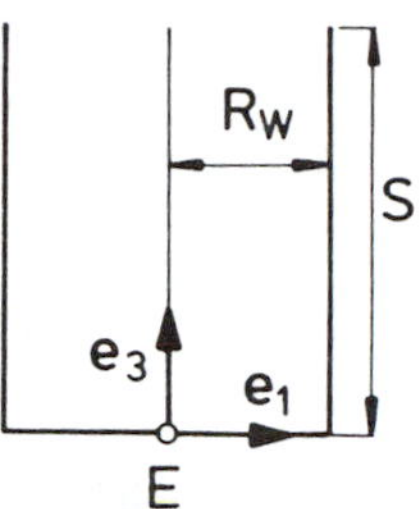

Fig. 16.6. Cylindrical-end cutter.

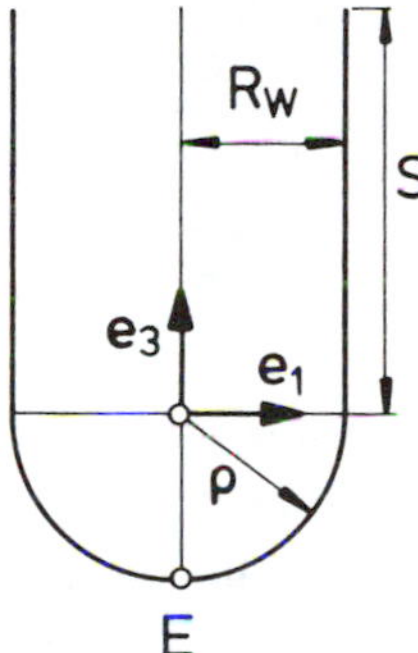

Fig. 16.7. Ball-end cutter.

while the straight line can be written in the form

$$\boldsymbol{k}_2(v) = (R_W, 0, sS)^T, \qquad s \in [0,1]. \tag{16.2b}$$

Combining these formulae, we get the representation

$$\boldsymbol{k}(v) = (R_W \sin t, 0, sS - R_W \cos t)^T. \tag{16.3}$$

For $0 \le t \le \pi/2$ and $s = 0$, $\boldsymbol{k}(v)$ reduces to the curve $\boldsymbol{k}_1(v)$, while for $t = \pi/2$ and $0 \le s \le 1$, it reduces to the straight line $\boldsymbol{k}_2(v)$. Coupling (16.3) with the parametric formulae in (16.2a,b), and allowing a rotation described by the parameter $r \in [0, 2\pi]$, the surface of the tool can be expressed as

$$\boldsymbol{Wk}(r,s,t) = (R_W \cos r \sin t, R_W \sin r \sin t, sS - R_W \cos t)^T. \tag{16.4}$$

In using this formula, we must be careful that s and t are never allowed to vary simultaneously over their respective parameter intervals. We either must fix $s = 0$, in which case $\boldsymbol{Wk}$ describes the bottom half of a sphere as r, t vary, or we must fix $t = \pi/2$, in which case we get the cylindrical surface as r, s vary.

We now turn to *toroidal shaped milling tools*, which are especially useful for four- and five-axis milling. Here the ingredients are a piece of a torus connected to a cylinder, and a plane which serves as the bottom surface, see Fig. 16.8.

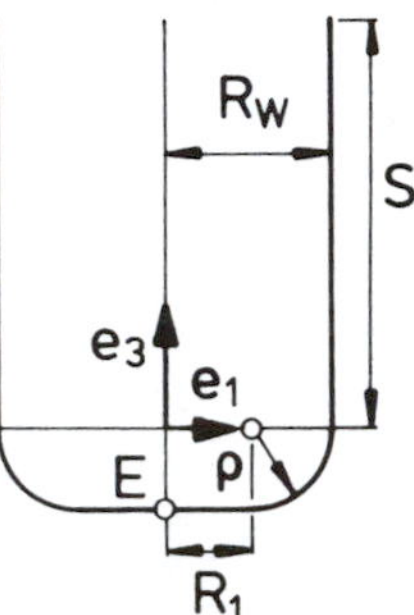

Fig. 16.8. Toroidal-shaped milling tool (BAFL).

The parametric formula for a torus does not allow a complete description of a toroidal shaped milling tool. Thus in addition to the corner radius ρ, we introduce two further parameters: q describes the tip of the cutter, and is taken over from formula (16.1) for a cylinder without corner radius, and s describes the length of the shaft. This results in the following parametric representation:

$$\boldsymbol{Wt}(q, r, s, t) = \begin{pmatrix} \cos r(qR_1 + \rho \sin t) \\ \sin r(qR_1 + \rho \sin t) \\ sS - \rho \cos t \end{pmatrix}, \tag{16.5}$$

where $q, s \in [0, 1]$, $r \in [0, 2\pi]$, and $t \in [0, \pi/2]$.

If we set $q = 0$ and $\rho = R_W$ in (16.5), then the formula reduces to that of the ball-end tool $\boldsymbol{Wk}$, while for $R_1 \neq 0$ and $\rho = 0$, it reduces to the parametric representation (16.1) of a cylindrical tool $\boldsymbol{Wz}$. The choice of constant ρ and $q = 1$ corresponds to the surface of a BAFL (ball-flat) milling tool.

We now move the tool coordinate system $(\boldsymbol{e}_1, \boldsymbol{e}_2, \boldsymbol{e}_3)$ in a direction parallel to the $\boldsymbol{e}_3$ axis. This corresponds to holding $s = 0$, $t = 0$, and $x_3 = 0$ fixed, so that

$$x_3 := sS + \rho(1 - \cos t).$$

Thus, the origin of the system lies at the end point E of the cutter.

The following tool formula is representative for all three tool forms:

$$\boldsymbol{W}(q, r, s, t) = \begin{pmatrix} \cos r(qR_1 + \rho \sin t) \\ \sin r(qR_1 + \rho \sin t) \\ sS + \rho(1 - \cos t) \end{pmatrix}. \tag{16.6}$$

We obtain a complete description of a tool by letting the parameters q, t, s run over their corresponding intervals (in that order) for all r. Once one parameter reaches its upper boundary, its value is held fixed. The parameters q, t, s never run over their respective intervals simultaneously, and so they can be combined to a common parameter $v = v(q, t, s)$ which takes values from the interval $[0, B_L]$, where

$$B_L := R_1 + \rho\frac{\pi}{2} + S.$$

For values of v along this parameter line, we have the following cases:

a) "bottom": $0 \leq v < R_1$, $\quad q = v/R_1,\ t = 0,\ s = 0,$

b) "circle": $R_1 \leq v < (\rho\frac{\pi}{2} + R_1)$, $\quad q = 1,\ s = 0,\ t = (v - R_1)/\rho,$

c) "shaft": $(\rho\frac{\pi}{2} + R_1) \leq v \leq B_L$, $\quad q = 1,\ t = \frac{\pi}{2},\ s = (v - (\rho\frac{\pi}{2} + R_1))/S.$

For five-axis milling, we can improve the operation of the cutting tool, and at the same time avoid collisions, by introducing an appropriate *angle of tilt* β. We can think of this angle as describing the rotation of the tool about the $\boldsymbol{e}_2$-axis, assuming that it is moving in the $\boldsymbol{e}_1$ direction.

If we multiply the matrix describing a rotation about the $\boldsymbol{e}_2$-axis,

$$R(\beta) = \begin{pmatrix} \cos\beta & 0 & \sin\beta \\ 0 & 1 & 0 \\ -\sin\beta & 0 & \cos\beta \end{pmatrix},$$

with the general tool formula $\boldsymbol{W}$ in (16.6), we get

$$\begin{aligned} \boldsymbol{W}(\beta, q, r, s, t) &= R(\beta) \cdot \boldsymbol{W}(q, r, s, t) \\ &= \begin{pmatrix} \cos\beta\cos r[qR_1 + \rho\sin t] + \sin\beta[sS + \rho(1-\cos t)] \\ \sin r[qR_1 + \rho\sin t] \\ -\sin\beta\cos r[qR_1 + \rho\sin t] + \cos\beta[sS + \rho(1-\cos t)] \end{pmatrix}, \end{aligned} \tag{16.7}$$

with respect to the basis $(\boldsymbol{e}_1, \boldsymbol{e}_2, \boldsymbol{e}_3)$. Fig. 16.9 shows the positions of the three types of cutters corresponding to an angle of tilt $\beta > 0$ in (16.7).

16.2. Cutting Paths

To construct tool cutting paths, we have to choose a strategy: we can either follow a zig-zag path (parallel milling) or follow contour lines (contour milling), see Fig. 16.10.

For milling with a two-axis tool of cylindrical shape, we have to guide the cutter along an offset curve at a distance equal to the radius of the cylinder,

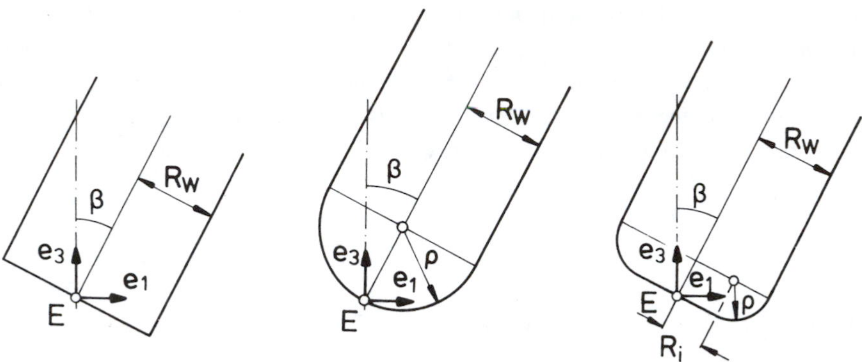

Fig. 16.9. Angle of tilt β for the three types of cutting tools.

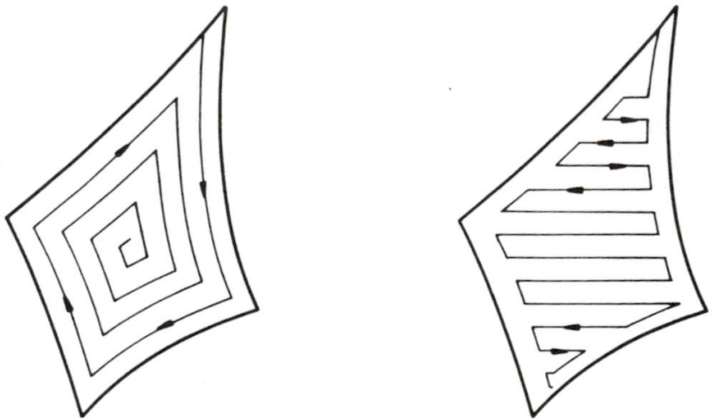

Fig. 16.10. Contour and parallel milling.

see Fig. 15.6. For milling a free-form surface with a three-axis tool of spherical shape, the center of the sphere has to move along an offset surface at a distance equal to the radius of the sphere, see Fig. 15.9. For a five-axis tool, there is no simple way to describe the guiding surface, and so the contact point of the tool has to be kept on the desired surface as we move along selected tool paths, see Fig. 16.11 (which was kindly provided by Mercedes-Benz, see also [Kla 91]).

- It is simplest to use a family of *parametric lines* for the tool paths. But if the parametric lines are not uniformly spaced, then we get varying widths for the cuts, see Fig. 16.11a.

- The problem of the spacing between parametric lines can be avoided if we always cut along the *lines of intersection* of the surface with a family of planes or some other appropriately selected family of surfaces (e.g., cylindrical surfaces), see Fig. 16.11b.
- Another strategy is based on *offset curves* corresponding to a given curve, say to one of the boundary curves, see Fig. 16.11c.
- Instead of using offset curves, we can also work with *curves of equal distance* corresponding to two boundary curves on opposite sides of the surface, see Fig. 16.11d.
- Another possibility is to project a given *network of curves* onto the surface, see Fig. 16.11e.
- A method which takes account of the geometry of the surface itself is to choose special curves defined in terms of differential geometric properties of the surface, e.g., *curvature lines*, see Chap. 2.

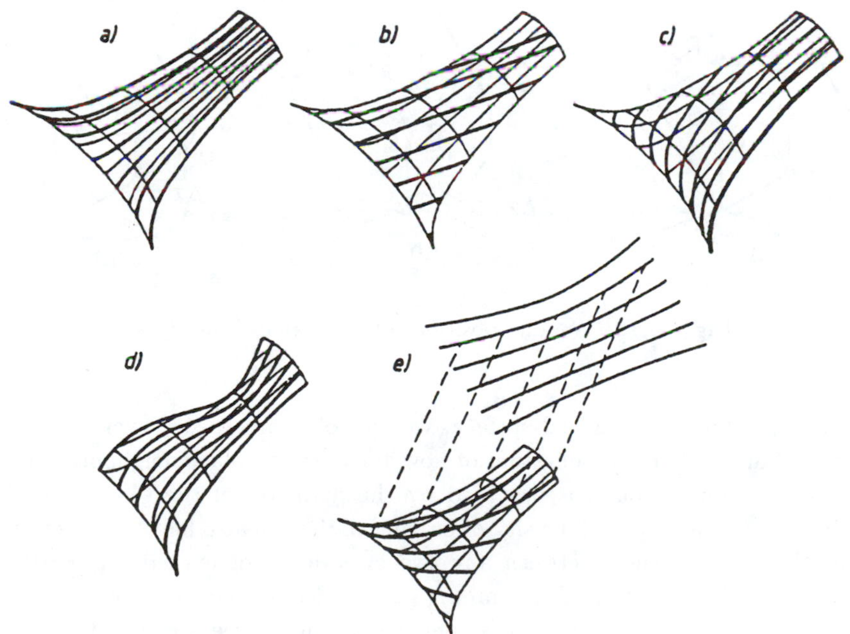

Fig. 16.11. Possible tool paths.

In order to steer the cutting tool along a surface curve $\boldsymbol{X}(t)$ in the direction $\dot{\boldsymbol{X}}(t)$, we introduce an orthonormal coordinate system $(\boldsymbol{T}, \boldsymbol{S}, \boldsymbol{N})$ which moves along the surface $\boldsymbol{X}(u, v)$, where $\boldsymbol{T}$ is the tangent vector to the selected cutting path, $\boldsymbol{N}$ is the normal vector to the surface, and $\boldsymbol{S} = \boldsymbol{N} \times \boldsymbol{T}$ is the binormal. In order to position the tool contact point on the desired surface, the coordinate system $(\boldsymbol{e}_1, \boldsymbol{e}_2, \boldsymbol{e}_3)$ of the tool (cf. (16.7) or Fig. 16.9) has to be converted to the coordinate system $(\boldsymbol{T}, \boldsymbol{S}, \boldsymbol{N})$ by translation. This transformation is described by the displacement of the origin of the coordinate system:

$$\boldsymbol{X} = \boldsymbol{E} + (\Delta h \boldsymbol{T} - \Delta z \boldsymbol{N}) =: \boldsymbol{E} + \boldsymbol{V}, \tag{16.8}$$

where $\boldsymbol{V}$ is the displacement vector, and $\boldsymbol{X}$ denotes a point on the cutting path. The path taken by the tool end point $\boldsymbol{E}$ along the cutting path $\boldsymbol{X}$ is called the *tool path*. The components of the displacement vector vary according to the type of cutter being used, see Fig. 16.12, and can be found from elementary geometry. After the transformation (16.8), the tool is in the correct position to begin cutting.

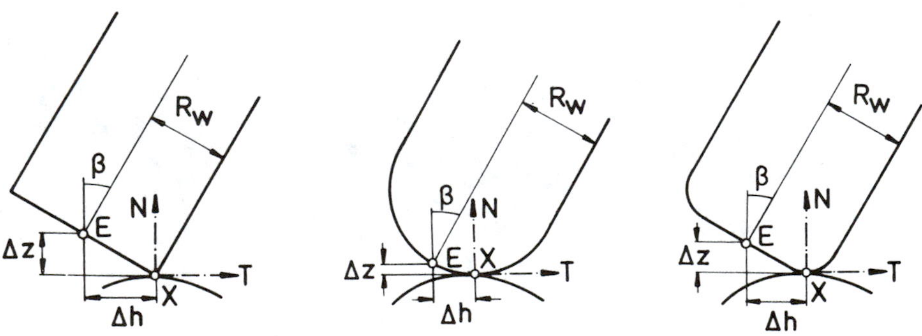

Fig. 16.12. Working positions of the various tool types.

To produce the finished surface, we will of course need several cutting paths. This leads to the question of how far apart to place neighboring tool paths. An upper bound is provided by the diameter of the tool shaft. In order to find the required density of cutting paths, we also have to make sure that the scallop height between any two cuts does not exceed a prescribed tolerance. The structure of the milling profile also depends on the tool type, and whether the milling process is concave or convex, see Fig. 16.13.

Several strategies for finding tool paths for three-axis cutting tools can be found in [Bob 85], [Lone 87], [Oli 90], [Lee 91a], and [Cui 91].

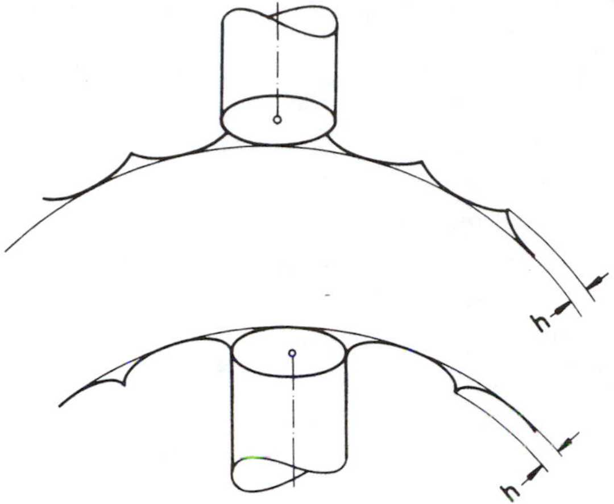

Fig. 16.13. Dependence of cutting paths on scallop height and surface curvature.

16.3. Milling Surfaces

The usual approach to milling a surface is to approximate the cutting path by a sequence of polygons and to report the position of each vertex of the polygons to the cutting tool controller. Then cutting algorithms built into the tool are used to produce the cutting paths themselves. The sequence of cutting profiles describes the desired surface, see Figs. 16.3 and 16.13. [Jer 88] presents some special cutting algorithms for three-axis tools based on approximating the cutting tool envelope.

Here we shall take a somewhat different approach. As the tool moves along the cutting path, it generates a second surface which consists of a family of curves whose hulls along the direction of motion always contain all points of the cutter. The resulting scalloped surface is the *envelope* of the cutting motion, see [Jer 89]. The collection of envelopes corresponding to cutting along various cutting paths describe a family of surfaces which enclose the desired surface. In the following we find this family of envelopes directly. This approach assumes in advance that the cutting paths are differentiable curves (e.g., splines). Solving the envelope equations then leads directly to points on the envelope of the moving tool. For simplicity, we assume now that the cutting path is a parametric line $\boldsymbol{X}(\mu)$ corresponding to $\mu = u$ or $\mu = v$ (for a general cutting path, $\mu = \mu(u, v)$). We first transform the tool representation (16.7) into the associated frame $\boldsymbol{T}, \boldsymbol{S}, \boldsymbol{N}$, and thus obtain the following formula for the family of cuts as we move along the cutting path:

$$\boldsymbol{M}(\mu) = \boldsymbol{X}(\mu) + (\boldsymbol{T}\ \boldsymbol{S}\ \boldsymbol{N})\boldsymbol{W}(\beta, q, r, s, t) + \boldsymbol{V}(\mu, \beta, q, r, s, t), \tag{16.9}$$

where $(\boldsymbol{T}\ \boldsymbol{S}\ \boldsymbol{N})$ is the matrix formed from the indicated column vectors, and $\boldsymbol{V}$ is the displacement vector of (16.8).

For a given angle of incidence β and a given parametric value μ, we get a complete representation for the tool as we move along the tool path $\boldsymbol{X}(\mu)$ as the parameters q, r, s, and t run over their intervals of definition, see Fig. 16.14. Formula (16.9) covers both face cutting and side cutting; in the latter case we take $\beta = \frac{\pi}{2}$.

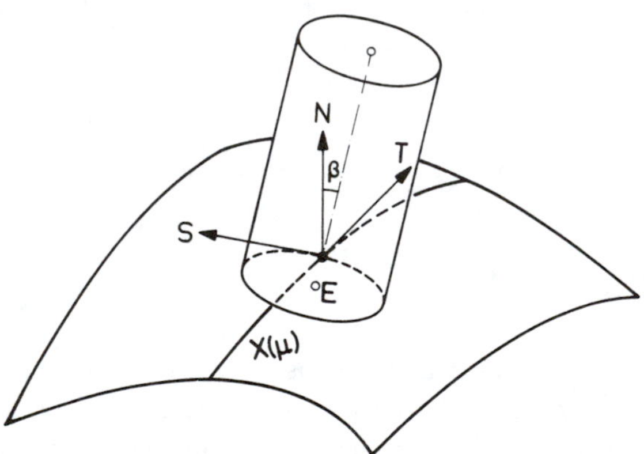

Fig. 16.14. Motion of a cylindrical-end tool along a cutting path $\boldsymbol{X}$.

We now find the points of the tool which lie on the envelope of the moving tool. The envelope touches the cutting surface tangentially, *i.e.*, the envelope points must be orthogonal to the associated surface normal. This leads to the following *envelope condition* (see Chap 2):

$$\boldsymbol{M}_\mu \cdot \boldsymbol{WN} = 0, \tag{16.10}$$

where $\boldsymbol{M}_\mu = \dfrac{\partial \boldsymbol{M}}{\partial \mu}$, and $\boldsymbol{WN}$ is the normal to the tool surface.

To compute $\boldsymbol{WN}$, we need to compute the partial derivatives of the function $\boldsymbol{W}(\beta, q, r, s, t)$ in (16.7) with respect to the parameters q, r, s, and t. Aside from insignificant factors, we get

$$\boldsymbol{W}_q = \frac{\partial \boldsymbol{W}}{\partial q} = (\cos\beta\cos r,\ \sin r,\ -\sin\beta\cos r)^T, \tag{16.11a}$$

$$\boldsymbol{W}_r = \frac{\partial \boldsymbol{W}}{\partial r} = (-\cos\beta\sin r,\ \cos r,\ \sin\beta\sin r)^T, \tag{16.11b}$$

$$\boldsymbol{W}_s = \frac{\partial \boldsymbol{W}}{\partial s} = (\sin\beta,\ 0,\ \cos\beta)^T, \tag{16.11c}$$

$$\boldsymbol{W}_t = \frac{\partial \boldsymbol{W}}{\partial t} = \begin{pmatrix} \cos\beta\cos r\cos t + \sin\beta\sin t \\ \sin r\ cost \\ \cos\beta\sin t - \sin\beta\cos r\cos t \end{pmatrix}. \tag{16.11d}$$

Now forming the three cross products $\boldsymbol{W}_r \times \boldsymbol{W}_q$, $\boldsymbol{W}_r \times \boldsymbol{W}_s$, and $\boldsymbol{W}_r \times \boldsymbol{W}_t$, we find that with respect to the frame $(\boldsymbol{T}\ \boldsymbol{S}\ \boldsymbol{N})$,

$$\boldsymbol{WN}_{rq} = \frac{\boldsymbol{W}_r \times \boldsymbol{W}_q}{|\boldsymbol{W}_r \times \boldsymbol{W}_q|} = -(\sin\beta)\boldsymbol{T} - (\cos\beta)\boldsymbol{N}, \tag{16.12a}$$

$$\begin{aligned} \boldsymbol{WN}_{rs} = \frac{\boldsymbol{W}_r \times \boldsymbol{W}_s}{|\boldsymbol{W}_r \times \boldsymbol{W}_s|} &= (\cos\beta\cos r)\boldsymbol{T} + (\sin r)\boldsymbol{S} \\ &- (\sin\beta\cos r)\boldsymbol{N}, \end{aligned} \tag{16.12b}$$

$$\begin{aligned} \boldsymbol{WN}_{rt} = \frac{\boldsymbol{W}_r \times \boldsymbol{W}_t}{|\boldsymbol{W}_r \times \boldsymbol{W}_t|} &= (\cos\beta\cos r\sin t - \sin\beta\cos t)\boldsymbol{T} \\ &+ (\sin r\sin t)\boldsymbol{S} - (\cos\beta\cos t + \sin\beta\cos r\sin t)\boldsymbol{N}. \end{aligned} \tag{16.12c}$$

If we examine the equations (16.12a) – (16.12c) more closely, we see that (16.12a) – (16.12b) are contained in (16.12c). Indeed, for $t = 0$, (16.12c) implies (16.12a), while for $t = \frac{\pi}{2}$ it implies (16.12b). As a consequence, we only need one vector formula to describe the tool normal, *i.e.*, we can take

$$\boldsymbol{WN} := \boldsymbol{WN}_{rt} =: \boldsymbol{WN}_1\boldsymbol{T} + \boldsymbol{WN}_2\boldsymbol{S} + \boldsymbol{WN}_3\boldsymbol{N}. \tag{16.13}$$

In addition, for constant β,

$$\boldsymbol{M}_\mu = \boldsymbol{X}_\mu + (\boldsymbol{T}_\mu\ \boldsymbol{S}_\mu\ \boldsymbol{N}_\mu)W(\beta, q, r, s, t) + \boldsymbol{V}_\mu, \tag{16.14}$$

with

$$\boldsymbol{T}_\mu = \frac{\partial \boldsymbol{T}}{\partial \mu}, \quad \boldsymbol{S}_\mu = \frac{\partial \boldsymbol{S}}{\partial \mu}, \quad \boldsymbol{N}_\mu = \frac{\partial \boldsymbol{N}}{\partial \mu}, \quad \boldsymbol{X}_\mu = \frac{\partial \boldsymbol{X}}{\partial \mu}, \quad \boldsymbol{V}_\mu = \frac{\partial \boldsymbol{V}}{\partial \mu}.$$

The derivatives of the vectors defining the frame satisfy

$$\frac{\partial \boldsymbol{T}}{\partial \mu} = \frac{\partial}{\partial \mu}\left(\frac{\dot{\boldsymbol{X}}}{|\dot{\boldsymbol{X}}|}\right), \tag{16.15a}$$

$$\frac{\partial \boldsymbol{S}}{\partial \mu} = \frac{\partial}{\partial \mu}(N \times \boldsymbol{T}) = \boldsymbol{N}_\mu \times \boldsymbol{T} + \boldsymbol{N} \times \boldsymbol{T}_\mu, \tag{16.15b}$$

$$\frac{\partial \boldsymbol{N}}{\partial \mu} = \frac{\partial}{\partial \mu}\left(\frac{\boldsymbol{X}_u \times \boldsymbol{X}_v}{|\boldsymbol{X}_u \times \boldsymbol{X}_v|}\right) = \frac{\partial}{\partial \mu}(\boldsymbol{X}_u \times \boldsymbol{X}_v) \cdot [(|\boldsymbol{X}_u \times \boldsymbol{X}_v|)^2]^{-1/2}$$

$$= (\boldsymbol{R}_u + \boldsymbol{R}_v)(\boldsymbol{R}^2)^{-1/2} - \boldsymbol{R} \cdot (\boldsymbol{R}^2)^{-3/2} \cdot (\boldsymbol{R} \cdot (\boldsymbol{R}_u + \boldsymbol{R}_v)), \quad (16.15c)$$

where

$$\boldsymbol{R} = \boldsymbol{X}_u \times \boldsymbol{X}_v, \qquad \boldsymbol{R}_u = \boldsymbol{X}_{u\mu} \times \boldsymbol{X}_v, \qquad \boldsymbol{R}_v = \boldsymbol{X}_u \times \boldsymbol{X}_{v\mu}.$$

Then

$$\boldsymbol{M}_\mu = \boldsymbol{X}_\mu + W_1 \boldsymbol{T}_\mu + W_2 \boldsymbol{S}_\mu + W_3 \boldsymbol{N}_\mu + \boldsymbol{V}_\mu, \quad (16.16)$$

where W_1, W_2, W_3 are the components of $\boldsymbol{W}$ in the above formula for $\boldsymbol{M}_\mu$.

Inserting (16.16) and (16.13) in the envelope condition (16.10) leads to

$$\boldsymbol{M}_\mu \cdot \boldsymbol{WN} = \boldsymbol{X}_\mu \cdot \boldsymbol{WN} + W_1(\boldsymbol{T}_\mu \cdot \boldsymbol{WN}) + W_2(\boldsymbol{S}_\mu \cdot \boldsymbol{WN}) + W_3(\boldsymbol{N}_\mu \cdot \boldsymbol{WN}) + \boldsymbol{V}_\mu \cdot \boldsymbol{WN} = 0. \quad (16.17)$$

Since $\boldsymbol{WN}$ contains the vectors of the accompanying frame $\boldsymbol{T}$, $\boldsymbol{S}$, $\boldsymbol{N}$, the equation (16.17) can be further simplified. First, in terms of the first fundamental form $g_{\mu\mu}$ in the parametric direction μ (cf. (2.13)),

$$\boldsymbol{X}_\mu \cdot \boldsymbol{T} = \sqrt{g_{\mu\mu}} \cdot \boldsymbol{T}^2 = \sqrt{g_{\mu\mu}}. \quad (16.18)$$

Since $|\boldsymbol{T}| = |\boldsymbol{S}| = |\boldsymbol{N}| = 1$, it follows that the derivatives of the unit vectors $\boldsymbol{T}$, $\boldsymbol{S}$, $\boldsymbol{N}$ with respect to μ satisfy

$$\boldsymbol{T} \cdot \boldsymbol{T}_\mu = 0, \qquad \boldsymbol{S} \cdot \boldsymbol{S}_\mu = 0, \qquad \boldsymbol{N} \cdot \boldsymbol{N}_\mu = 0. \quad (16.19)$$

Moreover,

$$\boldsymbol{T} \cdot \boldsymbol{S} = \boldsymbol{T} \cdot \boldsymbol{N} = \boldsymbol{S} \cdot \boldsymbol{N} = 0, \quad (16.20)$$

and thus

$$\boldsymbol{X}_\mu \cdot \boldsymbol{S} = \boldsymbol{X}_\mu \cdot \boldsymbol{N} = 0. \quad (16.21)$$

From this we see that the equation (16.17) reduces to

$$\begin{aligned} \boldsymbol{M}_\mu \cdot \boldsymbol{WN} = \sqrt{g_{\mu\mu}}\, WN_1 &+ W_1\, WN_2(\boldsymbol{T}_\mu \cdot \boldsymbol{S}) + W_1\, WN_3(\boldsymbol{T}_\mu \cdot \boldsymbol{N}) \\ &+ W_2\, WN_1(\boldsymbol{S}_\mu \cdot \boldsymbol{T}) + W_2\, WN_3(\boldsymbol{S}_\mu \cdot \boldsymbol{N}) \\ &+ W_3\, WN_1(\boldsymbol{N}_\mu \cdot \boldsymbol{T}) + W_3\, WN_2(\boldsymbol{N}_\mu \cdot \boldsymbol{S}) + \boldsymbol{V}_\mu \cdot \boldsymbol{WN} = 0, \end{aligned} \quad (16.22)$$

where WN_1, WN_2, and WN_3 are the components of the tool normal $\boldsymbol{WN}$, and W_1, W_2, and W_3 are the components of $\boldsymbol{W}$.

In this form, equation (16.22) cannot be solved to get an equation for the tool path without some additional work, depending on the particular type of tool. For example, for a cylindrical cutting tool without a corner radius, we have to solve the above equation twice, once for the normal to the bottom face of the cutter, and once for the normal to the shaft. For a cylindrical tool with corner radius, the equation (16.22) has to be solved three times, once for each of the three types of tool normals in (16.12a) – (16.12c).

We conclude with an example of finding the envelope of a ball-end tool of radius R_W being moved along a circular cylinder of radius R with an angle of incidence $\beta = 0$. We do not bother to take the shaft of the tool into consideration. Setting $\rho = R_W$, $R_1 = 0$, $s = 0$, and $\beta = 0$ in (16.16), we get the formula

$$\boldsymbol{W} = \begin{pmatrix} R_W \cos r \sin t \\ R_W \sin r \sin t \\ R_W(1 - \cos t) \end{pmatrix}$$

for the tool. We describe the cylinder by $\boldsymbol{X}(u, v) = (R\cos u, R\sin u, v)^T$, and the cutting path by the generator $u = u_0$, i.e., $\boldsymbol{X}(\mu) = (R\cos u_0, R\sin u_0, \mu)^T$ and $\boldsymbol{V} = 0$. This leads to the frame

$$\boldsymbol{X}_\mu = \boldsymbol{T} = (0, 0, 1)^T, \quad \boldsymbol{S} = (-\sin u_0, \cos u_0, 0)^T, \quad \boldsymbol{N} = (\cos u_0, \sin u_0, 0)^T,$$

which can be written in matrix form as

$$(\boldsymbol{T}\ \boldsymbol{S}\ \boldsymbol{N}) = \begin{pmatrix} 0 & -\sin u_0 & \cos u_0 \\ 0 & \cos u_0 & \sin u_0 \\ 1 & 0 & 0 \end{pmatrix}. \tag{$*$}$$

Then using (16.12a), we find that the tool normal is given by

$$\boldsymbol{WN} = (\cos r \sin t, \sin r \sin t, -\cos t)^T.$$

This normal must be transformed to the coordinate system $(*)$ via

$$\boldsymbol{WN} = (\boldsymbol{T}\ \boldsymbol{S}\ \boldsymbol{N})N_W = \begin{pmatrix} -\sin u_0 \sin r \sin t - \cos u_0 \cos t \\ \cos u_0 \sin r \sin t - \sin u_0 \cos t \\ \cos r \sin t \end{pmatrix}. \tag{16.23}$$

We now need the derivatives of the frame vectors:

$$\begin{aligned} \frac{\partial \boldsymbol{W}}{\partial \mu} &= (0, 0, 1) + (\boldsymbol{T}_\mu\ \boldsymbol{S}_\mu\ \boldsymbol{N}_\mu)\boldsymbol{W}, \\ \boldsymbol{T}_\mu &= \boldsymbol{T}_v = 0, \qquad \boldsymbol{S}_\mu = \boldsymbol{S}_v = 0, \qquad \boldsymbol{N}_\mu = \boldsymbol{N}_v = 0. \end{aligned} \tag{16.24}$$

The envelope condition (16.10), along with (16.23) and (16.24), implies that $\cos r \sin t = 0$, and it follows that $r = \frac{\pi}{2}$. The value $t = 0$ leads to the path of the tool end point only, which is also included in the solution $r = \frac{\pi}{2}$.

We now get the following equation for the envelope:

$$\begin{aligned} \boldsymbol{X} &= \begin{pmatrix} R\cos u_0 \\ R\sin u_0 \\ \mu \end{pmatrix} + \begin{pmatrix} -R_W \sin u_0 \sin t \\ R_W \cos u_0 \sin t \\ 0 \end{pmatrix} + \begin{pmatrix} R_W \cos u_0 \,(1-\cos t) \\ R_W \sin u_0 \,(1-\cos t) \\ 0 \end{pmatrix} \\ &= \begin{pmatrix} (R+R_W)\cos u_0 \\ (R+R_W)\sin u_0 \\ \mu \end{pmatrix} + R_W \begin{pmatrix} -\cos(t-u_0) \\ \sin(t-u_0) \\ 0 \end{pmatrix}. \end{aligned}$$

This means that the center of the sphere must move along a cylinder of radius $R + R_W$, and at each point we have to draw a circle of radius R_W, see Fig. 16.15.

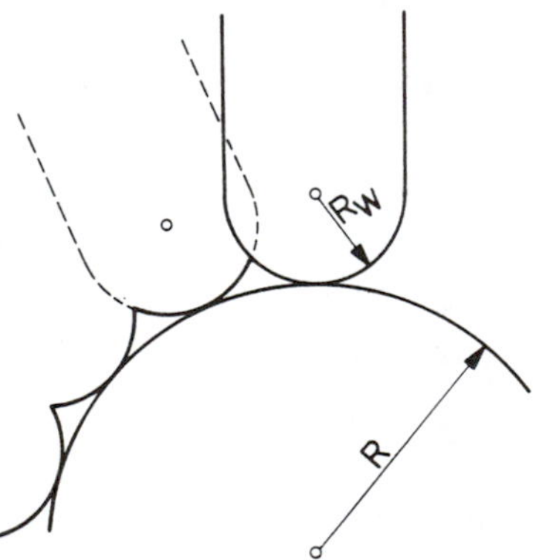

Fig. 16.15. Milling profile and tool position for a three-axis ball-end tool moving along a cylinder.

16.4. Collision Control

So far we have concentrated on the local tool geometry. We also have to look at the *global* situation in order to make sure that the tool does not collide with the surface being milled at some undesired point, see Fig. 16.4. In milling a concave surface, we also have to take into account that the tool is not larger than the minimal cross curvature (see (2.10a,b)), so that it can reach all desired points, see Fig. 16.5, [Bob 85], [Cho 89a], and [Kla 91].

Gouges as in Fig. 16.4 can be avoided by changing the angle of incidence β. [Hen 76] gives a simple algorithm for controlling gouging: we go through the

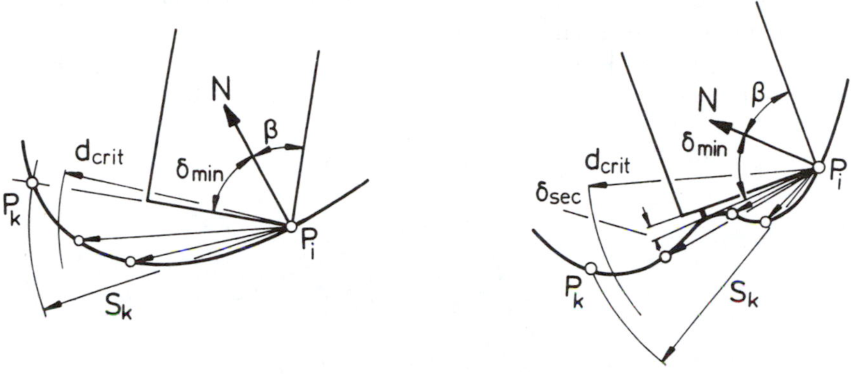

Fig. 16.16. Critical angles of incidence.

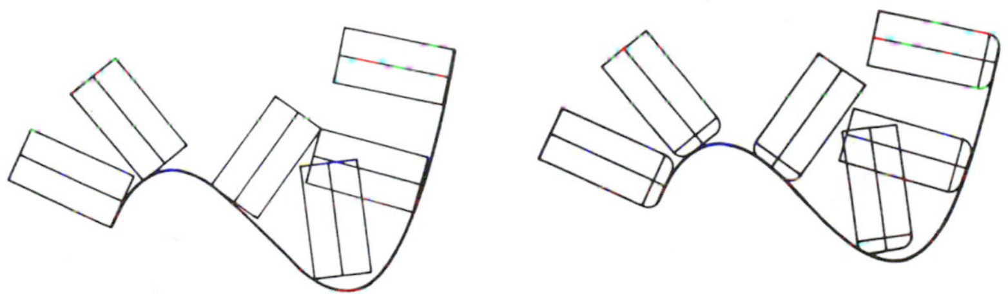

Fig. 16.17. Diagram of collision control for cylindrical and BAFL milling tools.

points $\boldsymbol{P}_k$ which have already been passed, one after another, and iteratively check the angle δ between the surface normal $\boldsymbol{N}$ at the present point $\boldsymbol{P}_i$ and the vector $\boldsymbol{S}_k = \boldsymbol{P}_i - \boldsymbol{P}_k$, until for some k, $|\boldsymbol{S}_k| \geq d_{crit}$. For a cylindrical tool, for example, we choose $d_{crit} = 2R_W + \epsilon$ where $\epsilon > 0$ and R_W is the tool radius. In general, the smallest angle δ between $\boldsymbol{P}_i$ and $\boldsymbol{P}_k$ will occur at $\boldsymbol{P}_k$ itself, see Fig. 16.16a. In this case we can take the angle of incidence to be $\beta = \frac{\pi}{2} - \delta_{min}$. If, however, the smallest angle δ_{min} occurs for a point between $\boldsymbol{P}_i$ and $\boldsymbol{P}_k$, then δ_{min} should be increased by δ_{sec}, and the angle of incidence is replaced by $\beta = \frac{\pi}{2} - \delta_{min} + \delta_{sec}$, see Fig. 16.16b.

Fig. 16.17 shows the movement of cylindrical and BAFL milling tools with collision control.

Fig. 16.18 shows another situation where the tool should not touch given bounding surfaces. To insure this, offset surfaces corresponding to a given tolerance are defined which restrict the path of the tool [Stei 90]. Collision control can then be accomplished using appropriate surface intersection algorithms.

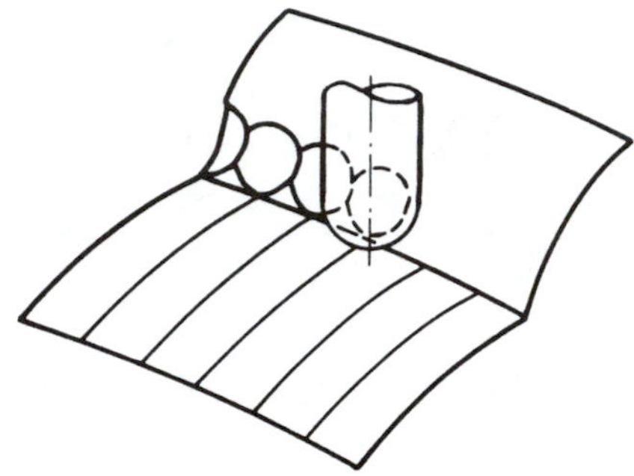

Fig. 16.18. Collision with a bounding surface.

Bibliography

Textbooks

[ABR 91] Abramowski, S.; Müller, H.: Geometrisches Modellieren. BI 1991.

[ADA 88] Adams, J. A.; Billow, L. M.: Descriptive Geometry and Geometric Modeling. A Basis for Design. Holt, Rinehart and Winston, Inc. 1988.

[AHL 67] Ahlberg, J. H.; Nilson, E. N.; Walsh, J. L.: The Theory of Splines and Their Applications. Academic Press 1967.

[ANG 81] Angell, O.: A Practical Introduction to Computer Graphics. McMillen Press 1981.

[BARN 74] Barnhill, R. E.; Riesenfeld, R. F.: Computer Aided Geometric Design. Academic Press 1974.

[BARN 83] Barnhill, R. E.; Boehm, W.: Surfaces in Computer Aided Geometric Design. North-Holland 1983.

[BARN 85] Barnhill, R. E.; Boehm, W.: Surfaces in CAGD '84. North-Holland 1985.

[BARR 87] Barr, A. H.; Barzel, R.; Haumann, D.; Kass, M.; Platt, J.; Terzopoulos, D.; Witkin, A.: Topics in physically-based modeling. ACM SIGGRAPH '87, Course Notes 17, 1987.

[BARS 88] Barsky, B. A.: Computer Graphics and Geometric Modeling Using Beta-Splines. Springer 1988.

[BART 87] Bartels, R. H.; Beatty, J. C.; Barsky, B. A.: An Introduction to Splines for Use in Computer Graphics & Geometric Modeling. Morgan Kaufmann Publishers 1987.

[BER 70] Berisin, I. S.; Shidkow, N. P.: Numerische Methoden I. VEB-Verlag der Wiss. 1970.

[BEZ 72] Bézier, P.: Numerical Control, Mathematics and Applications. Wiley 1972.

[BEZ 86] Bézier, P.: The mathematical basis of the UNISURF CAD system. Butterworth 1986.

[BLA 60] Blaschke, W.: Kinematik und Quaternionen. VEB-Verlag der Wiss. 1960.

[BOEH 85] Boehm, W.; Gose, G.; Kahmann, J.: Methoden der Numerischen Mathematik. 2nd ed. Vieweg 1985.

[BÖH 74] Böhmer, K.: Splinefunktionen. Teubner 1974.

[BOO 78] Boor de, C.: A Practical Guide to Splines. Springer 1978.

[BRA 77] Brauner, H.; Kickinger, W.: Baugeometrie I. Bauverlag Wiesbaden 1977.

[BRA 86] Brauner, H.: Lehrbuch der Konstruktiven Geometrie. Springer 1986.

[BRO 80] Brodlie, K. W.: Mathematical Methods in Computer Graphics and Design. Academic Press 1980.

[CAM 79] Campbell, S. L.; Meyer, C. D.: Generalized Inverses of Linear Transformations. Pittman 1979.

[CAR 76] Carmo do, M.: Differential Geometry of Curves and Surfaces. Prentice Hall 1976.

[CAS 86] Casteljau de, P.: Shape Mathematics and CAD. Kogan Page, London 1986

[CHAS 78] Chasen, S. H.: Geometric Principles and Procedures for Computer Graphic Applications. Prentice-Hall 1978.

[CHI 88] Chiyokura, H.: Solid Modelling with Designbase. AddisonWesley 1988.

[CHU 87] Chui, C. K.; Schumaker, L. L.; Utreras, F. I.: Topics in Multivariate Approximation. Academic Press 1987.

[CONT 72] Conte, S. D.; de Boor, C.: Elementary Numerical Analysis. McGraw Hill 1972.

[DAV 75] Davis, P.: Interpolation and Approximation. Dover 1975.

[EARN 85] Earnshaw, R. A.: Fundamental Algorithms for Computer Graphics. Springer 1985.

[EHR 64] Ehrenfeucht, A.: The cube made interesting. Pergamon Press 1964.

[ELS 70] Elsgolc, L. E.: Variationsrechnung. BI 1970.

[ENC 75] Encarnacao, J. L.: Computer Graphics. Oldenburg 1975.

[ENC 88] Encarnacao, J. L.; Straßer, W.: Computer Graphics. 3rd ed. Oldenburg 1988.

[END 89] Endl, K.; Endl, R.: Computergrafik 1. Würfel-Verlag 1989.

[ENG 85] Engeln-Müllges, G.; Reutter, E.: Numerische Mathematik für Ingenieure. 4th ed. BI 1985.

[FAR 87] Farin, G.: Geometric Modeling: Algorithms and New Trends. SIAM 1987.

[FAR 90] Farin, G.: Curves and Surfaces for Computer Aided Geometric Design. A Practical Guide. 2nd ed. Academic Press 1990.

[FAU 81] Faux, I. D.; Pratt, M. J.: Computational Geometry for Design and Manufacture. Ellis Horwood 1981.

[FEL 88] Fellner, W. D.: Computer Graphik. BI 1988.

[FIN 77] Finckenstein, F. von: Einführung in die Numerische Mathematik I. Hanser 1977.

[FLA 75] Fladt, K.; Bauer, A.: Analytische Geometrie spezieller Flächen und Raumkurven. Vieweg 1975.

[FOLE 82] Foley, J. D.; Van Dam, A.: Fundamentals of Interactive Computer Graphics. Addison-Wesley 1982.

[FORS 12] Forsyth, A. R.: Lectures on Differential Geometry of Curves and Surfaces, Cambridge University Press 1912.

[FORS 77] Forsythe, G. E.; Malcolm, M. A.; Moler, C. B.: Computer Methods for Mathematical Computations. Prentice Hall, Englewood Cliffs 1977.

[FOU 87] Fournier, A.; Bloomenthal, J.; Oppenheimer, P.; Reeves, W. T.; Smith, A. R.: The modeling of natural phenomena. ACM SIGGRAPH '87, Course Notes 16, 1987.

[FRÜ 91] Frühauf, M.; Göbel, M.: Visualisierung von Volumendaten. Springer 1991.

[GAN 85] Gander, W.: Computer Mathematik. Birkhäuser 1985.

[GLA 89] Glassner, A.: Introduction to Ray Tracing. Academic Press 1989.

[GOU 74] Goult, R. J.; Hoskins, R. F.; Millner, J. A.; Pratt, M. J.: Computational Methods in Linear Algebra. Wiley 1974.

[GRIE 87] Grieger, I.: Graphische Datenverarbeitung. Mathematische Methoden. Springer Hochschultext, Springer 1987.

[GRO 57] Grosche, G.: Projektive Geometrie I, II. Teubner 1957.

[HARTM 88] Hartmann, E.: Computerunterstützte Darstellende Geometrie. Teubner 1988.

[HAW 62] Hawk, M. C..: Descriptive Geometry. Schaums Outlines 1962.

[HEL 91] Held, M.: On the Computational Geometry of Pocket Machining. Lecture Notes in Computer Science 500, Springer 1991.

[HIL 56] Hildebrandt, F. B.: Introduction to Numerical Analysis. McGraw Hill 1956.

[HOFF 89] Hoffmann, Ch. M.: Geometric and Solid Modeling. An Introduction. Morgan Kaufmann Publishers 1989.

[HORS 79] Horst, R.: Nichtlineare Optimierung. Hanser 1979.

[HOS 87] Hoschek, J.; Lasser, D.: CAGD-Grundlagen der geometrischen Datenverarbeitung. Lehrbriefe der Fernuniversität Hagen 1-6, 1987.

[HOW 51] Howe, H. B.: Descriptive Geometry. Ronald 1951.

[KAUF 90] Kaufman, A.: Visualization '90. IEEE Computer Society Press 1990.

[KAUF 90a] Kaufman, A.: Volume Visualization. IEEE Computer Society Press 1990.

[KLI 78] Klingenberg, W.: A Course in Differential Geometry. Springer 1978.

[LAN 86] Lancaster, P.; Salkauskas, K.: Curve and Surface Fitting. Academic Press 1986.

[LAU 65] Laugwitz, D.: Differential and Riemannian Geometry. Academic Press 1965.

[LEI 91] Leister, W.; Müller, H.; Stößer, A.: Fotorealistische Computeranimation. Springer 1991.

[LIM 44] Liming, R. A.: Practical Analytical Geometry with Applications to Aircraft. Macmillan 1944.

[LORE 53] Lorentz, G.: Bernstein Polynomials. Toronto Press 1953.

[LOR 11] Loria, G.: Spezielle algebraische und transzendente ebene Kurven. Theorie und Geschichte II. Teubner 1911.

[LUT 89] Luther, W.; Ohsmann, M.: Mathematische Grundlagen der Computergraphik. Vieweg 1989.

[LYC 89] Lyche, T.; Schumaker, L. L.: Mathematical Methods in Computer Aided Geometric Design. Academic Press 1989.

[MÖB 67] Möbius, A. F.: Der barycentrische Calcul 1827. in Möbius, A. F.: Gesammelte Werke Band I. Sändig 1967.

[MOR 52] Morehead, J. C.: A Handbook of Perspective Drawing. Elsevier 1952.

[MORT 85] Mortenson, M. E.: Geometric Modeling. Wiley 1985.

[MÜL 80] Müller, G.: Rechnerorientierte Darstellung beliebig geformter Bauteile. Hanser 1980.

[MÜLL 88] Müller, H.: Realistische Computergraphik: Algorithmen, Datenstrukturen und Maschinen. Informatik-Fachberichte 163, Springer 1988.

[MYE 83] Myers, R. E.: Mikrocomputer Grafik. Pandasoft 1983.

[NEW 79] Newman, W. M.; Sproull, R. F.: Principles of Interactive Computer Graphics. McGraw Hill 1979.

[NIE 90] Nielson, G. M.; Shriver, B. D.: Visualization in Scientific Computing. IEEE Computer Society Press 1990.

[NIE 91] Nielson, G. M.; Rosenblum, L.: Visualization '91. IEEE Computer Society Press 1991.

[NOW 83] Nowacki, H.; Gnatz, R. (ed.): Geometrisches Modellieren. Informatik-Fachberichte 65, Springer 1983.

[NUT 88] Nutbourne, A. W.; Martin, R. R.: Differential Geometry applied to curve and surface design. Vol. 1: Foundations. Ellis Horwood 1988.

[OKO 76] Okoshi, T.: Three Dimensional Imaging Techniques. Academic Press 1976.

[PRE 75] Prenter, P. M.: Splines and Variational Methods. Wiley 1975.

[PREP 85] Preparata, F. P.; Shamos, M. I.: Computational Geometry, an Introduction. Springer 1985.

[QUI 87] Quilin, D.; Davis, B.: Surface Engineering Geometry for Computer Aided Design and Manufacture. Ellis Horwood 1987.

[ROG 76] Rogers, D. F.; Adams, J. A.: Mathematical Elements for Computer Graphics. McGraw Hill 1976.

[ROG 85] Rogers, D. F.: Procedural elements for computer graphics. McGraw Hill 1985.

[RUS 73] Rushing, T. B.: Topological Embeddings. Academic Press 1973.

[SABO 87] Sabonnadiere, J. C.; Coulomb, J. L.: Finite Element Methods in CAD: Electrical and Magnetic Fields. Springer 1987.

[SAL 85] Salmon, G.: Lessons Introducing to the Modern Higher Algebra. New York: Reprinted by Chelsea 1885.

[SCHEF 01] Scheffers, G.: Anwendungen der Differential und Integralrechnung auf die Geometrie. Bd. I: Einführung in die Theorie der Kurven in der Ebene und im Raume. von Veit 1901.

[SCHÖ 77] Schörner, E.: Darstellende Geometrie. Hanser 1977.

[SCHU 81] Schumaker, L. L.: Spline Functions: Basic Theory. Wiley 1981.

[SCHUL 87] Schulz, C.; Bielig-Schulz, G.: 3D-Grafik in PASCAL. Teubner 1987.

[SCHW 80] Schwarz, H. R.: Methode der finiten Elemente. Teubner 1980.

[SCHW 76] Schwidefsky, K.; Ackermann, F.: Photogrammetrie. Teubner 1976.

[SMI 78] Smith, G. D.: Numerical Solution of Partial Differential Equations. Oxford University Press 1978.

[SPÄ 83] Späth, H.: Spline Algorithmen zur Konstruktion glatter Kurven und Flächen. 3rd ed. Oldenburg 1983.

[SPÄ 90] Späth, H.: Eindimensionale Spline-Interpolations-Algorithmen. Oldenbourg 1990.

[SPÄ 91] Späth, H.: Zweidimensionale Spline-Interpolations-Algorithmen. Oldenbourg 1991.

[SPI 79] Spivak, M.: A Comprehensive Introduction to Differential Geometry 2. Publish or Perish 1979.

[STRA 73] Strang, G.; Fix, G.: An Analysis of the Finite Element Method. Prentice-Hall 1973.

[STR 50] Struik, D. J.: Differential Geometry. Addison-Wesley 1950.

[SU 89] Su Buchin; Liu Dingyuan: Computational Geometry. Academic Press 1989.

[THO 85] Thompson, J. F.; Warsi, Z. U.; Mastin, C. W.: Numerical Grid Generation, Foundations and Applications. North-Holland 1985.

[TIM 56] Timoshenko, S.: Strength of Materials II. 3rd ed. Van Nostrand 1956.

[VAR 62] Varga, R. S.: Matrix Iterative Analysis. Prentice-Hall 1962.

[WALK 50] Walker, R. J.: Algebraic Curves. Princeton University Press 1950.

[WER 79] Werner, H.; Schaback, R.: Praktische Mathematik II. 2nd ed. Springer 1979.

[WUN 76] Wunderlich, W.: Darstellende Geometrie II. BI 1976.

[YAM 88] Yamaguchi, F.: Curves and Surfaces in Computer Aided Geometric Design. Springer 1988.

[ZIE 77] Zienkiewicz, O. C.: The Finite Element Method. 3rd ed. McGraw Hill 1977.

Articles

[Abh 83] Abhyankar, S. S.: Desingularization of Plane Curves. American Mathematical Society, Providence, RI (1983) 1-45.

[Abh 87] Abhyankar, S. S.; Bajaj, C.: Automatic parametrization of rational curves and surfaces I: Conics and conicoids. Computer-Aided Design 19 (1987) 11-14.

[Abh 87a] Abhyankar, S. S.; Bajaj, C.: Automatic parametrization of rational curves and surfaces II: Cubics and cubicoids. Computer-Aided Design 19 (1987) 499-502.

[Abh 88] Abhyankar, S. S.; Bajaj, C.: Automatic parametrization of rational curves and surfaces III: Algebraic plane curves. Computer Aided Geometric Design 5 (1988) 309-321.

[Abh 89] Abhyankar, S. S.; Chandrasekar, S.; Chandru, V.: Degree Complexity Bounds on the Intersection of Algebraic Curves. Proceedings of the Fifth Annual Symposium on Computational Geometry. ACM Press (1989) 88-93.

[Abh 90] Abhyankar, S. S.; Chandrasekar, S.; Chandru, V.: Improper Intersection of Algebraic Curves. ACM Transactions on Graphics 9 (1990) 147-159.

[Ada 75] Adams, J. A.: The intrinsic method for curve definition. Computer-Aided Design 7 (1975) 243-249.

[Agi 91] Agishtein, M. E.; Migdal, A.: Smooth surface reconstruction from scattered data points. Computers & Graphics 15 (1991) 29-39.

[Aki 70] Akima, J. A.: A New Method of Interpolation and Smooth Curve Fitting Based on Local Procedures. Journal of the ACM 17 (1970) 589-602.

[Aki 78] Akima, H.: A method of bivariate interpolation and smooth surface fitting for irregularly distributed data points. ACM Transactions on Mathematical Software 4 (1978) 148-159.

[Aki 78a] Akima, H.: Algorithm 526: Bivariate interpolation and smooth surface fitting for irregularly distributed data points. ACM Transactions on Mathematical Software 4 (1978) 160-164.

[Aki 84] Akima, H.: On Estimating Partial Derivatives for Bivariate Interpolation of Scattered Data. Rocky Mountain Journal of Mathematics 14 (1984) 41-52.

[Alf 84] Alfeld, P.: A bivariate C^2 Clough-Tocher scheme. Computer Aided Geometric Design 1 (1984) 257-267.

[Alf 84a] Alfeld, P.: A trivariate Clough-Tocher scheme for tetrahedral data. Computer Aided Geometric Design 1 (1984) 169-181.

[Alf 84b] Alfeld, P.; Barnhill, R. E.: A Transfinite C^2 Interpolant over Triangles. Rocky Mountain Journal of Mathematics 14 (1984) 17-39.

[Alf 84c] Alfeld, P.: A Discrete C^1 Interpolant for Tetrahedral Data. Rocky Mountain Journal of Mathematics 14 (1984) 5-16.

[Alf 85] Alfeld, P.: Derivative generation from multivariate scattered data by functional minimization. Computer Aided Geometric Design 2 (1985) 281-296.

[Alf 85a] Alfeld, P.: Multivariate Perpendicular Interpolation. SIAM Journal on Numerical Analysis 22 (1985) 95-106.

[Alf 87] Alfeld, P.; Piper, B.; Schumaker, L. L.: Minimally supported bases for spaces of bivariate piecewise polynomials of smoothness r and degree $d \geq 4r + 1$. Computer Aided Geometric Design 4 (1987) 105-123.

[Alf 89] Alfeld, P.: Scattered Data Interpolation in Three or More Variables. in Lyche, T.; Schumaker, L. L. (ed.): Mathematical Methods in Computer Aided Geometric Design. Academic Press (1989) 1-33.

[All 91] Allgower, E. L.; Gnutzmann, S.: Simplicial pivoting for mesh generation of implicitly defined surfaces. Computer Aided Geometric Design 8 (1991) 305-325.

[And 88] Andersson, E.; Andersson, R.; Boman, M.; Elmroth, T.; Dahlberg, B.; Johansson, B.: Automatic construction of surfaces with prescribed shape. Computer-Aided Design 22 (1988) 317-324.

[Anj 87] Anjyo, K.; Ochi, T.; Usami, Y.; Kawashima, Y.: A practical method of constructing surfaces in three-dimensional digitized space. The Visual Computer 3 (1987) 4-12.

[Aom 90] Aomura, S.; Uehara, T.: Self-intersection of an offset surface. Computer-Aided Design 22 (1990) 417-422.

[Arn 86] Arnold, R.: Quadratische und kubische Offset-Bézierkurven. Diss. Dortmund 1986.

[Ast 88] Asteasu, C.: Intersection of arbitrary surfaces. Computer-Aided Design 20 (1988) 533-538.

[Ast 91] Asteasu, C.; Orbegozo, A.: Parametric Piecewise Surface Intersection. Computers & Graphics 15 (1991) 9-13.

[Avi 83] Avis, D.; Bhattacharya, B. K.: Algorithms for computing d-dimensional Voronoi diagrams and their duals. in Preparata, F. P. (ed.): Advances in computing research I, JAI Press Greenwich (1983) 159-180.

[Azi 90] Aziz, N. M.; Bata, R. M.; Bhat, S. P.: Bézier Surface/Surface Intersection. IEEE Computer Graphics & Applications 10 (1990) 50-58.

[Baa 84] Baass, K. G.: The use of clothoid templates in highway design. Transportation Forum 1 (1984), 47-52.

[Baj 88] Bajaj, C. L.; Hoffmann, C. M.; Lynch, R. E.; Hopcroft, J. E. H.: Tracing surface intersections. Computer Aided Geometric Design 5 (1988) 285-307.

[Baj 89] Bajaj, C. L.; Ihm, I.: Hermite Interpolation using Real Algebraic Surfaces. Proceedings of the Fifth Annual Symposium on Computational Geometry. ACM Press (1989) 94-103.

[Baj 90] Bajaj, C. L.: Rational Hypersurface Display. ACM (1990) 117-127.

[Baj 92] Bajaj, C. L.; Ihm, I.: Algebraic Surface Design with Hermite Interpolation. ACM Transactions on Graphics 11 (1992) 61-91.

[Ball 86] Ball, A. A.; Storry, D. J. T.: A matrix approach to recursively generated B-spline surfaces. Computer-Aided Design 18 (1986) 437-442.

[Ball 88] Ball, A. A.; Storry, D. J. T.: Conditions for tangent plane continuity over recursively generated B-spline surfaces. ACM Transactions on Graphics 7 (1988) 83-102.

[Ball 90] Ball, A. A.; Storry, D. J. T.: An Investigation of Curvature Variations Over Recursively Generated B-Spline Surfaces. ACM Transactions on Graphics 9 (1990) 424-437.

[Bal 81] Ballard, D. H.: Strip trees: a hierarchical representation for curves. Communications of the ACM 24 (1981) 310-321.

[Bär 77] Bär, G.: Parametrische Interpolation empirischer Raumkurven. ZAMM 57 (1977) 305-314.

[Bar 89] Bardis, L.; Patrikalakis, N. M.: Approximate Conversion of Rational B-Spline Patches. Computer Aided Geometric Design 6 (1989) 189-204.

[Bar 89a] Bardis, L.; Patrikalakis, N. M.: Blending Rational B-Spline Surfaces. Eurographics 89 (1989) 453-462.

[Barn 73] Barnhill, R. E.; Birkhoff, G.; Gordon, W. J.: Smooth interpolation in triangles. Journal of Approximation Theory 8 (1973) 114-128.

[Barn 75] Barnhill, R. E.; Gregory, J. A.: Polynomial Interpolation to Boundary Data on Triangles. Mathematics of Computation 29 (1975) 726-735.

[Barn 76] Barnhill, R. E.: Blending function interpolation: A survey and some new results. in Collatz, L.; Werner, H.; Meinardus, G. (ed.): Numerische Methoden der Approximationstheorie. Birkhäuser (1976) 43-89.

[Barn 77] Barnhill, R. E.: Representation and approximation of surfaces. in Rice, J. D. (ed.): Mathematical Software III. Academic Press (1977) 69-120.

[Barn 78] Barnhill, R. E.; Brown, J. H.; Klucewicz, I. M.: A New Twist in Computer Aided Geometric Design. Computers Graphics and Image Processing 8 (1978) 78-91.

[Barn 81] Barnhill, R. E.; Farin, G.: C^1 quintic interpolation over triangles: two explicit representations. International Journal for Numerical Methods in Engineering 17 (1981) 1763-1778.

[Barn 82] Barnhill, R. E.: Coons' Patches. Computers in Industry 3 (1982) 37-43.

[Barn 83] Barnhill, R. E.: Computer aided surface representation and design. in Barnhill, R. E.; Boehm, W. (ed.): Surfaces in Computer Aided Geometric Design. North-Holland (1983) 1-24.

[Barn 83a] Barnhill, R. E.; Dube, R. P.; Little, F. F.: Properties of Shepard's surfaces. Rocky Mountain Journal of Mathematics 13 (1983) 365-382.

[Barn 84] Barnhill, R. E.; Whelan, T.: A geometric interpretation of convexity conditions for surfaces. Computer Aided Geometric Design 1 (1984) 285-287.

[Barn 84a] Barnhill, R. E.; Little, F. F.: Three and Four-Dimensional Surfaces. Rocky Mountain Journal of Mathematics 14 (1984) 77-102.

[Barn 84b] Barnhill, R. E.; Stead, S.: Multistage trivariate surfaces. Rocky Mountain Journal of Mathematics 14 (1984) 103-118.

[Barn 84c] Barnhill, R. E.; Worsey, A. J.: Smooth interpolation over hypercubes. Computer Aided Geometric Design 1 (1984) 101-113.

[Barn 85] Barnhill, R. E.: Surfaces in computer-aided geometric design: a survey with new results. Computer Aided Geometric Design 2 (1985) 1-17.

[Barn 87] Barnhill, R. E.; Farin, G.; Jordan, M.; Piper, B. R.: Surface/surface intersection. Computer Aided Geometric Design 4 (1987) 3-16.

[Barn 87a] Barnhill, R. E.; Makatura, G. T.; Stead, S. E.: A New Look at Higher Dimensional Surfaces through Computer Graphics. in Farin, G. (ed.): Geometric Modeling, Algorithms and New Trends, SIAM (1987) 123-129.

[Barn 87b] Barnhill, R. E.; Piper, B. R.; Rescorla, K. L.: Interpolation to Arbitrary Data on a Surface. in Farin, G. (ed.): Geometric Modeling, Algorithms and New Trends, SIAM (1987) 281-289.

[Barn 88] Barnhill, R. E.; Farin, G.; Fayard, L.; Hagen, H.: Twists, curvatures and surface interrogation. Computer-Aided Design 20 (1988) 341-346.

[Barn 90] Barnhill, R. E.; Ou, H. S.: Surfaces defined on Surfaces. Computer Aided Geometric Design 7 (1990) 323-336.

[Barn 90a] Barnhill, R. E.; Kersey, S. N.: A marching method for parametric surface/surface intersection. Computer Aided Geometric Design 7 (1990) 257-280.

[Barn 91] Barnhill, R. E.; Foley, T. A.: Methods for Constructing Surfaces on Surfaces. in Hagen, H.; Roller, D. (ed.): Geometric Modeling. Methods and Applications. Springer (1991) 1-15.

[Barn 92] Barnhill, R. E.; Bloomquist, B.; Worsey, A. J.: Adaptive Contouring for Triangular Patches. in Barnhill, R. E. (ed.): Geometry Processing for Design and Manufacturing. SIAM (1992) 125–136.

[Barn 92a] Barnhill, R. E.; Bloomquist, B.; Worsey, A. J.: Adaptive Contouring for Trivariate Bézier Surfaces. (1992).

[Barr 84] Barr, A. H.: Global and local deformations of solid primitives. ACM Computer Graphics 18 (1984) 21-29.

[Barry 88] Barry, P. J.; Goldman, R. N.: de Casteljau-type subdivision is peculiar to Bézier curves. Computer-Aided Design 20 (1988) 114-116.

[Bars 81] Barsky, B. A.: The beta-spline: A local representation based on shape parameters and fundamental geometric measures. Thesis Univ. of Utah. Salt Lake City 1981.

[Bars 81a] Barsky, B. A.; Thomas, Sp. W.: TRANSPLINE - A system for representing curves using transformations among four spline formulations. The Computer Journal 24 (1981) 271-277.

[Bars 82] Barsky, B. A.: End conditions and boundary conditions for uniform B-spline curve and surface representations. Computers in Industry 3 (1982) 17-29.

[Bars 83] Barsky, B. A.; Beatty, J. C.: Local Control of Bias and Tension in Beta-splines. ACM Transactions on Graphics 2 (1983) 109-134.

[Bars 84] Barsky, B. A.: Exponential and Polynomial Methods for Applying Tension to an Interpolating Spline Curve. Computer Vision, Graphics, and Image Processing 27 (1984) 1-18.

[Bars 84a] Barsky, B. A.; DeRose, T. D.: Geometric Continuity of Parametric Curves. University of California, Berkeley, Computer Science Division, Technical Report UCB/CSD 84/205, 1984.

[Bars 87] Barsky, B. A.; DeRose, T. D.; Dippe, M. D.: An Adaptive Subdivision Method with Crack Prevention for Rendering Beta-spline Objects. University of California, Berkeley, Computer Science Division, Technical Report UCB/CSD 87/348, 1987.

[Bars 88] Barsky, B. A.: Introducing the Rational Beta-Spline. Proceedings of the Third International Conference on Engineering Graphics and Descriptive Geometry 1. Wien (1988) 16-27.

[Bars 88a] Barsky, B. A.; DeRose, T. D.: Three characterizations of geometric continuity for parametric curves. University of California, Berkeley, Computer Science Division, Technical Report UCB/CSD 88/417, 1988.

[Bars 89] Barsky, B. A.; DeRose, T. D.: Geometric Continuity of Parametric Curves: Three Equivalent Characterizations. IEEE Computer Graphics & Applications 9 (1989) 60-68.

[Bars 90] Barsky, B. A.; DeRose, T. D.: Geometric Continuity of Parametric Curves: Constructions of Geometrically Continuous Splines. IEEE Computer Graphics & Applications 10 (1990) 60-68.

[Bart 80] Bartel, W.; Volkmer, R.; Haubitz, I.: Thomsensche Minimalfläche - analytisch anschaulich. Resultate der Mathematik 3 (1980) 129-154.

[Bart 84] Bartels, R. H.; Beatty, J. C.: Beta-Splines with a Difference. University of Waterloo, Computer Science Department, Technical Report CS-83-40, 1984.

[Bee 86] Beeker, E.: Smoothing of shapes designed with free-form surfaces. Computer-Aided Design 18 (1986) 224-232.

[Beh 79] Behforooz, G. H.; Papamichael, N.: End Conditions for Cubic Spline Interpolation. Journal of the Institute of Mathematics and its Applications 23 (1979) 355-366.

[Beh 81] Behforooz, G. H.; Papamichael, N.: End Conditions for Interpolatory Quintic Splines. IMA Journal of Numerical Analysis 1 (1981) 81-93.

[Bei 92] Beier, P.: Implementation of a Highlight Band Algorithm on a Graphics Supercomputer. in Sapidis, N. (ed.): Designing Fair Curves and Surfaces. SIAM (1992).

[Bel 69] Bell, K.: A refined triangular plate bending element. International Journal on Numerical Methods Engineering 1 (1969) 101-122.

[Bez 74] Bézier, P.: Mathematical and practical possibilities of UNISURF. in Barnhill, R. E.; Riesenfeld, R. (ed.): Computer Aided Geometric Design. Academic Press (1974) 127-152.

[Bez 78] Bézier, P.: General distortion of an ensemble of biparametric surfaces. Computer-Aided Design 10 (1978) 116-120.

[Bid 92] Bidasaria, H. B.: Defining and rendering of textured objects through the use of exponential functions. CVGIP: Graphical Models and Image Processing 54 (1992) 97-102.

[Bie 91] Bien, A. P.; Cheng, F.: A Blending Model for Parametrically Defined Geometric Objects. in Rosignac, J.; Turner, J. (ed.): Proceedings Symposium on Solid Modeling Foundations and CAD/CAM Applications. Austin, Texas, June 1991. ACM Press (1991) 339-347.

[Bil 89] Billera, L. J.; Rose, L. L.: Gröbner Basis Methods for Multivariate Splines. in Lyche, T.; Schumaker, L. L. (ed.): Mathematical Methods in Computer Aided Geometric Design. Academic Press (1989) 93-103.

[Bir 74] Birkhoff, G.; Mansfield, L.: Compatible Triangular Finite Elements. Journal of Mathematical Analysis and Applications 47 (1974) 531-553.

[Bli 82] Blinn, J. F.: A generalization of algebraic surface drawing. ACM Transactions on Graphics 1 (1982) 235-256.

[Bli 87] Blinn, J. F.: How many Ways Can You Draw a Circle? IEEE Computer Graphics & Applications 7 (1987) 39-44.

[Bloo 88] Bloomenthal, J.: Polygonization of implicit surfaces. Computer Aided Geometric Design 5 (1988) 341-355.

[Bloo 90] Bloomenthal, J.: Interactive Techniques for Implicit Modeling. ACM Computer Graphics (1990) 109-116.

[Bloo 91] Bloomenthal, J.; Shoemake, K.: Convolution Surfaces. ACM Computer Graphics 25 (1991) 251-256.

[Blo 89] Bloor, M. I. G.; Wilson, M. J.: Generating blend surfaces using partial differential equations. Computer-Aided Design 21 (1989) 165-171.

[Blo 89a] Bloor, M. I. G.; Wilson, M. J.: Blend Design as a Boundary-Value Problem. in Straßer, W.; Seidel, H. P. (ed.): Theory and Practice of Geometric Modeling. Springer (1989) 221-234.

[Blo 89b] Bloor, M. I. G.; Wilson, M. J.: Generating N-sided patches with partial differential equations. in Earnshaw, R. A.; Wyvill, B. (ed.): Advances in Computer Graphics. Springer (1989) 129-145.

[Blo 90] Bloor, M. I. G.; Wilson, M. J.: Using partial differential equations to generate free-form surfaces. Computer-Aided Design 22 (1990) 202-212.

[Blo 90a] Bloor, M. I. G.; Wilson, M. J.: Representing PDE surfaces in terms of B-splines. Computer-Aided Design 22 (1990) 324-331.

[Bob 85] Bobrow, J. E.: NC machine tool path generation from CSG part representation. Computer-Aided Design 17 (1985) 69-76.

[Boeh 76] Boehm, W.: Darstellung und Korrektur symmetrischer Kurven und Flächen auf EDV-Anlagen. Computing 17 (1976) 79-85.

[Boeh 76a] Boehm, W.: Parameterdarstellung kubischer und bikubischer Splines. Computing 17 (1976) 87-92.

[Boeh 77] Boehm, W.: Über die Konstruktion von B-Spline Kurven. Computing 18 (1977) 161-166.

[Boeh 77a] Boehm, W.: Cubic B-Spline-Curves and Surfaces in Computer Aided Geometric Design. Computing 19 (1977) 29-34.

[Boeh 80] Boehm, W.: Inserting new knots into B-spline curves. Computer-Aided Design 12 (1980) 199-201.

[Boeh 81] Boehm, W.: Generating the Bézier Points of B-Spline Curves and Surfaces. Computer-Aided Design 13 (1981) 365-366.

[Boeh 82] Boehm, W.: On cubics, a survey. Computer Graphics and Image Processing 19 (1982) 201-226.

[Boeh 83] Boehm, W.: The de Boor Algorithm for triangular splines. in Barnhill, R. E.; Boehm, W. (ed.): Surfaces in Computer Aided Geometric Design, North-Holland (1983) 109-120.

[Boeh 83a] Boehm, W.: Subdividing multivariate splines. Computer-Aided Design 15 (1983) 345-352.

[Boeh 84] Boehm, W.; Farin, G.; Kahmann, J.: A survey of curve and surface methods in CAGD. Computer Aided Geometric Design 1 (1984) 1-60.

[Boeh 84a] Boehm, W.: Efficient evaluation of splines. Computing 33 (1984) 171-177.

[Boeh 85] Boehm, W.: Triangular spline algorithm. Computer Aided Geometric Design 2 (1985) 61-67.

[Boeh 85a] Boehm, W.: Curvature continuous curves and surfaces. Computer Aided Geometric Design 2 (1985) 313-323.

[Boeh 86] Boehm, W.: Multivariate spline methods in CAGD. Computer-Aided Design 18 (1986) 102-104.

[Boeh 86a] Boehm, W.: Multivariate spline algorithms. in Gregory, J. A. (ed.): The Mathematics of Surfaces. Clarendon Press (1986) 197-215.

[Boeh 87] Boehm, W.: Smooth curves and surfaces. in: Farin, G. E. (ed): Geometric Modeling, Algorithms and New Trends. SIAM (1987) 175-184.

[Boeh 87a] Boehm, W.: Rational geometric splines. Computer Aided Geometric Design 4 (1987) 67-77.

[Boeh 87b] Boehm, W.; Prautzsch, H.; Arner, P.: On Triangular Splines. Constructive Approximation 3 (1987) 157-167.

[Boeh 88] Boehm, W.: On de Boor-like algorithms blossoming. Computer Aided Geometric Design 5 (1988) 71-79.

[Boeh 88a] Boehm, W.: On the definition of geometric continuity. Letter to the Editor. Computer-Aided Design 20 (1988) 370-372.

[Boeh 88b] Boehm, W.: Visual Continuity. Computer-Aided Design 20 (1988) 307311.

[Boeh 89] Boehm, W.: Some Remarks on Cyclides in Solid Modeling. in Straßer, W.; Seidel, H. P. (ed.): Theory and Practice of Geometric Modeling. Springer (1989) 247-252.

[Boeh 90] Boehm, W.: Algebraic and Differential Geometric Methods in CAGD. in Dahmen, W.; Gasca, M.; Micchelli, C. A. (ed.): Computation of Curves and Surfaces. Kluwer (1990) 425-455.

[Boeh 90a] Boehm, W.: On cyclides in geometric modeling. Computer Aided Geometric Design 7 (1990) 243-255.

[Boeh 91] Boehm, W.; Hansford, D.: Bézier Patches on Quadrics. in Farin, G. (ed.): NURBS for Curve and Surface Design. SIAM (1991) 1-14.

[Boe 91] Boender, E.: A Survey of Intersection Algorithms for Curved Surfaces. Computers & Graphics 15 (1991) 109-115.

[Boi 84] Boissonnat, J. D.: Geometric Structures for Three-Dimensional Shape Representation. ACM Transactions on Graphics 3 (1984) 266-286.

[Boi 88] Boissonnat, J. D.: Shape reconstruction from planar cross sections. Computer Vision, Graphics, and Image Processing 44 (1988) 1-29.

[Bon 86] Bonfigliolo, L.: An algorithm for silhouette of curved surfaces based on graphical relations. Computer-Aided Design 18 (1986) 95-101.

[Boni 76] Bonitz, P.: Ein Beitrag zur Theorie des Entwurfes doppelt gekrümmter Flächen und differentialgeometrischen technischen Aspekten. Diss. Dresden 1976.

[Boo 62] Boor de, C.: Bicubic spline interpolation. Journal of Mathematics and Physics 41 (1962) 212-218.

[Boo 66] Boor de, C.; Lynch, R. E.: On Splines and Their Minimum Properties. Journal of Mathematics and Mechanics 15 (1966) 953-968.

[Boo 72] Boor de, C.: On calculating with B-Splines. Journal of Approximation Theory 6 (1972) 50-62.

[Boo 77] Boor de, C.: Efficient Computer Manipulation of Tensor Products. Univ. of Wisconsin, Mathematics Research Center, Report 1810, 1977.

[Boo 82] Boor de, C.; Höllig, K.: B-splines from parallelepipeds. Journal d'Analyse Mathématique 42 (1982) 99-115.

[Boo 87] Boor de, C.; Höllig, K.; Sabin, M.: High accuracy geometric Hermite interpolation. Computer Aided Geometric Design 4 (1987) 269-278.

[Boo 87a] Boor de, C.: B-form Basics. in Farin, G. (ed.): Geometric Modeling, Algorithms and New Trends, SIAM (1987) 131-148.

[Boo 88] Boor de, C.: What is a Multivariate Spline. ICIAM 87: Proceedings of the First International Conference on Industrial and Applied Mathematics. SIAM (1988) 90-101.

[Bou 85] Bouville, Ch.: Bounding Ellipsoids for Ray-Fractal Intersection. ACM Computer Graphics 19 (1985) 45-52.

[Bow 81] Bowyer, A.: Computing Dirichlet tessellations. The Computer Journal 24 (1981) 162-166.

[Bre 82] Breden, D.: Die Verwendung von bikubischen Splineflächen zur Darstellung von Tragflügeln und Propellern. Diss. Braunschweig 1982.

[Brew 77] Brewer, J. A.; Anderson, D. C.: Visual Interaction with Overhauser Curves and Surfaces. Computer Graphics 11 (1977) 132-137.

[Bri 91] Brickmann, J.; Heiden, W.; Goetze, T.; Marching Cube Algorithmen zur schnellen Generierung von Isoflächen auf der Basis dreidimensionaler Datenfelder. in Frühauf, M.; Göbel, M. (ed.): Visualisierung von Volumendaten. Springer (1991) 112-117.

[Brod 90] Brode, J.: Konvertieren von Polynomen in CAGD. Diplomarbeit Braunschweig 1990.

[Bro 91] Brodlie, K. W.; Butt, S.: Preserving convexity using piecewise cubic interpolation. Computer & Graphics 15 (1991) 15-23.

[Brow 91] Brown, J. L.: Vertex based data dependent triangulation. Computer Aided Geometric Design 8 (1991) 239-251.

[Brow 92] Brown, J. L.; Worsey, A. J.: Problems with Defining barycentric Coordinates for the sphere. Mathematical Modelling and Numerical Analysis 26 (1992) 37-49.

[Bru 84] Bruce, J. W.; Giblin, P. J.; Gibson, C. G.: Caustics through the looking glass. Mathematical Intelligencer 6 (1984) 47-58.

[Brü 80] Brückner, I.: Construction of Bézier points of quadrilaterals from those of triangles. Computer-Aided Design 12 (1980) 21-24.

[Brun 85] Brunet, P.: Increasing the smoothness of bicubic spline surfaces. Computer Aided Geometric Design 2 (1985) 157-164.

[Brun 88] Brunet, P.: Including shape handles in recursive subdivision surfaces. Computer Aided Geometric Design 5 (1988) 41-50.

[Brun 91] Brunet, P., Vinacua, A.: Surfaces in Solid Modeling. in Hagen, H.; Roller, D. (ed.): Geometric Modeling. Methods and Applications. Springer (1991) 17-34.

[Bruz 90] Bruzzone, E.; de Floriani, L.: Two data structures for building tetrahedralizations. The Visual Computer 6 (1990) 266-283.

[Buch 85] Buchberger, B.: Gröbner bases: An algorithmic method in polynomial ideal theory. in Bose, N. K. (ed.): Multidimensional Systems Theory. D. Reidel Publishing Company (1985) 184-232.

[Buch 89] Buchberger, B.: Applications of Gröbner Bases in Non-Linear Computational Geometry. in Kapur, D.; Mundy, J. L. (ed.): Geometric Reasoning. MIT Press, Elsevier Science Publisher (1989) 413-446.

[Butl 79] Butland, J.: Surface drawing made simple. Computer-Aided Design 11 (1979) 19-22.

[But 88] Butzer, P. L.; Schmidt, M.; Stark, E. L.: Observations on the History of Central B-Splines. Archive for History of Exact Science 39 (1988) 137-156.

[Cao 91] Cao, Y.; Hua, X.: The convexity of quadratic parametric triangular Bernstein-Bézier surfaces. Computer Aided Geometric Design 8 (1991) 1-6.

[Car 82] Carlson, W. E.: An algorithm and data structure for 3-D object synthesis using surface patch intersections. Computer Graphics 16 (1982) 255-263.

[Carl 91] Carlson, R. E.; Foley, Th. A.: The Parameter R^2 in Multiquadric Interpolation. Computers & Mathematics with Applications 21 (1991) 29-42.

[Casa 85] Casale, M. S.; Stanton, E. L.: An overview of analytic solid modeling. IEEE Computer Graphics & Applications 5 (1985) 45-56.

[Casa 87] Casale, M. S.: Free-form solid modeling with trimmed surface patches. IEEE Computer Graphics & Applications 7 (1987) 33-43.

[Casa 89] Casale, M. S.; Bobrow, J. E.: A set operation algorithm for sculptured solids modeled with trimmed patches. Computer Aided Geometric Design 6 (1989) 235-248.

[Casa 92] Casale, M. S.; Bobrow, J. E.; Underwood, R.: Trimmed-patch boundary elements: bridging the gap between solid modeling and engineering analysis. Computer-Aided Design 24 (1992) 193-199.

[Cas 59] Casteljau de, P.: Outillage méthodes calcul. Paris: André Citroen Automobiles S. A. 1959.

[Cat 74] Catmull, E. E.: A Subdivision Algorithm for Computer Display of Curved Surfaces. University of Utah, Salt Lake City, Department of Computer Science, Technical Report UTEC-CSc-74-133, 1974.

[Cat 78] Catmull, E. E.; Clark, J.: Recursively generated B-spline surfaces on arbitrary topological meshes. Computer-Aided Design 10 (1978) 350-355.

[Catl 87] Catley, D.; Whittle, C.; Thornton, P.: Applications of the General Surface Definition and Manipulation System, GENSURF. in Martin, R. R. (ed.): The Mathematics of Surfaces II. Oxford University Press (1987) 171-202.

[Cav 89] Cavaretta, A. S.; Micchelli, Ch. A.: The Design of Curves and Surfaces by Subdivision Algorithms. in Lyche, T.; Schumaker, L. L. (ed.): Mathematical Methods in Computer Aided Geometric Design. Academic Press (1989) 115-153.

[Cen 87] Cendes, Z. J.; Wong, St. H.: C^1 quadratic interpolation over arbitrary point sets. IEEE Computer Graphics & Applications 7 (1987) 8-16.

[Chad 89] Chadwick, J. E.; Haumann, D. R.; Parent, R. E.: Layered Construction for Deformable Animated Characters. ACM Computer Graphics 23 (1989) 243-252.

[Chai 74] Chaikin, G. M.: An Algorithm for High-Speed Curve Generation. Computer Graphics and Image Processing 3 (1974) 346-349.

[Chan 87] Chandru, V; Kochar, B. S.: Analytic Techniques for Geometric Intersection Problems. in Farin (ed.): Geometric Modeling: Algorithms and New Trends. SIAM (1987) 305-318.

[Chan 89] Chandru, V.; Dutta, D.; Hoffmann, Ch. M.: On the geometry of Dupin cyclides. The Visual Computer 5 (1989) 277-290.

[Chan 90] Chandru, V.; Dutta, D.: Variable Radius Blending Using Dupin Cyclides. in Wosny, M. J.; Turner, J. U.; Preiss, K. (ed.): Geometric Modeling for Product Engineering. North-Holland (1990) 39-57.

[Cha 82] Chang, G.: Matrix formulations of Bézier technique. Computer-Aided Design 14 (1982) 345-350.

[Cha 84] Chang, G.; Davis, Ph. J.: The convexity of Bernstein polynomials over triangles. Journal of Approximation Theory 40 (1984) 11-28.

[Cha 84a] Chang, G.; Feng, Y.: An improved condition for the convexity of Bernstein-Bézier surfaces over triangles. Computer Aided Geometric Design 1 (1984) 279-283.

[Cha 85] Chang, G.; Hoschek, J.: Convexity and variation diminishing property of Bernstein polynomials over triangles. in Schempp, W.; Zeller, K. (ed.): Multivariate Approximation Theory III, Birkhäuser (1985) 61-70.

[Chang 84] Chang, R. C.; Lee, R. C. T.: On the average length of Delaunay triangulations. BIT 24 (1984) 269-273.

[Char 84] Charrot, P.; Gregory, J. A.: A pentagonal surface patch for computer aided geometric design. Computer Aided Geometric Design 1 (1984) 87-94.

[Che 87] Chen, Y. J.; Ravani, B.: Offset surface generation and contouring in computer-aided design. ASME Journal of Mechanisms, Transmissions, and Automation in Design 109 (1987) 133-142.

[Che 88] Chen, J. J.; Ozsoy, T. M.: Predictor-corrector type of intersection algorithm for C^2 parametric surfaces. Computer-Aided Design 20 (1988) 347-352.

[Che 91] Chen, D. P.; Lin, T. L.; Hsu, C. C.: Complex-surface modelling with bias and tension. Computer-Aided Design 23 (1991) 189-194.

[Che 92] Chen, L. L.; Chou, S. Y.; Woo, T. C.: Separating and intersecting spherical polygons for computing visibility on 3-, 4-, and 5-axis machines. ASME Transactions Journal of Mechanical Design (1992).

[Chen 89] Cheng, K. P.: Using Plane Vector Fields to Obtain all the Intersection Curves of Two General Surfaces. in Straßer, W.; Seidel, H. P. (ed.): Theory and Practice of Geometric Modeling. Springer (1989) 187-204.

[Chio 91] Chionh, E. W.; Goldman, R. N.; Miller, J. R.: Using multivariate resultants to find the intersection of three quadric surfaces. ACM Transactions on Graphics 10 (1991) 378-400.

[Chio 92] Chionh, E. W.; Goldman, R. N.: Using multivariate resultants to find the implicit equation of a rational surface. The Visual Computer 8 (1992) 171-180.

[Chiu 90] Chiungtung Kao, T.; Knott, G. D.: An efficient and numerically correct algorithm for the 2D convex hull problem. BIT 30 (1990) 311-331.

[Chi 83] Chiyokura, H.; Kimura, F.: Design of solids with free-form surfaces. ACM Computer Graphics 17 (1983) 289-298.

[Chi 84] Chiyokura, H.; Kimura, F.: A new surface interpolation method for irregular curve models. Computer Graphics Forum 3 (1984) 209-218.

[Chi 86] Chiyokura, H.: Localized Surface Interpolation for Irregular Meshes. in Kunii, T. L. (ed.): Advanced Computer Graphics. Springer (1986) 3-19.

[Chi 87] Chiyokura, H.: An Extended Rounding Operation for Modelling Solids with Free-Form Surfaces. in Kunii, T. L. (ed.): Computer Graphics 1987. Springer (1987) 249-268.

[Chi 91] Chiyokura, H.; Takamura, T.; Konno, K.; Harada, T.: G^1 Surface Interpolation over Irregular Meshes with Rational Curves. in Farin, G. (ed.): NURBS for Curve and Surface Design. SIAM (1991) 15-34.

[Cho 88] Choi, B. K.; Shin, H. Y.; Yoon, Y. I.; Lee, J. W.: Triangulation of scattered data in 3D space. Computer-Aided Design 20 (1988) 239-248.

[Cho 88a] Choi, B. K.; Lee, C. S.; Hwang, J. S.; Jun, C. S.: Compound surface modelling and machining. Computer-Aided Design 20 (1988) 127-136.

[Cho 89] Choi, B. K.; Ju, S. Y.: Constant-radius blending in surface modeling. Computer-Aided Design 21 (1989) 213-220.

[Cho 89a] Choi, B. K.; Jun, C. S.: Ball-end cutter interference avoidance in NC machining of sculptured surfaces. Computer-Aided Design 21 (1989) 371-378.

[Cho 90] Choi, B. K.; Lee, C. S.: Sweep surfaces modelling via coordinate transformations and blending. Computer-Aided Design 22 (1990) 87-96.

[Cho 90a] Choi, B. K.; Yoo, W. S.; Lee, C. S.: Matrix representation for NURB curves and surfaces. Computer-Aided Design 22 (1990) 235-240.

[Chr 78] Christiansen, H. N.; Sederberg, Th. W.: Conversion of complex contour line definitions into polygonal element mosaics. ACM Computer Graphics 12 (1978) 187-192.

[Chu 90] Chui, C. K.; He, T. X.: Bivariate C^1 quadratic finite elements and vertex splines. Math. Comp. 54 (1990) 169-187.

[Cla 79] Clark, J. H.: A Fast Scan-line Algorithm for Rendering Parametric Surfaces. SIGGRAPH 79, ACM Computer Graphics 13 (1979).

[Cli 74] Cline, A. K.: Scalar- and Planar-Valued Curve Fitting Using Splines under Tension. Communications of the ACM 17 (1974) 218-220.

[Cli 84] Cline, A. K.; Renka, R. L.: A storage-efficient method for construction of a Thiessen triangulation. Rocky Mountain Journal of Mathematics 14 (1984) 119-139.

[Cli 90] Cline, A. K.; Renka, R. L.: A Constrained Two-Dimensional Triangulation and the Solution of Closest Node Problems in the Pressence of Barriers. SIAM Journal on Numerical Analysis 27 (1990) 1305-1321.

[Clo 65] Clough, R. W.; Tocher, J. L.: Finite element stiffness matrices for the analysis of plate bending. Proceedings of the 1st Conference on Matrix Methods in Structural Mechanics. Wright-Patterson AFB AFFDL TR 66-80 (1965) 515-545.

[Coh 80] Cohen, E.; Lyche, T; Riesenfeld, R. F.: Discrete B-splines and subdivision techniques in computer-aided geometric design and computer graphics. Computer Graphics and Image Processing 14 (1980) 87-111.

[Coh 82] Cohen, E.; Riesenfeld, R. F.: General matrix representations for Bézier and B-spline curves. Computers in Industry 3 (1982) 9-15.

[Coh 85] Cohen, E.; Schumaker, L. L.: Rates of convergence of control polygons. Computer Aided Geometric Design 2 (1985) 229-235.

[Coh 85a] Cohen, E.; Lyche, T.; Schumaker, L. L.: Algorithms for Degree-Raising of Splines. ACM Transactions on Graphics 4 (1985) 171-181.

[Coh 87] Cohen, E.: A new local basis for designing with tensioned splines. ACM Transactions on Graphics 6 (1987) 81-122.

[Coh 89] Cohen, E.; O'Dell, C. L.: A Data Dependent Parametrization for Spline Approximation. in Lyche, T.; Schumaker, L. L.: (ed.): Mathematical Methods in Computer Aided Geometric Design. Academic Press (1989) 155-166.

[Cohe 82] Cohen, S.: Ein Beitrag zur steuerbaren Interpolation von Kurven und Flächen. Diss. TU Dresden 1982.

[Com 68] Comba, P. G.: A Procedure for Detecting Intersections of Three-Dimensional Objects. Journal of the Association for Computing Machinery 15 (1968) 354-366.

[Cook 81] Cook, R. L.; Torrance, K. E.: A Reflectance Model for Computer Graphics. ACM Computer Graphics 15 (1981).

[Coo 64] Coons, S. A.: Surfaces for Computer Aided Design. MIT, Mechan Engin. Department, Design Division, 1964.

[Coo 67] Coons, S. A.: Surfaces for computer aided design of space forms. MIT Project MAC-TR-41, 1967.

[Coo 77] Coons, S. A.: Modification of the shape of piecewise curves. Computer-Aided Design 9 (1977) 178-180.

[Coq 87] Coquillart, S.: Computing offsets of B-spline curves. Computer-Aided Design 19 (1987) 305-309.

[Coq 90] Coquillart, S.: Extended Free-Form Deformation: A Sculpturing Tool for 3D Geometric Modeling. ACM Computer Graphics 24 (1990) 187-196.

[Coq 91] Coquillart, S.; Jancene, P.: Animated Free-Form Deformation: An Interactive Animation Technique. ACM Computer Graphics 25 (1991) 23-26.

[Cos 88] Costantini, P.: An algorithm for computing shape-preserving interpolating splines of arbitrary degree. Journal of Computational and Applied Mathematics 22 (1988) 89-136.

[Cos 90] Costantini, P.; Fontanella, F.: Shape-Preserving Bivariate Interpolation. SIAM Journal on Numerical Analysis 27 (1990) 488-506.

[Cox 71] Cox, M. G.: The numerical evaluation of B-splines. Nat. Phys. Lab. England: Teddington 1971.

[Cra 67] Crain, I. K.; Bhattacharya, B. K.: Treatment of non-equispaced two- dimensional data with a digital computer. Geoexploration 5 (1967) 173-194.

[Cre 59] Cressman, G. P.: An operational objective analysis system. Monthly Weather Review 87 (1959) 367-374.

[Cro 87] Crocker, G. A.; Reinke, W. F.: Boundary Evaluation of Non-Convex Primitives to Produce Parametric Trimmed Surfaces. ACM Computer Graphics (1987) 129-136.

[Cui 91] Cui, Zhu: Tool-path generation in manufacturing sculptured surfaces with a cylindrical end-milling cutter. Computers in Industry 17 (1991) 385-389.

[Dae 91] Daehlen, M.; Lyche, T.: Box Splines and Applications. in Hagen, H.; Roller, D. (ed.): Geometric Modeling. Methods and Applications. Springer (1991) 35-93.

[Dah 84] Dahmen, W.; Micchelli, C. A.: Subdivision algorithm for the generation of box spline surface. Computer Aided Geometric Design 1 (1984) 115-129.

[Dah 86] Dahmen, W.: Subdivision algorithms converge quadratically. Journal of Computational and Applied Mathematics 16 (1986) 145-158.

[Dah 89] Dahmen, W.: Smooth Piecewise Quadric Surfaces. in Lyche, T.; Schumaker, L. L. (ed.): Mathematical Methods in Computer Aided Geometric Design. Academic Press (1989) 181-193.

[Dah 91] Dahmen, W.: Convexity and Bernstein-Bézier Polynomials. in Laurent, P. J.; LeMéhauté, A.; Schumaker, L. L. (ed.): Curves and Surfaces. Academic Press (1991) 107-134.

[Dam 76] Damsohn, H.: Fünfachsiges NC-Fräsen. Beitrag zur Technologie, Teileprogrammierung und Postprozessorverarbeitung. Springer 1976.

[Dani 89] Daniel, M.; Daubisse, J. C.: The numerical problem of using Bézier curves and surfaces in the power basis. Computer Aided Geometric Design 6 (1989) 121-128.

[Dan 85] Dannenberg, L.; Nowacki, H.: Approximate conversion of surface representations with polynomial bases. Computer Aided Geometric Design 2 (1985) 123-132.

[Day 63] Dayhoff, M. I.: A Contour-Map Program for X-Ray Crystallography. Communications of the ACM 6 (1963) 620-622.

[Deg 88] Degen, W. L. F.: Some remarks on Bézier curves. Computer Aided Geometric Design 5 (1988) 259-268.

[Deg 90] Degen, W. L. F.: Explicit continuity conditions for adjacent Bézier surface patches. Computer Aided Geometric Design 7 (1990) 181-189.

[Deg 92] Degen, W. L. F.: Best approximations of parametric curves by splines. in Lyche, T.; Schumaker, L. L. (ed.): Mathematical Methods in Computer Aided Geometric Design II. Academic Press (1992) 171-184.

[Deh 91] DeHaemer, Jr., M. J.; Zyda, M. J.: Simplication of Objects Rendered by Polygonal Approximations. Computers & Graphics 15 (1991) 175–184.

[Dero 85] DeRose, T. D.: Geometric Continuity: A Parametrization Independent Measure of Continuity for Computer Aided Geometric Design. Thesis Berkeley 1985.

[Dero 87] DeRose, T. D.; Holman, T. J.: The Triangle: A Multiprocessor Architecture for Fast Curve and Surface Generation. University of Washington, Department of Computer Science, Technical Report No 87-08-07, 1987.

[Dero 88] DeRose, T. D.; Barsky, B. A.: Geometric Continuity, Shape Parameters, and Geometric Constructions for Catmull-Rom Splines. ACM Transactions on Graphics 7 (1988) 1-41.

[Dero 88a] DeRose, T. D.: Composing Bézier Simplices. ACM Transactions on Graphics 7 (1988) 198-221.

[Dero 89] DeRose, T. D.; Bailey, M. L.; Barnard, B.; Cypher, R.; Dobrikin, D.; Ebeling, C.; Konstantinidou, S.; McMurchie, L.; Mizrahi, H.; Yost, B.: Apex: two architectures for generating parametric curves and surfaces. The Visual Computer 5 (1989) 264-276.

[Dero 90] DeRose, T. D.: Necessary and sufficient conditions for adjacent Bézier surface patches. Computer Aided Geometric Design 7 (1990) 165-179.

[Dero 91] DeRose, T. D.: Rational Bézier Curves and Surfaces on Projective Domains. in Farin, G. (ed.): NURBS for Curve and Surface Design. SIAM (1991) 35-45.

[Dic 89] Dickinson, R. R.; Bartels, R. H.; Vermeulen, A. H.: The Interactive Editing and Contouring of Empirical Fields. IEEE Computer Graphics & Applications 9 (1989) 34-43.

[Die 81] Dierckx, P.: An Algorithm for Surface-Fitting with Spline Functions. IMA Journal of Numerical Analysis 1 (1981) 267-283.

[Die 89] Dierckx, P.; Tytgat, B.: Generating the Bézier points of a β-spline curve. Computer Aided Geometric Design 6 (1989) 279-291.

[Die 89a] Dierckx, P.; Tytgat, B.: Inserting new knots into beta-spline curves. in Lyche, T.; Schumaker, L. L. (ed.): Mathematical Methods in Computer Aided Geometric Design. Academic Press (1989) 195-206.

[Dob 90] Dobkin, D. P.; Levy, S. V. F.; Thurston, W. P.; Wilks, A. R.: Contour Tracing by Piecewise Linear Approximations. ACM Transactions on Graphics 9 (1990) 389-423.

[Dok 85] Dokken, T.: Finding intersections of B-spline represented geometries using recursive subdivision techniques. Computer Aided Geometric Design 2 (1985) 189-195.

[Dok 89] Dokken, T.; Skytt, V.; Ytrehus, A.-M.: Recursive Subdivision and Iteration in Intersections and Related Problems. in Lyche, T.; Schumaker, L. L. (ed.): Mathematical Methods in Computer Aided Geometric Design. Academic Press (1989) 207-214.

[Doo 78] Doo, D. W. H.; Sabin, M. A.: Behaviour of Recursive Division Surfaces Near Extraordinary Points. Computer-Aided Design 10 (1978) 356-360.

[Doo 78a] Doo, D. W. H.: A Subdivision Algorithm for Smoothing Down Irregularly Shaped Polyhedrons. in Proceedings of an International Conference on Interactive Techniques in Computer Aided Design. Bologna (1978) 157-165.

[Dre 88] Drebin, R. A.; Carpenter, L.; Hanrahan, P.: Volume Rendering. ACM Computer Graphics 22 (1988) 65-74.

[Drex 77] Drexler, F. J.: Eine Methode zur Berechnung sämtlicher Lösungen von Polynomgleichungssystemen. Numerische Mathematik 29 (1977) 45-58.

[Du 88] Du, W. H.; Schmitt, F. J. M.: New Results for the smooth connection between Tensor Product Bézier Patches. in Magnenat-Thalmann, N.; Thalmann, D. (ed.): New Trends in Computer Graphics. Proceedings of CG International '88. Springer (1988) 351-363.

[Du 90] Du, W. H.; Schmitt, F. J. M.: On the G^1 continuity of piecewise Bézier surfaces: a review with new results. Computer-Aided Design 22 (1990) 556-573.

[Duc 77] Duchon, J.: Splines minimizing rotation invariant semi-norms in Sobolev spaces. in Schempp, W.; Zeller, K. (ed.): Constructive Theory of Functions of Several Variables. Lecture Notes in Mathematics Vol. 571, Springer (1977) 85-100.

[Dür 88] Dürst, M. J.: Additional reference to "marching cubes". ACM Computer Graphics 22, 2 (1988) Letters.

[Dwy 87] Dwyer, R. A.: A Faster Divide-and-Conquer Algorithm for Constructing Delaunay Triangulations. Algoritmica 2 (1987) 137-151.

[Dyn 85] Dyn, N.; Micchelli, Ch. A.: Piecewise polynomial spaces and geometric continuity of curves. IBM Thomas J. Watson Research Center, Yorktown Heights, New York 1985.

[Dyn 86] Dyn, N.; Levin, D.; Rippa, S.: Numerical Procedures for surface fitting of scattered data by radial functions. SIAM Journal on Scientific and Statistical Computing 7 (1986) 639-659.

[Dyn 87] Dyn, N.; Edelman, A.; Micchelli, Ch. A.: On locally supported basis functions for the representation of geometrically continuous curves. Analysis 7 (1987) 313-341.

[Dyn 87a] Dyn, N.: Interpolation of Scattered Data by Radial Functions. in Chui, C. K.; Schumaker, L. L.; Utreras, F. I. (ed.): Topics in Multivariate Approximation. Academic Press (1987) 47-61.

[Dyn 87b] Dyn, N.; Gregory, J. A.; Levin, D.: A four-point interpolatory subdivision scheme for curve design. Computer Aided Geometric Design 4 (1987) 257-268.

[Dyn 89] Dyn, N.: Interpolation and Approximation by Radial and Related Functions. in Chui, C. K.; Schumaker, L. L.; Ward, J. D. (ed.): Approximation Theory VI: Volume 1. Academic Press (1989) 211-234.

[Dyn 90] Dyn, N.; Levin, D.; Rippa, S.: Data dependent triangulations for piecewise linear interpolation. IMA Journal of Numerical Analysis 10 (1990) 137-154.

[Dyn 90a] Dyn, N.; Levin, D.; Gregory, J. A.: A Butterfly Subdivision Scheme for Surface Interpolation with Tension Control. ACM Transactions on Graphics 9 (1990) 160-169.

[Dyn 90b] Dyn, N.; Levin, D.; Rippa, S.: Algorithms for the construction of data dependent triangulations. in Mason, J. C.; Cox, M. G. (ed.): Algorithms for Approximation II. Chapman and Hall (1990) 185-192.

[Dyn 90c] Dyn, N.; Levin, D.; Micchelli, C. A.: Using parameters to increase smoothness of curves and surfaces generated by subdivision. Computer Aided Geometric Design 7 (1990) 129-140.

[Dyn 91] Dyn, N.; Gregory, J. A.; Levin, D.: Analysis of Uniform Binary Subdivision Schemes for Curve Design. Constructive Approximation 7 (1991) 127-147.

[Dyn 92] Dyn, N.: Subdivision Schemes in CAGD. School of Mathematical Sciences. Sackler Faculty of Exact Sciences. Tel Aviv University 1992.

[Dyn 92a] Dyn, N.; Levin, D.; Liu, D.: Interpolatory convexity-preserving subdivision schemes for curves and surfaces. Computer-Aided Design 24 (1992) 211-216.

[Eck 87] Eck, M.: Allgemeine Konzepte geometrischer Bézier- und B-Spline-kurven. Diplomarbeit Darmstadt 1987.

[Eck 89] Eck, M.; Lasser, D.: B-Spline-Bézier Representation of Geometric Spline Curves. Quartics and Quintics. Preprint 1234 Fachbereich Mathematik, TH-Darmstadt 1989.

[Eck 90] Eck, M.: Geometrische Verfahren zur dreidimensionalen Osteotomieplanung. Diss. Darmstadt 1990.

[Edd 77] Eddy, W. F.: A new convex hull algorithm for planar sets. ACM Transactions on Mathematical Software 3 (1977) 398-403.

[Eis 92] Eisele, E. F.: Chebyshev approximation of planar curves by splines. Computer Aided Geometric Design 9 (1992).

[Elb 90] Elber, G.; Cohen, E.: Hidden Curve Removal for Free Form Surfaces. ACM Computer Graphics 24 (1990) 95-104.

[Eps 76] Epstein, M. P.: On the influence of parametrization in parametric interpolation. SIAM Journal on Numerical Analysis 13 (1976) 261-268.

[Eva 87] Evans, B. M.: View from practice. Computer-Aided Design 19 (1987) 203-211.

[Far 79] Farin, G.: Subsplines über Dreiecken. Diss. Braunschweig 1979.

[Far 82] Farin, G.: A Construction for Visual C^1 Continuity of Polynomial Surface Patches. Computer Graphics and Image Processing 20 (1982) 272-282.

[Far 82a] Farin, G.: Visually C^2 cubic splines. Computer-Aided Design 14 (1982) 137-139.

[Far 83] Farin, G.: Algorithms for rational Bézier curves. Computer-Aided Design 15 (1983) 73-77.

[Far 83a] Farin, G.: Smooth Interpolation to Scattered 3D Data. in Barnhill, R. E.; Boehm, W. (ed.): Surfaces in Computer Aided Geometric Design. North-Holland (1983) 43-63.

[Far 85] Farin, G.: Some remarks on V^2 splines. Computer Aided Geometric Design 2 (1985) 325-328.

[Far 85a] Farin, G.: A modified Clough-Tocher interpolant. Computer Aided Geometric Design 2 (1985) 19-27.

[Far 86] Farin, G.: Triangular Bernstein-Bézier patches. Computer Aided Geometric Design 3 (1986) 83-127.

[Far 87] Farin, G.; Rein, G.; Sapidis, N.; Worsey, A. J.: Fairing cubic B-spline curves. Computer Aided Geometric Design 4 (1987) 91-103.

[Far 87a] Farin, G.; Piper, B.; Worsey, A. J.: The octant of a sphere as a nondegenerate triangular Bézier patch. Short Communication, Computer Aided Geometric Design 4 (1987) 329-332.

[Far 89] Farin, G.: Rational Curves and Surfaces. in Lyche, T.; Schumaker, L. L. (ed.): Mathematical Methods in Computer Aided Geometric Design. Academic Press (1989) 215-238.

[Far 89a] Farin, G.: Curvature Continuity and Offset for Piecewise Conics. ACM Transactions on Graphics 8 (1989) 89-99.

[Far 90] Farin, G.: Surfaces over Dirichlet tessellations. Computer Aided Geometric Design 7 (1990) 281-292.

[Far 90a] Farin, G.; Hansford, D.; Worsey, A.: The singular cases for γ-spline interpolation. Computer Aided Geometric Design 7 (1990) 533-546.

[Far 92] Farin, G.; Hagen, H.: A Local Twist Estimator. in Hagen, H. (ed.): Topics in Surface Modeling. SIAM (1992) 79-84.

[Faro 85] Farouki, R. T.: Exact offset procedures for simple solids. Computer Aided Geometric Design 2 (1985) 257-279.

[Faro 85a] Farouki, R. T.; Hinds, J. K.: A hierarchy of geometric forms. IEEE Computer Graphics & Applications 5 (1985) 51-78.

[Faro 86] Farouki, R. T.: The Characterization of Parametric Surface Sections. Computer Vision, Graphics, and Image Processing 33 (1986) 209-236.

[Faro 86a] Farouki, R. T.: The approximation of non-degenerate offset surfaces. Computer Aided Geometric Design 3 (1986) 15-43.

[Faro 87] Farouki, R. T.; Rajan, V. T.: On the numerical condition of polynomials in Bernstein form. Computer Aided Geometric Design 4 (1987). 191-216.

[Faro 87a] Farouki, R. T.: Direct surface section evaluation. in Farin, G. (ed.): Geometric Modeling, Algorithms and New Trends. SIAM (1987) 319-334.

[Faro 88] Farouki, R. T.; Rajan, V. T.: Algorithms for polynomials in Bernstein form. Computer Aided Geometric Design 5 (1988) 1-26.

[Faro 88a] Farouki, R. T.; Rajan, V. T.: On the numerical conditions of algebraic curves and surfaces, 1. Implicit equations. Computer Aided Geometric Design 5 (1988) 215-252.

[Faro 89] Farouki, R. T.: Hierarchical Segmentations of Algebraic Curves and Some Applications. in Lyche, T.; Schumaker, L. L. (ed.): Mathematical Methods in Computer Aided Geometric Design. Academic Press (1989) 207-214.

[Faro 89a] Farouki, R. T.; Neff, C. A.; O'Connor, M. A.: Automatic Parsing of Degenerate Quadric-Surface Intersection. ACM Transactions on Graphics 8 (1989) 174-203.

[Faro 90a] Farouki, R. T.; Neff, C. A.: Analytic properties of plane offset curves. Computer Aided Geometric Design 7 (1990) 83-100.

[Faro 90b] Farouki, R. T.; Neff, C. A.: Algebraic properties of plane offset curves. Computer Aided Geometric Design 7 (1990) 101-127.

[Faro 92] Farouki, R. T.: Pythagorean-Hodograph Curves in Practical Use. in Barnhill, R. E. (ed.): Geometry Processing for Design and Manufacturing. SIAM (1992) 3-33.

[Farw 86] Farwig, R.: Multivariate interpolation of arbitrarily spaced data by moving least squares methods. Journal of Computational and Applied Mathematics 16 (1986) 79-93.

[Farw 86a] Farwig, R.: Rate of Convergence of Shepard's Global Interpolation Formula. Mathematics of Computation 46 (1986) 577-590.

[Faw 86] Fawzy, T.; Schumaker, L. L.: A Piecewise Polynomial Lacunary Interpolation Method. Journal of Approximation Theory 48 (1986) 407-426.

[Fels 91] Felsenstein, L.: Berechnung von Fräsbahnen mittels räumlicher Parallelkurven auf Parallelflächen. Diplomarbeit Darmstadt 1991.

[Fen 88] Feng, Y. Y.; Kozak, J.: An approach to the interpolation of nonuniformly spaced data. Journal of Computational and Applied Mathematics 23 (1988) 169-178.

[Fer 64] Ferguson, J. C.: Multivariable curve interpolation. Journal of the ACM 11 (1964) 221-228.

[Ferg 86] Ferguson, D. R.: Construction of curves and surfaces using numerical optimization techniques. Computer-Aided Design 18 (1986) 15-21.

[Fil 86] Filip, D. J.: Adaptive subdivision algorithms for a set of Bézier triangles. Computer-Aided Design 18 (1986) 74-78.

[Fil 86a] Filip, D. J.; Magedson, R.; Markot, R.: Surface algorithms using bounds on derivatives. Computer Aided Geometric Design 3 (1986) 295-311.

[Fil 89] Filip, D. J.: Blending Parametric Surfaces. ACM Transactions on Graphics 8 (1989) 164-173.

[Fil 89a] Filip, D. J.; Ball, Th. W.: Procedurally Representing Lofted Surfaces. IEEE Computer Graphics & Applications 9 (1989) 27-33.

[Fja 86] Fjallström, P. O.: Smoothing of Polyhedral Models. in ACM Proceedings of the 2nd Symposium on Computational Geometry. Yorktown Heights (1986) 226-235.

[Fle 86] Fletcher, G. Y.; McAllister, D. F.: Natural bias approach to shape preserving curves. Computer-Aided Design 18 (1986) 48-52.

[Fle 89] Fletcher, G. Y.; McAllister, D. F.: A Tension-Compatible Patch for Shape-Preserving Surface Interpolation. IEEE Computer Graphics & Applications 5 (1989) 45-55.

[Flo 87] Floriani de, L.: Surface representations based on triangular grids. The Visual Computer 3 (1987) 27-50.

[Fol 80] Foley, Th. A.; Nielson, G. M.: Multivariate interpolation to scattered data using delta iteration. in Cheney, E. W. (ed.): Multivariate Approximation Theory III. Academic Press (1980) 419-424.

[Fol 84] Foley, Th. A.: Three-stage Interpolation to Scattered Data. Rocky Mountain Journal of Mathematics 14 (1984) 141-149.

[Fol 86] Foley, Th. A.: Scattered data interpolation and approximation with error bound. Computer Aided Geometric Design 3 (1986) 163-177.

[Fol 86a] Foley, Th. A.: Local control of interval tension using weighted splines. Computer Aided Geometric Design 3 (1986) 281-294.

[Fol 87] Foley, Th. A.: Interpolation with Interval and Point Tension Controls Using Cubic Weighted ν-Splines. ACM Transaction on Mathematical Software 13 (1987) 68-96.

[Fol 87a] Foley, Th. A.: Weighted bicubic spline interpolation to rapidly varying data. ACM Transactions on Graphics 6 (1987) 1-18.

[Fol 87b] Foley, Th. A.: Interpolation and approximation of 3-D and 4-D scattered data. Computers and Mathematics with Applications 13 (1987) 711-740.

[Fol 88] Foley, Th. A.: A shape preserving interpolant with tension controls. Computer Aided Geometric Design 5 (1988) 105-118.

[Fol 89] Foley, Th. A.; Nielson, G. M.: Knot Selection for Parametric Spline Interpolation. in Lyche, T.; Schumaker, L. L.; (ed.): Mathematical Methods in Computer Aided Geometric Design. Academic Press (1989) 261-272.

[Fol 90] Foley, Th. A.; Lane, D. A.; Nielson, G. M.; Ramaraj, R.: Visualizing functions over a sphere. IEEE Computer Graphics & Applications 10 (1990) 32-40.

[Fol 90a] Foley, Th. A.; Lane, D. A.: Visualization of Irregular Multivariate Data. in Kaufman, A. (ed.): Visualization '90. IEEE Computer Society Press (1990) 247-254.

[Fol 90b] Foley, Th. A.; Lane, D. A.; Nielson, G. M.; Franke, R.; Hagen, H.: Interpolation of scattered data on closed surfaces. Computer Aided Geometric Design 7 (1990) 303-312.

[Fol 90c] Foley, Th. A.: Interpolation to scattered data on a spherical domain. in Cox, M.; Mason, J. (ed.): Algorithms for Approximation II. Chapman and Hall (1990) 303-310.

[Fol 91] Foley, Th. A.; Lane, D. A.: Multi-Valued Volumetric Visualization. in Nielson, G. M.; Rosenblum, L. (ed.): Visualization '91. IEEE Computer Society Press (1991) 218-225.

[Fol 92] Foley, Th. A.; K. Opitz: Hybrid Cubic Bézier Triangle Patches. in Lyche, T.; Schumaker, L. L. (ed.): Mathematical Methods in Computer Aided Geometric Design II. Academic Press (1992) 275-286.

[Fon 87] Fontanella, F.: Shape Preserving Surface Interpolation. in Chui, C. K.; Schumaker, L. L.; Utreras, F. (ed.): Topics in Multivariate Approximation. Academic Press (1987) 63-78.

[For 72] Forrest, A. R.: Interactive interpolation and approximation by Bézier polynomials. Computer Journal 15 (1972) 71-79.

[For 72a] Forrest, A. R.: On Coons and other methods for the representation of curved surfaces. Computer Graphics and Image Processing 1 (1972) 341-359.

[Fow 66] Fowler, A. H.; Wilson, C. W.: Cubic spline, a curve fitting routine, Y-12 Plant Report Y-1400 (Revision 1), Union Carbide Corporation, Oak Ridge, TN 1966.

[Fra 77] Franke, R.: Locally determined smooth interpolation at irregularly spaced points in several variables. Journal of the Institute of Mathematics and its Applications 19 (1977) 471-482.

[Fra 80] Franke, R.; Nielson, G. M.: Smooth interpolation of large sets of scattered data. International Journal for Numerical Methods in Engineering 15 (1980) 1691-1704.

[Fra 82] Franke, R.: Scattered data interpolation: Tests of some methods. Mathematics of Computation 38 (1982) 181-200.

[Fra 82a] Franke, R.: Smooth interpolation of scattered data by local thin plate splines. Computers and Mathematics with Applications 8 (1982) 273-281.

[Fra 83] Franke, R.; Nielson, G. M.: Surface approximation with imposed conditions. in Barnhill, R. E.; Boehm, W. (ed.): Surfaces in Computer Aided Geometric Design. North-Holland (1983) 135-146.

[Fra 85] Franke, R.: Thin plate splines with tension. in Barnhill, R. E.; Boehm, W. (ed.): Surfaces in CAGD '84. North-Holland (1985) 87-95.

[Fra 87] Franke, R.: Recent advances in the approximation of surfaces from scattered data. in Chui, C. K.; Schumaker, L. L.; Utreras, F. I. (ed.): Topics in Multivariate Approximation, Academic Press (1987) 79-97.

[Fra 87a] Franke, R.; Schumaker, L. L.: A Bibliography of Multivariate Approximation. in Chui, C. K.; Schumaker, L. L.; Utreras, F. I. (ed.): Topics in Multivariate Approximation. Academic Press (1987) 275-335.

[Fra 91] Franke, R.; Nielson, G. M.: Scattered Data Interpolation and Applications: A Tutorial and Survey. in Hagen, H.; Roller, D. (ed.): Geometric Modeling. Springer (1991) 131-160.

[Fre 81] Freeden, W.: On spherical spline interpolation and approximation. Mathematical Methods in the Applied Sciences 3 (1981) 551-575.

[Fre 84] Freeden, W.: Spherical spline interpolation: Basic theory and computational aspects. Journal Computational and Applied Mathematics 11 (1984) 367-375.

[Fri 80] Fritsch, F. N.; Carlson, R. E.: Monotone piecewise cubic interpolation. SIAM Journal on Numerical Analysis 17 (1980) 238-246.

[Fri 86] Fritsch, F. N.: The Wilson-Fowler Spline is a ν-spline. Computer Aided Geometric Design 3 (1986) 155-162.

[Fri 86a] Fritsch, F. N.: History of the Wilson-Fowler Spline. Preprint UCID-20746, Lawrence Livermore National Laboratory 1986.

[Fri 87] Fritsch, F. N.: Energy Comparison of Wilson-Fowler Splines with other Interpolating Splines. in Farin, G. (ed.): Geometric Modeling: Algorithms and New Trends. SIAM (1987) 185-201.

[Fri 92] Fritsch, F. N.; Nielson, G. M.: On the Problem of Determining the Distance between Parametric Curves. in Hagen, H. (ed.): Curve and Surface Design. SIAM (1992) 123-141.

[Fu 90] Fu, Q.: The intersection of a bicubic Bézier patch and a plane. Computer Aided Geometric Design 7 (1990) 475-488.

[Gal 89] Gallagher, R. S.; Nagtegaal, J. C.: An efficient 3-D visualization technique for finite element models and other coarse volumes. ACM Computer Graphics 23 (1989) 185-194.

[Gar 79] Garcia, C. B.; Zangwill, W. I.: Finding all Solutions to Polynomial Systems and other Systems of Equations. Mathematical Programming 16 (1979) 159-176.

[Garr 89] Garrity, Th.; Warren, J.: On computing the intersection of a pair of algebraic surfaces. Computer Aided Geometric Design 6 (1989) 137-154.

[Garr 91] Garrity, Th.; Warren, J.: Geometric Continuity. Computer Aided Geometric Design 8 (1991) 51-65.

[Gas 90] Gassmann, V.; Tolle, B.: Algorithmen zur Approximation von ebenen Kurven mit Kreisbögen gleicher Tangente in den Anschlußpunkten. VDI-Bericht Nr. 847, VDI-Verlag Düsseldorf 1990.

[Geig 91] Geiger, B.; Müller, H.: Interpolation und Visualisierung von Körpern aus ebenen Schnitten. in Frühauf, M.; Göbel, M. (ed.): Visualisierung von Volumendaten. Springer (1991) 92-111.

[Gei 62] Geise, G.: Über berührende Kegelschnitte einer ebenen Kurve. ZAMM 42 (1962) 297-304.

[Gei 84] Geise, G.; Stammler, L.; Harms, S.; Uhlig, A.: Krümmungsabhängige Schrittweitensteuerung ohne höhere als erste Ableitung durch Nutzung eines kreisnahen Polygonstreifenmodells. Beiträge zur Algebra und Geometrie 17 (1984) 83-109.

[Gei 88] Geise, G.: Projektion des Schnittes zweier Quadriken. Beiträge zur Algebra und Geometrie 27 (1988) 21-32.

[Gei 90] Geise, G.; Langbecker, U.: Finite quadric segments with four conic boundary curves. Computer Aided Geometric Design 7 (1990) 141-150.

[Gig 89] Giger, Ch.: Ray Tracing Polynomial Tensor Product Surfaces. in Hansmann, W.; Hopgood, F. R. A.; Straßer, W. (ed.): EUROGRAPHICS '89. North-Holland (1989) 125-136.

[Gla 88] Glaeser, G.: Problemangepaßte schnelle Algorithmen zur Bestimmung spezieller Flächenkurven. CAD und Computergraphik 11 (1988) 119-127.

[Glass 66] Glass, J. M.: Smooth-curve interpolation: A generalized spline-fit procedure. BIT 6 (1966) 277-293.

[Gme 90] Gmelig Meyling, R. H. J.; Pfluger, R. R.: Smooth interpolation to scattered data by bivariate piecewise polynomials of odd degree. Computer Aided Geometric Design 7 (1990) 435-458.

[Gold 77] Gold, C. M.: Automated contour mapping using triangular element data structures and an interpolant over each irregular triangular domain. ACM Computer Graphics 11 (1977) 170-175.

[Gol 82] Goldman, R. N.: Using degenerate Bézier triangles and tetrahedra to subdivide Bézier curves. Computer-Aided Design 14 (1982) 307-311.

[Gol 83] Goldman, R. N.: Subdivision algorithms for Bézier triangles. Computer-Aided Design 15 (1983) 159-166.

[Gol 84] Goldman, R. N.: Linear subdivision is strictly a polynomial phenomenon. Computer Aided Geometric Design 1 (1984) 269-278.

[Gol 85] Goldman, R. N.: The method of resolvents. A technique for the implicitization, inversion and intersection of non-planar, parametric, rational cubic curves. Computer Aided Geometric Design 2 (1985) 237-256.

[Gol 86] Goldman, R. N.; DeRose, T. D.: Recursive subdivision without the convex hull property. Computer Aided Geometric Design 3 (1986) 247-265.

[Gol 87] Goldman, R. N.; Filip, D. J.: Conversion from Bézier rectangles to Bézier triangles. Computer-Aided Design 19 (1987) 25-27.

[Gol 87a] Goldman, R. N.; Sederberg, Th. W.: Analytic approach to intersection of all piecewise parametric rational cubic curves. Computer-Aided Design 19 (1987) 282-292.

[Gol 88] Goldman, R. N.: Urn Models and B-splines. Constructive Approximation 4 (1988) 265-288.

[Gol 88a] Goldman, R. N.: Urn Models, Approximations, and Splines. Journal of Approximation Theory 54 (1988) 1-66.

[Gol 89] Goldman, R. N.; Barsky, B. A.: On Beta-continuous Functions and their Application to the Construction of Geometrically Continuous Curves and Surfaces. in Lyche, T.; Schumaker, L. L. (ed.): Mathematical Methods in Computer Aided Geometric Design. Academic Press (1989) 299-311.

[Gol 89a] Goldman, R. N.; Micchelli, C. A.: Algebraic Aspects of Geometric Continuity. in Lyche, T.; Schumaker, L. L. (ed.): Mathematical Methods in Computer Aided Geometric Design. Academic Press (1989) 313-332.

[Gol 91] Goldman, R. N.; Miller, J. R.: Combining Algebraic Rigor with Geometric Robustness for the Detection and Calculation of Conic Sections in the Intersection of Two Natural Quadric Surfaces. ACM (1991) 221-231.

[Gon 83] Gonska, H.; Meier, J.: A bibliography on approximation of functions by Bernstein type operators. in Chui, C. K.; Schumaker, L. L.; Ward, J. (ed.): Approximation Theory IV. Academic Press (1983) 739-785.

[Goo 85] Goodman, T. N. T.; Unsworth, K.: Generation of β-spline curves using a recurrence relation. in Earnshaw, R. A. (ed.): Fundamental Algorithms for Computer Graphics. Springer (1985) 325-357.

[Goo 85a] Goodman, T. N. T.: Properties of beta splines. Journal of Approximation Theory 44 (1985) 132-153.

[Goo 86] Goodman, T. N. T.; Unsworth, K.: Manipulating Shape and Producing Geometric Continuity in β-spline Curves. IEEE Computer Graphics & Applications 6 (1986) 50-56.

[Goo 87] Goodman, T. N. T.; Lee, S. L.: Geometrically continuous surfaces defined parametrically from piecewise polynomials. in Martin, R. (ed.): The Mathematics of Surfaces II. Oxford University Press (1987) 343-361.

[Goo 88] Goodman, T. N. T.; Unsworth, K.: Shape preserving interpolation by curvature continuous parametric curves. Computer Aided Geometric Design 5 (1988) 323-340.

[Goo 89] Goodman, T. N. T.: Shape preserving representations. in Lyche, T.; Schumaker, L. L. (ed.): Mathematical Methods in Computer Aided Geometric Design. Academic Press (1989) 333-351.

[Goo 90] Goodman, T. N. T.: Constructing piecewise rational curves with Frenet frame continuity. Computer Aided Geometric Design 7 (1990) 15-31.

[Goo 91] Goodman, T. N. T.: Closed biquadratic surfaces. Constructive Approximation 7 (1991) 149-160.

[Goo 91a] Goodman, T. N. T.: Convexity of Bézier nets on triangulations. Computer Aided Geometric Design 8 (1991) 175-180.

[Gop 91] Gopalsamy, S.; Khandekar, D.; Mudur, S. P.: A new method of evaluating compact geometric bounds for use in subdivision algorithms. Computer Aided Geometric Design 8 (1991) 337-356.

[Gor 69] Gordon, W. J.: Distributive lattices and the approximation of multivariate functions. in Schoenberg, I. J. (ed.): Approximation with Special Emphasis on Spline Functions. Academic Press (1969) 223-277.

[Gor 71] Gordon, W. J.: Blending function methods of bivariate and multivariate interpolation and approximation. SIAM Journal on Numerical Analysis 8 (1971) 158-177.

[Gor 74] Gordon, W. J.; Riesenfeld, R. F.: Bernstein-Bézier Methods for Computer Aided Design of Free-Form Curves and Surfaces. Journal of the ACM 21 (1974) 293-310.

[Gor 74a] Gordon, W. J.; Riesenfeld, R. F.: B-Spline Curves and Surfaces. in Barnhill R. E.; Riesenfeld, R. F. (ed.): Computer Aided Geometric Design, Academic Press (1974) 95-126.

[Gor 78] Gordon, W. J.; Wixom, J. A.: Shepard's method of "metric interpolation" to bivariate and multivariate interpolation. Mathematics of Computation 32 (1978) 253-264.

[Gra 91] Grabowski, H.; Li, X.: General Matrix Representation for NURBS Curves and Surfaces for Interface. in Hoschek, J. (ed.): Freeform Tools in CAD-Systems – A Comparison. Teubner (1991) 219-232.

[Gree 78] Green, P. J.; Sibson, R.: Computing Dirichlet tesselations in the plane. Computer Journal 21 (1978) 168-173.

[Gre 74] Gregory, J. A.: Smooth interpolation without twist constraints. in Barnhill, R. E.; Riesenfeld, R. F. (ed.): Computer Aided Geometric Design, Academic Press (1974) 71-87.

[Gre 75] Gregory, J. A.: Error bounds for linear interpolation on triangles. in Whiteman, J. (ed.): The Mathematics of Finite Elements and Application II. Academic Press (1975) 163-170.

[Gre 78] Gregory, J. A.: A blending function interpolant for triangles. in Handscomb, D. G. (ed.): Multivariate Approximation, Academic Press (1978) 279-287.

[Gre 80] Gregory, J. A.; Charrot, P.: A C^1 triangular interpolation patch for computer-aided geometric design. Computer Graphics and Image Processing 13 (1980) 80-87.

[Gre 83] Gregory, J. A.: C^1 rectangular and non-rectangular surface patches. in Barnhill, R. E.; Boehm, W. (ed.): Surfaces in Computer Aided Geometric Design, North-Holland (1983) 25-33.

[Gre 85] Gregory, J. A.: Interpolation to boundary data on the simplex. in Barnhill, R. E.; Boehm, W. (ed.): Surfaces in CAGD '84. North-Holland (1985) 43-52.

[Gre 86] Gregory, J. A.: Shape preserving spline interpolation. Computer-Aided Design 18 (1986) 53-57.

[Gre 86a] Gregory, J. A.: N-sided surface patches. in Gregory, J. A. (ed.): The mathematics of surfaces. Clarendon Press (1986) 217-232.

[Gre 87] Gregory, J. A.; Hahn, J.: Geometric Continuity and Convex Combination Patches. Computer Aided Geometric Design 4 (1987) 79-89.

[Gre 89] Gregory, J. A.: Geometric Continuity. in Lyche, T.; Schumaker, L. L. (ed.): Mathematical Methods in Computer Aided Geometric Design. Academic Press (1989) 353-371.

[Gre 89a] Gregory, J. A.; Hahn, J.: A C^2 polygonal surface patch. Computer Aided Geometric Design 6 (1989) 69-75.

[Gre 91] Gregory, J. A.; Zhou, J.: The weighted ν-spline as a double knot B-spline. Brunel University, Department of Mathematics and Statistics, Technical Report No. TR/03/91, 1991.

[Gre 91a] Gregory, J. A.; Zhou, J.: Convexity of Bézier nets on subtriangles. Computer Aided Geometric Design 8 (1991) 207-211.

[Grie 85] Grieger, I.: Geometry cells and surface definition by finite elements. Computer Aided Geometric Design 2 (1985) 213-222.

[Gries 89] Griessmair, J.; Purgathofer, W.: Deformation of Solids with Trivariate B-Splines. in Hansmann, W.; Hopgood, F. R. A.; Straßer, W. (ed.): EUROGRAPHICS '89. North-Holland (1989) 137-148.

[Grif 75] Griffiths, J. G.: A data structure for the elimination of hidden surfaces by patch subdivision. Computer-Aided Design 7 (1975) 171-178.

[Grif 78] Griffiths, J. G.: Bibliography of hidden-line and hidden-surface algorithms. Computer-Aided Design 10 (1978) 203-206.

[Grif 79] Griffiths, J. G.: Eliminating hidden edges in line drawings. Computer-Aided Design 11 (1979) 71-78.

[Grif 81] Griffiths, J. G.: Tape-oriented hidden-line algorithm. Computer-Aided Design 13 (1981) 19-26.

[Gui 85] Guibas, L.; Stolfi, J.: Primitives for the Manipulation of General Subdivisions and the Computation of Voronoi Diagrams. ACM Transactions on Graphics 4 (1985) 74-123.

[Guj 88] Gujar, U. G.; Bhavsar, V. C.; Datar, N. N.: Interpolation Techniques for 3-D Object Generation. Computer & Graphics 12 (1988) 541-555.

[Hag 85] Hagen, H.: Geometric spline curves. Computer Aided Geometric Design 2 (1985) 223-227. Detailed Version: Technical Report TR-85-011, Department of Computer Science, Arizona State University, Tempe 1985.

[Hag 86] Hagen, H.: Bézier-curves with curvature and torsion continuity. Rocky Mountain Journal of Mathematics 16 (1986) 629-638.

[Hag 86a] Hagen, H.: Geometric surface patches without twist constraints. Computer Aided Geometric Design 3 (1986) 179-184.

[Hag 87] Hagen, H.; Schulze, G.: Automatic smoothing with geometric surface patches. Computer Aided Geometric Design 4 (1987) 231-236.

[Hag 88] Hagen, H.: Generalized Coons' Patches. Computers in Industry 10 (1988) 267-269.

[Hag 89] Hagen, H.; Pottmann, H.: Curvature Continuous Triangular Interpolants. in Lyche, T.; Schumaker, L. L. (ed.): Mathematical Methods in Computer Aided Geometric Design. Academic Press (1989) 373-384.

[Hag 90] Hagen, H.; Schulze, G.: Extremalprinzipien im Kurven- und Flächen-design. in Encarnacao, J. L.; Hoschek, J.; Rix, J. (ed.): Geometrische Verfahren der Graphischen Datenverarbeitung. Springer (1990) 46-60.

[Hag 91] Hagen, H.; Bonneau, G. P.: Variational design of smooth rational Bézier curves. Computer Aided Geometric Design 8 (1991) 393-399.

[Hag 92] Hagen, H.; Santarelli, P.: Variational Design of Smooth B-Spline Surfaces. in Hagen, H. (ed.): Topics in Surface Modeling. SIAM (1992) 85-94.

[Hah 89] Hahn, J. M.: Filling polygonal holes with rectangular patches. in Straßer, W.; Seidel, H. P. (ed.): Theory and Practice of Geometric Modeling. Springer (1989) 81-91.

[Hah 89a] Hahn, J. M.: Geometric continuous patch complexes. Computer Aided Geometric Design 6 (1989) 55-67.

[Ham 91] Hamann, B.: Visualization and Modeling Contours of Trivariate Functions. Thesis, Arizona State University 1991.

[Ham 91a] Hamann, B.; Farin, G.; Nielson, G. M.: A Parametric Triangular Patch Based on Generalized Conics. in Farin, G. (ed.): NURBS for Curve and Surface Design. SIAM (1991) 75-85.

[Ham 92] Hamann, B.: Modeling Contours of Trivariate Data. Mathematical Modelling and Numerical Analysis 26 (1992) 51-76.

[Han 87] Handscomb, D.: Knot-Elimination: Reversal of the Oslo Algorithm. in Collatz, L.; Meinardus, G.; Nürnberger, G. (ed.): Numerische Methoden der Approximationstheorie, Band 8. Birkhäuser Verlag (1987) 103-111.

[Hann 83] Hanna, S. L.; Abel, J. F.; Greenberg, D. P.: Intersection of parametric surfaces by means of look-up tables. Computer Graphics & Applications 3 (1983) 39-48.

[Hans 90] Hansford, D.: The neutral case for the min-max triangulation. Computer Aided Geometric Design 7 (1990) 431-438.

[Hans 91] Hansford, D.: Boundary Curves with Quadric Precision for a Tangent Continuous Scattered Data Interpolant. Thesis, Arizona State University 1991.

[Hansm 87] Hansmann, W.: Interactive Design and Geometric Description of Smooth Transitions Between Curved Surfaces. in Computers in Offshore and Arctic Engineering, 6th International Symposium on Offshore Mechanics and Engineering, Houston (1987) 19-26.

[Hansm 88] Hansmann, W.: Benutzerfreundliche interaktive Erfassung von Kurven auf Flächen höherer Ordnung. in GI - 18. Jahrestagung, Springer (1988) 743-751.

[Har 82] Harada, K.; Nakamae, E.: Application of the Bézier curve to data interpolation. Computer-Aided Design 14 (1982) 55-59.

[Har 84] Harada, K.; Kaneda, K.; Nakamae, E.: A further investigation of segmented Bézier interpolants. Computer-Aided Design 16 (1984) 186-190.

[Har 90] Harada, T.; Toriya, H.; Chiyokura, H.: An Enhanced Rounding Operation Between Curved Surfaces in Solid Modeling. in Chua, T. S.; Kunii, T. L. (ed.): CG International '90. Springer (1990) 563-588.

[Hard 72] Harder, R. L.; Desmarais, R. N.: Interpolation using surface splines. Journal of Aircraft 9 (1972) 189-197.

[Hardy 71] Hardy, R. L.: Multiquadric equations of topography and other irregular surfaces. Journal Geophysical Research 76 (1971) 1905-1915.

[Hardy 75] Hardy, R. L.; Göpfert, W. M.: Least squares prediction of gravity anomalies, geoidal undulations, and deflections of the vertical with multiquadric harmonic functions. Geophysical Research Letters 2 (1975) 423-426.

[Hardy 90] Hardy, R. L.: Theory and Applications of the multiquadric-biharmonic method. Computers and Mathematics with Applications 19 (1990) 163-208.

[Hartl 80] Hartley, P. J.; Judd, C. J.: Parametrization and shape of B-spline curves for CAD. Computer-Aided Design 12 (1980) 235-238.

[Hartm 90] Hartmann, E.: Blending of implicit surfaces with functional splines. Computer-Aided Design 22 (1990) 500-506.

[Hartm 92] Hartmann, E.: On the convexity of functional splines. Computer Aided Geometric Design 9 (1992).

[Hartw 83] Hartwig, R.; Nowacki, H.: Isolinien und Schnitte in Coonsschen Flächen. in Nowacki, H.; Gnatz, R. (ed.): Geometrisches Modellieren. Informatik-Fachberichte 65. Springer (1983) 239-344.

[Hau 77] Haubitz, I.: Programm zum Zeichnen von allgemeinen Flächenstücken. Computing 18 (1977) 295-315.

[Hauc 88] Hauck, R.: Glätten von Bézierflächen über Variation von Flächen-punkten. Diplomarbeit Darmstadt 1988.

[Hay 74] Hayes, J. G.; Halliday, J.: The least-squares fitting of cubic spline surfaces to general data sets. Journal Institute Mathematics and Applications 14 (1974) 89-103.

[Hei 86] Heidemann, U.: Linearer Ausgleich mit Exponentialsplines bei automatischer Bestimmung der Intervallteilungspunkte. Computing 36 (1986) 217-227.

[Hein 79] Heindl, G.: Interpolation and approximation by piecewise quadratic C^1 functions of two variables. in Schempp, W.; Zeller, K. (ed.): Multivariate Approximation Theory. Birkhäuser (1979) 146-161.

[Hein 85] Heindl, G.: Construction and applications of Hermite interpolating quadratic spline functions of two and three variables. in Schempp, W.; Zeller, K. (ed.): Multivariate Approximation Theory III. Birkhäuser (1985) 232-252.

[Hel 91] Held, M.: On the Computational Geometry of Pocket Machining. Lecture Notes in Computer Science 500, Springer 1991.

[Hen 76] Henning, H.: Fünfachsiges NC-Fräsen gekrümmter Flächen. Beitrag zur numerischen Flächendarstellung, Programmierung und Fertigung. Springer 1976.

[Hen 76a] Henning, H.; Sanzenbacher, M.: Neuere Erkenntnisse beim fünfachsigen Fräsen. wt. -Z. ind. Fert. 66 (1976) 259-264.

[Her 83] Hering, L.: Closed (C^2 and C^3) continuous Bézier and B-spline curves with given tangent polygons. Computer-Aided Design 15 (1983) 3-6.

[Herm 89] Hermann, T.; Renner, G.: Subdivision of N-sided regions into four-sided patches. in Martin, R. (ed.): The Mathematics of Surfaces III. Clarendon Press Oxford (1989) 347-357.

[Herm 92] Hermann, T.; Renner, G.; Varady, T.: Surface interpolation based on a curve network of general topology. Computer-Aided Design 24 (1992).

[Herr 85] Herron, G.: Smooth closed surfaces with discrete triangular interpolants. Computer Aided Geometric Design 2 (1985) 297-306.

[Herr 87] Herron, G.: Techniques for Visual Continuity. in Farin, G. (ed.): Geometric Modeling: Algorithms and New Trends. SIAM (1987) 163-174.

[Herr 89] Herron, G.: Polynomial bases for quadratic and cubic polynomials which yield control points with small convex hulls. Computer Aided Geometric Design 6 (1989) 1-9.

[Herz 87] Herzen von, B.; Barr, A. H.: Accurate Triangulations of Deformed, Intersecting Surfaces. ACM Computer Graphics 21 (1987) 103-110.

[Herz 90] Herzen von, B.; Barr, A. H.; Zatz, H. R.: Geometric Collisions for Time-Dependent Parametric Surfaces. ACM Computer Graphics 24 (1990) 39-48.

[Heß 86] Heß, W.; Schmidt, J. W.: Convexity Preserving Interpolation with Exponential Splines. Computing 36 (1986) 335-342.

[Hob 91] Hobby, J. D.: Numerically Stable Implicitization of Cubic Curves. ACM Transactions on Graphics 10 (1991) 255-296.

[Höll 86] Höllig, K.: Geometric Continuity of Spline Curves and Surfaces. University of Wisconsin, Madison, Computer Science Technical Report 645, also SIGGRAPH 86, Course 5: Extension of B-spline Curve Algorithms to Surfaces (1986).

[Höll 90] Höllig, K.; Mögerle, H.: *G*-splines. Computer Aided Geometric Design 7 (1990) 197-207.

[Hölz 83] Hölzle, G. E.: Knot placement for piecewise polynomial approximation of curves. Computer-Aided Design 15 (1983) 295-296.

[Hoff 85] Hoffmann, Ch. M.; Hopcroft, J.: Automatic surface generation in computer aided design. The Visual Computer 1 (1985) 92-100.

[Hoff 86] Hoffmann, Ch. M.; Hopcroft, J.: Quadratic blending surfaces. Computer-Aided Design 18 (1986) 301-306.

[Hoff 87] Hoffmann, Ch. M.; Hopscroft, J.: The potential method for blending surfaces and corners. in Farin, G. (ed.): Geometric Modeling, Algorithms and New Trends. SIAM (1987) 347-365.

[Hoff 88] Hoffmann, Ch. M.; Hopcroft, J.: The Geometry of Projective Blending Surfaces. in Kapur, D.; Mundy, L. (ed.): Geometric Reasoning. MIT Press (1988) 357-376, see also Artificial Intelligence 37 (1988) 357-376.

[Hoff 90] Hoffmann, Ch. M.: Algebraic and Numerical Techniques for Offsets and Blends. in Dahmen, W.; Gasca, M.; Micchelli, C. A. (ed.): Computation of Curves and Surfaces. Kluwer Academic Publishers (1990) 499-528.

[Hoff 90a] Hoffmann, Ch. M.: A dimensionality paradigm for surface interrogations. Computer Aided Geometric Design 7 (1990) 517-532.

[Hoh 89] Hohmeyer, M. E.; Barsky, B. A.: Rational Continuity: Parametric, Geometric, and Frenet Frame Continuity of Rational Curves. ACM Transactions on Graphics 8 (1989) 335-359.

[Hoh 91] Hohmeyer, M. E.: A Surface Intersection Algorithm Based on Loop Detection. in Rosignac, J.; Turner, J. (ed.): Proceedings Symposium on Solid Modeling Foundations and CAD/CAM Applications. Austin, Texas, June 1991. ACM Press (1991) 197-207.

[Hol 87] Holmström, L.: Piecewise quadratic blending of implicitly defined surfaces. Computer Aided Geometric Design 4 (1987) 171-189.

[Hol 88] Holmström, L.; Laakko, T.: Rounding facility for solid modelling of mechanical parts. Computer-Aided Design 20 (1988) 605-614.

[Hol 89] Holmström, L.; Laakko, T., Mäntylä, M., Rekola, R.: Ray Tracing of Boundary Models with Implicit Blend Surfaces. in Straßer, W.; Seidel, H. P. (ed.): Theory and Practice of Geometric Modeling. Springer (1989) 253-271.

[Horn 85] Hornung C.; Lellek, W.; Rehwald, P.; Straßer, W.: An area-oriented analytical visibility method for displaying parametrically defined tensor-product surfaces. Computer Aided Geometric Design 2 (1985) 197-206.

[Hosa 69] Hosaka, M.: Theory of curves and surface synthesis and their smooth fitting. Information Processing in Japan 9 (1969) 60-68.

[Hosa 78] Hosaka, M.; Kimura, F.: Synthesis methods of curves and surfaces in interactive CAD. Proceedings: Interactive Techniques in Computer- Aided Design, Bologna (1978) 151-156.

[Hosa 80] Hosaka, M.; Kimura, F.: A theory and methods for three dimensional free-form shape construction. Journal of Information Processing 3 (1980) 140-156.

[Hosa 84] Hosaka, M.; Kimura, F.: Non-four-sided patch expressions with control points. Computer Aided Geometric Design 1 (1984) 75-86.

[Hos 77] Hoschek, J.: Zur Berechnung der Hauptkrümmungen und Hauptkrümmungsrichtungen bei empirisch vorgegebenen Flächenstücken. Abhandlungen der Braunschweigischen Wissenschaftlichen Gesellschaft 28 (1977) 107-117.

[Hos 83] Hoschek, J.: Dual Bézier-Curves and Surfaces. in Barnhill, R. E.; Boehm, W. (ed.): Surfaces in Computer Aided Geometric Design, North-Holland (1983) 147-156.

[Hos 84] Hoschek, J.: Detecting regions with undesirable curvature. Computer Aided Geometric Design 1 (1984) 183-192.

[Hos 85] Hoschek, J.: Offset curves in the plane. Computer-Aided Design 17 (1985) 77-82.

[Hos 85a] Hoschek, J.: Smoothing of curves and surfaces. in Barnhill, R. E.; Boehm, W. (ed.): Surfaces in CAGD '84. North-Holland (1985) 97-105.

[Hos 87] Hoschek, J.: Approximate conversion of spline curves. Computer Aided Geometric Design 4 (1987) 59-66.

[Hos 87a] Hoschek, J.: Algebraische Methoden zum Glätten und Schneiden von Splineflächen. Results in Mathematics 12 (1987) 119-133.

[Hos 88] Hoschek, J.: Intrinsic parametrization for approximation. Computer Aided Geometric Design 5 (1988) 27-31.

[Hos 88a] Hoschek, J.: Spline approximation of offset curves. Computer Aided Geometric Design 5 (1988) 33-40.

[Hos 88b] Hoschek, J.; Wissel, N.: Optimal approximate conversion of spline curves and spline approximation of offset curves. Computer-Aided Design 20 (1988) 475-483.

[Hos 88c] Hoschek, J.; Wassum, P.; Schneider, F.-J.: Optimal approximate conversion of spline curves and spline surfaces, spline approximation of offset curves. Proceedings of the Third International Conference on Engineering Graphics and Descriptive Geometry 1. Vienna (1988) 246-249.

[Hos 89] Hoschek, J.; Schneider, F.-J.; Wassum, P.: Optimal Approximate Conversion of Spline Surfaces. Computer Aided Geometric Design 6 (1989) 293-306.

[Hos 90] Hoschek, J.; Schneider, F.-J.: Spline conversion for trimmed rational Bézier- and B-spline surfaces. Computer-Aided Design 22 (1990) 580-590.

[Hos 90a] Hoschek, J.; Schneider, F.-J.: Spline approximation of offset curves and offset surfaces. in Manley, J. et al. (ed.): Proceedings of the Third European Conference on Mathematics in Industry. (1990) 383-389.

[Hos 91] Hoschek, J.; Schneider, F.-J.: Approximate Conversion and Merging of Spline Surface Patches. in Hoschek, J. (ed.): Freeform Tools in CAD Systems – A Comparison. Teubner (1991) 233-245.

[Hos 91a] Hoschek, J.; Hartmann, E.: G^{n-1} functional Splines for Modeling. in Hagen, H.; Roller, D. (ed.): Geometric Modeling. Methods and Applications. Springer (1991) 185-211.

[Hos 92] Hoschek, J.: Spline curves and surface patches on quadrics. in Lyche, T.; Schumaker, L. L. (ed.): Mathematical Methods in CAGD II. Academic Press (1992) 331-342.

[Hos 92a] Hoschek, J.; Schneider, F.-J.: Approximate Spline Conversion for Integral and Rational Bézier- and B-Spline Surfaces. in Barnhill, R. E. (ed.): Geometry Processing for Design and Manufacturing. SIAM (1992) 45-85.

[Hos 92b] Hoschek, J.; Seemann, G.: Spherical Splines. Mathematical Modelling and Numerical Analysis 26 (1992) 1-22.

[Hos 92c] Hoschek, J.: Circular splines. Computer-Aided Design 24 (1992) 611-618.

[Hou 85] Houghton, E. G.; Emnett, R. F.; Factor, J. D.; Sabharwal, C. L.: Implementation of a divide-and-conquer method for intersection of parametric surfaces. Computer Aided Geometric Design 2 (1985) 173-183.

[Hu 86] Hu, Ch. L.; Schumaker, L. L.: Complete Spline Smoothing. Numerische Mathematik 49 (1986) 1-10.

[Hu 91] Hu, J.; Pavlidis, Th.: Function Plotting Using Conic Splines. IEEE Computer Graphics & Applications 11 (1991) 89-94.

[Hua 89] Huang, Y.: Triangular Irregular Network Generation and Topographical Modeling. Computers in Industry 12 (1989) 203-213.

[Hus 89] Hussain, F.: Conic Rescue of Bézier Founts. Proceedings Computer Graphics International. Springer (1989) 97-119.

[Jac 88] Jackson, I. R. H.: Convergence properties of radial basis functions. Constructive Approximation 4 (1988) 243-264.

[Jen 87] Jensen, Th.: Assembling Triangular and Rectangular Patches and Multivariate Splines. in Farin, G. (ed.): Geometric Modeling, Algorithms and New Trends, SIAM (1987) 203-220.

[Jer 89] Jerard, R. B.; Hussaini, S. Z.; Drysdale, R. L.; Schaudt, B.: Approximate methods for simulation and verification of numerically controlled machining programs. The Visual Computer 5 (1989) 329-348.

[Jie 90] Jie, T.: A Geometric Condition for Smoothness between Adjacent Rational Bézier Surfaces. Computers in Industry (1990) 355-360.

[Joe 86] Joe, B.: Delaunay triangular meshes in convex polygons. SIAM Journal on Scientific and Statistical Computing 7 (1986) 514-539.

[Joe 87] Joe, B.: Discrete Beta-Splines. SIGGRAPH '87, ACM Computer Graphics 21 (1987) 137-144.

[Joe 89] Joe, B.: Multiple knot and rational cubic β-splines. ACM Transactions on Graphics 8 (1989) 100-120.

[Joe 90] Joe, B.: Knot insertion for Beta-spline curves and surfaces. ACM Transactions on Graphics 9 (1990) 41-65.

[Joe 90a] Joe, B.: Quartic Beta-Splines. ACM Transactions on Graphics 9 (1990) 301-337.

[Joe 91] Joe, B.: Construction of three-dimensional Delaunay triangulations using local transformations. Computer Aided Geometric Design 8 (1991) 123-142.

[Joh 90] Johnstone, J. K.; Bajaj, Ch. L.: Sorting points along an algebraic curve. SIAM Journal on Computing 19 (1990) 925-967.

[Joh 91] Johnstone, J. K.; Goodrich, M. T.: A localized method for intersecting plane algebraic curve segments. The Visual Computer 7 (1991) 60-71.

[Jon 88] Jones, A. K.: Nonrectangular surface patches with curvature continuity. Computer-Aided Design 20 (1988) 325-335.

[Jor 84] Jordan, M. C.; Schindler, F.: Curves under Tension. Computer Aided Geometric Design 1 (1984) 291-300.

[Jou 92] Jou, E. D.; Han, W.: Minimal-Energy Splines with Various End Constraints. in Hagen, H. (ed.): Curve and Surface Design. SIAM (1992) 23-40.

[Joy 86] Joy, K.; Bhetanabhotla, M.: Ray Tracing Parametric Patches Utilizing Numerical Techniques and Ray Coherence. ACM Computer Graphics 20 (1986) 279-284.

[Kah 82] Kahmann, J.: Krümmungsübergänge zusammengesetzter Kurven und Flächen. Diss. Braunschweig 1982.

[Kah 83] Kahmann, J.: Continuity of Curvature between Adjacent Bézier Patches. in Barnhill, R. E.; Boehm, W. (ed.): Surfaces in Computer Aided Geometric Design. North-Holland (1983) 65-75.

[Kaj 82] Kajiya, J. T.: Ray tracing parametric patches. Computer Graphics 16 (1982) 245-254.

[Kal 87] Kallay, M.: Approximating a composite cubic curve by one with fewer pieces. Computer-Aided Design 19 (1987) 539-543.

[Kal 90] Kallay, M.; Ravani, B.: Optimal twist vectors as a tool for interpolating a network of curves with a minimum energy surface. Computer Aided Geometric Design 7 (1990) 465-473.

[Kalr 89] Kalra, D.; Barr, A. H.: Guaranteed Ray Intersections with Implicit Surfaces. ACM Computer Graphics 23 (1989) 297-306.

[Kan 86] Kansa, E. J.: Application of Hardy's multiquadric interpolation to hydrodynamics. Computer Simulation 4 (1986) 111-116.

[Kan 90] Kansa, E. J.: Multiquadrics – A Scattered Data Approximation Scheme with Applications to Computational Fluid-Dynamics. Computers and Mathematics with Applications 19 (1990) Part I: 127-145, Part II: 147-161.

[Kao 90] Kao, Th. Ch.; Knott, G. D.: An Efficient and Numerically Correct Algorithm for the 2D Convex Hull Problem. BIT 30 (1990) 311-331.

[Kat 88] Katz, Sh.: Genus of the intersection curve of two rational surface patches. Computer Aided Geometric Design 5 (1988) 253-258.

[Kau 88] Kaufmann, E.; Klass, R.: Smoothing surfaces using reflection lines for families of splines. Computer-Aided Design 20 (1988) 312-316.

[Kauf 89] Kaufmann, A.: Parallelization of the Subdivision Algorithm for Intersection of Bézier Curves on the FPS T20. in Lyche, T.; Schumaker, L. L. (ed.): Mathematical Methods in Computer Aided Geometric Design. Academic Press (1989) 403-411.

[Kay 86] Kay, T. L.; Kajiya, J. T.: Ray Tracing Complex Scenes. Computer Graphics 20 (1986) 269-278.

[Kep 75] Keppel, E.: Approximating Complex Surfaces by Triangulation of Contour Lines. IBM Journal of Research and Development 19 (1975) 2-11.

[Kim 84] Kimura, F.: Geomap III: Designing solids with free-form surfaces. IEEE Computer Graphics & Applications 4 (1984) 58-72.

[Kje 83] Kjellander, J. A. P.: Smoothing of cubic parametric splines. Computer-Aided Design 15 (1983) 175-178.

[Kje 83a] Kjellander, J. A. P.: Smoothing of bicubic parametric surfaces. Computer-Aided Design 15 (1983) 288-293.

[Kla 80] Klass, R.: Correction of local surface irregularities using reflection lines. Computer-Aided Design 12 (1980) 73-77.

[Kla 83] Klass, R.: An offset spline approximation for plane cubic splines. Computer-Aided Design 15 (1983) 297-299.

[Kla 91] Klass, R.; Schramm, P.: Numerically-controlled milling of CAD surface data. in Hagen, H.; Roller, D. (ed.): Geometric Modeling. Methods and Applications. Springer (1991) 213-226.

[Kla 92] Klass, R.; Kuhn, B.: Fillet and Surface Intersections Defined by Rolling Balls. Computer Aided Geometric Design 9 (1992) 185-193.

[Klo 86] Klok, F.: Two moving coordinate frames for sweeping along a 3D trajectory. Computer Aided Geometric Design 3 (1986) 217-229.

[Klu 78] Klucewicz, I. M.: A piecewise C^1 interpolant to arbitrarily spaced data. Computer Graphics and Image Processing 8 (1978) 92-112.

[Kob 86] Kobori, K.; Iwazu, M.; Jones, K. M.; Nishioka, I.: Polygonal Subdivision of Parametric Surfaces. in Kunii, T. L. (ed.): Advanced Computer Graphics. Springer (1986) 50-59.

[Kop 83] Koparkar, P. A.; Mudur, S. P.: A new class of algorithms for the processing of parametric curves. Computer-Aided Design 15 (1983) 41-45.

[Kop 86] Koparkar, P. A.; Mudur, S. P.: Generation of continuous smooth curves resulting from operations on parametric surface patches. Computer-Aided Design 18 (1986) 193-206.

[Kop 91] Koparkar, P. A.: Designing parametric blends: surface model and geometric correspondence. The Visual Computer 7 (1991) 39-58.

[Kop 91a] Koparkar, P. A.: Surface intersection by switching from recursive subdivision to iterative refinement. The Visual Computer 8 (1991) 47-63.

[Kop 91b] Koparkar, P. A.: Parametric Blending Using Fanout Surfaces. in Rosignac, J.; Turner, J. (ed.): Proceedings Symposium on Solid Modeling Foundations and CAD/CAM Applications. Austin, Texas, 5–7. June 1991. ACM Press (1991) 317-327.

[Kos 89] Kosters, M.: Quadratic blending surfaces for complex corners. The Visual Computer 5 (1989) 134-146.

[Kos 91] Kosters, M.: An Extension of the Potential Method to Higher Order Blendings. in Rosignac, J.; Turner, J. (ed.): Proceedings Symposium on Solid Modeling Foundations and CAD/CAM Applications. Austin, Texas, 5–7. June 1991. ACM Press (1991) 329-337.

[Koz 86] Kozak, J.: Shape Preserving Approximation. Computers in Industry 7 (1986) 435-440.

[Kre 72] Kreiling, W.: Ein neuer Stereobetrachter. Bildmessung und Luftbildwesen 40 (1972) 206.

[Kri 90] Kriezis, G. A.; Prakash, P. V.; Patrikalakis, N. M.: A method for intersecting algebraic surfaces with rational polynomial patches. Computer-Aided Design 22 (1990) 645-654.

[Kri 92] Kriezis, G. A.; Patrikalakis, N. M.; Wolter, F. E.: Topological and differential-equation methods for surface intersections. Computer-Aided Design 24 (1992) 41-55.

[Krip 85] Kripac, J.: Classification of edges and its application in determining visibility. Computer-Aided Design 17 (1985) 30-36.

[Krü 90] Krüger, W.: The Application of Transport Theory to Visualization of 3D Scalar Data Fields. in Kaufman, A. (ed.): Visualization '90. IEEE Computer Society Press (1990) 273-279.

[Kul 91] Kulkarni, R.; Laurent, P. J.: Q-Splines. University Joseph Fourier, LMC-IMAG Grenoble 1991.

[Lac 88] Lachance, M. A.: Chebyshev economization for parametric surfaces. Computer Aided Geometric Design 5 (1988) 195-208.

[Lan 79] Lancaster, P.: Moving weighted least-squares methods. in Sahney, B. N. (ed.): Polynomial and Spline Approximation. Reidel Publishing Company (1979) 103-120.

[Lane 77] Lane, J. M.: Shape Operators for Computer Aided Geometric Design. Diss. University of Utah 1977.

[Lane 79] Lane, J. M.; Carpenter, L.: A generalized scan line algorithm for the computer display of parametrically defined surfaces. Computer Vision, Graphics, and Image Processing 11 (1979) 290-297.

[Lane 80] Lane, J. M.; Riesenfeld, R. F.: A theoretical development for the computer generation of piecewise polynomial surfaces. IEEE Transaction on Pattern Analysis and Machine Intelligence PAMI 2 (1980) 35-45.

[Lane 80a] Lane, J. M.; Carpenter, L. C.; Whitted, J. T.; Blinn, J. F.: Scan Line Methods for Displaying Parametrically Defined Surfaces. Communications of the ACM, Vol. 23 (1980) 23-34.

[Lane 83] Lane, J. M.; Riesenfeld, R. F.: A geometric proof for the variation diminishing property of B-spline approximation. Journal of Approximation Theory 37 (1983) 1-4.

[Lang 84] Lang, J.: Zur Konstruktion von Isophoten im Computer-aided Design. CAD-Computergraphik und Konstruktion, Heft 34 (1984) 1-7.

[Langr 84] Langridge, D. J.: Detection of Discontinuities in the first derivatives of surfaces. Computer Vision Graphics and Image Processing 27 (1984) 291-308.

[Las 84] Lasser, D.: Ein neuer Aspekt des de Casteljau Algorithmus. Preprint 853, Fachbereich Mathematik, Technische Hochschule Darmstadt 1984.

[Las 85] Lasser, D.: Das Interpolations- und Spline-Interpolationsproblem in drei Variablen. Preprint 948, Fachbereich Mathematik, Technische Hochschule Darmstadt 1985.

[Las 85a] Lasser, D.: Bernstein-Bézier Representation of Volumes. Computer Aided Geometric Design 2 (1985) 145-150.

[Las 86] Lasser, D.: Intersection of parametric surfaces in the Bernstein-Bézier representation. Computer-Aided Design 18 (1986) 186-192.

[Las 87] Lasser, D.: Bernstein-Bézier-Darstellung trivariater Splines. Diss. Darmstadt 1987.

[Las 88] Lasser, D.: Self-Intersections of Parametric Surfaces. Proceedings of the Third International Conference on Engineering Graphics and Descriptive Geometry: Volume 1. Vienna (1988) 322-331.

[Las 88a] Lasser, D.; Eck, M.: Bézier Representation of Geometric Spline Curves. A general concept and the quintic case. Technical Report NPS 53-88-004, Naval Postgraduate School, Monterey 1988.

[Las 89] Lasser, D.: Calculating the Self-Intersections of Bézier Curves. Computers in Industry 12 (1989) 259-268.

[Las 90] Lasser, D.: Two Remarks on Tau-Splines. ACM Transactions on Graphics 2 (1990) 198-211.

[Las 90a] Lasser, D.: Visually Continous Quartics and Quintics. Computing 45 (1990) 119-129.

[Las 90b] Lasser, D.: Schnittalgorithmus für Freiform-Volumina. Informationstechnik 2 (1990) 135-141.

[Las 91] Lasser, D.; Purucker, A.: B-Spline-Bézier Representation of Rational Geometric Spline Curves: Quartics and Quintics. in Farin, G. (ed.): NURBS for Curve and Surface Design. SIAM (1991) 115-130.

[Las 91a] Lasser, D.; Kirchgeßner, P.: Rationale Bézier Volumina. in Frühauf, M.; Göbel, M. (ed.): Visualisierung von Volumendaten. Springer (1991) 68-91.

[Las 92] Lasser, D.; Hagen, H.: Interval-Weighted Tau-Splines. in Hagen, H. (ed.): Curve and Surface Design. SIAM (1992) 41–53.

[Las 92a] Lasser, D.: Composition of Tensor Product Bézier Representations. Report 213/91, Fachbereich Informatik, Universität Kaiserslautern 1991, to appear in Computing 47 (1992).

[Law 72] Lawson, C. L.: Generation of a triangular grid with application to contour plotting. Internal Technical Memorandum 299, Jet Propulsion Laboratory, Pasadena 1972.

[Law 77] Lawson, C. L.: Software for C^1 surface interpolation. in Rice, J. R. (ed.): Mathematical Software III, Academic Press (1977) 161-194.

[Law 84] Lawson, C. L.: C^1 surface interpolation for scattered data on a sphere. Rocky Mountain Journal of Mathematics 14 (1984) 177-202.

[Law 86] Lawson, C. L.: Properties of n-dimensional triangulations. Computer Aided Geometric Design 3 (1986) 231-247.

[Lee 80] Lee, D. T.; Schachter, B. J.: Two algorithms for constructing Delaunay triangulations. International Journal of Computer and Information Sciences 9 (1980) 219-242.

[Lee 73] Lee, E. H.; Forsythe, G. E.: Variational Study of nonlinear spline curves. SIAM Review 15 (1973) 120-133.

[Lee 82] Lee, E. T. Y.: A simplified B-spline computation routine. Computing 29 (1982) 365-373.

[Lee 86] Lee, E. T. Y.: Comments on some B-Spline-Algorithms. Computing 36 (1986) 229-238.

[Lee 87] Lee, E. T. Y.: The rational Bézier representation for conics. in Farin, G. (ed.): Geometric Modeling: Algorithms and New Trends. SIAM (1987) 3-19.

[Lee 89] Lee, E. T. Y.: Choosing nodes in parametric curve interpolation. Computer-Aided Design 21 (1989) 363-370.

[Lee 84] Lee, R. B.; Fredericks, D. A.: Intersection of parametric surfaces and a plane. IEEE Computer Graphics & Applications 4 (1984) 48-51.

[Lee 91] Lee, S. L.; Majid, A. A.: Closed Smooth Piecewise Bicubic Surfaces. ACM Transactions on Graphics 10 (1991) 342-365.

[Lee 91a] Lee, Y. S.; Chang, T. C.: CASCAM - An automated system for sculptered surface cavity machining. Computers in Industry 16 (1991) 321-342.

[Lev 76] Levin, J. Z.: A Parametric Algorithm for drawing pictures of Solid Objects composed of Quadric Surfaces. Communications of the ACM 19 (1976) 555-563.

[Lev 79] Levin, J. Z.: Mathematical models for determining the intersection of quadric surfaces. Computer Graphics and Image Processing 11 (1979) 73-87.

[Levo 88] Levoy, M.: Display of Surfaces from Volume Data. IEEE Computer Graphics & Applications 8 (1988) 29-37.

[Levo 90] Levoy, M.: Volume Rendering. IEEE Computer Graphics & Applications 10 (1990) 33-40.

[Levo 90a] Levoy, M.: Efficient Ray Tracing of Volume Data. ACM Transactions on Graphics 9 (1990) 245-261.

[Lewi 78] Lewis, B. A.; Robinson, J. S.: Triangulation of planar regions with applications. The Computer Journal 21 (1978) 324-332.

[Lew 75] Lewis, J.: B-spline bases for splines under tension, ν-splines and fractional order splines. Presented at the SIAM-SIGNUM-meeting, San Fransisco 1975.

[Li 88] Li, L.: Hidden-line algorithm for curved surfaces. Computer-Aided Design 20 (1988) 466-470.

[Li 90] Li, J.; Hoschek, J.; Hartmann, E.: G^1 functional splines for interpolation and approximation of curves, surfaces and solids. Computer Aided Geometric Design 7 (1990) 209-220.

[Lia 88] Liang, Y. Ye X.; Feng, S.: G^1 smoothing solid objects by bicubic Bézier patches. Proceedings of EUROGRAPHICS '88 (1988) 343-355.

[Lin 89] Lin, W. C.; Chen, S. Y.; Chen, C. T.: A New Surface Interpolation Technique for Reconstructing 3D Objects from Serial Cross-Sections. Computer Vision, Graphics, and Image Processing 48 (1989) 124-143.

[Lit 83] Little, F.: Convex Combination Surfaces. in Barnhill, R. E.; Boehm, W. (ed.): Surfaces in Computer Aided Geometric Design, North-Holland (1983) 99-108.

[Liu 86] Liu, D.: A geometric condition for smoothness between adjacent rectangular Bézier patches. Acta Math. Appl. Sinica 9 (1986) 432-442.

[Liu 89] Liu, D.; Hoschek, J.: GC^1 continuity conditions between adjacent rectangular and triangular Bézier surface patches. Computer-Aided Design 21 (1989) 194-200.

[Liu 90] Liu, D.: GC^1 Continuity Conditions between two Adjacent Rational Bézier Surface Patches. Computer Aided Geometric Design 7 (1990) 151-163.

[Lod 90] Lodha, S.; Warren, J.: Bézier representation for quadric surface patches. Computer-Aided Design 22 (1990) 574-579.

[Loh 81] Loh, R.: Convex B-spline surfaces. Computer-Aided Design 13 (1981) 145-149.

[Lone 87] Loney, G. C.; Ozsoy, T. M.: NC machining of free form surfaces. Computer-Aided Design 19 (1987) 85-90.

[Long 87] Long, C.: Special Bézier quartics in three dimensional curve design and interpolation. Computer-Aided Design 19 (1987) 77-84.

[Lon 87] Longhi, L.: Interpolating patches between cubic boundaries. University of California, Berkeley, Technical Report UCB/CSD 87/313, 1987.

[Loo 89] Loop, Ch. T.; DeRose, T. D.: A Multisided Generalization of Bézier Surfaces. ACM Transactions on Graphics 8 (1989) 204-234.

[Loo 90] Loop, Ch. T.; DeRose, T. D.: Generalized B-spline Surfaces of Arbitrary Topology. ACM Computer Graphics 24 (1990) 347-356.

[Lor 87] Lorensen, W. E.; Cline, H. E.: Marching Cubes: A High Resolution 3D Surface Construction Algorithm. ACM Computer Graphics 21 (1987) 163-169.

[Lot 88] Lott, N. J.; Pullin, D. I.: Method for fairing B-Spline surfaces. Computer-Aided Design 22 (1988) 597-604.

[Low 90] Lowe, T. W.; Bloor, M. I. G.; Wilson, M. J.: Functionality in blend design. Computer-Aided Design 22 (1990) 655-665.

[Luk 89] Lukacs, G.: The generalized inverse matrix and the surface-surface intersection problem. in Straßer, W.; Seidel, H. P. (ed.): Theory and Practice of Geometric Modeling. Springer (1989) 167-185.

[Lus 87] Luscher, N.: Die Bernstein-Bézier-Technik in der Methode der finiten Elemente. Diss. Braunschweig 1987.

[Lyc 87] Lyche, T.; Morken, K.: Knot removal for parametric B-spline curves and surfaces. Computer Aided Geometric Design 4 (1987) 217-230.

[Lyc 88] Lyche, T.; Morken, K.: A data-reduction strategy for splines with applications to the approximation of functions and data. IMA Journal of Numerical Analysis 8 (1988) 185-208.

[Lyn 82] Lynch, R. W.: A method for choosing a tension factor for spline under tension interpolation. Thesis, University of Texas, Austin 1982.

[Mad 90] Madych, W. R.; Nelson, S. A.: Error Bounds for Multiquadric Interpolation. in Chui, C. K.; Schumaker, L. L.; Ward, J. D. (ed.): Approximation Theory VI: Volume 2 (1990) 413-416.

[Mal 77] Malcolm, M. A.: On the Computation of Nonlinear Spline Functions. SIAM Journal on Numerical Analysis 14 (1977) 254-282.

[Malo 91] Malosse, J. J.: Search of intersection loops by solving algebraic equations. Intergraph Corporation 1991.

[Mann 92] Mann, St.; Loop, Ch.; Lounsbery, M.; Meyers, D.; Painter, J.; DeRose, T. D.; Sloan, K.: A Survey of Parametric Scattered Data Fitting Using Triangular Interpolants. in Hagen, H. (ed.): Curve and Surface Design. SIAM (1992) 145-172.

[Man 74] Manning, J. R.: Continuity conditions for spline curves. Computer Journal 17 (1974) 181-186.

[Mano 90] Manocha, D.; Barsky, B. A.: Basis Functions for Rational Continuity. in Chua, T. S.; Kunii, T. L. (ed.): CG International '90. Springer (1990) 521-541.

[Mano 91] Manocha, D.; Canny, J. F.: A New Approach for Surface Intersection. in Rosignac, J.; Turner, J. (ed.): Proceedings Symposium on Solid Modeling Foundations and CAD/CAM Applications. Austin, Texas, 5–7. June 1991. ACM Press (1991) 209-219.

[Mano 91a] Manocha, D.; Canny, J. F.: Rational curves with polynomial parameterization. Computer-Aided Design 23 (1991) 645-652.

[Mans 74] Mansfield, L.: Higher Order Compatible Triangular Finite Elements. Numerische Mathematik 22 (1974) 89-97.

[Marc 84] Marciniak, K.; Putz, B.: Approximation of spirals by piecewise curves of fewest circular arc segments. Computer-Aided Design 16 (1984) 87-90.

[Mark 89] Markot, R. P.; Magedson, R. L.: Solutions of tangential surface and curve intersections. Computer-Aided Design 21 (1989) 421-429.

[Mark 91] Markot, R. P.; Magedson, R. L.: Procedural method for evaluating the intersection curves of two parametric surfaces. Computer-Aided Design 23 (1991) 395-404.

[Mar 86] Martin, R. R.; de Pont, J.; Sharrock, T. J.: Cyclide surfaces in computer aided design. in Gregory, J. A. (ed.): The mathematics of surfaces. Clarendon Press (1986) 253-267.

[Mar 90] Martin, R. R.; Stephenson, P. C.: Sweeping of three-dimensional objects. Computer-Aided Design 22 (1990) 223-234.

[Mas 86] Mastin, C. W.: Parameterization in grid generation. Computer-Aided Design 18 (1986) 22-24.

[Max 68] Maxwell, J. C.: On the Cyclide. Quarterly Journal of Pure and Ap- plied Mathematics IX (1868) 111-126.

[Mcc 70] McConalogue, D. J.: A quasi-intrinsic scheme for passing a smooth curve through a discrete set of points. The Computer Journal 13 (1970) 392-396.

[Mcco 92] McCormick, M.; Davies, N.: 3-D Worlds. Physics World 5 (1992) 42-46.

[Mcl 83] McLaughlin, H. W.: Shape-Preserving Planar Interpolation: an Algorithm. IEEE Computer Graphics & Applications 3 (1983) 58-67.

[Mcla 74] McLain, D. H.: Drawing contours from arbitrary data points. The Computer Journal 17 (1974) 318-324.

[Mcla 76] McLain, D. H.: Two dimensional interpolation from random data. The Computer Journal 19 (1976) 178-181.

[Mcla 84] McLean, D. H.: A method for generating surfaces as a composite of cyclide patches. The Computer Journal 28 (1984) 433-438.

[Mcm 87] McMahon, J. R.; Franke, R.: Knot selection for least squares thin plate splines. Technical Report NPS-53-87-005, Naval Postgraduate School, Monterey 1987.

[Mee 89] Meek, D. S.; Walton, D. J.: The use of Cornu spirals in drawing planar curves of controlled curvature. Journal on Computational and Applied Mathematics 25 (1989) 69-78.

[Mee 90] Meek, D. S.; Walton, D. J.: Offset curves of clothoidal splines. Computer-Aided Design 22 (1990) 199-201.

[Meh 74] Mehlum, E.: Nonlinear Splines. in Barnhill, R. E.; Riesenfeld, R. F. (ed.): Computer Aided Geometric Design, Academic Press (1974) 173-208.

[Mei 87] Meier, H.: Der differentialgeometrische Entwurf und die analytische Darstellung krümmungsstetiger Schiffsoberflächen und ähnlicher Freiformflächen. Fortschritt-Berichte VDI, Reihe 20, Heft 5. VDI-Verlag 1987.

[Mein 79] Meinguet, J.: Multivariate Interpolation at Arbitrary Points Made Simple. Journal of Applied Mathematics and Physics (ZAMP) 30 (1979) 292-304.

[Mein 79a] Meinguet, J.: An intrinsic approach to multivariate spline interpolation at arbitrary points. in Sakney, B. N. (ed.): Polynomial and Spline Approximation. Reidel Publishing Company (1979) 163-190.

[Mic 86] Micchelli, C. A.: Interpolation of scattered data: distance matrices and conditionally positive definite functions. Constructive Approximation 2 (1986) 11-22.

[Mic 87] Micchelli, C. A.; Prautzsch, H.: Computing surfaces invariant under subdivision. Computer Aided Geometric Design 4 (1987) 321-328.

[Mic 89] Micchelli, C. A.; Prautzsch, H.: Uniform refinement of curves. Linear Algebra and Applications 114/115 (1989) 841-870.

[Mid 85] Middleditch, A. E.; Sears, K. H.: Blend Surfaces for Set Theoretic Volume Modelling Systems. SIGGRAPH Computer Graphics 19 (1985) 161-170.

[Mil 86] Miller, J. R.: Sculptured Surfaces in Solid Models: Issues and Alternative Approaches. IEEE Computer Graphics & Applications 6 (1986) 37-48.

[Mil 87] Miller, J. R.: Geometric Approaches to Nonplanar Quadric Surface Intersection Curves. ACM Transactions on Graphics 6 (1987) 274-307.

[Mir 82] Mirante, A.; Weingarten, N.: The radial sweep algorithm for constructing triangulated irregular networks. IEEE Computer Graphics & Applications 2 (1982) 11-21.

[Mon 86] Montaudouin, Y. de; Tiller, W.; Vold, H.: Applications of power series in computational geometry. Computer-Aided Design 18 (1986) 514-524.

[Mon 91] Montaudouin, Y. de: Resolution of $P(x, y) = 0$. Computer-Aided Design 23 (1991) 653-654.

[Mont 87] Montefusco, L. B.: An Interactive Procedure for Shape Preserving Cubic Spline Interpolation. Computer & Graphics 11 (1987) 389-392.

[Mont 89] Montefusco, L. B.; Casciola, G.: Algorithm 677: C^1 surface interpolation. ACM Transactions on Mathematical Software 15 (1989) 365-374.

[Morg 83] Morgan, A. P.: A Method for Computing all Solutions to Systems of Polynomial Equations. ACM Transactions on Mathematical Software 9 (1983) 1-17.

[Mülle 90] Müllenheim, G.: Convergence of a surface/surface intersection algorithm. Computer Aided Geometric Design 7 (1990) 415-423.

[Mülle 91] Müllenheim, G.: On determining start points for a surface/surface intersection algorithm. Computer Aided Geometric Design 8 (1991) 401-408.

[Mur 91] Muraki, S.: Volumetric Shape Description of Range Data using "Blobby Model". ACM Computer Graphics 25 (1991) 227-235.

[Nac 82] Nackman, L. R.: Curvature Relations in Three-Dimensional Symmetric Axes. Computer Graphics & Image Processing 20 (1982) 43-57.

[Nas 87] Nasri, A. H.: Polyhedral Subdivision Methods for Free-Form Surfaces. ACM Transactions on Graphics 6 (1987) 29-73.

[Nas 91] Nasri, A. H.: Surface interpolation on irregular networks with normal conditions. Computer Aided Geometric Design 8 (1991) 89-96.

[Nas 91a] Nasri, A. H.: Boundary-corner control in recursive-subdivision surfaces. Computer-Aided Design 23 (1991) 405-410.

[Nat 90] Natarajan, B. K.: On Computing the Intersection of B-Splines. Proceedings of the sixth Annual Symposium on Computational Geometry. ACM Press (1990) 157-167.

[Nef 89] Neff, C. A.: Decomposing algebraic sets using Gröbner bases. Computer Aided Geometric Design 6 (1989) 249-263.

[Neu 92] Neuser, D. A.: Curve and surface interpolation using quintic weighted tau-splines. in Hagen, H. (ed.): Curve and Surface Design. SIAM (1992) 55-85.

[Nie 74] Nielson, G. M.: Some piecewise polynomial alternatives to splines under tension. in Barnhill, R. E.; Riesenfeld, R. F. (ed.): Computer Aided Geometric Design. Academic Press (1974) 209-235.

[Nie 79] Nielson, G. M.: The side-vertex method for interpolation in triangles. Journal of Approximation Theory 25 (1979) 318-336.

[Nie 80] Nielson, G. M.: Minimum norm interpolation in triangles. SIAM Journal on Numerical Analysis 17 (1980) 44-62.

[Nie 83] Nielson, G. M.; Franke, R.: Surface construction based upon triangulations. in Barnhill, R. E.; Boehm, W. (ed.): Surfaces in Computer Aided Geometric Design. North-Holland (1983) 163-177.

[Nie 83a] Nielson, G. M.: A method for interpolating scattered data based upon a minimum norm network. Mathematics of Computation 40 (1983) 253-271.

[Nie 84] Nielson, G. M.: A locally controllable spline with tension for interactive curve design. Computer Aided Geometric Design 1 (1984) 199-205.

[Nie 84a] Nielson, G. M.; Franke, R.: A method for construction of surfaces under tension. Rocky Mountain Journal of Mathematics 14 (1984) 203-222.

[Nie 86] Nielson, G. M.: Rectangular ν-splines. IEEE Computer Graphics & Applications 6 (1986) 35-40.

[Nie 87] Nielson, G. M.: Coordinate free scattered data interpolation. in Chui, C. K.; Schumaker, L. L.; Utreras, F. I. (ed.): Topics in Multivariate Approximation. Academic Press (1987) 175-184.

[Nie 87a] Nielson, G. M.: An example with a local minimum for the minimax ordering of triangulations. Report TR-87-014, Computer Science Department, Arizona State University, Tempe 1987.

[Nie 87b] Nielson, G. M.; Ramaraj, R.: Interpolation over a sphere based upon a minimum norm network. Computer Aided Geometric Design 4 (1987) 41-57.

[Nie 87c] Nielson, G. M.: A Transfinite, Visually Continuous Triangular Interpolant. in Farin, G. (ed.): Geometric Modeling, Algorithms and New Trends, SIAM (1987) 235-246.

[Nie 88] Nielson, G. M.: Interactive Surface Design Using Triangular Network Splines. Proceedings of the Third International Conference on Engineering Graphics and Descriptive Geometry 2. Vienna (1988) 70-77.

[Nie 89] Nielson, G. M.; Foley, Th. A.: A survey of applications of an affine invariant metric. in Lyche, T.; Schumaker, L. L. (ed.): Mathematical Methods in Computer Aided Geometric Design. Academic Press (1989) 445-468.

[Nie 90] Nielson, G. M.; Hamann, B.: Techniques for the Visualization of Volumetric Data. in Kaufman, A. (ed.): Visualization '90. IEEE Computer Society Press (1990) 45-50.

[Nie 91] Nielson, G. M.; Hamann, B.: The Asymptotic Decider: Resolving the Ambiguity in Marching Cubes. in Nielson, G. M.; Rosenblum, L. (ed.): Visualization '91. IEEE Computer Society Press (1991) 83-91.

[Nie 91a] Nielson, G. M.; Foley, Th. A.; Hamann, B.; Lane, D.: Visualizing and Modeling Scattered Multivariate Data. IEEE Computer Graphics & Applications 11 (1991) 47-55.

[Nie 91b] Nielson, G. M.: Visualization in Scientific and Engineering Computation. IEEE Computer 24 (1991) 58-66.

[Nis 90] Nishita, T.; Sederberg, Th. W.; Kakimoto, M.: Ray Tracing Trimmed Rational Surface Patches. ACM Computer Graphics 24 (1990) 337-345.

[Now 83] Nowacki, H.; Reese, D.: Design and fairing of ship surfaces. in Barnhill, R. E.; Boehm, W. (ed.): Surfaces in Computer Aided Geometric Design. North-Holland (1983) 121-134.

[Now 89] Nowacki, H.; Liu, D.; Lü, X.: Mesh Fairing GC^1 Surface Generation Method. in Straßer, W.; Seidel, H. P. (ed.): Theory and Practice of Geometric Modeling. Springer (1989) 93-108.

[Now 90] Nowacki, H.: Mathematische Verfahren zum Glätten von Kurven und Flächen. in Encarnacao, J. L.; Hoschek, J.; Rix, J. (ed.): Geometrische Verfahren der Graphischen Datenverarbeitung. Springer (1990) 22-45.

[Now 90a] Nowacki, H.; Liu, D.; Lü, X.: Fairing Bézier-curves with constraints. Computer Aided Geometric Design 7 (1990) 43-55.

[Now 92] Nowacki, H.; Kaklis, P.; Weber, J.: Curve Mesh Fairing and GC^2 Surface Interpolation. Mathematical Modelling and Numerical Analysis 26 (1992) 113-136.

[Nut 72] Nutbourne, A. W.; McLellan, P. M.; Kensit, R. M. L.: Curvature profiles for plane curves. Computer-Aided Design 4 (1972) 176-184.

[Nyd 72] Nydegger, R. W.: A Data Minimization Algorithm of Analytical Models for Computer Graphics. University of Utah, Salt Lake City, Utah (1972).

[Ock 84] Ocken, S.; Schwartz, J. T.: Precise implementation of CAD primitives using rational parametrizations of standard surfaces. in Pickett, M. S.; Boyse, J. W. (ed.): Solid Modeling by Computers. From Theory to Applications. Plenum Press (1984) 259-273.

[Oli 90] Oliver, J. H.; Goodman, E. D.: Direct dimensional NC verification. Computer-Aided Design 22 (1990) 3-9.

[Owe 87] Owen, J. C.; Rockwood , A. P.: Intersection of General Implicit Surfaces. in Farin, G. (ed.): Geometric Modeling: Algorithms and New Trends. SIAM (1987) 335-345.

[Paj 90] Pajon, J. L.; Tran, V. B.: Visualization of scalar data defined on a structured grid. in Kaufman, A. (ed.): Visualization '90. IEEE Computer Society Press (1990) 281-287.

[Pal 77] Pal, T. K.; Nutbourne, A. W.: Two-dimensional curve synthesis using linear curvature elements. Computer-Aided Design 9 (1977) 121-134.

[Pal 78] Pal, T. K.: Intrinsic spline curve with local control. Computer-Aided Design 10 (1978) 19-29.

[Pal 78a] Pal, T. K.: Mean tangent rotational angles and curvature integration. Computer-Aided Design 10 (1978) 30-34.

[Patr 89] Patrikalakis, N. M.: Approximate conversion of rational splines. Computer Aided Geometric Design 6 (1989) 155-166.

[Patr 89a] Patrikalakis, N. M.; Kriezis, G. A.: Representation of piecewise continuous algebraic surfaces in terms of B-splines. The Visual Computer 5 (1989) 360-374.

[Patr 89b] Patrikalakis, N. M.; Bardis, L.: Offsets of Curves on Rational B-Spline Surfaces. Engineering with Computers 5 (1989) 39-46.

[Patr 90] Patrikalakis, N. M.; Kriezis, G. A.: Piecewise continuous algebraic surfaces in terms of B-splines. in Wozny, M. J.; Turner, J. U.; Preiss, K. (ed.): Geometric Modeling for Product Engineering. North-Holland (1990) 3-19.

[Patt 78] Patterson, M. R.: Contur: A Subroutine to Draw Contour Lines for Randomly Located Data. Technical Report ORNL/CSD TM-59. Oak Ridge National Laboratory 1978.

[Pat 85] Patterson, R. R.: Projective Transformations of the Parameter of a Bernstein-Bézier Curve. ACM Transactions on Graphics 4 (1985) 276-290.

[Pat 88] Patterson, R. R.: Parametric cubics as algebraic curves. Computer Aided Geometric Design 5 (1988) 139-159.

[Pau 88] Paukowitsch, P.: Fundamental ideas for computer-supported descriptive geometry. Computers & Graphics 12 (1988) 3-14.

[Pav 83] Pavlidis, Th.: Curve Fitting with Conic Splines. ACM Transactions on Graphics 2 (1983) 1-31.

[Peck 85] Peckham, R. J.: Shading evaluations with general three-dimensional models. Computer-Aided Design 17 (1985) 305-310.

[Peg 87] Pegna, J.: Variable Sweep Geometric Modeling. Thesis Stanford 1987.

[Peg 90] Pegna, J.; Wilde, D. J.: Spherical and Circular Blending of Functional Surfaces. Transactions of ASME, Journal of Offshore Mechanics and Arctic Engineering 112 (1990) 134-142.

[Peg 90a] Pegna, J.; Wolter, F. E.: Designing and Mapping Trimming Curves on Surfaces Using Orthogonal Projection. Advances in Design Automation, Vol. I (1990) 235-245.

[Peg 92] Pegna, J.; Wolter, F. E.: Geometrical Criteria to Guarantee Curvature Continuity of Blend Surfaces. Transactions of ASME, Journal of Mechanical Design 114 (1992) 201-210.

[Pen 84] Peng, Q. S.: An algorithm for finding the intersection lines between two B-Spline surfaces. Computer-Aided Design 16 (1984) 191-196.

[Perc 76] Percell, P.: On cubic and quartic Clough-Tocher elements. SIAM Journal on Numerical Analysis 13 (1976) 100-103.

[Per 78] Persson, H.: NC machining of arbitrarily shaped pockets. Computer-Aided Design 10 (1978) 169-174.

[Pet 90] Peters, J.: Local smooth surface interpolation: a classification. Computer Aided Geometric Design 7 (1990) 191-195.

[Pet 90a] Peters, J.: Smooth mesh interpolation with cubic patches. Computer-Aided Design 22 (1990) 109-120.

[Pet 90b] Peters, J.: Local cubic and bicubic C^1 surface interpolation with linearly varying boundary normal. Computer Aided Geometric Design 7 (1990) 499-516.

[Pet 91] Peters, J.: Smooth Interpolation of a Mesh of Curves. Constructive Approximation 7 (1991) 221-246.

[Pete 84] Petersen, C. S.: Adaptive contouring of three-dimensional surfaces. Computer Aided Geometric Design 1 (1984) 61-74.

[Pete 87] Petersen, C. S.; Piper, B. R.; Worsey, A. J.: Adaptive Contouring of a Trivariate Interpolant. in Farin, G. (ed.): Geometric Modeling, Algorithms and New Trends. SIAM (1987) 385-395.

[Petr 87] Petrie, G.; Kennie, T. J. M.: Terrain modelling in surveying and civil engineering. Computer-Aided Design 19 (1987) 171-187.

[Pfe 85] Pfeifer, H. U.: Methods used for intersecting geometrical entities in the GPM module for volume geometry. Computer-Aided Design 17 (1985) 311-318.

[Pfl 89] Pfluger, P. R.; Gmelig Meyling, R. H. J.: An algorithm for smooth interpolation to scattered data in $\mathbb{R}^2$. in Lyche, T.; Schumaker, L. L. (ed.): Mathematical Methods in Computer Aided Geometric Design. Academic Press (1989) 469-480.

[Pha 88] Pham, B.: Offset approximation of uniform B-splines. Computer-Aided Design 20 (1988) 471-474.

[Pha 92] Pham, B.: Offset curves and surfaces: a brief survey. Computer-Aided Design 24 (1992) 223-229.

[Phi 84] Phillips, M. B.; Odell, G. M.: An Algorithm for Locating and Displaying the Intersection of Two Arbitrary Surfaces. IEEE Computer Graphics & Applications 4 (1984) 48-58.

[Pho 75] Phong, Bui-Tuong: Illumination of Computer Generated Pictures. Communications of the ACM 18 (1975) 311-317.

[Pie 86] Piegl, L.: A Geometric Investigation of the Rational Bézier Scheme of Computer Aided Design. Computers in Industry 7 (1986) 401-410.

[Pie 86a] Piegl, L.: Curve fitting algorithm for rough cutting. Computer-Aided Design 18 (1986) 79-82.

[Pie 87] Piegl, L.: On the use of infinite control points in CAGD. Computer Aided Geometric Design 4 (1987) 155-166.

[Pie 87a] Piegl, L.: Interactive Data Interpolation by Rational Bézier Curves. IEEE Computer Graphics & Applications 7 (1987) 45-58.

[Pie 87b] Piegl, L.; Tiller, W.: Curve and surface constructions using rational B-splines. Computer-Aided Design 19 (1987) 485-498.

[Pie 88] Piegl, L.: Hermite and Coons-like interpolants using rational Bézier approximation form with infinite control points. Computer-Aided Design 20 (1988) 2-10.

[Pie 89] Piegl, L.: Geometric method of intersecting natural quadrics represented in trimmed surface form. Computer-Aided Design 21 (1989) 201-212.

[Pie 89a] Piegl, L.: Modifying the shape of rational B-splines. Part 1: curves. Computer-Aided Design 21 (1989) 509-518.

[Pie 89b] Piegl. L.: Modifying the shape of rational B-splines. Part 2: surfaces. Computer-Aided Design 21 (1989) 538-546.

[Pie 89c] Piegl, L.; Tiller, W.: A Menagerie of Rational B-Spline Circles. IEEE Computer Graphics & Applications 9 (1989) 48-56.

[Pie 91] Piegl, L.: On NURBS: A Survey. IEEE Computer Graphics & Applications 11 (1991) 55-71.

[Pie 92] Piegl, L.: Constructive Geometric Approach to Surface-Surface Intersection. in Barnhill, R. E. (ed.): Geometry Processing for Design and Manufacturing. SIAM (1992) 137-159.

[Pip 87] Piper, B. R.: Visually Smooth Interpolation with Triangular Bézier Patches. in Farin, G. (ed.): Geometric Modeling, Algorithms and New Trends, SIAM (1987) 221-233.

[Pit 67] Pitteway, M. L. V.: Algorithm for Drawing Ellipses or Hyperbolas with a Digital Plotter. Computer Journal 10 (1967) 282-289.

[Pla 83] Plass, M.; Stone, M.: Curve-Fitting with Piecewise Parametric Cubics. ACM Computer Graphics 17 (1983) 229-238.

[Ple 89] Pletinckx, D.: Quaternion calculus as a basic tool in Computer graphics. The Visual Computer 5 (1989) 2-13.

[Plu 85] Plunkett, D. J.; Bailey, M. J.: The Vectorization of a Ray-Tracing Algorithm for Improved Execution Speed. IEEE Computer Graphics & Applications 5 (1985) 52-60.

[Pob 91] Pobegailo, A. P.: Local interpolation with weight functions for variable smoothness curve design. Computer-Aided Design 23 (1991) 579-582.

[Pös 84] Pöschl, Th.: Detecting surface irregularities using isophotes. Computer Aided Geometric Design 1 (1984) 163-168.

[Pot 88] Pottmann, H.: Curves and Tensor Product Surfaces with Third Order Geometric Continuity. Proceedings of the Third International Conference on Engineering Graphics and Descriptive Geometry 2. Vienna (1988) 107-116.

[Pot 88a] Pottmann, H.: Eine Verfeinerung der Isophotenmethode zur Qualitätsanalyse von Freiformflächen. CAD und Computergraphik 11 (1988) 99-109.

[Pot 89] Pottmann, H.: Projectively invariant classes of geometric continuity for CAGD. Computer Aided Geometric Design 6 (1989) 307-322.

[Pot 90] Pottmann, H.: Smooth curves under tension. Computer-Aided Design 22 (1990) 241-245.

[Pot 90a] Pottmann, H.; Eck, M.: Modified multiquadric methods for scattered data interpolation over a sphere. Computer Aided Geometric Design 7 (1990) 313-321.

[Pot 90b] Pottmann, H.; Divivier, A.: Interpolation von Meßdaten auf Flächen. in Encarnacao, J. L.; Hoschek, J.; Rix, J. (ed.): Geometrische Verfahren der Graphischen Datenverarbeitung. Springer (1990) 104-120.

[Pot 91] Pottmann, H.: Scattered Data Interpolation Based upon Generalized Minimum Norm Networks. Constructive Approximation 7 (1991) 247-256.

[Pot 91a] Pottmann, H.; Hagen, H.; Divivier, A.: Visualizing Functions on a Surface. The Journal of Visualization and Computer Animation 2 (1991) 52-58.

[Pot 92] Pottmann, H.: Interpolation on surfaces using minimum norm networks. Computer Aided Geometric Design 9 (1992) 51-67.

[Pow 72] Powell, M. J. D.: Problems Related to Unconstrained Optimisation. in Murray, W. (ed.): Numerical Methods for Unconstrained Optimisation. Academic Press (1972).

[Pow 74] Powell, M. J. D.: Piecewise quadratic surface fitting for contour plotting. in Evans, D. J. (ed.): Software for Numerical Mathematics. Academic Press (1974) 253-271.

[Pow 77] Powell, M. J. D.; Sabin, M. A.: Piecewise quadratic approximations on triangles. ACM Transactions on Mathematical Software 3 (1977) 316-325.

[Pow 87] Powell, M. J. D.: Radial basis functions for multivariable interpolation: a review. in Griffiths, D. F.; Watson, G. A. (ed.): Numerical Analysis 1987. Longman Scientific & Technical, Harlow (1987) 223-241.

[Prat 86] Pratt, M. J.; Geisow, A. D.: Surface/surface intersection problems. in Gregory, J. A. (ed.): The Mathematics of Surfaces. Clarendon Press (1986) 117-142.

[Prat 88] Pratt, M. J.: Applications of Cyclide Surfaces in Geometric Modelling. in Handscomb, D. C. (ed.): The Mathematics of Surfaces III. Oxford University Press (1988) 405-428.

[Prat 89] Pratt, M. J.: Cyclide Blending in Solid Modelling. in Straßer, W.; Seidel, H. P. (ed.): Theory and Practice of Geometric Modeling. Springer (1989) 235-245.

[Prat 90] Pratt, M. J.: Cyclides in computer aided geometric design. Computer Aided Geometric Design 7 (1990) 221-242.

[Pratt 85] Pratt, V.: Techniques for Conic Splines. ACM Computer Graphics 19 (1985) 151-159.

[Pratt 87] Pratt, V.: Direct least square fitting of algebraic surfaces. ACM Computer Graphics 21 (1987) 145-152.

[Pra 84] Prautzsch, H.: Unterteilungsalgorithmen für multivariate Splines, ein geometrischer Zugang. Diss., Braunschweig 1984.

[Pra 84a] Prautzsch, H.: Degree elevation of B-spline curves. Computer Aided Geometric Design 1 (1984) 193-198.

[Pra 85] Prautzsch, H.: Generalized subdivision and convergence. Computer Aided Geometric Design 2 (1985) 69-75.

[Pra 87] Prautzsch, H.; Micchelli, C. A.: Computing curves invariant under halving. Computer Aided Geometric Design 4 (1987) 133-140.

[Pra 91] Prautzsch, H.; Piper, B.: A fast algorithm to raise the degree of spline curves. Computer Aided Geometric Design 8 (1991) 253-265.

[Prep 77] Preparata, F. P.; Hong, S. J.: Convex hulls of finite sets of points in two and three dimensions. Communications of the ACM 20 (1977) 87-93.

[Prep 79] Preparata, F. P.: An optimal real-time algorithm for planar convex hulls. Communications of the ACM 22 (1979) 402-405.

[Preu 84] Preusser, A.: Computing Contours by Successive Solution of Quintic Polynomial Equations. ACM Transactions on Mathematical Software 10 (1984) 464-472.

[Preu 84a] Preusser, A.: Algorithm 626. TRICP: A contour plot program for triangular meshes. ACM Transactions on Mathematical Software 10 (1984) 473-475.

[Preu 85] Preusser, A: A Remark on Algorithm 526. ACM Transactions on Mathematical Software 11 (1985) 186-187.

[Preu 86] Preusser, A.: Computing area filling contours for surfaces defined by piecewise polynomials. Computer Aided Geometric Design 3 (1986) 267-279.

[Preu 89] Preusser, A.: Algorithm 671 FARB-E-2D: Fill Area with Bicubics on Rectangles - A Contour Plot Program. ACM Transactions on Mathematical Software 15 (1989) 79-89.

[Preu 90] Preusser, A.: Efficient Formulation of a Bivariate Nonic C^1 Hermite Polynomial on Triangles. ACM Transactions on Mathematical Software 16 (1990) 246-252.

[Preu 90a] Preusser, A.: Algorithm 684. C^1 and C^2 Interpolation on Triangles with Quintic and Nonic Bivariate Polynomials. ACM Transactions on Mathematical Software 16 (1990) 253-257.

[Pri 86] Pritschow, G.; Viefhaus, R.: Mehrachsiges NC-Fräsen von Blechumformwerkzeugen. wt. -Z. ind. Fert. 76 (1986) 619-623.

[Pru 76] Pruess, S.: Properties of splines in tension. Journal of Approximation Theory 17 (1976) 86-96.

[Pue 87] Pueyo, X.; Brunet, P.: A parametric-space-based scan-line algorithm for rendering bicubic surfaces. IEEE Computer Graphics & Applications 7 (1987) 17-25.

[Qua 89] Quak, E.; Schumaker, L. L.: C^1 surface fitting using data dependent triangulations. in Chui, C. K.; Schumaker, L. L.; Ward, J. D. (ed.): Approximation Theory VI: Volume 2. Academic Press (1989) 545-548.

[Qua 90] Quak, E.; Schumaker, L. L.: Cubic spline fitting using data dependent triangulations. Computer Aided Geometric Design 7 (1990) 293-301.

[Qua 90a] Quak, E.; Schumaker, L. L.: Calculation of the energy of a piecewise polynomial surface. in Cox, M.; Mason, J. (ed.): Algorithms for Approximation II. Chapman and Hall (1990) 134-143.

[Quar 86] Quarendon, P.; Woodwark, J. R.: The Model for Graphics. Proceedings of International Summer Institute, Stirling (1986).

[Raj 91] Raj, R. N.; Pegna, J.: Surface Blending Using Variable Sweep Geometric Modeling. Rensselaer Polytechnic Institute, Design Research Center Technical Report No. 90022, 1991.

[Ram 87] Ramshaw, L.: Blossoming: A Connect-the-Dots Approach to Splines. Research Report 19, Digital Systems Research Center, Palo Alto 1987.

[Ram 88] Ramshaw, L.: Béziers and B-splines as Multiaffine Maps. in Earnshaw, R. A. (ed.): Theoretical Foundations of Computer Graphics and CAD. Springer (1988) 757-776.

[Ram 89] Ramshaw, L.: Blossoms are polar forms. Computer Aided Geometric Design 6 (1989) 323-358.

[Ran 91] Rando, T.; Roulier, J. A.: Designing faired parametric surfaces. Computer-Aided Design 23 (1991) 492-497.

[Rat 88] Rath, W.: Computergestützte Darstellung von Hyperflächen des $\mathbb{R}^4$ und deren Anwendungsmöglichkeiten im CAGD. CAD und Computergraphik 11 (1988) 111-117.

[Ree 83] Reese, D.; Riedger, M.; Lang, R.: Flächenhaftes Glätten und Verändern von Schiffsoberflächen. Bericht des Institutes für Schiffs- und Meerestechnik Berlin 14, 1983.

[Rei 67] Reinsch, Ch. H.: Smoothing by Spline Functions. Numerische Mathematik 10 (1967) 177-183.

[Rei 71] Reinsch, Ch. H.: Smoothing by Spline Functions II. Numerische Mathematik 16 (1971) 451-454.

[Ren 84] Renka, R. J.; Cline, A. K.: A triangle-based C^1 interpolation method. Rocky Mountain Journal of Mathematics 14 (1984) 223-237.

[Ren 84a] Renka, R. J.: Interpolation of data on the surface of a sphere. ACM Transactions on Mathematical Software 10 (1984) 417-436.

[Ren 84b] Renka, R. J.: Algorithm 624: Triangulation and interpolation at arbitrarily distributed points in the plane. ACM Transactions on Mathematical Software 10 (1984) 440-442.

[Ren 87] Renka, R. J.: Interpolatory tension splines with automatic selection of tension factors. SIAM Journal on Scientific and Statistical Computing 8 (1987) 393-415.

[Ren 88] Renka, R. J.: Multivariate interpolation of large sets of scattered data. ACM Transactions on Mathematical Software 14 (1988) 139-148.

[Ren 88a] Renka, R. J.: Algorithm 660: QSHEP2D: Quadratic Shepard method for bivariate interpolation of scattered data. ACM Transactions on Mathematical Software 14 (1988) 149-150.

[Ren 88b] Renka, R. J.: Algorithm 661: QSHEP3D: Quadratic Shepard method for trivariate interpolation of scattered data. ACM Transactions on Mathematical Software 14 (1988) 151-152.

[Renn 82] Renner, G.: A method of shape description for mechanical engineering practice. Computers in Industry 3 (1982) 137-142.

[Renn 91] Renner, G.: Designing complex surfaces based on curve networks with irregular topology. Computers in Industry 16 (1991) 189-196.

[Rent 80] Rentrop, P.: An algorithm for the computation of the exponential spline. Numerische Mathematik 35 (1980) 81-83.

[Renz 82] Renz, W.: Interactive smoothing of digitized point data. Computer-Aided Design 14 (1982) 267-269.

[Req 82] Requicha, A. A. G.; Voelcker, H. B.: Solid Modelling: A Historical Summary & Contemporary Assessment. IEEE Computer Graphics & Applications 2 (1982) 9-24.

[Res 86] Rescorla, K. L.: Cardinal interpolation: A bivariate polynomial example. Computer Aided Geometric Design 3 (1986) 313-321.

[Res 87] Rescorla, K. L.: C^1 trivariate polynomial interpolation. Computer Aided Geometric Design 4 (1987) 237-244.

[Reu 89] Reuding, Th.: Bézier patches on cubic grid curves – An application to the preliminary design of a yacht hull surface. Computer Aided Geometric Design 6 (1989) 11-21.

[Ric 73] Ricci, A.: A constructive geometry for computer graphics. The Computer Journal 16 (1973) 157-160.

[Rie 73] Riesenfeld, R. F.: Applications of B-spline approximation to geometric problems of computer-aided design. Thesis, Syracuse 1973.

[Rie 75] Riesenfeld, R. F.: On Chaikin's Algorithm. Computer Graphics and Image Processing 4 (1975) 304-310.

[Rip 90] Rippa, S.: Minimal roughness property of the Delaunay triangulation. Computer Aided Geometric Design 7 (1990) 489-497.

[Rip 90a] Rippa, S.; Schiff, B.: Minimum Energy Triangulations for Elliptic Problems. Computer Methods in Applied Mechanics and Engineering 84 (1990) 257-274.

[Rock 84] Rockwood, A. P.: Introducing sculptured surfaces into a geometric modeler. in Pickett, M. S.; Boyse, J. W. (ed.): Solid Modeling by Computers. From Theory to Applications. Plenum Press (1984) 237-258.

[Rock 87] Rockwood, A. P.; Owen, J.: Blending surfaces in solid modeling. in Farin, G. (ed.): Geometric Modeling: Algorithms and New Trends. SIAM (1987) 367-384.

[Rock 89] Rockwood, A. P.; Heaton, K.; Davis, T.: Real Time Rendering of Trimmed Surfaces. ACM Computer Graphics 23 (1989) 107-116.

[Rock 90] Rockwood, A. P.: Accurate Display of Tensor Product Isosurfaces. in Kaufman, A. (ed.): Visualization '90. IEEE Computer Society Press (1990) 353-360.

[Rog 89] Rogers, D. F.; Fog, N. G.: Constrained B-spline curve and surface fitting. Computer-Aided Design 21 (1989) 641-648.

[Rös 37] Rössler, F.: Geometrische Grundlagen der Konstruktion von Helligkeitsgleichungen für eine neue Helligkeitshypothese. Monatshefte für Mathematik 46 (1937) 157-171.

[Ross 84] Rossignac, J. R.; Requicha, A. A. G.: Constant-radius blending in solid modeling. Computers in Mechanical Engineering, July (1984) 65-73.

[Ross 86] Rossignac, J. R.; Requicha, A. A. G.: Offsetting operations in solid modelling. Computer Aided Geometric Design 3 (1986) 129-148.

[Rot 82] Roth, S. D.: Ray Casting for Modeling Solids. Computer Graphics and Image Processing 18 (1982) 109-144.

[Rou 88] Roulier, J. A.: Bézier curves of positive curvature. Computer Aided Geometric Design 5 (1988) 59-70.

[Rud 58] Rudin, M. E.: An unshellable triangulation of a tetrahedron. Bulletin of the American Mathematical Society 64 (1958) 90-91.

[Run 01] Runge, C.: Über empirische Funktionen und die Interpolation zwischen äquidistanten Ordinaten. ZAMM 46 (1901) 224-243.

[Sabe 88] Sabella, P.: A rendering algorithm for visualizing 3D scalar fields. ACM Computer Graphics 22 (1988) 51-58.

[Sab 68] Sabin, M. A.: Conditions for continuity of surface normals between adjacent parametric surfaces. British Aircraft Corporation Ltd. Technical Report 1968.

[Sab 76] Sabin, M. A.: A method for displaying the intersection curve of two quadric surfaces. Computer Journal 19 (1976) 336-338.

[Sab 76a] Sabin, M. A.: The use of piecewise forms for the numerical representation of shape. Diss. Hungarian Academy of Science, Budapest 1976.

[Sab 80] Sabin, M. A.: Contouring – A review of methods for scattered data. in Brodlie, K. W. (ed.): Mathematical Methods in Computer Graphics and Design. Academic Press (1980) 63-85.

[Sab 83] Sabin, M. A.: Non-rectangular patches suitable for inclusion in a B-spline surface. Ten Hagen (ed.): Proceedings Eurographics. North-Holland (1983) 57-69.

[Sab 85] Sabin, M. A.: Non-four-sided patch expressions with control points. Computer Aided Geometric Design 1 (1985) 289-290.

[Sab 85a] Sabin, M. A.: Contouring – the State of the Art. in Earnshaw, R. A. (ed.): Fundamental Algorithms for Computer Graphics. NATO ASI series F 17. Springer (1985) 411-482.

[Sab 86] Sabin, M. A.: Some negative results in N-sided patches. Computer-Aided Design 18 (1986) 38-44.

[Sab 86a] Sabin, M. A.: Recursive Division. in Gregory, J. A. (ed.): The Mathematics of Surfaces. Clarendon Press (1986) 269-282.

[Sabl 78] Sablonniere, P.: Spline and Bézier polygons associated with a polynomial spline curve. Computer-Aided Design 10 (1978) 257-261.

[Sabl 85] Sablonniere, P.: Bernstein-Bézier methods for the construction of bivariate spline approximants. Computer Aided Geometric Design 2 (1985) 29-36.

[Sabl 85a] Sablonniere, P.: Composite finite elements of class C^k. Journal of Computational and Applied Mathematics 12 & 13 (1985) 541-550.

[Sabl 87] Sablonniere, P.: Composite finite elements of Class C^1. in Chui, C. K.; Schumaker, L. L.; Utreras, F. I. (ed.): Topics in Multivariate Approximation. Academic Press (1987) 207-217.

[Sae 88] Saeed, S. E. O.; de Pennington, A.; Dodsworth, J. R.: Offsetting in geometric modelling. Computer-Aided Design 20 (1988) 67-74.

[Sai 87] Saia, A.; Bloor, M. S.; de Pennington, A.: Sculptured Solids in a CSG Based Geometric Modelling System. in Martin, R. R. (ed.): The Mathematics of Surfaces II. Oxford University Press (1987) 321-341.

[Sait 90] Saitoh, T.; Hosaka, M.: Interpolating Curve Networks with new Blending Patches. in Vandoni, C. E.; Duce, D. A. (ed.): EUROGRAPHICS '90. North-Holland (1990) 137-146.

[Sak 88] Sakai, M.; Usmani, R. A.: A shape preserving area true approximation of histograms by rational splines. BIT 28 (1988) 329-339.

[Salk 74] Salkauskas, K.: C^1 splines for interpolation of rapidly varying data. Rocky Mountain Journal of Mathematics 14 (1974) 239-250.

[Sanc 91] Sanchez-Reyes, J.: Single-valued surfaces in cylindrical coordinates. Computer-Aided Design 23 (1991) 561-568.

[San 90] Sanglikar, M. A.; Koparkar, P.; Joshi, V. N.: Modelling rolling ball blends for computer aided geometric design. Computer Aided Geometric Design 7 (1990) 399-414.

[Sap 88] Sapidis, N. S.; Kaklis, P. D.: An algorithm for constructing convexity and monotonicity-preserving splines in tension. Computer Aided Geometric Design 5 (1988) 127-137.

[Sap 90] Sapidis, N. S.; Farin, G.: Automatic fairing algorithm for B-Spline curves. Computer-Aided Design 22 (1990) 121-129.

[Sar 83] Sarraga, R. F.: Algebraic Methods for Intersections of Quadric Surfaces in GMSOLID. Computer Vision, Graphics and Image Processing 22 (1983) 222-238.

[Sar 87] Sarraga, R. F.: G^1 interpolation of generally unrestricted cubic Bézier curves. Computer Aided Geometric Design 4 (1987) 23-39.

[Sar 89] Sarraga, R. F.: Errata: G^1 interpolation of generally unrestricted cubic Bézier curves. Computer Aided Geometric Design 6 (1989) 167-172.

[Sar 90] Sarraga, R. F.: Computer Modeling of Surfaces with Arbitrary Shapes. IEEE Computer Graphics & Applications (1990) 67-77.

[Sark 91] Sarkar, B.; Menq, C. H.: Parameter optimization in approximating curves and surfaces to measured data. Computer Aided Geometric Design 8 (1991) 267-290.

[Sark 91a] Sarkar, B.; Menq, C. H.: Smooth-surface approximation and reverse engineering. Computer-Aided Design 23 (1991) 623-628.

[Sat 85] Satterfield, St. G.; Rogers, D. F.: A procedure for generating contour lines from a B-spline surface. IEEE Computer Graphics & Applications 5 (1985) 71-75.

[Sch 88] Schaal, H.: Computer Graphical Treatments of Perspective Pictures. Computers & graphics 12 (1988) 15-32.

[Schab 89] Schaback, R.: Interpolation with piecewise quadratic visually C^2 Bézier polynomials. Computer Aided Geometric Design 6 (1989) 219-233.

[Schab 92] Schaback, R.: Rational Curve Interpolation. in Lyche, T.; Schumaker, L. L. (ed.): Mathematical Methods in CAGD II. Academic Press (1992) 517-536.

[Schag 82] Schagen, I. P.: Automatic Contouring from Scattered Data Points. The Computer Journal 25 (1982) 7-11.

[Sche 78] Schechter, A.: Synthesis of 2D curves by blending piecewise linear curvature profiles. Computer-Aided Design 10 (1978) 8-18.

[Sche 78a] Schechter, A.: Linear blending of curvature profiles. Computer-Aided Design 10 (1978) 101-109.

[Schel 84] Schelske, H. J.: Lokale Glättung segmentierter Bézierkurven und Bézierflächen. Diss., Darmstadt 1984.

[Scher 78] Scherrer, P. K.; Hillberry, B. M.: Determining distance to a surface represented in piecewise fashion with surface patches. Computer-Aided Design 10 (1978) 320-324.

[Schm 89] Schmidt, J. W.: Results and problems in shape preserving interpolation and approximation with polynomial splines. in Schmidt, J. W.; Späth, H. (ed.): Splines in Numerical Analysis – Mathematical Research 52, Akademie-Verlag (1989) 159-170.

[Schm 79] Schmidt, R. M.: Einstufige Verfahren zur Flächenapproximation unregelmäßig verteilter Daten durch Tensor-Produkt B-Splines. Berlin Hahn-Meitner-Institutsbericht 268, 1979.

[Schm 83] Schmidt, R. M.: Fitting scattered surface data with large gaps. in Barnhill, R. E.; Boehm, W. (ed.): Surfaces in Computer Aided Geometric Design. North-Holland (1983) 185-189.

[Schm 85] Schmidt, R. M.: Ein Beitrag zur Flächenapproximation über unregelmäßig verteilten Daten. in Schempp, W.; Zeller, K.: Multivariate Approximations Theory III. Birkhäuser (1985) 363-369.

[Schm 92] Schmidt, M.: Modellierung und Approximation von Kurven und Flächen in implizierter Darstellung. Diss., Darmstadt 1992.

[Schn 92] Schneider, F.-J.: Interpolation, Approximation und Konvertierung mit rationalen B-Spline-Kurven und Flächen. Diss., Darmstadt 1992.

[Scho 67] Schoenberg, I. J.; Greville, T. N. E.: On spline functions. in Shisha, O. (ed.): Inequalities. Academic Press (1967) 255-291.

[Schul 90] Schulze, G.: Elastische Wege und nichtlineare Splines im CAGD. Diss., Kaiserslautern 1990.

[Schu 76] Schumaker, L. L.: Fitting surfaces to scattered data. in Lorentz, G. G.; Chui, C. K.; Schumaker, L. L. (ed.): Approximation Theory II. Academic Press (1976) 203-268.

[Schu 79] Schumaker, L. L.: Two-stage methods for fitting surfaces to scattered data. Berlin: Hahn-Meitner-Institutsbericht 314, 1979.

[Schu 79a] Schumaker, L. L.: On the dimension of spaces of piecewise polynomials in two variables. in Schemp, W.; Zeller, K. (ed.): Multivariate Approximation Theory. Birkhäuser (1979) 396-412.

[Schu 83] Schumaker, L. L.: On shape preserving quadratic spline interpolation. SIAM Journal on Numerical Analysis 20 (1983) 854-864.

[Schu 84] Schumaker, L. L.: Bounds on the dimension of spaces of multivariate piecewise polynomials. Rocky Mountain Journal of Mathematics 14 (1984) 251-264.

[Schu 86] Schumaker, L. L.; Volk, W.: Efficient evaluation of multivariate polynomials. Computer Aided Geometric Design 3 (1986) 149-154.

[Schu 87] Schumaker, L. L.: Triangulation methods. in Chui, C. K.; Schumaker, L. L.; Utreras, F. I. (ed.): Topics in Multivariate Approximation, Academic Press (1987) 219-232.

[Schu 87a] Schumaker, L. L.: Numerical aspects of spaces of piecewise polynomials on triangulations. in Mason, J. C.; Cox, M. G. (ed.): Algorithms for the Approximation of Functions and Data. Oxford University Press (1987) 373-406.

[Schu 88] Schumaker, L. L.: Dual bases for spline spaces on cells. Computer Aided Geometric Design 4 (1988) 277-284.

[Schu 90] Schumaker, L. L.: Reconstructing 3D objects from cross-sections. in Dahmen, W.; Gasca, M.; Micchelli, C. A. (ed.): Computation of Curves and Surfaces. Kluwer (1990) 275-309.

[Schw 66] Schweikert, D. G.: An Interpolation Curve Using a Spline In Tension. Journal Mathematics and Physics 45 (1966) 312-317.

[Sed 83] Sederberg, Th. W.: Implicit and parametric curves and surfaces for Computer Aided Geometric Design. Thesis, Purdue University 1983.

[Sed 84] Sederberg, Th. W.; Anderson, D. C.; Goldman, R. N.: Implicit Representation of Parametric Curves and Surfaces. Computer Vision, Graphics, and Image Processing 28 (1984) 72-84.

[Sed 84a] Sederberg, Th. W.: Planar piecewise algebraic curves. Computer Aided Geometric Design 1 (1984) 241-255.

[Sed 85] Sederberg, Th. W.; Anderson, D. C.: Steiner surface patches. IEEE Computer Graphics & Applications 5 (1985) 23-36.

[Sed 85a] Sederberg, Th. W.; Anderson, D. C.; Goldman, R. N.: Implicitization, inversion, and intersection of planar rational cubic curves. Computer Vision, Graphics, and Image Processing 31 (1985) 89-102.

[Sed 85b] Sederberg, Th. W.: Piecewise algebraic surface patches. Computer Aided Geometric Design 2 (1985) 53-59.

[Sed 86] Sederberg, Th. W.; Parry, S. R.: Comparison of three curve intersection algorithms. Computer-Aided Design 18 (1986) 58-63.

[Sed 86a] Sederberg, Th. W.; Parry, S. R.: Free-form deformation of solid geometric models. ACM 20 (1986) 151-160.

[Sed 87] Sederberg, Th. W.: Algebraic geometry for surface and solid modeling. in Farin, G. (ed.): Geometric Modeling: Algorithms and New Trends. SIAM (1987) 28-42.

[Sed 87a] Sederberg, Th. W.; Wang, X.: Rational hodographs. Computer Aided Geometric Design 4 (1987) 333-335.

[Sed 87b] Sederberg, Th. W.; Snively, J. P.: Parametrization of Cubic Algebraic Surfaces. in Martin, R. R. (ed.): The Mathematics of Surfaces II. Oxford University Press (1987) 299-319.

[Sed 88] Sederberg, Th. W.; Meyers, R. J.: Loop detection in surface patch intersections. Computer Aided Geometric Design 5 (1988) 161-171.

[Sed 89] Sederberg, Th. W.; White, S. C.; Zundel, A. K.: Fat arcs: A bounding region with cubic convergence. Computer Aided Geometric Design 6 (1989) 205-218.

[Sed 89a] Sederberg, Th. W.; Zhao, J.; Zundel, A. K.: Approximate Parametrization of Algebraic Curves. in Straßer, W.; Seidel, H. P. (ed.): Theory and Practice of Geometric Modeling. Springer (1989) 33-54.

[Sed 89b] Sederberg, Th. W.; Christiansen, H. N.; Katz, S.: Improved test for closed loops in surface intersections. Computer-Aided Design 21 (1989) 505-508.

[Sed 90] Sederberg, Th. W.; Nishita, T.: Curve intersection using Bézier clipping. Computer-Aided Design 22 (1990) 538-549.

[Sed 90a] Sederberg, Th. W.: Techniques for Cubic Algebraic Surfaces. Part One. IEEE Computer Graphics & Applications 10 (1990) 14-25.

[Sed 90b] Sederberg, Th. W.: Techniques for Cubic Algebraic Surfaces. Part Two. IEEE Computer Graphics & Applications 10 (1990) 12-21.

[Sed 91] Sederberg, Th. W.; Nishita, T.: Geometric Hermite approximation of surface patch intersection curves. Computer Aided Geometric Design 8 (1991) 97-114.

[Sei 88] Seidel, H. P.: Ein neuer Zugang zu B-Splines in der graphischen Datenverarbeitung. CAD und Computergraphik 11 (1988) 189-198.

[Sei 88a] Seidel, H. P.: Knot insertion from a blossoming point of view. Computer Aided Geometric Design 5 (1988) 81-86.

[Sei 89] Seidel, H. P.: A new multiaffine approach to B-splines. Computer Aided Geometric Design 6 (1989) 23-32.

[Sei 89a] Seidel, H. P.: Computing B-Spline Control Points. in Straßer, W.; Seidel, H. P. (ed.): Theory and Practice of Geometric Modeling. Springer (1989) 17-32.

[Sei 90] Seidel, H. P.: Quaternionen in Computergraphik und Robotik. Informationstechnik 32 (1990) 266-275.

[Sew 88] Sewell, G.: Plotting Contour Surfaces of a Function of Three Variables. Algorithm 657. ACM Transactions on Mathematical Software 14 (1988) 33-41.

[Shan 88] Shantz, M.; Chang, S.: Rendering trimmed NURBS with adaptive forward differencing. ACM Computer Graphics 22 (1988) 189-198.

[Sha 82] Sharpe, R. J.; Thorne, R. W.: Numerical method for extracting an arc length parameterization from parametric curves. Computer-Aided Design 14 (1982) 79-81.

[She 91] Shene, Ch. K.; Johnstone, J. K.: On the Planar Intersection of Natural Quadrics. in Rosignac, J.; Turner, J. (ed.): Proceedings Symposium on Solid Modeling Foundations and CAD/CAM Applications. Austin, Texas, 5–7. June 1991. ACM Press (1991) 233-242.

[Shep 68] Shepard, D.: A two dimensional interpolation function for irregular spaced data. Proceedings 23rd ACM National Conference (1968) 517-524.

[Shet 91] Shetty, S.; White, P. R.: Curvature-continuous extensions for rational B-spline curves and surfaces. Computer-Aided Design 23 (1991) 484-491.

[Shi 87] Shirman, L. A.; Séquin, C. H.: Local surface interpolation with Bézier patches. Computer Aided Geometric Design 4 (1987) 279-295.

[Shi 90] Shirman, L. A.; Séquin, C. H.: Local surface interpolation with shape parameters between adjoining Gregory patches. Computer Aided Geometric Design 7 (1990) 375-388.

[Shi 91] Shirman, L. A.; Séquin, C. H.: Local surface interpolation with Bézier patches: errata and improvements. Computer Aided Geometric Design 8 (1991) 217-221.

[Sib 78] Sibson, R.: Locally equiangular triangulations. The Computer Journal 21 (1978) 243-245.

[Sib 78a] Sibson, R.; Green, P. J.: Computing Dirichlet tessellations in the plane. The Computer Journal 21 (1978) 168-173.

[Sie 81] Sielaff, W.: Fünfachsiges NC-Umfangfräsen von Werkstücken mit verwundenen Regelflächen. Ein Beitrag zur Technologie und Teileprogrammierung. Springer 1981.

[Sny 87] Snyder, J. M., Barr, A. H.: Ray Tracing Complex Models Containing Surface Tessellations. ACM Computer Graphics 21 (1987) 119-128.

[Snyd 78] Snyder, W. V.: Algorithm 531, Contour Plotting [J6]. ACM Transactions on Mathematical Software 4 (1978) 290-294.

[Spä 69] Späth, H.: Exponential spline interpolation. Computing 4 (1969) 225-233.

[Spä 71] Späth, H.: Algorithm 16: Two-Dimensional Exponential Splines. Computing 7 (1971) 364-369.

[Spe 90] Speray, D.; Kennon, St.: Volume Probes: Interactive Data Exploration on Arbitrary Grids. Computer Graphics 24 (1990) 5-12.

[Stä 76] Stärk, E.: Mehrfach differenzierbare Bézier-Kurven und Bézier-Flächen. Diss., Braunschweig 1976.

[Stan 88] Standerski, N. B.: Die Generierung und Verzerrung von Schiffsoberflächen beschrieben mit globalen Tensorproduktflächen. Diss., TU Berlin 1988.

[Sta 81] Stark, E. L.: Bernstein-Polynome, 1912-1955. in Butzer, P. L.; Nagy, B. Sz.; Gorlich, E. (ed.): Functional Analysis and Approximation. Birkhäuser (1981) 443-461.

[Ste 84] Stead, S. E.: Estimation of Gradients from Scattered Data. Rocky Mountain Journal of Mathematics 14 (1984) 265-279.

[Stei 90] Steiner, S.: Äquidistante Kurven und Flächen – Anwendungen einer klassischen mathematischen Theorie auf Probleme in CAGD. in Encarnacao, J. L.; Hoschek, J.; Rix, J. (ed.): Geometrische Verfahren der Graphischen Datenverarbeitung. Springer (1990) 88-103.

[Sto 89] Storry, D. J. T.; Ball, A. A.: Design of an n-sided surface patch from Hermite boundary data. Computer Aided Geometric Design 6 (1989) 111-120.

[Suf 84] Suffern, K. G.: Contouring Functions of two Variables. The Australian Computer Journal 16 (1984) 102-106.

[Suh 90] Suh, Y. S.; Lee, K.: NC milling tool path generation for arbitrary pockets defined by sculptured surfaces. Computer-Aided Design 22 (1990) 273-284.

[Sut 76] Sutcliffe, D. C.: A remark on a contouring algorithm. The Computer Journal 19 (1976) 333-335.

[Sut 76a] Sutcliffe, D. C.: An algorithm for drawing the curve $f(x, y) = 0$. The Computer Journal 19 (1976) 246-249.

[Sut 80] Sutcliffe, D. C.: Contouring over rectangular and skewed rectangular grids – an introduction. in Brodlie, K. (ed.): Mathematical Methods in Computer Graphics and Design. Academic Press (1980) 39-62.

[Suth 74] Sutherland, I.; Sproull, R.; Schumaker, R.: A Characterization of ten Hidden-Surface Algorithms. Computer Surveys 6 (1974) 1-55.

[Swe 86] Sweeney, M.; Bartels, R.: Ray Tracing Free-Form B-Spline Surfaces. IEEE Computer Graphics & Applications 6 (1986) 41-49.

[Szi 91] Szilvasi-Nagy, M.: Flexible rounding operation for polyhedra. Computer-Aided Design 23 (1991) 629-633.

[Tak 88] Takai, M. K.: Free-form Surface Generation System for Camera Design. Proceedings of the USA-Japan Symposium on Flexible Automation (1988) 939-946.

[Tak 91] Takai, M. K.; Wang, K. K.: C^2 Gregory Patch. Proceedings EUROGRAPHICS '91 (1991) 481-492.

[Taka 90] Takamura, T.; Ohta, M.; Toriya, H.; Chiyokura, H.: A Method to Convert a Gregory Patch and a Rational Boundary Gregory Patch to a Rational Bézier Patch and its Applications. in Chua, T. S.; Kunii, T. L. (ed.): CG International '90. Springer (1990) 543-562.

[Tan 86] Tan, S. T.; Chan, K. C.: Generation of high order surfaces over arbitrary polyhedral meshes. Computer-Aided Design 18 (1986) 411-423.

[Tan 87] Tan, S. T.; Chan, K. C.: Biquadratic B-spline surfaces generated from arbitrary polyhedral meshes: A constructive approach. Computer Vision, Graphic, and Image Processing 39 (1987) 144-166.

[Ter 88] Terzopoulos, D.; Fleischer, K.: Deformable models. The Visual Computer 4 (1988) 306-331.

[Til 83] Tiller, W.: Rational B-splines for curve and surface representation. IEEE Computer Graphics & Applications 3 (1983) 61-69.

[Til 84] Tiller, W.; Hanson, E. G.: Offsets of Two-Dimensional Profiles. IEEE Computer Graphics & Applications 4 (1984) 36-46.

[Tim 77] Timmer, H. G.: Analytical background for computation of surface intersections. Douglas Aircraft Company Technical Memorandum C1-250-CAT-77-036, 1977.

[Tod 86] Todd, P. H.; McLeod, R. J. Y.: Numerical estimation of the curvature of surfaces. Computer-Aided Design 18 (1986) 33-37.

[Töp 82] Töpfer, H. J.: Models for smooth curve fitting. in Numerical Methods of Approximation Theory 6, Birkhäuser Verlag (1982) 209-224.

[Tot 85] Toth, D.: On Ray Tracing Parametric Surfaces. ACM Computer Graphics 19 (1985) 171-179.

[Tse 91] Tseng, Y. J.; Joshi, S.: Determining feasible tool-approach directions for machining Bézier curves and surfaces. Computer-Aided Design 23 (1991) 367-379.

[Tur 88] Turner, J. U.: Accurate Solid Modeling Using Polyhedral Approximations. IEEE Computer Graphics & Applications 4 (1988) 14-28.

[Unb 85] Unbehaun, K.: Drucken in drei Dimensionen - Analyse der alternativen verfügbaren 3-D-Verfahren. Deutscher Drucker (1985) 142-160.

[Ups 88] Upson, C.; Keeler, M.: The v-buffer: Visible volume rendering. ACM Computer Graphics 22 (1988) 59-64.

[Var 87] Varady, T.: Survey and new results in N-sided patch generation. in Martin, R. (ed.): The Mathematics of Surfaces II. Oxford University Press (1987) 203-235.

[Var 89] Varady, T.; Martin, R. R.; Vida, J.: Topological considerations in blending boundary representation solid models. in Straßer, W.; Seidel, H. P. (ed.): Theory and Practice of Geometric Modeling. Springer (1989) 205-220.

[Var 91] Varady, T.: Overlap patches: a new scheme for interpolating curve networks with n-sided regions. Computer Aided Geometric Design 8 (1991) 7-27.

[Vee 82] Veenman, P. R.: The Design of Sculptured Surfaces Using Recursive Division Techniques. in Proceedings of a Conference on CAD/CAM Technology in Mechanical Engineering. MIT-Press (1982).

[Vero 76] Veron, M.; Ris, G.; Musse, J. P.: Continuity of biparametric surface patches. Computer-Aided Design 8 (1976) 267-273.

[Vers 75] Verspille, K. J.: Computer-aided design applications of the rational B-spline approximation form. Thesis, Syracuse University 1975.

[Vie 89] Viefhaus, R.: Fräsergeometriekorrektur in numerischen Steuerungen für das fünfachsige Fräsen. ISW-Forschung und Praxis 79, Springer 1989.

[Vin 89] Vinacua, A.; Brunet, P.: A construction for VC^1 continuity of rational Bézier patches. in Lyche, T.; Schumaker, L. L. (ed.): Mathematical Methods in Computer Aided Geometric Design. Academic Press (1989) 601-611.

[Vri 91] Vries-Baayens, A.: CAD product data exchange: conversions for curves and surfaces. Thesis, Delft University Press 1991.

[Wag 86] Waggenspack, W. N.; Anderson, D. C.: Converting standard bivariate polynomials to Bernstein form over arbitrary triangular regions. Computer-Aided Design 18 (1986) 529-532.

[Wag 89] Waggenspack, W. N.; Anderson, D. C.: Piecewise parametric approximations for algebraic curves. Computer Aided Geometric Design 6 (1989) 33-53.

[Wah 81] Wahba, G.: Spline interpolation and smoothing on a sphere. SIAM Journal on Scientific and Statistical Computing 2 (1981) 5-16.

[Wah 84] Wahba, G.: Surface fitting with scattered noisy data on Euclidean D-space and on the sphere. Rocky Mountain Journal of Mathematics 14 (1984) 281-299.

[Wal 71] Walter, H.: Numerische Darstellung von Oberflächen unter Verwendung eines Optimalprinzips. Diss., TU München 1971.

[Wal 82] Walter, W.: Interaktive NC-Programmierung von Werkstücken mit gekrümmten Flächen. Springer 1982.

[Wan 84] Wang, G.: The subdivision method for finding the intersection between two Bézier curves or surfaces. Zhejiang University Journal. Special Issue on Computational Geometry (1984).

[Wan 91] Wang, G. J.: Rational cubic circular arcs and their application in CAD. Computers in Industry 16 (1991) 283-288.

[Wan 91a] Wang. G.: The Termination Criterion for Subdivision of the Rational Bézier Curves. CVGIP: Graphical Models and Image Processing 53 (1991) 93-96.

[War 89] Warren, J.: Blending algebraic surfaces. ACM Transactions on Graphics 8 (1989) 263-278.

[Was 91] Wassum, P.: Bedingungen und Konstruktionen zur geometrischen Stetigkeit und Anwendungen auf approximative Basistransformationen. Diss., Darmstadt 1991.

[Watk 88] Watkins, M. A.; Worsey, A. J.: Degree reduction of Bézier curves. Computer-Aided Design 20 (1988) 398-405.

[Watk 88a] Watkins, M. A.: Problems in geometric continuity. Computer-Aided Design 20 (1988) 499-502.

[Wat 81] Watson, D. F.: Computing the n-dimensional Delaunay tessellation with application to Voronoi polytopes. The Computer Journal 20 (1981) 167-172.

[Wat 84] Watson, D. F.; Philip, G. M.: Survey: Systematic triangulations. Computer Vision, Graphics, and Image Processing 26 (1984) 217-223.

[Web 90] Weber, J.: Methoden zur Konstruktion krümmungsstetiger Flächen. Diss. TU Berlin 1990.

[Web 90a] Weber, J.: Constructing a Boolean-sum curvature-continuous surface. in Krause, F. L.; Jansen, H. (ed.): Advanced Geometric Modelling for Engineering Applications. IFIP/GI. North-Holland (1990) 115-128.

[Wei 66] Weiss, R. A.: Be VISION, a package of IBM 7090 FORTRAN programs to draw orthographic views of combinations of plane and quadric surfaces. Journal of the ACM 13 (1966) 194-204.

[Wel 84] Welbourn, D. B.: Full Three-Dimensional CAD/CAM. CAE Journal 1 (1984) 54-60, 189-192.

[Wer 79] Werner, H.: An introduction to non-linear splines. in Salney, B. N. (ed.): Polynomial and Spline Approximation. Reidel (1979) 247-306.

[Wev 88] Wever, U. A.: Darstellung von Kurven und Flächen mittels datenreduzierender Algorithmen. Diss., TU München 1988.

[Wev 88a] Wever, U. A.: Non-negative exponential splines. Computer-Aided Design 20 (1988) 11-16.

[Wev 91] Wever, U. A.: Global and local data reduction strategies for cubic splines. Computer-Aided Design 23 (1991) 127-132.

[Whe 86] Whelan, T.: A representation of a C^2 interpolant over triangles. Computer Aided Geometric Design 3 (1986) 53-66.

[Whi 78] Whitted, T.: A Scan-line Algorithm for Computer Display of Curved Surfaces. ACM Computer Graphics 12 (1978) 26, see [Lane 80a].

[Whi 80] Whitted, T.: An Improved Illumination Model for Shaded Display. Communications of the ACM 23 (1980) 343-349.

[Wij 86] Wijk van, J. J.: Bicubic patches for approximating non-rectangular control-point meshes. Computer Aided Geometric Design 3 (1986) 1-13.

[Wil 90] Wilhelms, J.; Gelder, A. V.: Topological Considerations in Isosurface Generation. ACM Computer Graphics 24 (1990) 79-86.

[Wils 87] Wilson, P. R.: Conic Representations for Shape Description. IEEE Computer Graphics & Applications 7 (1987) 23-30.

[Wit 90] Witkin, A.; Welch, W.: Fast Animation and Control of Nonrigid Structures. ACM Computer Graphics 24 (1990) 243-252.

[Witt 81] Wittram, M.: Hidden-line algorithm for scenes of high complexity. Computer-Aided Design 13 (1981) 187-192.

[Wör 91] Wördenweber, B.; Santarelli, P.: Digitizing Sculptured Surfaces. in Hoschek, J. (ed.): Freeform Tools in CAD Systems. A Comparison. Teubner (1991) 37-44.

[Wol 92] Wolter, F. E.; Tuohy, S. T.: Approximation of High-Degree and Procedural Curves. Engineering with Computers 8 (1992) 61-80.

[Woo 87] Woodward, Ch. D.: Cross-Sectional Design of B-Spline Surfaces. Computer & Graphics 11 (1987) 193-201.

[Woo 89] Woodward, Ch. D.: Ray Tracing Parametric Surfaces by Subdivision in Viewing Plane. in Straßer, W.; Seidel, H. P. (ed.): Theory and Practice of Geometric Modeling. Springer (1989) 273-290.

[Wood 87] Woodwark, J. R.: Blends in Geometric Modelling. in Martin, R. R. (ed.): The Mathematics of Surfaces II. Oxford University Press (1987) 255-297.

[Woon 71] Woon, P. Y.; Freeman, H.: A procedure for generating visible-line projections of solids bounded by quadric surfaces. in Proceedings 1971 IFIP Congress, North-Holland (1971) 1120-1125.

[Wor 85] Worsey, A. J.: C^2 interpolation over hypercubes. Computer Aided Geometric Design 2 (1985) 107-115.

[Wor 87] Worsey, A. J.; Farin, G.: An n-Dimensional Clough-Tocher Interpolant. Constructive Approximation 3 (1987) 99-110.

[Wor 88] Worsey, A. J.; Piper, B.: A trivariate Powell-Sabin interpolant. Computer Aided Geometric Design 5 (1988) 177-186.

[Wor 90] Worsey, A. J.; Farin, G.: Contouring a bivariate quadratic polynomial over a triangle. Computer Aided Geometric Design 7 (1990) 337-351.

[Wri 85] Wright, A. H.: Finding All Solutions to a System of Polynomial Equations. Mathematics of Computation 44 (1985) 125-133.

[Wrig 79] Wright, T.; Humbrecht, J.: Isosurf – an algorithm for plotting iso-valued surfaces of a function of three variables. ACM Computer Graphics 13 (1979) 182-189.

[Wun 50] Wunderlich, W.: Zur Geometrie gewisser Glanzerscheinungen. Monatshefte für Mathematik 54 (1950) 330-344.

[Wun 77] Wunderlich, W.: Über die gefährlichen Örter bei zwei Achtpunkt-problemen und einem Fünfpunktproblem. Österr. Z. Vermessungswesen und Photogrammetrie 64 (1977) 119-128.

[Wyv 86] Wyvill, G. M.; McPheeters, C.; Wyvill, B.: Data structures for soft objects. The Visual Computer 2 (1986) 227-234.

[Wyv 86a] Wyvill, B.; McPheeters, C.; Wyvill, G. M.: Animating soft objects. The Visual Computer 2 (1986) 235-242.

[Wyv 89] Wyvill, B.; Wyvill, G. M.: Field functions for implicit surfaces. The Visual Computer 5 (1989) 75-82.

[Yan 87] Yang, C. G.: On speeding up ray tracing of B-spline surfaces. Computer-aided Design 19 (1987) 122-130.

[Yan 87a] Yang, C. G.: Illumination models for generating images of curved surfaces. Computer-Aided Design 20 (1987) 544-554.

[Yen 91] Yen, J.; Spach, S.; Smith, M.; Pulleyblank, R.: Parallel Boxing in B-Spline Intersection. IEEE Computer Graphics & Applications 11 (1991) 72-79.

[Zen 70] Zenisek, A.: Interpolation Polynomials on the Triangle. Numerische Mathematik 15 (1970) 283-296.

[Zen 73] Zenisek, A.: Polynomial Approximation on tetrahedrons in the finite element method. Journal of Approximation Theory 7 (1973) 334-351.

[Zha 86] Zhang, D. Y.; Bowyer, A.: CSG Set-Theoretic Solid Modelling and NC Machining of Blend Surfaces. Proceedings of the ACM 2nd Symposium on Computational Geometry, Yorktown Heights (1986) 236-245.

[Zho 90] Zhou, C. Z.: On the convexity of parametric Bézier triangular surfaces. Computer Aided Geometric Design 7 (1990) 459-464.

[Zhou 91] Zhou, X.; Straßer, W.: A NURBS Representation for Cyclides. in Kunii, T. L. (ed.): Modeling in Computer Graphics. Springer (1991) 77-92.

[Zir 88] Zirbs, J.: Konzept zur Kollisionsüberwachung bei der NC-Programmierung komplexer Oberflächen. HFG-Kurzberichte. Industrie Anzeiger 39 (1988) 32-33.

[Zir 89] Zirbs, J.: Fertigungsgerechte Aufbereitung von Flächenverbänden bei der NC-Programmierung im Formenbau. ISW-Forschung und Praxis 80, Springer 1989.

[Zyd 87] Zyda, M. A.; Jones, A. R.; Hogan, P. G.: Surface construction from planar contours. Computer & Graphics 11 (1987) 393-408.

Index

A

B

D

E

F

G

H

I

O

P

Q

R

S

T

Other Titles of Interest

from

A K PETERS, LTD

Barnsley/Hurd
Fractal Image Compression
ISBN 1-56881-000-8

Birmingham et al.
Automating the Design of Computer Systems
ISBN 0-86720-241-6

Boehm/Prautzsch
Geometric Concepts for Geometric Design
ISBN 1-56881-004-0

Brown/Reynolds
Computer Graphics File Formats
ISBN 1-56881-008-3

Cowan
Colour Principles for Computer Graphics
ISBN 1-56881-009-1

Iterated Systems Inc.
Snapshots: True-Color Photo Images Using the Fractal Formatter
ISBN 0-86720-299-8

Klinker
A Physical Approach to Color Image Understanding
ISBN 1-56881-013-X

Parke/Waters
Computer Facial Animation
ISBN 1-56881-014-8

Whitman
Multiprocessor Methods for Computer Graphics Rendering
ISBN 0-86720-293-9

289 Linden Street
Wellesley, Massachusetts
(617) 235-2210
Fax (617) 235-2404